Integrated Drought Management, Volume 2

The second volume of this comprehensive global perspective on Integrated Drought Management is focused on drought modeling, meteorological prediction, and the use of remote sensing in assessing, analyzing, and monitoring drought. It discusses risk management, planning, policy, and societal impacts of drought such as water pricing, water transfer, water quality, and crop insurance. Through numerous case studies from India, Iran, Brazil, the US, Nepal, and other countries that cover a broad range of topics and geographical regions, this volume serves as a valuable resource for all professionals, researchers, and academics who want to advance their knowledge about droughts.

Features

- Provides a global perspective on drought prediction and management, and a synthesis of the recent state of knowledge.
- Covers a wide range of topics from essential concepts and advanced techniques for forecasting and modeling drought to societal impacts, consequences, and planning.
- Presents numerous case studies with different management approaches from different regions and countries.
- Addresses how climate change impacts drought, the increasing challenges associated with managing drought, decision-making, and policy implications.
- Includes contributions from hundreds of experts from around the world.

Professionals, researchers, academics, and postgraduate students with knowledge in environmental sciences, ecology, agriculture, forestry, hydrology, water resources engineering, and earth sciences, as well as those interested in how climate change impacts drought management, will gain new insights from the experts featured in this two-volume handbook.

Drought and Water Crises
Series Editor: Donald A. Wilhite

Published Titles:

For more information about this series visit https://www.routledge.com/Drought-and-Water-Crises/book-series/CRCDROANDWATCRI

Integrated Drought Management, Volume 2

Forecasting, Monitoring, and Managing Risk

Edited by

Vijay P. Singh
Deepak Jhajharia
Rasoul Mirabbasi
Rohitashw Kumar

CRC Press
Taylor & Francis Group
Boca Raton London New York

CRC Press is an imprint of the
Taylor & Francis Group, an **informa** business

Designed cover image: © Shutterstock

First edition published 2024
by CRC Press
6000 Broken Sound Parkway NW, Suite 300, Boca Raton, FL 33487-2742

and by CRC Press
4 Park Square, Milton Park, Abingdon, Oxon, OX14 4RN

CRC Press is an imprint of Taylor & Francis Group, LLC

ISBN: 9781032231686 (hbk)
ISBN: 9781032232782 (pbk)
ISBN: 9781003276548 (ebk)

DOI: 10.1201/9781003276548

Typeset in Times
by Deanta Global Publishing Services, Chennai, India

Contents

Editors

Vijay P. Singh, PhD, DSc, is a University Distinguished Professor, a Regents Professor, and Caroline and William N. Lehrer Distinguished Chair in Water Engineering at Texas A&M University, USA. He earned his BS, MS, PhD, and DSc in engineering. He is a registered professional engineer, a registered professional hydrologist, and an honorary diplomat of ASCE-AAWRE. He is a distinguished member of ASCE; an honorary member of IWRA; a distinguished fellow of AGGS; and an Honorary Member of AWRA; and a fellow of EWRI-ASCE, IAH, ISAE, IWRS, and IASWC. He has published extensively in the areas of hydrology, irrigation engineering, hydraulics, groundwater, water quality, and water resources with more than 1370 journal articles; 32 textbooks; 75 edited reference books, including *Handbook of Applied Hydrology* and *Encyclopedia of Snow, Ice and Glaciers*; 115 book chapters; and 315 conference papers. He has received more than 100 national and international awards, including three honorary doctorates. He is a member of 12 international science/engineering academies. He has served as president of the American Institute of Hydrology (AIH); president of the American Academy of Water Resources Engineers; president of the International Association for Water, Environment, Energy, and Society; and chair of the Watershed Council of the American Society of Civil Engineers. He has served as editor-in-chief of five journals and two book series and serves on the editorial boards of more than 25 journals and three book series. His Google Scholar citations number 72,000; h-index of 119; and I10-index of 1025.

Deepak Jhajharia, PhD, is currently working as a Professor in the Department of Soil & Water Conservation Engineering, College of Agricultural Engineering & Post Harvest Technology (Central Agricultural University), Ranipool, Gangtok, Sikkim, India. He is also acting as principal investigator of the All India Coordinated Research Project on Plastic Engineering in Agriculture Structures and Environment Management (CAEPHT Centre) funded by the Indian Council of Agricultural Research-Central Institute of Post Harvest Engineering and Technology, Ludhiana, Punjab, India. He studied agricultural engineering, graduating in 1998 from the College of Technology and Engineering (MPUAT), Udaipur, Rajasthan, India; and completed his postgrad studies in water resources engineering at the Department of Civil Engineering, Indian Institute of Technology Delhi, India. He earned his PhD from the Department of Hydrology, Indian Institute of Technology Roorkee, Uttarakhand, India. He is the recipient of the prestigious Science Without Borders scholarship – Young Talent of CNPq (Brazil) as a research collaborator at the Department of Agronomy, Universidade Federal Rural De Pernambuco (UFRPE), Ministry of Education, Brazil. He has 22 years of academic experience and has published more than 85 papers in peer-reviewed journals, books, reports, or extension bulletins. He has guided one PhD and 13 MTech theses in the field of soil and water conservation engineering along with many undergraduate theses in the field of agricultural engineering. He is coeditor of *Agricultural Impacts of Climate Change* and *Applied Agricultural Practices for Mitigating Climate Change*, published by CRC Press/Taylor & Francis Group. He also conducted a 21-day summer school for scientists from ICAR and faculty members from different universities of India and one 90-day skill development training program on Greenhouse Technology for school drop-outs and unemployed rural youth from six states of northeast India. He was awarded the CSIRO Land and Water Publication Award 2013, CSIRO Australia, for a global review paper published in the *Journal of Hydrology*. He was also judged the best extension scientist (2017–2018) of the AICRP on PET in recognition of his outstanding contribution to the extension and popularization of plasticulture technologies in Sikkim. He is the recipient of the Distinguished Alumni Award (in 2016) by the College of Technology and Engineering Alumni Society, CTAE (MPUAT), Udaipur. He was elected as a Fellow of the Indian Association of Hydrologists, Roorkee (in 2015); and Indian Water Resources Society, Roorkee (2019). He is also a life member of 14 different professional societies from India and abroad. His Google Scholar citations are 3042 and h-index of 18.

Rasoul Mirabbasi, PhD, is an Associate Professor of Hydrology and Water Resources Engineering at Shahrekord University, Iran. He is also head of the Water Resources Center of Shahrekord University. His research focuses mainly on Statistical and Environmental Hydrology and Climate Change. In particular, he is working on modeling natural hazards including flood, drought, wind, and pollution toward a sustainable environment. Formerly, he was a visiting researcher at the University of Connecticut. He has contributed to more than 150 publications in journals, books, or technical reports. Dr. Mirabbasi is the reviewer of about 30 Web of Science (ISI) journals. His Google Scholar citations are 1916 and h-index of 22.

Rohitashw Kumar, PhD, is a Professor at the College of Agricultural Engineering and Technology, Sher-e-Kashmir University of Agricultural Sciences and Technology of Kashmir, Srinagar, India. He is also holding additional charge of Associate Dean, College of Agricultural Engineering and Technology and SKUAST- Kashmir, Srinagar (India). He is also Professor Water Chair (Sheikkul Alam Shiekh Nuruddin Water Chair), Ministry of Water Resources, Government of India, at National Institute of Technology, Srinagar (J&K). He earned his PhD in water resources engineering from NIT, Hamirpur; and Master's of Engineering in irrigation water management engineering from MPUAT, Udaipur. He received the Special Research Award in 2017 and Student Incentive Award 2015 (PhD research) from the Soil Conservation Society of India, New Delhi. He also earned the first prize in India for best MTech thesis in agricultural engineering in 2001. He graduated from Maharana Pratap University of Agricultural and Technology, Udaipur, India, in agricultural engineering. He has published more than 80 papers in peer-reviewed journals, four practical manuals, and 20 chapters in books. He has guided ten postgraduate students in soil and water engineering. He has led more than ten research projects as a principal or co-principal investigator. Since 2011, he has been the principal investigator of the All India Coordinated Research Project on Plastic Engineering in Agriculture Structures and Environment Management (Srinagar Centre) funded by the Indian Council of Agricultural Research–Central Institute of Post Harvest Engineering and Technology, Ludhiana, Punjab, India. His Google Scholar citations are 1148 and h-index of 18.

Contributors

Khodayar Abdollahi
Shahrekord University
Shahrekord, Iran

Wanny K. Adidarma
Parahyangan Catholic University
Bandung, Indonesia

Raimunda Adlany Dias da Silva
Federal University of Rio Grande do Norte
Natal, Brazil

Nand Kishor Agrawal
International Centre for Integrated Mountain
    Development (ICIMOD)
Kathmandu, Nepal

Mohamed Al Mulla
Ministry of Energy and Industry
Dubai, United Arab Emirates

Rawshan Ali
Department of Petroleum, Koya Technical
    Institute, Erbil Polytechnic University
Erbil, Iraq

Mohammad Ali Ghorbani
University of Tabriz
Tabriz, Iran

Mohammad Arab Amiri
K.N. Toosi University of Technology
Tehran, Iran

Abdel Azim Ebraheem
United Arab Emirates University
Al Ain, United Arab Emirates

D.K. Bastia
All India Coordinated Research Project for
    Dryland Agriculture
Odisha University of Agriculture and
    Technology
Kandhamal, Odisha, India

Deborah J. Bathke
University of Nebraska-Lincoln
Lincoln, Nebraska, USA

S.K. Behera
All India Coordinated Research Project for
    Dryland Agriculture
Odisha University of Agriculture and
    Technology
Kandhamal, Odisha, India

Sita Ram Bhakar
Maharana Pratap University of Agriculture and
    Technology
Udaipur, India

A. Bhattacharyya
Birbal Sahni Institute of Palaeosciences
Lucknow, Uttar Pradesh, India

Sanjeev Bhuchar
International Centre for Integrated Mountain
    Development (ICIMOD)
Kathmandu, Nepal

Chandrashekhar Bhuiyan
Sikkim Manipal Institute of Technology
Sikkim, India

Jimmy Byakatonda
Gulu University
Gulu, Uganda

Tommaso Caloiero
National Research Council of Italy (CNR)
Institute for Agricultural and Forest Systems in
    the Mediterranean (ISAFOM)

Virgínia Maria Cavalari Henriques
Federal University of Rio Grande do Norte
Natal, Brazil

Arun Chakraborty
Indian Institute of Technology Kharagpur
Kharagpur, India

David Bency
Institute of Geography
Friedrich-Alexander-Universität
    Erlangen-Nürnberg
Erlangen, Germany
and
Birbal Sahni Institute of Palaeosciences
Lucknow, Uttar Pradesh, India

Majid Dehghani
Vali-e-Asr University of Rafsanjan
Rafsanjan, Iran

Madhav Prasad Dhakal
International Centre for Integrated Mountain
    Development (ICIMOD)
Kathmandu, Nepal

Zahra Eslami
Shahrekord University
Shahrekord, Iran

Saeid Eslamian
Department of Water Science and Engineering
College of Agriculture
and
Center of Excellence on Risk Management and
    Natural Hazards
Isfahan University of Technology
Isfahan, Iran

Juliana Espada Lichston
Federal University of Rio Grande do Norte
Natal, Brazil

Flavia D. Frederick
Parahyangan Catholic University
Bandung, Indonesia

R.V. Galkate
CIHRC-Bhopal, National Institute of Hydrology
Roorkee, India

Magda Maria Guilhermino
Federal University of Rio Grande do Norte
Natal, Brazil

Shivam Gupta
Acharya Narendra Dev University of
    Agriculture and Technology
Ayodhya, India

Vivek Gupta
Indian Institute of Technology Roorkee
Roorkee, India

Abu Reza Md. Towfiqul Islam
Begum Rokeya University
Rangpur, Bangladesh

Manoj Kumar Jain
Indian Institute of Technology Roorkee
Roorkee, India

R.K. Jaiswal
CIHRC-Bhopal, National Institute of
    Hydrology
Roorkee, India

Deepak Jhajharia
Central Agricultural University
Imphal, India

Shanhu Jiang
Hohai University
Nanjing, China

Piet K. Kenabatho
University of Botswana
Gaborone, Botswana

Ercan Kahya
Istanbul Technical University
Istanbul, Turkey
and
Tashkent University of Architecture and Civil
    Engineering
Tashkent, Uzbekistan

P. Kanthavel
ICAR-Central Institute of Agricultural
    Engineering
Bhopal, India

Mahshid Karimi
Sari Agricultural Science and Natural
    Resources University
Sari, Iran

Karishma Khadka
International Centre for Integrated Mountain
    Development (ICIMOD)
Kathmandu, Nepal

Subha Khanal
International Centre for Integrated Mountain
    Development (ICIMOD)
Kathmandu, Nepal

N. Kodandapani
Center for Advanced Spatial and
    Environmental Research (CASER)
Bengaluru, India

Anil Kumar
Department of Soil and Water Conservation
    Engineering, College of Technology,
    G.B. Pant University of Agriculture and
    Technology
Pantnagar, Uttarakhand, India

Navsal Kumar
Shoolini University
Solan, Himachal Pradesh, India

Rohitashw Kumar
Sher-e-Kashmir University of Agricultural
    Sciences and Technology
Kashmir, India

Alban Kuriqi
CERIS, Instituto Superior Técnico,
    Universidade de Lisboa
Lisbon, Portugal

Yi Liu
Hohai University
Nanjing, China

A.K. Lohani
National Institute of Hydrology
Roorkee, India

T. Loidang Chanu
Central Agricultural University
Imphal, India

Rebecca Luna Lucena
Federal University of Rio Grande do Norte
Natal, Brazil

Aribam Priya Mahanta Sharma
Central Agricultural University
Imphal, India

Mousa Maleki
Islamic Azad University of Isfahan
Isfahan, Iran

Anurag Malik
Punjab Agricultural University, Regional
    Research Station, Bathinda
Punjab, India

Rasoul Mirabbasi
Shahrekord University
Shahrekord, Iran

Mirhassan Miryaghoubzadeh
Urmia University
Urmia, Iran

D.B. Moalafhi
Botswana University of Agriculture and
    Natural Sciences
Gaborone, Botswana

G. Krishna Mohan
International Institute of Information
    Technology
Hyderabad, India

Ehsan Moradi
University of Tehran, Iran

Diptimayee Nayak
Indian Institute of Technology Roorkee
Roorkee, India

R.K. Panda
Siksha 'O' Anusandhan University
Bhubaneswar, India

B.P. Parida
Botswana International University of Science
    and Technology
Palapye, Botswana

Saida Parvizi
Isfahan University of Technology
Isfahan, Iran

Ghanshyam T. Patle
Central Agricultural University
Imphal, India

Arunava Poddar
Shoolini University, Solan
Himachal Pradesh, India

Swayam Prava Singh
Soil Conservation Office-Cum-Project
    Director, Watershed, Naupada
Odisha, India

Nejem Raheem
Emerson College
Boston, Massachusetts, USA

M. Rajesh
International Institute of Information
    Technology
Hyderabad, India

Parminder S. Ranhotra
Birbal Sahni Institute of Palaeosciences
Lucknow, Uttar Pradesh, India

Md Mamunur Rashid
University of Southern Mississippi
Ocean Springs, Mississippi, USA

Tayeb Raziei
Soil Conservation and Watershed Management
    Research Institute
Agricultural Research Education and Extension
    Organization
Tehran, Iran

GVS Reddy
University of Agricultural Sciences
Raichur, Karnataka, India

S. Rehana
International Institute of Information
    Technology
Hyderabad, India

Liliang Ren
Hohai University
Nanjing, China

Ipsita Roy
Birbal Sahni Institute of Palaeosciences
Lucknow, Uttar Pradesh, India

Roquia Salam
Begum Rokeya University
Rangpur, Bangladesh

Ahmed Sefelnasr
United Arab Emirates University
Al Ain, United Arab Emirates

Kaka Shahedi
Sari Agricultural Science and Natural
    Resources University
Sari, Iran

F.A. Shaheen
Sher-e-Kashmir University of Agricultural
    Sciences & Technology of Kashmir
Shalimar-Srinagar, Jammu and
    Kashmir, India

Vijay Shankar
National Institute of Technology Hamirpur
Hamirpur, India

Priyank J. Sharma
Indian Institute of Technology Indore
Indore, India

Ashutosh Sharma
Indian Institute of Technology Roorkee
Roorkee, India

Mayank Shekhar
Birbal Sahni Institute of Palaeosciences
Lucknow, Uttar Pradesh, India

Mohsen Sherif
United Arab Emirates University
Al Ain, United Arab Emirates

Ayushi Singh
Birbal Sahni Institute of Palaeosciences
Lucknow, Uttar Pradesh, India

Gurjeet Singh
Indian Institute of Technology Bhubaneswar
Bhubaneswar, India
and
Michigan State University
East Lansing, Michigan, USA

Deepak Singh Bisht
WHRC - Jammu, National Institute of
   Hydrology
Roorkee, India

Ghanashyam Singh Yurembam
Central Agricultural University
Imphal, India

R.K. Srivastava
University of Hohenheim
Stuttgart, Germany

Yazid Tikhamarine
Department of Science and Technology,
   University of Tamanrasset, Sersouf
Tamanrasset, Algeria
and
Southern Public Works Laboratory,
   Tamanrasset Unity
Tamanrasset, Algeria

Nidhi Tomar
Birbal Sahni Institute of Palaeosciences
Lucknow, Uttar Pradesh, India

Fatih Tosunoglu
Erzurum Technical University
Erzurum, Turkey

Donald A. Wilhite
University of Nebraska
Lincoln, Nebraska, USA

Nilufa Yeasmin
Begum Rokeya University
Rangpur, Bangladesh

Francisco Zambrano Bigiarini
Universidad Mayor
Santiago, Chile

Mansoor Zargar
Vali-e-Asr University of Rafsanjan
Rafsanjan, Iran

1 Spatial and Temporal Linkages between Large-Scale Atmospheric Oscillations and Hydrologic Drought Indices in Turkey

Fatih Tosunoglu, Ercan Kahya, and Mohammad Ali Ghorbani

CONTENTS

1.1 INTRODUCTION

Drought is known as a complex phenomenon due to the subjectivity function of the field of study and the development of the system. Drought can be characterized by a prolonged deviation from the normal conditions of variables, such as precipitation, streamflow, groundwater, and soil moisture. Among the common drought categories, the meteorological drought, which is often considered the precursor to other types, generally occurs when precipitation is significantly lower than normal. It is always an important task to identify the onset and termination of a drought (Tosunoglu et al. 2018). Hence, drought status is generally defined using various indicators for analysis (Mishra and Singh

DOI: 10.1201/9781003276548-1

1

2011). Among these drought indicators, the Standardized Precipitation Index (SPI), proposed by McKee et al. (1993), is the most popular definition because it has many advantages.

However, if streamflow (or lake, reservoir level, and groundwater level) is of primary interest, it is called as hydrologic drought, which is mainly related to the streamflow deficit with respect to normal conditions (Talaee et al. 2014). Among all the drought types, hydrological drought is considered the most important type requiring further scientific attention, as human life is vitally dependent on water resources. Frequently used hydrological drought indices are the Standardized Streamflow (Runoff) Index (SSI), Streamflow Drought Index (SDI), Palmer Hydrological Drought Index (PHDI), Surface Water Supply Index (SWSI), and index-based low flows. Over the last decade, several researchers have applied these indices to evaluate the spatial and temporal variability of meteorological and hydrological droughts in different regions of the world (Lorenzo-Lacruz et al. 2010; Mo 2008; Tosunoglu and Can 2016; Vicente-Serrano et al. 2012). Analysis of the linkages between droughts and atmospheric oscillations helps us understand the climate and hydrological mechanisms driving the development of droughts. Recent studies have indicated that many droughts, which occurred over different parts of the world, are influenced by large-scale atmospheric circulations (e.g., Southern Oscillation (SO), North Atlantic Oscillation (NAO), and Atlantic Multidecadal Oscillation (AMO)). For instance, Sarlak et al. (2009) analyzed the impact of the NAO on the probability distribution functions of critical droughts in Göksu River, Turkey. They revealed the NAO has remarkable impacts on the transition probabilities and expectation of the critical drought duration. Mo (2008) investigated the impacts of the ENSO (El Niño–Southern Oscillation) and AMO on drought over the United States. They used the six-month SPI, SSI, and soil moisture anomalies to represent drought events. Talaee et al. (2014) analyzed the relationship between hydrological droughts in Iran and the NAO and SO. They showed that the connection between the SSI and SOI (Southern Oscillation Index) was more powerful than the corresponding relations with NAO indices. Tosunoglu et al. (2018) investigated teleconnections between meteorological droughts and the large-scale atmospheric circulation indices in Turkey. The linkages between large-scale atmospheric patterns and drought indices that are extracted from climate and hydrological variables (precipitation, streamflow, temperature, etc.) have been widely studied in the world (Kalayci et al. 2004; Karabork and Kahya 2003; Marti et al. 2010; Turkes and Erlat 2003). In this chapter, the spatial and temporal linkages between large-scale atmospheric oscillations and hydrologic drought indices are of primary interest.

1.2   LARGE-SCALE ATMOSPHERIC OSCILLATIONS

1.2.1   North Sea–Caspian Pattern (NCP)

The North Sea–Caspian Pattern (NCP), which was defined by Kutiel and Benaroch (2002), is an upper-level (500 hPa) atmospheric teleconnection between the North Sea (0°–10° E, 55° N) and Caspian Sea (50°–60° E, 45° N). The authors also reported an index (the North Sea–Caspian Pattern Index (NCPI)) that measures the geopotential height difference between the two aforementioned locations (Figure 1.1). The NCPI value below −0.5 is defined as belonging to the negative phase, which implies an increased counterclockwise anomaly circulation around its western pole and an increased clockwise anomaly circulation around its eastern pole. The opposite conditions occur when the NCPI is larger than 0.5. In the negative phase, it is expected to have an increased westerly anomaly circulation toward central Europe, and an increased easterly anomaly circulation toward Georgia, Armenia, and eastern Turkey causing an increased southwesterly anomaly circulation toward the Balkans and western Turkey. In the positive phase, there is an increased northwesterly circulation toward Eastern Europe, and an increased northeasterly circulation toward the Black Sea that results in an increased northeasterly anomaly circulation toward the Balkans (Ghasemi and Khalili 2008; Kutiel 2011). Kutiel and Benaroch (2002) pointed out that the NCPI has a slightly better capability than the other large-scale atmospheric teleconnection indices (such as the North

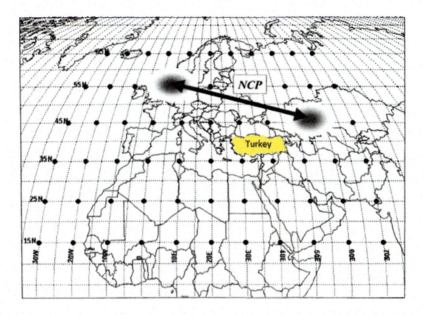

FIGURE 1.1 Location map for poles of the North Sea–Caspian Pattern (Kutiel and Benaroch 2002).

Atlantic Oscillation Index (NAOI) and Southern Oscillation Index (SOI)) to differentiate between lower or above normal temperatures. Therefore, the use of the NCPI in forecasting studies might be preferable over the NAOI and SOI. Kutiel and Turkes (2005) also pointed out that the Anatolian Peninsula and the Middle East are largely affected by the NCP, which is mainly active during October–April.

1.2.2 ARCTIC OSCILLATION (AO)

The Arctic Oscillation (AO) is a large-scale atmospheric oscillation characterized by atmospheric pressure changes between the Arctic region and middle latitudes in the Northern Hemisphere. The AO is known as one of the most important controllers of long- and short-term climatic variability in the North Atlantic, European, and Mediterranean basins, such as the North Atlantic Oscillation. The AO has positive and negative phases (Figure 1.2). When it is in its positive phase, a ring of strong winds circulating the North Pole acts to confine colder air across polar regions. This belt of winds becomes weaker and more distorted in the negative phase of the AO, which allows easier southward penetration of colder, Arctic air masses and increased storminess into the midlatitudes (https://www.ncdc.noaa.gov/teleconnections/ao/).

1.2.3 NORTH ATLANTIC OSCILLATION (NAO)

The NAO is the most dominant large-scale atmospheric oscillation pattern affecting the climate of Europe, especially in the winter season. The NAOI is generally considered as the difference in the normalized sea level atmospheric pressures between the stations located near the Azores and near Iceland. There are several NAO indices available in the literature and provided by researchers that have implemented different methodological approaches (e.g., Barnston and Livezey 1987; Hurrell 1995; Jones et al. 1997). The NAOI is positive (negative) when the pressure difference between the two locations is large (small), as shown in Figure 1.3. The NAO index used in this study is calculated from the difference between the standardized pressure anomalies measured at Gibraltar (south of Spain) and Reykjavík (Iceland). The advantage of using these stations is discussed in depth by

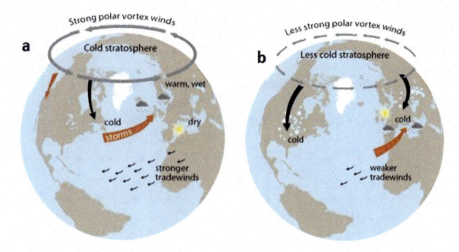

FIGURE 1.2   The phases of the Arctic Oscillation: (a) positive and (b) negative (https://www.amap.no/docu-ments/doc/the-arctic-oscillation-and-circulation-positive-and-negative-phase/947).

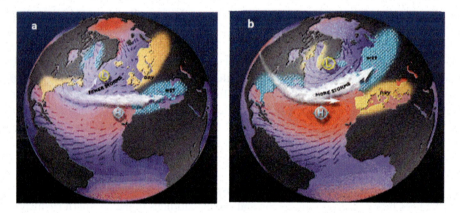

FIGURE 1.3   (a) Negative and (b) positive phases of the North Atlantic Oscillation (https://www.ldeo.colum-bia.edu/res/pi/NAO).

Jones et al. (1997), who noted that Gibraltar appears to better represent the southern part of the NAO dipole than other commonly used stations, such as Lisbon or Ponta Delgada in the Azores (Lorenzo-Lacruz et al. 2011)**.** The monthly NAO indices were supplied by the Climatic Research Unit (CRU), University of East Anglia (http://www.cru.uea.ac.uk/). When this index is positive, it causes stronger winter storms over the Atlantic Ocean on a more northerly track and yields warm and wet winters in northern Europe while the Mediterranean experiences dry and cold winters. On the other hand, during the negative phase, the opposite conditions dominate over northern Europe and the Mediterranean (Karabork et al. 2005).

1.2.4   SOUTHERN OSCILLATION (SO)

The SO is a natural part of the global climate system and results from large-scale interactions between the ocean and atmosphere that occur mainly across its core region in the tropical-subtrop-ical Pacific Ocean basin (Allan 2000). The SOI is considered a good representative of the atmo-spheric state of the SO phenomena and it is simply defined as the difference in the normalized sea level atmospheric pressure between Tahiti and Darwin, Australia (Karabork and Kahya 2009)

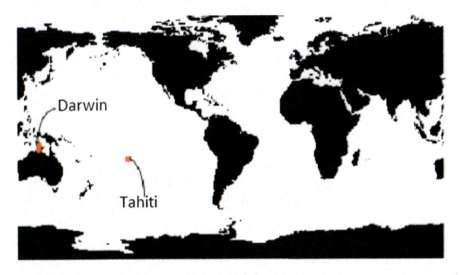

FIGURE 1.4 Location map of Tahiti and Darwin where SO mainly occurs.

(Figure 1.4). The extreme phases of the SO, called the El Niño and La Niña, have been shown to have significant impacts all over the world. The former, which is also known as the warm phase, is characterized by low atmospheric pressure throughout the eastern Pacific with positive sea surface temperature anomalies, while the La Niña (cold phase) has the opposite conditions (Tosunoglu et al. 2018). Although the SO has severe impacts on the tropic latitudes, in particular, for countries that are affected by monsoons (such as India, Indonesia, Africa, and Australia), its influence has been detected in different locations all over the globe (Bhuvaneswari et al. 2013).

1.2.5 COMPUTATION OF HYDROLOGIC DROUGHT INDICES

In recent years, two hydrological drought indices, namely the Streamflow Drought Index (SDI) and the Standardized Streamflow (Runoff) Index (SSI), have been proposed for hydrological drought studies. The computing of these indices is very similar to each other as the computation of the hydrological drought index requires first transforming monthly streamflow values into z-scores. To compute more precise SDI or SSI values, the best suitable probability distribution for streamflow data is precisely identified. Then, the SDI, which is used in this study, can be computed as follows:

$$SSI = -\left(t - \frac{C_0 + C_1 t + C_2 t^2}{1 + d_1 t + d_2 t^2 + d_3 t^3}\right) for \rightarrow 0 < H(x) \le 0,5 \tag{1.1}$$

$$SSI = +\left(t - \frac{C_0 + C_1 t + C_2 t^2}{1 + d_1 t + d_2 t^2 + d_3 t^3}\right) for \rightarrow 0,5 < H(x) \le 1 \tag{1.2}$$

where

$$t = \left(\sqrt{\ln\left(\frac{1}{(H(x))^2}\right)}\right) for \rightarrow 0 < H(x) \le 0,5 \tag{1.3}$$

$$t = \left(\sqrt{\ln\left(\frac{1}{1 - (H(x))^2}\right)}\right) for \rightarrow 0,5 < H(x) \le 1 \tag{1.4}$$

in which $C_0 = 2,515517$, $C_1 = 0,802853$, $C_1 = 0,010328$, $d_1 = 1,432788$ $d_2 = 0,189269$, $d_3 = 0,001308$, and $H(x)$ is the cumulative distribution function of the best-fitted distribution. If the selected probability distribution is correct, the mean and standard deviation value of the SSI time series must be 0 and 1, respectively.

1.3 RESULTS

1.3.1 DATA ANALYSIS

To illustrate a numerical example, the geographic location of Turkey, which is situated in the Northern Hemisphere with bounding coordinates 25° E–42° N and 45° E–36° N, was selected as a case study. In conducting a hydrological drought analysis, monthly streamflow records of 96 gauge stations were provided by the General Directorate of State Hydraulic Works, Turkey. The period 1969–2004 was selected as the study period in this research. Three stations having missing stream-flow records with less than 2.5% were estimated using simple and multivariate linear regression models. The data of the nearest neighboring stations having a strong correlation with the troubled station were used to obtain precise estimation. The determination (R^2) and root mean-squared error (RMSE) coefficients, which are the widely used performance criteria, were employed to evaluate the estimation results. To ensure that the streamflow data can be reliably used for further analysis, the homogeneity of 96 stations data series was tested by the standard normal homogeneity test (SNHT), Buishand test, and Pettitt test. It should be noted that each homogeneity test includes a different ability to catch inhomogeneity (breaks) in the considered time series. For instance, the SNHT can detect breaks at the beginning and the end of the series, while the Pettitt and Buishand tests can catch inhomogeneity years in the middle of the series (Tosunoglu et al. 2018; Arıkan and Kahya 2019). The result of each test was analyzed at a significance level of 95% and the years causing inhomogeneous were defined, and all results are given in Table 1.1. Out of 96 stations, 19 were found to be inhomogeneous, as data of these stations failed according to at least one of the three tests. Subsequently, the double-mass curve technique was implemented in each of the 19 inhomogeneous stations to see if there is any possibility to get their data homogeneous. The double-mass curve for each station was drawn by putting the annual mean streamflow of inhomogeneous stations (on the ordinate axis) against the corresponding average of the annual mean streamflow of neighboring homogeneous stations (on the horizontal axis). A minimum of four neighboring stations were used in this analysis.

The results indicated that 7 out of 19 inhomogeneous stations passed all three tests after being applied to correction based on the double-mass curve method and the geographical distribution of these stations is illustrated in Figure 1.5. One example (station 1413) of the double-mass curve-based correction is given in Figure 1.6a, b. As can be seen from Figure 1.6a, there is a sudden change (jump) in the mean after the year 1993 and this implies an anthropogenic influence. The detected anthropogenic influence in streamflow data of this station was successfully corrected using the double-curve mass technique (Figure 1.6b). The homogeneity of the new data series was tested by the SNHT, Buishand, and Pettitt tests, and the new data series were found to be homogeneous at the 5% significance level. In this example, streamflow data of the four neighboring stations (1401, 1402, 1418, and 1424) were used for the correction method. Finally, streamflow data of 84 gauge stations were found to be homogeneous and the locations of these stations with corresponding identification numbers are presented in Figure 1.7.

1.3.2 COMPUTING SSI SERIES

In this part of the study, hydrological drought indices (SSI) are computed using monthly streamflow data of the selected stations. To obtain more precise SSI values, a probability density function that best fits monthly streamflow data at each station was first defined. For this purpose, eight commonly

TABLE 1.1

Homogeneity Test Results Obtained by SNHT, Buishand, and Pettitt Tests

Station No.	Test Statistic (Break Year)		
	Pettitt Test (K_T)	SNHT (T_0)	Buishand Test (Q)
311	253 (1987)	15.524 (1987)	11.969 (1987)
510	259 (1987)	13.770 (1984)	11.220 (1984)
514	235 (1987)	12.362 (1987)	10.697 (1986)
518	289 (1987)	17.802 (1984)	12.578 (1984)
601	260 (1984)	14.608 (1984)	11.547 (1984)
706	289 (1987)	17.661 (1985)	12.767 (1985)
713	266 (1988)	15.176 (1971)	11.449 (1987)
808	294 (1988)	18.035 (1985)	12.901 (1985)
1203	261 (1987)	15.163 (1985)	11.829 (1985)
1218	277 (1987)	—	11.469 (1985)
1224	310 (1986)	17.991 (1985)	12.885 (1985)
1413	209 (1993)	12.785 (1993)	10.023 (1993)
1414	249 (1989)	16.704 (1989)	12.261 (1989)
1501	—	9.460 (2000)	—
1517	159 (1995)	—	8.319 (1995)
1622	222 (1990)	12.656 (2000)	10.280 (1990)
1714	206 (1968)	9.231 (1968)	9.187 (1968)
1719	192 (1982)	9.316 (1996)	8.446 (1989)
1907	310 (1988)	20.625 (1988)	13.272 (1988)

Source: Tosunoglu (2014).

*Critical values at the 95% confidence level are 8.009, 8.04, and 157 for the SNHT, Buishand, and Pettitt tests, respectively (Buishand 1982; Wijngaard et al. 2003).

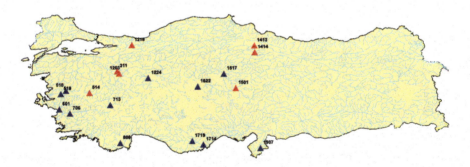

FIGURE 1.5 Geographical distribution of inhomogeneous gauge stations indicated by SNHT, Buishand, and Pettitt tests (red points indicate adjusted stations using the double-mass curve method) (Tosunoglu 2014).

applied distributions in hydrology, namely, the normal (NORM), lognormal (LN2), Pearson type III (P3), logistic (LOGS), gamma (G2), the general extreme value (GEV), generalized Pareto (GPA), and Weibull (WBL) were fitted to streamflow data. To define the most suitable one, Akaike information criterion (AIC) was employed, and the distributions, which provided minimum AIC value, were selected as the best-fitted distributions for monthly streamflow series. The spatial and temporal distribution of the probability distributions providing the best fit to the monthly streamflow of

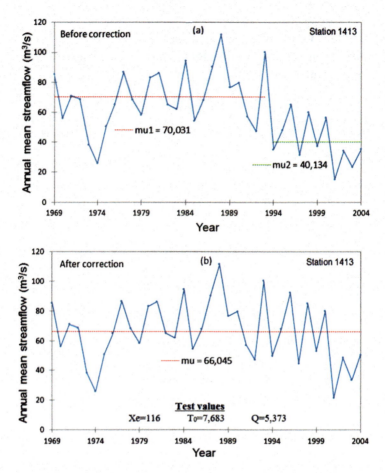

FIGURE 1.6 Homogeneity test graphs (a) before and (b) after the application of the double-mass curve method.

FIGURE 1.7 Geographical distribution of homogeneous gauge stations with the related numbers (Tosunoglu 2014).

each station is presented in Figure 1.8. It is seen from the monthly maps that there is no specific distribution providing a better overall fit for a particular month or region. Figure 1.8 moreover demonstrates no clear spatial pattern in the probability distributions provided the most suitable distribution to the streamflow time series. Even gauge stations along the same river course follow different

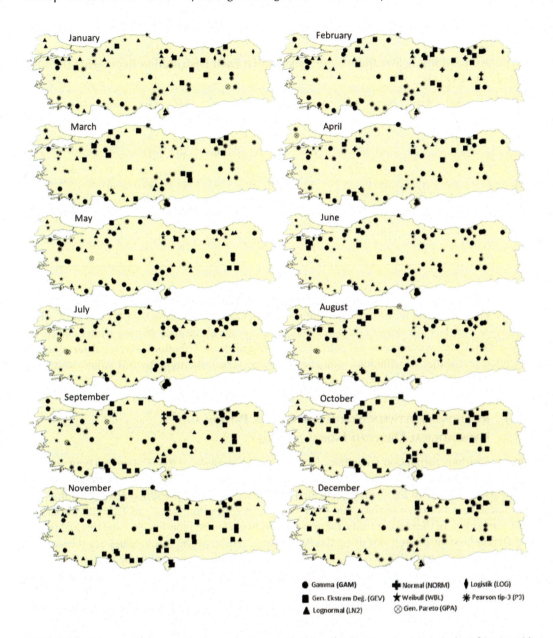

FIGURE 1.8 Spatial and temporal evaluation of the probability distributions providing the best fit to monthly streamflow data (Tosunoglu 2014).

distributions. The main reason behind this can be attributed to the physical characteristics, climate, hydrological regimes, and hierarchy of Turkey's river basins. Table 1.2 represents the frequency with which each distribution has been chosen for each month by the AIC. The numbers show the percentage of the monthly streamflow series in which the AIC values are the lowest. According to Table 1.2, the gamma distribution provided the best fit for 24% of the series; the lognormal distribution showed the best fit for 22% of the series; and the GEV, WBL, and P3 distributions showed the best fit for 20%, 17%, and 10% of the series, respectively. The three least suitable were the NORM, LOGS, and GPA, which provided the best fit for 2%, 1%, and 1% of the series, respectively. Some seasonal differences were also observed. For instance, the gamma distribution provided the best fit

TABLE 1.2

Percentage of Monthly Streamflow Series in Which Each Distribution Represents the Best Fit According to AIC

Distribution	Jan	Feb	Mar	Apr	May	Jun	Jul	Aug	Sept	Oct	Nov	Dec	Annual
Gamma	11	23	14	24	35	39	37	34	27	18	17	11	24
GEV	10	23	17	18	6	10	7	11	26	56	43	18	20
LN2	39	24	18	10	24	13	22	8	19	14	29	42	22
NORM	5	4	0	5	2	2	2	6	7	2	0	0	3
WBL	15	10	25	32	23	30	26	20	11	10	4	2	17
GPA	1	0	0	0	0	0	5	4	2	0	0	0	1
LOGS.	0	1	0	0	1	1	1	0	1	0	0	0	0
LLOGS	0	1	8	0	1	1	0	8	1	0	0	0	2
P3	19	14	18	11	8	4	0	9	6	0	7	27	10

in the highest percentage of stations in the summer; the GEV distribution showed the most suitable fit in the highest percentage of stations in autumn (especially late autumn); the LN2 and WBL distributions showed the most suitable fit in the highest percentage of the station in the winter and spring, respectively. There was no seasonal pattern for the rest of the distributions. Having determined the best-fitted probability distributions of streamflow time series, the monthly SSI values were computed by using the procedure explained in the previous sections.

1.3.3 RELATIONSHIP BETWEEN NORTH SEA–CASPIAN PATTERN AND HYDROLOGICAL DROUGHT INDICES IN TURKEY

There are various correlation techniques for defining the strength of the relationship between variables. Among these methods, the Pearson, Kendall, and Spearman's rank correlation are the most widely used methods in the hydrology field. The Pearson correlation method always requires normal distribution of both variables, but there is no such restriction for use of Kendall/Spearman's rank correlation. In the present study, the correlations between the monthly NCP indices and the SSI in different lags were computed using the Pearson correlation as the index values are the standard normal variables. The monthly NCP indices used in this study were supplied by the CRU.

Figure 1.9 and Figure 1.10 show the computed lag-zero and lag-one correlations, respectively, between the monthly SSI and the monthly NCPI at each station for the period 1969–2004. The Student's t-test was used to evaluate the statistical significance of the calculated correlation values (at the significance level of 0.05). Here, monthly correlation maps were drawn by ArcGIS software, and dark blue/dark red circle points were used to define significant negative and positive correlations, respectively. It is evident in Figure 1.9 that significant negative correlations (lag zero) occurred during January, February, and March, while significant positive relationships mostly occurred in April. Moreover, a few significant positive correlations were detected during August, October, and November. In January, the west part of Turkey was strongly affected by the NCP and the lag-zero negative correlations between the NCPI and SSI took the value of 0.50–0.55 especially the stations located in the southwest part. In February, a dominant spatial and temporal pattern with significant negative correlations mostly occurred in the middle and east parts of Turkey. During March, the significant negative correlations were more regional and poorer. In April, statistically positive correlations were observed between the NCPI and SSI values of the stations mostly located in northwestern Turkey.

Based on the lag-one correlations indicated in Figure 1.10, significant negative relations appeared between the January NCPI–February SSI, February NCPI–March SSI, and May NCPI–June SSI

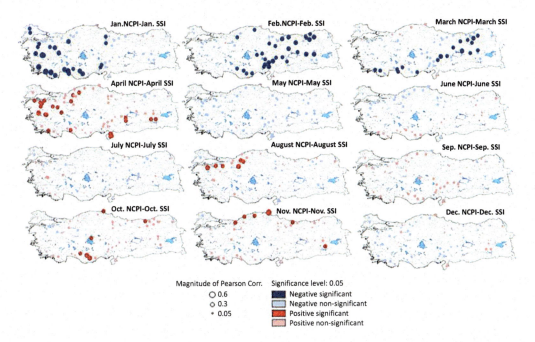

FIGURE 1.9 Spatial and temporal distribution of the lag-zero correlations between the monthly NCPI and SSI (Tosunoglu 2014).

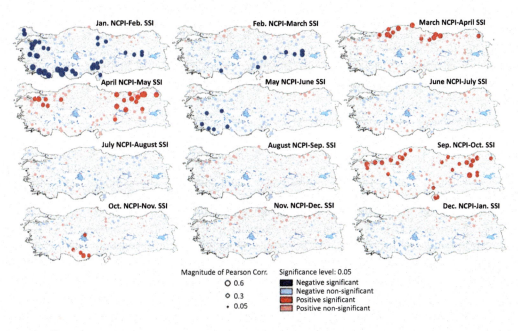

FIGURE 1.10 Spatial and temporal distribution of the lag-one correlations between the monthly NCPI and SSI (Tosunoglu 2014).

values. The computed Pearson correlations between the January NCPI and February SSI were more dominant in the western and southwestern parts of Turkey, and the correlations take on remarkable values of 0.65 for the station located on the southwestern coast. We point out that 43% of the variance of SSI values can be explained by NCP for this region. Moreover, significant positive

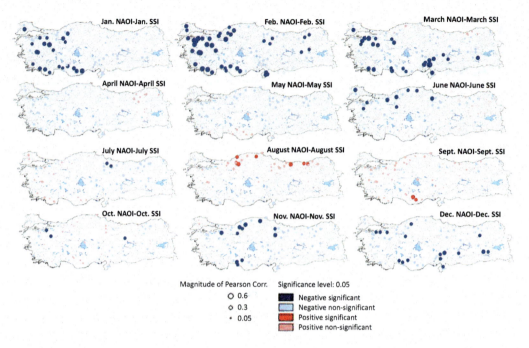

FIGURE 1.11 Spatial and temporal distribution of the lag-zero correlations between the monthly NAOI and SSI (Tosunoglu 2014).

connections mostly occurred between March, April, and September NCPI and one-month ahead SSI values. Between April NCPI and May SSI, the computed positive correlations take on remarkable values of 0.50–0.55. Significant correlations between the February NCPI–March SSI values occurred in the stations that take place in the basins near the northern coasts.

1.3.4 RELATIONSHIP BETWEEN NORTH ATLANTIC OSCILLATION AND HYDROLOGICAL DROUGHT INDICES IN TURKEY

The spatial and temporal distribution of the Pearson correlations between the monthly NAOI and SSI at each station for the period 1969–2004 is shown in Figure 1.11. The significant negative correlations occurred during the months of December, January, February, and March, in which the NAO is more dynamically active. Very strong and significant correlations of the NAOI with the SSI were mostly detected at the stations located in the western parts of Turkey. These correlations for January exceed −0.55 at the stations located in the northwestern regions. The correlations with higher lags (e.g., lag-one, lag-two) were also computed, but the maps are not shown here as the results did not contain striking outcomes.

1.3.5 RELATIONSHIP BETWEEN ARCTIC OSCILLATION AND HYDROLOGICAL DROUGHT INDICES IN TURKEY

Figure 1.12 shows the spatial and temporal distribution of lag-zero correlations between monthly SSI values and AO indices for the period 1969–2004. The monthly AO indices used in this study were supplied by Climate Prediction Center, USA (https://www.cpc.ncep.noaa.gov/). It can be seen from the figure that significant relationships, which were negative, were mostly dominant during January, February, and March. For these months, significant negative correlations were mainly dominant in the west and southwest parts of Turkey. The computed correlations, for some stations, have a remarkably high value of around −0.60, which indicates a strong relationship.

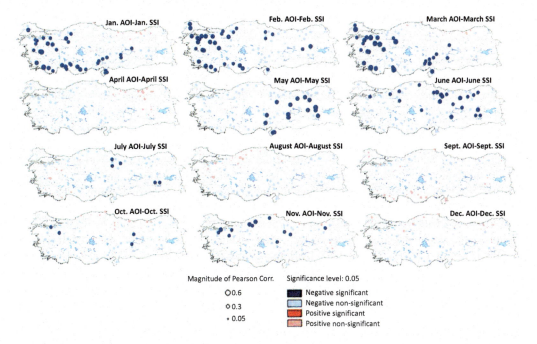

FIGURE 1.12 Spatial and temporal distribution of the lag-zero correlations between the monthly AOI and SSI (Tosunoglu 2014).

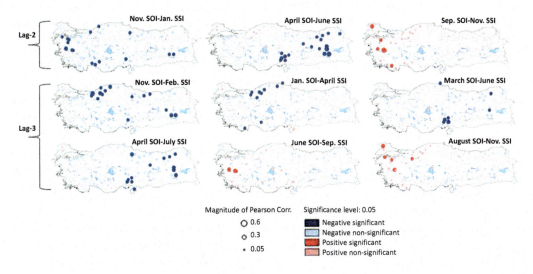

FIGURE 1.13 Spatial and temporal distribution of the computed correlations between the SOI and SSI (Tosunoglu 2014).

1.3.6 RELATIONSHIP BETWEEN SOUTHERN OSCILLATION AND HYDROLOGICAL DROUGHT INDICES IN TURKEY

To define spatial and temporal connections between the SO and hydrologic droughts, the simultaneous and lagged correlations for each month were computed. The monthly SO indices used in this study were supplied from the CRU. Among various types of lag correlations (from zero to five), lag-two and lag-three were found to be the most suitable lags that can demonstrate the impact of SOI on drought indices. Because of the scarcity of space, only monthly maps having significant correlations are presented in Figure 1.13. Negative relationships were dominant, along with no strong seasonality.

1.4 DISCUSSION AND CONCLUSION

Spatial and temporal distributions of the linkages between large-scale atmospheric oscillations (namely, the North Sea–Caspian Pattern, North Atlantic Oscillation, Arctic Oscillation, and Southern Oscillation) and hydrological droughts in Turkey are investigated in this study. Each monthly time indicator of the oscillations are described in an index format, the so-called NCPI, NAOI, AOI, and SOI, while the SSI is employed to define the hydrological drought indicator. The monthly streamflow records covering the period 1969–2004 from 96 stations uniformly distributed across Turkey are used for the analysis. Unlike previous studies that have often fitted unique probability distribution to calculate SSI values, the suitability of different distributions was evaluated in this study. The results showed that there is no evidence that the unique mode of distribution performed better than other candidates for any certain month or area. Moreover, the monthly streamflow records were best fitted by gamma, GEV, and LN2 distributions in many cases. Having the advantage of these distributions, more reliable values of the drought indices were generated. Monthly and one-month lag correlations were then calculated to evaluate the spatial and temporal distribution of the oscillation impact on hydrological drought in Turkey. The lag-zero monthly correlation maps obtained from ArcGIS software showed that the linkages between NCPI and SSI are more significant during the months of January/February/March and April. Significant negative correlations were mostly observed in January/February and March, while significant positive relations appeared in April, August, October, and November. From the monthly lag-one correlation maps, it was observed that the significant negative correlations between January NCPI and February SSI take remarkable values. These high correlations are seen at the stations mainly located in the Aegean and Mediterranean regions of Turkey. Moreover, significant positive connections reached notable values between April NCPI and May SSI. These connections are mostly observed for the stations located on the east Black Sea coast and northeast part of Turkey. The northeast part of Turkey has recently gained a growing and greater interest in the country economy. During the last decade, many large hydroelectric power plants (HEPPs) and dams have already been constructed (e.g., Deriner Dam and HEPP) or are under construction (e.g., Yusufeli Dam and HEPP) in this region. A key recommendation to researchers, who are acting in planning and managing water resources projects, agriculture, and land-use planning, should also consider the influence of the NCP to obtain more precise and reliable results. From the monthly correlation maps between NAOI and SSI, it was observed that the high correlations (negative) mostly occurred during the winter months and the impacts of the NAO were found to be more dominant in the west, southwest, and northwest parts of the country. Similar relationships were observed for the AOI and SSI analyses. The results indicated that there exists significant negative correlations between the two variables especially in the months January, February, March, May, and June. From the monthly correlation maps of the SOI–SSI linkage, we discovered that there is not a strong seasonality, although the relationships are mostly negative. A key finding of the SOI–SSI linkage is that the impacts of the SO are more detectable in higher lags (lag-two and lag-three), which has much importance in drought monitoring, modeling, and forecasting studies in Turkey. The indications of this study are consistent with those of Kahya (2011).

ACKNOWLEDGMENTS

The authors sincerely thank the General Directorate of State Hydraulic Works, Turkey, for providing the streamflow data used in the study.

REFERENCES

Allan RJ (2000) *ENSO and Climatic Variability in the Past 150 Years.* Cambridge, UK: Cambridge University Press.

Arıkan BB, Kahya E (2019) Homogeneity revisited: Analysis of updated precipitation series in Turkey. *Theor Appl Climatol* 135(1–2):211–220. https://doi.org/10.1007/s00704-018-2368-x

Barnston AG, Livezey RE (1987) Classification, seasonality and persistence of low-frequency atmospheric circulation patterns. *Mon Wea Rev* 115:1083–1126.

Bhuvaneswari K, Geethalakshmi V, Lakshmanan A, Srinivasan R, Sekhar NU (2013) The impact of El Niño/Southern oscillation on hydrology and rice productivity in the Cauvery Basin, India: Application of the soil and water assessment tool. *Weather Clim Extremes* 2:39–47.

Buishand TA (1982) Some methods for testing the homogeneity of rainfall records. *J Hydrol* 58:11–27.

Ghasemi AR, Khalili D (2008) The effect of the North Sea-Caspian pattern (NCP) on winter temperatures in Iran. *Theor Appl Climatol* 92:59–74.

Hurrell JW (1995) Decadal trends in the North Atlantic Oscillation: Regional temperatures and precipitation. *Science* 269:676–679.

Jones PD, Jonsson T, Wheeler, DA (1997) *Monthly Values of the North Atlantic Oscillation Index from 1821 to 2000.* PANGAEA.

Kahya E (2011) Impacts of the NAO on the hydrology of the Eastern Mediterranean in "hydrological, socio-economic and ecological impacts of the North Atlantic oscillation in the mediterranean region". Eds: SM Vicente-Serrano and RM Trigo. *Advances in Global Change Research*, Vol. 46, pp. 57–71, Springer.1st Edition, 2011, VIII, 236 p. https://doi.org/10.1007/978-94-007-1372-7.

Kalayci S, Karabork MC, Kahya E (2004) Analysis of El Nino signals on Turkish streamflow and precipitation patterns using spectral analysis. *Fresen Environ Bull* 13:719–725.

Karabork MC, Kahya E (2003) The teleconnections between the extreme phases of the southern oscillation and precipitation patterns over Turkey. *Int J Climatol* 23:1607–1625.

Karabork MC, Kahya E (2009) The links between the categorised Southern Oscillation indicators and climate and hydrologic variables in Turkey. *Hydrol Process* 23:1927–1936.

Karabork MC, Kahya E, Karaca M (2005) The influences of the Southern and North Atlantic Oscillations on climatic surface variables in Turkey. *Hydrol Process* 19(6):1185–1211. https://doi.org/10.1002/hyp.5560

Kutiel H (2011) A review on the impact of the North Sea-Caspian Pattern (NCP) on temperature and precipitation regimes in the Middle East. *Env Earth Sci-Ser* 11:1301–1312.

Kutiel H, Benaroch Y (2002) North Sea–Caspian Pattern (NCP)—An upper level atmospheric teleconnection affecting the eastern Mediterranean: Identification and definition. *Theor Appl Climatol* 71(1–2):17–28. https://doi.org/10.1007/s704-002-8205-x

Kutiel H, Turkes M (2005) New evidence for the role of the North Sea - Caspian Pattern on the temperature and precipitation regimes in continental Central Turkey. *Geogr Ann A* 87a:501–513.

Lorenzo-Lacruz J, Vicente-Serrano SM, Lopez-Moreno JI, Begueria S, Garcia-Ruiz JM, Cuadrat JM (2010) The impact of droughts and water management on various hydrological systems in the headwaters of the Tagus River (central Spain). *J Hydrol* 386:13–26.

Lorenzo-Lacruz J, Vicente-Serrano SM, López-Moreno JI, Gonzalez-Hidalgo JC, Moran-Tejeda E (2011) The response of Iberian Rivers to the North Atlantic Oscillation. *Hydrol Earth Syst Sci* 15:2581–2597. https://doi.org/10.5194/hess-15-2581-2011

Marti AI, Yerdelen C, Kahya E (2010) Enso modulations on streamflow characteristics. *Earth Sci Res J* 14:31–43.

McKee TB, Doesken N, Kleist J (1993) The relationship of drought frequency and duration to time scales. In: *Proceedings of the IX Conference on Applied Climatology.* Boston, MA: American Meteorological Society, pp. 179–184.

Mishra AK, Singh VP (2011) Drought modeling: A review. *J Hydrol* 403(1–2):157–175. https://doi.org/10.1016/j.jhydrol.2011.03.049.

Mo KC (2008) Model-based drought indices over the United States. *J Hydrometeorol* 9:1212–1230.

Sarlak N, Kahya E, Beg OA (2009) Critical drought analysis: Case study of Goumlksu River (Turkey) and North Atlantic oscillation influences. *J Hydrol Eng* 14:795–802.

Talaee PH, Tabari H, Ardakani SS (2014) Hydrological drought in the west of Iran and possible association with large-scale atmospheric circulation patterns. *Hydrol Process* 28:764–773.

Tosunoglu F (2014) *Investigating the Relationship Between Atmospheric Oscillations and Meteorological and Hydrological Droughts in Turkey*, PhD thesis, Atatürk University, Turkey.

Tosunoglu F, Can I (2016) Application of copulas for regional bivariate frequency analysis of meteorological droughts in Turkey. *Nat Hazards* 82:1457–1477.

Tosunoglu F, Can I, Kahya E (2018) Evaluation of spatial and temporal relationships between large-scale atmospheric oscillations and meteorological drought indexes in Turkey. *Int J Climatol* 38:4579–4596.

Turkes M, Erlat E (2003) Precipitation changes and variability in turkey linked to the North Atlantic oscillation during the period 1930–2000. *Int J Climatol* 23:1771–1796.

Vicente-Serrano SM, Lopez-Moreno JI, Begueria S, Lorenzo-Lacruz J, Azorin-Molina C, Moran-Tejeda E (2012) Accurate computation of a streamflow drought index. *J Hydrol Eng* 17:318–332.

Wijngaard JB, Kleink Tank AMG, Konnen GP (2003) Homogeneity of 20th century European daily temperature and precipitation series. *Int J Climatol* 23:679–692. https://doi.org/10.1002/joc.906

2 Spatiotemporal Drought Analysis

Priyank J. Sharma and Ashutosh Sharma

CONTENTS

2.1 BACKGROUND

2.1.1 DROUGHTS

Drought is a natural and recurring climatic phenomenon characterized by a severe reduction in water availability compared to normal conditions, which extends for a long time over a large area (Rossi, 2000). In comparison to other hydrometeorological hazards, such as floods and cyclones, the spatial extent of droughts is usually larger, but the impacts of droughts are nonstructural and difficult to quantify (Obasi, 1994). The effects of drought often accumulate slowly over a considerable period that may remain for years after the termination of the event. But the onset and end of a drought are difficult to determine. Because of this, drought is often referred to as a creeping phenomenon (Tannehill, 1947). If more than 20% of areas in a country is affected by droughts in that year, it is defined as a "drought year" (Sharma and Goyal, 2020).

Droughts have been classified into four major categories, viz., meteorological, hydrological, agricultural, and socioeconomic (Wilhite and Glantz, 1985). The first three drought categories are represented by a physical phenomenon, while the socioeconomic drought deals with the demand and supply of water as tracked within a socioeconomic system. Meteorological drought is defined based on the degree of dryness (i.e., deviation from normal conditions) and the length of the dry period. Precipitation in a region is commonly adopted for the analysis of meteorological droughts. Hydrological drought is defined based on the deficient surface or subsurface water availability affected by reduced amounts of precipitation for prolonged periods. Streamflows in a watershed or basin are usually analyzed for determining hydrological droughts. Agricultural droughts are defined by linking several hydrometeorological characteristics, such as precipitation deficits, soil moisture deficits, deviations in actual and potential evapotranspiration, and the decline in groundwater and

DOI: 10.1201/9781003276548-2

reservoir levels. Socioeconomic droughts are not directly dependent on meteorological factors, but rather the space–time variability of demand and supply of water. The present chapter focuses on exploring the spatiotemporal variability in meteorological droughts.

2.1.2 DROUGHT INDICES

A suite of drought indices has been developed and reported in the literature for the evaluation of drought characteristics (Keyantash and Dracup, 2002; Svoboda and Fuchs, 2016). The details of widely used drought indices are presented in Table 2.1. The Standardized Precipitation Index (SPI) is a common indicator of drought that does not require information about land surface conditions and needs only precipitation data to compute drought properties. It is a normalized score and represents an event departure from the mean, expressed in units of standard deviation. The SPI is simple, spatially invariant, and probabilistic in nature, and is used to estimate the effect of droughts on various water supplies at different timescales such as 3, 6, 9, or 12 months. The SPI represents the number of standard deviations (following a statistical distribution transformed to a normal distribution) above or below that an event happens to be from the long-run mean (Sims et al., 2002). To estimate the SPI, at a n-month timescale (hence, SPI-n), an accumulation window of n-months is applied to a given monthly precipitation time series, following which a statistical distribution is fitted. This facilitates the temporal analysis of drought events. The 6-month timescale SPI may be useful for seasonal drought identification, 12-month SPI for medium-term droughts, and the 24-month SPI for long-term drought analysis (Łabędzki, 2007). In the present study, a nonparametric procedure is adopted to estimate the SPI (Farahmand and AghaKouchak, 2015). In this procedure, instead of a two-parameter gamma distribution (McKee et al., 1993), the empirical probability is adopted to derive nonparametric SPI. The marginal probability of precipitation is derived using the empirical Gringorten plotting position:

$$p(x_i) = \frac{i - 0.44}{n + 0.12} \tag{2.1}$$

where i is the rank of non-zero precipitation data from the smallest, n represents the total number of samples, and $p(x_i)$ is the associated empirical probability. The empirical probabilities are transformed using the standard normal distribution function (ϕ) to derive the SPI:

$$SPI = \phi^{-1}(p) \tag{2.2}$$

2.1.3 DROUGHT INDICES DERIVED USING SPI

The schematic representation of drought characteristics derived from the SPI is shown in Figure 2.1. The drought event (for example, event 1) starts at month t_i, when the SPI value drops below the threshold limit, has a deficit volume or severity S_i (run- sum) that lasts over the deficit duration D (run length), and ends on month t_e (where $t = 1$ starts with the first month of the drought event and continues until the end of that event (over the duration, D). Yevjevich (1967) proposed the theory of runs to define drought characteristics, as shown in Figure 2.1. A run is defined as a portion of time series of drought parameter X_t, in which all values are either below or above the selected truncation level of X_0; accordingly, it is called either a negative run or a positive run. A drought period is assumed as a consecutive number of months where SPI values remain below a threshold (X_0) of −0.8 (Hao and AghaKouchak, 2013). Based on the SPI range, a drought period can be classified as moderate drought (−0.8 to −1.2), severe drought (−1.3 to −1.5), extreme drought (−1.6 to −1.9), and exceptional drought (−2 or less) (Svoboda et al., 2002).

TABLE 2.1

Details of Widely Adopted Meteorological Drought Indices

Drought Index	Developer	Input Parameter(s) Required	Characteristics	Merits	Demerits
Aridity Anomaly Index (AAI)	India Meteorological Department	AET, PET	• Takes water balance into account • Negative (positive) values denote moisture excess (stress)	• Ease in computation • Specific to agriculture applications	• Not applicable for long-term analysis
Deciles Index (DI)	Gibbs and Maher (1967)	P	• Uses entire period of record • Flexibility to compute for different time steps and timescales	• Simple and flexible approach • Useful in both dry and wet situations	• Impact of temperature is not considered • Longer period of record is required for better results
Percent of Normal Precipitation (PNP)	—	P	• Simple to calculate for any time period at different time steps • Minimum 30-year data length is desired	• Quick and easy to apply	• Difficult to make comparisons across different climatic regimes
Standardized Precipitation Index (SPI)	Mckee et al. (1993)	P	• Does not have specific data length requirements • Negative values of SPI signify droughts • Flexibility to calculate for various timescales	• Data requirements are less, hence, easy to calculate • Suitable for all climatic regimes • Can handle missing data records	• Does not account for the temperature component, which is important from the water balance perspective
Weighted Anomaly Standardized Precipitation Index (WASP)	Lyon (2004)	P (monthly and annual)	• Adopts grid-based monthly precipitation at 0.5° × 0.5° resolution • Used mainly in wet tropical regions	• Simpler in computation	• Not suitable for arid regions
Aridity Index (AI)	De Martonne (1925)	T, P	• Used for classification of various climatic regimes • Suited to identify droughts at shorter timescales	• Easy to compute • Flexible for various time steps	• Does not account for the carryover of dryness chronologically

(Continued)

TABLE 2.1 (CONTINUED)
Details of Widely Adopted Meteorological Drought Indices

Drought Index	Developer	Input Parameter(s) Required	Characteristics	Merits	Demerits
Drought Area Index (DAI)	Bhalme and Mooley (1980)	P (monsoon)	• For understanding the monsoon rainfall in India • Contribution of monthly precipitation to monsoon precipitation is quantified	• Suitable for Indian monsoon season	• Lack of universal applicability
Drought Reconnaissance Index (DRI)	Tsakiris and Vangelis (2005)	T, P (monthly)	• Represents simplified water balance equation considering P and PET	• More representative than SPI • Flexibility to calculate for multiple time steps	• PET estimates are solely based on temperature, while other factors are not considered
Palmer Drought Severity Index (PDSI)	Palmer (1965)	T, P (monthly)	• Considers the moisture received and moisture stored in the soil • Better identification of agricultural droughts	• Inclusion of soil data and total water balance approach makes it robust	• Requires continuous data • Drawback in identifying short-term droughts
Rainfall Anomaly Index	Van Rooy (1965)	P	• Addresses agriculture droughts	• Ease in calculation • Can be analyzed for multiple timescales	• Requires continuous data • Data variability should be small
Standardized Precipitation Evapotranspiration Index (SPEI)	Vicente-Serrano et al. (2010)	T, P	• Accounts for temperature influences on drought through water balance computations	• Applicable for all climatic regimes • Derive impact of climate change in model output	• Requires continuous data • Rapidly developing drought situations are not easily identified

Source: Svoboda and Fuchs (2016).

Notes: P – precipitation, T – temperature, AET – actual evapotranspiration, PET – potential evapotranspiration.

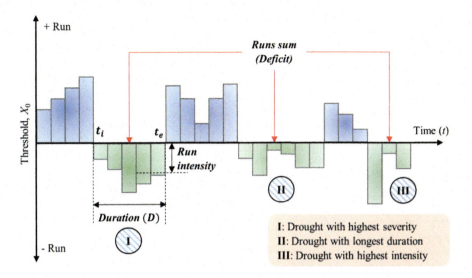

FIGURE 2.1 Schematic representation of drought characteristics estimated using the SPI.

The components of Figure 2.1 are described as follows:

a. Drought initiation time (t_i): It is the start of the water shortage period, which indicates the beginning of the drought event.

b. Drought termination time (t_e): It is the time when the water shortage became sufficiently small so that drought conditions no longer persist.

c. Drought duration (D): It is defined as the number of consecutive intervals (months) where SPI remains below the specified threshold value (SPI <-0.8). Since the drought event is defined at aggregation of the monthly timescale, the minimum duration of drought is one month. In other words, it is time between the initiation and termination of a drought event.

$$D_i = t_i - t_e \qquad \forall SPI < -0.8 \tag{2.3}$$

d. Drought severity (S): It indicates cumulative values of SPI within the drought duration. For convenience, the severity of drought event i, $S_i (i = 1,2,\ldots,n)$ is taken to be positive, which is given by (McKee et al., 1993):

$$S_i = -\sum_{i=1}^{D} SPI_{i,t} \tag{2.4}$$

e. Drought intensity (I): It is the average value of the SPI within the drought duration. It is measured as the drought severity divided by the duration.

$$I_i = \frac{S_i}{D_i} \tag{2.5}$$

The frequency analysis of drought occurrence is important as it quantifies the drought severity, duration, and frequency for an area, especially those regions where droughts are common. The spatiotemporal variability of droughts can be analyzed through the drought severity–area–frequency (SAF), severity–duration–frequency (SDF), and severity–area–duration (SAD) curves. The schematic representation of SAF and SDF curves are shown in Figure 2.2. More details regarding development of SAF and SDF curves are discussed in the following sections.

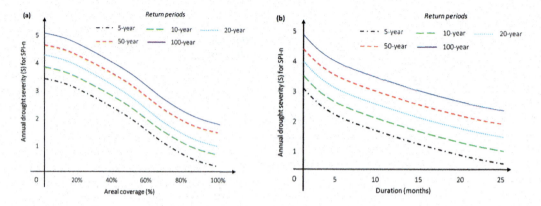

FIGURE 2.2 Schematic representation of (a) severity–area–frequency and (b) severity–duration–frequency relationships.

2.2 SEVERITY–AREA–FREQUENCY CURVES

A regional drought is assumed when a significant fraction of the total area of the region is under drought conditions. Frequency of drought occurrence for a region is aptly characterized by considering the severity, area, and frequency of droughts, which led to the development of drought SAF curves. A regional drought may be defined by its duration, cumulated areal deficit, intensity, and mean regional coverage. The drought SAF curves depict drought severity and drought area with respect to the drought return period, and characterizes the spatial and recurrence patterns of droughts. The series of steps involved in deriving SAF curves is illustrated in Figure 2.3.

Several studies have analyzed SAF curves to characterize the severity and spatial extent of droughts. SAF curves have been used to estimate the recurrent characteristics and areas affected by drought (Henriques and Santos, 1999). Mishra and Desai (2005) analyzed the spatiotemporal variations in drought occurrence for the Kangsabati River basin in India using monthly rainfall data from five rain gauges. The drought severity was estimated from SPI-n derived at different timescales (i.e., $n = 1, 3, 6, 9, 12$, and 24). The extreme value type I (EV-I) distribution was the best-fit distribution for modeling annual values of drought severity associated with different areal coverage for various return periods and SAF curves were constructed. The results indicated that short-term droughts (based on SPI-1 and SPI-3) were more frequent in the study region, while the medium- and long-term droughts were prevalent during the 1980s.

Later, Mishra and Singh (2009) compared historical SAF curves developed earlier Mishra and Desai (2005) with SAF curves developed based on projected rainfall using a general circulation model (GCM). The projected rainfall data was extracted from six GCMs and downscaled using artificial and Bayesian neural networks. The SAF curves were analyzed to compare the severity of historical vis-à-vis projected droughts for various climate change scenarios. The projected short- and long-term droughts were likely to become more severe during 2001–2050 across the basin. Akhtari et al. (2008) analyzed drought characteristics in Iran by deriving an SAF curve using the SPI. Several past studies (Mishra and Cherkauer, 2010; Bonaccorso et al., 2015; Goyal and Sharma, 2016; Ahmed et al., 2019) have reported the investigation of drought severity using SAF curves. Alamgir et al. (2020) evaluated the changes in the SAF relationship for seasonal droughts across Bangladesh using station-based rainfall data. The seasonal droughts were computed for pre-monsoon, monsoon, post-monsoon, and winter seasons using SPI for the last month of previous season, while SPI-6 was used for cropping seasons of rabi and kharif. The SAF curves derived from historical rainfall data were compared to those derived from the rainfall from 13 GCMs to assess the influence of climate change on droughts. The results indicated that kharif droughts had higher severity and larger areal extent, while the seasonal droughts are projected to become more severe due to climate change.

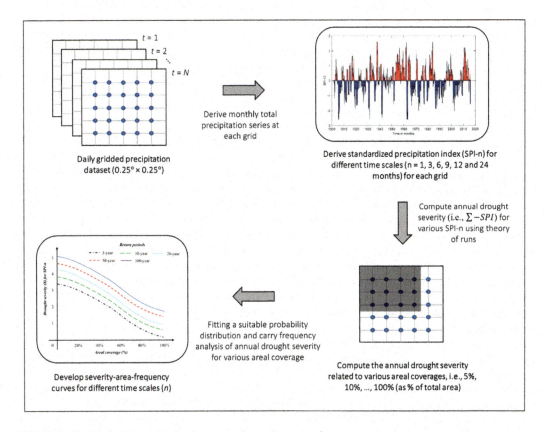

FIGURE 2.3   Methodology for development of severity–area–frequency curves.

2.3   SEVERITY–DURATION–FREQUENCY CURVES

The SDF curve of a drought relates drought severity, duration, and frequency of occurrence (in terms of recurrence interval) in a single diagram. Dalezios et al. (2000) developed the SDF relationships of drought episodes in Greece using extreme value distributions, which were used to prepare iso-severity maps of various return periods over the region. Saghafian et al. (2003) analyzed droughts in Iran using the theory of runs to derive the SDF curves and iso-severity maps of the region. In both the studies, empirical or semi-empirical relationships based on plotting position formulas were adopted for the extreme value analysis. Janga Reddy and Ganguli (2012) presented a copula-based multivariate probabilistic approach to model the SDF relationships of drought events in western India. Rahmat et al. (2015) analyzed drought risk by assessment of SDF curves using point rainfall observations in Victoria, Australia. The study used precipitation thresholds instead of SPI in computing drought severity. Also, SDF curves were developed using the partial duration series approach, wherein, log Pearson type III distribution was chosen to best fit the precipitation deficit for various drought durations. Adarsh et al. (2018) presented the bivariate copula approach to develop the SDF curves for Kerala in India using fine resolution (0.25° × 0.25°) gridded rainfall data. The three-month SPI was estimated to represent short-term drought severity and duration. The joint association between drought severity and duration was modeled by various copulas of the Archimedean family and SDF curves were derived. Sahana et al. (2020) employed the Multivariate Standardized Drought Index (MSDI) for simultaneous evaluation of precipitation and soil moisture deficient conditions for deriving copula-based SDF relationships in India. The SDF relationships are helpful to identify the regions with different degrees of drought severity, which would be helpful in implementing suitable mitigation strategies.

2.4 CASE STUDY

2.4.1 INTRODUCTION

The identification and characterization of droughts is vital for water resources management. The spatial and temporal variations of the droughts are explored by analyzing SAF and SDF curves using the SPI. The SPI is useful in quantifying the meteorological droughts for various durations. Thus, short-term and long-term drought occurrences can be derived from SPI analysis. These SAF curves would serve as a useful tool for planning sustainable water resources projects in the region.

2.4.2 STUDY AREA AND DATA

The development of SAF and SDF curves are presented for the study region comprising of three states of western India, namely, Rajasthan, Gujarat, and Maharashtra. The states collectively encompass a geographical area of 845,976 km², which accounts for approximately 26% of the total area of the country. Rajasthan and Gujarat states are categorized as severely affected by droughts (Jain et al., 2010), whereas the Marathwada region of Maharashtra is also severely drought prone (Sahana et al., 2020). Most parts of the study area experience semiarid and arid climates as per the Koppen–Geigger climate classification (Rubel and Kottek, 2010). The daily gridded rainfall data at 0.25° × 0.25° (Pai et al., 2014) for the 1172 grids falling in the study region are collected from the India Meteorological Department (IMD), Pune, India (see Figure 2.4). The monthly total rainfall at each of these grids is derived for the computation of the SPI. The spatial variation in the mean annual rainfall across the study region for the period 1901–2015 is shown in Figure 2.5. The

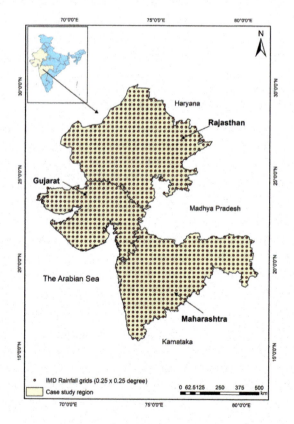

FIGURE 2.4 Locations of rainfall grid points in the study region.

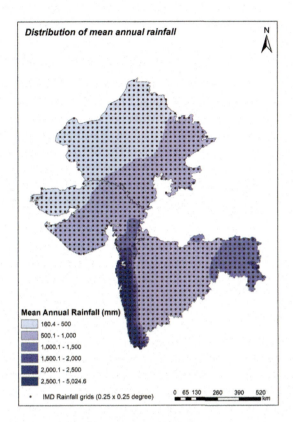

FIGURE 2.5 Spatial variation in the mean annual rainfall for the period 1901–2015 in the study region.

study area shows vast spatial variation in rainfall which is primarily driven by topographic variations. The western coast of Maharashtra, which is located on the windward side of the Western Ghats, received annual rainfall more than 2500 mm. The higher rainfall in this region is due to the orographic influences of the Western Ghats. On the other hand, the leeward side of the Western Ghats receives relatively lower rainfall, i.e., within the range 500–1000 mm. The westernmost part of the Gujarat and Rajasthan, which is the arid desert region, receives scanty rainfall, i.e., less than 500 mm. Also, the parts of Gujarat and Rajasthan exhibiting a semiarid climate show rainfall variability in the range 500–1000 mm. Thus, these regions that receive lower rainfall are more susceptible to droughts.

2.4.3 METHODOLOGY

The monthly rainfall at each grid is analyzed to derive the SPI using a nonparametric approach. The SPI is derived for different timescales, viz., SPI-1, SPI-3, SPI-6, SPI-9, SPI-12, and SPI-24. In the present study, a drought period is defined when the number of months where SPI values remain below a threshold of −0.8 is noticed. The annual drought severity for each SPI is derived at each grid point. Drought severity is thereafter estimated for different areal thresholds (i.e., percentage of total area under drought). The areal thresholds at every 5% increment are selected for the analysis. Thus, the time series for the drought severity for different areal thresholds and SPI of various time lengths are fitted with various probability distributions. The best-fit probability distribution is chosen, and their corresponding parameters are estimated. Using the parameter values, SAF curves are constructed for different return periods, viz., 5-, 10-, 20-, 50-, 80-, and 100-year return periods for various SPI-n.

2.4.4 RESULTS AND DISCUSSION

In this study, the annual drought severity values computed from the nonparametric SPI estimator are used for frequency analysis. The frequency analysis is carried out by fitting several candidate probability distributions to the annual drought severity time series corresponding to different areal thresholds and SPI-n. The distributions are ranked based on the loglikelihood estimator (LLE) and Akaike information criterion (AIC). The distribution showing maximum LLE and AIC values is chosen as the best-fit distribution. From the analysis, a two-parameter gamma distribution was found to best fit the annual drought severity series. The SPI-1 and SPI-3 are used for short-term drought analysis, whereas the SPI-6 and SPI-9 are used for medium-term drought analysis, and the SPI-12 and SPI-24 are used for long-term drought analysis (Mishra and Desai, 2005). The SAF curves for all SPI timescales for various return periods are shown in Figure 2.6. From the analysis, it is seen that the short-term droughts are less severe than the medium- and long-term droughts. The long-term droughts represented by SPI-24 show rather steeper SAF curves compared to other SPI durations. Thus, the droughts of higher durations would have severe impact on water scarcity across the region. The medium- and long-term droughts were noticed in the region when the annual rainfall was lower than the normal annual rainfall in that year and for two consecutive years, respectively (Mishra and Desai, 2005). Similar observations are also noted in the present study.

Further analysis is also carried out to determine the changes in SAF characteristics over the time. In this case, the entire duration (1901–2015) is divided into three equal periods, viz., initial period (1901–1938), mid-period (1939–1976), and recent period (1977–2015). For each subseries, the two-parameter gamma distribution is fitted and frequency analysis is carried out. The changes in the SAF characteristics for the short-term (SPI-3) and long-term (SPI-12) droughts are shown in Figures 2.7 and 2.8, respectively. From Figure 2.7a, the short-term droughts are found to be more frequent in the recent period (1977–2015) compared to the other periods. On the other hand, parts of the basin (having areal thresholds less than 20%) have shown an increase in severity of short-term droughts in the recent period, whereas other parts of the basin have shown a reducing tendency of drought severity in the recent and mid-periods as compared to the initial period. Further, for lower return periods (T = 5 years), the long-term drought (SPI-12) shows higher severity across most parts of the basin (Figure 2.8a). For higher return periods, the severity of long-term droughts has decreased

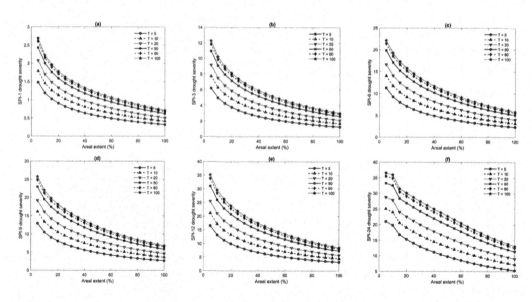

FIGURE 2.6 Drought severity–area–frequency curves for (a) SPI-1, (b) SPI-3, (c) SPI-6, (d) SPI-9, (e) SPI-12, and (f) SPI-24.

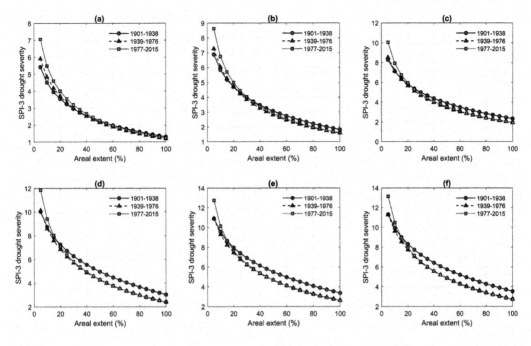

FIGURE 2.7 Changes in SPI-3 based SAF relationships across three temporal windows for (a) 5-, (b) 10-, (c) 20-, (d) 50-, (e) 80-, and (f) 100-year return periods.

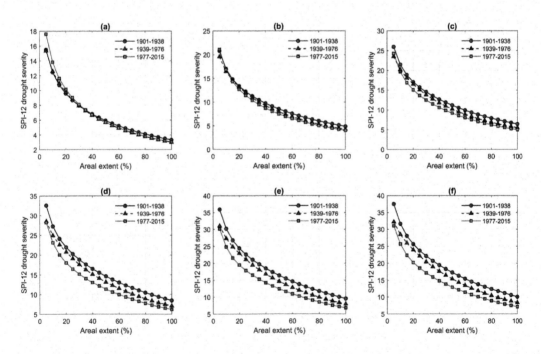

FIGURE 2.8 Changes in SPI-12 based SAF relationships across three temporal windows for (a) 5-, (b) 10-, (c) 20-, (d) 50-, (e) 80-, and (f) 100-year return periods.

in the recent period as compared to the initial and mid-periods. Thus, the SAF curves provide an exhaustive evaluation of changes in the drought characteristics across different timescales.

2.4.5 Conclusions from the Case Study

The analysis of drought severity was carried out for a period of 115 years using grid-based rainfall data for western India. The gridded data at $0.25° \times 0.25°$ at 1172 grid points in the study region were analyzed. The study region exhibited vast spatial variability in rainfall occurrence, wherein the mean annual rainfall varied from 160 to 5024 mm. The rainfall variability in the region was primarily driven by topographic characteristics. The SPI-based annual drought severity was computed at all the grids using theory of runs. The annual drought severity corresponding to various areal thresholds was evaluated and frequency analysis was carried out. A two-parameter gamma distribution was fitted to the time series and SAF curves were constructed. The short-term droughts were found to be less severe than the medium- and long-term droughts. The SAF curves for the long-term droughts (represented by SPI-24) showed steep characteristics as compared to other SPI durations, which could induce water scarcity in the region. The temporal variations in the SAF curves across three equal periods – initial period (1901–1938), mid-period (1939–1976), and recent period (1977–2015) – were also evaluated. The short-term droughts (SPI-3) were found to be more frequent in the recent period (1977–2015) compared to the other periods. On the other hand, the long-term drought (SPI-12) showed higher (lower) severity across most parts of the basin for lower (higher) return periods in the recent period as compared to the initial and mid-periods.

2.5 SUMMARY

The present study deals with the analysis of identification and characterization of spatiotemporal variations in droughts using SAF and SDF curves. Among various indicators used for the analysis of meteorological drought, such as the Standardized Precipitation Evapotranspiration Index (SPEI), Drought Reconnaissance Index (DRI), Palmer Drought Severity Index (PDSI), and Rainfall Anomaly Index (RAI), the Standardized Precipitation Index (SPI) is widely adopted and preferred for estimation of drought due to its simplicity and less data requirement. In the present study, a nonparametric estimator of SPI is adopted to overcome the bias associated with distributional assumptions. The drought severity is estimated for SPI at different timescales to analyze the short-, medium- and long-term droughts. The drought severity is also computed for different areal thresholds and frequency distribution is carried out. The SAF curves are thereafter constructed for the region and their variations are analyzed. Thus, SAF curves prove to be a useful tool for studying annual drought severity and its future occurrences, covering the percentage of the area in the basin.

REFERENCES

Adarsh, S., Karthik, S., Shyma, M., Prem, G. D., Parveen, A. S., & Sruthi, N. (2018). Developing short term drought severity-duration-frequency curves for Kerala meteorological subdivision, India using bivariate copulas. *KSCE Journal of Civil Engineering*, 22(3), 962–973.

Ahmed, K., Shahid, S., Chung, E. S., Wang, X. J., & Harun, S. B. (2019). Climate change uncertainties in seasonal drought severity-area-frequency curves: Case of arid region of Pakistan. *Journal of Hydrology*, 570, 473–485.

Akhtari, R., Bandarabadi, S. R., & Saghafian, B. (2008). Spatio-temporal pattern of drought in Northeast of Iran. In *International Conference on Drought management: Scientific and Technological Innovations, Zaragoza, Spain*, 71–77.

Alamgir, M., Khan, N., Shahid, S., Yaseen, Z. M., Dewan, A., Hassan, Q., & Rasheed, B. (2020). Evaluating severity–area–frequency (SAF) of seasonal droughts in Bangladesh under climate change scenarios. *Stochastic Environmental Research and Risk Assessment*, 34, 447–464.

Bhalme, H. N., & Mooley, D. A. (1980). Large-scale droughts/floods and monsoon circulation. *Monthly Weather Review*, 108(8), 1197–1211.

Bonaccorso, B., Peres, D. J., Castano, A., & Cancelliere, A. (2015). SPI-based probabilistic analysis of drought areal extent in Sicily. *Water Resources Management*, 29(2), 459–470.

Dalezios, N. R., Loukas, A., Vasiliades, L., & Liakopoulos, E. (2000). Severity-duration-frequency analysis of droughts and wet periods in Greece. *Hydrological Sciences Journal*, 45(5), 751–769.

De Martonne, E. (1925). *Traité de Géographie Physique*, 11, Colin, Paris.

Farahmand, A., & AghaKouchak, A. (2015). A generalized framework for deriving nonparametric standardized drought indicators. *Advances in Water Resources*, 76, 140–145.

Gibbs, W. J., & Maher, J. V. (1967). *Rainfall Deciles as Drought Indicators, Bureau of Meteorology Bulletin no. 48*. Commonwealth of Australia, Melbourne, 29.

Goyal, M. K., & Sharma, A. (2016). A fuzzy c-means approach regionalization for analysis of meteorological drought homogeneous regions in western India. *Natural Hazards*, 84(3), 1831–1847.

Hao, Z., & AghaKouchak, A. (2013). Multivariate standardized drought index: A parametric multi-index model. *Advances in Water Resources*, 57, 12–18.

Henriques, A. G., & Santos, M. J. J. (1999). Regional drought distribution model. *Physics and Chemistry of the Earth, Part B: Hydrology, Oceans and Atmosphere*, 24(1–2), 19–22.

Jain, S. K., Keshri, R., Goswami, A., & Sarkar, A. (2010). Application of meteorological and vegetation indices for evaluation of drought impact: A case study for Rajasthan, India. *Natural hazards*, 54(3), 643–656.

Janga Reddy, M., & Ganguli, P. (2012). Application of copulas for derivation of drought severity–duration–frequency curves. *Hydrological Processes*, 26(11), 1672–1685.

Keyantash, J., & Dracup, J. A. (2002). The quantification of drought: An evaluation of drought indices. *Bulletin of the American Meteorological Society*, 83(8), 1167–1180.

Łabędzki, L. (2007). Estimation of local drought frequency in central Poland using the standardized precipitation index SPI. *Irrigation and Drainage*, 56(1), 67–77.

Lyon, B. (2004). The strength of El Niño and the spatial extent of tropical drought. *Geophysical Research Letters*, 31(21), L21204 1–4.

McKee, T. B., Doesken, N. J., & Kleist, J. (1993, January). The relationship of drought frequency and duration to time scales. In *Proceedings of the 8th Conference on Applied Climatology* (Vol. 17, No. 22, pp. 179–183).

Mishra, A. K., & Desai, V. R. (2005). Spatial and temporal drought analysis in the Kansabati river basin, India. *International Journal of River Basin Management*, 3(1), 31–41.

Mishra, A. K., & Singh, V. P. (2009). Analysis of drought severity-area-frequency curves using a general circulation model and scenario uncertainty. *Journal of Geophysical Research: Atmospheres*, 114(D6), D06120 1–18.

Mishra, V., & Cherkauer, K. A. (2010). Retrospective droughts in the crop growing season: Implications to corn and soybean yield in the Midwestern United States. *Agricultural and Forest Meteorology*, 150(7–8), 1030–1045.

Obasi, G. O. P. (1994). WMO's role in the international decade for natural disaster reduction. *Bulletin of the American Meteorological Society*, 75(9), 1655–1661.

Pai, D. S., Sridhar, L., Rajeevan, M., Sreejith, O. P., Satbhai, N. S., & Mukhopadhyay, B. (2014). Development of a new high spatial resolution (0.25 × 0.25) long period (1901–2010) daily gridded rainfall data set over India and its comparison with existing data sets over the region. *Mausam*, 65(1), 1–18.

Palmer, W. C. (1965). *Meteorological drought (Research paper no. 45)*. US Weather Bureau, Washington, DC.

Rahmat, S. N., Jayasuriya, N., & Bhuiyan, M. (2015). Development of drought severity-duration-frequency curves in Victoria, Australia. *Australasian Journal of Water Resources*, 19(1), 31–42.

Rossi, G. (2000). Drought Mitigation Measures: A Comprehensive Framework. In: Vogt, J.V., Somma, F. (eds) *Drought and Drought Mitigation in Europe*. Advances in Natural and Technological Hazards Research, vol 14. Springer, Dordrecht, pp. 233–246.

Rubel, F., & Kottek, M. (2010). Observed and projected climate shifts 1901–2100 depicted by world maps of the Köppen-Geiger climate classification. *Meteorologische Zeitschrift*, 19(2), 135.

Saghafian, B., Shokoohi, A., & Raziei, T. (2003). Drought spatial analysis and development of severity-duration-frequency curves for an arid region. *International Association of Hydrological Sciences Publication*, 278, 305–311.

Sahana, V., Sreekumar, P., Mondal, A., & Rajsekhar, D. (2020). On the rarity of the 2015 drought in India: A country-wide drought atlas using the multivariate standardized drought index and copula-based severity-duration-frequency curves. *Journal of Hydrology: Regional Studies*, 31, 100727.

Sharma, A., & Goyal, M. K. (2020). Assessment of drought trend and variability in India using wavelet transform. *Hydrological Sciences Journal*, 65(9), 1539–1554.

Sims, A. P., Niyogi, D. D. S., & Raman, S. (2002). Adopting drought indices for estimating soil moisture: A North Carolina case study. *Geophysical Research Letters*, 29(8), 24-1–24-4.

Svoboda, M., & Fuchs, B. A. (2016). *Handbook of drought indicators and indices. Integrated drought management programme (IDMP), integrated drought management tools and guidelines series 2.* World Meteorological Organization and Global Water Partnership, Geneva, Switzerland, 52.

Svoboda, M., LeComte, D., Hayes, M., Heim, R., Gleason, K., Angel, J., Rippey, B., Tinker, R., Palecki, M., Stooksbury, D., Miskus, D., & Stephens, S. (2002). The drought monitor. *Bulletin of the American Meteorological Society*, 83(8), 1181–1190.

Tannehill, I. R. (1947). *Drought: Its Causes and Effects.* Princeton, NJ: Princeton University Press.

Tsakiris, G., & Vangelis, H. J. E. W. (2005). Establishing a drought index incorporating evapotranspiration. *European Water*, 9(10), 3–11.

van Rooy, M. P. (1965). A rainfall anomaly index independent of time and space. *Notos*, 14, 43–48.

Vicente-Serrano, S. M., Beguería, S., & López-Moreno, J. I. (2010). A multiscalar drought index sensitive to global warming: The standardized precipitation evapotranspiration index. *Journal of Climate*, 23(7), 1696–1718.

Wilhite, D. A., & Glantz, M. H. (1985). Understanding: The drought phenomenon: The role of definitions. *Water International*, 10(3), 111–120.

Yevjevich, V. (1967). *An Objective Approach to Definitions and Investigations of Continental Hydrologic Droughts.* Hydrology Paper No. 23, Colorado State University, Fort Collins.

3 Analysis of Spatial Variability and Patterns of Drought
A Case Study for Serbia

Milan Gocic and Mohammad Arab Amiri

CONTENTS

3.1 INTRODUCTION

Drought, as one of the natural hazards, has caused a number of direct or indirect damages all over the world affecting different sectors (e.g., agriculture, water management, and forestry) (Wilhite, 2000; Fleig et al., 2006; Smakhtin and Hughes, 2006; Solomon et al., 2007; Hanel et al., 2018). Drought is different from aridity. Aridity is a permanent rainfall deficiency, whereas drought is a temporary period of lower than average rainfall. Drought is categorized into four types by the American Meteorological Society: meteorological drought, agricultural drought, hydrological drought, and socioeconomic drought. All the mentioned drought types share the deficiency of precipitation as a common phenomenon (Mekonen et al., 2020). Meteorological drought refers to a precipitation deficit and the duration of a period with a precipitation deficit.

Meteorological drought events are difficult to detect and have different aspects and characteristics from one region to another due to different climates. Droughts can last for weeks, months, or even years, and its spatial extent is usually larger than that of other natural hazards (Obasi, 1994). Drought mitigation depends on drought monitoring, and its monitoring is usually performed using drought indices (Morid et al., 2006). To quantify drought, many different drought indices have been developed and evaluated (Wilhite and Glantz, 1985; Guttman, 1999; Keyantash and Dracup, 2002; Vicente-Serrano et al., 2010; Mishra and Singh, 2010).

This study analyzes the spatial distribution of drought and extreme precipitation at a yearly timescale using the Rainfall Anomaly Index (RAI), the Standardized Anomaly Index (SAI), and the Percent of Normal Precipitation (PNP), and annual precipitation data from Serbia between 1946 and 2019. The annual precipitation data were obtained through 28 well-distributed synoptic stations over Serbia. The used drought indices in this study are easy to calculate, with a single input (precipitation) that can be analyzed on different timescales.

DOI: 10.1201/9781003276548-3

Oladipo (1985) and Hänsel et al. (2016) compared the RAI with other similar drought indices requiring only rainfall as an input parameter and concluded that it can be determined on the same timescales and is not more complicated in calculation than the Standardized Precipitation Index (McKee et al., 1993). In Brazil, the RAI was used for drought monitoring in a decision support system (Freitas, 1998), while in Ethiopia it was used to analyze a rainfall climate (Tilahun, 2006).

In order to analyze the year-to-year fluctuations, the SAI can be considered (Valero et al., 2004). Also, in Southern Togo (Koudahe et al., 2017) and India (Siddharam Kambale et al., 2020), this index was used to investigate dry and wet periods.

In Serbia, drought has been analyzed using different drought indices finding different drought subregions and making recommendations to the local authorities (Mihajlovic, 2006; Stricevic et al., 2011; Tosic and Unkasevic, 2014; Gocic and Trajkovic, 2015; Trajkovic et al., 2020).

The main objectives of the present study are to (1) investigate the spatial patterns of drought and extreme precipitation at the most extreme years using RAI, SAI, and PNP in Serbia; (2) compare the spatial patterns of drought and extreme precipitation at the most extreme years obtained through the three aforementioned indices over Serbia; (3) propose a classification for SAI and PNP; and (4) analyze the temporal distribution of drought and extreme precipitation in different subperiods over Serbia.

3.2   ANALYZED STUDY AREA

As a case study area, Serbia, a country in Southeast Europe and a part of the Western Balkan region, was chosen. It is located in the Balkan Peninsula with a moderate continental climate. Analyzing precipitation patterns, Serbia can be divided into three subregions: the first subregion located in the north part of the country is driest, the second in the western part is wettest, and the third located in the south part of Serbia has precipitation values under the average (Gocic and Trajkovic, 2014a). The least amount of rain falls in February, whereas the months with the highest amount of rain are May and June (Gocic and Trajkovic, 2013).

A series of monthly precipitation data with no missing values and quality controlled by the Republic Hydrometeorological Service of Serbia (http://www.hidmet.gov.rs) were used for drought analysis from 28 synoptic stations for the period of 1946–2019. The spatial distribution of selected stations is presented in Figure 3.1.

3.3   APPLIED DROUGHT INDICES

3.3.1   Rainfall Anomaly Index

The RAI was derived by Van Rooy (1965) as a flexible drought index that can be calculated on different timescales (monthly, seasonal, and annual). It uses normalized precipitation values based upon the long-term rainfall data. The RAI is easy to calculate and only uses precipitation data as an input. However, it requires complete rainfall data sets with no missing values. To calculate the RAI, variations within a year should be small in comparison with temporal variations. The RAI can be formulated as

$$RAI = 3\left[\frac{R - \bar{R}}{\bar{M} - \bar{R}}\right]$$

(3.1)

for positive anomalies, and

$$RAI = -3\left[\frac{R - \bar{R}}{\bar{X} - \bar{R}}\right]$$

(3.2)

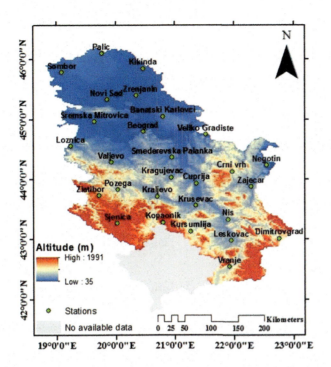

FIGURE 3.1 The spatial distribution of the synoptic stations in Serbia.

for negative anomalies, where R is the annual total rainfall, $\bar{R}$ is the long-term average annual rainfall, $\bar{M}$ is the average of the ten highest annual rainfall of the historical series, and $\bar{X}$ is the average of the ten lowest annual rainfall of the historical series. Years with annual precipitation above average have positive anomalies and should be calculated with Equation 3.1, whereas years with annual precipitation below average or years with negative anomalies should be calculated with Equation 3.2.

3.3.2 STANDARDIZED ANOMALY INDEX

The SAI introduced by Kraus (1977) is the difference between the annual total rainfall of a particular year and the long-term average rainfall divided by the standard deviation of the long-term rainfall data. The SAI can also be calculated on monthly and seasonal timescales. The SAI only uses rainfall data as input and can be calculated for any defined period. But, calculations are strongly dependent on the quality of the data. The SAI formula is given as

$$SAI_t = \frac{1}{N} \sum_{i=1}^{N} \frac{x_{it} - \mu_i}{\sigma_i} \tag{3.3}$$

where x_{it} is the total rainfall of a particular period t; and μ_i and σ_i are the mean and the standard deviation of the total rainfall at the ith station, respectively.

3.3.3 PERCENT OF NORMAL PRECIPITATION

The PNP is one of the most used drought indices that calculates the deviation of rainfall from its long-term average. It can be computed on daily, weekly, monthly, seasonal, and annual timescales. The PNP is calculated by dividing actual precipitation by long-term average precipitation for the

TABLE 3.1

RAI, SAI, and PNP Classification According to Their Original Definition, Definitions Used by Previous Research, and Definitions Proposed in This Study

Original Definition by Van Rooy (1965)		Classification Used in This Study		Classification Used by Morid et al. (2006)		Classification Used in This Study	
RAI	Description	SAI	Description	PNP (%)	Description	PNP (%)	Description
≥4	Extremely humid	≥2	Extremely humid	≥110	Wet	≥140	Extremely humid
2 to 4	Very humid	1 to 2	Very humid	80 to 110	Normal	120 to 140	Very humid
0 to 2	Humid	0 to 1	Humid	55 to 80	Moderately dry	100 to 120	Humid
–2 to 0	Dry	–1 to 0	Dry	40 to 55	Severely dry	80 to 100	Dry
–4 to –2	Very dry	–2 to –1	Very dry	≤40	Extremely dry	60 to 80	Very dry
≤–4	Extremely dry	≤–2	Extremely dry			≤60	Extremely dry

time being considered and multiplying by 100. So, the normal is considered as a long-term mean precipitation value at a specific location and at least 30 years of rainfall data is needed to compute the normal. It should be noted that the normal may be perceived differently in different regions with different climate regimes.

3.3.4 CLASSIFICATION OF DROUGHT INDICES USED IN THIS STUDY

There are some studies in the literature that categorized the used drought indices into different drought classes to interpret drought events. The RAI was classified originally by Van Rooy (1965). Since there is no classification for the SAI, we proposed a classification in this chapter. Moreover, there is a classification for the PNP that was proposed by Morid et al. (2006). We also proposed some thresholds for categorization of the PNP in the present chapter. We used both classifications of PNP to compare the results. Table 3.1 represents some classifications for the used drought indices (i.e., RAI, SAI, and PNP) that were defined originally, used in previous research, or proposed in the present study.

3.4 ANALYSIS OF SPATIAL VARIABILITY OF DROUGHT

Calculating the RAI using yearly precipitation data sets, the calculated RAI values of all the stations for every year from 1946 to 2019 are plotted on a point chart (Figure 3.2). From Figure 3.2, it can be realized that the years 2000 and 2014 were the most extreme years over the whole period, where the year 2000 was the driest year and the year 2014 was the most humid one. There were also several other extreme years, such as 1946, 1990, and 2012 as the driest years, and 1954, 1955, 1999, and 2016 as the wettest years. But, most of the calculated values fell between –2 and 2 that can be categorized as dry to humid according to Table 3.1.

The calculated RAI values of all the stations in two of the most extreme years were interpolated using the kriging interpolation method to show the spatial distribution of the RAI throughout the study region (Figure 3.3). The interpolated map of the RAI in the year 2000 classified the country

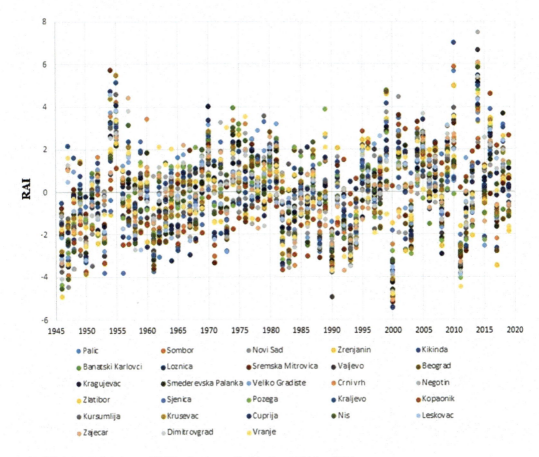

FIGURE 3.2 Point chart of RAI values over Serbia from 1946 to 2019.

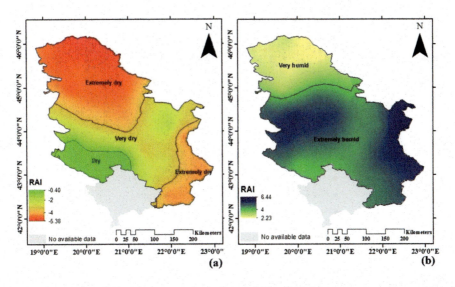

FIGURE 3.3 Spatial distribution of RAI values over Serbia in two of the most extreme years: (a) year 2000 and (b) year 2014.

into three classes – extremely dry in the north and southeast, very dry in the center, and dry in the southwest – which is in line with the results presented in Gocic and Trajkovic (2013, 2014b). The southwestern corner of Serbia that was classified as dry is a mountainous region with the highest precipitation in Serbia. On the other hand, the interpolated map of the RAI in the wettest year, i.e., the year 2014, classified the region into two subregions: very humid in the north and extremely humid in the remaining part. The northern part of Serbia that was categorized as very humid is a flat region that receives the lowest amount of precipitation over Serbia. The produced maps for both years are in accordance with the topography map of the country.

The SAI was developed based on the RAI, and the RAI is a component of the SAI. These indices are similar, but both are unique drought indices. Having computed the SAI for every station from 1946 to 2019, the SAI values were plotted on a point chart (Figure 3.4). The produced point chart is pretty similar to the point chart of the RAI; and it also determined the years 2000 and 2014 as the driest and wettest years, respectively. Both charts indicate that extreme years have been increasing in the last 20 years.

The SAI values in two of the most extreme years (i.e., the years 2000 and 2014) were then interpolated to show the spatial distribution of the SAI over Serbia (Figure 3.5). The spatial patterns of both maps in Figure 3.5 are pretty similar to the patterns observed in Figure 3.3. Also, the topography map of Serbia confirms the recognized classes. So, the classification of the SAI that is proposed in this study works well, and since the SAI is easier to compute than the RAI, calculating

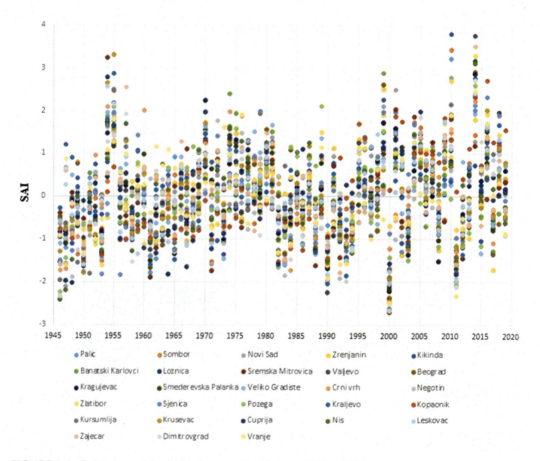

FIGURE 3.4   Point chart of the SAI values over Serbia from 1946 to 2019.

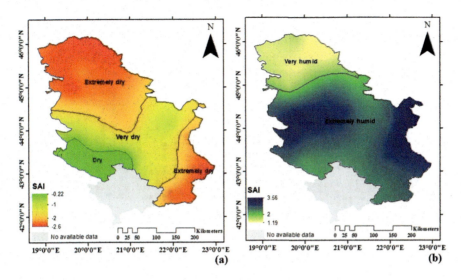

FIGURE 3.5 Spatial distribution of the SAI values over Serbia in two of the most extreme years: (a) year 2000 and (b) year 2014.

the SAI and classifying it based on the proposed categorization could be a better choice or a more convenient method than the RAI in drought studies.

The PNP is another drought index that was calculated in this study. The computed PNP values were plotted in a point chart in Figure 3.6. In this study, the long-term average of yearly total precipitation at each station was considered as the normal. So, the normal is 100%, and extreme years could be recognized by comparing their PNP values to the normal (i.e., 100%). The general pattern of the point chart in Figure 3.6 is also similar to Figure 3.2 and Figure 3.4. This index confirms the results of the RAI and SAI; as 2000 and 2014 are the driest and wettest years, respectively. But, the point chart of the PNP is more compact in comparison to the point charts of the RAI and SAI. More near-normal values are seen in the point chart of the PNP.

The interpolate maps of PNP in the years 2000 and 2014 are shown in Figure 3.7. To interpret the degree of drought, we should compare the PNP values to the normal (i.e., 100%). So, if the PNP in a region is more than 100%, the region is wet; and the higher the PNP, the wetter the region. But, if the PNP in a region is lower than 100%, the region is dry; and the lower the PNP, the drier the region. The general patterns of both maps for the years 2000 and 2014 that are classified by the categorization proposed in this study and by Morid et al. (2006) are shown in Figure 3.7. The results of classification of PNP based on the proposed classification are in close agreement with the subregions found in Figure 3.3 and Figure 3.5. But, the classified maps produced by classification presented in Morid et al. (2006) don't confirm the classes found by the RAI and SAI. So, the proposed classification for the PNP in the present study works better than the classification proposed by Morid et al. (2006), at least in the study region.

Every point in the point charts (Figure 3.2, Figure 3.4, and Figure 3.6) is a calculated index in a specific year. Table 3.2 counts the number of points that have greater and less values than the defined threshold values. The threshold values are defined here as having RAI values greater than 2 and less than –2, SAI values larger than 1 and smaller than –1, and PNP values greater than 120 and less than 80 as the highest and lowest values, respectively. The percentages of the highest and lowest values to the total points were also calculated. Similar results are seen through all three indices. About 30% of the total points show extreme conditions, of which half (15%) are very dry and the other half (15%) are very humid. In other words, 30% of the recorded precipitation values by all the stations during the whole period show extreme conditions (i.e., very dry or very humid).

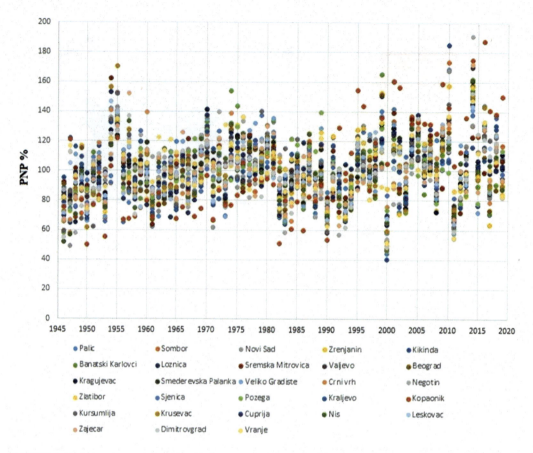

FIGURE 3.6   Point chart of the PNP values in percent over Serbia from 1946 to 2019.

TABLE 3.2

Total Number and Percentage of Points with Extreme Precipitation

Index Extreme Statistics	RAI	SAI	PNP
Number of points with fewer values than the defined thresholds	321	313	323
Percentage of points with fewer values than the defined thresholds	15.5	15.1	15.6
Number of points with greater values than the defined thresholds	268	320	318
Percentage of points with greater values than the defined thresholds	12.9	15.4	15.3

The number of points with greater and less values than the defined thresholds are plotted in four subperiods (i.e., 1946–1959, 1960–1979, 1980–1999, and 2000–2019) (Figure 3.8). The line graph shows that the number of points with the computed values less than the defined thresholds using all three indices are higher in the last two subperiods (i.e., from 1980 to 2019); and the number of points with values greater than the defined thresholds is by far higher in the last subperiod (i.e., between 2000 and 2019). Hence, more drought events took place in the last 40 years and more floods occurred in the last 20 years in the country.

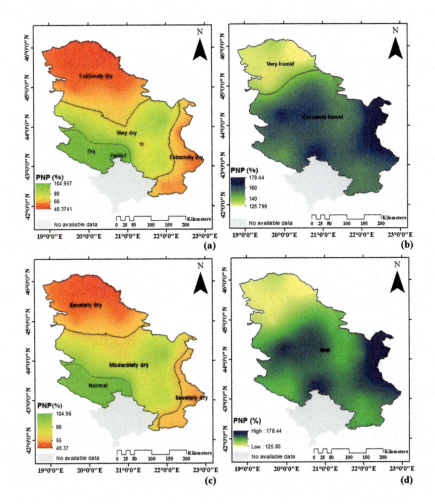

FIGURE 3.7 Spatial distribution of PNP values over Serbia in two of the most extreme years: (a) year 2000; (b) year 2014, using the classification proposed in this study; and (c) year 2000; (d) year 2014, using the classification presented in Morid et al. (2006).

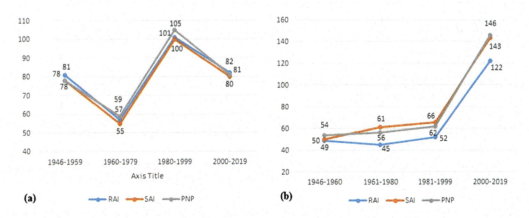

FIGURE 3.8 Number of points with the calculated index values (a) less and (b) greater than the defined thresholds in the four considered subperiods using three indices.

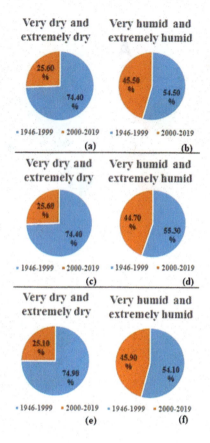

FIGURE 3.9 Percentages of points with values less and greater than the threshold values occurred in the two subperiods (i.e., 1946–1999 and 2000–2019) using the (a and b) RAI index, (c and d) SAI index, and (e and f) PNP index.

Percentages of points with the computed values less and greater than the defined thresholds in two subperiods, from 1946–1999 and 2000–2019, are shown in Figure 3.9. This graph shows the temporal distribution of extreme precipitation in the two subperiods. The very dry and extremely dry drought events are monotonically distributed in comparison to the very humid and extremely humid events. Around half of the very humid and extremely humid points fell in the past 20 years. Thus, planning for agricultural activities needs more attention in Serbia.

3.5 CONCLUSIONS

In this study, monthly precipitation data from 28 meteorological stations from Serbia for the period 1946–2019 were used. Also, the three drought indices were applied, i.e., the RAI, SAI, and PNP.

All selected indices showed that the years 2000 and 2014 were the driest and wettest years, respectively. There were also several other extreme years, such as 1946, 1990, and 2012 as the driest years, and 1954, 1955, 1999, and 2016 as the wettest years. In total, 30 percent of the recorded precipitation values by all the stations during the whole period show extreme conditions (i.e., very dry or very humid).

The point chart of the PNP is more compact in comparison to the point charts of the RAI and SAI. More near-normal values are seen in the point chart of the PNP. The proposed classification for the PNP in the present study works better than the classification proposed by Morid et al. (2006), at least in the study region.

The obtained results can be used in planning water resources, irrigation systems, and agricultural production as well as crop scheduling in the future.

CONFLICT OF INTEREST

The authors declare that they have no conflict of interest.

REFERENCES

Fleig, A.K., Tallaksen, L.M., Hisdal, H. & Demuth, S. (2006). A global evaluation of streamflow drought characteristics. *Hydrology and Earth System Sciences* 10, 535–552.

Freitas, M.A.S. (1998). Um sistema de suporte à decisão para o monitoramento de secas meteorológicas em regiões semi-áridas. *Revista Tecnologia* 19, 84–95.

Gocic, M. & Trajkovic, S. (2013). Analysis of precipitation and drought data in Serbia over the period 1980–2010. *Journal of Hydrology* 494, 32–42.

Gocic, M. & Trajkovic, S. (2014a). Spatio-temporal patterns of precipitation in Serbia. *Theoretical and Applied Climatology* 117(3–4), 419–431.

Gocic, M. & Trajkovic, S. (2014b). Spatiotemporal characteristics of drought in Serbia. *Journal of Hydrology* 510, 110–123.

Gocic, M. & Trajkovic, S. (2015). Water Surplus Variability Index as an indicator of drought. *Journal of Hydrologic Engineering* 20(2), 04014038.

Guttman, N.B. (1999). Accepting the standardized precipitation index: A calculation algorithm. *Journal of the American Water Resources Association* 35, 311–322.

Hanel, M., Rakovec, O., Markonis, Z., Maca, P., Samaniego, L., Kzselz, J. & Kumar, R. (2018). Revisiting the recent European droughts from a long-term perspective. *Scientific Reports* 8:9499.

Hänsel, S., Schucknecht, A. & Matschullat, J. (2016). The Modified Rainfall Anomaly Index (mRAI)—Is this an alternative to the Standardised Precipitation Index (SPI) in evaluating future extreme precipitation characteristics?. *Theoretical and Applied Climatology* 123, 827–844.

Keyantash, J. & Dracup, J.A. (2002). The quantification of drought: An evaluation of drought indices. *Bulletin of the American Meteorological Society* 83(8), 1167–1180.

Koudahe, K., Kayode, A., Samson, A., Adebola, A. & Djaman, K. (2017). Trend analysis in standardized precipitation index and standardized anomaly index in the context of climate change in Southern Togo. *Atmospheric and Climate Sciences* 7, 401–423.

Kraus, E. (1977). Subtropical droughts and cross-equatorial energy transports. *Monthly Weather Review* 105(8), 1009–1018.

McKee, T.B., Doesken, N.J. & Kleist, J. (1993). The relationship of drought frequency and duration to time scales. Proceedings of the Eighth Conference on Applied Climatology, American Meteorological Society, 179–184.

Mekonen, A.A., Berlie, A.B. & Ferede, M.B. (2020). Spatial and temporal drought incidence analysis in the northeastern highlands of Ethiopia. *Geoenvironmental Disasters* 7(1). https://doi.org/10.1186/s40677-020-0146-4

Mihajlovic, D. (2006). Monitoring the 2003–2004 meteorological' drought over Pannonian part of Croatia. *International Journal of Climatology* 26, 2213–2225.

Mishra, A.K. & Singh, V.P. (2010). A review of drought concepts. *Journal of Hydrology* 354 (1–2), 202–216.

Morid, S., Smakhtin, V. & Moghaddasi, M. (2006). Comparison of seven meteorological indices for drought monitoring in Iran. *International Journal of Climatology* 26, 971–985.

Obasi, G.O.P. (1994). WMO's role in the international decade for natural disaster reduction. *Bulletin of the American Meteorological Society* 75(9), 1655–1661.

Oladipo, E.O. (1985). A comparative performance analysis of three meteorological drought indices. *Journal of Climatology* 5, 655–664.

Siddharam Kambale, J.B., Basavaraja, D., Nemichandrappa, M. & Dandekar, A.T. (2020). Assessment of long term spatio-temporal variability and Standardized Anomaly Index of rainfall of Northeastern region, Karnataka, India. *Climate Change* 6(21), 1–11.

Smakhtin, V.U. & Hughes, D.A. (2006). Automated estimation and analyses of meteorological drought characteristics of monthly rainfall data. *Environmental Modelling and Software* 22(6), 880–890.

Solomon, S., et al. (2007). Technical summary. In: Solomon, S. et al., (ed.) *Climate Change 2007: The Physical Science Basis. Contribution of Working Group I to the Fourth Assessment Report of the Intergovernmental Panel on Climate Change*, Cambridge University Press, New York, pp. 19–91.

Stricevic, R., Djurovic, N. & Djurovic, Z. (2011). Drought classification in Northern Serbia based on SPI and statistical pattern recognition. *Meteorological Application* 18 (1), 60–69.

Tilahun, K. (2006). Analysis of rainfall climate and evapotranspiration in arid and semi-arid regions of Ethiopia using data over the last half a century. *Journal of Arid Environments* 64, 474–487.

Tosic, I. & Unkasevic, M. (2014). Analysis of wet and dry periods in Serbia. *International Journal of Climatology* 34, 1357–1368.

Trajkovic, S., Gocic, M., Misic, D. & Mladenovic, M. (2020). Spatio-temporal distribution of hydrological and meteorological droughts in the South Morava Basin. In: Gocic, M., Aronica, G., Stavroulakis, G., Trajkovic, S., (eds) *Natural Risk Management and Engineering NatRisk Project, Springer Tracts in Civil Engineering*, Springer, Cham, pp. 225–242.

Valero, F., Luna, M.Y., Martin, M.L., Morata, A. & Gonzalez-Rouco, F. (2004). Coupled modes of large-scale climatic variables and regional precipitation in the western Mediterranean in autumn. *Climate Dynamics* 22, 307–323.

Van Rooy, M.P. (1965). A rainfall anomaly index independent of time and space. *Notos* 14, 43–48.

Vicente-Serrano, S.M., Begueria, S. & Lopez-Moreno, J.I. (2010). A multiscalar drought index sensitive to global warming: The standardized precipitation evapotranspiration index. *Journal of Climate* 23, 1696–1718.

Wilhite, D.A. (2000). Drought as a natural hazard: Concepts and definitions. In: Wilhite, D.A., (ed.) *Drought: A Global Assessment*, Routledge, London, 3–18.

Wilhite, D.A. & Glantz, M.H. (1985). Understanding the drought phenomenon: The role of definitions. *Water International* 10(3), 111–120.

4 Spatial and Temporal Trend Patterns of Drought in Bangladesh

Abu Reza Md. Towfiqul Islam, Nilufa Yeasmin, and Roquia Salam

CONTENTS

4.1 INTRODUCTION

Drought is one of the most geoenvironmental disasters that has notably influenced socioeconomic progress in recent eras. There has been considerable attention to drought appraisal at various spatiotemporal scales varying from regional to continental levels (Sheffield et al. 2012; Peña-Gallardo et al. 2019). In the context of climate change, an increase in global temperature may change the water cycle via elevated evaporation demand (Vicente-Serrano et al. 2020). As a result, precipitation decreases integrated with high evaporation demand can substantially enhance drought severity and frequency worldwide (Islam et al. 2017; Vicente-Serrano et al. 2020). The effects of drought may differ significantly from one area to another because of the existing climatic conditions, adaptation plans, and prevailing infrastructure (Dai et al. 2018).

In South Asia, many cited works have appraised drought variability at various spatial and temporal scales (Trenberth et al. 2014; Bonaccorso et al. 2013; Dina and Islam 2020). Most work has used drought monitoring tools like the Standardized Precipitation Index (SPI) and the Standardized Precipitation Evapotranspiration Index (SPEI). The review of these drought appraisals reveals increased occurrences of droughts in the southeast region, including Bangladesh, in the last decade.

Like other Southeast Asian regions, Bangladesh is more susceptible to climate change (Ali 1999; IPCC 2014; Rahman and Islam 2019). Droughts have been recognized as one of the most

DOI: 10.1201/9781003276548-4

detrimental natural hazards in Bangladesh, affecting agricultural crop production, water resource practices, ecological balance, and human health, among many other fields. Bangladesh, especially northern Bangladesh, has already experienced severe drought events that triggered huge socio-economic and infrastructure damages (Islam et al. 2014; 2020). Climate change has exaggerated the frequency and intensity of drought events (Selvaraju et al. 2006). In addition, over exploration of groundwater puts extra pressure on the groundwater table daily (Shahid and Behrawan 2008). Climate change adds a new dimension to drought risk and vulnerability in most disaster-induced Bangladesh (Ahmed 2006; Islam et al. 2019). Due to the limited surface water resources, agriculture in the country profoundly relies on irrigation, groundwater abstraction, and seasonal rainfall (Ruane et al. 2013; Parvin et al. 2015; Zannat et al. 2019). Drought mainly occurs during the dry season from October to April, when seasonal rainfall is deficient (Umma et al. 2011). Bangladesh faced several extreme drought phenomena in the late 1980s, mid-1990s, and mid-2010s (Miah et al. 2017). Thus, it is essential to examine statistical analysis to determine Bangladesh's most suitable tool for drought appraisal.

Drought phenomena are frequently identified by more than a hundred indices developed by many global research scholars (Niemeyer 2008; Miah et al. 2017). These indices have been successfully utilized in hydrometeorology, climatology, and agriculture (Zargar et al. 2011; Prodhan et al. 2020). The SPI has become the most widespread drought index, measured from only the rainfall data set (McKee et al.1993). The SPEI, another popular drought index in recent times, which was developed by Vicente-Serrano et al. (2010), requires rainfall and temperature data sets to identify drought conditions. Of these, some drought indices like the Palmer Drought Severity Index (PDSI) (Palmer 1965), Surface Water Supply Index (SWSI) (Shafer 1982), and Standardized Streamflow Index (SSI) are also widely used for drought monitoring systems (Kao and Rao 2010). However, in this study, we have used the SPI, SPEI, and precipitation anomaly (PA) for meteorological drought appraisal, because of their easiness, popularity, and easy computation. They require only one or two parameters, easily available at various timescales. Current literature associated with drought-related studies (Shahid and Behrawan 2008; Alamgir et al. 2015; Rahman and Lateh 2016; Miah et al. 2017; Islam et al. 2017; Uddin et al., 2020; Prodhan et al. 2020; Zinat et al. 2020; Mondol et al. 2021; Kamruzzaman et al. 2022) commonly concentrated on SPI-based techniques in the country. However, a few papers also utilized remote-sensing tools for drought evaluation in Bangladesh (Alamgir et al. 2015; Rahman and Lateh 2016; Prodhan et al. 2020). Nevertheless, due to the lack of proper scientific investigation, a single index cannot manage the drought hazard. Thus, multi-index tools are required to monitor drought conditions appropriately. To narrow the gap in the literature, we have used two indices to calculate the actual state of drought and also explore the causal relationship between them. Hence, the main goal of this study is to evaluate a long-term spatiotemporal trend in drought events over Bangladesh and to show the relationship between the SPEI, SPI, PET, and precipitation anomaly.

4.2 MATERIAL AND METHODS

4.2.1 STUDY AREA DESCRIPTION

Bangladesh, a deltaic riverine sediment-induced land, is located in Southeast Asia (Rahman and Islam 2019). In most regions of the nation, drought is a sporadic occurrence, but the northern and western zones are more prone to drought events than other zones due to irregular rainfall patterns. In addition, sandy soil has a low capacity for holding moisture and a high infiltration rate. We divided the whole of Bangladesh into four regions (see Figure 4.1) – (i) northeastern (region 1); (ii) northwestern (region 2); (iii) central (region 3); and (iv) southern (region 4) –according to their geographical locations and similar kinds of hydrological settings, climatic variation, and soil type (Banglapedia 2014). Region 1 consists of Sylhet; region 2, Rangpur, Rajshahi, and Bogura; region 3, Dhaka, Faridpur, Madaripur, and Mymensingh; and region 4, Barishal, Bhola, Cox's Bazar, Feni,

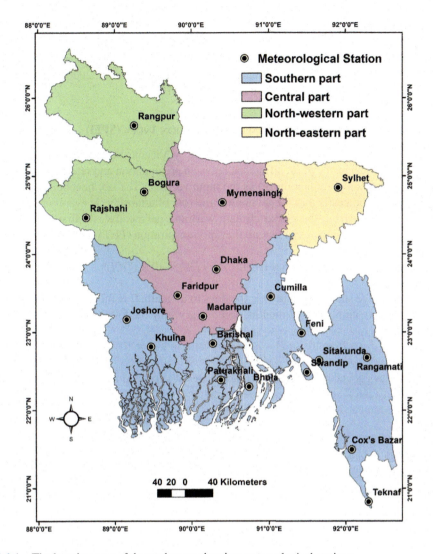

FIGURE 4.1 The location map of the study area showing meteorological stations.

Joshore, Patuakhali, Swandip, Chottogram, Teknaf, and Khulna. the average minimum temperature is 10°C and the average maximum temperature hardly reaches more than 32°C in Bangladesh. In this region, fog is a common phenomenon in winter. The northwestern part is known as an area of extremes. The summer is dry and hot, but the rainy monsoon season is wet with heavy precipitation. The central part's rainfall is abundant, being above 190 cm. The southern region includes Chittagong and a slip of country north to Comilla. The remaining parts of the country have a slight temperature change, scarcely below a mean of 12°C and over a mean of 32°C. Rainfall is about 2540 mm annually in Bangladesh.

4.2.2 DATA SOURCES

For estimating drought indices, monthly rainfall and temperature data sets from 20 meteorological stations were used in the period 1980–2017. The data sets were provided by the Bangladesh Metrological Department (BMD). However, missing data sets were observed in almost all 20 stations. From 1985 to 2015, the missing data from the 20 stations was less than 5%. The past data records filled in the missing data for each station for the respective days from the neighboring stations. The BMD follows the guidelines of the World Meteorological Organization (WMO) for weather data set acquisition and record archiving. The homogeneity test and autocorrelation of the data set were performed to show any anomaly in the data set (Islam et al. 2020). All of the data sets were passed through quality control by the staff of the BMD.

4.2.3 STANDARDIZED PRECIPITATION EVAPOTRANSPIRATION INDEX (SPEI)

The SPEI, an advanced version of the SPI, pioneered by Vicente-Serrano et al. (2010), computes the climatic water balance, the main difference between precipitation and reference evapotranspiration (P-ETo), rather than precipitation (P) as the input variable. The key benefit of SPEI over other extensively utilized indices is that it can monitor the role of atmospheric evaporative demand and temperature variability from the global climate warming perspective. At first, the Thornthwaite model was employed to compute the potential evapotranspiration (PET) by the following equation:

$$\text{PET} = 16 \times (N/12) \times (m/30) \times 10 \times (Ti/I)^a \tag{4.1}$$

where N is the average sunshine hour, m is the days in a month, Ti is the average monthly temperature, I is a cumulative number of 12 months, and a is given by the following Equation 4.2:

$$a = 6.75 \times 10^{-7} \times I^3 - 7.71 \times 10^{-5} \times I^2 + 1.79 \times 10^{-2} \times I + 0.49 \tag{4.2}$$

The shortage or excess accumulation of climatic water at various timescales (D_i) is determined by the difference between P and PET for a specific day:

$$D_i = P_i - PET_i \tag{4.3}$$

The computed D_i values are accumulated at various time scales, following a similar process as in the SPI for given month j and year i relies on the specific time scale k (months). For instance, the aggregated difference for one day in a specific year i with a 90-day timescale is computed using Equations 4.4 and 4.5:

$$X_{ij}^k = \sum_{l=j-k+1}^{90} D_{i-1,l} + \sum_{l=1}^{j} D_{i,j} \text{ ,if } j < k \text{ and} \tag{4.4}$$

$$X_{ij}^k = \sum_{l=j-k+1}^{j} D_{i,l} \text{ ,if } j \geq k \tag{4.5}$$

Next, normalize the water balance into a log–logistic distribution to get the SPEI series. To get the SPEI, the log–logistic distribution was chosen for standardizing the D time series by following Equation 4.6:

$$f(x) = \frac{\beta}{\alpha} \left(\frac{x - \lambda}{\alpha} \right) \left[1 + \left(\frac{x - \lambda}{\alpha} \right) \right]^{-2} \tag{4.6}$$

where α, β, and γ are the scale, shape, and origin parameters, respectively, for D values in the range ($\gamma > D < \infty$). Therefore, the PDF (probability density function) of the D series is provided by Equation 4.7:

$$F(x) = \left[1 + \left(\frac{\alpha}{x - \gamma} \right)^{\beta} \right]^{-1} \tag{4.7}$$

With $F(x)$, the SPEI can get the standardized values of $F(x)$ using the classical estimate by Equation 4.8:

$$SPEI = W - \frac{C_0 + C_1 W + C_2 W^2}{1 + d_1 W + d_2 W^2 + d_3 W^3} \qquad (4.8)$$

where $W = V-2 \ln (P)$ for $P \leq 0.5$ and P is the likelihood of exceeding a measured D value, $P = 1-F(x)$. If $P > 0.5$, then P is substituted by $1-P$ and the sign of the subsequent SPEI is inverted. The constants are $C_0 = 2.515517$, $C_1 = 0.802853$, $C_2 = 0.010328$, $d1 = 1.432788$, $d_2 = 0.189269$, and $d_3 = 0.001308$.

4.2.4 STANDARDIZED PRECIPITATION INDEX (SPI)

The SPI, the most popular, extensively applied index, is utilized to monitor drought events based on only the precipitation data set (McKee et al. 1993; McKee et al. 1995; Guttman 1998). Though the SPI is not a drought forecasting technique, the SPI has been employed to recognize dry and wet situations thoroughly and to assess their effects on water resource practices.

Statistically, the SPI tool was based on the cumulative probability distribution of a specific precipitation phenomenon at a particular station. At first, a two-parameter gamma density distribution function (GDDF) was used to fit the precipitation frequency of each given month for a site. The GDDF is calculated by

$$f(x) = \frac{1}{\beta^\alpha \Gamma(\alpha)} x^{\alpha-1} e^{-x/\beta} \qquad (4.9)$$

where α, β is the shape and scale parameter. x denotes the precipitation at a monthly scale. The two parameters were determined from the Thom method (Thom 1958). Next, the cumulative probability $G(x)$ at x can be calculated using the GDDF. Ultimately, $G(x)$ is altered into the SPI value with the inverse of the cumulative standard GDDF. The detailed computation technique of the SPI can be found elsewhere (Zhou and Liu 2016).

4.2.5 COMPUTATION OF THE PRECIPITATION ANOMALY

The Pa percentage is measured by using the following equation:

$$Pa = P-P_o / P \times 100\% \qquad (4.10)$$

where P is the mean precipitation at the monthly and annual scales.

4.2.6 NONPARAMETRIC MANN–KENDALL TEST

The Mann–Kendall (MK), a robust nonparametric tool, is the most applied test in hydrometeorological studies (Mann 1945; Kendall1 1975; Praveen et al. 2020). The main benefit of the MK test is that it did not require any specific sample distribution, thus ignoring the potential interference of a given outlier. This study utilized the MK test for identifying a continual variation in a time series data set. The MK test statistic is computed by

$$S = \sum_{k=0}^{n} \binom{n}{k} x^k a^{n-k} \qquad (4.11)$$

where x represents the sequential time series data values, and n is the length of the data sets by Equation 4.12:

$$sgn\left(x_j - x_i\right) = \begin{cases} 1 & \text{if } x_j > x_i \\ 0 & \text{if } x_j = x_i \\ -1 & \text{if } x_j < x_i \end{cases} \tag{4.12}$$

When $n \geq 8$, the statistic S is nearly normally distributed, with the variance as given by Equation 4.13:

$$V\left(S\right) = \frac{n\left(n-1\right)\left(2n+5\right) - \sum_{m=1}^{n} t_m m\left(m-1\right)\left(2m+5\right)}{18} \tag{4.13}$$

where tm is the number of extent m. The standardized test statistic Z is calculated by

$$Z = \begin{cases} \dfrac{S-1}{\sqrt{V\left(S\right)}} & S > 0 \\ 0 & S = 0 \\ \dfrac{S+1}{\sqrt{V\left(S\right)}} & S < 0 \end{cases} \tag{4.14}$$

It is noted that the confidence levels are set at 0.01, 0.05, and 0.1 in this study.

4.2.7 LINEAR REGRESSION MODEL

To find out the association among the SPEI, SPI, Pa, and PET, linear correlation was used to understand how the variables are linked with each other. Linear correlation can be signified by R, and the value will be between –1 and 1. The elements denote a strong relationship if the value is 1. Likewise, if the coefficient comes close to –1, it is considered a negative relationship. If the linear coefficient is zero, there is no relation between the data given. The R technique estimates the causal factors from the many parameters involved. It provides the significance of the estimating parameters (Thompson 1995).

4.3 RESULTS

4.3.1 THE SPATIAL TREND PATTERN OF DROUGHT MONITORING INDICES

This study assessed the past drought scenarios of Bangladesh using both short-term (3 and 6 months) and long-term (12 and 24 months) observations by calculating the SPEI and SPI for 1980–2017. The past drought trend is presented in Figure 4.2 and Figure 4.3 using MK-Z statistics. The more the positive value of Z statistics, the more the drought trend. Short-term observation of SPEI-3 revealed that Jashore and Rangpur showed a significant ($P < 0.05$) increasing trend of drought, followed by Bogura and Rajshahi (these two showed no significant trend). MK-Z values of SPEI-6 showed that the Jashore, Bogura, Rajshahi, and Rangpur stations showed a 95% significant increasing trend ($3.17 > Z > 1.94$) of drought. Both SPEI-12 and SPEI-24 showed that a 99% significant ($P < 0.01$) increasing trend was demonstrated by the Jashore and Rangpur stations; and a 95% significant ($P < 0.05$) increasing trend was shown by the Rajshahi and Bogura stations (Figure 4.2). Mymensingh, Dhaka, and Faridpur also showed an increasing trend of drought found by the SPEI-3, SPEI-6, SPEI-12, and SPEI-24 (Figure 4.2). MK-Z values of the SPI-3, SPI-6, SPI-12, and SPI-24 are presented in Figure 4.3. In both short-term and long-term observations, Jashore showed a 99% significant increasing trend (Z value greater than 3.17); Rajshahi and Bogura showed a 95% significant growing drought

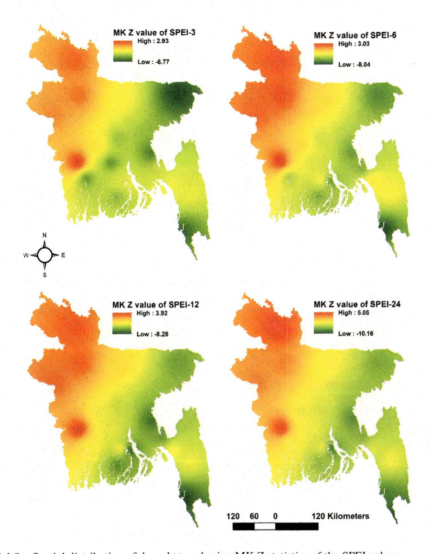

FIGURE 4.2   Spatial distribution of drought trend using MK-Z statistics of the SPEI values.

trend found by SPI (Figure 4.3). Rangpur showed a 95% increasing trend seen by the SPI-3 and SPI-6, while the SPI-12 and SPI-24 revealed a 99% significant growing drought trend in Rangpur (Figure 4.3). Like the SPEI, Mymensingh, Dhaka, and Faridpur also showed an increasing trend of drought as found by the SPI.

4.3.2   CORRELATIONS OF DROUGHT MONITORING INDICES WITH PET AND PA

The correlation coefficient between the SPEI and PET, and SPEI and Pa is presented in Table 4.1. The SPEI and PET showed a positive correlation in all the subregions, in which the SPEI-3 and PET of the northwestern region showed a significant ($P < 0.05$) correlation (0.524). The SPEI-3 and Pa, and SPEI-6 and Pa showed a significant ($P < 0.05$) positive correlation in all the subregions in Bangladesh, except SPI-3 versus Pa in the southern region. No significant correlation was found between the SPEI-24 and Pa. The SPEI-12 showed a significant positive correlation with Pa in the central and southern areas. Overall, the SPEI-3, SPEI-6, and SPEI-12 showed a significant positive

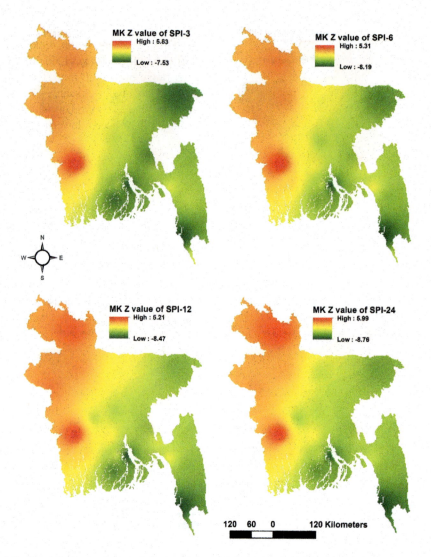

FIGURE 4.3   Spatial distribution of drought trend using MK-Z statistics of the SPI values.

TABLE 4.1

Correlation Coefficient (R) of SPEI with PET and Pa at the 3-, 6-, 12-, 24-Month Timescales for Different Regions of Bangladesh

Region	SPEI vs. PET				SPEI vs. Pa			
	3 Months	6 Months	12 Months	24 Months	3 Months	6 Months	12 Months	24 Months
1	.299	.386	.369	.386	.725*	.619*	.494	.407
2	.524*	.298	.291	.127	.697*	.502*	.351	.107
3	.125	.363	.184	.047	.744*	.654*	.500*	.286
4	.186	.192	.192	.173	.494	.585*	.580*	.139
Whole Country	.320	.188	.189	.194	.543*	.528*	.585*	.131

* Correlation is significant at the 0.05 level.

TABLE 4.2

Correlation Coefficient (R) of SPI with PET and Pa at the 3-, 6-, 12-, 24-Month Timescales for Different Regions of Bangladesh

Region	SPI vs. PET				SPI vs. Pa			
	3 Months	6 Months	12 Months	24 Months	3 Months	6 Months	12 Months	24 Months
1	.442	.278	.428	.368	.716*	.382	.473	.452
2	.533*	.134	.279	.312	.790*	.392	.324	.199
3	.408	.136	.109	.010	.810*	.628*	.523*	.249
4	.329	.114	.143	.064	.729*	.472	.441	.056
Whole Country	.375	.148	.188	.064	.774*	.509*	.511*	.056

*Correlation is significant at the 0.05 level.

correlation with Pa throughout the country (Table 4.1). Like the SPEI, a positive correlation was found between the PET and SPI-3, SPI-6, SPI-12, and SPI-24 in all the regions (Table 4.2). The SPI-3 and PET showed a significant positive correlation in the northwestern area. Overall, the SPI and PET showed an insignificant correlation for the whole country. In contrast, the SPI-3 and Pa showed a significant positive correlation in all the subregions. In the central region, the SPI-6 and SPI-12 significantly positively correlated with Pa. Overall, the SPI-3, SPI-6, and SPI-12 significantly positively correlated with Pa for the whole country (Table 4.2). The SPI-24 showed an insignificant positive correlation in all the subregions and the entire country.

4.3.3 CORRELATION BETWEEN SPEI AND SPI

Both the SPI and SPEI are effective and the most used tools for assessing drought. This study employed both tools for determining drought. A Pearson correlation technique was used to estimate the correlation between the SPEI and SPI (Figure 4.4). A very strong correlation (the value of R is very close to 0.9) was explored in subregion number 3 (central region). Then, in the southern part, the correlation between the SPEI and SPI was strong (R = 0.78). Correlation in the northeastern area was also strong (R = 0.75). Only in the northwestern part does correlation drop down to the high stage. But, in all the subregions, a significant correlation was found, which means the results of the SPEI and SPI are reliable and usable.

4.4 DISCUSSION

Our study assessed both short-term and long-term drought conditions from 1980 to 2017. Both the SPEI and SPI at 3-, 6-, 12-, and 24-month timescales showed a significantly (P < 0.01 and P < 0.05) increasing trend of drought in Jashore, Rangpur, Rajshahi, and Bogura (northwestern region). An insignificant increasing trend of the drought was found in Mymensingh, Dhaka, and Faridpur (Central region) stations. Miah et al. (2017) found by using the SPEI with similar results as the present study that the frequency and increased drought trend were higher in the northwestern part of Bangladesh, and extreme drought was explored in Jashore. Alamgir et al. (2019) examined that the north and northwest regions showed extreme drought events and a significant increasing trend returning to 100 years, and the southern part showed a low drought increasing trend revealed by calculating the Drought Risk Index (DRI), which is similar to the findings of this study. Hoque et al. (2020) demonstrated by using an analytical hierarchy process that 77% of northwestern Bangladesh's land is vulnerable to moderate to extreme drought hazards, and drought in this region

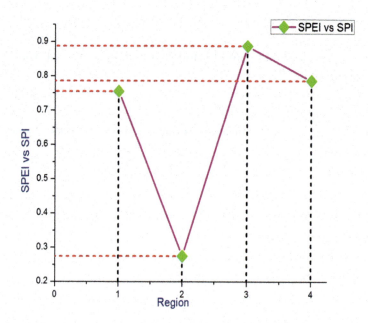

FIGURE 4.4 The correlation trend graph between the SPI and SPEI at different timescales in the selected four regions.

significantly increased, which is analogous to this study. Mondol et al. (2018) found similar findings to the present study: the northwestern part showed an increasing trend of meteorological drought, followed by the central region.

An insignificant positive correlation was found between both the SPEI and PET, and SPI and PET. The SPI and SPEI at 3, 6, and 12 months showed a significant positive correlation with Pa all over the country. Yao et al. (2018) reported that the drought severity calculated by the SPEI varied with the variation of PET in Xinjiang, which is similar to the findings of this study in that SPEI-3 in the northwestern region was significantly correlated with the variation of PET. Merabti et al. (2018) reported that SPI hardly showed a significant correlation with PET as this index uses only the precipitation data, and the present study demonstrated the findings of Merabti et al. (2018) except in one region on a three-month timescale, where all the correlation results between the SPI and PET showed insignificant relations. The previous study rarely employs correlation between the SPI and Pa, or SPEI and Pa As the SPI and SPEI indices are mainly based on the variation of precipitation data, the correlation of these drought indices with Pa is significant in understanding the actual scenarios. The present study revealed that drought intensity and frequency significantly increased with the increase of Pa and vice versa.

The SPEI and SPI showed a significant correlation in all the subregions. But in the central region, the SPEI and SPI showed a very strong correlation. Labudova et al. (2014) revealed a very low correlation between the SPI-3 and SPEI-3 in southern Slovakia. Stagge et al. (2014) found a significant moderate-to-high correlation between SPEI and SPI for calculating moderate drought, which is analogous to this study. Yao et al. (2018) revealed a significant correlation between SPI and SPEI in four river basins of China for assessing drought, which is almost the same as the present study's findings. In addition, Banimahd and Khalili (2013) found a significant correlation between the SPI and SPEI at three- and six-month timescales in several climatic zones, which is analogous to the present study.

4.5 CONCLUSIONS

The present study assessed the drought condition of Bangladesh through both short-term and long-term observations using the SPI and SPEI from 1980 to 2017. The northwestern part of the country showed a significantly increasing drought trend as found by the MK test of both the SPI and SPEI at 3-, 6-, 12-, and 24-month timescales. The central region also showed an increasing drought trend in the study period. The present study found an insignificant positive correlation between PET and the other two drought monitoring indices (SPI and SPEI). In contrast, the SPI and SPEI at 3-, 6-, and 12-month timescales showed a significant positive trend with Pa, as these two indices heavily depended on rainfall variation. The SPEI and SPI showed a significant correlation in all the subregions, and a very strong correlation was found in the central region. The findings of this study will help planners think sustainably to reduce the impacts of drought in the northwestern and central areas of this country. As this study was involved in the spatial and temporal variation of different timescales, it is anticipated that it will be a helpful guide toward understanding drought features and assist in formulating comprehensive management approaches to overcome the drought problem effectively in Bangladesh.

REFERENCES

Ahmed, AU (2006). Bangladesh climate change impacts and vulnerability: A synthesis. *Climate Change.* Cell, Department of Environment, Dhaka.

Alamgir, M, Shahid, S, Hazarika, MK et al. (2015). Analysis of meteorological drought pattern during different climatic and cropping seasons in Bangladesh. *J Amer Water Resour Assoc* 51(3):794–806. https://doi.org/10.1111/jawr.12276.

Alamgir, M, Mohsenipour, M, Homsi, R, Wang, X, Shahid, S, Shiru, MS, Alias, NE, Yuzir, A (2019). Parametric assessment of seasonal drought risk to crop production in Bangladesh. *Sustainability* 11(5):1442, 1–17.

Ali, A (1999). Climate change impacts and adaptation assessment in Bangladesh. *Clim Res* 12:109–116. American Meteorological Society 91:1351–1354.

Banglapedia. (2014). Drought in Bangladesh. Accessed: Feb. 12, 2019. [Online]. Available: http://en.banglapedia.org/index.php?.

Banimahd, SA, Khalili, D (2013). Factors influencing Markov chains predictability characteristics, utilizing SPI, RDI, EDI and SPEI drought indices in different climatic zones. *Water Resour Manage* 27(11):3911–3928.

Bonaccorso, B, Peres, D, Cancelliere, A, Rossi, G (2013). Large scale probabilistic drought characterization over europe. *Water Resour Manag* 27(6):1675–1692.

Dai, AG, Zhao, T, Chen, J (2018). Climate change and drought: A precipitation and evaporation perspective. *Curr Clim Change Rep* 4:301–312.

Dina, RA, Islam, ARMT (2020). Assessment of drought disaster risk in Boro rice cultivated areas of northwestern Bangladesh. *Eur J Geosci, EURAASS* 2(1):19–29.

Guttman, NB (1998). Comparing the palmer drought index and the standardized precipitation index. *J Am Water Resour Assoc* 34:113–121.

Hoque, MAA, Pradhan, B, Ahmed, N (2020). Assessing drought vulnerability using geospatial techniques in northwestern part of Bangladesh. *Sci Total Environ* 705:135957.

IPCC (2014). *Climate Change 2014: IPCC Fifth Assessment Synthesis Report-Summary for Policymakers: An Assessment Of Inter-governmental Panel on Climate Change.* Cambridge University Press, Cambridge.

Islam, ARMT, Rahman, MS, Khatun, R, Hu, Z (2020). Spatiotemporal trends in the frequency of daily rainfall in Bangladesh during 1975–2017. *Theor Appl Climatol* 141(3–4):869–887. https://doi.org/10.1007/s00704-020-03244-x.

Islam, ARMT, Shen, S, Hu, Z, Rahman, MA (2017). Drought hazard evaluation in boro paddy cultivated areas of western Bangladesh at current and future climate change conditions. *Adv Meteorol* 2017:3514381, 12 pages. https://doi.org/10.1155/2017/3514381.

Islam, ARMT, Shen, S, Yang, SB, Hu, Z, Chu, R (2019). Assessing recent impacts of climate change on design water requirement of Boro rice season in Bangladesh. *Theor Appl Climatol* 138:97–113. https://doi.org/10.1007/s00704-019-02818-8.

Islam, ARMT, Tasnuva, T, Sarker, SC, Rahman, M, et al. (2014). Drought in Northern Bangladesh: Social, agro-ecological impact and local perception. *Int J Ecosyst* 4(3):150–158. https://doi.org/10.5923.j.ije .20140403.07.

Kamruzzaman, M, Almazroui, M, Salam, MA, Mondol, MAH, Rahman, MM, Deb, L, Kundu, PK, Zaman, MAU, Islam, ARMT (2022). Spatiotemporal drought analysis in Bangladesh using the standardized precipitation index (SPI) and standardized precipitation evapotranspiration index (SPEI). *Sci Rep* 12:20694. https://doi.org/10.1038/s41598-022-24146-0.

Kao, SC, Rao, SG (2010). A copula-based joint deficit index for droughts. *J Hydrol* 380 (1–2):121–134.

Kendall, MG (1975). *Rank Correlation Methods*, 4th edn. Charles Griffin, London.

Labudova, L, Schefczyk, L, Heinemann, G (2014). The comparison of the SPI and the SPEI using COSMO model data in two selected Slovakian river basins.   EGU2014-4357-2, 16, 1p.

Mann, HB (1945). Nonparametric tests against trend. *Econometrica* 13:245–259.

McKee, TB, Doesken, NJ, Kleist, J (1993). The relationship of drought frequency and duration to time scales. In Preprints Eighth Conference, on Applied Climatology. American Meteorology Society, Anaheim, pp. 179–184.

McKee, TB, Doesken, NJ, Kleist, L (1995). Drought monitoring with multiple scales. In: Proceedings of the 9th Conference on Applied Climatology, Boston, MA, pp. 233–236.

Merabti, A, Meddi, M, Martins, DS, & Pereira, LS (2018). Comparing SPI and RDI applied at local scale as influenced by climate. *Water Resour Manage* 32(3):1071–1085.

Miah, MG, Abdullah, HM, Jeong, C (2017). Exploring standardized precipitation evapotranspiration index for drought assessment in Bangladesh. *Environ Monit Assess* 189(11):547.

Mondol, MAH, Al-Mamun, IM, Jang D (2018). Precipitation concentration in Bangladesh over different temporal periods. *Adv in Meteorology* 2018:1849050. https://doi.org/10.1155/2018/1849050.

Mondol, MAH, Zhu, X, Dunkerley, D, Henley, BJ (2021). Observed meteorological drought trends in Bangladesh identified with the Effective Drought Index (EDI). *Agric Water Manag* 255:107001.

Niemeyer, S (2008). New drought indices. *Options Méditerranéennes* 80(80):267–274. https://doi.org/10.1017 /CBO9781107415324.004.

Palmer, WC (1965). Meteorological drought. US Department of Commerce, Weather Bureau. *Res Paper* 45:58.

Parvin, GA, Fujita, K, Matsuyama, A, Shaw, R, Sakamoto, M (2015). Climate change, flood. *Food Secur* 13:235–254.

Peña-Gallardo, M et al. (2019). Complex influences of meteorological drought time-scales on hydrological droughts in natural basins of the contiguous Unites States. *J Hydrol* 568:611–625.

Praveen, B, Talukdar, S, Shahfahad, MS, Mondal, J, Sharma, P, Islam, ARMT, Rahman, A (2020). Analyzing trend and forecasting of rainfall changes in India using non-parametrical and machine learning approaches. *Sci Rep* 10:10342. https://doi.org/10.1038/s41598-020-67228-7.

Prodhan, FA, Zhan, J et al. (2020). Monitoring of drought condition and risk in Bangladesh combined data from satellite and ground meteorological observations. *IEEE Acess* 8:93264–93282. https://doi.org/10 .1109/ACCESS.2020.2993025.

Rahman, MR, Lateh, H (2016). Meteorological drought in Bangladesh: Assessing, analysing and hazard mapping using SPI, GIS and monthly rainfall data. *Environ Earth Sci* 75(12):1026. https://doi.org/10.1007/ s12665-016-5829-5.

Rahman, MS, Islam, ARMT (2019). Are precipitation concentration and intensity changing in Bangladesh overtimes? Analysis of the possible causes of changes in precipitation systems, *Sci Total Environ* 690:370–387. https://doi.org/10.1016/j.scitotenv.2019.06.529.

Ruane, AC, Major, DC, Winston, HY, Alam, M, Hussain, SG, Khan, AS, Rosenzweig, C (2013). Multi-factor impact analysis of agricultural production in Bangladesh with climate change. *Glob Environ Change* 23(1):338–350.

Selvaraju, R, Subbiah, AR, Baas, S, Juergens, I (2006). *Livelihood Adaptation to Climate Variability and Change*. Rome, Italy: FAO.

Shafer, BA (1982). Development of a surface water supply index (SWSI) to assess the severity of drought conditions in snowpack runoff areas. In Proceedings of the 50th Annual Western Snow Conference. Fort Collins, CO: Colorado State University, 1982.

Shahid, S, Behrawan, H (2008). Drought risk assessment in the western part of Bangladesh. *Nat Hazard* 46:391–413. https://doi.org/10.1007/s11069-007-9191-5.

Sheffield, J, Wood, E, Roderick, M (2012). Little change in global drought over the past 60 years. *Nature* 491(7424):435–438.

Stagge, JH, Tallaksen, LM, Xu, CY, Van Lanen, HA (2014). Standardized precipitation-evapotranspiration index (SPEI): Sensitivity to potential evapotranspiration model and parameters. In *Hydrology in a Changing World*. IAHS Publ, (Vol. 363, pp. 367–373), In: *7th World Flow Regimes from International and Experimental Network Data-Water Conference*, FRIEND-Water 2014, Montpellier, 2014-10-07/2014-10-10.

Thom, HCS (1958). A note on the gamma distribution. *Mon Weather Rev* 86:117–122.

Thompson, B (1995). Stepwise regression and stepwise discriminant analysis need not apply. *Educ Psychol Meas* 55(4):525–534.

Trenberth, K, Dai, A, Schrier, G, Jones, P, Barichivich, J, Briffa, KR (2014). Global warming and changes in drought. *Nat Clim Chang* 4(1):17–22.

Uddin, MJ, Hu, J, Islam, ARMT, Eibek, KU, Zahan, MN (2020). A comprehensive statistical assessment of drought indices to monitor drought status in Bangladesh. *Arab J Geosci* 13:323. https://doi.org/10.1007/s12517-020-05302-0.

Umma, H, Rajib, S, Yukiko, T (2011). Socioeconomic impact of droughts in Bangladesh. *Community Environ Disaster Risk Manag* 8:25–48. https://doi.org/10.1108/S2040-7262(2011)0000008008.

Vicente-Serrano, SM, Beguería, S, López-Moreno, J (2010). A multiscalar drought index sensitive to global warming: The standardized precipitation evapotranspiration index. *J Clim* 23:1696–1718.

Vicente-Serrano, SM, Domínguez Castro, F, McVicar, TR, et al. (2020). Global characterization of hydrological and meteorological droughts under future climate change: The importance of timescales, vegetation-CO2 feedbacks and changes to distribution functions. *Int J Climatol* 40:2557–2567.

Yao, J, Zhao, Y, Chen, Y, Yu, X, & Zhang, R (2018). Multi-scale assessments of droughts: A case study in Xinjiang, China. *Sci Total Environ* 630:444–452.

Zannat, F, Islam, ARMT, Rahman, MA (2019). Spatiotemporal variability of rainfall linked to ground water level under changing climate in northwestern region, Bangladesh. *Eur J Geosci, EURAASS* 1(1):35–58,

Zargar, A, Sadiq, R, Naser, B, Khan, FI. (2011). A review of drought indices. *Environ Rev* 19(1):333–349, Dec. 2011. https://doi.org/10.1139/a11-013.

Zhou, H, Liu, Y (2016). SPI based meteorological drought assessment over a Humid Basin: Effects of processing schemes. *Water* 8(9):373. https://doi.org/10.3390/w8090373.

Zinat, MRM, Salam, R, Badhan, MA, Islam, ARMT (2020). Appraising drought hazard during Boro rice growing period in western Bangladesh. *Int J Biometeorol*. https://doi.org/10.1007/s00484-020-01949-2.

5 Spatiotemporal Analysis of Meteorological Drought in Tripura

*Aribam Priya Mahanta Sharma, Deepak Jhajharia,
Ghanashyam Singh Yurembam, Shivam Gupta,
Ghanshyam T. Patle, and T. Loidang Chanu*

CONTENTS

5.1 INTRODUCTION

Climate change poses a serious threat to global water regimes, altering seasonal and interannual precipitation patterns. Trend analyses of various climatic parameters in different regions of the world have conclusively proved that climate change may intensify with natural extremes, i.e., dry regions to drier and wet regions to wetter (Mahajan and Dodamani, 2015; Sharma and Goyal, 2020). Drought is one such natural extreme that occurs slowly and is caused mainly due to the deficiency of availability of water in a region. It is expected that the severity, duration, and intensity of droughts may increase, which in turn may be a significant concern in regions throughout the globe pertaining to the availability of water for various users or sectors, such as municipal and industrial use, irrigation, navigation support, hydropower, and environmental flows. Drought also provides a heuristic opportunity for the scientific community, water resource managers, and policymakers (Xu et al., 2005; Field et al., 2012). Drought is a slow phenomenon that has long-lasting impacts on the well-being of society and the environment, and also reduces gross primary productivity resulting in distorting the food security and welfare of about 55 million people globally (WHO, 2020; Dai et al., 2004).

Drought is classified as meteorological, agricultural, hydrological, or socioeconomic (Khan et al., 2017; Kwartang et al., 2017). A deficit of precipitation over a region in a specific period results in meteorological drought and its continuous persistence can cause other types of drought (Djebou, 2017). India is among the most drought-prone countries in the world and is highlighted among the top ten high socioeconomic-risk countries in the world due to ever-increasing drought

DOI: 10.1201/9781003276548-5

57

in the climate change scenario (Liu and Chen, 2020). The current major concern is that the northeastern part of India, which receives an average annual rainfall of 2000 mm to 5000 mm (Dikshit and Dikshit, 2014), has encountered more unprecedented drought than the western parts of India in recent years due to scanty and uneven precipitation over the region (Parida and Oinam, 2015). Tripura, one of the states among the eight states of NE India, has witnessed significant and drastic changes in the environment with the state getting dryer, leading to frequent drought-like conditions in the state in particular and the NE region in general (TSPCB, ENVIS-Centre). High-risk droughts are projected for regions with recurring drought episodes with increased spatial extent because of climate change (Kundzewicz et al., 2008; Alamgir et al., 2020). Lessening of drought impact is possible by timely identification, characterization, and mitigation for attaining water security through different drought-proofing systems. The meteorological drought is characterized by its severity, duration, and intensity. Conventional drought monitoring includes analysis of the rainfall data from rain gauge observations, which is limited to northeastern India with its vast hilly terrain. However, as an alternative, using gridded precipitation data allows a better assessment of drought events.

Drought studies conducted all over the world use the Standardized Precipitation Index (SPI) for describing and comparing droughts (Piccarreta et al., 2004; Xu et al., 2011; Nalbantis and Tsakiris, 2009; Paulo et al., 2005; Khalili et al., 2011). The World Meteorological Organization (WMO) has recommended the use of the SPI for the determination of meteorological drought (Hayes, 2003). McKee et al. (1993) developed the SPI to show the effect of precipitation deficits for both short periods (i.e., impact on agriculture) and long periods (i.e., impact on water resources). Trend detection is performed to detect the increase (or decrease) of the values of a random variable over some period of time in statistical terms (Helsel and Hirsch, 2002). Trend identification methods can further be classified as parametric and nonparametric methods (Tabari et al., 2011). The present study adopted the Mann–Kendall (MK) non–parametric test for the detection of trends (Yue et al., 2002; Drápela and Drápelová, 2011; Paulo et al., 2012), which has the advantage of not fitting any particular probability distribution.

In this study, we present an SPI-based drought analysis for Tripura, identifying trends in the SPI values and assessing the drought variability in the state. The analysis is performed using long-term gridded precipitation data over the period 1980–2013 of four stations in Tripura. The MK test was used for trend analysis and identification of the seasonality of the droughts, and the slope of the trend line was determined using Sen's slope estimator. This study will help in improving our understanding of the variability of drought occurrence and the drought trend in the region.

5.2   METHODS AND MATERIALS

5.2.1   STUDY AREA AND DATA

The state of Tripura in northeastern India is characterized by a warm and humid tropical climate with annual rainfall ranging from 1922 mm to 2855 mm, which increases from southwest to northeast. Tripura has a total geographical area of 10,492 km^2 with 58% of the total area under forest, followed by agricultural land. The location map of the study area and details of rainfall stations representing the four districts are presented in Figure 5.1 and Table 5.1, respectively. The 0.3° × 0.3° spatial resolution gridded rainfall data for the years 1980 to 2013 were acquired for all four districts of Tripura from Global Weather Data for SWAT (Soil and Water Assessment Tool). Figure 5.2 shows the monthwise details of monthly rainfall based on the gridded rainfall data of all four districts of Tripura state. Figure 5.2 reveals that the occurrence of the lowest (the highest) monthly rainfall values varied from about 50 mm (150 mm) in March to 300 mm (650 mm) in May over the South (North) Tripura district in the pre-monsoon season. On the other hand, in the four months of June to September, monsoon season, the monthly rainfall values were found to vary from more than 300 mm over South Tripura to more than 650 mm over the North Tripura district. There was

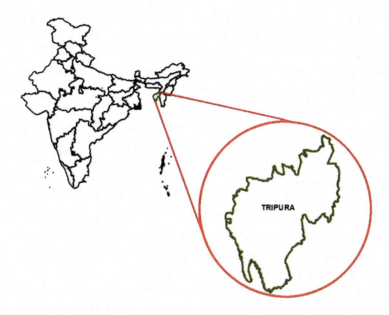

FIGURE 5.1   Location of the study area.

TABLE 5.1
Details of Districts of Tripura, Northeast India

District	Latitude °N	Longitude °E	Elevation (m)
North Tripura	24.19	92.18	56
West Tripura	23.88	91.56	74
Dhalai	23.88	91.87	97
South Tripura	23.57	91.56	97

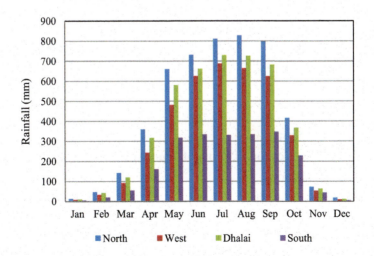

FIGURE 5.2   Monthly rainfall in all four districts of Tripura, Northeast India.

hardly any rainfall during the months of December and January, and to some extent in November and February over the whole state of Tripura.

5.2.2 STANDARDIZED PRECIPITATION INDEX (SPI)

The SPI was designed by McKee and his associates (1993) to quantify precipitation deficit for multiple timescales. SPI analysis involved assembling the precipitation time series over the time period of interest for several timescales to monitor drought Different SPI timescales possess different drought indicator qualities, e.g., short-term SPI scales (3 and 6 months) are effective for detecting meteorological drought, however, long-term SPI scales (e.g., 12 months) are effective for droughts affecting vegetation and agriculture, and water-resources-related (hydrological) drought (Edwards and McKee, 1997; Wu et al., 2001). The major advantages of the SPI are that it gives an indication for the initiation and termination of drought, and it requires minimum input data, i.e., precipitation for drought analysis (Buttafuoco et al., 2015). The first step of SPI calculation is to fit rainfall values with the gamma distribution. Next is to transform the fitted gamma distribution values to the standard normal distribution with mean 0 and standard deviation 1. The normalized format thus obtained is the SPI. For the standard normal distribution curve, the right-hand side of the curve indicates wet conditions and the left-hand side indicates dry conditions. McKee et al. (1993) classified the climate type from extremely wet to extremely dry based on the SPI values, as given in Table 5.2.

5.2.3 TREND ANALYSIS

The principle of trend analysis is to detect increasing or decreasing pattern changes over time for a series of observations of a random variable. Nonparametric trend tests are advantageous over the powerful parametric test in the fact that it takes independent data and can accommodate outliers in the data (Tabari et al., 2011). The MK test, based on ranking, is used for determining monotonic trends (Helsel and Hirsch, 2002) and has a major advantage in that it is free from statistical distributions, which are required for the parametric method. The WMO has recommended the MK test for the assessment of monotonic trends in hydrometeorological time series (Tian et al., 2012). In this study, trend analysis has been performed on the SPI values for 34 years (1980–2013). The rank-based nonparametric test or MK test trend analysis was used to detect the significant trends in the SPI. The MK statistic S is given as

$$S = \sum_{i=1}^{n-1} \sum_{j=i+1}^{n} sgn\left(x_j - x_i\right) \tag{5.1}$$

where S and sgn represent the MK statistic and signum function, respectively. The application of the trend test is done to a time series x_i that is ranked from $i = 1, 2, \ldots, (n-1)$ and x_j that is ranked from

---

TABLE 5.2

Classification of Drought Type Based on the SPI

Drought Class	SPI Value
Moderate drought	−1.00 to −1.49
Severe drought	−1.50 to −1.99
Extreme drought	−2.00 and less

Source: McKee et al. (1993).

$j = i + 1, 2, \ldots, n$. x_i for each data point is taken as a reference point and is compared with the rest of the data points x_j so that

$$sgn\left(x_j - x_i\right) = \begin{cases} 1 \, if \left(x_j - x_i\right) > 0 \\ 0 \, if \left(x_j - x_i\right) = 0 \\ -1 \, if \left(x_j - x_i\right) < 0 \end{cases} \tag{5.2}$$

The variance statistic is given as

$$Var\left(S\right) = \frac{n\left(n-1\right)\left(2n+5\right) - \sum_{i=1}^{m} t_i\left(i\right)\left(i-1\right)\left(2i+5\right)}{18} \tag{5.3}$$

where t_i is considered as the number of ties up to sample i. The presence of a statistically significant trend is evaluated using the Z_c value. The Z_c test statistic is given by

$$Z_c = \begin{cases} \dfrac{S-1}{\sqrt{Var\left(S\right)}} \, if \, S > 0 \\ 0 \, if \, S = 0 \\ \dfrac{S+1}{\sqrt{Var\left(S\right)}} \, if \, S < 0 \end{cases} \tag{5.4}$$

The presence of a statistically significant trend is evaluated by using the Z_c value. Positive values of Z_c indicate increasing trends, while negative values show decreasing trends. To test for an increasing or decreasing monotonic trend (a two-tailed test) at α level of significance, the H_0 (null hypothesis of no trend in data) should be rejected if $| Z_c | > Z_{1-\alpha/2}$, where $Z_{1-\alpha/2}$ is obtained from the standard normal cumulative distribution tables. For example, the null hypothesis was rejected at the 5% significance level if $| Z_c | > 1.96$. A higher Z_c value indicates that the trend is more statistically significant.

Sen's slope estimator is used to predict the magnitude of the trend and is given by

$$Q_i = \frac{\left(x_p - x_q\right)}{p - q} \, for \, i = 1, 2, \ldots N, p > q \tag{5.5}$$

where x_p and x_q are data values at times p and q. Sen's slope estimator is signified as the median of these N values of Q_i. The Sen's slope estimator is computed by

$$Q = \begin{cases} Q_{\frac{N+1}{2}} & if \, N \, is \, odd \\ \dfrac{\left(Q_{\frac{N}{2}} + Q_{\frac{N+2}{2}}\right)}{2} & if \, N \, is \, even \end{cases} \tag{5.6}$$

5.3 RESULTS AND DISCUSSION

5.3.1 STANDARDIZED PRECIPITATION INDEX

The precipitation data from the period 1980 to 2013 was used for the spatiotemporal evaluation of the SPI at three different monthly scales (3, 6, and 12 months) for all four districts of Tripura. The time series of the SPI for the study area at 3, 6, and 12 monthly scales are shown in Figure 5.3 and the positive values represent wet periods and negative values represent dry periods.

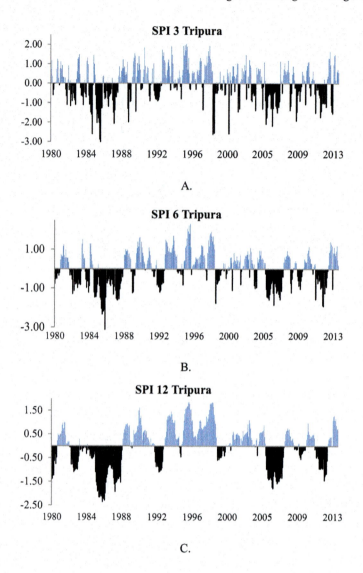

FIGURE 5.3 Temporal evaluation of drought over Tripura.

The minimum timescale considered (SPI-3) detected 26 drought events in the Tripura region, which has a maximum duration of 9 months exhibiting a severity of –17.18. In the intermediate timescale (SPI-6), 16 drought events were detected with a maximum drought duration of 9 months and maximum severity of –18.51. The maximum timescale (SPI-12) detected 7 drought events and had a maximum drought duration of 12 months and a severity of –33.86. The temporal evaluation of different SPI timescales over the Tripura region is represented in Figure 5.3.

Mainly, the short timescale of SPI values (3 months) recorded more numbers of drought, followed by the intermediate, then longest timescale recorded. The early warning of meteorological drought along with interim soil moisture and crop yield deficit, i.e., agricultural drought, can be described by shorter timescales (3 and 6 months) (Ajaz et al., 2019; Svoboda et al., 2012; Guo et al., 2017). On the other hand, the identification of wet and dry periods becomes much more evident and reflects a long-term drought pattern (hydrological drought) that is signified by a long SPI timescale.

5.3.2 Drought Identification and Characterization

SPI-6, being the best timescale for the identification of droughts (Fluixá-Sanmartín et al., 2018), was used to define drought event characteristics and perform run theory for the identification of drought duration, severity, and intensity in all four districts during the study period. The majority of the districts encountered 20 drought events; the South Tripura district experienced the highest number of drought events at 23, and the North and West Tripura district experienced the fewest drought events (20) during the entire study period. Figure 5.4 illustrated the occurrence of drought events in different districts of the study regions.

For the entire study period, from 1980 to 2013, the longest drought duration occurred for 23 months and the maximum severity recorded was –33.91. Upon district-wise analysis, North Tripura recorded three major droughts, one with maximum severity of –33.91 prevailing for 23 months from May 2005 until March 2007. In the case of West Tripura, three major droughts were also recorded with the worse scenario of –22.90 severity for 10 months from June 1985 until March 1986. For the Dhalai district, five major droughts were recorded with the worse event of –20.26 severity magnitude prevailing for 10 months from June 2011 until March 2012. South Tripura recorded three major droughts having the most severe event of magnitude –23.62 for 10 months of drought duration from June 1985 to March 1986, shown in Figure 5.5.

District-wise drought events for the Tripura region listed in Table 5.3. Almost 20 drought events were recorded in all four districts. The South Tripura district recorded the highest number of drought events (23). The longest drought duration was observed in North Tripura for 23 months. West Tripura, Dhalai, and South Tripura had the same maximum drought duration, i.e., 10 months. The worst drought severity magnitude (–33.91) occurred in North Tripura during the study period. The average drought characteristics of all the districts exhibited familiar drought conditions over the entire study area.

The spatial representation of three major drought events recorded over the Tripura region with SPI-6 during 1980–2013 is presented in Figure 5.6. The spatial pattern of SPI-6 for these drought events showed variability with the temporal pattern. Drought event 1 occurred from June 1985 to March 1986, and had severe conditions prevailing in the major portions of South and West Tripura. The severity of drought events decreases as it moves from the northeastern portion of the state. Contrary to the previous drought event, drought events 2 (1987) and 3 (2005) indicated severe drought conditions, mostly in the North Tripura district.

5.3.3 Trend Analysis

The MK test of trends was performed for the Tripura region on a district-wise basis at annual and seasonal scales during 1980–2013, and is shown in Table 5.4. All the districts showed a statistically significant trend in one or more timescales and also in many of the seasons. The negative trends were observed particularly during the pre-monsoon and winter season at a short timescale, i.e., three months SPI, in almost all the districts. The decreasing trend in the SPI for the pre-monsoon and winter seasons may be attributed to the precipitation deficit over the study region. However, in the monsoon and post-monsoon season, a positive trend is observed in three districts at all SPI timescales, except for the Dhalai district. On an annual basis, the Dhalai district showed a negative trend at all timescales, with sparse occurences in West Sikkim. The result of the trend analysis revealed that the magnitude of negative SPI trends is high during the pre-monsoon and winter seasons, which indicates an increasing trend of meteorological drought that could lead to soil moisture depletion and ultimately crop failure. This finding may be beneficial in regard to early information of drought onset for the different growing seasons (Pandey et al., 2020), which enables supplemental irrigation requirement analysis for a dry spell (Thomas et al., 2015) facilitating timely renovation of the traditional water harvesting structures, such as farm ponds, haveli, building check dams, and field bunding (Garg et al., 2020).

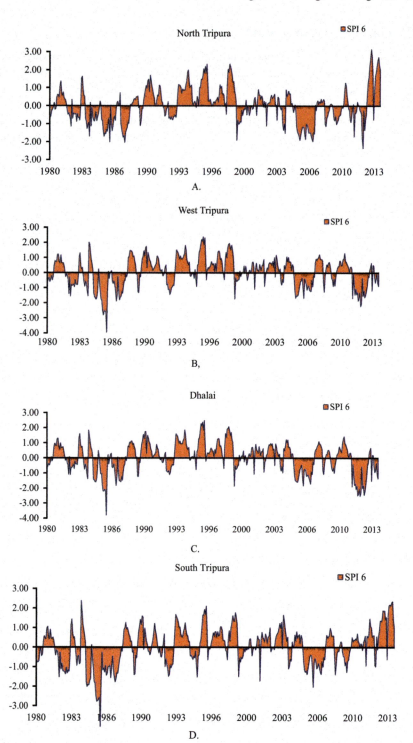

FIGURE 5.4 Temporal drought evaluation over four districts of Tripura.

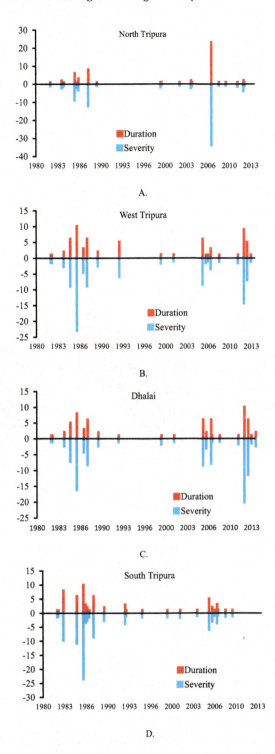

FIGURE 5.5 District-wise drought duration and drought severity for the drought events recorded by SPI-6.

TABLE 5.3

District-Wise Drought Characteristics by SPI at a Six-Month Timescale

SPI	North Tripura	West Tripura	Dhalai	South Tripura
Number of droughts	20	20	21	23
Max drought severity	−33.91	−22.90	−20.26	−23.62
Max drought duration	23	10	10	10
Max drought intensity	−1.47	−2.29	−2.02	−2.36
Min drought severity	−1.00	−2.58	−1.07	−1.00
Min drought duration	1	2	1	1
Min drought intensity	−1.00	−1.29	−1.07	−1.00
Avg drought severity	−4.26	−5.17	−5.09	−4.11
Avg drought duration	2.95	3.3	3.19	2.69
Avg drought intensity	−1.44	−1.59	−1.59	−1.52

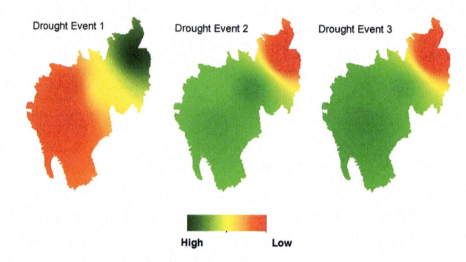

FIGURE 5.6   Spatial drought analysis by SPI-6 over Tripura, Northeast India.

5.4   SUMMARY AND CONCLUSIONS

Gridded precipitation data has proven an effective data source for analyzing spatiotemporal meteorological drought and estimating the characteristics of drought over a region, especially in a meteorological data-scarce region, such as northeast that has the last few remaining rainforests on the Indian subcontinent. The district-wide meteorological drought characteristics were analyzed by using the SPI over Tripura, one of the eight sister states of Northeast India, during the period from 1980 to 2013. All four districts experienced an average of 20 drought events during the 34-year study period. The highest drought intensity was observed in West Tripura (−2.29) and the highest drought severity in North Tripura (−33.91) district. Overall, the district of North Tripura exhibited the longest and the most severe drought, whereas intense drought was found in West Tripura. A significant decreasing trend at 5% significance level was observed for the SPI-3 during the pre-monsoon and winter seasons in all four districts. The falling trend indicates the increasing severity of meteorological drought in the region. The current scenario of climate change could escalate the severity level, which needs effective and solid contingency planning. The findings of the present study has ample scope in improving the water resources management and policy development

TABLE 5.4

Trends in SPI Values of Four Districts of Tripura by Using the MK Test and Sen's Slope Estimator

District	SPI	MK	Annual	Pre- Monsoon	Monsoon	Post- Monsoon	Winter
North Tripura	3	Z	0.00	−1.51	1.30	0.56	**−2.28**
		Slope	0.00	−0.02	0.02	0.01	−0.04
	6	Z	0.30	−1.60	0.89	0.89	0.15
		Slope	0.00	−0.02	0.01	0.02	0.01
	12	Z	0.42	0.50	0.39	0.50	0.42
		Slope	0.01	0.01	0.01	0.01	0.01
West Tripura	3	Z	−0.44	**−2.22**	0.33	1.10	**−2.05**
		Slope	−0.01	−0.35	0.01	0.02	−0.03
	6	Z	−0.27	−1.16	0.06	0.06	0.86
		Slope	0.00	−0.02	0.00	0.00	0.02
	12	Z	0.21	0.62	−0.03	−0.03	0.59
		Slope	0.00	0.02	0.00	0.00	0.01
Dhalai	3	Z	−1.13	**−2.67**	−0.09	0.86	**−2.16**
		Slope	−0.01	−0.04	0.00	0.02	−0.03
	6	Z	−0.77	**−2.11**	−0.65	0.03	0.86
		Slope	−0.01	−0.03	−0.01	0.00	0.02
	12	Z	−0.56	0.18	−0.53	−0.42	0.09
		Slope	−0.01	0.00	−0.01	−0.01	0.00
South Tripura	3	Z	1.45	−1.40	**2.68**	1.93	−1.57
		Slope	0.01	−0.02	0.04	0.04	−0.02
	6	Z	1.87	−0.27	1.90	**2.61**	1.78
		Slope	0.02	0.00	0.03	0.05	0.03
	12	Z	**2.16**	1.87	**2.13**	**2.19**	**2.13**
		Slope	0.03	0.03	0.03	0.03	0.03

for an economical drought-proof system in the northeastern region of India. This study will help in projecting the upcoming scenarios to facilitate drought contingency planning, climate adaptive mitigation planning, and the development of the most feasible policies.

ACKNOWLEDGMENTS

The first author gratefully acknowledges the competent authority of Central Agricultural University (CAU), Imphal, India, for three years of financial support through a university scholarship during PhD study in agricultural engineering with specialization in Soil and Water Conservation Engineering at College of Agricultural Engineering and Post Harvest Technology, CAU, Ranipool, Gangtok, Sikkim.

REFERENCES

Ajaz, A.; Taghvaeian, S.; Khand, K.; Gowda, P.; Moorhead, J.E. Development and Evaluation of an Agricultural Drought Index by Harnessing Soil Moisture and Weather Data. *Water* 2019, 11, 1375.
Alamgir, M.; Khan, N.; Shahid, S.; Yaseen, Z.M.; Dewan, A.; Hassan, Q.K.; Rasheed, B. Evaluating severity–area–frequency (SAF) of seasonal droughts in Bangladesh under climate change scenarios. *Stoch. Environ. Res. Risk Assess.* 2020, 34, 447–464.

Buttafuoco, G.; Caloiero, T.; Coscarelli, R. Analyses of drought events in Calabria (Southern Italy) using standardized precipitation index. *Water Resour. Manag.* 2015, 29, 557–573.

Dai, A.; Lamb, P.J.; Trenberth, K.E.; Hulme, M.; Jones, P.D.; Xie, P. The recent Sahel drought is real. *Int. J. Clim.* 2004, 24, 1323–1331.

Dikshit, K.R.; Dikshit, J.K. Weather and climate of north-east India. In *North-East India: Land, People and Economy* (pp. 149–173). Springer, Dordrecht, 2014.

Djebou, D.C.S. Bridging drought and climate aridity. *J. Arid Environ.* 2017, 144, 170–180.

Drápela, K.; Drápelová, I. Application of Mann-Kendall test and the Sen's slope estimates for trend detection in deposition data from Bílý Kř íž (Beskydy Mts., the Czech Republic) 1997–2010. *Beskydy* 2011, 4, 133–146.

Edwards, D.; Mckee, T. Characteristics of 20th century drought in the United States at multiple time scales. Atmospheric Science Paper No. 634, Climatology Report No. 97-2, Department of Atmospheric Science, Colorado State University, Fort Collins, CO 80523-1371, United States, pp. 155, 1997.

Field, C.B.; Barros, V.; Stocker, T.F.; Dahe, Q. *Managing the Risks of Extreme Events and Disasters to Advance Climate Change Adaptation: Special Report of the Intergovernmental Panel on Climate Change*; Cambridge University Press: Cambridge, UK, 2012.

Fluixá-Sanmartín, J.; Pan, D.; Fischer, L.; Orlowsky, B.; García-Hernández, J.; Jordan, F.; Haemmig, C.; Zhang, F., Xu, J. Searching for the optimal drought index and timescale combination to detect drought: A case study from the lower Jinsha River basin, China. *Hydrology and Earth System Sciences* 2018 Feb 1, 22(1), 889–910.

Garg, K.K.; Singh, R.; Anantha, K.; Singh, A.K.; Akuraju, V.R.; Barron, J.; Dev, I.; Tewari, R.; Wani, S.P.; Dhyani, S.; et al. Building climate resilience in degraded agricultural landscapes through water management: A case study of Bundelkhand region, Central India. *J. Hydrol.* 2020, 591, 125592.

Guo, H.; Bao, A.; Liu, T.; Ndayisaba, F.; He, D.; Kurban, A.; De Maeyer, P. Meteorological Drought Analysis in the Lower Mekong Basin Using Satellite-Based Long-Term CHIRPS Product. *Sustainability* 2017, 9, 901.

Hayes, M.J. *Drought Indices.* 2003. [Accessed on 2 May 2022].

Helsel, D.R.; Hirsch, R.M. Statistical methods in water resources techniques of water resources investigations. U.S. Geological Survey. 2002.

Khalili, D.; Farnoud, T.; Jamshidi, H.; Kamgar-Haghighi, A.; Zand-Parsa, S. Comparability analyses of the SPI and RDI meteorological drought indices in different climatic zones. *Water Resour. Manag.* 2011, 25, 1737–1757.

Khan, M.I.; Liu, D.; Fu, Q.; Saddique, Q.; Faiz, M.A.; Li, T.; Qamar, M.U.; Cui, S.; Cheng, C. Projected changes of future extreme drought events under numerous drought indices in the Heilongjiang province of China. *Water Resour. Manage.* 2017, 31, 3921–3937.

Kundzewicz, Z.W.; Mata, L.J.; Arnell, N.; Döll, P.; Jiménez, B.; Miller, K.; Oki, T.; ̦Sen, Z.; Shiklomanov, I. The implications of projected climate change for freshwater resources and their management. *Hydrol. Sci. J.* 2008, 53, 3–10.

Kwarteng, F; Shwetha, G.; Rahul, P. Reconnaissance drought index as potential drought monitoring tool in a Deccan Plateau, hot semi-arid climatic zone. *Int. J. Agric. Sci.* 2017, 9(1), 2183–2186.

Liu, Y.; Chen, J. Future global socioeconomic risk to droughts based on estimates of hazard, exposure, and vulnerability in a changing climate. *Sci. Total Environ.* 2020, 751, 142159.

Mahajan, D.; Dodamani, B. Trend analysis of drought events over upper Krishna basin in Maharashtra. *Aquat. Procedia* 2015, 4, 1250–1257.

Mckee, T.B.; Doesken, N.J.; Kleist, J. (1993). The relationship of drought frequency and duration to time scales. In *8th Conference on Applied Climatology*, 17–22 January, American Meteorological Society, 179–184.

Nalbantis, I.; Tsakiris, G. Assessment of hydrological drought revisited. *Water Resour. Manag.* 2009, 23, 881–897.

Pandey, V.; Srivastava, P.K.; Mall, R.; Munoz-Arriola, F.; Han, D. Multi-satellite precipitation products for meteorological drought assessment and forecasting in Central India. *Geocarto Int.* 2020, 1–20.

Parida, B.R.; Oinam, B. Unprecedented drought in North East India compared to Western India. *Curr. Sci.* 2015 Dec 10, 2121–2126.

Sharma, A.; Goyal, M.K. Assessment of drought trend and variability in India using wavelet transform. *Hydrol. Sci. J.* 2020, 65, 1539–1554.

Paulo, A.A.; Ferreira, E.; Coelho, C.; Pereira, L.S. Drought class transition analysis through Markov and Loglinear models, an approach to early warning. *Agric. Water Manag.* 2005, 77, 59–81.

Paulo, A.A.; Rosa, R.D.; Pereira, L.S. Climate trends and behaviour of drought indices based on precipitation and evapotranspiration in Portugal. *Nat. Hazards Earth Syst. Sci.* 2012, 12, 1481–1491.

Piccarreta, M.; Capolongo, D.; Boenzi, F. Trend analysis of precipitation and drought in Basilicata from 1923 to 2000 within a southern Italy context. *Int. J. Climatol.* 2004, 24, 907–922.

Svoboda, M.; Hayes, M.; Wood, D. *Standardized Precipitation Index User Guide*; World Meteorological Organization, Geneva, Switzerland, 2012.

Tabari, H.; Marofi, S.; Aeini, A.; Talaee, P.H.; Mohammadi, K. Trend analysis of reference evapotranspiration in the western half of Iran. *Agric. for Meteorol.* 2011, 151, 128–136.

Tian, Q.; Wang, Q.; Zhan, C.; Li, X.; Liu, X. Analysis of climate change in the coastal zone of eastern China, against the background of global climate change over the last fifty years: Case study of Shandong peninsula, China. *Int. J. Geosci.* 2012, 3, 379–90.

WHO. Drought, 2020. Available online: https://www.who.int/health-topics/drought#tab=tab_1 (accessed on 20 August 2020).

Wu, H.; Hayes, M.J.; Weiss, A.; Hu, Q. An evaluation of the Standardized Precipitation Index, the China-Z Index and the statistical Z-Score. *Int. J. Climatol.* 2001, 21, 745–758.

Xu, Y.-P.; Lin, S.-J.; Huang, Y.; Zhang, Q.-Q.; Ran,Q.-H. Drought analysis using multi-scale standardized precipitation index in the Han River Basin, China. *J. Zhejiang Univ. Sci.* 2011, 12, 483–494.

Xu, C.-Y.; Widén, E.; Halldin, S. Modelling hydrological consequences of climate change—Progress and challenges. *Adv. Atmos. Sci.* 2005, 22, 789–797.

Yue, S.; Pilon, P.; Cavadias, G. Power of the Mann–Kendall and Spearman's rho tests for detecting monotonic trends in hydrological series. *J. Hydrol.* 2002, 259, 254–271.

6 Drought Assessments in the Nonstationary Domain

Md Mamunur Rashid

CONTENTS

6.1 INTRODUCTION

Drought is one of the most prominent, costly natural disasters posing significant effects on infrastructure and socioenvironmental systems. With its devastating impacts on widespread sectors such as agriculture, energy, water resources, environment, and human health, it has turned into a potential multidisciplinary research topic. Drought is the persistent deficiency of water for a prolonged period compared to the climatological mean over a region. There are mainly four types of droughts, commonly known as meteorological, agricultural, hydrological, and socioeconomic droughts. Meteorological drought is characterized by a persistent deficit in precipitation compared to the long-term mean. Agricultural drought is attributed to a sustained decline in soil moisture, whereas a prolonged reduction in surface and/or groundwater causes hydrological drought. Socioeconomic drought is often the combined effect of all other types of droughts and has wider spatial, temporal, and sectoral (e.g., water resources, agriculture, and energy) coverage. Meteorological drought may occur frequently; the other droughts (in particular hydrological and socioeconomic droughts) are less frequent and often require months to years to take place because they occur when the precipitation deficit is enough to significantly reduce soil moisture, streamflows, and groundwater levels.

Drought occurrences have extended to many regions of the world, including both wet and humid, with different durations and severities (Dai, 2011). The risk of drought has intensified worldwide over the second half of the 20th century (Dai, 2013), and the frequency and intensity of drought will likely increase over the 21st century according to global climate model simulations (Touma et al., 2015). In addition to regions in the Northern Hemisphere, Australia is known to be a drought-prone region in the Southern Hemisphere. Several devastating prolonged droughts of various severity have occurred in Australia since 1900. The most recent devastating drought in Australia occurred from 1997 to 2009. According to the Australian Bureau of Agricultural and Resource Economics, this event, termed the Millennium Drought, reduced the national winter cereal crop production by 36% and cost rural Australia around AUD$3.5 billion (Wong et al., 2009). It was comparable in severity with the other historical long-term droughts such as the Federation Drought (1895–1902) and the Second War Drought (1936–1945) (Murphy and Timbal, 2008). Other remarkable Australian

DOI: 10.1201/9781003276548-6

droughts in the last century occurred in 1963–1968, 1972–1973, 1982–1983, and 1991–1995 (ABS, 2016).

Over the years, in different studies, spatial and temporal variability of droughts have been assessed in various ways using climate data from different sources such as satellite images, *in situ* historical records, and reanalysis and climate model hindcasts (Dai, 2011). However, historical precipitation data obtained from *in situ* gauge records are most widely used to quantify droughts in terms of various drought statistics. A simple statistic for drought quantification is the coefficient of variation of multiyear rainfall sequences, for example, two years and five years. Another simple statistic is the count of total annual rainfall events below the median, which are also common in drought quantification. Besides these, the well-known "run length" approach, which is defined as the number of consecutive months when precipitation is below a threshold, has been adopted in many studies in recent times. Peel et al. (2005) employ run length and run magnitude of consecutive years below or equal to the median of precipitation and runoff to quantify droughts. However, irrespective of the drought statistics often used in drought studies, short historical records often limited our ability to assess multiyear drought variability. Nevertheless, this can be resolved using stochastic rainfall models (Chowdhury et al., 2017; Chowdhury et al., 2019; Rashid et al., 2014c).

Droughts becoming more frequent and severe with time due to accelerated anthropogenic changes (Alexander et al., 2009; Dai, 2011; Mishra and Singh, 2009, 2010) warrants a suitable and effective drought monitoring system for water resource planning and management. Generally, droughts are monitored using different indices. Drought indices are often used to identify drought characteristics, such as severity, duration, peak, geographical extent, and frequency or return period (Hao et al., 2015; Rashid and Beecham, 2019a; Xu et al., 2015). To date, a number of drought indices have been developed to characterize meteorological, agricultural, and hydrological droughts (Mishra and Singh, 2010; Zargar et al., 2011). The most commonly used drought indices are the Standardized Precipitation Index (SPI) (Edwards, 1997; McKee et al., 1993), Palmer Drought Severity Index (PSDI) (Palmer, 1965), Soil Moisture Drought Index (SMDI) (Welford et al., 1993), Soil Moisture Deficit Index (SMDI) (Narasimhan and Srinivasan, 2005), and Standardized Runoff Index (SRI) (Shukla and Wood, 2008). While the indices mentioned are univariate, several recent studies have also proposed multivariate drought indices, such as the Standardized Precipitation Evapotranspiration Index (SPEI) (Vicente-Serrano et al., 2010), Aggregated Drought Index (ADI) (Keyantash and Dracup, 2004), Linearly Combined Drought Index (LDI) (Mo and Lettenmaier, 2014), and Multivariate Standardized Drought Index (MSDI) (Hao and AghaKouchak, 2014). Nevertheless, the SPI is the most widely used drought index and is recommended by the World Meteorological Organization (WMO) (Svoboda et al., 2012). The SPI is estimated based on the probability of precipitation. The main strength of the SPI is that it can be estimated for multiple timescales, which permits its use for monitoring both short-term (soil moisture deficit) and long-term (reduction of groundwater and streamflow) droughts (Rahmat et al., 2015; Rahmat et al., 2017; Szalai et al., 2000; Zhu et al., 2016). However, because the SPI is a standardized index derived by transforming precipitation probability to normal distribution, estimation of the SPI from biased precipitation data from model hindcasts (e.g., general circulation model) may misestimate the drought quantification if precipitation is not adequately bias corrected.

While many earlier studies have developed various drought indices, most of those assumed that the hydrometeorological time series is stationary. But hydrometeorological variables for example precipitation and temperature often show significant trends (Rashid et al., 2013; Rashid et al., 2015a) and plausible future climate change indicate that the stationary assumption in formulating drought indices is not valid. Recently, a few studies have characterized droughts from a nonstationary perspective. Wang et al. (2015) fitted a nonstationary gamma model to observed rainfall using a GAMLSS where the location parameter of the distribution varied only with time to derive a time-dependent SPI. Although their model was able to capture the long-term temporal trend in the precipitation series, the variability was not adequately reproduced because only time was considered as a covariate. It is expected that inclusion of climate indices as covariates would better represent the variability because large-scale climate often modulates the interannual to multidecadal variability of precipitation (Cai et al., 2009; Cai et al., 2011; Evans et al., 2009). For example, Sun et al. (2016) used the SPEI and SPI to assess the relationship between wet/dry conditions and atmospheric patterns. They found that approximately 38% of the global land surface area is dominated by El Niño–Southern Oscillation events, whereas western North America, northern South America, and eastern Russia are influenced more by the Pacific Decadal Oscillation. Another performed an comprehensive study and identified significant climatic connections of sustained drought and wet anomalies across Australia (Rashid et al., 2018).

Several studies have proposed different approaches to develop nonstationary SPI, for example, Li et al. (2015) proposed a model considering climate indices as external covariates for modeling precipitation and thereby estimating the nonstationary SPI. However, linear relationships between precipitation and various climate indices were considered in this study. Studies have claimed that these relationships of precipitation with climatic variations are not necessarily linear in a changing climate (Montazerolghaem et al., 2016; Rashid et al., 2014a). Additionally, for nonstationary SPI, most of the existing studies allow that the location parameter of the selected distribution varies with time and covariates, but in the case of precipitation with high temporal and spatial variability in a semiarid region like Australia (Rashid et al., 2014c), considering shape parameter as a function of explanatory variables in addition to the location parameter would provide more predictive capability (Beecham et al., 2014; Rashid and Beecham, 2019b). Rashid and Beecham (2019b) were the first to propose more flexible nonstationary SPI allowing both scale and shape parameters of gamma distribution that vary with selected climate indices and then derive the SPI from the fitted probability distribution of precipitation.

This chapter will first discuss various characteristics of drought based on the run theory and show how droughts have been changed over the historical period considering South Australia as a case study. Then, this chapter will focus on the formulation of a Nonstationary Standardized Precipitation Index (NSPI) where both location and scale parameters of gamma distribution are allowed to vary with time considering large-scale climate indices as external covariates. NSPI will be tested on high-quality 46 rainfall stations across South Australia (SA). The rest of the chapter will flow as follows: description of study area and data, explanation of methods, and finally a focus on the results and conclusions.

6.2 STUDY AREA AND DATA

South Australia is one of the drought-prone regions in the Southern Hemisphere and has a temperate climate with dry summer in the south, whereas the northern part of SA has a persistently dry hot desert-type climate. Although South Australia's climate is highly variable, a warming trend is observed after the 1970s (Rashid et al., 2014b; Rashid et al., 2015b). Across SA rainfall varies strongly in space and time, with higher rainfall in the south along the coast to very low rainfall in the north (Rashid et al., 2015c). Mean annual rainfall decreases from 667 mm in the south to 162 mm in the north, with some exceptionally high rainfall in some mountain areas. There are eight Natural Resource Management (NRM) hydrological regions (http://nrmregionsaustralia.com.au/nrm-regions-map/) in SA, as shown in Figure 6.1. The South Australian Arid Lands region (538,000 km^2) covers approximately 55% of SA. The Murray Darling Basin NRM region includes a portion of Australia's largest surface water catchment and covers approximately 70,000 km^2. The Adelaide and Mount Lofty Ranges region includes almost 80% of SA's total population. Diverse spatial and temporal variability of rainfall across SA makes this a suitable case study area to examine the advantages of nonstationary SPI over stationary SPI. The Australian Bureau of Meteorology (BOM) provides long-term, high-quality monthly rainfall data across Australia. Of the 379 BOM high-quality rainfall stations, 56 stations are in SA. Eight rainfall stations were excluded from this study because of a large number of missing rainfall values at these stations, and thus 46 stations were considered during the analysis in the study. The spatial locations of these selected rainfall stations over SA are shown in Figure 6.1.

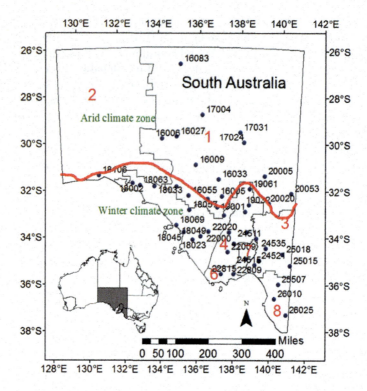

FIGURE 6.1 High-quality rainfall stations with corresponding BOM station IDs in the following Natural Resource Management (NRM) regions across South Australia: (1) South Australia Arid Lands, (2) Alinytiara Wilurara, (3) South Australia Murray Darling Basin, (4) Northern and Yorke, (5) the Eyre Peninsula, (6) Kangaroo Island, (7) Adelaide and Mount Lofty Ranges, and (8) South East.

There are many climate indices available in the literature often used for understanding and modeling the remote connections with hydrologic variables. Among those, five prominent climate indices that often show strong connections with rainfall variability across Australia are considered as external covariates for nonstationary modeling of rainfall. Table 6.1 lists selected climate indices with their corresponding ocean basin and sources of data.

6.3 METHODS

6.3.1 STANDARDIZED PRECIPITATION INDEX

The SPI is the most widely used index for monitoring and quantification of meteorological droughts that was first proposed by McKee et al. (1995). The SPI can be estimated for different accumulation periods from month to years. The SPI is generally derived using a stationary model by fitting a distribution to the observed rainfall. The gamma probability distribution is often considered as the best candidate, whereas other parametric and empirical distributions are also used. In SPI derivation, the first gamma distribution is fitted to the total precipitation for any accumulation period and cumulative distribution functions (CDFs) are estimated. Then, the CDFs are transformed into standard normal values to obtain corresponding SPIs.

Let y be the monthly precipitation measured at any time t, then the precipitation for an accumulation period of M months (y_m) is calculated as

$$y_m(t) = \sum_{i=t-1-m}^{t} y_i \tag{6.1}$$

TABLE 6.1

Climate Indices Used in the Study

Climate index	Description	Ocean Basin	Source of Data
Southern Oscillation Index (SOI)	Standardized sea level pressure differences between Tahiti and Darwin	Pacific	Bureau of Meteorology, Australia
Niño3.4	Average SSTA over 5° N to 5° S and 170° W to 120° W	Pacific	NOAA Earth System Research Laboratory
Pacific Decadal Oscillation (PDO)	Leading principal component of monthly sea surface temperature anomalies in the North Pacific poleward of 20° N	Pacific	Joint Institute for the study of the Atmosphere and Ocean
Southern Annular Model (SAM)	Monthly mean Antarctic Oscillation Index described by the north–south movement of the westerly wind belt that circles Antarctica	Antarctica	NOAA Earth System Research Laboratory
Indian Ocean Dipole (IOD)	Coupled ocean–atmosphere phenomenon in the Indian Ocean (Saji et al., 1999) that is measured in terms of DMI, which is the anomaly of the sea surface temperature gradient between the western (50° E–70° E and 10° S–10° N) and southeastern (90° E–110° E and 10° S–0° N) Indian Ocean	Indian	Japan Agency for Marine-Earth Science and Technology

A two-parameter gamma distribution is fitted to the aggregated precipitation series y_m. CDFs of precipitation data are transformed into standard normal values using Equation 6.2. Here, SPI are estimated using a stationary gamma distribution is termed as the Stationary Standardized Precipitation Index (SSPI).

$$SPI(t) = \Phi^{-1}\left[F\left(y_m(t)\right)\right] \tag{6.2}$$

where $y_m(t)$ is the aggregated precipitation at any time t, F are the CDFs of the precipitation data, and Φ^{-1} is the inverse of standard normal CDFs.

6.3.2 NONSTATIONARY STANDARDIZED PRECIPITATION INDEX (NSPI)

To develop a nonstationary drought index (i.e., nonstationary SPI), one needs to first develop a nonstationary model by fitting a nonstationary probability distribution to the aggregate rainfall of the target timescale. GAMLSS modeling framework was adopted to developing nonstationary models for aggregated rainfall series at each station. Although different aggregation levels could be considered, the six-month aggregation level is considered as a case study. Five large-scale climate indices – namely, the Southern Oscillation Index (SOI), Niño3.4, Pacific Decadal Oscillation (PDO), Southern Annular Model (SAM), and Dipole Mode Index (DMI) – were used as external covariates to incorporate the low frequency variability in the nonstationary modeling framework. Low frequency variability is important to capture the persistence in rainfall time series. In order to reduce the monthly random variability, running means (over a period similar to the aggregation level of rainfall, i.e., at the SPI scale) of climate indices were considered. A cross-correlation analysis was used to identify the lag relationships between aggregated rainfall and running mean climate indices. Lag series of climate indices corresponding to the maximum lag correlations at the 5% significance were used as potential covariates.

Details of the GAMLSS are available in Stasinopoulos and Rigby (2007) and Rigby and Stasinopoulos (2005). While GAMLSS has several different possible models, a semiparametric additive model formulation was used in this study as follows:

$$g_k\left(\theta_k\right) = X_k\beta_k + \sum_{j=1}^{j_k} h_{jk}\left(x_{jk}\right) \tag{6.3}$$

where θ_k is a vector representing the parameters of the distribution, X_k is a matrix of covariates of order $n \times j_k$, $\beta_k(\beta_{1k},...,\beta_{jkk})$ is a parameter vector of length j_k, and h_{jk} (.) represents the dependence function of the distribution parameters on the covariates. The dependence could be linear, or a smoothing term is included to allow more flexibility for modeling the dependence of the distribution parameters on the covariates. Significant covariates for each distribution parameter were identified following the stepwise model fitting approach proposed by Rigby and Stasinopoulos (2005). Gamma distribution functions were considered for nonstationary modeling of six-month rainfall data because gamma distribution is commonly used for SPI estimation. The Akaike information criterion (AIC) and the Schwarz Bayesian information criterion (SBIC) were used for the selection of significant covariates. In addition, normality and independence of the residuals were assessed by examining the first four moments of the residuals and by visual inspection of a diagnostic plot of the residuals including Q-Q and worm plots. Finally, the fitted nonstationary model used to estimate the cumulative probabilities and convert to standard normal values is the NSPI. The NSPI is analogous to the SSPI since in both cases cumulative probabilities are converted into standard normal values. So, positive NSPI values represent wet conditions and negative NSPI values indicate dry conditions.

6.3.3 Drought Characteristics

The run theory is generally used for identifying and estimating different drought characteristics from SPI time series. SPI values less than −1 and greater than +1 are considered as the drought and wet conditions, respectively. A drought event is defined as a period for which the SPI values are continuously negative starting from a neutral condition (i.e., SPI value is 0) and reach a threshold value of −1.0 or less before they return to a neutral condition. For a drought event, the duration is quantified as the total number of months in the event and the severity is the cumulative sum of all SPI values over the duration. Each drought event has its own severity and duration. For each rainfall station, drought characteristics such as the total number of drought events (i.e., drought frequency) and the corresponding drought durations and severities are estimated from stationary and nonstationary SPI series over the period 1960 to 2010 and compared. We have considered the mean of these durations and severities obtained from each drought event to represent the spatial variability.

6.3.4 Multivariate Frequency Analysis

In case of drought monitoring and risk assessment, univariate frequency analysis of droughts in terms of various drought characteristics (e.g., severity and duration) is often employed. However, due to the significant dependence among drought indicators, multivariate frequency analysis is warrant. Multivariate modeling based on the copula theory has been widely used in this regard. For a two-dimensional random vector, $X = \left(X_1, X_2\right)$, with continuous marginal CDFs F_1 and F_2, according to Sklar's theorem (Sklar, 1959), the multivariate joint CDF of X can be expressed as

$$H\left(x\right) = C\left[F_1\left(x_1\right), F_2\left(x_2\right)\right] \tag{6.4}$$

where the unique function $C : \left[1,0\right]^d \rightarrow \left[0,1\right]$ is called a copula. The copula C represents the dependence information between the components (X_1, X_2), and the marginal cumulative distributions

(F_1, F_2) contain all information on the marginal distributions. Construction of a multivariate dependence model consists of estimation of marginal CDFs and the copula C. For the marginal, seven commonly used univariate probability distributions were used. These are the beta, exponential, gamma, logistic, lognormal, normal, and Weibull distributions. Widely used copula functions such as the Gumbel, Frank, Clayton, normal, and t copula were considered as the candidate copulas. The L-moment method was used to estimate the parameters of the marginal distribution. The maximum likelihood (ML) method was used to fit the copulas. The best performing marginal distribution and copula function were selected based on the root mean square error (RMSE) and the AIC.

In a bivariate setting, the recurrence interval of drought exceeding any x_1 and x_2 values of interest can be calculated as follows:

$$T = \frac{\mu}{1 - F_1(x_1) - F_2(x_2) + C_{12}(F_1(x_1), F_2(x_2))} \qquad (6.5)$$

where F_1 and F_2 are the marginal CDFs of any threshold values of interest for x_1 and x_2, respectively, and μ is the average interarrival time of the drought events and is defined as the ratio of the number of years in the analysis period (N_{year}) to the number of drought events ($N_{drought}$). While for flood frequency analysis μ is always 1, in the case of drought it is different because drought durations can be longer than one year, or several drought events may occur within a single year.

6.4 RESULTS

Severity and duration are two common attributes often used to characterize droughts. Figures 6.2 and 6.3 show the spatial variability of drought severity and duration across SA. Drought severity and duration vary with timescales of the SPI. Drought severity ranges from 5 to 7 and 29 to 52 for the 3-month and 24-month SPIs, respectively. Drought duration varies from 5 to 8 months and 31 to

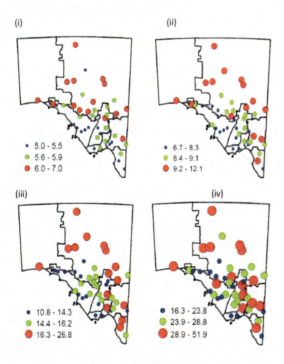

FIGURE 6.2 Mean drought severity over the period 1960 to 2010 based on SPI at timescales of (i) 3 months, (ii) 6 months, (iii) 12 months, and (iv) 24 months.

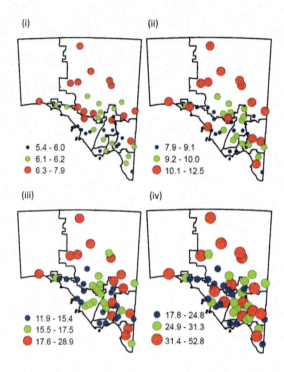

FIGURE 6.3 Mean drought duration (months) over the period 1960 to 2010 based on SPI at timescales of (i) 3 months, (ii) 6 months, (iii) 12 months, and (iv) 24 months.

53 months for the 3-month and 24-month SPI, respectively. Generally, the higher the timescale of the SPI, the higher the drought severity and duration. SA often experiences considerably longer dry spells due to sustained low or no rainfall leading to prolonged drought with higher severity that is captured when the SPI is estimated at higher timescales.

The winter climate zone of SA generally experienced droughts with lower magnitude of severity and duration, whereas droughts are comparatively higher across the arid region. However, for a few instances, drought severity and duration of stations in the winter climate zone are as high as that of the arid regions at 24-month SPI. For example, the southeast coast and Murray Darling Basin show drought severity and duration as high as the arid region, particularly for higher timescales (i.e., 24-month) of the SPI. This reveals that some regions in the winter-dominating climate zone have been suffering from prolonged droughts like the dry regions.

Many stations across SA show significant trends of SPI whether positive or negative. Positive, (or increasing) trends of the SPI represent the intensification of floods, whereas negative (or decreasing) trends of the SPI indicate intensification of droughts. Results show that a number of stations with significant negative trends increase with longer SPI timescales, revealing that long-term prolonged droughts are more prominent than the short-term droughts across SA. The recent long-term drought in Australia known as the Millennium Drought (1997 to 2009) is an example of intensification of long-term droughts. In general, the stations showing significant negative trends of SPI are often found in the winter-dominating climate zone. Stations within the arid climate zone do not show significant trends. Results indicate that with significant negative trends in SPI intensify the drought over the winter-dominating zone of SA where most of the population reside, hence posing significant threats to the society and environment. This intensification of drought in SA is related to variability observed in the large-scale climatology (Rashid and Beecham, 2019a).

Discussion in the earlier section indicates that droughts vary significantly both in space and time and require nonstationary modeling for appropriate characterization of droughts (e.g., drought severity and duration). For drought prediction and monitoring, standardized indices, for example the SPI, are often derived by fitting suitable (generally gamma) distribution based on the stationary assumption where parameters of the distribution are considered constant over time. However, in a changing climate, this assumption in not valid because hydrometeorological variables exhibit nonstationarity in their statistical characteristics. Hence, traditional SPI, estimated based on the stationary assumption, is unreliable to use in drought prediction and monitoring. The nonstationary attributes and spatiotemporal variability of rainfall is often connected to the large-scale climate variability (Kenyon and Hegerl, 2010; Rashid et al., 2014c; Rashid et al., 2015a). Hence, large-scale climate indices are considered as potential external covariates for developing nonstationary SPI by fitting a nonstationary gamma distribution to observed rainfall. Comparing the ability of stationary and nonstationary models to reproduce rainfall variability is a quick check if the nonstationary modeling approach outperforms the stationary one in deriving the SPI. Figure 6.4 represents prediction errors in terms of mean absolute error (MAE) for the stationary and nonstationary models. The MAEs of the nonstationary model range from 39.4 to 115.5, whereas for the stationary model they range from 60.2 to 165.9. Results indicate that the perdition errors are significantly lower for the nonstationary model compared to that of the stationary model for almost all stations. This reveals that the nonstationary model better captures the observed rainfall variability, and eventually will quantify droughts with higher accuracy compared to the stationary model. Additionally, nonstationary attributions are conditioned to the large-scale climate variability and allowed to outperform the proposed nonstationary model even in a changing climate. In contrast, SPI derived from the stationary model is susceptible to erroneous estimation of droughts if there are significant changes in the rainfall statistics (e.g., mean and standard deviation) over time. The limiting ability of predicting historical droughts by stationary SPI are evident in earlier studies (Rashid and Beecham, 2019b; Wang et al., 2015).

Figure 6.5 shows the spatial plots of different drought characteristics such as interannual time, severity, duration, and peak obtained by interpolating station-based drought characteristics using inverse weighted distance. Results show that drought characteristics are substantially different for the SSPI and NSPI. Strong differences in the drought characteristics are observed over the arid

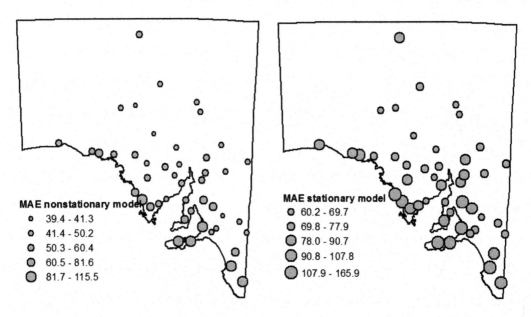

FIGURE 6.4 Mean absolute error (MAE) estimated from (left) nonstationary and (right) stationary models for 46 rainfall stations across South Australia.

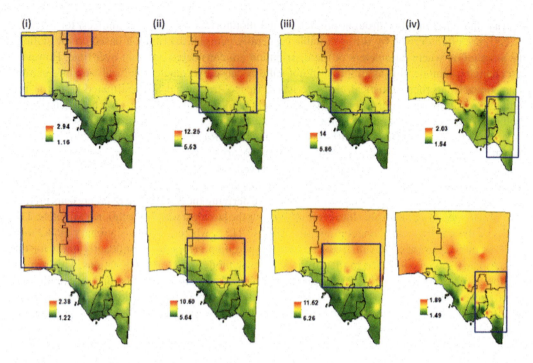

FIGURE 6.5 Drought characteristics estimated by the SSPI (upper row) and NSPI (lower row): (i) interarrival time, (ii) severity, (iii) duration, and (iv) peak (shown as absolute value of drought index). Blues boxes highlight areas where drought characteristics estimated from the SSPI and NSPI are different.

region of SA and indicate that the stationary model was unable to adequately capture the persistent dry anomalies often observed in an arid region with consistently low rainfall. Significant differences are observed in the interarrival time and peak of droughts. The interarrival times of droughts estimated from the SSPI series are lower than those estimated from the NSPI series and drought peaks estimated from the SSPI series are higher than those estimated from the NSPI series. Additionally, significant differences are observed for the spatial extent of droughts in terms of various drought characteristics for the SSPI and NSPI.

Joint return periods of drought events exceeding certain severity and duration of interest are quantified from the NSPI and SSPI time series. Figure 6.6 represents the joint return periods for selected stations. Results show significant differences in the joint return period between NSPI and SSPI, originating due to the nonstationary behavior of rainfall. For example, at station ID 16006 a drought event with severity of 22 and duration of 23 months is expected to occur once in 100 years when estimated from the SSPI, whereas droughts become more frequent events with a recurrence interval of less than 50 years when nonstationary behavior of rainfall is considered to derive the SPI (i.e., NSPI). Similar differences are observed at other stations mainly caused by the inability of stationary models to capture the nonstationary attributes of observed rainfall. On the other hand, with the ability of the nonstationary model to capture the statistical changes of rainfall over time it allows the characterization of drought in terms of joint return period of severity and duration well with higher confidence. Thus, in the context of climate change, application of nonstationary models to derive the SPI for monitoring and prediction of drought is crucial. Hence, we should be aware and vigilant of nonstationary behavior of rainfall and many other hydrometeorological variables when used for deriving an index for drought monitoring and predictions.

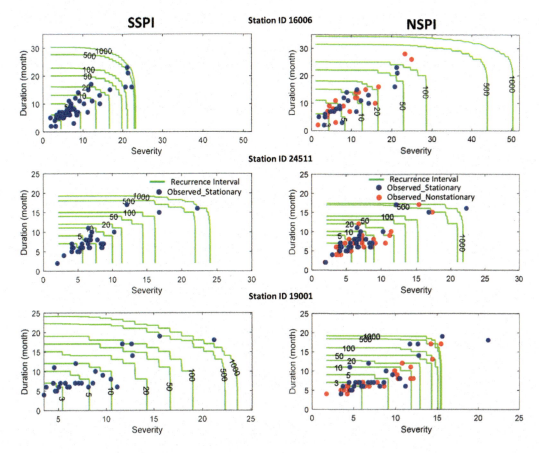

FIGURE 6.6 Joint return period T (in years) of drought events of any severity and duration (in months) for selected stations for the SSPI and NSPI.

6.5 CONCLUSIONS

Drought is one of the most devastating natural hazards, often causing significant damage to society and the environment along with sectoral impacts such as energy, agriculture, water resources, and human health. The biggest impact of drought is generally observed in the agriculture sector threatening global food security. Over time, drought has become more prominent and devastating due to the increase in global temperature accelerated by anthropogenic changes. Therefore, prediction and monitoring of drought are of high importance for sustainable development. This chapter discussed different drought characteristics in terms of severity, duration, peak, and interarrival time, as well as developed a nonstationary drought index (i.e., NSPI) to capture and include the nonstationary behavior of hydrometeorological variables (here rainfall) when used to derive drought indices. This is particularly important from the viewpoint of continuing climate change. It is found that the NSPI has more potential for long-term prediction and monitoring of drought in a future warming climate possibly downscaling from the general circulation models compared to the traditional SPI (i.e., SSPI).

REFERENCES

ABS, 2016. Retrieved Sep 11, 2016. Available at http://www.abs.gov.au/AUSSTATS/abs@.nsf/lookup/1301 .0Feature%20Article151988.

Alexander, L.V. et al., 2009. Climate extremes: Progress and future directions. *International Journal of Climatology*, 29(3): 317–319.

Beecham, S., Rashid, M., Chowdhury, R.K., 2014. Statistical downscaling of multi-site daily rainfall in a South Australian catchment using a generalized linear model. *International Journal of Climatology, Royal Meteorological Society*. https://doi.org/10.1002/joc.3933

Cai, W., Cowan, T., Sullivan, A., 2009. Recent unprecedented skewness towards positive Indian Ocean Dipole occurrences and its impact on Australian rainfall. *Geophysical Research Letters*, 36(11): L11705. https://doi.org/10.1029/2009GL037604

Cai, W., van Rensch, P., Cowan, T., Hendon, H.H., 2011. Teleconnection pathways of ENSO and the IOD and the mechanisms for impacts on Australian rainfall. *Journal of Climate*, 24(15): 3910–3923.

Chowdhury, A. et al., 2017. Development and evaluation of a stochastic daily rainfall model with long-term variability. *Hydrology and Earth System Sciences*, 21(12): 6541–6558.

Chowdhury, A.K., Kar, K.K., Shahid, S., Chowdhury, R., Rashid, M.M., 2019. Evaluation of spatio-temporal rainfall variability and performance of a stochastic rainfall model in Bangladesh. *International Journal of Climatology*, 39(11): 4256–4273.

Dai, A., 2011. Drought under global warming: A review. *Wiley Interdisciplinary Reviews: Climate Change*, 2(1): 45–65.

Dai, A., 2013. Increasing drought under global warming in observations and models. *Nature Climate Change*, 3(1): 52–58.

Edwards, D.C., 1997. Characteristics of 20th century drought in the United States at multiple time scales. *Atmospheric Science*, Paper No. 634, May 1–30.

Evans, A.D., Bennett, J.M., Ewenz, C.M., 2009. South Australian rainfall variability and climate extremes. *Climate Dynamics*, 33(4): 477–493. https://doi.org/10.1007/s00382-008-0461-z

Hao, C., Zhang, J., Yao, F., 2015. Multivariate drought frequency estimation using copula method in Southwest China. *Theoretical and Applied Climatology*: 1–15.

Hao, Z., AghaKouchak, A., 2014. A nonparametric multivariate multi-index drought monitoring framework. *Journal of Hydrometeorology*, 15(1): 89–101.

Kenyon, J., Hegerl, G.C., 2010. Influence of modes of climate variability on global precipitation extremes. *Journal of Climate*, 23(23): 6248–6262.

Keyantash, J.A., Dracup, J.A., 2004. An aggregate drought index: Assessing drought severity based on fluctuations in the hydrologic cycle and surface water storage. *Water Resources Research*, 40(9): W09304 (1-3). https://doi.org/10.1029/2003WR002610

Li, J., Wang, Y., Li, S., Hu, R., 2015. A Nonstationary Standardized Precipitation Index incorporating climate indices as covariates. *Journal of Geophysical Research: Atmospheres*, 120(12). https://doi.org/10.1002/2015JD023920

McKee, T.B., Doesken, N.J., Kleist, J., 1993. The relationship of drought frequency and duration to time scales, *Proceedings of the 8th Conference on Applied Climatology*. American Meteorological Society, Boston, MA, pp. 179–183.

McKee, T. B., Doesken, N., Kleist, J., 1995. Drought monitoring with multiple time scales. Proceedings of the 9th Conference on Applied Climatology, American Meteorological Society Dallas, Boston, MA: 233–236.

Mishra, A., Singh, V.P., 2009. Analysis of drought severity-area-frequency curves using a general circulation model and scenario uncertainty. *Journal of Geophysical Research: Atmospheres*, 114(D6). https://doi.org/10.1029/2008JD010986

Mishra, A.K., Singh, V.P., 2010. A review of drought concepts. *Journal of Hydrology*, 391(1): 202–216.

Mo, K.C., Lettenmaier, D.P., 2014. Objective drought classification using multiple land surface models. *Journal of Hydrometeorology*, 15(3): 990–1010.

Montazerolghaem, M., Vervoort, W., Minasny, B., McBratney, A., 2016. Long-term variability of the leading seasonal modes of rainfall in south-eastern Australia. *Weather and Climate Extremes*, 13(2016): 1–14. 10.1016/j.wace.2016.04.001

Murphy, B.F., Timbal, B., 2008. A review of recent climate variability and climate change in southeastern Australia. *International Journal of Climatology*, 28(7): 859–879. https://doi.org/10.1002/joc.1627

Narasimhan, B., Srinivasan, R., 2005. Development and evaluation of Soil Moisture Deficit Index (SMDI) and Evapotranspiration Deficit Index (ETDI) for agricultural drought monitoring. *Agricultural and Forest Meteorology*, 133(1): 69–88.

Palmer, W., 1965. *Meteorological Drought*. US Department of Commerce, Washington, DC. Weather Bureau Research Paper, 45.

Peel, M.C., McMahon, T.A., Pegram, G.G., 2005. Global analysis of runs of annual precipitation and runoff equal to or below the median: Run magnitude and severity. *International Journal of Climatology*, 25(5): 549–568.

Rahmat, S.N., Jayasuriya, N., Bhuiyan, M.A., 2015. Assessing droughts using meteorological drought indices in Victoria, Australia. *Hydrology Research*, 46(3): 463–476.

Rahmat, S.N., Jayasuriya, N., Bhuiyan, M.A., 2017. Short-term droughts forecast using Markov chain model in Victoria, Australia. *Theoretical and Applied Climatology*, 129: 445–457.

Rashid, M.M., Beecham, S., 2019a. Characterization of meteorological droughts across South Australia. *Meteorological Applications*, 26(4): 556–568.

Rashid, M.M., Beecham, S., 2019b. Development of a non-stationary Standardized Precipitation Index and its application to a South Australian climate. *Science of the Total Environment*, 657: 882–892.

Rashid, M.M., Beecham, S., Chowdhury, R., 2013. Assessment of statistical characteristics of point rainfall in the Onkaparinga catchment in South Australia. *Hydrology and Earth System Sciences Discussions*, 10(5): 5975–6017.

Rashid, M.M., Beecham, S., Chowdhury, R.K., 2014a. Influence of climate drivers on variability and trends in seasonal rainfall in the Onkaparinga catchment in South Australia: A wavelet approach. 13th International Conference on Urban Drainage (ICUD) 2014, 7–12 September, Kuching, Sarawak, Malaysia.

Rashid, M.M., Beecham, S., Chowdhury, R.K., 2014b. Statistical characteristics of rainfall in the Onkaparinga catchment in South Australia. *Journal of Water and Climate Change*, IWA Publishing. https://doi.org /10.2166/wcc.2014.031

Rashid, M.M., Beecham, S., Chowdhury, R.K., 2015a. Assessment of trends in point rainfall using Continuous Wavelet Transforms. *Advances in Water Resources*, Elsevier, 82: 1–15.

Rashid, M.M., Beecham, S., Chowdhury, R.K., 2015b. Assessment of trends in point rainfall using continuous wavelet transforms. *Advances in Water Resources*, 82: 1–15.

Rashid, M.M., Beecham, S., Chowdhury, R.K., 2015c. Statistical characteristics of rainfall in the Onkaparinga catchment in South Australia. *Journal of Water and Climate Change*, 6(2): 352–373.

Rashid, M.M., Johnson, F., Sharma, A., 2018. Identifying sustained drought anomalies in hydrological records: A wavelet approach. *Journal of Geophysical Research: Atmospheres*, 123(14): 7416–7432.

Rigby, R.A., Stasinopoulos, D.M., 2005. Generalized additive models for location, scale and shape, (with discussion). *Applied Statistics*, 54: 507–554.

Saji, N., Goswami, B., Vinayachandran, P., Yamagata, T., 1999. A dipole mode in the tropical Indian Ocean. *Nature*, 401(6751): 360–363.

Shukla, S., Wood, A.W., 2008. Use of a standardized runoff index for characterizing hydrologic drought. *Geophysical Research Letters*, 35(2): L02405. https://doi.org/10.1029/2007GL032487

Sklar, M., 1959. Fonctions de répartition à n dimensions et leurs marges. *Publications de l'Institut de Statistique de L'Université de Paris*, 8: 229–231.

Stasinopoulos, D.M., Rigby, R.A., 2007. Generalized additive models for location scale and shape (GAMLSS) in R. *Journal of Statistical Software*, 23: 1–46. DOI:10.18637/jss.v023.i07.

Sun, Q., Miao, C., AghaKouchak, A., Duan, Q., 2016. Century-scale causal relationships between global dry/ wet conditions and the state of the Pacific and Atlantic Oceans. *Geophysical Research Letters*, 43(12): 6528–6537.

Svoboda, M., Hayes, M., Wood, D., 2012. *Standardized Precipitation Index User Guide*. World Meteorological Organization, Geneva, Switzerland.

Szalai, S., Szinell, C., Zoboki, J., 2000. Drought monitoring in Hungary. In: *Early Warning Systems for Drought Preparedness and Drought Management*, WMO, Geneva, pp.161–176.

Touma, D., Ashfaq, M., Nayak, M.A., Kao, S.-C., Diffenbaugh, N.S., 2015. A multi-model and multi-index evaluation of drought characteristics in the 21st century. *Journal of Hydrology*, 526: 196–207.

Vicente-Serrano, S.M., Beguería, S., López-Moreno, J.I., 2010. A multiscalar drought index sensitive to global warming: The standardized precipitation evapotranspiration index. *Journal of Climate*, 23(7): 1696–1718.

Wang, Y., Li, J., Feng, P., Hu, R., 2015. A time-dependent drought index for non-stationary precipitation series. *Water Resources Management*, 29(15): 5631–5647.

Welford, M.R., Hollinger, S.E., Isard, S.A., 1993. A new soil moisture drought index for predicting crop yields. In: *Preprints, Eighth Conf. on Applied Climatology, Anaheim, CA, American Meteorological Society*, pp. 187–190.

Wong, G., Lambert, M., Leonard, M., Metcalfe, A., 2009. Drought analysis using trivariate copulas conditional on climatic states. *Journal of Hydrologic Engineering*, 15(2): 129–141.

Xu, K., Yang, D., Xu, X., Lei, H., 2015. Copula based drought frequency analysis considering the spatio-temporal variability in Southwest China. *Journal of Hydrology*, 527: 630–640.

Zargar, A., Sadiq, R., Naser, B., Khan, F.I., 2011. A review of drought indices. *Environmental Reviews*, 19: 333–349.

Zhu, Y., Wang, W., Singh, V.P., Liu, Y., 2016. Combined use of meteorological drought indices at multi-time scales for improving hydrological drought detection. *Science of the Total Environment*, 571: 1058–1068.

7 Drought Monitoring in Arid and Semiarid Environments Using Aridity Indices and Artificial Neural Networks

Jimmy Byakatonda, B.P. Parida, D.B. Moalafhi, and Piet K. Kenabatho

CONTENTS

7.1   INTRODUCTION

Hydroclimatic extremes in the form of floods and droughts have become more frequent in the last half of the 20th century and the first half of the 21st century (Zarch et al., 2015; Zhang et al., 2009). These extremes have shown variant effects on water resources and in turn crop productivity. It has also been reported that more than 2 billion of the global population live in water-scarce areas, which are mainly located in arid/semiarid zones (Ortigara et al., 2018). It is anticipated that if no actions are taken, these hydroclimatic extremes are likely to worsen and cause more damage. Global circulation models (GCMs) paint even a more oblique picture with some of their projections indicating a continued increase in global temperatures under various emission scenarios (Akinsanola et al., 2018; Dyderski et al., 2018; Garreaud et al., 2020; PaiMazumder et al., 2013). This rise in temperature could have great influence on other meteorological variables such as precipitation and humidity. In yet another study by Kenabatho et al. (2012), it was revealed that indeed global warming could alter precipitation regimes in many locations in southern Africa. This calls for periodic appraisal

DOI: 10.1201/9781003276548-7

of the global climate to avert possible catastrophes that may result from these perturbations. In the midlatitudes, hydroclimatic extremes in the form of droughts is the main challenge facing livelihoods in lieu of increasing population pressure and rising food demand. Due to their "creeping" nature, droughts have caused far much more damage than floods (Byakatonda et al., 2018a; IPCC, 2012; Van Loon, 2013). Droughts often cover more areas spatially compared to river basins, hence they affect a larger population compared to floods (Khan et al., 2016; Lloyd-Hughes, 2012). It is for this reason that this chapter focuses on drought monitoring rather than floods.

Until recently, methods for direct drought measurement did not exist and for this reason a number of drought indices have been applied over time (Mishra and Singh, 2010; Nalbantis and Tsakiris, 2009). Drought indices are considered as proxies of the hydroclimate (Sheffield et al., 2012; Sivakumar et al., 2011). Some of such indices include the aridity index (AI) (Costa and Soares, 2009; Croitoru et al., 2013; Moalafhi et al., 2017), Standardized Precipitation Evaporation Index (SPEI) (Vicente-Serrano et al., 2010), Standardized Precipitation Index (SPI) (McKee et al., 1993), Standardized Flow Index (SFI) (Nalbantis and Tsakiris, 2009), and Standardized Soil Moisture Index (SSMI) (Homdee et al., 2016). Even currently there is no blueprint on the exact index to use in drought monitoring. A host of these indices are location specific, in that there is no global acceptance. The aridity index and the SPEI stand out from the numerous indices available and have been applied in a number of hydroclimatic studies all over the world (Byakatonda et al., 2018d; Dai, 2013; Huang et al., 2016; Zarch et al., 2015). The aridity index has been since recommended by United Nations Education and Scientific Organization (UNESCO) since 1979 to classify global climates (Byakatonda et al., 2016b; Huo et al., 2013; Nastos et al., 2013). On the other hand, use of the SPEI is a recent preference in the advent of the global warming phenomenon (Byakatonda et al., 2016a; Vicente-Serrano et al., 2010) and is recommended by the World Meteorological Organization (WMO). Both these indices use the interrelations between temperature and precipitation with respect to climate variability and change. Due to the significance of drought planning and monitoring, various attempts have been made through studies globally that have applied some of these drought indices, such as the aridity index, SPI, and SPEI. To enable us arrive at a general agreement, critical examination of the findings from some of the related studies is required. For example, Zhang et al. (2009), using climatic data observed in the Pearl River basin of China over a long term (1960–2005), computed the SPI and aridity index to study the changes in droughts and found that the river basin was getting drier, particularly during the winter season. In Romania, Croitoru et al. (2013) used the aridity index to investigate changes in drought over a period between 1961 and 2007. In their findings, they reported semiarid tendencies in the southern region of Romania. In the quest to trace the climate change signal in China, Huo et al. (2013), Zhao et al. (2014) and Wang et al. (2015) combined the reference evapotranspiration (ET_{ref}) and aridity index in the semiarid regions of China for a period between 1955 and 2012. From their studies they reported significantly deceasing trends in both the ET_{ref} and aridity index during the 50-year period of analysis. Trends in the ET_{ref} and aridity index were detected using Mann–Kendall statistic in these studies. Using a global gridded data set, aridity changes have been investigated by Zarch et al. (2015) while applying the Mann–Kendall test statistic over a period of 1960–2009. They observed a global reversal in trend around 1980, which was attributed to abrupt changes in global temperatures. A more interesting finding was that aridity was being reversed in arid and hyperarid locations. However, gridded data has limitations due to their coarse resolution. They may not reveal local attributes associated with local topography and vegetation influence. Interdecadal variability in aridity in Mongolia China was also investigated over a period of 1961 to 2010 by Tong et al. (2017). Their findings revealed moderate aridity, which increased from west to east. The study also reported a decrease in humid spatial extent in Inner Mongolia. In yet another study covering the Limpopo river basin shared among southern African countries including Botswana, Moalafhi et al. (2017) used precipitation and temperature dynamically downscaled at a high resolution of 10 km between 1980 and 2010 to investigate drought using the SPI and aridity index. In their study, it was revealed that rainfed agriculture can on average thrive only during the summer months of December, January, and February with moisture availability becoming increasingly limited over time. In this era of global warming, the SPI has been found inadequate in monitoring droughts, which is rather a shortcoming for this method as the SPI does not consider temperature in its formulation (Byakatonda et al., 2018d). Most recent studies in other semiarid areas such as Botswana by Byakatonda et al. (2018d) used both the SPEI and SPI to monitor droughts. Their findings revealed that the SPEI performed better in tracking drought events. This study, however, did not consider the aridity index as another tool for drought monitoring.

To aid drought planning and mitigation of possible impacts of climate change, there have been several attempts in making futuristic projections of aridity. For instance, Yin et al. (2015) made projections using five GCMs and four representative concentration pathway (RCP) scenarios over China for the mid-21st century. In their findings, they projected negative anomalies in the aridity index. In the same spirit, Ramachandran et al. (2015) studied future aridity in India using RCP 4.5 for a period between 2006 and 2100. In their findings they reported increasing aridity toward the end of the 21st century. At a global scale Zarch et al. (2017) in their recent study used GCMs to project the future AI for a period of 2006–2100. Their findings projected an increase in arid and semiarid areas by 7% and 6%, respectively, by the end of the 21st century. Most recently, Greve et al. (2019) projected the AI under global warming using the Coupled Model Intercomparison Project 5 (CMIP5). In their study, they discovered an increasing AI with an increase in carbon emission by the end of the 21st century. All these projections of the AI have been performed by GCMs, however, there are inherent uncertainties partly as a result of the structure of the applied models, the emission scenarios selected themselves at times, and the choice of initial assumptions (Stonevičius et al., 2018; Zarch et al., 2017). To mitigate these uncertainties, in this chapter we demonstrate the ability of artificial neural networks (ANNs) in the prediction of the AI by developing and applying a multistep ahead model. ANN models are a category of artificial intelligence that requires no initial assumptions to learn the interrelations between the input variables. We hope this will give a new paradigm in AI future projections and studies in climate variability and change. Therefore, the main objective of this research is to investigate aridity changes in arid and semiarid environments, assessing the relationship between SPEI and AI in drought monitoring and development of a multistep ahead ANN model that will be used to perform five-years ahead predictions of AI.

In this chapter, Botswana, a classic case of a typical arid/semiarid country, is selected to help in understanding aridity changes in arid zones. The study is based on long-term climatic data observed over years between 1960 and 2019. Botswana mostly depends on rainfall for all its water demands to support livelihoods. Most of the rainfall is harvested and stored in surface and groundwater reservoirs for use in domestic, agricultural and industrial sectors. Therefore, any climatic shocks greatly affect livelihoods and the economy as a whole. It is expected that findings from this study can aid in drought monitoring and devise mitigation measures for any diverse effects arising from the current global warming tendencies.

7.2 ARIDITY INDEX (AI) IN DROUGHT MONITORING

Aridity is defined in this chapter as a water deficit in a given allocation. It is a common phenomenon in water-scarce areas, especially in the midlatitudes that experience frequent anticyclone weather behaviors (Byakatonda et al., 2016b; Galarneau et al., 2008). In other wards, locations where precipitation is unable to meet the evaporation demand are prone to aridity. Aridity and drought are synonymous, but in effect they differ. Whereas drought can be a short form of moisture shortfall, aridity is a long-term condition of moisture deficit. It is a climatic classification often applied in land and water development. In effect long-term drought may lead to aridity. From the foregoing operational definition, it can be observed that aridity is a function of both precipitation and potential evapotranspiration.

Just like drought, aridity has been categorized using a host of aridity indices. These aridity indices in essence measure the degree of dryness of a particular location. These indices have the great

TABLE 7.1

De Martonne Climate Classification

Climate Classification	Aridity Index (AI)
Arid	$AI < 10$
Semiarid	$10 \leq AI < 20$
Mediterranean	$20 \leq AI < 24$
Semihumid	$24 \leq AI < 28$
Humid	$28 \leq AI < 35$
Very humid	$35 \leq AI < 55$
Extremely humid	$AI > 55$

ability of revealing the effect of climate change on regional and local water resources, starting with the initial classification of Koppen in 1900 (Alvares et al., 2013; Byakatonda et al., 2019a; Zarch et al., 2015). This initial formulation of aridity considered the two most important parameters of precipitation and temperature. As an improvement to the Koppen aridity index, UNESCO in 1979 introduced another aridity index based on the ratio of annual precipitation to potential evapotranspiration. The formulation is presented next, as suggested by UNESCO and applied in Alvares et al. (2013) and Zarch et al. (2015):

$$AI = \frac{P_a}{ET_0} \tag{7.1}$$

where P_a is the annual precipitation and ET_0 is the potential evapotranspiration.

The formulation of the index is simple and easy to use, especially in warm climates. The only shortcoming of this index is that it can only be computed at the annual scale (Croitoru et al., 2013), moreover the degree of dryness is a function of timescale. To alleviate this shortcoming, the De Martonne formula was introduced. This aridity index (Table 7.1) is flexible and can be computed at monthly, seasonal, and annual timescales (Byakatonda, 2018e; Tong et al., 2017). This helps in quantifying dryness at different time horizons. The formulation of the De Martonne aridity index is presented next. The aridity or humidity of a particular location for a given month is given by

$$AI_m = \frac{12P_m}{\left(T_m + 10\right)} \tag{7.2}$$

where m denotes the number of months less than five is used.

The AI was computed at monthly, seasonal, and annual scales from which a relationship was established with the SPEI at a 12-month timescale for the annual series and SPEI-6 for the seasonal series using Spearman's rank correlation.

7.3 RELATING STANDARDIZED PRECIPITATION EVAPORATION INDEX (SPEI) AND ARIDITY

As indicated earlier, many drought indices exist that can quantify drought in terms of its magnitude, duration, and intensity. In this chapter, we select the SPEI as it has been recommended by the WMO to be applied in drought monitoring (Byakatonda et al., 2016a; Vicente-Serrano et al., 2012). The other advantage of using the SPEI in drought monitoring is that just like the aridity index, it uses rainfall and temperature in its formulation. This way it is easy to compare one index against the other. The SPEI is multiscalar, which allows it to be compared at various timescales. The various timescales allow the study of effects of droughts on various components of the hydrological cycle.

At low timescales of 1–3 months, drought can be classified as a meteorological drought; at 3–6 months, an agricultural drought; and at 12–24 months, a hydrological drought. In this chapter we computed the SPEI at timescales on 1, 6, and 12 months. At timescales of 6 and 12 months, a relationship was established between the SPEI and aridity index. This relationship allows identification of the climatic zones where drought severity can be used to approximate the extent of aridity in a given area. The computation steps of SPEI are summarized:

1. Determination of potential evapotranspiration (ET_0).
2. Generation of the climatic water balance (D), which is the difference between monthly rainfall amounts (P) and potential evapotranspiration (ET_0).
3. Aggregation of the climatic water balance (D) from step 2 over a period of n months, where n is the drought timescale.
4. The aggregated climatic water balance time series obtained in step 3 are then standardized through a Gaussian transformation.

7.4 SPATIOTEMPORAL TREND ANALYSIS OF ARIDITY TIME SERIES

Trend analysis is crucial in studying time series, as it reveals the inherent behavior of the data over the period of record (Byakatonda et al., 2018b). Trend analysis has been applied in climatological studies over time as part of extreme events monitoring (Byakatonda et al., 2018c; Costa and Soares, 2009; Oguntunde et al., 2012). There exists various techniques of establishing trends in time series. These techniques are classified as parametric and nonparametric. Nonparametric methods are preferred over parametric ones since they do not require time series to be normally distributed. Nonparametric methods also perform better with data containing outliers. It is for these reasons that in this chapter we use the nonparametric Mann–Kendall (MK) method to establish the direction of the trend. The formulation of the MK test is presented next.

The MK statistic as presented by Mann (1945) and Kendall (1975) is given by

$$S = \sum_{i}^{n-1} \sum_{k=i+1}^{n} \text{Sgn}\left(AI_k - AI_i\right) \qquad \text{for} < I \quad (7.3)$$

where Sgn is the sign function given by

$$\text{Sgn}\left(AI_k - AI_i\right) = \begin{cases} +1 & \text{if}\left(AI_k - AI_i\right) > 0 \\ 0 & \text{if}\left(AI_k - AI_i\right) = 0 \\ -1 & \text{if}\left(AI_k - AI_i\right) < 0 \end{cases} \qquad (7.4)$$

The statistical significance of the trend is tested using the Z-statistic obtained from

$$Z_w = \begin{cases} \dfrac{S-1}{\sqrt{\text{Var}(S)}} & if\ S > 0 \\ 0\ if\ S = 0 \\ \dfrac{S+1}{\sqrt{\text{Var}(S)}} & if\ S < 0 \end{cases} \qquad (7.5)$$

where Var(S) is the variance of the S statistic obtained from Equation 7.6:

$$\text{Var}(S) = \frac{n(n-1)(2n+5) - \sum_{i=1}^{q} p_i(p_i-1)(2p_i+5)}{18} \qquad (7.6)$$

where q and p_i are the total number of tied groups and ties of extend i, respectively. Positive values of Z_w designate upward trends, and negative values otherwise. The null hypothesis assumes that no trend exists, and the alternative hypothesis assumes otherwise. The significance of the trend was tested at 95% confidence level. Negative trends in aridity designate an increase in aridity while positive trends indicate recovery from arid to humid conditions.

7.4.1 DETERMINATION OF TREND MAGNITUDE

The MK trend is only able to reveal the direction of the trend but not the magnitude. For this reason, the Sen slope estimator is used to determine the magnitude of this trend. It is important to know this magnitude to inform the scale of the mitigation measures of any negative impacts. The formulation of the Sen slope estimator is

$$Q_i = \frac{\left(AI_k - AI_i\right)}{k - i} \qquad \text{for i=1,2,...,N (7.7)}$$

The median slope of Q_i arranged in ascending order gives Sen's slope and is computed from

$$Q_{med} = \begin{cases} Q_{\frac{(N+1)}{2}} & \textit{if N is odd} \\ \dfrac{Q_{\left(\frac{N}{2}\right)} + Q_{\left[\frac{N+2}{2}\right]}}{2} \textit{if N is even} \end{cases} \qquad (7.8)$$

7.5 MODELING ARIDITY USING ARTIFICIAL NEURAL NETWORKS (ANNS)

To aid planning and even mitigate impacts of climate variability and change on the population, it is necessary to carry out futuristic projections. As demonstrated in the literature, projections of aridity have been made in recent times using GCMs. In this chapter we present the use of ANNs in making five-years ahead predictions. ANNs are a form of artificial intelligence that mimics neurobiological information to learn interrelations between inputs to generate targets (Byakatonda et al., 2019a; Byakatonda et al., 2016a). ANNs do not require initial conditions to generate targets. Different forms of ANNs exist, but it has been demonstrated that dynamic networks are more robust in learning interlinkages of processes. In this chapter we present the nonlinear autoregressive with exogenous inputs (NARX) model, which is a class of the recurrent neural network (RNN). This category of RNN model is equipped with tapped delay neurons that provides more model memory enhancing predictions (Chang et al., 2014; Gao and Meng Joo, 2005; Menezes and Barreto, 2008). The model construction steps of the NARX are summarized in the following:

1. Data is mapped to be in the same range between −1 and 1 to simplify the learning of process nodes.
2. The data available is then segregated into exogenous inputs and autoregressive target nodes.
3. As part of the model architecture, the number of hidden layers capable of learning the input–output relations is selected.
4. Training of the model in open loop and eventually predictions in the closed-loop mode.

7.5.1 MODEL ARCHITECTURE, TRAINING, AND LEARNING

The NARX model comprises of a hidden layer, and input and output regressors. The hidden layer neurons are chosen based on the procedure suggested by Hecht-Nielsen (1987) and applied in

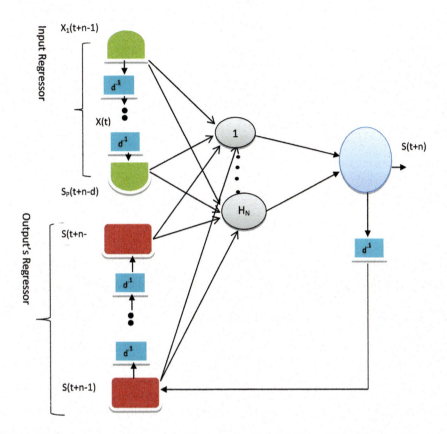

FIGURE 7.1 Nonlinear autoregressive with exogenous inputs (NARX) model architecture.

Byakatonda et al. (2019). The relationship between neurons and the number of input variables is given by

$$H_N \le 2X_{iN} + 1 \tag{7.9}$$

where H_N is the number of neurons and X_{iN} are input variables.

The NARX uses a feedforward with a backpropagation that uses a sigmoid transfer function. A Levenberg–Marquardt optimization function is used to update the weights during the training period while keeping the errors to the minimum. The model architecture is presented in Figure 7.1.

The NARX inputs with n steps ahead is given by

$$\hat{S}(t+n) = f\left[\hat{S}(t+n-1),....,\hat{S}(t+n-d); X_p(t)\right] \tag{7.10}$$

where $X_p(t)$ is the input vector, $P = 1, ..., m$, and m is the number of input variables. $\hat{S}(t+n)$ is the output at time step t, and n is the number of times of multisteps ahead. The two regressors of the model are $X_p(t)$, which is the implicit exogenous variables, and S(t+n), which is the autoregressive component of the model.

The training is performed in the open-loop connection where the output regressor is only comprised of actual values. During this period, the network performs a one-step ahead prediction. In the training mode, Equation 7.10 becomes

$$\hat{S}(t+1) = f\left[S(t),....,S(t+1-d); X_p(t)\right] \tag{7.11}$$

7.5.2   Nonlinear Autoregressive with Exogenous Inputs (NARX) Model Performance

Model performance is evaluated based on errors from training, validation, and testing phases. In this chapter, the mean squared error (MSE) and correlation coefficient (r) between the targets and model simulations are used in the performance evaluation as follows:

$$\text{MSE} = \frac{1}{n}\sum_{i=1}^{n}e_i^2 = \frac{1}{n}\sum_{i=1}^{n}\left(S_i - \hat{S}_i\right)^2 \tag{7.12}$$

$$r_{y\hat{y}} = \frac{\sum_{i=i}^{N}\left(S_i - \overline{y}_i\right)\left(\hat{S}_i - \overline{\hat{y}}_i\right)}{\sqrt{\sum_{i=1}^{N}\left(S_i - \overline{y}_i\right)^2}\sqrt{\sum_{i=1}^{N}\left(\hat{S}_i - \overline{\hat{S}}_i\right)^2}} \tag{7.13}$$

where y_i and $\hat{S}$ are the observed and predicted values, respectively, and n is the period of prediction (in this case five years). $\overline{S}_i$ and $\overline{\hat{S}}_i$ are average values of observed and model outputs, respectively.

7.6   BOTSWANA AS A CASE STUDY

Botswana, which is located in the multitudes (16° S–28° S and 19° E–30° E) in southern Africa and classified as semiarid, has been selected as a case study. Synonymous to its climate classification, Botswana has always been affected by recurrent droughts. The most recent drought events recorded in Botswana were in 1961/62–1965/66, 1980/81–1986/87, 1991/92, 2001/02–2005/06, 2009/10–2011/12, and 2014/15–2015/16. The duration and intensity has been increasing with the latest becoming the most devastating (Byakatonda et al., 2019b). This drought event caused drying up of the Gaborone Dam, which is the main water source of the country's capital. Botswana's water resources are largely dependent on rainfall, which only comes during the summer months of November to April (Byakatonda et al., 2019a). The country is water scarce, which has a bearing on the national economy. Water is an input in most of the sectors including human consumption, agriculture, and mining. This scarcity could partly be attributed to the presence of the Kalahari Desert, which occupies a large area of the western region of Botswana and largely moderates the regional climate. Conversely, Botswana is a host of the iconic Okavango Delta in the north. The delta has been declared a World Heritage Site by UNESCO. With projections indicating increasing aridity at a global scale, it is therefore feasible to use Botswana as a case. The Botswana case aids understanding of the interaction between aridity and the natural environment specifically for drought-prone areas (Figure 7.2).

7.7   RESULTS AND DISCUSSION FROM THE BOTSWANA CASE STUDY

7.7.1   Historical Aridity Trends in Botswana

Long-term historical (1960–2019) time series of aridity were generated for annual, summer, and winter seasons for Botswana using Equations 7.1 and 7.2. The aridity time series were determined at different timescales to aid understanding of the effect of drought across seasons. The summer season is particularly important in Botswana since 90% of the rainfall is received during this season. It is the period when most agricultural activities are carried out and surface reservoirs replenished for eventual future use. Winters are largely cold and dry across the case study.

   In Figure 7.3 annual aridity time series are presented across the regions, including Ghanzi (Figure 7.3a) in the west, Shakawe (Figure 7.3b) and Maun (Figure 7.3d) in the north, and Francistown (Figure 7.3c) and Mahalapye (Figure 7.3f) located in the east. Tsabong Figure 7.3e) located in the

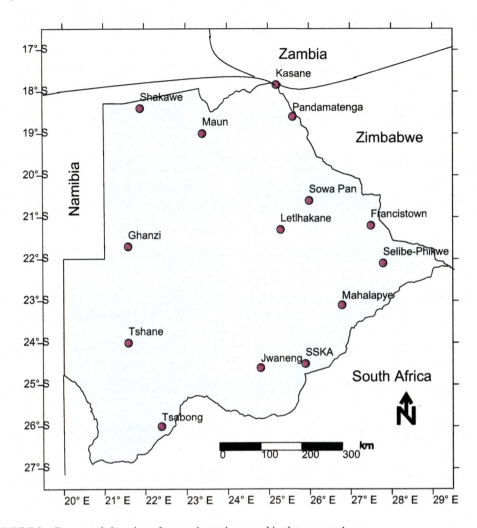

FIGURE 7.2 Botswana's location of synoptic stations used in the case study.

south is also included in this presentation. From the linear trends presented in Figure 7.3, deceasing trends, which depict increasing aridity, are registered in all regions except in the east at Francistown. In the north at Shakawe and Maun together with Mahalapye in the southeast, aridity is increasing at a rate of 9% annually. Also, in the west at Ghanzi and south at Tsabong, aridity is increasing but at a slower rate of approximately 3% per annum. Francistown presents with increasing trends, which is an indication of reducing aridity that could imply that this location is becoming relatively humid at a rate of 1%.

The relatively slower rate of aridity at Tsabong and Ghanzi could be attributed to the fact that they are located in the most arid zone of the Kalahari desert. Being an arid region there may be limited moisture supply for evaporation to occur. The fact that these arid locations are still showing tendencies of aridity poses more concern, as these locations could become hyperarid. In another study across the study area by Byakatonda et al. (2018d), it was revealed that at the locations investigated here were experiencing increasing temperature and decreasing precipitation. An exception to this was Francistown, where it was observed that whereas temperature was rising, precipitation was also on the increase. This may explain perhaps why Francistown in the east is becoming humid. It has also been reported that Botswana's climate is influenced by other external factors such as the El

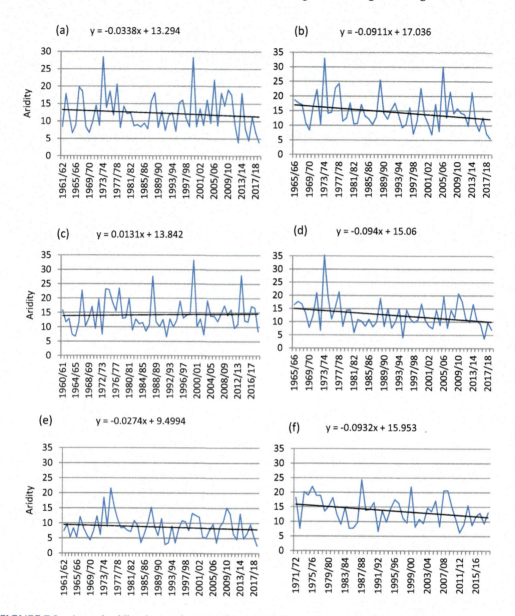

FIGURE 7.3 Annual aridity time series at (a) Ghanzi, (b) Shakawe, (c) Francistown, (d) Maun, (e) Tsabong, and (f) Mahalapye.

Niño–Southern Oscillation (Byakatonda et al., 2019b), which may be associated with these isolated behaviors in some locations. Due to these discrepancies from location to location, it was therefore necessary to investigate location-specific climatic attributes with a more robust MK nonparametric test statistic. To further understand these climatic behaviors even better, seasonal trends were also investigated.

7.7.2 MANN–KENDALL TREND TEST RESULTS

At each of the 14 stations, a trend analysis using the Mann–Kendall statistic and the Sen slope estimator was conducted to establish the direction and magnitude of the trend, respectively. The results

TABLE 7.2

MK and Sen Slope Estimator

Station	Annual		Summer		Winter	
	MK- Z	mm/°C/yr	MK- Z	mm/°C/yr	MK- Z	mm/°C/yr
Francistown	0.52	0.020	0.85	0.052	−2.05*	−0.028
Ghanzi	−0.72	−0.022	−0.80	−0.057	0.03	0.000
Jwaneng	−0.71	−0.072	−0.93	−0.157	−0.79	−0.109
Kasane	−1.21	−0.064	−1.70	−0.162	0.12	0.006
Letlhakane	−1.28	−0.148	−1.99*	−0.237	−0.35	−0.023
Mahalapye	−1.88	−0.102	−1.50	−0.152	−1.13	−0.046
Maun	−1.79	−0.073	−1.82	−0.116	−1.25	−0.047
Pandamatenga	−1.60	−0.287	−1.48	−0.345	−1.24	−0.235
Selibe–Phikwe	−0.70	−0.131	−0.56	−0.117	−0.91	−0.113
Shakawe	−2.06*	−0.093	−2.37*	−0.143	−1.22	−0.047
SSKA	−1.16	−0.104	−1.57	−0.229	−0.09	−0.015
Sowa Pan	0.00	0.001	−1.63	−0.320	2.04*	0.237
Tsabong	−0.51	−0.017	0.24	0.009	−1.52	−0.056
Tshane	−1.22	−0.043	−0.48	−0.024	−1.54	−0.046

* indicates the MK estimates that exceed the 95% confidence limits

from this analysis are presented in Table 7.2. From the annual trends, increasing aridity (decreasing trends) is observed at all stations except Francistown just as it was the case with linear trends. Significantly increasing aridity was reported at Shakawe in the north at a rate of −0.093 mm/°C/yr. At Sowa Pan also in the north, there was no trend recorded at the annual scale.

During the summer season, significant drying trends were again observed at Shakawe and Letlhakane at rates of −0.143 mm/°C/yr and −0.237 mm/°C/yr, respectively. In the summer season it can be observed that the rate of aridity increases, which may be attributed to the fact that summer is the hottest season in Botswana and hence increased evaporative demand. As reported in the linear trends, locations near the Kalahari desert such as Tshane and Ghanzi reported the lowest increase in aridity. Tsabong also located at the heart of the Kalahari desert presented with increasing trends in aridity.

In the winter season, incidentally, Francistown, which reported reducing aridity, now records a reversal with significantly increasing aridity at a rate of −0.028 mm/°C/yr. Similarly Sowa Pan, which recorded no trend at the annual scale, now registers a significantly increasing trend (decrease in aridity) during the winter season of −0.237 mm/°C/yr. Equally Ghanzi and Kasane recorded reducing aridity. It may be observed that the rate of aridity is low during winter as compared to the rest of the seasons. This is due to the fact that the winter seasons are the coldest and hence low evaporative demand. To further understand the countrywide distribution of aridity trends, spatial representations were made. These results are shown in Figures 7.4, 7.5, and 7.6 for the annual, summer, and winter trends, respectively. In Figure 7.4, it is observed that highly negative aridity trends are mostly registered in the northwest and north transiting to the southeast. The biggest percentage of the study area is under negative trend apart from only 3.3% that experiences positive trends around Francistown in the east.

For the summer spatial representation in Figure 7.5, positive trends are only observed in the southwest around Tsabong and Francistown in the east. The northeast to southeast gradient is still observed as was the case in annual trends. The area under increasing trends has now increased to 8.6% compared to annual trends. This is mostly owing to the fact that during summer, there is increased moisture circulation across the study area.

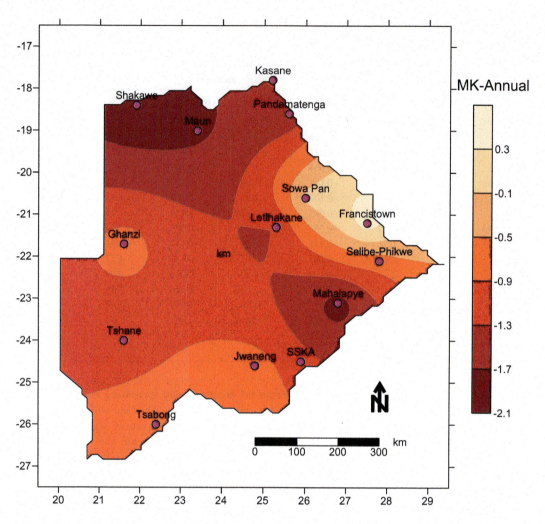

FIGURE 7.4   Spatial representation of annual Mann–Kendall (MK) trends.

During the winter season (Figure 7.6), positive trends are now observed in a larger spatial extent spreading to locations such as Kasane, Sowa Pan, and Ghanzi. The area under reducing aridity has now increased to 24% during the winter season. The northwest, south, and east are still experiencing drying episodes. There is no spatial pattern that can be observed from the aridity trend spatial representation in Figure 7.6. The winter season is largely dry and this could be the reason for an increase in the areal extent of the positive trends in aridity since there is less solar radiation to effect evapotranspiration. With these seasonal variations in aridity trends, it was hence necessary to classify Botswana's climate across the three timescales of annual, summer, and winter seasons.

7.7.3   BOTSWANA'S CLIMATE CLASSIFICATION USING ARIDITY INDEX

Climate classification of any location is important for planning and mitigation of any impacts of climate disasters. In this chapter we demonstrate how aridity index classification according to Table 7.1 was used to characterize Botswana's climatic zones. As shown in Section 7.7.2, climate classification varies between seasons. Classification using an annual aridity series, presented in Figure 7.7a, reveals that only the southwest is classified as arid with the rest of the country categorized

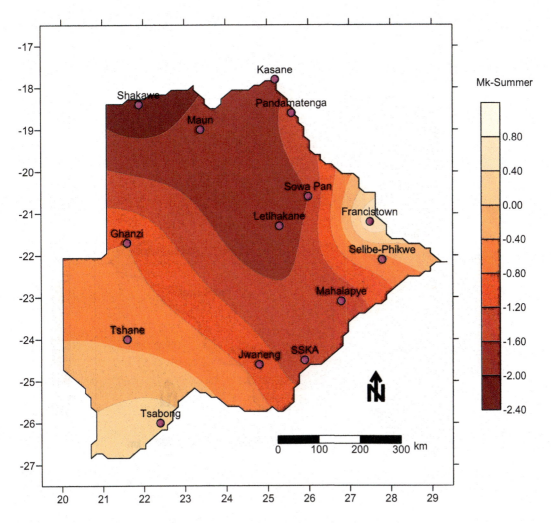

FIGURE 7.5 Spatial representation of summer Mann–Kendall (MK) trends.

as semiarid. The areas of northeast and east around Francistown are semiarid tending toward a Mediterranean type of climate. From Figure 7.7a, the area classified as Mediterranean climate is approximately 4.5% of Botswana's land area.

The locations classified as Mediterranean climate are where intense agricultural activities currently take place. In a recent study of dry spell analysis by Byakatonda et al. (2019a), which corroborates this finding, it was revealed that locations in the east around Francistown and the north had the lowest dry spell making them suitable for agricultural production.

During the summer season when most of the study area is moist, climatic reclassification can be observed, as shown in Figure 7.7b. In this season the area under arid climate reduces tremendously, while some locations in the northeast and east that had been classified as Mediterranean now tends toward semihumid. The Mediterranean to semihumid area increases to 52.5% during summer, while the rest of the area remains semiarid. This reclassification is no surprise as this is the season when the country receives the majority of its precipitation. During the winter season, when Botswana is devoid of moisture supply, the entire study area is classified as arid. During this season not much agricultural activity takes place, as the country depends on water harvested during the summer rainy season.

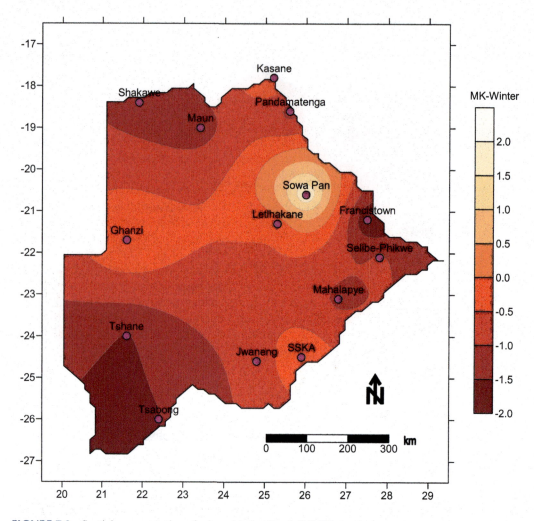

FIGURE 7.6   Spatial representation of winter Mann–Kendall (MK) trends.

7.8   RELATIONSHIP BETWEEN ARIDITY INDEX AND STANDARDIZED PRECIPITATION EVAPORATION INDEX (SPEI)

The SPEI has been used in drought monitoring in Botswana and has demonstrated its ability to track drought events. In other semiarid locations such as China and Romania (Croitoru et al., 2013; Tong et al., 2017), the aridity index has been used in drought monitoring, which is not the case for Botswana. In an attempt to use aridity in drought monitoring, a relationship with the SPEI was established using Spearman's rank correlation. The results of this analysis are presented in Figure 7.8a for the annual series and Figure 7.8b for the summer series. For the annual aridity series, the correlations were carried out with SPEI-12. Their results reveal positive and significant correlations for the entire study area at 95% confidence level. Whereas the lowest correlations were recorded in locations classified as subhumid, the highest correlations were recorded in locations classified as arid and semiarid. Ghanzi, Jwaneng, and SSKA, all located in the south, recorded the highest correlation of 0.84. This implies that up to 84% of the variations in aridity can be explained by the SPEI. While the SPEI accumulates a moisture deficit over a time period, the aridity index does not.

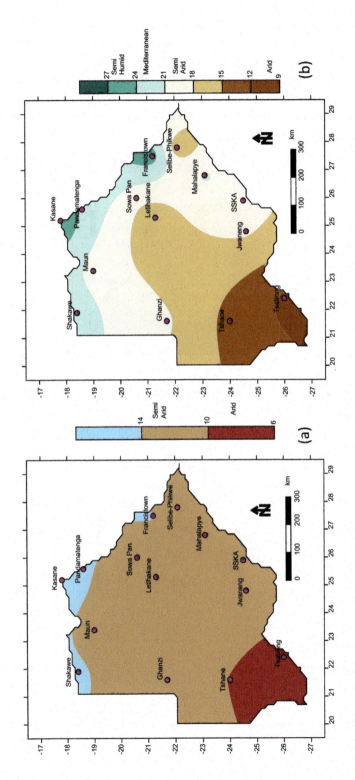

FIGURE 7.7 Climate classification using annual aridity

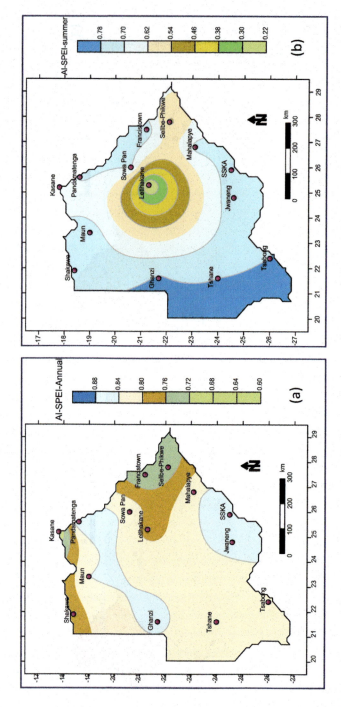

FIGURE 7.8   Correlations between (a) annual aridity and SPEI-12 and (b) summer aridity and SPEI-6.

This may explain why in locations with adequate moisture supply lower correlations are recorded compared to arid and semiarid locations.

Correlations for summer aridity time series with SPEI-6 are presented in Figure 7.8b. SPEI-6 was used here because six months is a typical length of the summer season in Botswana that occurs between November and April. It is observed that correlations take the same pattern as those observed with the annual series, although they are lower during the summer. From Figure 7.8b, correlations in arid and semiarid at the western axis were highest at 0.78. Similarly, during the summer season, up to 78% variations in aridity can be explained by the SPEI. This clearly demonstrates that in arid to semiarid climates, the aridity index can be used in drought monitoring.

7.9   ARIDITY PREDICTIONS USING ARTIFICIAL NEURAL NETWORK (ANNS)

We have demonstrated in this chapter the ability of ANNs in predicting annual aridity, which is crucial in planning and drought disaster preparedness. The model performance and predictions are presented in Figure 7.9 and Table 7.3, respectively. As observed in Table 7.3, the errors during training (one-step ahead predictions) are kept to a minimum. This implies the model was in position to learn the interrelations between the inputs, which included, rainfall, SPEI, and minimum and maximum temperature. The number of neurons used was 12 with a tapped delay of 2, which showed sufficient ability in minimizing errors compared to any other combination. The model errors during training were kept lower than those during the prediction phase in all cases. The correlations between observed and simulated values are all significant at 95% confidence level. The highest correlations of 0.88 and 0.85 were recorded at Letlhakane in the central and Jwaneng in the south, respectively.

From Table 7.3, based on the long-term historical mean, the model does not indicate much variation in the next five years. At most locations the five-year predictions are above the historical mean, with exceptions at Jwaneng, Letlhakane, and Selibe-Phikwe. There is no evidence that arid areas

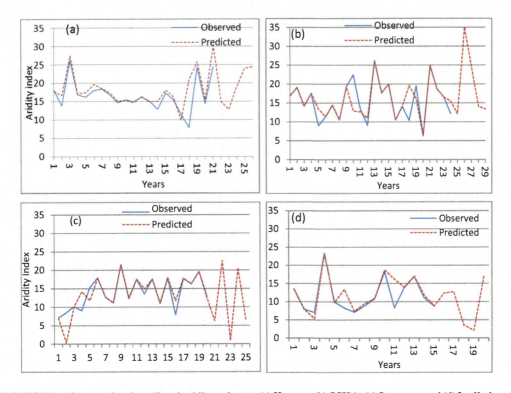

FIGURE 7.9   Observed and predicted aridity values at (a) Kasane, (b) SSKA, (c) Jwaneng, and (d) Letlhakane.

TABLE 7.3

Annual Aridity Predictions and Historic Mean

Station	Mean Square Errors		Correlation (R)	Historical Mean Annual AI	5-Year ANN Predictions				
	One-Step Ahead	Multisteps Ahead			2020/21	2021/22	2022/23	2023/24	2024/25
Francistown	0.14	0.71	0.71	14.2	17.3	14.7	19.5	17.5	14.9
Ghanzi	0.25	0.10	0.81	12.3	15.2	10.7	21.0	20.8	13.5
Jwaneng	0.07	0.12	0.85	12.6	6.2	22.5	0.9	20.6	6.2
Kasane	0.13	0.61	0.72	16.1	14.8	12.8	18.9	24.0	24.3
Letlhakane	0.09	0.53	0.88	11.2	12.3	12.7	3.5	2.1	16.8
Mahalapye	0.20	0.31	0.62	13.7	5.4	29.9	27.2	4.7	20.7
Maun	0.06	0.14	0.76	12.5	13.2	17.3	11.7	19.1	12.0
Pandamatenga	0.08	0.48	0.80	14.5	11.0	20.6	15.1	9.2	18.8
Selibe-Phikwe	0.28	0.61	0.71	10.5	11.0	9.6	10.5	9.0	12.1
Shakawe	0.12	0.52	0.75	14.5	11.7	27.8	14.3	19.0	8.0
SSKA	0.11	0.18	0.80	13.9	12.1	35.0	24.2	14.0	13.3
Sowa Pan	0.46	0.59	0.74	12.5	12.2	14.0	23.6	10.4	19.6
Tsabong	0.10	0.33	0.72	8.7	10.7	7.6	1.6	13.3	11.4
Tshane	0.10	0.33	0.74	9.9	10.2	7.9	16.7	11.8	20.0

will slip into hyperarid in the next five years. This is rather not in agreement with what has been projected using GCMs reported in Zarch et al. (2017) and Stonevičius et al. (2018). This discrepancy could be due to the various assumptions applied to GCMs before projections are made, while the ANN predictions require no assumptions to be generated.

To further demonstrate performance of the model simulations and predictions, a plot of the fit between observed and predicted values is provided in Figure 7.9. At all four locations presented (Kasane, Jwaneng, Letlhakane, and SSKA), it can be observed that the model was able to learn the various input combinations. The simulations were able to fit to the troughs and peaks with a few exceptions at SSKA (Figure 7.9b). This is a clear demonstration that ANNs have the ability to provide credible predictions.

7.10   CONCLUSIONS FROM THE CASE STUDY

In this case study, spatiotemporal trends of aridity were analyzed and predictions made using ANNs to aid in drought planning, preparedness, and management. Based on the results and discussions presented herein, the following conclusions can be drawn.

1. Aridity was observed to increase across all regions at a rate of 9% annually with an exception of the eastern region around Francistown. It was also observed that arid areas were drying up at a rate less than the humid areas. Significantly increasing aridity at a rate of −0.143 mm/°C/yr is mainly observed in the north at Shakawe.
2. There is variation in climate classification across seasons with a semihumid climate only available during summer, while in the winter season the whole country turns into an arid region.
3. It was demonstrated that 84% of the variation in aridity can be explained by the SPEI, especially in the arid zones. Correlations lower to 78% during the summer season due to increased moisture supply and reduced aridity.
4. Model prediction does not show any major changes in aridity in the next five years with the highest model performance of R = 0.88 recorded in the study area. All predictions were within the order of the historical mean.

It is anticipated that information generated here will be useful to governments and climate scientists, especially those working on drought monitoring and climate disaster reduction strategies.

REFERENCES

Akinsanola, A.A., Ajayi, V.O., Adejare, A.T., Adeyeri, O.E., Gbode, I.E., Ogunjobi, K.O., Nikulin, G., Abolude, A.T., 2018. Evaluation of rainfall simulations over West Africa in dynamically downscaled CMIP5 global circulation models. *Theor. Appl. Climatol.* 132, 437–450.

Alvares, C.A., Stape, J.L., Sentelhas, P.C., de Moraes, G., Leonardo, J., Sparovek, G., 2013. Köppen's climate classification map for Brazil. *Meteorol. Zeitschrift* 22, 711–728.

Baltas, E., 2007. Spatial distribution of climatic indices in northern Greece. *Meteorol. Appl. A J. Forecast. Pract. Appl. Train. Tech. Model.* 14, 69–78.

Byakatonda, J., 2018. Climatological droughts and their implications on water resources and agricultural production in semiarid regions: The case of Botswana. www.ub.bw.

Byakatonda, J., Parida, B.P., Kenabatho, P.K., Moalafhi, D.B., 2016a. Modeling dryness severity using artificial neural network at the Okavango Delta, Botswana. *Glob. Nest J.* 18, 463–481.

Byakatonda, J., Parida, B.P., Kenabathob, P.K., Books, R., OER, R., SCARDA, R., Tenders, R., 2016b. Aridity changes and its association with drought severity in Botswana, in: Fifth African Higher Education Week and RUFORUM Biennial Conference 2016, "Linking Agricultural Universities with Civil Society, the Private Sector, Governments and Other Stakeholders in Support of Agricultural Development in Africa", Cape Town, South Africa. pp. 325–333.

Byakatonda, J., Parida, B.P., Kenabatho, P.K., 2018a. Relating the dynamics of climatological and hydrological droughts in semiarid Botswana. *Phys. Chem. Earth* 1–13. doi:10.1016/j.pce.2018.02.004

Byakatonda, J., Parida, B.P., Kenabatho, P.K., Moalafhi, D.B., 2018b. Analysis of rainfall and temperature time series to detect long-term climatic trends and variability over semi-arid Botswana. *J. Earth Syst. Sci.* 127, 25. doi:10.1007/s12040-018-0926-3

Byakatonda, J., Parida, B.P., Kenabatho, P.K., Moalafhi, D.B., 2018c. Analysis of rainfall and temperature time series to detect long-term climatic trends and variability over semi-arid Botswana. *J. Earth Syst. Sci. Indian Acad. Sci.* 127. doi:10.1007/s12040-018-0926-3

Byakatonda, J., Parida, B.P., Moalafhi, D.B., Kenabatho, P.K., 2018d. Analysis of long term drought severity characteristics and trends across semiarid Botswana using two drought indices. doi:10.1016/j.atmosres.2018.07.002

Byakatonda, J., Parida, B.P., Kenabatho, P.K., Moalafhi, D.B., 2019a. Prediction of onset and cessation of austral summer rainfall and dry spell frequency analysis in semiarid Botswana. *Theor. Appl. Climatol.* 135, 101–117.

Byakatonda, J., Parida, B.P., Moalafhi, D.B., Kenabatho, P.K., Lesolle, D., 2019b. Investigating relationship between drought severity in Botswana and ENSO. *Nat. Hazards* 1–24.

Chang, F.-J., Chen, P.-A., Lu, Y.-R., Huang, E., Chang, K.-Y., 2014. Real-time multi-step-ahead water level forecasting by recurrent neural networks for urban flood control. *J. Hydrol.* 517, 836–846. doi:10.1016/j.jhydrol.2014.06.013

Costa, A.C., Soares, A., 2009. Trends in extreme precipitation indices derived from a daily rainfall database for the South of Portugal. *Int. J. Climatol.* 29, 1956–1975.

Croitoru, A.-E., Piticar, A., Imbroane, A.M., Burada, D.C., 2013. Spatiotemporal distribution of aridity indices based on temperature and precipitation in the extra-Carpathian regions of Romania. *Theor. Appl. Climatol.* 112, 597–607.

Dai, A., 2013. Increasing drought under global warming in observations and models. *Nat. Clim. Chang.* 3, 52–58.

Dyderski, M.K., Paź, S., Frelich, L.E., Jagodziński, A.M., 2018. How much does climate change threaten European forest tree species distributions? *Glob. Chang. Biol.* 24, 1150–1163.

Galarneau, T.J., Bosart, L.F., Aiyyer, A.R., 2008. Closed anticyclones of the subtropics and midlatitudes: A 54-yr climatology (1950–2003) and three case studies, in: *Synoptic Dynamic Meteorology and Weather Analysis and Forecasting. Springer*, pp. 349–392.

Gao, Y., Meng Joo, E., 2005. NARMAX time series model prediction: Feedforward and recurrent fuzzy neural network approaches. *Fuzzy Sets Syst.* 150, 331–350.

Garreaud, R.D., Boisier, J.P., Rondanelli, R., Montecinos, A., Sepúlveda, H.H., Veloso-Aguila, D., 2020. The central Chile mega drought (2010–2018): A climate dynamics perspective. *Int. J. Climatol.* 40, 421–439.

Greve, P., Roderick, M.L., Ukkola, A.M., Wada, Y., 2019. The aridity Index under global warming. *Environ. Res. Lett.* 14, 124006.

Hecht-Nielsen, R., 1987. Kolmogorov's mapping neural network existence theorem, in: Proceedings of the International Conference on Neural Networks. pp. 11–13.

Homdee, T., Pongput, K., Kanae, S., 2016. A comparative performance analysis of three standardized climatic drought indices in the Chi River basin, Thailand. *Agric. Nat. Resour.* 50, 211–219.

Huang, J., Ji, M., Xie, Y., Wang, S., He, Y., Ran, J., 2016. Global semi-arid climate change over last 60 years. *Clim. Dyn.* 46, 1131–1150.

Huo, Z., Dai, X., Feng, S., Kang, S., Huang, G., 2013. Effect of climate change on reference evapotranspiration and aridity index in arid region of China. *J. Hydrol.* 492, 24–34.

IPCC, 2012. *Managing the Risks of Extreme Events and Disasters to Advance Climate Change Adaption.* Cambridge University Press.

Kenabatho, P.K., Parida, B.P., Moalafhi, D.B., 2012. The value of large-scale climate variables in climate change assessment: The case of Botswana's rainfall. *Phys. Chem. Earth* 50–52, 64–71. doi:10.1016/j.pce.2012.08.006

Khan, M.I., Liu, D., Fu, Q., Dong, S., Liaqat, U.W., Faiz, M.A., Hu, Y., Saddique, Q., 2016. Recent climate trends and drought behavioral assessment based on precipitation and temperature data series in the Songhua River Basin of China. *Water Resour. Manag.* 30, 4839–4859. doi:10.1007/s11269-016-1456-x

Lloyd-Hughes, B., 2012. A spatio-temporal structure-based approach to drought characterisation. *Int. J. Climatol.* 32, 406–418.

McKee, T.B., Doesken, N.J., Kleist, J., others, 1993. The relationship of drought frequency and duration to time scales. In: Proceedings of the 8th Conference on Applied Climatology. pp. 179–183.

Menezes, J.M.P., Barreto, G.A., 2008. Long-term time series prediction with the NARX network: An empirical evaluation. *Neurocomputing* 71, 3335–3343. doi:10.1016/j.neucom.2008.01.030

Mishra, A.K., Singh, V.P., 2010. A review of drought concepts. *J. Hydrol.* 391, 202–216.

Moalafhi, D.B., Sharma, A., Evans, J.P., 2017. Reconstructing hydro-climatological data using dynamical downscaling of reanalysis products in data-sparse regions: Application to the Limpopo catchment in southern Africa. *J. Hydrol. Reg. Stud.* 12, 378–395.

Nalbantis, I., Tsakiris, G., 2009. Assessment of hydrological drought revisited. *Water Resour. Manag.* 23, 881–897.

Nastos, P.T., Politi, N., Kapsomenakis, J., 2013. Spatial and temporal variability of the Aridity Index in Greece. *Atmos. Res.* 119, 140–152.

Oguntunde, P.G., Abiodun, B.J., Lischeid, G., 2012. Spatial and temporal temperature trends in Nigeria, 1901–2000. *Meteorol. Atmos. Phys.* 118, 95–105. doi:10.1007/s00703-012-0199-3

Ortigara, A.R.C., Kay, M., Uhlenbrook, S., 2018. A review of the SDG 6 synthesis report 2018 from an education, training, and research perspective. *Water* 10, 1353.

PaiMazumder, D., Sushama, L., Laprise, R., Khaliq, M.N., Sauchyn, D., 2013. Canadian RCM projected changes to short-and long-term drought characteristics over the Canadian Prairies. *Int. J. Climatol.* 33, 1409–1423.

Ramachandran, A., Praveen, D., Jaganathan, R., Palanivelu, K., 2015. Projected and observed aridity and climate change in the east coast of south India under RCP 4.5. *Sci. World J.* 2015, 11.

Sheffield, J., Wood, E.F., Roderick, M.L., 2012. Little change in global drought over the past 60 years. *Nature* 491, 435–438.

Sivakumar, M.V.K., Motha, R., Wilhite, D., Wood, D., 2011. Agricultural drought indices. Proceedings of an Expert Meeting: 2-4 June, 2010, Murcia, Spain. WMO.

Stonevičius, E., Rimkus, E., Bukantis, A., Kriaučiuiene, J., Akstinas, V., Jakimavičius, D., Povilaitis, A., Kesminas, V., Virbickas, T., Pliuraite, V., others, 2018. Recent aridity trends and future projections in the Nemunas River basin. *Clim. Res.* 75, 143–154.

Tong, S., Zhang, J., Bao, Y., 2017. Inter-decadal spatiotemporal variations of aridity based on temperature and precipitation in Inner Mongolia, China. *Polish J. Environ. Stud.* 26, 819–826.

Van Loon, A.F., 2013. *On the Propagation of Drought: How Climate and Catchment Characteristics Influence Hydrological Drought Development and Recovery*. Wageningen University.

Vicente-Serrano, S.M., Begueria, S., López-Moreno, J.I., 2010. A multiscalar drought index sensitive to global warming: The standardized precipitation evapotranspiration index. *J. Clim.* 23, 1696–1718.

Vicente-Serrano, S.M., Begueria, S., Lorenzo-Lacruz, J., Camarero, J.J., López-Moreno, J.I., Azorin-Molina, C., Revuelto, J., Morán-Tejeda, E., Sanchez-Lorenzo, A., 2012. Performance of drought indices for ecological, agricultural, and hydrological applications. *Earth Interact.* 16, 1–27.

Wang, W., Zhu, Y., Xu, R., Liu, J., 2015. Drought severity change in China during 1961–2012 indicated by SPI and SPEI. *Nat. Hazards* 75, 2437–2451.

Yin, Y., Ma, D., Wu, S., Pan, T., 2015. Projections of aridity and its regional variability over China in the mid-21st century. *Int. J. Climatol.* 35, 4387–4398.

Zarch, M.A.A., Sivakumar, B., Sharma, A., 2015. Assessment of global aridity change. *J. Hydrol.* 520, 300–313.

Zarch, M.A.A., Sivakumar, B., Malekinezhad, H., Sharma, A., 2017. Future aridity under conditions of global climate change. *J. Hydrol.* 554, 451–469.

Zhang, Q., Xu, C.-Y., Zhang, Z., 2009. Observed changes of drought/wetness episodes in the Pearl River basin, China, using the standardized precipitation index and aridity index. *Theor. Appl. Climatol.* 98, 89–99.

Zhao, Y., Zou, X., Zhang, J., Cao, L., Xu, X., Zhang, K., Chen, Y., 2014. Spatio-temporal variation of reference evapotranspiration and aridity index in the Loess Plateau Region of China, during 1961–2012. *Quat. Int.* 349, 196–206.

8 Soil Moisture Drought Estimation Using Hydrological Modeling Approach for a River Basin of Eastern India

Shivam Gupta and Deepak Jhajharia

CONTENTS

8.1 INTRODUCTION

Drought is a severe climatic hazard that affects all sectors, and agriculture is among the worst affected sectors (Narasimhan and Srinivasan, 2005). Sometimes drought is taken as synonymous with aridity, but aridity is the climatic condition that is defined as the area having a low rainfall and high evapotranspiration rate, whereas drought is a temporary phenomenon that could last for months or years (Wilhite, 1992), hence, drought is a temporary aberration. Drought is a climatic phenomenon that varies from region to region depending on the climatic condition and cannot be defined by a single value for all regions. Variation in hydroclimatic conditions and socioeconomic structure in different regions of the world is a major challenge in precisely defining the drought (Mishra and Singh, 2010). There are several types of droughts to define their severity, such as agricultural drought, meteorological drought, hydrological drought, and socioeconomic drought. Prolonged meteorological and hydrological drought causes the soil moisture deficit along with surface water and groundwater shortage, which ultimately leads to a decrease in agricultural production (Funk et

DOI: 10.1201/9781003276548-8

al., 2008; Panda et al., 2007). Drought events affect the surface water as well as groundwater, which subsequently alters the water availability and supply, causes crop failure, and affects socioeconomic ventures (Riebsame et al., 1991).

Defining the drought universally is a challenging task for researchers, as the hydroclimatic and socioeconomic conditions of the region vary from place to place, and hence the severity of drought also varies (Mishra and Singh, 2010). Wilhite and Glantz (1987) suggested defining the drought based on conceptual and operational basis; operational definitions address the drought severity, frequency, and duration, whereas conceptual definitions focuse on relative terms such as longevity of the spell and its dryness. Gumbel (1963), Palmer (1965), and Linseley et al. (1959) defined drought events in terms of deficit in meteorological or hydrological parameters compared to the average of a long period. The Food and Agriculture Organization (FAO, 1983) defined drought events as the duration of a year when crops fail due to lack of soil moisture, whereas the World Meteorological Department (WMO, 1986) defined drought as a long spell of precipitation deficit.

Drought indices are the numerical representation of drought events, which give an insight into the severity, duration, frequency, and other parameters of drought. Several drought indices have been proposed by researchers from time to time, which have their own strengths and weakness. Following are some of the popular drought indices, along with their strength and weakness.

8.1.1 PALMER DROUGHT SEVERITY INDEX (PDSI)

Using precipitation and temperature for estimating moisture supply and demand within a two-layer soil model, Palmer (1965) formulated what is now referred to as the Palmer Drought Severity Index (PDSI). This was the first comprehensive effort to assess the total moisture status of a region. Since its inception, some modified versions of the PDSI have evolved. For example, Karl (1986) described a modified version known as the Palmer Hydrological Drought Index (PHDI), which is used for water supply monitoring. For operational purposes, a real-time version of the PDSI, called the modified PDSI ,was introduced by Heddinghaus and Sahol (1991). The PDSI is perhaps the most widely used regional drought index for monitoring droughts. The index has been used to illustrate the areal extent and severity of various drought episodes, investigate the spatial and temporal drought characteristics (Lawson et al., 1971; Soule, 1993), explore the periodic behavior of droughts, monitor hydrologic trends, forecast crops, assess potential fire severity (Heddinghaus and Sahol, 1991), monitor droughts over large geographic areas, and forecast droughts (Johnson and Kohne, 1993).

8.1.1.1 Limitations of PDSI

Some of the rules used to establish the PDSI are arbitrary, and the limitations of the PDSI have been documented in several studies (Karl and Knight, 1985). Limitations of the PDSI include (1) an inherent timescale making the PDSI more suitable for agricultural impacts and not so much for hydrologic droughts. (2) The PDSI assumes that all precipitation is rain, thus making values during winter months and at high elevations are often questionable. The PDSI also assumes that runoff only occurs after all soil layers have become saturated, leading to an underestimation of runoff. (3) The PDSI can be slow to respond to developing and diminishing droughts (Hayes et al., 1999). While there are criticisms of PDSI, there are positive aspects as well. It has been in use for a long time, and has been well tested and verified in many cases. It accounts for temperature and soil characteristics and is standardized so comparisons of different climatic zones are possible. The PDSI is also sensitive to precipitation and temperature, which is discussed in the following section.

8.1.2 STANDARDIZED PRECIPITATION INDEX (SPI)

The Standardized Precipitation Index (SPI) for any location is calculated based on the long-term precipitation record for the desired period. This long-term record is fitted to a probability distribution, which is then transformed to a normal distribution so that the mean SPI for the location and desired period is zero (McKee et al., 1993). The fundamental strength of the SPI is that it can be calculated for a variety of timescales. This versatility allows the SPI to monitor short-term water supplies, such as soil moisture, which is important for agricultural production and long-term water resources, such as groundwater supplies, streamflow, and lake and reservoir levels. Soil moisture conditions respond to precipitation anomalies on a relatively short scale. Groundwater, streamflow, and reservoir storage reflect the long-term precipitation anomalies. For example, Szalai and Szinell (2000) examined how strong the connection of SPI is with hydrological features, such as streamflow and groundwater level at stations in Hungary. The correlation of the SPI with streamflow was the highest on a two-month timescale, while for groundwater levels the best correlations were found at widely different timescales. They also concluded that agricultural drought (proxied by soil moisture content) was replicated best by the SPI on a scale of two–three months. The SPI has been used for studying different aspects of droughts, for example, forecasting (Mishra et al., 2007), frequency analysis, spatiotemporal analysis, and climate impact studies.

8.1.3 SURFACE WATER SUPPLY INDEX (SWSI)

The Surface Water Supply Index (SWSI) was primarily developed as a hydrological drought index and it is calculated based on monthly nonexceedance probability from available historical records of reservoir storage, streamflow, snowpack, and precipitation (Shafer and Dezman, 1982). The purpose of the SWSI is primarily to monitor abnormalities in surface water supply sources. Hence, it is a good measure to monitor the impact of hydrologic droughts on urban and industrial water supplies, irrigation, and hydroelectric power generation. Four inputs are required within the SWSI: snowpack, streamflow, precipitation, and reservoir storage (Wilhite and Glantz, 1987; Garen, 1993). Because it is dependent on the season, the SWSI is computed with only snowpack, precipitation, and reservoir storage in winter. During summer months, streamflow replaces snowpack as a component within the SWSI equation.

8.1.3.1 Limitations of SWSI

The definition of surface water supply and the factor weights vary with spatial scale (one watershed to another) as well as temporal scale (season or month) due to differences in hydroclimatic variability resulting in SWSIs with different statistical properties. For example, the hydroclimatic differences that characterize river basins in the western United States result in SWSIs that do not have the same meaning and significance in all areas and at all times (Doesken et al., 1991).

8.1.4 SOIL MOISTURE DEFICIT INDEX (SMDI)

Narasimhan and Srinivasan (2005) developed the Soil Moisture Deficit Index (SMDI) and Evapotranspiration Deficit Index (ETDI) based on weekly soil moisture and evapotranspiration simulated by a calibrated hydrologic model, respectively. The drought indices were derived from the soil moisture deficit and evapotranspiration deficit, and scaled between −4 and +4 for spatial comparison of droughts, irrespective of climatic conditions.

India's vulnerability to drought is very high, as the large population thrives on agriculture, which is mostly dependent on rainfall (Mishra, 2020). The major proportion of the annual rainfall in India occurs in monsoon months, which also increases the vulnerability of the agriculture sector in India (Mishra, 2020). India has been suffering from drought for a long time, which had caused a loss of life and assets (Mishra and Singh, 2010; Mooley and Pant, 1981). In India, a major concern is the increasing frequency of drought events. It is observed that a drought event has been reported every 3 years in the past 50 years (FAO, 2002; Mishra and Singh, 2010). In this article, a case study of drought estimation using the SMDI has been presented using the hydrological modeling and water balance approach.

8.2 MATERIAL AND METHODS

8.2.1 STUDY AREA AND DATA

The Sabari River is the major tributary of the Godavari River basin in the eastern part of India, and a major part of its basin extends into Odisha state. It is also known as the Kolab River. It also covers some portion of Chhattisgarh state and in the end, it joins the Godavari River in Andhra Pradesh state. Figure 8.1 shows the geographical location of the Sabari River basin, and elevation variation within the basin varies from 19 m to 1700 m above mean sea level. The average annual rainfall in the Sabari River basin is 1250 mm, and major rainfall is received in the monsoon season. This river is important for the agricultural and hydropower projects in the region, and the Upper Kolab project is one of the main hydropower projects on the river that is also used for irrigation projects.

In this study, the India Meteorological Department (IMD) gridded precipitation (0.25°×0.25°) and temperature (1°×1°) data were used for the hydrological modeling. Other meteorological data sets, i.e., relative humidity, solar radiation, and wind speed, were taken from the reanalysis data sets available on the Soil and Water Assessment Tool (SWAT) website. A land use/land cover map (Figure 8.2a) provided by MODIS, soil map (Figure 8.2b) from the Harmonized World Soil Database (HWSD) provided by the Food and Agriculture Organization (FAO), and Satellite Radar Topographic Mission (SRTM) digital elevation model data (Figure 8.1) with 90 m × 90 m spatial resolution were used for the hydrological modeling.

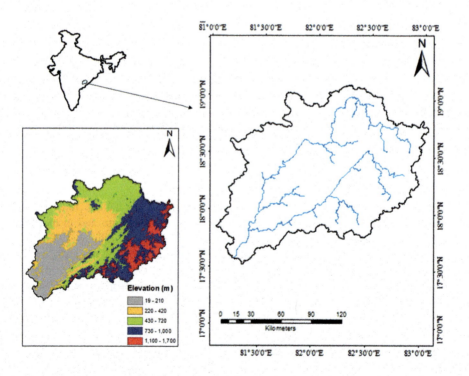

FIGURE 8.1   Location of the Sabari River basin in the peninsular region of India.

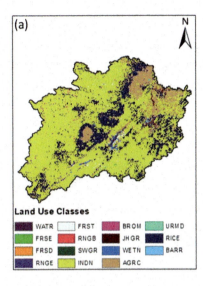
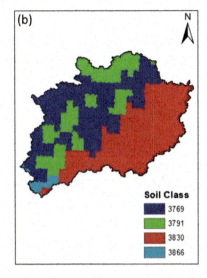

FIGURE 8.2   (a) Land use/land cover map and (b) soil map of the Sabari River basin.

8.2.2   METHODOLOGY

8.2.2.1   Hydrological Modeling

Hydrological modeling of the Sabari River basin was performed by using the SWAT. SWAT is a semidistributed physical-based model that works on the water balance principle and uses the Soil Conservation Service Curve Number method (SCS-CN) approach for simulating the surface run-off. It changes the curve number on a daily basis based on the change in soil moisture. SWAT is a physical-based model that requires a digital elevation model, slope map, soil map, and land use/land cover map for simulating the hydrological parameters such as surface runoff, evapotranspiration, and infiltration. SWAT uses the following water balance equation for calculating all the hydrological components.

$$SW_t = SW_0 + \sum_{i=1}^{n} (R_{day} + Q_{sur} + E_a + W_{seep} + Q_{gw}) \tag{8.1}$$

where SW_t is the final soil water content (mm H_2O), SW_0 the initial soil water content (mm H_2O), t is time in days, R_{day} is the amount of precipitation on day i (mm H_2O), Q_{surf} is the amount of surface runoff on day i (mm H_2O), E_a is the amount of evapotranspiration on day i (mm H_2O), W_{seep} is the amount of percolation and bypass exiting the soil profile bottom on day i (mm H_2O), and Q_{gw} is the amount of return flow on day i (mm H_2O).

8.2.2.2   Soil Moisture Deficit Index

The SWAT model can efficiently simulate the soil moisture, which can be used for the calculation of the Soil Moisture Deficit Index and Crop Moisture Index for estimating the drought index. DeLiberty and Legates (2003) carried out a study of soil moisture variation using the SWAT model simulation for Oklahoma and found the SWAT model efficient to simulate soil moisture. In the present study, soil moisture simulated from the calibrated and validated SWAT model was used for calculating the Soil Moisture Deficit Index to find out the soil moisture drought situation in crops.

The daily model output of available soil water in the root zone was averaged over a 7-day period to get weekly soil water for each of the 52 weeks in a year for each sub-basin. The long-term soil moisture for each week in a year was obtained by taking the median of the available soil water for that week during a 70-year period (1911–1980). The median was chosen over the mean as a measure of "normal" available soil water because the median is more stable and is not influenced by a few outliers. The maximum and minimum soil water for each week was also obtained from the 70-year data. Using this long-term median, maximum, and minimum soil water, the weekly percentage soil moisture deficit (SD) or excess for 98 years (1901–1998) was calculated as

$$SD_{i,j} = \frac{SW_{i,j} - MSW_j}{MSW_j - minSW_j} \times 100 \quad if \quad SW_{i,j} = MSW_j \tag{8.2}$$

$$SD_{i,j} = \frac{SW_{i,j} - MSW_j}{maxMSW_j - MSW_j} \times 100 \quad if \quad SW_{i,j} > MSW_j \tag{8.3}$$

where $SD_{i,j}$ is the soil water deficit (%), $SW_{i,j}$ is the mean weekly soil water available in the soil profile (mm), MSW_j is the long-term median available soil water in the soil profile (mm), $maxMSW_j$ the long-term maximum available soil water in the soil profile (mm), and $minMSW_j$ is the long-term minimum available soil water in the soil profile (mm).

By using Equation 8.1, the seasonality inherent in soil water was removed. Hence, the deficit values can be compared across seasons. The SD values during a week range from –100 to +100 indicating very dry to very wet conditions. As the SD values for all the sub-basins were scaled between –100 and +100, they are also spatially comparable across different climatic zones (humid or arid).

The SD value during any week gives the dryness (wetness) during that week when compared to long-term historical data. Drought occurs only when the dryness continues for a prolonged period of time that can affect crop growth. As the limits of SD values were between –100 and +100, the worst drought can be represented by a straight line with the equation:

$$\sum\nolimits_{t=1}^{j} Z_t = -100t - 100 \tag{8.4}$$

where t is the time in weeks. If this line defines the worst drought (i.e., 4 for the drought index to be comparable with PDSI), then SMDI for any given week can be calculated by

$$SMDI_j = \frac{\Sigma_{t=1}^{j} SD_t}{(25t + 25)} \tag{8.5}$$

8.3   RESULTS AND DISCUSSION

8.3.1   HYDROLOGICAL MODELING OF RIVER BASIN

A spatially distributed hydrologic model is essential for developing the drought index. In this study, the hydrologic model SWAT was used. SWAT is a physically based basin scale, continuous-time distributed parameter hydrologic model that uses spatially distributed data on soil, land cover, digital elevation model, and weather data for hydrologic modeling, and operates on a daily time step. Major model components include weather, hydrology, soil temperature, plant growth, nutrients, pesticides, and land management. A complete description of the SWAT model components can be found in Arnold et al. (1998) and Neitsch et al. (2002).

For spatially explicit parameterization, SWAT subdivides watersheds into sub-basins based on topography, which are further subdivided into hydrologic response units based on unique soil and land cover characteristics. Four storage volumes represent the water balance in each hydrologic response unit in the watershed: snow, soil profile (0–2 m), shallow aquifer (2–20 m), and deep aquifer (>20 m). The soil profile can be subdivided into multiple layers. Soil water processes include surface runoff, infiltration, evaporation, plant water uptake, interflow, and percolation to shallow and deep aquifers.

SWAT simulates surface runoff using the modified SCS-CN method (USDA Soil Conservation Service, 1972) with daily precipitation data. Based on the soil hydrologic group, vegetation type, and land management practice, initial CN values are assigned from the SCS hydrology handbook (USDA Soil Conservation Service, 1972). SWAT updates the CN values daily based on changes in soil moisture. The excess water available after accounting for initial abstractions and surface runoff, using the SCS-CN method, infiltrates into the soil. A storage routing technique is used to simulate the flow through each soil layer. SWAT directly simulates saturated flow only and assumes that water is uniformly distributed within a given layer. Unsaturated flow between layers is indirectly modeled using depth distribution functions for plant water uptake and soil water evaporation. Downward flow occurs when the soil water in the layer exceeds field capacity and the layer below is not saturated. The rate of downward flow is governed by the saturated hydraulic conductivity.

8.3.2 Calibration and Validation of the Model

Calibration and validation of SWAT model is performed by optimizing the hydrological parameters such as curve number, baseflow factor, and groundwater delay (Figure 8.3).

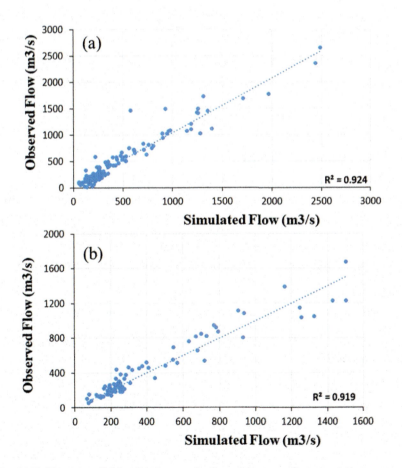

FIGURE 8.3 (a) Calibration and (b) validation result of monthly streamflow simulation.

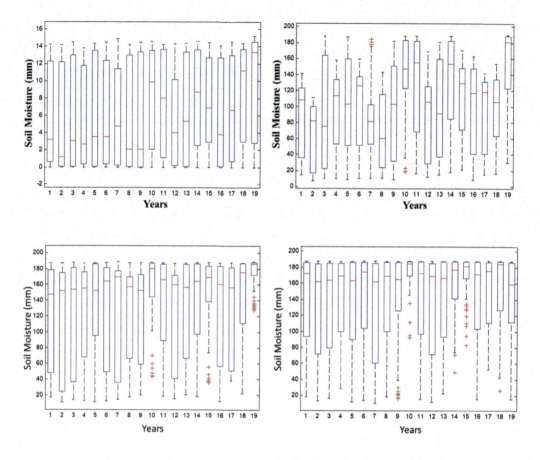

FIGURE 8.4   Boxplot representation of soil moisture (mm) variation in four stations.

8.3.3   Time Series Variability of SMDI

A spatial variability analysis was done to study the effect of the spatially distributed model on soils moisture availability and weather variables like precipitation and temperature on the drought index. Standard deviation of the drought index, calculated from the drought index of sub-basins within the watershed during each time step, was used as a measure of spatial variability of the drought index. The spatial standard deviation was calculated for each week of simulated year. Considering that the range of drought indices is from −4 to +4, a standard deviation of 1.0 indicates that the spatial variability of the drought index is high. The distribution of standard deviation for 52 weeks during the 20-year period is shown in Figures 8.4 and 8.5.

The mean and standard deviation of weekly precipitation and reference crop ET for each watershed were also analyzed to determine the reason for spatial variability in the drought index. Analysis of 98 years of precipitation data showed that the precipitation distribution in a year was bimodal. For both watersheds, with high precipitation occurring during late spring and mid-fall seasons. Precipitation was the highly variable component both spatially and temporally during different years for the same season. Reference crop ET also showed some spatial variability with high variability occurring during the summer season. The spatial variability (standard deviation) of the drought indices, especially ETDI, during different seasons closely followed the variability in precipitation and evapotranspiration across seasons. In order to get a sense of how the standard deviation reflects the spatial distribution of drought indices, SMDI derived during the 46th week

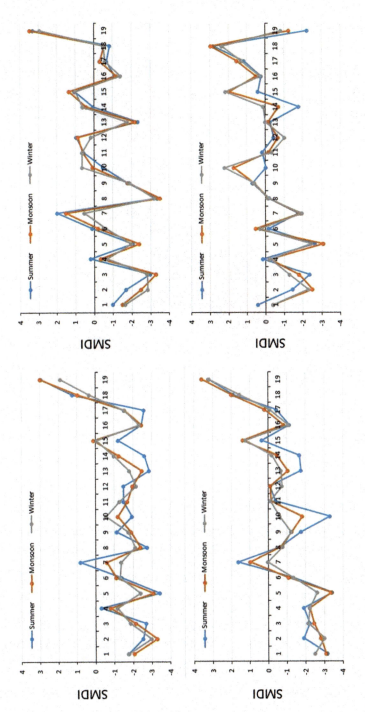

FIGURE 8.5 Soil Moisture Deficit Index time series at monthly timescale.

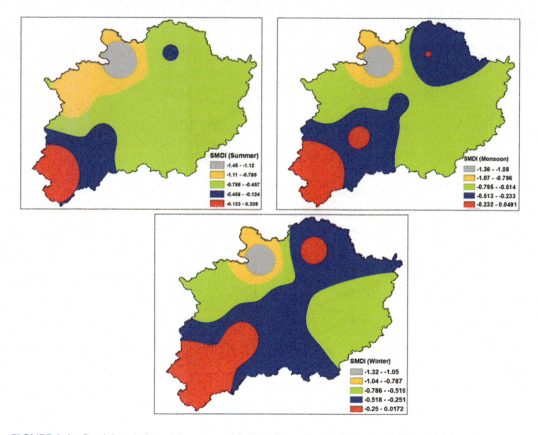

FIGURE 8.6 Spatial variation of the seasonal Soil Moisture Deficit Index in the Sabari River basin.

of 1988 and 24th week of 1990 with standard deviations of 1.0 and 1.5, respectively, are shown in Figures 8.6a, b. As the standard deviation increased, the spatial variability of the drought index also increased considerably.

8.3.4 SPATIAL VARIATION OF SMDI

The spatial standard deviation of the SMDI increased from 0.75 during the spring season to as much as 1.5 at the end of the summer season (see Figures 8.4 and 8.5). This high value of the SMDI was recorded as the ET during summer season was very high, which is followed by a high precipitation season that recharged the soil moisture to an extent depending on the physical properties of soil and precipitation magnitude. This affects actual evapotranspiration that depends on the amount of water already in the soil profile, soil physical properties, and land cover characteristics.

The spatial variability of the SMDI as shown in Figure 8.6 indicates that all soil regions had values from −0.78 to −0.457 during the summer and this changes to −1.32 to −1.05 during the winter. The SMDI decreased during the summer because regions are characterized by a reduction in rainfall (less spatial variability) and increased ET values (less moisture availability in soils) and thus, the spatial variability of the SMDI reduced during the summer. The spatial analysis of the SMDI indicates considerable spatial variability in drought across the region depending on the weather, soil, and land use characteristics.

8.4 SUMMARY AND CONCLUSIONS

Weekly soil moisture simulated by the hydrologic model was used to develop drought indices. The drought index was derived from soil moisture deficit and scaled between −4 and +4 for spatial comparison of drought extent, irrespective of climatic conditions. The SMDI very efficiently depicts the moisture dynamics and its correlation with the evapotranspiration from the available soil layer. Therefore, the SMDI could be a good indicator of short-term agricultural droughts. The spatial variability of the drought indices was high with a standard deviation greater than 1.0 during most weeks of a year, which was not apparent in the existing drought indices due to large-scale spatial lumping. The high spatial variability in the drought indices was mainly due to high spatial variability in rainfall distribution. The spatial variability (standard deviation) of the drought indices during different seasons closely followed the variability in precipitation and evapotranspiration across seasons.

REFERENCES

Arnold, J.G., Srinivasan, R., Muttiah, R.S., Williams, J.R., 1998. Large area hydrologic modeling and assessment. Part 1. Model development. *J. Am. Soc. Water Res. Assoc.* 34(1), 73–89.

DeLiberty, T.L., Legates, D.R., 2003. Interannual and seasonal variability of modelled soil moisture in Oklahoma. *Int. J. Climatol.* 23, 1057–1086.

Doesken, N.J., McKee, T.B., Kleist, J., 1991. *Development of a Surface Water Supply Index for the Western United States.* Climatology Report Number 91-3, Colorado State University, Fort Collins, Colorado.

FAO, 2002. *Report of FAO-CRIDA Expert Group Consultation on Farming System and Best Practices for Drought-prone Areas of Asia and the Pacific Region. Food and Agricultural Organisation of United Nations.* Published by Central Research Institute for Dryland Agriculture, Hyderabad, India.

Food and Agriculture Organization, 1983. *Guidelines: Land Evaluation for Rainfed Agriculture.* FAO Soils Bulletin 52, Rome.

Funk, C., Dettinger, M.D., Michaelsen, J.C., Verdin, J.P., Brown, M.E., Barlow, M., Hoell, A., 2008. Warming of the Indian Ocean threatens eastern and southern African food security but could be mitigated by agricultural development. *Proc. Natl. Acad Sci.* 105(32), 11081–11086. https://doi.org/10.1073/pnas.0708196105.

Garen, D.C., 1993. Revised surface-water supply index for western United States. *J. Water Resour. Plan. Manage.* 119(4), 437–454.

Gumbel, E.J., 1963. Statistical forecast of droughts. *Bull. Int. Assoc. Sci. Hydrol.* 8(1), 5.23.

Hayes, M.J., Svoboda, M.D., Wilhite, D.A., Vanyarkho, O.V., 1999. Monitoring the 1996 drought using the standardized precipitation index. *Bull. Am. Meteorol. Soc.* 80(3), 429–438.

Heddinghaus, T.B., Sahol, P., 1991. A review of the palmer drought severity index and where do we go from here? In: Proceedings of the 7th Conference on Applied Climatology, September 10-13, 1991, American Meteorological Society, Boston, MA, pp. 242–246.

Johnson, W.K., Kohne, R.W., 1993. Susceptibility of reservoirs to drought using Palmer index. *J. Water Resour. Plann. Manage.* 119(3), 367–387.

Karl, T.R., 1986. The sensitivity of the Palmer drought severity index and Palmer's Z index to their calibration coefficients including potential evapotranspiration. *J. Clim. Appl. Meteorol.* 25, 77–86.

Karl, T., Knight, R.W., 1985. *Atlas of Monthly Palmer Drought Severity Index (1931–1983) for the Contiguous United States* (Vol. 3). National Climatic Data Center.

Lawson, M.P., Reiss, A., Phillips, R., Livingston, K., 1971. *Nebraska Droughts A Study of Their Past Chronological and Spatial Extent with Implications for the Future.*

Linsely Jr., R.K., Kohler, M.A., Paulhus, J.L.H., 1959. *Applied Hydrology.* McGraw Hill, New York.

McKee, T.B., Doesken, N.J., Kleist, J., 1993. The Relationship of Drought Frequency and Duration to Time Scales, Paper Presented at 8th Conference on Applied Climatology. American Meteorological Society, Anaheim, CA.

Mishra, V. (2020). Long-term (1870–2018) drought reconstruction in context of surface water security in India. *J. Hydrol.* 580, 124228.

Mishra, A.K., Desai, V.R., Singh, V.P., 2007. Drought forecasting using a hybrid stochastic and neural network model. *J. Hydrologic Eng. ASCE* 12(6), 626–638.

Mishra, A.K., Singh, V.P., 2010. A review of drought concepts. *J. Hydrol.* https://doi.org/10.1016/j.jhydrol.2010 .07.012.

Mooley, D.A., Pant, G.B., 1981. Droughts in India over the last 200 years, their socioeconomic impacts and remedial measures for them. Climate and history. pp. 465–478. http://moeseprints.incois.gov.in/1331/.

Narasimhan, B., Srinivasan, R., 2005. Development and evaluation of soil moisture deficit index (SMDI) and evapotranspiration deficit index (ETDI) for agricultural drought monitoring. *Agric. For. Meteorol.* 133, 69–88.

Neitsch, S.L., Arnold, J.G., Kiniry, J.R., Williams, J.R., King, K.W., 2002. *Soil and Water Assessment Tool, Theoretical Documentation: Version 2000.* TWRI TR-191. Texas Water Resources Institute, College Station, TX.

Palmer, W.C., 1965. Meteorologic drought. US Department of Commerce, Weather Bureau. *Res. Paper* 45, 58.

Panda, D.K., Mishra, A., Jena, S.K., James, B.K., Kumar, A., 2007. The influence of drought and anthropogenic effects on groundwater levels in Orissa, India. *J. Hydrol.* https://doi.org/10.1016/j.jhydrol.2007.06 .007.

Riebsame, W.E., Changnon, S.A., Karl, T.R., 1991. *Drought and Natural Resource Management in the United States: Impacts and Implications of the 1987–1989 Drought.* Westview Press, Boulder, CO, p. 174.

Shafer, B.A., Dezman, L.E. (1982, January). Development of surface water supply index (SWSI) to assess the severity of drought condition in snowpack runoff areas. Proceeding of the Western Snow Conference.

Soulé, P.T., 1993. Hydrologic drought in the contiguous United States, 1900–1989: Spatial patterns and multiple comparison of means. *Geophys. Res. Lett.* 20(21), 2367–2370.

Szalai, S., Szinell, C.S., 2000. Comparison of two drought indices for drought monitoring in Hungary: A case study. In: *Drought and Drought Mitigation in Europe* (pp. 161–166). Springer, Dordrecht.

USDA Soil Conservation Service, 1972. *National Engineering Handbook, Hydrology, Section 4*, Chapters 4–10. GPO, Washington, DC.

Wilhite, D.A., 1992. *Preparing for Drought: A Guidebook for Developing Countries*, Climate Unit, United Nations Environment Program, Nairobi, Kenya.

Wilhite, D.A., Glantz, M,H., 1987. Understanding the drought phenomena: The role of definitions. In: Donald, A., Wilhite, Easterling Willam, E., Deobarah, A., (Eds.), *Planning of Drought: Towards a Reduction of Societal Vulnerability*, Westview Press, Wood, Boulder, CO, pp. 11–27.

World Meteorological Organization (WMO), 1986. *Report on Drought and Countries Affected by Drought During 1974–1985*, WMO, Geneva, p. 118.

9 Meteorological Drought Prediction Using Hybrid Machine Learning Models
Ant Lion Optimizer versus Multi-Verse Optimizer

Anurag Malik, Yazid Tikhamarine, Rawshan Ali, Alban Kuriqi, and Anil Kumar

CONTENTS

9.1 INTRODUCTION

Drought is a recurring phenomenon that generally starts with a significant deficit in precipitation, resulting in subsequent deficits in soil moisture and decreased stream discharge, affecting water resources and agricultural production, particularly in rainfed farming systems (Rahmati et al. 2020). Droughts influence both surface and groundwater resources, and result in deteriorated water quality, reduced water supply, crop failure, and diminished range productivity, along with economic and social activities (Riebsame et al. 1991). In the last two decades, due to climate change, extreme weather events have occurred more severely and frequently on Earth (Dai 2011a, 2011b, 2013; Ali et al. 2020). The prediction and early warning signs of drought are especially crucial in managing and preparing farm resources before drought begins.

Droughts generally fall into four categories: hydrological, agricultural, meteorological, and socioeconomic (Wilhite and Glantz 1985). Meteorological drought is commonly characterized as lower than average precipitation for a given area over a given period and is related to the deficiency in precipitation. Agricultural drought occurs when crop production needs and soil humidity is insufficient to satisfy plant growth (Mishra and Singh 2010). Meteorological and agricultural droughts typically have a reasonably short time lag depending on the antecedent soil moisture and soil composition. Hydrological drought occurs when a precipitation shortage reduces the levels of streamflow, groundwater, or reservoir, leading to a reduction in the water supply. Socioeconomic drought is a major failure of water resources systems to meet people's water demands and their activities, resulting in many socioeconomic consequences, such as impacts on income and lifestyles (Mishra and Singh 2010; Rahmati et al. 2020; Ali et al. 2020).

The first thing dealt with in the prediction of drought is the type of drought, i.e., meteorological, hydrological, agricultural, or socioeconomic, and then selection of a suitable index for monitoring and prediction. Several meteorological drought indices have been proposed over the past few decades based on climatic variables, such as the Palmer Drought Severity Index (PDSI) (Palmer 1965), Standardized Precipitation Index (SPI) (McKee et al. 1993), Effective Drought Index (EDI) (Byun and Wilhite 1999), Reconnaissance Drought Index (RDI) (Tsakiris et al. 2007), and Standardized Precipitation Evapotranspiration Index (SPEI) (Vicente-Serrano et al. 2010). Among these, the EDI has been considered a more accurate index for meteorological drought characteristics (i.e., duration, severity, intensity) investigation (Byun and Wilhite 1999; Deo et al. 2017a). Thus, researchers from different disciplines (e.g., hydrology, meteorology, agriculture, economics, and environment) concentrate on drought prediction with multiple aspects like risk management, drought preparedness, and policymaking (Wilhite 1986; Mishra and Singh 2011; Kisi et al. 2019).

Another challenge in droughts prediction is the selection of a robust and reliable model. Several models have been previously applied to predict the occurrence and severity of droughts by employing autoregressive integrated moving average (ARIMA) and seasonal ARIMA (SARIMA), artificial neural networks (ANNs), and hybrid models (Mishra and Desai 2005, 2006; Mishra et al. 2007; Morid et al. 2007; Malik et al. 2019c, 2020a; Malik and Kumar 2020). The time series–based models (i.e., ARIMA/SARIMA) have been frequently utilized for hydrometeorological droughts prediction, but found not to be more reliable because of the complex nature of droughts (Zhang 2003; Elkiran et al. 2019). While machine learning (ML)–based models predict complex drought events more accurately (Danandeh Mehr et al. 2014; Ali et al. 2017, 2018; Kisi et al. 2019). Deo and Şahin (2015) stated that ML models are found potentially viable for evaluating the complex relationships between historical drought events and other climatic variables. Zhang et al. (2019) employed Extreme Gradient Boosting (XGBoost), ANNs, and the distributed nonlinear lag model (DLNM) to forecast the multiscale SPEI in China. Analysis of results shows that the XGBoost model forecast multiscale SPEI more accurately than the other models. Das et al. (2020) predicted meteorological drought based on the SPI in Karnataka state, India, by using standalone ANN, SVR, and hybrid wavelet (W-ANN and W-SVR) models. They found a hybrid W-ANN model outperformed the other models for SPI-1 and SPI-6 prediction in the study region.

Recently, ML-based models integrated with several nature-inspired algorithms have been successfully applied in modeling various hydrological processes (Tikhamarine et al. 2019a, 2019b, 2020a, 2020b, 2020c; Banadkooki et al. 2020; Fadaee et al. 2020; Guan et al. 2020; Majumder and Eldho 2020). They are more potent than the simple ML models and can handle the big data with higher computational efficiency. So, by considering the aforementioned literature, we found that the application of hybrid ML models is limited for meteorological drought prediction in the study region. Thus, the motivation of research was to examine the comparative potential of hybrid support vector regression (SVR) integrated with the Ant Lion Optimizer (SVR-ALO) and Multi-Verse Optimizer (SVR-MVO) algorithms to predict the meteorological drought at Almora, Champawat, Pauri Garhwal, and Pithoragarh stations based on the EDI. Monthly rainfall data from 115 years (1901–2015) at these stations were used for EDI computation. The outcomes of models were evaluated using performance indicators and by visual inspection to designate the most appropriate model for EDI prediction in the study region.

9.2 MATERIALS AND METHODS

9.2.1 STUDY LOCATION AND DATA COLLECTION

The present study was conducted at four meteorological stations – Almora, Champawat, Pauri Garhwal, and Pithoragarh – in the state of Uttarakhand (India). The locations of stations are illustrated in Figure 9.1. Information about the geographical coordinates and data of the study locations are listed in Table 9.1. The state of Uttarakhand contains 13 districts divided into two administrative divisions: Garhwal (Uttarkashi, Tehri Garhwal, Rudraprayag, Pauri Garhwal, Chamoli, Dehradun, and Haridwar) and Kumaon (Udham Singh Nagar, Nainital, Pithoragarh, Champawat, Bageshwar, and Almora). Most of the rainfall is received in the rainy season (June–September), with yearly variation from 260 to 3955 mm over the study region. The climate varies from temperate to tropical with subzero to 43°C in the state. Rainfall data was gathered on a monthly scale for 115 years (1901–2015) from the study locations, courtesy of the Indian Meteorological Department (IMD), Pune.

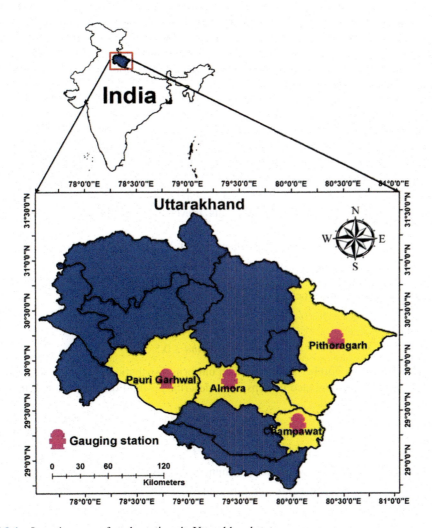

FIGURE 9.1 Location map of study stations in Uttarakhand state.

TABLE 9.1

Geographical Specifications and Rainfall Data Accessibility at Study Stations

Station	Longitude	Latitude	Elevation	Data of Rainfall (years)
Pithoragarh	80° 21′ 54″ E	30° 11′ 31″ N	3669 m	1901–2015
Champawat	80° 04′ 26″ E	29° 21′ 54″ N	1791 m	1901–2015
Almora	79° 26′ 13″ E	29° 48′ 40″ N	1759 m	1901–2015
Pauri Garhwal	78° 43′ 44″ E	29° 50′ 28″ N	1134 m	1901–2015

9.2.2 Effective Drought Index

The EDI was initially developed by Byun and Wilhite (1999) to investigate meteorological drought. Computation of the EDI includes (i) effective precipitation (EP), the summed value of daily precipitation with a time-dependent reduction function; (ii) means of EP (MEP), which shows climatological characteristics of precipitation as a water resource for a station or region; (iii) deviation of EP (DEP); and (iv) standardized value of DEP. Over the year, many applications of the EDI have been found in the literature for meteorological drought quantification on daily and monthly scales (Smakhtin and Hughes 2007; Morid et al. 2007; Kim et al. 2009; Dogan et al. 2012; Jain et al. 2015; Deo et al. 2017a). For profound information about the EDI computation procedure, readers can refer to Byun and Wilhite (1999).

9.2.3 Ant Lion Optimizer

Mirjalili (2015) first proposed the concept of a novel Ant Lion Optimizer (ALO) algorithm, which mimics the hunting mechanism of antlions inspired by nature. The hunting mechanism of prey includes five main steps: (i) random walk of ants, (ii) building traps, (iii) entrapment of ants in traps, (iv) catching prey, and (v) rebuilding traps. When searching for food, ants move stochastically in nature; a random walk is elected for modeling ants' movement using Equation 9.1:

$$X(t) = \left[0, cumsum\left(2r(t_1)-1\right), cumsum\left(2r(t_2)-1\right), \ldots, cumsum\left(2r(t_n)-1\right) \right] \tag{9.1}$$

in which *cumsum* is the cumulative sum, r is the maximum number of iterations, t is a step of random walk, and $r(t)$ is the stochastic function and defined as

$$r(t) = \begin{cases} 1 & \text{if } rand > 0.5 \\ 0 & \text{if } rand \leq 0.5 \end{cases} \tag{9.2}$$

where *rand* is a random number generated with uniform distribution in the interval of [0, 1]. The hunting behavior of ants mentioned in steps i–v is described using Equations 9.3 to 9.10:

$$X_i^t = \frac{\left(X_i^t - a_i\right) \times \left(d_i - C_i^t\right)}{\left(d_i^t - a_i\right)} + C_i \tag{9.3}$$

in which a_i and d_i are, respectively, the minimum and maximum of random walk of the ith variable; and C_i^t and d_i^t are, respectively, the minimum and maximum of the ith variable at the tth iteration and defined as

$$C_i^t = Antlion_j^t + C^t \tag{9.4}$$

$$d_i^t = Antlion_j^t + d^t \tag{9.5}$$

where $Antlion_j^t$ represents the position of the nominated jth antlion at the tth iteration, and C^t and d^t are the minimum and maximum vectors, respectively, of all the variables at the tth iteration, and expressed as

$$C^t = \frac{C^t}{I} \tag{9.6}$$

$$d^t = \frac{d^t}{I} \tag{9.7}$$

in which I is a ratio and written as

$$I = 10^w \frac{t}{T} \tag{9.8}$$

where T is the maximum number of iterations, t is the current iteration, and w is a constant defined based on the current iteration, which adjusts the accuracy level of exploitation. The ant reaches the bottom of the pit and is caught in the antlion's jaw, pulls the ant inside the sand, and consumes its body; this process is called catching prey. After that, the antlion updates its position to the latest position of the hunted, and to repeat the process for catching new prey, rebuilding of the pit occurs, and is expressed using Equation 9.9:

$$Antlion_j^t = Ant_j^t \quad if \; f\left(Ant_j^t\right) > f\left(Antlion_j^t\right) \tag{9.9}$$

in which Ant_j^t is the position of the ith ant at the tth iteration. Finally, at any stage of optimization, the best solutions were obtained using the elitism mechanism as follows:

$$Ant_j^t = \frac{R_A^t + R_E^t}{2} \tag{9.10}$$

in which R_A^t is the random walk around the antlion elected by the roulette wheel at the tth iteration, and R_E^t is the random walk around the elite at the tth iteration. Over the years, ALO algorithms have been effectively applied to solve numerous optimization problems (Mouassa et al. 2017; Tharwat and Hassanien 2018; Kose 2018; Dinkar and Deep 2019). Readers can refer to Mirjalili (2015) for details about the ALO algorithm.

9.2.4 Multi-Verse Optimizer

The Multi-Verse Optimizer (MVO) is an advanced nature-inspired algorithm, first introduced by Mirjalili et al. (2016). The MVO is based on three concepts in cosmology: white hole, black hole, and wormhole. These three concepts are compiled in one mathematical model to perform the exploration, exploitation, and local search, respectively, according to two principal coefficients: WEP (wormhole existence probability) and TDR (traveling distance rate). The WEP and TDR are expressed as

$$WEP = minimum + l \times \left(\frac{maximum - minimum}{L}\right) \tag{9.11}$$

$$TDR = 1 - \frac{l^{1/p}}{L^{1/p}} \tag{9.12}$$

in which l is the current iteration, L is the maximum iterations (minimum, 0.2; maximum, 1.0), and p is the exploitation accuracy over the iterations ($p = 6$). The sooner and more accurate the exploitation/local search, the higher value of p. After that, the roulette wheel mechanism is utilized to create the best solution position using Equation 9.13:

$$X_i^j = \begin{cases} X_j + TDR + \left(\left(ub_j - lb_j \right) \times r_4 + lb_j \right) & \text{if } r_2 < WEP \text{ and } r_3 < 0.5 \\ X_j - TDR + \left(\left(ub_j - lb_j \right) \times r_4 + lb_j \right) & \text{if } r_2 < WEP \text{ and } r_3 \geq 0.5 \\ X_i^j & \text{if } r_2 \geq WEP \end{cases} \tag{9.13}$$

in which X_i^j is the jth parameter of the ith universe; ub_j and lb_j are the upper and lower bounds, respectively, of jth parameters; and r_2, r_3, and r_4 are arbitrary statistics [0, 1]. In recent times, the MVO algorithm has been successfully applied in diverse fields of engineering (Hu et al. 2016; Trivedi et al. 2016; Peng et al. 2017; Fathy and Rezk 2018; Tikhamarine et al. 2019a, 2020a; Mohammadi et al. 2020). For more information about the MVO algorithms, readers can refer to Mirjalili et al. (2016).

9.2.5 HYBRID SUPPORT VECTOR REGRESSION

SVR is a popular ML technique developed by Vapnik (1995) to solve complex engineering problems. SVR works based on statistical learning theory and the principle of structural risk minimization (SRM) to identify the association between input and output parameters (Su et al. 2018; Tikhamarine et al. 2019b):

$$y = f(x) = w \times \phi(x) + b \tag{9.14}$$

in which y is the output, x is the input, w is the weight vector, ϕ is the mapping function, and b is the bias term. The w and b parameters can be defined by employing the principle of SRM (Su et al. 2018; Tikhamarine et al. 2020c) using Equations 9.15 and 9.16:

$$Minimize : \left[\frac{1}{2} w^2 + C \sum_{i=1}^{N} \left(\xi_i + \xi_i^* \right) \right] \tag{9.15}$$

$$Subject\ to : \begin{cases} y_i - \left(w \times \phi(x_i) - b \right) \leq \varepsilon + \xi_i \\ \left(w \times \phi(x_i) + b \right) - y_i \leq \varepsilon + \xi_i^* \\ \xi_i, \xi_i^* \geq 0,\ i = 1,2,\dots,N \end{cases} \tag{9.16}$$

where C is the penalty constant that balances the empirical risk and model flatness ($C > 0$), ξ_i and ξ_i^* are the slack variables, and ε is a constant named the tube size that characterizes the optimization performance. Thus, the dual optimization problem can be solved by using Lagrangian multipliers, and the regression function of SVR was calculated (Su et al. 2018; Panahi et al. 2020) using Equation 9.17:

$$f(x) = \sum_{i=1}^{N} \left(a_i - a_i^* \right) K(x, x_i) + b \tag{9.17}$$

in which a_i and a_i^* are the Lagrange multipliers, and $K(x, x_i) = \left\{ \phi(x),\ \phi(x_i) \right\}$ denotes the kernel function. A variety of kernel functions exist, such as linear, polynomial, sigmoid, and radial basis function (RBF) (Ansari and Gholami 2015; Zhang et al. 2017; Tikhamarine et al. 2020a). The selection of a suitable kernel function is essential to solve the nonlinear regression problem into a linear function. The following RBF kernel was used in this study:

$$K\left(x, x_i\right) = exp\left(-\gamma x - x_i^2\right) \tag{9.18}$$

where γ is the kernel parameters. The optimization of SVR in the training phase largely depends on C, γ, and ε parameters. In this study, two nature-inspired algorithms, i.e., ALO and MVO, were used to find the best SVR parameter (i.e., C, γ, and ε). Also, the SVR coupled with these algorithms is called hybrid SVR-ALO and SVR-MVO models. Figure 9.2 demonstrates the flowchart of proposed hybrid SVR models used for EDI prediction in this study.

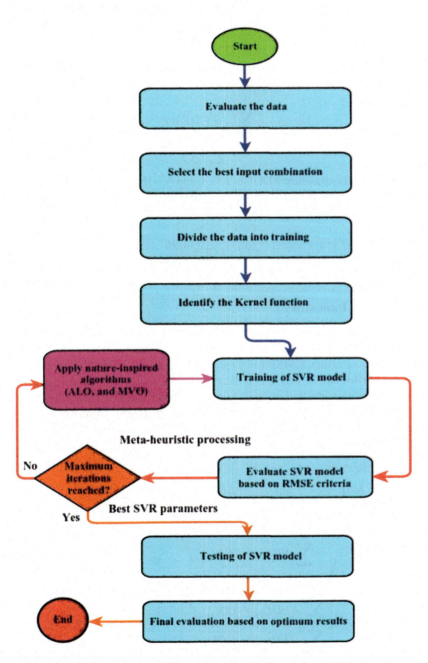

FIGURE 9.2   Flowchart of the proposed hybrid SVR coupled with ALO and MVO algorithms.

9.2.6 OPTIMAL INPUT SELECTION

In this study, after the computation of the EDI, the partial autocorrelation function (PACF) was utilized to decide the optimal inputs (lags) at a 5% significance level for Almora, Champawat, Pauri Garhwal, and Pithoragarh stations. The PACF explores the association of correlation of the residuals with the next lag value. Such information is beneficial to develop the EDI prediction model based on selected significant inputs (lags). The PACF was calculated as (Tiwari and Adamowski 2013; Deo et al. 2017b; Malik et al. 2019c)

$$PACF_{k,k} = \frac{ACF - \sum_{j=1}^{k-1} PACF_{k-1,j} \, ACF_{k-1}}{1 - \sum_{j=1}^{k-1} PACF_{k-1,j} \, ACF_{k-1}} \tag{9.19}$$

in which k is the lag through the sequence Y_t, and ACF is the autocorrelation function and computed as (Tiwari and Adamowski 2013; Deo et al. 2017b; Malik et al. 2019c)

$$ACF_k = \frac{\sum_{t=1}^{N-k} (Y_t - \bar{Y})(Y_{t+k} - \bar{Y})}{\sum_{t=1}^{N} (Y_t - \bar{Y})} \tag{9.20}$$

where $\bar{Y}$ is the average of the entire EDI data series, and N is the number of observations in the EDI series. Subsequently, the PACF values for kth lags were verified at a 5% significance level by drawing the upper and lower critical limits (UCL and LCL) using Equation (9.21):

$$UCL\,/\,LCL = \frac{1.96}{\sqrt{N}} \tag{9.21}$$

9.2.7 PERFORMANCE INDICATORS

Both quantitative and qualitative analyses were used to assess the prediction efficacy of hybrid SVR-ALO and SVR-MVO models. The quantitative analysis includes five performance indicators – MAE (mean absolute error), RMSE (root mean square error), NSE (Nash–Sutcliffe efficiency), COC (coefficient of correlation), and WI (Willmott index) – while the qualitative analysis comprises thorough visual interpretation (radar chart, temporal variation and scatterplots, Taylor diagram). The MAE (Tikhamarine et al. 2019a; Elbeltagi et al. 2020b), RMSE (Malik et al. 2019b; Elbeltagi et al. 2020a), NSE (Nash and Sutcliffe 1970), COC (Malik et al. 2018, 2020b), and WI (Willmott 1981; Malik et al. 2019a) are expressed as

$$MAE = \frac{1}{N} \sum_{i=1}^{N} \left| EDI_{pre,i} - EDI_{com,i} \right| \qquad (0 < MAE < \infty) \tag{9.22}$$

$$RMSE = \sqrt{\frac{1}{N} \sum_{i=1}^{N} (EDI_{com,\,i} - EDI_{pre,i})^2} \qquad \left(0 < RMSE < \infty\right) \tag{9.23}$$

$$NSE = 1 - \left[\frac{\sum_{i=1}^{N} (EDI_{com,i} - EDI_{pre,i})^2}{\sum_{i=1}^{N} (EDI_{com,i} - \overline{EDI_{com}})^2} \right] \qquad \left(-\infty < NSE < 1\right) \tag{9.24}$$

$$COC = \frac{\sum_{i=1}^{N}\left(EDI_{com,i} - \overline{EDI_{com}}\right)\left(EDI_{pre,i} - \overline{EDI_{pre}}\right)}{\sqrt{\sum_{i=1}^{N}(EDI_{com,i} - \overline{EDI_{com}})^2} \; \sqrt{\sum_{i=1}^{N}(EDI_{pre,i} - \overline{EDI_{pre}})^2}} \qquad (-1 < COC < 1) \qquad (9.25)$$

$$WI = 1 - \left[\frac{\sum_{i=1}^{N}(EDI_{pre,i} - EDI_{com,i})^2}{\sum_{i=1}^{N}(|EDI_{pre,i} - \overline{EDI_{com}}| + |EDI_{com,i} - \overline{EDI_{com}}|)^2}\right] \qquad (0 < WI \le 1) \qquad (9.26)$$

where N is the number of observations in the ith data set; EDI_{com} and EDI_{pre} are the computed (observed) and predicted EDI values, respectively; and $\overline{EDI_{com}}$ and $\overline{EDI_{pre}}$ are the mean of computed and predicted EDI values, respectively. If a model demonstrates NSE > 0.50, COC > 0.75, and WI > 0.70, and lower values of MAE and RMSE, then it is designated as an optimal model (Moriasi et al. 2012; Kouchi et al. 2017; Paul and Negahban-Azar 2018) for meteorological prediction at the study stations.

9.3 RESULTS AND DISCUSSION

9.3.1 MODEL INPUT NOMINATION

The PACF analysis was performed to determine the optimal data entry (at different intervals) for the desired output regarding the EDI data series at all stations considered in this study. The significant input variables (lags) used for EDI prediction by building the SVR-ALO and SVR-MVO models at the study stations are given in Table 9.2. Figure 9.3a–d shows the PACF findings for the EDI time series at Almora, Champawat, Pauri Garhwal, and Pithoragarh stations, respectively. The blue dash line in each respective figure designates the upper (UCL) and lower (LCL) confidence limits at 5% significance level; PACF values above these limits are considered statistically significant. As results show, spikes 1, 2, 5, and 6 are considered statistically significant at the 5% level of significance (as being above the upper and below lower confidence limits) and inserted as inputs (intervals) to predict EDI time series for the Almora station (Figure 9.3a); spikes 1, 2, 3, 5, and 6 for the Champawat station (Figure 9.3b); spikes 1, 2, 5, 10 for the Pauri Garhwal station (Figure 9.3c); and spikes 1 and 9 for the Pithoragarh station (Figure 9.3d). After constructing the hybrid SVR-ALO and SVR-MVO models, 70% (1901–1981) of the data of computed EDI was used for training, and the remaining 30% (1982–2015) was used for testing of these models.

TABLE 9.2

Optimized Hybrid SVR Models for EDI Prediction at Study Stations

Station	Output = Input Variables	Nature-Inspired Algorithms	
		ALO	MVO
Almora	$EDI = f(EDI_{t-1}, EDI_{t-2}, EDI_{t-5}, EDI_{t-6})$	√	√
Champawat	$EDI = f(EDI_{t-1}, EDI_{t-2}, EDI_{t-5}, EDI_{t-6})$	√	√
Pauri Garhwal	$EDI = f(EDI_{t-1}, EDI_{t-2}, EDI_{t-5}, EDI_{t-10})$	√	√
Pithoragarh	$EDI = f(EDI_{t-1}, EDI_{t-9})$	√	√

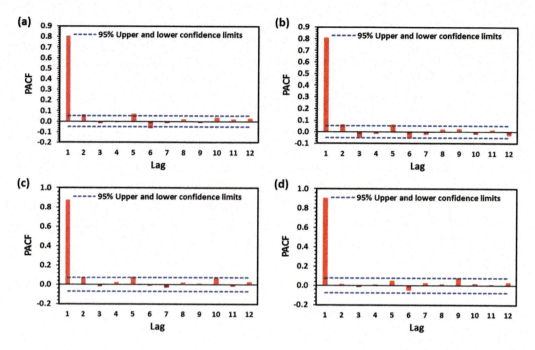

FIGURE 9.3 PACF of EDI time series at (a) Almora, (b) Champawat, (c) Pauri Garhwal, and (d) Pithoragarh stations.

9.3.2 Drought Prediction Using Hybrid SVR Models

The values of the performance indicators, i.e., MAE, RMSE, NSE, COC, and WI, through the testing stage generated by SVR-ALO (MAE = 0.419, 0.423, 0.375, 0.366; RMSE = 0.604, 0.660, 0.591, 0.539; NSE = 0.574, 0.645, 0.804, 0.859; COC = 0.763, 0.817, 0.898, 0.930; WI = 0.864, 0.889, 0.942, 0.958) and SVR-MVO (MAE = 0.420, 0.426, 0.391, 0.379; RMSE = 0.607, 0.662, 0.603, 0.552; NSE = 0.571, 0.643, 0.796, 0.852; COC = 0.761, 0.817, 0.896, 0.929; WI = 0.864, 0.887, 0.937, 0.954) models at Almora, Champawat, Pauri Garhwal, and Pithoragarh stations are presented in Table 9.3. Observations show that the performance of the SVR-ALO model revealed to be better through the testing stage at all stations. However, as mentioned earlier at Pithoragarh station, performance of the SVR-ALO model (MAE = 0.366, RMSE = 0.539, NSE = 0.859, COC = 0.930, WI = 0.958) was notably better (satisfactory) regarding all performance evaluation metrics and following the criteria of lower values of MAE and RMSE, and NSE > 0.50, COC > 0.75, and WI > 0.70 (Moriasi et al. 2012; Kouchi et al. 2017; Paul and Negahban-Azar 2018).

The temporal difference among predicted and computed EDI time series by the SVR-ALO and SVR-MVO models through the testing stage for all four study stations are plotted through a line diagram (left) and scatterplots (right) in Figure 9.4. The regression line in scatterplots shows the coefficient of determination (R^2) for both SVR-ALO and SVR-MVO models applied at the four stations. Although with no significant differences, the regression line was revealed to be much closer to the best fit (1:1) line for the SVR-ALO model at the four stations. It is worth noting that at the Almora and Champawat stations the coefficient of determination exhibits quite low values ($R^2 = 0.582$ and $R^2 = 0.579$) and ($R^2 = 0.668$ and $R^2 = 0.667$), respectively. In contrast, at two other stations, the coefficient of determination shows much higher values, which indicate better performance of both models at those two stations. Specifically, for the Pauri Garhwal station, both models demonstrate higher values ($R^2 = 0.807$ and $R^2 = 0.802$), and for the Pithoragarh station ($R^2 = 0.865$ and $R^2 = $

TABLE 9.3

Performance Indicators Results during the Testing Period at Study Stations

Station	Models	Optimal Parameters			Performance Indicators				
		γ	C	ε	MAE	RMSE	NSE	COC	WI
Almora	SVR-ALO	5.031E–04	29.803	0.028	0.419	0.604	0.574	0.763	0.864
	SVR-MVO	3.105E–03	13.282	0.061	0.420	0.607	0.571	0.761	0.863
Champawat	SVR-ALO	8.708E–04	17.379	0.273	0.423	0.660	0.645	0.817	0.889
	SVR-MVO	1.601E–03	7.757	1.712	0.426	0.662	0.643	0.817	0.887
Pauri Garhwal	SVR-ALO	5.233E–04	21.461	1.394	0.375	0.591	0.804	0.898	0.942
	SVR-MVO	7.037E–04	5.254	1.375	0.391	0.603	0.796	0.896	0.937
Pithoragarh	SVR-ALO	9.874E–04	32.524	0.045	0.366	0.539	0.859	0.930	0.958
	SVR-MVO	4.557E–04	98.996	1.761	0.379	0.552	0.852	0.929	0.954

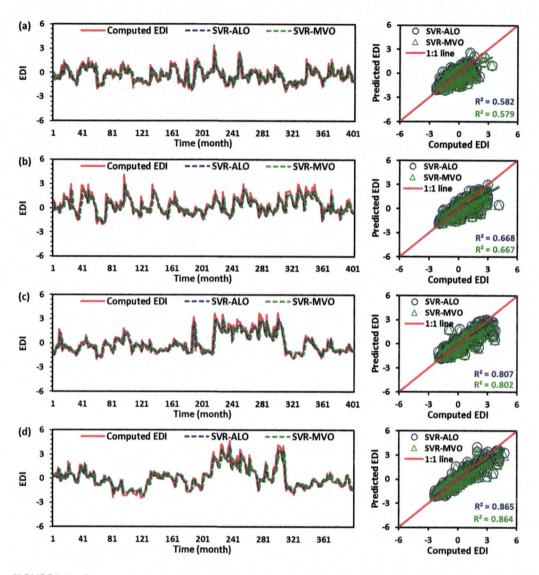

FIGURE 9.4 Computed versus predicted EDI values by hybrid SVR models during testing at (a) Almora, (b) Champawat, (c) Pauri Garhwal, and (d) Pithoragarh stations.

0.864). As mention earlier, in the case of Pithoragarh station, both models demonstrated a higher predictive performance compared to the other stations.

Figure 9.5 shows the spatial variation of predicted and calculated (i.e., observed) EDI time series by using the SVR-ALO and SVR-MVO models through the testing stage at the four study stations using a Taylor diagram (Taylor 2001) as a polar plot for acquiring a graphic evaluation of model performance based on COC, standard deviation (SD), MAE, and RMSE. The Taylor diagram confirmed that particularly the SVR-ALO model estimates are close to the computed, which has lower values of MAE, RMSE, and SD, and higher values of COC at all stations, but with a notable difference in the Pithoragarh station. Thus, although the SVR-ALO and SVR-MVO models show a slight difference among them from station to station, the difference was not so significant. Therefore, both models with selected intervals (i.e., inputs) are suitable to be utilized for prediction EDI on monthly timescales at the four stations and/or other similar case studies.

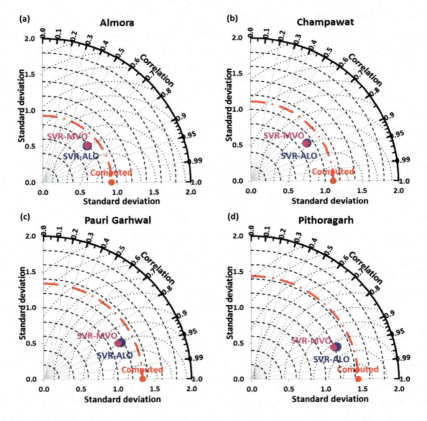

FIGURE 9.5 Taylor diagram of SVR-ALO and SVR-MVO models during testing at (a) Almora, (b) Champawat, (c) Pauri Garhwal, and (d) Pithoragarh stations.

9.3.3 Discussion

This study aimed to evaluate the capabilities of two hybrid ML models for the prediction of an EDI at four different stations. Droughts most commonly originate as a result of a precipitation deficit which leads to a reduction of runoff, and as a result tremendous socioeconomic consequences (Xu et al. 2018). Therefore, accurate prediction and quantification of droughts are crucial for sustainable and effective planning and water resources management. Droughts are a complex phenomenon that is hardly predictable both at the temporal and spatial scales. Nevertheless, utilization of the ML approaches has shown to be robust in the prediction of meteorological drought (Kisi et al. 2019; Malik and Kumar 2020; Malik et al. 2020a) and hydrological drought (Malik et al. 2019c; Shamshirband et al. 2020). In general ML models have shown to be more robust than stochastic models in many cases (Reddy and Singh 2014). Our results showed that two hybrid ML models applied in this study could provide robust results. Both models, SVR-ALO and SVR-MVO, provided quite similar results among them with regards to several evaluation performance metrics. In this regard, our findings are in line with other authors' findings. Xu et al. (2018) applied several ML models over China for drought prediction, and they achieved consistent results. Similar results concerning ML model performance for droughts prediction were reported by several authors (Dayal et al. 2017; Kousari et al. 2017; Soh et al. 2018; Das et al. 2020). One crucial element particularly in droughts prediction is the data quality, which consists of the length of the time series, resolution (e.g., daily, monthly), and gaps in the data records.

Therefore, it should be highlighted that the quality of the data records may significantly affect the performance of the models (Xu et al. 2018). As observed at the Pithoragarh station, COC was much higher than other stations for both models. Therefore, we suggest a careful pretreatment of the data before the computations. Notably, in those cases where data records are scarce or with many gaps, special attention should be given to the data treatment and reconstruction to get close to the natural pattern of the data. Inappropriate data records with numerous gaps may produce misleading information for water administrators. The main findings from this study demonstrate that hybrid SVR-ALO and SVR-MVO models stand to support meteorological drought estimation on a monthly scale in the given study region.

9.4 CONCLUSIONS

Precise and early prediction of drought is crucial for the effective utilization of water resources. It reduces drought-related impacts by formulating drought mitigation strategies. The present study explores the comparative potential of the ALO and MVO algorithms hybridized with SVR to predict the monthly EDI at Almora, Champawat, Pauri Garhwal, and Pithoragarh stations situated in Uttarakhand state, India. Comparison of results showed better performance of the SVR-ALO model over the SVR-MVO model at all study locations, conferring with performance indicators (i.e., MAE, RMSE, NSE, COC, WI) and visual examination (radar chart, temporal variation and scatterplots, Taylor diagram). Similarly, the SVR-ALO exhibits the best prediction of EDI at the Pithoragarh station. The results confirmed the better feasibility of the hybrid SVR-ALO model for EDI prediction, which could be of great assistance in the designing of the hydrologic system for sustainable development and management of water resources to combat the drought consequences in the given study region.

CONFLICT OF INTEREST

None.

REFERENCES

Ali M, Deo RC, Downs NJ, Maraseni T (2018) An ensemble-ANFIS based uncertainty assessment model for forecasting multi-scalar standardized precipitation index. *Atmos Res* 207:155–180. https://doi.org/10.1016/j.atmosres.2018.02.024

Ali R, Kuriqi A, Kisi O (2020) Human–environment natural disasters interconnection in China: A review. *Climate* 8:48. https://doi.org/10.3390/cli8040048

Ali Z, Hussain I, Faisal M, et al (2017) Forecasting drought using multilayer perceptron artificial neural network model. *Adv Meteorol* 2017:1–9. https://doi.org/10.1155/2017/5681308

Ansari HR, Gholami A (2015) An improved support vector regression model for estimation of saturation pressure of crude oils. *Fluid Phase Equilib* 402:124–132. https://doi.org/10.1016/j.fluid.2015.05.037

Banadkooki FB, Ehteram M, Ahmed AN, et al (2020) Suspended sediment load prediction using artificial neural network and ant lion optimization algorithm. *Environ Sci Pollut Res*. https://doi.org/10.1007/s11356-020-09876-w

Byun H-R, Wilhite DA (1999) Objective quantification of drought severity and duration. *J Clim* 12:2747–2756. https://doi.org/10.1175/1520-0442(1999)012<2747:OQODSA>2.0.CO;2

Dai A (2011a) Characteristics and trends in various forms of the Palmer Drought Severity Index during 1900–2008. *J Geophys Res* 116:D12115. https://doi.org/10.1029/2010JD015541

Dai A (2011b) Drought under global warming: A review. *Wiley Interdiscip Rev Clim Chang* 2:45–65. https://doi.org/10.1002/wcc.81

Dai A (2013) Increasing drought under global warming in observations and models. *Nat Clim Chang*. https://doi.org/10.1038/nclimate1633

Danandeh Mehr A, Kahya E, Özger M (2014) A gene–wavelet model for long lead time drought forecasting. *J Hydrol* 517:691–699. https://doi.org/10.1016/j.jhydrol.2014.06.012

Das P, Naganna SR, Deka PC, Pushparaj J (2020) Hybrid wavelet packet machine learning approaches for drought modeling. *Environ Earth Sci* 79:221. https://doi.org/10.1007/s12665-020-08971-y

Dayal K, Deo R, Apan AA (2017) Drought modelling based on artificial intelligence and neural network algorithms: A case study in Queensland, Australia. In: Leal Filho, W. (eds) *Climate Change Adaptation in Pacific Countries*. Climate Change Management. Springer, Cham. https://doi.org/10.1007/978-3-319 -50094-2_11.

Deo RC, Byun H-R, Adamowski JF, Begum K (2017a) Application of effective drought index for quantification of meteorological drought events: A case study in Australia. *Theor Appl Climatol* 128:359–379. https://doi.org/10.1007/s00704-015-1706-5

Deo RC, Şahin M (2015) Application of the extreme learning machine algorithm for the prediction of monthly Effective Drought Index in eastern Australia. *Atmos Res* 153:512–525. https://doi.org/10.1016/j.atmosres .2014.10.016

Deo RC, Tiwari MK, Adamowski JF, Quilty JM (2017b) Forecasting effective drought index using a wavelet extreme learning machine (W-ELM) model. *Stoch Environ Res Risk Assess* 31:1211–1240. https://doi .org/10.1007/s00477-016-1265-z

Dinkar SK, Deep K (2019) Accelerated opposition-based antlion optimizer with application to order reduction of linear time-invariant systems. *Arab J Sci Eng*. https://doi.org/10.1007/s13369-018-3370-4

Dogan S, Berktay A, Singh VP (2012) Comparison of multi-monthly rainfall-based drought severity indices, with application to semi-arid Konya closed basin, Turkey. *J Hydrol* 470–471:255–268. https://doi.org/10 .1016/j.jhydrol.2012.09.003

Elbeltagi A, Aslam MR, Malik A, et al (2020a) The impact of climate changes on the water footprint of wheat and maize production in the Nile Delta, Egypt. *Sci Total Environ*. https://doi.org/10.1016/j.scitotenv .2020.140770

Elbeltagi A, Deng J, Wang K, et al (2020b) Modeling long-term dynamics of crop evapotranspiration using deep learning in a semi-arid environment. *Agric Water Manag*. https://doi.org/10.1016/j.agwat.2020 .106334

Elkiran G, Nourani V, Abba SI (2019) Multi-step ahead modelling of river water quality parameters using ensemble artificial intelligence-based approach. *J Hydrol*. https://doi.org/10.1016/j.jhydrol.2019.123962

Fadaee M, Mahdavi-Meymand A, Zounemat-Kermani M (2020) Suspended sediment prediction using integrative soft computing models: On the analogy between the butterfly optimization and genetic algorithms. *Geocarto Int*. https://doi.org/10.1080/10106049.2020.1753821

Fathy A, Rezk H (2018) Multi-verse optimizer for identifying the optimal parameters of PEMFC model. *Energy*. https://doi.org/10.1016/j.energy.2017.11.014

Guan Y, Mohammadi B, Pham QB, et al (2020) A novel approach for predicting daily pan evaporation in the coastal regions of Iran using support vector regression coupled with krill herd algorithm model. *Theor Appl Climatol*. https://doi.org/10.1007/s00704-020-03283-4

Hu C, Li Z, Zhou T, et al (2016) A multi-verse optimizer with levy flights for numerical optimization and its application in test scheduling for network-on-chip. *PLoS One*. https://doi.org/10.1371/journal.pone .0167341

Jain VK, Pandey RP, Jain MK, Byun H-R (2015) Comparison of drought indices for appraisal of drought characteristics in the Ken River Basin. *Weather Clim Extrem* 8:1–11. https://doi.org/10.1016/j.wace.2015 .05.002

Kim D-W, Byun H-R, Choi K-S (2009) Evaluation, modification, and application of the effective drought index to 200-Year drought climatology of Seoul, Korea. *J Hydrol* 378:1–12. https://doi.org/10.1016/j .jhydrol.2009.08.021

Kisi O, Docheshmeh Gorgij A, Zounemat-Kermani M, et al (2019) Drought forecasting using novel heuristic methods in a semi-arid environment. *J Hydrol* 578:124053. https://doi.org/10.1016/j.jhydrol.2019.124053

Kose U (2018) An ant-lion optimizer-trained artificial neural network system for chaotic electroencephalogram (EEG) prediction. *Appl Sci* 8:1613. https://doi.org/10.3390/app8091613

Kouchi DH, Esmaili K, Faridhosseini A, et al (2017) Sensitivity of calibrated parameters and water resource estimates on different objective functions and optimization algorithms. *Water* 9:384. https://doi.org/10 .3390/w9060384

Kousari MR, Hosseini ME, Ahani H, Hakimelahi H (2017) Introducing an operational method to forecast long-term regional drought based on the application of artificial intelligence capabilities. *Theor Appl Climatol*. https://doi.org/10.1007/s00704-015-1624-6

Majumder P, Eldho TI (2020) Artificial neural network and grey wolf optimizer based surrogate simulation-optimization model for groundwater remediation. *Water Resour Manag*. https://doi.org/10.1007/s11269 -019-02472-9

Malik A, Kumar A (2020) Meteorological drought prediction using heuristic approaches based on effective drought index: A case study in Uttarakhand. *Arab J Geosci* 13:276. https://doi.org/10.1007/s12517-020 -5239-6

Malik A, Kumar A, Ghorbani MA, et al (2019a) The viability of co-active fuzzy inference system model for monthly reference evapotranspiration estimation: Case study of Uttarakhand State. *Hydrol Res* 50:1623–1644. https://doi.org/10.2166/nh.2019.059

Malik A, Kumar A, Kisi O (2018) Daily pan evaporation estimation using heuristic methods with gamma test. *J Irrig Drain Eng* 144:04018023. https://doi.org/10.1061/(ASCE)IR.1943-4774.0001336

Malik A, Kumar A, Kisi O, Shiri J (2019b) Evaluating the performance of four different heuristic approaches with Gamma test for daily suspended sediment concentration modeling. *Environ Sci Pollut Res* 26:22670–22687. https://doi.org/10.1007/s11356-019-05553-9

Malik A, Kumar A, Salih SQ, et al (2020a) Drought index prediction using advanced fuzzy logic model: Regional case study over Kumaon in India. *PLoS One* 15:e0233280. https://doi.org/10.1371/journal.pone.0233280

Malik A, Rai P, Heddam S, et al (2020b) Pan evaporation estimation in Uttarakhand and Uttar Pradesh States, India: Validity of an integrative data intelligence model. *Atmosphere (Basel)* 11:553. https://doi.org/10.3390/atmos11060553

Malik A, Kumar A, Singh RP (2019c) Application of Heuristic Approaches for Prediction of Hydrological Drought Using Multi-scalar Streamflow Drought Index. *Water Resour Manag* 33:3985–4006. https://doi.org/10.1007/s11269-019-02350-4

Mckee TB, Doesken NJ, Kleist J (1993) The relationship of drought frequency and duration to time scales. In: Eighth Conference on Applied Climatology, 17–22 January 1993, Anaheim, CA.

Mirjalili S (2015) The ant lion optimizer. *Adv Eng Softw.* https://doi.org/10.1016/j.advengsoft.2015.01.010

Mirjalili S, Mirjalili SM, Hatamlou A (2016) Multi-verse optimizer: A nature-inspired algorithm for global optimization. *Neural Comput Appl.* https://doi.org/10.1007/s00521-015-1870-7

Mishra AK, Desai VR (2005) Drought forecasting using stochastic models. *Stoch Environ Res Risk Assess* 19:326–339. https://doi.org/10.1007/s00477-005-0238-4

Mishra AK, Desai VR (2006) Drought forecasting using feed-forward recursive neural network. *Ecol Modell* 198:127–138. https://doi.org/10.1016/j.ecolmodel.2006.04.017

Mishra AK, Singh VP (2010) A review of drought concepts. *J Hydrol* 391:202–216. https://doi.org/10.1016/j.jhydrol.2010.07.012

Mishra AK, Singh VP (2011) Drought modeling: A review. *J Hydrol* 403:157–175. https://doi.org/10.1016/j.jhydrol.2011.03.049

Mishra AK, Desai VR, Singh VP (2007) Drought forecasting using a hybrid stochastic and neural network model. *J Hydrol Eng* 12:626–638. https://doi.org/10.1061/(ASCE)1084-0699(2007)12:6(626)

Mohammadi B, Ahmadi F, Mehdizadeh S, et al (2020) Developing novel robust models to improve the accuracy of daily streamflow modeling. *Water Resour Manag* 34:3387–3409. https://doi.org/10.1007/s11269-020-02619-z

Moriasi DN, Wilson BN, Douglas-Mankin KR, et al (2012) Hydrologic and water quality models: Use, calibration, and validation. *Trans ASABE* 55:1241–1247. https://doi.org/10.13031/2013.42265

Morid S, Smakhtin V, Bagherzadeh K (2007) Drought forecasting using artificial neural networks and time series of drought indices. *Int J Climatol* 27:2103–2111. https://doi.org/10.1002/joc.1498

Mouassa S, Bouktir T, Salhi A (2017) Ant lion optimizer for solving optimal reactive power dispatch problem in power systems. *Eng Sci Technol an Int J.* https://doi.org/10.1016/j.jestch.2017.03.006

Nash JE, Sutcliffe JV (1970) River flow forecasting through conceptual models part I: A discussion of principles. *J Hydrol* 10:282–290. https://doi.org/10.1016/0022-1694(70)90255-6

Palmer WC (1965) Meteorological drought. Office of Climatology, U.S. weather bur, Washington, DC. *Res Pap* No. 45: pp. 1–65.

Panahi M, Sadhasivam N, Pourghasemi HR, et al. (2020) Spatial prediction of groundwater potential mapping based on convolutional neural network (CNN) and support vector regression (SVR). *J Hydrol* 588:125033. https://doi.org/10.1016/j.jhydrol.2020.125033

Paul M, Negahban-Azar M (2018) Sensitivity and uncertainty analysis for streamflow prediction using multiple optimization algorithms and objective functions: San Joaquin Watershed, California. *Model Earth Syst Environ* 4:1509–1525. https://doi.org/10.1007/s40808-018-0483-4

Peng T, Zhou J, Zhang C, Fu W (2017) Streamflow forecasting using empirical wavelet transform and artificial neural networks. *Water.* https://doi.org/10.3390/w9060406

Rahmati O, Falah F, Dayal KS, et al (2020) Machine learning approaches for spatial modeling of agricultural droughts in the south-east region of Queensland Australia. *Sci Total Environ* 699:134230. https://doi.org/10.1016/j.scitotenv.2019.134230

Reddy MJ, Singh VP (2014) Multivariate modeling of droughts using copulas and meta-heuristic methods. *Stoch Environ Res Risk Assess.* https://doi.org/10.1007/s00477-013-0766-2

Riebsame WE, Changnon SA, Karl TR (1991) *Drought and Natural Resource Management in the United States: Impacts and Implications of the 1987–1989 Drought.* Westview Press, Boulder, 174

Shamshirband S, Hashemi S, Salimi H, et al (2020) Predicting Standardized Streamflow index for hydrological drought using machine learning models. *Eng Appl Comput Fluid Mech.* https://doi.org/10.1080/19942060.2020.1715844

Smakhtin V, Hughes D (2007) Automated estimation and analyses of meteorological drought characteristics from monthly rainfall data. *Environ Model Softw* 22:880–890. https://doi.org/10.1016/j.envsoft.2006.05.013

Soh YW, Koo CH, Huang YF, Fung KF (2018) Application of artificial intelligence models for the prediction of standardized precipitation evapotranspiration index (SPEI) at Langat River Basin, Malaysia. *Comput Electron Agric* 144:164–173. https://doi.org/10.1016/j.compag.2017.12.002

Su H, Li X, Yang B, Wen Z (2018) Wavelet support vector machine-based prediction model of dam deformation. *Mech Syst Signal Process* 110:412–427. https://doi.org/10.1016/j.ymssp.2018.03.022

Taylor KE (2001) Summarizing multiple aspects of model performance in a single diagram. *J Geophys Res Atmos* 106:7183–7192. https://doi.org/10.1029/2000JD900719

Tharwat A, Hassanien AE (2018) Chaotic antlion algorithm for parameter optimization of support vector machine. *Appl Intell* 48:670–686. https://doi.org/10.1007/s10489-017-0994-0

Tikhamarine Y, Malik A, Kumar A et al (2019a) Estimation of monthly reference evapotranspiration using novel hybrid machine learning approaches. *Hydrol Sci J* 64:1824–1842. https://doi.org/10.1080/02626667.2019.1678750

Tikhamarine Y, Souag-Gamane D, Kisi O (2019b) A new intelligent method for monthly streamflow prediction: Hybrid wavelet support vector regression based on grey wolf optimizer (WSVR–GWO). *Arab J Geosci* 12:540. https://doi.org/10.1007/s12517-019-4697-1

Tikhamarine Y, Malik A, Souag-Gamane D, Kisi O (2020a) Artificial intelligence models versus empirical equations for modeling monthly reference evapotranspiration. *Environ Sci Pollut Res* 27:30001–30019. https://doi.org/10.1007/s11356-020-08792-3

Tikhamarine Y, Souag-Gamane D, Ahmed AN, et al (2020b) Rainfall-runoff modelling using improved machine learning methods: Harris hawks optimizer vs. particle swarm optimization. *J Hydrol* 589:125133. https://doi.org/10.1016/j.jhydrol.2020.125133

Tikhamarine Y, Souag-Gamane D, Najah Ahmed A, et al (2020c) Improving artificial intelligence models accuracy for monthly streamflow forecasting using grey Wolf optimization (GWO) algorithm. *J Hydrol* 582:124435. https://doi.org/10.1016/j.jhydrol.2019.124435

Tiwari MK, Adamowski J (2013) Urban water demand forecasting and uncertainty assessment using ensemble wavelet-bootstrap-neural network models. *Water Resour Res* 49:6486–6507. https://doi.org/10.1002/wrcr.20517

Trivedi IN, Parmar SA, Bhesdadiya RH, Jangir P (2016) Voltage stability enhancement and Voltage Deviation Minimization using ant-lion optimizer algorithm. In: Proceeding of IEEE - 2nd International Conference on Advances in Electrical, Electronics, Information, Communication and Bio-Informatics, IEEE - AEEICB 2016

Tsakiris G, Pangalou D, Vangelis H (2007) Regional drought assessment based on the Reconnaissance Drought Index (RDI). *Water Resour Manag* 21:821–833. https://doi.org/10.1007/s11269-006-9105-4

Vicente-Serrano SM, Beguería S, López-Moreno JI (2010) A multiscalar drought index sensitive to global warming: The standardized precipitation evapotranspiration index. *J Clim* 23:1696–1718. https://doi.org/10.1175/2009JCLI2909.1

Wilhite DA (1986) Drought policy in the U.S. and Australia: A comparative analysis. *J Am Water Resour Assoc* 22:425–438. https://doi.org/10.1111/j.1752-1688.1986.tb01897.x

Wilhite DA, Glantz MH (1985) Understanding the drought phenomenon: The role of definitions. *Water Int* 10:111–120.

Willmott CJ (1981) On the validation of models. *Phys Geogr* 2:184–194. https://doi.org/10.1080/02723646.1981.10642213

Xu L, Chen N, Zhang X, Chen Z (2018) An evaluation of statistical, NMME and hybrid models for drought prediction in China. *J Hydrol.* https://doi.org/10.1016/j.jhydrol.2018.09.020

Zhang PG (2003) Time series forecasting using a hybrid ARIMA and neural network model. *Neurocomputing.* https://doi.org/10.1016/S0925-2312(01)00702-0

Zhang R, Chen Z-Y, Xu L-J, Ou C-Q (2019) Meteorological drought forecasting based on a statistical model with machine learning techniques in Shaanxi province, China. *Sci Total Environ* 665:338–346. https://doi.org/10.1016/j.scitotenv.2019.01.431

Zhang X, Wang J, Zhang K (2017) Short-term electric load forecasting based on singular spectrum analysis and support vector machine optimized by Cuckoo search algorithm. *Electr Power Syst Res* 146:270–285. https://doi.org/10.1016/j.jpgr.2017.01.035

10 Uncertainty Analysis of Bivariate Modeling of Hydrological Drought Severity and Duration Using Copula

Mansoor Zargar and Majid Dehghani

CONTENTS

10.1 INTRODUCTION

Drought defines as a period of time with a considerable shortage in available water compared to the average water availability in the same period (Zhang et al., 2017; Dehghani et al., 2019a). Drought as a temporary phenomenon affects the agricultural, industrial, and even the domestic water availability, and happens in all climates. Due to its creeping nature, determining the onset, end, and the affected area is difficult (Yoo et al., 2016; Dehghani et al., 2014). Drought was classified into four types by Wilhite and Glantz (1985) – meteorological, hydrological, agricultural, and socioeconomic – while Mishra and Singh (2010) proposed groundwater drought as the fifth drought type. Among these types, hydrological drought more directly relates to human activities (Zhang et al., 2017; Liu et al., 2017; Long et al., 2013). Hydrological drought is defined as a period of time in which river streamflow, lake volume, and dam reservoir storage are below normal conditions. So, it is possible to say that hydrological drought threatens human water availability. As drought is a dynamic phenomenon, it is important to assess drought characteristics such as duration, severity, and magnitude simultaneously for drought risk management (Shiau and Modarres, 2009). Several studies have been carried out to explore drought characteristics using probabilistic approaches (Liu et al., 2016; Dodangeh et al., 2017; Ayantobo et al., 2017; Samaniego et al., 2017; Yang et al., 2018; Tijdeman et al., 2018; Ahmed et al., 2018; Huang et al., 2018; Rudd et al., 2019). As different distributions fitted to drought characteristics, copula functions have the capability of constructing the joint multivariate distribution by linking the marginals (Dehghani et al., 2020). In recent decades, copula functions reached high popularity in water resources analysis and modeling (Sohn and Tom, 2016; Dehghani et al., 2019a). Also, various copula functions have been used for drought characteristics analysis (Zhang et al., 2017; Adarsh et al., 2018; Dehghani et al., 2019a; Nguyen-Huy et al., 2019; Nabaei et al. 2019). In this field several studfies have been carried out to find the relation between duration, severity, and frequency of drought (Shiau and Modarres, 2009; Zhang et al., 2013; Yoo et al., 2016; Kwon and Lall, 2016; Tosunoglu and Can, 2016; Hao et al., 2017; She and Xia et al., 2018; Nabaei et al., 2019). All these studies reported promising results. However, there are some considerations. As, the length of the records are somehow short, the number of drought events are limited for analysis. For example, Shiau and Modarres (2009) used 83 drought events and Adarsh et al. (2018) reported 90 drought events. So, the reliability of the results may be doubtful. An uncertainty analysis is needed to show the reliability of the the results of such frequency analysis. There is a very limited number of uncertainty analysis in this field. Zhang et al. (2015) studied the uncertainty of hydrological drought by copula in the East River basin in China. They used a Bayesian framework and reported considerable uncertainty in the results. Yin et al. (2018) investigated the uncertainty in design flood estimation by copula. Also, Dodangeh et al. (2019) evaluated the bivariate uncertainty of rainfall–runoff modeling using a copula. They used a bootstrapping method to resample from a p-level curve. Overall, there are a few studies in the field of uncertainty analysis of copula modeling, especially in drought analysis. So, in this research, a novel method was used for uncertainty analysis of bivariate hydrological drought characteristics modeling. For this purpose, first, the Standardized Hydrological Drought Index (SHDI) was calculated in three aggregated months. Then, the drought events during the period of 1963 to 2017 were determined and the corresponding drought duration and severity were extracted. Several univariate probability distributions wrer fitted to the duration and severity series to find the best fit. Also, four Archimedean copula functions were fitted to the joint of duration and severity. In the next step, 3000 samples were generated based on the best marginals and best-fitted copula to explore the uncertainty associated with the drought severity–duration–frequency (SDF) analysis. The rest of the chapter is organized as follows. In Section 10.2, the case study and data, the copula theory, SHDI, and uncertainty analysis are presented. In Section 10.3, the results of the frequency analysis of drought characteristics and uncertainty analysis are described. Finally, the conclusions of this study are drawn in Section 10.4.

10.2 METHODOLOGY

10.2.1 Case Study

The Dez Dam is a double-arch dam located in the southwest of Iran on the Dez River (Figure 10.1). It was constructed in 1963. The dam height is 203 m, and the total capacity and effective capacity are 3460 and 2600 mm^3, respectively. The Dez basin is located between 48° to 50° east and 31° to 34° north. The basin area is 17,163 km^2 with a mean elevation of 1915 m above sea level (Dehghani et al., 2019b). Tale-Zang is the hydrometric station just upstream of the dam, where the inflow to the dam is measured (Figure 10.1).

The monthly streamflow data, which covers 1963 to 2017, was gathered from Iran's Water Resources Management Company (http://www.wrm.ir/index.php?l=EN). The dam mainly controls floods, as it is the only reservoir on the Dez River. A 520 MW power plant is installed on the dam and the reservoir supplies the irrigation water for 125,000 ha of downstream farmlands.

10.2.2 Standardized Hydrological Drought Index (SHDI)

For drought characterization, the drought must be quantified. For this purpose, several drought indices were proposed for each drought type mentioned in the introduction. A suitable drought index is the one that has the capability of comparison in both time and space, and needs few input

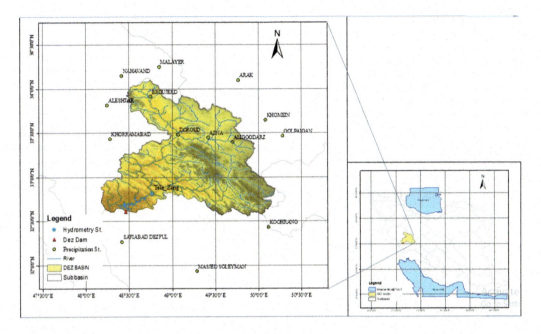

FIGURE 10.1   Location of the Dez basin.

parameters. MacKee et al. (1993) proposed the Standardized Precipitation Index (SPI), which is based on a single input: precipitation. Dehghani et al. (2014) proposed the SHDI based on the procedure that is used for SPI calculation, replacing the precipitation with the streamflow. For this, several probability distributions are fitted to the streamflow time series. Then, the best-fitted distribution will transform to the normal distribution with mean zero and standard deviation of unity using an equiprobability transformation. For more details on SHDI calculation, one may refer to Dehghani et al. (2014) and Dehghani et al. (2017).

The classification of the SHDI is presented in Table 10.1. In this research, the SHDI in three aggregated months was calculated. Drought is defined as a period in which the SHDI values are below −0.5. So, the period with SHDI < −0.5 defines the drought duration, which is denoted by D.

TABLE 10.1

SHDI Classification

Class	Value
Extremely wet	2 and above
Severely wet	1.5 to 1.99
Moderately wet	1 to 1.49
Slightly wet	0.5 to 0.99
Near normal	−0.49 to 0.49
Slightly drought	−0.5 to −1
Moderately dry	−1 to −1.49
Severely dry	−1.5 to −1.99
Extremely dry	−2 and less

Source: Dehghani et al. (2019a).

Also, the cumulative SHDI values during the drought period are defined as drought severity and denoted by S. For convenience, drought severity is considered a positive value.

10.2.3 COPULA FUNCTION

To model the dependence structure of two dependent variables, it is possible to use copula functions. In the case when the joint probability distribution is known, it is possible to determine the marginals. However, knowing the marginal distributions may not lead to defining the joint probability distribution directly. In this case, a copula can detect the joint probability by linking the marginal distributions. A copula considers the dependence structure between the variables without considering the marginal effects.

A copula is a function $C(u,v)$ that satisfies the following conditions:

1. $C(u,0) = 0 = C(0,v)$, $C(u,1) = u$ and $C(1,v) = v$
2. For every u_1, u_2, v_1, v_2 such that $0 \le u_1 < u_2 \le 1$ and $0 \le v_1 < v_2 \le 1$,

$$C(u_2, v_2) - C(u_2, v_1) - C(u_1, v_2) + C(u_1, v_1) \ge 0 \qquad (10.1)$$

As a copula is capable of linking the marginal distributions and detecting the joint probability distribution, it is possible to construct bivariate modeling between S and D variables using copula. In a mathematical view, suppose $F_S(s)$ and $G_D(d)$ are the marginal cumulative distribution functions (CDFs) of the random variables S and D, respectively, and $H_{S,D}(s,d)$ is a joint cumulative distribution function (JCDF) with marginals of $F_S(s)$ and $G_D(d)$ (Zhang et al., 2013). Then, based on the Sklar's (1959) theorem, there exists a copula C such that

$$H_{S,D}(s,d) = C(F_S(s), G_D(d)), \qquad \forall s,d \in R \qquad (10.2)$$

As $F_S(d)$ and $G_D(d)$ are continuous, then the copula C is unique.

There are several copula families. Among them, Archimedean copulas are most often used in drought analysis. So in this research we consider four Archimedean copulas: Clayton, Frank, Gumbel–Hougaard, and Joe (Table 10.2). An Archimedean copula is given by

$$C_{\phi_\theta}(u,v) = \phi_\theta^{-1}(\phi_\theta(u) + \phi_\theta(v)) \qquad (10.3)$$

where the function ϕ_θ is defineds as (Genest et al., 2006)

$$\phi_\theta(1) = 0, \qquad (-1)^i \frac{d^i}{dt^i} \phi_\theta^{-1}(t) > 0, \qquad i = 1,2 \qquad (10.4)$$

It is needed to find the best copula among the representatives. So, a robust goodness-of-fit test must be carried out. For this purpose, the procedure proposed by Genest et al. (2006) was used for the goodness of fit. A detailed description of this procedure is presented in Dehghani et al. (2019a) and Dehghani et al. (2020).

10.2.4 BIVARIATE DROUGHT SEVERITY–DURATION–FREQUENCY ANALYSIS

As presented earlier, the drought events are considered by the random variables S and D. It is possible to model the relationship between drought severity, drought duration, and frequency based on the conditional recurrence interval as follows:

TABLE 10.2

Families of Archimedean Copulas

Copula	$C(u,v)$	θ	$\varphi_\theta(t)$
Clayton	$\left[\max\left(u^{-\theta}+v^{-\theta}-1,0\right)\right]^{\frac{-1}{\theta}}$	$[-1,\infty)-\{0\}$	$\dfrac{1}{\theta}(t^{-\theta}-1)$
Frank	$-\dfrac{1}{\theta}\log\left(1+\dfrac{\left(e^{-\theta u}-1\right)\left(e^{-\theta v}-1\right)}{e^{-\theta}-1}\right)$	$(-\infty,\infty)-\{0\}$	$\log\left(\dfrac{1-e^{-\theta}}{1-e^{-\theta t}}\right)$
Gumbel–Hougaard	$\exp\left(-\left[\left(-\log u\right)^{\theta}+\left(-\log v\right)^{\theta}\right]^{\frac{1}{\theta}}\right)$	$[-1,\infty)-\{0\}$	$(-\log t)^{\theta}$
Joe	$1-\left[\left(1-u\right)^{\theta}+\left(1-v\right)^{\theta}-\left(1-u\right)^{\theta}\left(1-v\right)^{\theta}\right]^{\frac{1}{\theta}}$	$[-1,\infty)-\{0\}$	$-\log\left(1-(1-t)^{\theta}\right)$

$$T_{S|D}(s\mid d)=\frac{1}{\gamma\left(1-H_{S|D}(s\mid d)\right)} \tag{10.5}$$

where $H_{S|D}(s\mid d)=P(S\le s\mid D=d)$ is the conditional CDF of variable S given $D=d$ and γ is the arriving rate of the drought events.

The $H_{S|D}(s|d)$ for continuous variables S and D is given by

$$H_{S|D}(s\mid d)=\int_{-\infty}^{s}\frac{h_{S,D}(t,d)}{g_D(d)}\,dt \qquad \forall t\in R$$

$$=\frac{1}{g_D(d)}\int_{-\infty}^{s}\frac{\partial^2 H_{S,D}(t,d)}{\partial t\partial d}\,dt$$

$$=\frac{1}{g_D(d)}\int_{-\infty}^{s}\frac{\partial}{\partial t}\left(\frac{\partial H_{S,D}(t,d)}{\partial d}\right)dt$$

$$=\frac{1}{g_D(d)}\frac{\partial H_{S,D}(s,d)}{\partial d}$$

$$=\frac{\partial H_{S,D}(s,d)}{\partial G_D(d)} \tag{10.6}$$

where $h_{S,D}(s,d)$ and $H_{S,D}(s,d)$ are the JPDF (joint probability density function) and the JCDF of the bivariate variable (S,D), respectively, and $g_D(d)$ and $G_D(d)$ are the PDF (probability density function) and the CDF of variable D, respectively.

According to Equation 10.2, it is possible to rearrange Equation 10.6 as follows:

$$H_{S|D}(s \mid d) = \frac{\partial H_{S,D}(s,d)}{\partial G_D(d)} = \frac{\partial C\big(F_S(s), G_D(d)\big)}{\partial G_D(d)} \tag{10.7}$$

where C is a copula function. By placing Equation 10.7 into Equation 10.5, the conditional T formulation is given as follows:

$$T_{S|D}(s \mid d) = \frac{1}{\gamma\left(1 - \dfrac{\partial C\big(F_S(s), G_D(d)\big)}{\partial G_D(d)}\right)} \tag{10.8}$$

By considering a constant value for T and γ, one can calculate the drought severity and duration relationship as follows:

$$s = F_{S|D}^{-1}\left(\frac{T\gamma - 1}{T\gamma}\right) \tag{10.9}$$

Based on Equation 10.8, the CDFs of S and D are needed. So, in this research, six commonly used univariate distributions in drought analysis, presented in Table 10.3, were fitted to the series of S and D. To find the best-fitted univariate marginal distribution, several statistical tests and criteria,

---

TABLE 10.3

Univariate CDFs Fitted to the Drought Severity and Duration

Distribution	CDF	Parameters
Exponential	$F(x) = 1 - \exp\left\{-\dfrac{x}{\Lambda}\right\}, \quad x \geq 0$	$\Lambda > 0,$ scale
Generalized extreme value	$F(x) = \exp\left\{-\left(1 + k\dfrac{x-\mu}{\sigma}\right)^{-\frac{1}{k}}\right\},$ $1 + k\dfrac{x-\mu}{\sigma} > 0, \quad k \neq 0$	$k \in R,$ shape $\sigma > 0,$ scale $\mu \in R,$ location
Johnson SB	$F(x) = \Phi\left(\gamma + \delta\ln\left(\dfrac{z}{1-z}\right)\right), \quad z = \dfrac{x-\xi}{\lambda}, \; \xi \leq x \leq \xi + \lambda$	$\delta > 0, \gamma \in R,$ shape $\lambda > 0,$ scale $\xi \in R,$ location
Log-logistic	$F(x) = \left[1 + \left(\dfrac{\beta}{x}\right)^{\alpha}\right]^{-1}, \quad x > 0$	$\alpha > 0,$ shape $\beta > 0,$ scale
Lognormal	$F(x) = \Phi\left(\dfrac{\ln x - \mu}{\sigma}\right), \quad x \geq 0$	$\sigma > 0,$ scale $\mu \in R,$ location
Weibull	$F(x) = 1 - \exp\left\{-\left(\dfrac{x}{\beta}\right)^{\alpha}\right\}, \quad x \geq 0$	$\alpha > 0,$ shape $\beta > 0,$ scale

including the Kolmogorov–Smirnov (K-S), Anderson–Darling (A-D), log-likelihood, and Akaike information criterion (AIC) measures are used.

10.2.5 UNCERTAINTY ANALYSIS

In order to quantify the uncertainty in the bivariate modeling of drought severity, duration, and frequency, Monte Carlo simulation is used. For this purpose, by considering the best-fitted copula $C(u,v)$ to the joint of (S,D), 3000 samples were generated using a parametric bootstrap sampling procedure as follows (Nelsen, 2006):

1. Generate two independent uniform $(0,1)$ variates u and t according to the number of drought events (sample size = 41).
2. Calculate the conditional distribution function for V given $U = u$, which is the differentiation of $C(u,v)$ with respect to u:

$$C_u(v) = \frac{\partial C(u,v)}{\partial u} \tag{10.10}$$

3. Compute the inverse of $C_u(v)$ denoted by C_u^{-1}.
4. By setting t from step 1 to C_u^{-1}, $v = C_u^{-1}(t)$.
5. The desired pair is (u,v), which follows $C(u,v)$.
6. The inverse $F_S(s)$ and $G_D(d)$ are denoted by $F_S^{-1}(s)$ and $G_D^{-1}(d)$.
7. Set u and v from step 5 into $F_S^{-1}(s)$ and $G_D^{-1}(d)$, respectively. So, two vectors of $s = F_S^{-1}(u)$ and $d = G_D^{-1}(v)$ will be generated, which is the desired pair of (s,d).
8. This procedure is repeated 3000 times to generate 3000 pair samples.
9. By considering the best-fitted copula, the dependence parameter θ was estimated using 3000 pair samples of (s,d).
10. The drought severity is calculated using equation (10.9) by setting the generated vector of d and constant values of γ and T. T was considered to be 2, 5, 10, 20, 30, 40, 60, and 100 years.
11. The 95 percent estimated uncertainties (95 PEU) are calculated by extracting the 2.5th (X_L) and 97.5th (X_U) percentiles of each simulated vector of s in step 10.
12. The average distance $\bar{d}_X$ is determined by Equation 10.11, which shows the degree of uncertainty (Abbaspour et al., 2007):

$$\bar{d}_X = \frac{1}{n} \sum_{i=1}^{n} (X_U - X_L), \tag{10.11}$$

where n is the number of drought events. The best situation is the case when all observations are bracketed by the 95 PEU and $\bar{d}_X = 0$. However, due to the associated uncertainty in modeling, it is impossible to achieve these values. So, Abbaspour et al. (2007) proposed the d-factor as follows:

$$d-factor = \frac{\bar{d}_X}{\sigma_X} \tag{10.12}$$

where σ_X is the standard deviation of the measured variable. The d-factor is the average thickness of the 95 PEU band divided by the standard deviation of the observed data. The desired d-factor value is about zero, which almost never achieved due to the associated uncertainty in fitting the marginal and copula distributions, and in estimation of the dependence parameter of a copula. So, a value around zero is desired for the d-factor (Abbaspour et al., 2007).

10.3   RESULTS

In this research, the uncertainty associated with the bivariate analysis of drought duration, severity, and frequency was evaluated. For this purpose, first, the marginal distributions of S and D were determined. Then, the best-fitted copula to the joint of (S, D) was determined and the dependence parameter of copula was estimated. In the next step, the relationship between S, D and T was determined, and, finally, the uncertainty in bivariate modeling was evaluated using a parametric bootstrap sampling. The results of this procedure are presented in the following sections.

10.3.1   MARGINAL CDFs FOR DROUGHT SEVERITY AND DURATION

According to Equation 10.7, it is needed to fit the marginal CDFs to the drought severity and drought duration. In Table 10.3, six competitor univariate CDFs along with their parameters are presented. The parameters of these CDFs are estimated by the maximum likelihood procedure. The K-S and A-D goodness-of-fit tests, the log-likelihood, and AIC measures are used for choosing the best-fitted distributions to S and D variables. The estimation of parameters, the test statistics and P-values of the K-S and A-D tests, the log-likelihood and the AIC for each of competitor univariate distributions are reported in Table 10.4. At a 5% significance level, Table 10.4 indicates that all univariate CDFs fitted to the S and D appropriately. Based on the statistical tests, the exponential and the Johnson SB distributions were the best fits to the S and D, respectively. These two distributions were considered for subsequent analyses.

According to estimated parameters in Table 10.4, $F_S(s)$ and $G_D(d)$ could be shown as

$$F_S(s) = 1 - e^{-0.193s}, \qquad s \geq 0 \tag{10.13}$$

$$F_D(d) = \Phi\left(0.323 + 0.394\ln\left(\frac{z}{1-z}\right)\right), \qquad 0.805 \leq d \leq 11.115 \tag{10.14}$$

where $z = \dfrac{d - 0.805}{10.31}$, and Φ is the standard normal CDF.

10.3.2   COPULA FUNCTIONS FOR THE BIVARIATE RANDOM VARIABLE (S, D)

Since there is a strong dependence between drought severity and drought duration, copula functions can be used to construct the joint relationship. For this perpuse, four Archimedean copulas, presented in Table 10.2, were utilized. Results are presented in Table 10.5. In Table 10.5, the goodness of fit of four copula functions are considered based on S_n and T_n statistics. Table 10.5 indicates that the Kendall correlation coefficient between S and D is 0.871. According to the results, the Joe copula was rejected based on both the S_n and T_n statistics, while the Frank and Gumbel–Hougaard copulas were rejected and accepted based on the S_n and T_n, respectively. However, the Clayton copula function with $\theta = 13.492$ was accepted based on both statistics and it is an appropriate copula for modeling S and D jointly.

The 3D perspective plot of the Clayton copula function with $\theta = 13.492$ and the scatterplot of the joint variable (S, D) are shown in Figure 10.2, confirming a suitable fit in general but inappropriate fit in tails. Therefore, the fitted copula function to (S, D) is given by

$$C(u, v) = \left(u^{-13.492} + v^{-13.492} - 1\right)^{-\frac{1}{13.492}}, \qquad 0 < u, v < 1 \tag{10.15}$$

TABLE 10.4

Results of Univariate CDF Fit to the S and D

Variable	Margin	Parameters	Log-Likelihood	AIC	K-S Test Statistics	K-S Test p-Value	A-D Test Statistics	A-D Test p-Value
Duration	Generalized extreme value	$(k,\sigma,\mu)=(0.279, 2.363, 2.959)$	−103.766	213.532	0.156	0.256	1.429	0.194
	Johnson SB	$(\delta,\lambda,\xi,\gamma)=(0.394, 10.31, 0.805, 0.323)$	**−86.865**	**181.745**	**0.117**	**0.605**	**0.651**	**0.600**
	Log-logistic	$(\alpha,\beta)=(1.935, 3.813)$	−103.841	211.681	0.139	0.386	1.357	0.214
	Lognormal	$(\sigma,\mu)=(0.845, 1.297)$	−101.888	207.776	0.154	0.272	1.463	0.186
	Weibull	$(\alpha,\beta)=(446, 5.492)$	−100.636	205.272	0.143	0.354	1.243	0.252
Severity	**Exponential**	$\Lambda = 0.193$	**−105.583**	**213.565**	**0.102**	**0.757**	**0.497**	**0.749**
	Generalized extreme value	$(k,\sigma,\mu)=(0.889, 1.925, 1.933)$	−107.721	221.443	0.124	0.532	1.019	0.347
	Log-logistic	$(\alpha,\beta)=(1.528, 3.255)$	−107.417	218.833	0.106	0.722	0.745	0.522
	Lognormal	$(\sigma,\mu)=(1.076, 1.147)$	−105.783	215.167	0.115	0.625	0.751	0.517
	Weibull	$(\alpha,\beta)=(1.074, 5.329)$	−105.621	215.243	0.106	0.721	0.602	0.645

TABLE 10.5

Goodness of Fit of Bivariate Copulas Based on S_n and T_n Statistics

Copula	θ	τ	n	S_{n0}	Critical value of S_n	p-Value of S_n	T_{n0}	Critical Value of T_n	p-Value of T_n
Clayton	13.492	0.871	41	0.043	0.054	0.209	0.488	0.592	0.543
Frank	29.242	0.871	41	0.054	0.047	0.019	0.525	0.562	0.100
Gumbel–Hougaard	0.068	0.871	41	0.068	0.048	0.005	0.500	0.548	0.101
Joe	14.162	0.871	41	1.062	1.026	0.002	0.668	0.581	0.017

Duration – Severity

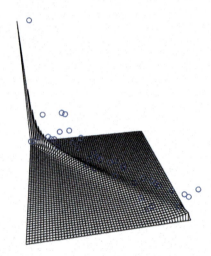

FIGURE 10.2   Scatterplot of (S,D) and the perspective plot of the Clayton copula function with $\theta = 13.492$.

Then, based on Equation 10.2, the JCDF of (S,D) is given by

$$H_{S,D}(s,d) = C\big(F_S(s), G_D(d)\big) = \big(F_S(s)^{-13.492} + G_D(d)^{-13.492} - 1\big)^{-\frac{1}{13.492}} \tag{10.16}$$

where $F_S(s)$ and $G_D(d)$ are the CDFs of S and D, respectively.

10.3.3 DROUGHT SDF CURVE

The conditional distribution function $H_{S|D}(s\,|\,d)$ defined in Equation 10.7 is given by

$$H_{S|D}(s\,|\,d) = \frac{\partial C\big(F_S(s), G_D(d)\big)}{\partial F_D(d)} = \left\{1 + G_D(d)^{13.492}\big(F_S(s)^{-13.492} - 1\big)\right\}^{-1.07} \tag{10.17}$$

According to Equation 10.5, the conditional T is calculated as

$$T_{SID}(s \mid d) = \cfrac{1}{\gamma \left(1 - \left\{ 1 + G_D(d)^{13.492} \left(F_S(s)^{-13.492} - 1 \right) \right\}^{-1.07} \right)} \tag{10.18}$$

Now, the drought severity for a given T can be derived from Equation 10.18 as follows:

$$s = F_{SID}^{-1}\left(\frac{T\gamma - 1}{T\gamma} \right) = F_S^{-1}\left\{ \left[\frac{\left(\frac{T\gamma - 1}{T\gamma} \right)^{-0.931} - 1}{G_D(d)^{13.492}} + 1 \right]^{-\frac{1}{13.492}} \right\} \tag{10.19}$$

Since the best-fitted CDF for the S is the exponential distribution, according to Table 10.4, the distribution function of variable S is $F_S\left(s \right) = 1 - e^{-\lambda s}$ and its inverse function is $F_S^{-1}\left(s \right) = -\frac{1}{\Lambda}\ln\left(1 - s \right)$ for every $s > 0$. By considering $\gamma = \frac{41}{55} = 0.745$, $\theta = 13.492$, and $\Lambda = 0.193$, it is possible to calculate S based on the following equation:

$$s = -\frac{1}{0.193}\ln\left\{ 1 - \left[\frac{\left(\frac{0.745T - 1}{0.745T} \right)^{-0.931} - 1}{G_D(d)^{13.492}} + 1 \right]^{-\frac{1}{13.492}} \right\} \tag{10.20}$$

In the final step, the SDF curves were plotted considering $T = 2, 5, 10, 20, 30, 40, 60,$ and 100 years as shown in Figure 10.3. In Figure 10.3, the SDF curves show a convexity model. Based on these curves and for all T, as the drought duration increased, the drought severity increased. The drought

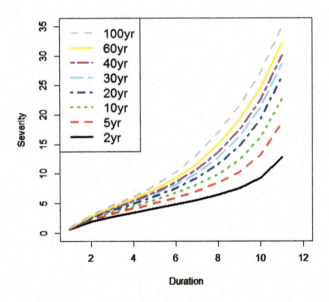

FIGURE 10.3 Drought SDF curves.

severity increases slowly until the drought duration reaches six months, then the drought severity happens faster. Also, the difference between drought severity longer than six months is more distinctive than between one and six months. It must be noted that for all T, by increasing the drought duration, the drought severity increases.

10.3.4 UNCERTAINTY ANALYSIS

The main purpose of this research was analyzing the associated uncertainty in bivariate modeling of drought severity–duration–frequency. So, based on the Clayton copula with dependence parameter $\theta = 13.492$ and Johnson SB and exponential univariate marginal distributions, 3000 samples of (s,d) were generated. Then, 3000 dependence parameters θ were estimated based on the Clayton copula. Using these dependence parameters and the generated D, 3000 drought severities were calculated based on Equation 10.9. In the final step, the 2.5th and 97.5th percentiles of calculated drought severities, the $\bar{d}_x$ and the d-factor were calculated. Results ar3e presented in Table 10.6.

According to Table 10.6, it is obvious that for $T = 2, 5$ the d-factor is less than 1, meaning that the bivariate modeling was appropriate and the associated uncertainty in the modeling is acceptable. Also, 73 and 85 of the observed data were bracketed by the 95 PEU, which is more than 50%, and show satisfactory results. However, for $T > 5$, the d-factors are much greater than 1, which shows a great uncertainty in bivariate modeling. Based on the results, as the recurrence interval increased, the d-factor and, as a result, the associated uncertainty in the modeling increased considerably. Although the bracketed observed data by the 95 PEU is greater than 50% in $T > 5$, as the d-factor is greater than 1, the results of bivariate modeling in these cases were not be reliable. It means that, in these cases $(T > 5)$, the average distance between the upper and lower bands of 95 PEU is large enough to consist a of large number of observations, which leads to a high d-factor. So, as a result, just the results of bivariate modeling of drought severity–duration–frequency for $T \leq 5$ are reliable and can be used for water resources decision-making.

Based on Figure 10.4, for $T = 2$, most of the observed values laid in the 95% of confidence intervals. The majority of the observed data that were out of the 95 PEU were the ones with duration more than five months. In this case, the results show an underestimation in the resampled data. However, as T increased the observed data got closer to the lower bound of 95 PEU, and in some cases outside of the lower bound. This means that in small recurrence intervals, the bivariate modeling has underestimation, while in large recurrence intervals, the copula-based modeling has overestimation. As it is apparent, most of the observed values with duration less than five months laid in the confidence intervals of all the recurrence intervals. However, as the duration increased,

TABLE 10.6
Results of Uncertainty Analysis of Bivariate Modeling of S and D

Return Period (T)	d-Factor	Observed Bracket
2	0.629	73.17
5	0.946	85.36
10	1.220	78.04
20	1.507	68.29
30	1.651	65.85
40	1.804	65.85
60	2.022	63.41
100	2.249	63.41

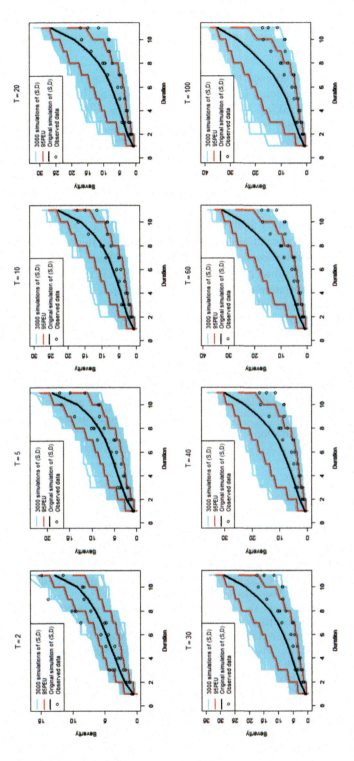

FIGURE 10.4   Plots of observed and simulated values and 95 PEU for different *T*.

the number of observed values that laid outside the confidence intervals increased. Overall, it is possible to conclude that the bivariate modeling in $T \leq 5$ leads to reliable results, while for $T > 5$, the results were associated with a high uncertainty.

10.4 CONCLUSIONS

In this research, the associated uncertainty in bivariate modeling of drought severity–duration–frequency, which mainly neglected in previous research, was analyzed. As the SDF curves are highly usesd for decision-making in drought conditions, the reliability of these curves has a substantial role in the accuracy of decisions made by water resources policy makers. So, in this research, the uncertainty in copula-based modeling of (S, D) was analyzed using a parametric bootstrap sampling technique. For this purpose, the SHDI in a three-months timescale was calculated in the Dez River basin, and the drought events were extracted. In the next step, the joint relationship between drought duration and severity was modeled by copulas considering different recurrence intervals. Based on the results, the Clayton copula was the best fit for the joint of (S, D). Finally, the uncertainty analysis of bivariate modeling was carried out by generating 3000 samples of (S, D). Results indicated that for $T \leq 5$, there is an acceptable uncertainty in the modeling and about 75% of the observed data laid in the confidence intervals. However, for $T > 5$, a high uncertainty is associated with the modeling, which makes the results unreliable. Also, the results showed that the bivariate modeling has overestimation, especially of long durations. For $T \geq 30$, most, and in some cases all, of the observed values with duration more than eight months laid outside the confidence intervals. It means that the bivariate modeling is overestimated especially in extreme values. The results of this study show that there is a high uncertainty embedded in the bivariate modeling of drought severity and duration ,especially in high recurrence intervals. This uncertainty may be due to the few number of drought events, which is common in severity–duration modeling as mentioned in the introduction. The few numbers of drought events led to improper fitting of the copulas and marginal distributions, especially in extreme events. So, the SDF curves must be used with caution by water resources decision makers, especially in sustained droughts of long duration.

REFERENCES

Abbaspour, K. C., Yang, J., Maximov, I., Siber, R., Bogner, K., Mieleitner, J., ... & Srinivasan, R. (2007). Modelling hydrology and water quality in the pre-alpine/alpine Thur watershed using SWAT. *Journal of Hydrology*, 333(2–4), 413–430.

Adarsh, S., Karthik, S., Shyma, M., Prem, G. D., Parveen, A. S., & Sruthi, N. (2018). Developing short term drought severity-duration-frequency curves for Kerala meteorological subdivision, India using bivariate copulas. *KSCE Journal of Civil Engineering*, 22(3), 962–973.

Ahmed, K., Shahid, S., & Nawaz, N. (2018). Impacts of climate variability and change on seasonal drought characteristics of Pakistan. *Atmospheric Research*, 214, 364–374.

Ayantobo, O. O., Li, Y., Song, S., & Yao, N. (2017). Spatial comparability of drought characteristics and related return periods in mainland China over 1961–2013. *Journal of Hydrology*, 550, 549–567.

Dehghani, M., Saghafian, B., Nasiri Saleh, F., Farokhnia, A., & Noori, R. (2014). Uncertainty analysis of streamflow drought forecast using artificial neural networks and Monte Carlo simulation. *International Journal of Climatology*, 34(4), 1169–1180.

Dehghani, M., Saghafian, B., Rivaz, F., & Khodadadi, A. (2017). Evaluation of dynamic regression and artificial neural networks models for real-time hydrological drought forecasting. *Arabian Journal of Geosciences*, 10(12), 1–13.

Dehghani, M., Saghafian, B., & Zargar, M. (2019a). Probabilistic hydrological drought index forecasting based on meteorological drought index using Archimedean copulas. *Hydrology Research*, 50(5), 1230–1250.

Dehghani, M., Riahi-Madvar, H., Hooshyaripor, F., Mosavi, A., Shamshirband, S., Zavadskas, E. K., & Chau, K. W. (2019b). Prediction of hydropower generation using grey wolf optimization adaptive neuro-fuzzy inference system. *Energies*, 12(2), 289.

Dehghani, M., Zargar, M., Riahi-Madvar, H., & Memarzadeh, R. (2020). A novel approach for longitudinal dispersion coefficient estimation via tri-variate Archimedean copulas. *Journal of Hydrology*, 584, 124662.

Dodangeh, E., Shahedi, K., Shiau, J. T., & MirAkbari, M. (2017). Spatial hydrological drought characteristics in Karkheh River basin, southwest Iran using copulas. *Journal of Earth System Science*, 126(6), 80.

Dodangeh, E., Shahedi, K., Solaimani, K., Shiau, J. T., & Abraham, J. (2019). Data-based bivariate uncertainty assessment of extreme rainfall-runoff using copulas: Comparison between annual maximum series (AMS) and peaks over threshold (POT). *Environmental Monitoring and Assessment*, 191(2), 67.

Genest, C., Quessy, J. F., & Rmillard, B. (2006). Goodness-of-fit procedures for copula models based on the probability integral transformation. *Scandinavian Journal of Statistics*, 33(2), 337–366.

Hao, C., Zhang, J., & Yao, F. (2017). Multivariate drought frequency estimation using copula method in Southwest China. *Theoretical and Applied Climatology*, 127(3–4), 977–991.

Huang, J., Zhai, J., Jiang, T., Wang, Y., Li, X., Wang, R., ... & Fischer, T. (2018). Analysis of future drought characteristics in China using the regional climate model CCLM. *Climate Dynamics*, 50(1–2), 507–525.

Kwon, H. H., & Lall, U. (2016). A copula-based nonstationary frequency analysis for the 2012–2015 drought in California. *Water Resources Research*, 52(7), 5662–5675.

Liu, X., Yang, T., Hsu, K., Liu, C., & Sorooshian, S. (2017). Evaluating the streamflow simulation capability of PERSIANN-CDR daily rainfall products in two river basins on the Tibetan Plateau. *Hydrology and Earth System Sciences (Online)*, 21(1), 169–181.

Liu, Z., Wang, Y., Shao, M., Jia, X., & Li, X. (2016). Spatiotemporal analysis of multiscalar drought characteristics across the Loess Plateau of China. *Journal of Hydrology*, 534, 281–299.

Long, D., Scanlon, B. R., Longuevergne, L., Sun, A. Y., Fernando, D. N., & Save, H. (2013). GRACE satellite monitoring of large depletion in water storage in response to the 2011 drought in Texas. *Geophysical Research Letters*, 40(13), 3395–3401.

McKee, T. B., Doesken, N. J., & Kleist, J. (1993, January). The relationship of drought frequency and duration to time scales. In Proceedings of the 8th Conference on Applied Climatology (Vol. 17, No. 22, pp. 179–183).

Mishra, A. K., & Singh, V. P. (2010). A review of drought concepts. *Journal of Hydrology*, 391(1–2), 202–216.

Nabaei, S., Sharafati, A., Yaseen, Z. M., & Shahid, S. (2019). Copula based assessment of meteorological drought characteristics: Regional investigation of Iran. *Agricultural and Forest Meteorology*, 276, 107611.

Nelsen, R. B. (2006). *An Introduction to Copulas*. Springer Science & Business Media.

Nguyen-Huy, T., Deo, R. C., Mushtaq, S., Kath, J., & Khan, S. (2019). Copula statistical models for analyzing stochastic dependencies of systemic drought risk and potential adaptation strategies. *Stochastic Environmental Research and Risk Assessment*, 33(3), 779–799.

Rudd, A. C., Kay, A. L., & Bell, V. A. (2019). National-scale analysis of future river flow and soil moisture droughts: Potential changes in drought characteristics. *Climatic Change*, 156(3), 323–340.

Samaniego, L., Kumar, R., Breuer, L., Chamorro, A., Flrke, M., Pechlivanidis, I. G., ... & Zeng, X. (2017). Propagation of forcing and model uncertainties on to hydrological drought characteristics in a multi-model century-long experiment in large river basins. *Climatic Change*, 141(3), 435–449.

She, D., & Xia, J. (2018). Copulas-based drought characteristics analysis and risk assessment across the loess plateau of China. *Water Resources Management*, 32(2), 547–564.

Shiau, J. T., & Modarres, R. (2009). Copula-based drought severity-duration-frequency analysis in Iran. *Meteorological Applications: A Journal of Forecasting, Practical Applications, Training Techniques and Modelling*, 16(4), 481–489.

Sklar, A. (1959). Fonctions de rpartitions n dimensions et leurs marges Paris Distriburtion functions with n dimensions and their margins). *Publications de LInstitut Statistiques de LU-niversite de Paris* 8, 229–231.

Sohn, S. J., & Tam, C. Y. (2016). Long-lead station-scale prediction of hydrological droughts in South Korea based on bivariate pattern-based downscaling. *Climate Dynamics*, 46(9–10), 3305–3321.

Tijdeman, E., Hannaford, J., & Stahl, K. (2018). Human influences on streamflow drought characteristics in England and Wales. *Hydrology and Earth System Sciences*, 22(2), 1051–1064.

Tosunoglu, F., & Can, I. (2016). Application of copulas for regional bivariate frequency analysis of meteorological droughts in Turkey. *Natural Hazards*, 82(3), 1457–1477.

Wilhite, D. A., & Glantz, M. H. (1985). Understanding: The drought phenomenon: The role of definitions. *Water International*, 10(3), 111–120.

Yang, J., Chang, J., Wang, Y., Li, Y., Hu, H., Chen, Y. & Yao, J. (2018). Comprehensive drought characteristics analysis based on a nonlinear multivariate drought index. *Journal of Hydrology*, 557, 651–667.

Yin, J., Guo, S., Liu, Z., Yang, G., Zhong, Y., & Liu, D. (2018). Uncertainty analysis of bivariate design flood estimation and its impacts on reservoir routing. *Water Resources Management*, 32(5), 1795–1809.

Yoo, J., Kim, D., Kim, H., & Kim, T. W. (2016). Application of copula functions to construct confidence intervals of bivariate drought frequency curve. *Journal of Hydro-environment Research*, 11, 113–122.

Zhang, D., Chen, P., Zhang, Q., & Li, X. (2017). Copula-based probability of concurrent hydrological drought in the Poyang lake-catchment-river system (China) from 1960 to 2013. *Journal of Hydrology*, 553, 773–784.

Zhang, Q., Xiao, M., Singh, V. P., & Chen, X. (2013). Copula-based risk evaluation of hydrological droughts in the East River basin, China. *Stochastic Environmental Research and Risk Assessment*, 27(6), 1397–1406.

Zhang, Q., Xiao, M., & Singh, V. P. (2015). Uncertainty evaluation of copula analysis of hydrological droughts in the East River basin, China. *Global and Planetary Change*, 129, 1–9.

11 Copula-Based Bivariate Frequency Analysis of Drought Characteristics over India

Vivek Gupta, Manoj Kumar Jain, and Shivam Gupta

CONTENTS

11.1 INTRODUCTION

Drought is a creeping natural hazard that occurs due to deficient water availability as compared to normal. Drought virtually occurs in all climates irrespective of the amount of rainfall an area receives. Almost every year, some parts of the world suffer from drought, which translates into a huge economic loss. Understanding the changing behavior of the distribution and occurrence of drought has remained significant to hydrologists and climatologists worldwide for the past few decades. India is among the countries that is are most vulnerable to drought, thus it is important to characterize the drought frequency and distribution to lay out an effective management strategy. Drought frequency analysis is a widely used technique to characterize the probability and return period of severe drought events.

Many drought indices have been developed to characterize different types of droughts. The Standard Precipitation Index (SPI), developed by McKee et al. (1993), is the most commonly used meteorological drought indicator. The SPI is the standard normal deviant of the cumulative probability series that is obtained by fitting different probability density functions to historical series of precipitation. Various univariate and multivariate techniques have been developed and widely used for probabilistic analysis of droughts since various drought characteristics are random in nature (Song and Singh, 2010; Ayantobo et al., 2019). Since drought characteristics are not completely independent of one another and generally have a correlation with other drought characteristics, it is recommended to use multivariate analysis for characterizing dependence between different drought characteristics. In traditional multivariate probability analysis, different drought characteristics are treated as independent and are expected to follow the same marginal distribution (Kwon and Lall, 2016). However, in reality, drought characteristics have been found to have certain correlation among the characteristics of the same drought event. Copula functions help in overcoming this difficulty

DOI: 10.1201/9781003276548-11

153

by allowing each drought characteristic to preserve its own probability distribution and dependence structure among various drought characteristics. In recent years, copulas have been widely used for modeling multivariate dependence in various domains of hydrology and climatology, for example, rainfall frequency analysis (Michele and Salvadori, 2003; Zhang and Singh, 2007a, 2007b; Kao and Govindaraju, 2007; Zhang et al., 2013, Joo et al., 2016), flood frequency analysis (Grimaldi and Serinaldi, 2006; Wang et al., 2009; Chowdhary et al., 2011; Fu and Butler, 2014; Szolgay et al., 2015; Ozga-Zielinski et al., 2016), and drought frequency analysis (González and Valdés, 2003, Shiau, 2006; Song and Singh, 2010; Wu et al., 2015; Ge et al., 2016; Yu et al., 2016).

In this chapter, we present the univariate and bivariate frequency analyses by testing different univariate distributions and many bivariate copulas to find the best-fit copula for different combinations of drought characteristics using four different scales of the SPI, i.e., SPI3, SPI6, SPI9, and SPI12. Further, the computation for return periods has been done using different threshold values of drought characteristics over India.

11.2   STANDARDIZED PRECIPITATION INDEX

McKee et al. (1993) developed the SPI to assess characteristics of meteorological drought. The SPI is one of the most widely used parametric indices to assess meteorological drought characteristics. The scale of the SPI may range from 1 month to as long as 48 months (McKee et al., 1993). Longer timescales, such as 12 or 24 months, represent generally a deficit in water storage, such as streamflows, reservoir levels, and groundwater levels . A gamma probability density function has been widely used (Diani et al., 2019; Zakhem and Kattaa, 2016; Hosseini-Moghari et al., 2019), therefore, the same is used in this study to obtain individual cumulative probabilities of various monthly cumulative precipitation events. Then, cumulative probabilities were normalized to a zero mean and unit variance using the inverse Gaussian function. MATLAB's statistical toolbox was used to fit the probability density function and calculate SPI values. McKee et al. (1993) suggested the SPI intervals to classify drought events as shown in Table 11.1.

11.3   DEFINITIONS OF DROUGHT CHARACTERISTICS

This study uses the theory of runs for defining numerous characteristics of droughts considering the SPI-12 series. The theory of runs chooses a run (drought event) based on chosen values of thresholds. In this study, a value of SPI equal to -1 has been considered as a threshold value for the theory of runs, and a drought event is considered when the SPI value falls below -1 consecutively. Also, only drought events that last more than three months are considered in this study.

Various drought characteristics definitions used in this study are as follows:

TABLE 11.1

SPI Intervals for Drought Class Identification Based on SPI Values

SPI Values	Drought Class
$SPI \geq 0$	Non-drought
$-1 < SPI \leq 0$	Near normal
$-1.5 < SPI \leq -1$	Moderate
$SPI \leq -1.5$	Severe/extreme

Source: McKee et al. (1993).

(i) Length of the time period for which SPI values remain below the threshold can be defined as drought duration (D).

(ii) Drought magnitude can be defined as the cumulative SPI value for a particular drought event.

$$M = \sum_{i=1}^{D} SPI_i \tag{11.1}$$

where D is drought duration and SPI_i is the ith SPI value.

(iii) Drought interval (I) can be defined as the time period between the starting point of a drought event and the starting point of the next drought event.

(iv) Intensity (S) can be considered as the ratio between the magnitude (M) and duration (D) of a particular drought event.

$$S = M / D \tag{11.2}$$

11.4 UNIVARIATE MARGINAL DISTRIBUTIONS OF DROUGHT CHARACTERISTICS

Best-fit marginal distribution may differ for different characteristics based on different drought indices (Liu et al., 2016). Thus, the literature suggests testing marginal distributions for the best fit. In this analysis, four two-parameter distribution functions (Gaussian, exponential, gamma, and Weibull) and two three-parameter distribution functions (generalized Pareto distribution (GPD) and generalized extreme value distribution (GEV)) have been used. The maximum likelihood method was used to assess the parameters of the marginal distributions. The Bayesian information criterion (BIC) is generally utilized to test the goodness of fit of a distribution function. BIC introduces a penalty based on a number of parameters thus reducing the chances of overfitting. The minimum BIC corresponds to the best-fit marginal distribution:

$$BIC = \ln(n)k - 2\ln(\hat{L}) \tag{11.3}$$

where $\hat{L} = p(x \mid \hat{\theta}, M)$ is the maximized value of likelihood function where $\hat{\theta}$ is the parameter at which the likelihood function reaches its maximum. x is the observed data and, n and k are number of data points and number of parameters, respectively.

11.5 BIVARIATE FREQUENCY ANALYSIS USING COPULAS

In recent years, copulas have been widely used for multivariate joint distribution analysis and multivariate frequency analysis due to their capability of modeling dependence between random variables and describing marginal distributions for different variables independently. Among the different copula families, elliptical copulas (normal copula and t-copula) and Archimedean copulas (Clayton copula, and Gumbel–Hougaard copula, and Frank copula) are most widely used copulas for droughts and hydrology. Archimedean copulas are simple in form and known for their generation properties, however elliptical copulas prefer to simulate arbitrary pairwise dependencies between variables through a correlation matrix. The brief descriptions of copulas used in this study are shown in Table 11.2.

TABLE 11.2

Copula Descriptions Used in the Study

Name of Copula	Expression for Copula
Gumbel	$C_{Gumbel}(M) = \exp\left\{-\left[\sum_{i=1}^{n}(-\ln M_i)^{\theta}\right]\right\}$
Frank	$C_{frank}(M) = \dfrac{-1}{\theta}\left\{1 + \dfrac{\prod_{i=}^{n}\left(e^{-\theta M_i} - 1\right)}{\left(e^{-\theta} - 1\right)^{n}}\right\}$
Normal	$C_{Normal}(M, \pounds) = \displaystyle\int_{-\infty}^{\phi^{-1}(M_1)}\int_{-\infty}^{\phi^{-1}(M_2)}\ldots\int_{-\infty}^{\phi^{-1}(M_n)} \dfrac{1}{\sqrt{(2\pi)^{n}\lvert\pounds\rvert}}\exp\left(-\dfrac{1}{2}x'\pounds^{-1}x\right)dx$
Clayton	$C_{clayton}(M) = \left(M_1^{-\theta} + M_2^{-\theta} + \ldots + M_n^{-\theta} - 3\right)^{\frac{-1}{\theta}}$

Note: n is the number of drought characteristics, M_i is the marginal distribution for ith drought characteristics, and θ is the copula dependence parameter.

TABLE 11.3

Kendall's Tau Values for Lagged-One Drought Characteristics

SPI Scale	Peak	Magnitude	Duration
3 months	0.024	0.028	0.018
6 months	0.022	0.022	0.027
9 months	0.019	0.019	0.017
12 months	0.009	0.012	0.002

11.5.1 EXAMINATION OF INDEPENDENCE IN TIME

Kendall's tau values were calculated to test the time independence of drought characteristics. Further, tau values were spatially averaged to obtain a single value for drought characteristics. As it can be observed from Table 11.3, all drought characteristics have their Kendall's tau absolute values less than 0.2 ,which indicates their independence in time.

11.6 UNIVARIATE MARGINAL DISTRIBUTION

Nine different marginal distributions, namely, exponential, EV, gamma, GEV, GPD, inverse Gaussian, lognormal, normal, and Weibull, were used to find the best-fit marginal distribution at different grid points over India. It is important to note that discrete drought events were considered as continuous time series. BIC were used to estimate the best fit, and the parameters of the distributions were estimated using the maximum likelihood estimation technique. The best marginal obtained from using the BIC is shown in Figure 11.1.

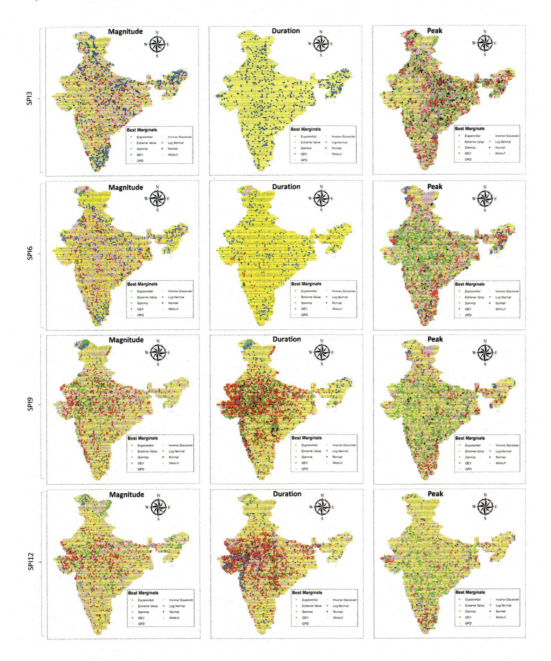

FIGURE 11.1 Spatial distribution of best-fit marginal distributions obtained using the BIC for different drought characteristics.

For drought magnitude, mostly GPD was found to fit best for almost all the country, except for Himalayan and south India regions where GEV is found to dominate best fit for drought magnitudes for short-scale SPI such as SPI3 and SPI6. For the western and central regions, lognormal distribution is found be the best fit.

For drought duration, GPD is found to fit best for almost all part of the country for SPI3 and SPI6, except few points fitting GEV distributed uniformly across all the country for SPI3, and a very few points fitting normal and lognormal distribution for SPI6. For longer SPI, lognormal and gamma distribution have aggregation in western and central India and Western Ghats regions.

For drought peaks, gamma and GPD mainly dominate the country for all scales of SPI. For SPI3, aggregation of lognormal distribution can be observed in the Bihar and Jharkhand regions of India. However, for SPI6 and longer scales, some aggregation of lognormal distribution can be observed in southern India. Unlike previous studies, we have not considered the unified approach for calculating return periods in this work. Cumulative probability for different distribution as shown in Figure 11.2 were obtained for different values of drought characteristics.

11.7 MULTIVARIATE JOINT DISTRIBUTION

Kendall's tau were estimated to examine the interdependence of drought characteristics. Results are illustrated in Figure 11.2. The correlation between peak and duration is found to be much lesser than the correlation of peak–magnitude and magnitude–duration. Substantial correlation between drought features indicates the likelihood of joint modeling of different drought characteristics.

Spatial distribution of Kendall's tau differs for different parts of the country and for different scales of the SPI. Generally, an increase in scale of SPI value was found among all drought characteristics. For peak–duration the correlation is found to be much less for SPI3, however, with an increase in scale of the SPI, the correlation increases, especially in the Himalayan and south Indian region. Whereas for the peak–magnitude relationship, the correlation is found to be high especially for western states such as Gujarat and Rajasthan and for the northeastern region for SPI3. For SPI12, the correlation was found to be very high for almost all over the country. Magnitude and duration were found to be very well correlated for all scales of the SPI. Spatial distribution of correlation between magnitude and duration were found to be very high.

Four commonly used copulas, namely, Gaussian, Gumbel, Frank, and Clayton, were used to model the dependence among the drought variables (P-D, P-M, and M-D). The maximum likelihood approach was utilized to estimate the parameters of all copulas. The best-fit copula was assessed by using the BIC. Selected copulas for different scales of SPI are shown in Figure 11.3.

The spatial distribution of all copulas is different for different regions in India and also for different scales of the SPI. For P-D, the Frank and Clayton copulas mainly dominate the best fit for SPI3 with 38.38% and 30.95%, respectively. Howeve,r for higher scales of the SPI, the Frank copula mainly aggregates in the country with 60.15%, 63.7%, and 63.46% for SPI6, SPI9, and SPI12, respectively. For P-M, the Frank copula mainly dominates the best fit for bivariate distribution for all SPIs with 48.3%, 47.3%, 50.5%, and 63.9% for SPI3, SPI6, SPI9, and SPI12, respectively. For S-D, the Frank copula also dominates the country with the highest percentage of area, but as the scale of SPI increases, the Gumbel copula also starts contributing to some grid points of the country with 35.6%, 31.56%, and 33.78% for SPI6, SPI9, and SPI12, respectively. Other than one copula dominating the country, some spatial pattern can also be observed in the best-fit copulas. For P-D and P-M, the Clayton copula aggregates in Jammu and Kashmir for SPI3, and for SPI9 and SPI12, the Gumbel copula aggregates. However, for M-D, the Gumbel copula aggregates in north India and the eastern coastal regions of India. Best-fit copulas for different grid points were chosen to calculate bivariate return periods at each grid point.

11.8 SPATIAL DISTRIBUTION OF UNIVARIATE RETURN PERIODS

Figures 11.4, 11.5, 11.6, and 11.7 present the univariate return periods for duration (D), magnitude (M), and peak (P) for different values of severity. A very high spatial heterogeneity exists between in return period values across India for all scales. However, a common spatial pattern among all the return periods can be observed. The values of return periods were found to be higher in the western coastal states of India such as Gujarat, Goa, Maharashtra, and Karnataka, which indicates that in these regions drought occurs less frequently as compared to other regions. For smaller values of duration (6 and 9 months) and magnitude (9 and 13), the Jammu and Kashmir regions and the northeastern region were found to have the most frequent droughts with very low values of return

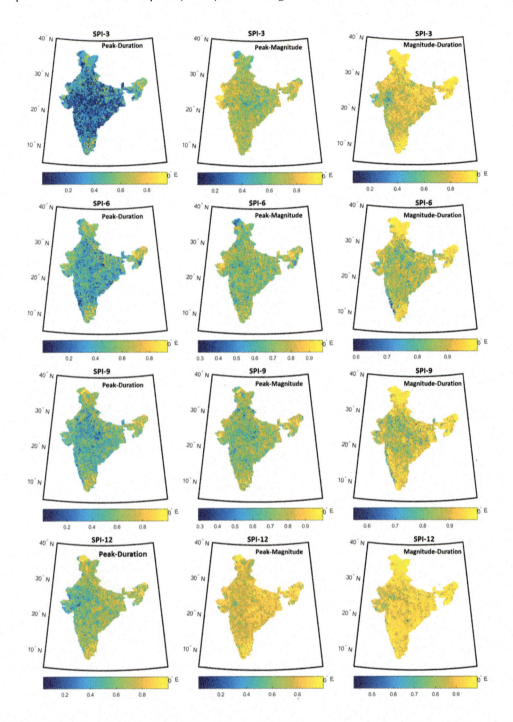

FIGURE 11.2 Spatial distribution of Kendall's tau for different drought characteristics and different SPI scales.

periods. However, very long droughts of more than 24 months were less frequent in the Jammu and Kashmir regions. For central India, including the Ganga basin and parts of southern central India, a very high heterogeneity was observed in observed return periods, which indicates that for some points droughts are very frequent and for other points less frequent.

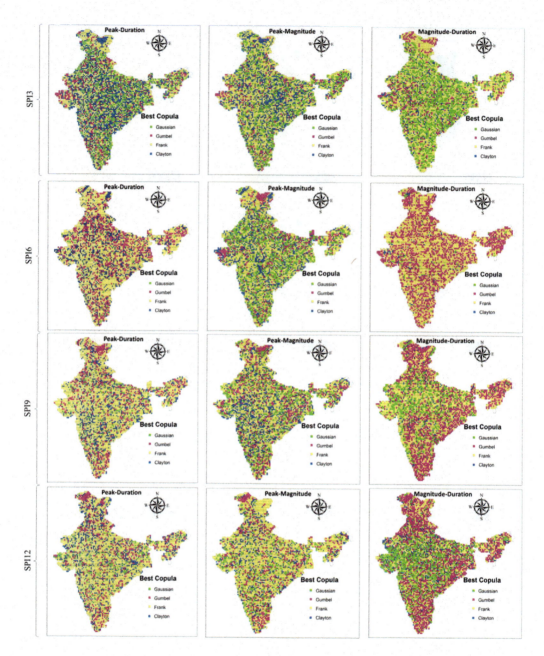

FIGURE 11.3 Spatial distribution of best-fit bivariate copula obtained using the BIC for different pairs of drought characteristics.

When comparing different scales of the drought index, it was observed that with drought events greater than 24 months the return periods values were very high. For most parts of the country, the values were higher ranging from 10 to 100 years for SPI3 and SPI12. However, for SPI6 and SPI9, the return periods for most parts of India were more than 100 years except for extreme southern India and northeastern India. Similarly, for M > 36, the occurrence frequency was found to be very rare with very high values of return periods for all scales of the SPI. As magnitude and duration are better correlated as compared with the peak values, the return period patterns for magnitude and

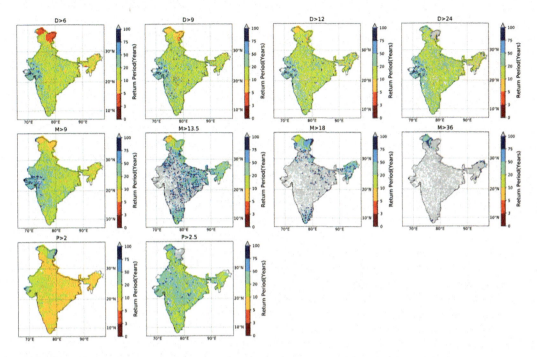

FIGURE 11.4   Spatial distribution of univariate return periods for SPI3 over India.

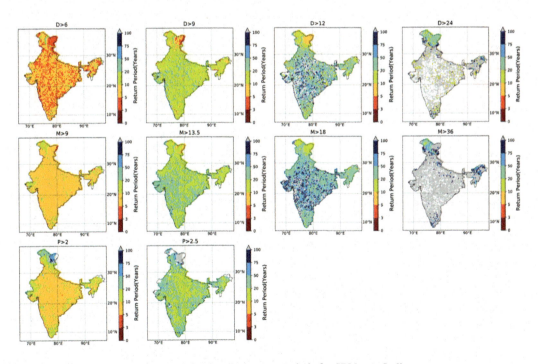

FIGURE 11.5   Spatial distribution of univariate return periods for SPI6 over India.

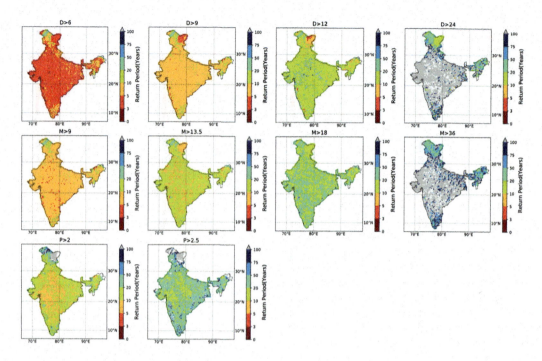

FIGURE 11.6   Spatial distribution of univariate return periods for SPI9 over India.

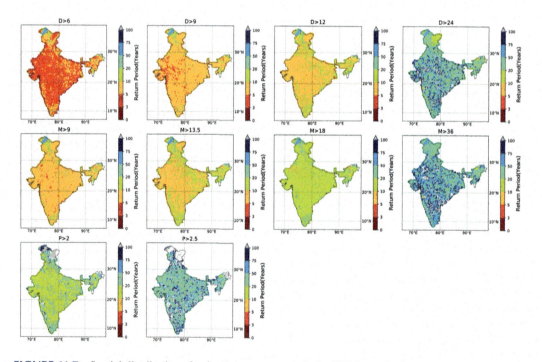

FIGURE 11.7   Spatial distribution of univariate return periods for SPI12 over India.

duration were similar. However, return periods for peak values greater than 2.5 had different patterns. For P > 2.5, more parts of southern India had small values of return periods, which indicated that south India has more probability of getting a drought, for which the peak is more than 2.5 (SPI < 2.5).

11.9 SPATIAL DISTRIBUTION OF BIVARIATE RETURN PERIODS

After analyses of univariate return periods, three different combinations of drought characteristics, i.e., M-D, P-D, and P-M, were selected for bivariate return period analysis. The spatial distribution of the bivariate return period also follows similar patterns as the univariate return periods. Western parts of country were found to be having higher return period values as compared to other parts of the country.

Joint return periods for all combinations of drought characteristics were found to be higher than their individual exceedance values of drought characteristics, which indicates that the two characteristics very rarely exceed their respective values together as compared to individually. For most combinations, the return period values for Jammu and Kashmir, Ladakh, northeast India, and the southern states of south India were found be less as compared to other regions, which reveals that the drought exceedance probability of two characteristics exceeding their threshold values together is higher as compared to other regions. Further, for M > 9 and D > 6, the return values (for SPI3) for most parts of the country were found to be in range of 20 to 50 years. For this combination, in some parts of Jammu and Kashmir and northeastern India, the return values ranged between 5 and 10 years. However, for Gujarat and parts of Maharashtra, the values ranged between 100 and 500 years. For M > 9 and with an increase in the value of duration (for D > 9, 12, and 24), the return period values slightly increase, but remain in the range of 50 to 100 years. For M > 9 and D > 24, the return values for Ladakh region were between 100 and 500 years, which represents that these values exceed very rarely. In Figures 11.8 to 11.11, the white pixels inside the country boundary

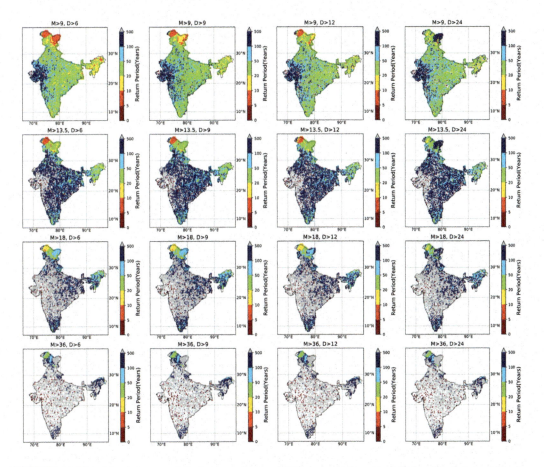

FIGURE 11.8A Joint return periods for exceedance of different combinations of drought characteristics for SPI3.

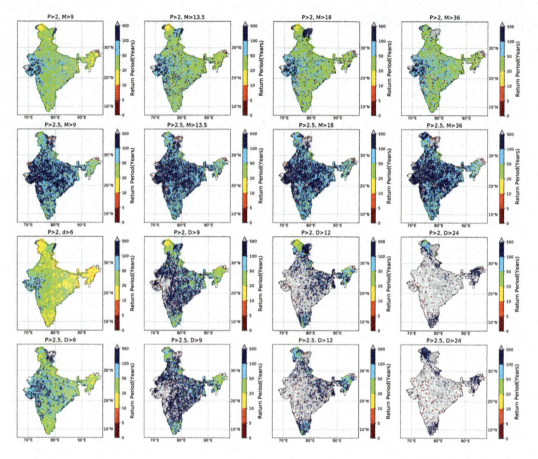

FIGURE 11.8B Joint return periods for exceedance of different combinations of drought characteristics for SPI3.

show grids for which computations were not possible or return values were infinite. Gray pixels represent all values that exceed the color bar. For most of the combinations, which has M > 13.5, the return values for most parts of India were found to more than 100 years for SPI3.

For SPI6, for most parts of India the return period values were between 5 and 10 years for M > 9 and D > 6, which are not in the range of their respective univariate return periods, indicating that the droughts with M > 9 and D > 6 occur frequently similar to their individual occurrence. Similarly, for the combination of M > 9 and D > 9, the return values were in range (but slightly higher) of both their individual values for more parts of India, which represents that these variables exceed their threshold together for most parts of India. However, as the values of drought characteristics start increasing, the occurrence of two variable exceeding their values together start becoming less frequent. Further, it is interesting to observe that the spatial patterns in return period are similar to the patterns of individual variables with higher value. As it can be observed in Figure 11.9a, M > 9 and D > 12, the joint return period follows the pattern of the univariate return period pattern of duration, which signifies that for drought events exceeding 12-months duration the magnitude generally crosses 9. Similar phenomenon can also be observed in the combination M > 13.5 and D

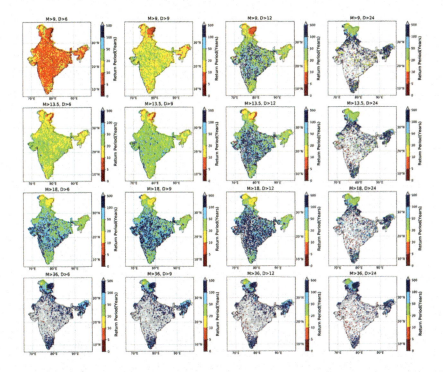

FIGURE 11.9A Joint return periods for exceedance of different combinations of drought characteristics for SPI6.

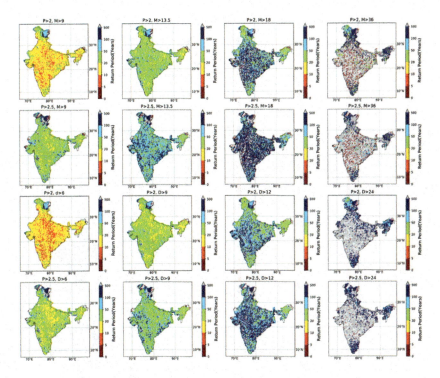

FIGURE 11.9B Joint return periods for exceedance of different combinations of drought characteristics for SPI6.

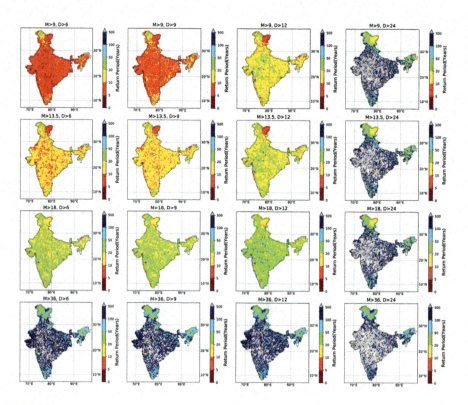

FIGURE 11.10A   Joint return periods for exceedance of different combinations of drought characteristics for SPI9.

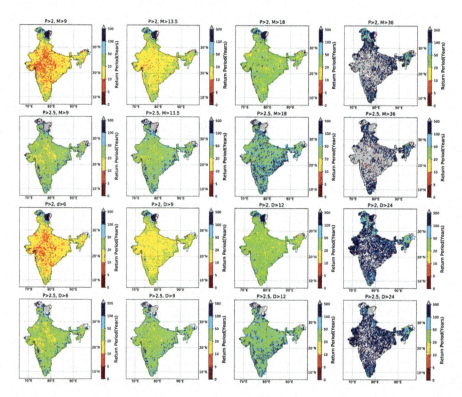

FIGURE 11.10B   Joint return periods for exceedance of different combinations of drought characteristics for SPI9.

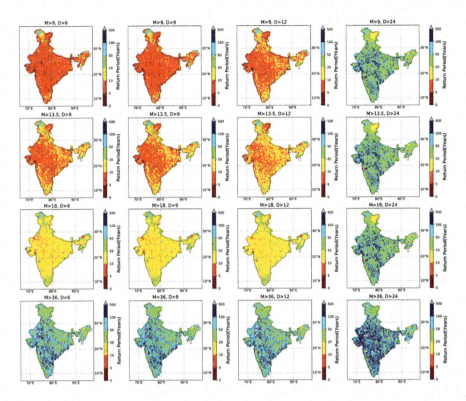

FIGURE 11.11A Joint return periods for exceedance of different combinations of drought characteristics for SPI12.

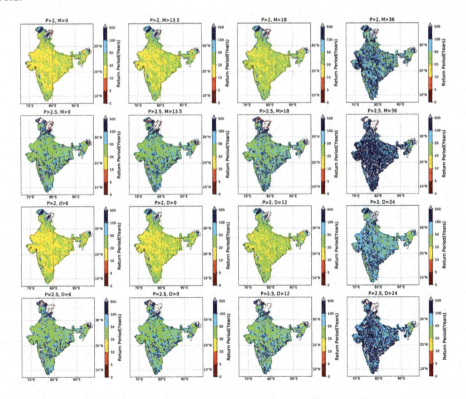

FIGURE 11.11B Joint return periods for exceedance of different combinations of drought characteristics for SPI12.

> 6 where the joint return period follows the spatial pattern of M > 13.5, which again represents that for most of the time when the magnitude crosses 13.5, the droughts are longer than six months.

When comparing different scales of the SPI, for smaller scales, the joint return periods had very high values, but with the increase in the scale of the SPI, the joint return periods for most parts of India were found to be decreasing significantly for all combinations of drought characteristics. The reason for the decrease can be explained by the nature of SPI at different scales. Since SPI3 is computed using a three-month running series, it contains higher fluctuations as compared to other large-scale drought indices. A higher fluctuation in the slight increase in rainfall pushed the value of the SPI toward the positive side; therefore, the length of drought events in the SPI do not generally get longer until there has been really severe dry conditions for a prolonged period of time. However, drought indices with longer timescales move slowly as compared to shorter-scale SPI. Thus, they result in longer drought events, since short-term rainfall doesn't affect the drought index values by a large amount. Since any particular drought event is larger for the SPI with large timescales, then accumulation of the magnitude also results in drought events with higher magnitude. Therefore, the drought events with longer duration and higher magnitude occur more frequently in drought indices with large temporal scales than drought indices with shorter scales.

11.10 CONCLUSIONS

Multivariate drought modeling was conducted using four different scales of the SPI – SPI3, SPI6, SPI9, and SPI12 – over India. Drought events were identified using the theory of runs, and drought characteristics were computed for each event at every grid point across India. Univariate modeling of drought characteristics, such as magnitude, duration, and peak, were done by testing nine different univariate probability distributions selecting the best-fit distribution. Further, the return periods for magnitude exceeding 9, 13.5, 18, and 36; duration exceeding 6, 9, 12, and 24 months; and peak exceeding 2 and 2.5 were computed for India. Further, for bivariate modeling, four different copulas were tested and the best-fit copula was chosen for each grid point over India. Bivariate return periods for different drought characteristics exceeding the aforementioned values were computed throughout India. Results show that the copulas are an excellent tool to model the drought characteristics. Further, a high correlation among drought variables was found. Also, it was also revealed that the SPI12 shows the highest probability of occurrence of longer and more severe droughts as compared to the other three scales. Overall, these results can be used for the planning and design of water resources in India to combat droughts and make India more drought resilient.

REFERENCES

Ayantobo, O. O., Li, Y., & Song, S. (2019). Multivariate drought frequency analysis using four-variate symmetric and asymmetric Archimedean copula functions. *Water Resources Management, 33*(1), 103–127.

Chowdhary, H., Escobar, L. A., & Singh, V. P. (2011). Identification of suitable copulas for bivariate frequency analysis of flood peak and flood volume data. *Hydrology Research, 42*(2–3), 193–216.

De Michele, C., & Salvadori, G. (2003). A generalized Pareto intensity-duration model of storm rainfall exploiting 2-copulas. *Journal of Geophysical Research: Atmospheres, 108*(D2), 1–11.

Diani, K., Kacimi, I., Zemzami, M., Tabyaoui, H., & Haghighi, A. T. (2019). Evaluation of meteorological drought using the Standardized Precipitation Index (SPI) in the High Ziz River basin, Morocco. *Limnological Review, 19*(3), 125–135.

Fu, G., & Butler, D. (2014). Copula-based frequency analysis of overflow and flooding in urban drainage systems. *Journal of Hydrology, 510*, 49–58.

Ge, Y., Cai, X., Zhu, T., & Ringler, C. (2016). Drought frequency change: An assessment in northern India plains. *Agricultural Water Management, 176*, 111–121.

González, J., & Valdés, J. B. (2003). Bivariate drought recurrence analysis using tree ring reconstructions. *Journal of Hydrologic Engineering, 8*(5), 247–258.

Grimaldi, S., & Serinaldi, F. (2006). Asymmetric copula in multivariate flood frequency analysis. *Advances in Water Resources, 29*(8), 1155–1167.

Hosseini-Moghari, S. M., Araghinejad, S., & Ebrahimi, K. (2019). A comparison of parametric and non-parametric methods of standardized precipitation index (SPI) in drought monitoring (Case study: Gorganroud basin). *Iranian Journal of Rainwater Catchment Systems*, 6(4), 25–34.

Joo, K., Kim, S., Kim, H., Ahn, H., &Heo, J. H. (2016, April). Bivariate frequency analysis with nonstationary Gumbel/GEV marginal distributions for rainfall event. In EGU General Assembly Conference Abstracts (Vol. 18, p. 11427).

Kao, S. C., & Govindaraju, R. S. (2007). A bivariate frequency analysis of extreme rainfall with implications for design. *Journal of Geophysical Research: Atmospheres*, 112(D13).

Kwon, H. H., & Lall, U. (2016). A copula-based nonstationary frequency analysis for the 2012–2015 drought in California. *Water Resources Research*, 52(7), 5662–5675.

Liu, X. F., Wang, S. X., Zhou, Y., Wang, F. T., Yang, G., & Liu, W. L. (2016). Spatial analysis of meteorological drought return periods in China using Copulas. *Natural Hazards*, 80(1), 367–388.

McKee, T. B., Doesken, N. J., & Kliest, J. (1993). The relationship of drought frequency and duration time scales. In *Proceedings of the Eighth International Conference on Applied Climatology* (pp. 179–184). American Meteorological Society, Boston.

Ozga-Zielinski, B., Ciupak, M., Adamowski, J., Khalil, B., & Malard, J. (2016). Snow-melt flood frequency analysis by means of copula based 2D probability distributions for the Narew River in Poland. *Journal of Hydrology: Regional Studies*, 6, 26–51.

Shiau, J. T. (2006). Fitting drought duration and severity with two-dimensional copulas. *Water Resources Management*, 20(5), 795–815.

Song, S., & Singh, V. P. (2010). Meta-elliptical copulas for drought frequency analysis of periodic hydrologic data. *Stochastic Environmental Research and Risk Assessment*, 24(3), 425–444.

Song, S., & Singh, V. P. (2010). Meta-elliptical copulas for drought frequency analysis of periodic hydrologic data. *Stochastic Environmental Research and Risk Assessment*, 24(3), 425–444.

Szolgay, J., Gaál, L., Kohnová, S., Hlavcová, K., Výleta, R., Bacigál, T., &Blöschl, G. (2015). A process-based analysis of the suitability of copula types for peak-volume flood relationships. *Proceedings of the International Association of Hydrological Sciences*, 370, 183.

Wang, C., Chang, N. B., &Yeh, G. T. (2009). Copula-based flood frequency (COFF) analysis at the confluences of river systems. *Hydrological Processes*, 23(10), 1471.

Wu, Z., Lin, Q., Lu, G., He, H., & Qu, J. J. (2015). Analysis of hydrological drought frequency for the Xijiang River Basin in South China using observed streamflow data. *Natural Hazards*, 77(3), 1655–1677.

Yu, J. S., Yoo, J. Y., Lee, J. H., & Kim, T. W. (2016). Estimation of drought risk through the bivariate drought frequency analysis using copula functions. *Journal of Korea Water Resources Association*, 49(3), 217–225.

Zakhem, B. A., & Kattaa, B. (2016). Investigation of hydrological drought using cumulative Standardized Precipitation Index (SPI 30) in the eastern Mediterranean region (Damascus, Syria). *Journal of Earth System Science*, 125(5), 969–984.

Zhang, L., & Singh, V. P. (2007a). Bivariate rainfall frequency distributions using Archimedean copulas. *Journal of Hydrology*, 332(1), 93–109.

Zhang, L., & Singh, V. P. (2007b). Gumbel–Hougaard copula for trivariate rainfall frequency analysis. *Journal of Hydrologic Engineering*, 12(4), 409–419.

Zhang, Q., Li, J., Singh, V. P., & Xu, C. Y. (2013). Copula-based spatio-temporal patterns of precipitation extremes in China. *International Journal of Climatology*, 33(5), 1140–1152.

12 Application of Fuzzy Rule–Based Model for Forecasting Drought

A.K. Lohani, R.K. Jaiswal, and R.V. Galkate

CONTENTS

12.1 INTRODUCTION

Due to the spatial and temporal heterogeneity observed in precipitation, different areas on the Earth suffer from drought and floods intermittently. Drought is caused when precipitation conditions are drier than normal and thus the deficit rainfall results in major economic loss. Drought is defined as a slow-onset, insidious hazard that is well established for months or years before it is recognized as a threat (Pongracz et al., 1999). Throughout human existence, drought has been considered a major threat to society. Drought leads to famine, migration of populations, and wars. Despite massive developmental activities, drought still distresses the global community. Researchers are trying to discover the interrelationships between society and drought so that mitigation studies can be planned to reduce the impact of drought. Droughts cannot be avoided, however, preparedness and mitigation strategies for dealing with then can be developed. These preparedness and mitigation strategies solely depend upon how well the droughts and their characteristics are defined.

DOI: 10.1201/9781003276548-12

Several classifications exist based on different base periods and truncation levels to define rainfall deficiency/drought. Any rainless period of six days or more in Bali is considered a drought, however, in Libya, it is recognized only after two consecutive years without rain (Dracup et al., 1980). Subramanyam (1964) developed the aridity index as the percentage ratio of water deficiency to water need at an annual scale. Palmer (1965) applied the water balance technique and incorporated current and antecedent rainfall, evapotranspiration, and soil moisture in his drought index. Appa Rao (1981) applied the aridity index extensively to classify drought. Furthermore, he identified moderate and severe droughts from 1875 to 1980 and affected areas in India (Appa Rao, 1987). The rainfall criterion is suitable for continuous monitoring of the monsoon season. The total seasonal rainfall is used as the basis for defining a region under moderate or severe drought. If more than 50% of a country's area is facing moderate or severe drought, then it is defined as a severe drought, and it is a moderate drought when 26%–50% of the area is affected (Singh, 2000). The Drought Prone Area Programme (DPAP, 1973) defines drought-prone areas as areas receiving less than 750 mm rainfall per annum. Areas receiving annual rainfall between 750 mm to 800 mm are defined as vulnerable to drought. According to the India Meteorological Department (IMD), an area is classified as drought-hit when annual rainfall is less than 75% of the normal rainfall, (i.e., the rainfall deficit is 25% to 50%) and severe drought when the deficiency is above 50%. The Irrigation Commission assumes an area is drought-hit if the irrigated area falls short by 70% of the irrigable area. So, when the irrigated area is less than 30% of the irrigable area, the areas are to be called drought-prone. Areas with more than a 20% probability of experiencing a drought are defined as drought-prone areas, while areas with more than a 40% probability of experiencing a drought are defined as a chronic drought-prone area (Singh, 2000). The Standardized Precipitation Index (SPI) was introduced by McKee et al. (1993). The SPI is computed for various timescales to measure the shortage in precipitation. According to the IMD (2002), when the monsoon rainfall deficiency in an area is within 10% of the long-term average, then it is defined as a normal rainfall year, and when it is more than 10% it is defined as a drought year. Drought always starts with a rainfall deficit and it may further affect soil moisture, streamflow, groundwater, agriculture, and the ecosystem. Therefore, drought has been classified in different ways, e.g., meteorological drought, hydrological drought, and agricultural drought, describing water shortage in different sectors. When the precipitation of a place is lower than its normal precipitation, it gives an indication of the meteorological drought conditions. A reduced amount of rainfall causes soil moisture reduction and also reduces surface and groundwater resources. The spatially extensive and temporally long periods of water deficit may result in the occurrence of different types of droughts. Initially, it may reflect on the surface and groundwater availability, leading to hydrological drought and later, with respect to agricultural operations, becomes an agricultural drought. Any external (interbasin transfers) source of water may mitigate the effect on an area without hydrological and agricultural droughts.

Due to the inherent spatial and temporal variability in rainfall, drought is observed in many parts of the world. However, the readiness of the government and society for such situations can help in reducing the overall impact of drought. The success of both depends, among others, on how well the droughts are defined and the drought characteristics quantified. In order to manage water resources during a drought, it is important to predict the drought conditions more precisely. In order to find workable answers to water management in drought-prone areas, forecasting of drought events is very important (Bordi and Sutera, 2007). A number of methods ranging from conceptual to data driven to physical are available for the forecasting of drought. Conceptual and physical models provide insight into the processes, but sometimes their implementation is tough in forecasting due to large data requirements (Beven, 2006). The usefulness of the stochastic models in drought forecasting studies has been demonstrated by various researchers (Mishra and Desai, 2005, 2006; Mishra et al., 2007; Han et al., 2010). As the hydrological processes are nonlinear, stochastic models have limitations due to their linear nature. In forecasting, soft computing–based models are very useful due to their less data requirement, easy implementation, very fast development time, and better forecasting accuracy (Adamowski, 2008;). Furthermore, the soft computing models, e.g. artificial neural networks (ANNs) and fuzzy logic, are very effective models in forecasting nonlinear data therefore, they have been successfully used in drought forecasting (Mishra and Desai, 2006; Morid et al., 2007; Bacanli et al., 2008; Barros and Bowden, 2008; Cutore et al., 2009; Karamouz et al., 2009; Marj and Meijerink, 2011). ANNs and fuzzy logic have been applied to model rainfall–runoff relationships (Lohani et al., 2011), rating curve analysis, sediment rating curve, and flood forecasting (Lohani et al., 2005a, 2005b, 2014). The objective of this study is to compare ANNs and fuzzy logic–based models in drought forecasting. These models were compared for a six-month lead period SPI 3 forecast, as the SPI is a good indicator for meteorological drought.

12.2 STANDARDIZED PRECIPITATION INDEX (SPI)

The SPI is the most commonly used indicator for forecasting drought. The primary reason for using the SPI indicator, which was developed by McKee et al. (1993), is that the SPI is based on rainfall alone, so drought assessment is possible even if other hydrometeorological measurements are not available. According to Edwards and McKee (1997), the SPI indicator measures precipitation anomalies at a given location. As the SPI is defined from the rainfall values, it is independent of topography. For developing the SPI, different timescales are used so that the developed SPI can define the conditions of the meteorological, hydrological, and agricultural droughts. A probability distribution (gamma distribution) is fitted on the historical rainfall data, and after this it is transformed into normal distribution so that the mean value of the SPI for that location and period is zero. The SPI confirms the consistency of the extreme events frequencies at a location and on any timescale. The SPI also promptly identifies a deficiency in moisture. The results of the SPI for a 12-month timescale detect long-duration dry periods related to the global impact of drought (Paulo et al., 2005).

The SPI was used in this study to quantify the precipitation deficit. As mentioned earlier, the SPI indicates the deviation of observed values of rainfall from the mean rainfall in terms of the number of standard deviations, considering the rainfall as normally distributed. Because rainfall is not normally distributed, a transformation function is applied to the rainfall data, so that after the transformation, rainfall values follow a normal distribution (Rouault and Richard, 2003). On the long-term rainfall data, a gamma distribution is fitted, then it is transformed into normal distribution for the calculation of the SPI. The criteria for drought classification as reported by Hayes et al. (1999) and Steinemann (2003) for the meteorological drought was assessed using the categories presented in Table 12.1. On the other hand, Table 12.2 presents the main strengths and weaknesses of the SPI. The SPI is calculated for accumulated rainfall of different periods (1–48 months). These SPI values of the different accumulation periods represent different impacts of the meteorological drought:

- SPI 1 to SPI 3 indicate immediate impacts such as reduced soil moisture, snowpack, and flow in smaller creeks.
- SPI 3 to SPI 12 indicate reduced streamflow and reservoir storage.
- SPI 12 to SPI 48 indicate reduced reservoir and groundwater recharge.

TABLE 12.1
Standardized Precipitation Index Categories

Drought Category	SPI Values
Extremely wet	≥ 2
Very wet	Between 1.5 and 1.99
Moderately wet	Between 1.0 and 1.49
Near normal	Between −0.99 and 0.99
Moderately dry	Between −1.00 and −1.49
Severely dry	Between −1.50 and −1.99
Extremely dry	≤ -2.0

TABLE 12.2

Strength and Weakness of the SPI

Strengths	Weaknesses
The SPI can be compared for any geographic location and for any number of timescales.	Estimation of the SPI indicator should be done carefully in the areas having a very high probability of zero values of rainfall.
The SPI is very effective in analyzing dry periods and cycles.	As the SPI indicator is computed only from rainfall, it does not describe the impact of high temperatures on drought conditions. For this, Vicente-Serrano et al. (2010) developed a new indicator named the Standardized Precipitation and
The SPI is computed from only the rainfall values and thus it is less complex.	Evapotranspiration Index (SPEI).

12.3 DROUGHT FORECASTING

Forecasting drought is a primary requirement to reduce the detrimental effects of droughts in a region. Accurate drought forecasts enable decision makers to prepare for optimal operation of irrigation systems. Various stochastic models are used for forecasting of drought. Stochastic models are not able to model the nonlinear behavior of hydrological processes. These models also have limited ability to model nonstationarities in the data. Hydrological variables such as monthly and annual rainfall and streamflows have been widely simulated using autoregressive (AR) and autoregressive moving average (ARMA) models. However, it is necessary for hydrologists to consider alternative models when nonlinearity and nonstationarity play a significant role in forecasting.

In the case of droughts, it is observed that there are vagueness in definition, imprecision in measurements, and uncertainty in parameter estimation. Vagueness, impression, and uncertainty can be handled by applying the fuzzy logic theory. Errors in the measurement of hydrologic data and noise can be taken care of by the fuzzy membership functions and rules. The creation of the fuzzy rules and membership functions from the historical data establishes an inherent consistency between the model and the hydrologic behavior observed in the basin. Furthermore, the transparency of the fuzzy rules provides explicit qualitative and quantitative insights into the physical behavior of the system.

12.4 STUDY AREA AND DATA USED

The Mahanadi River, the lifeline of Chhattisgarh, covers more than half of the geographical area of the state and is a major source for meeting different sectorial demands including irrigation, and industrial and domestic uses. The river originates from the Sihawan forest in the Dhamtari district of Chhattisgarh traversing nearly 851 km in length in Chhattisgarh and Odisha before merging with the Bay of Bengal. The Hasdeo, Seonath, Mand, Ib, Ong, Tel, and Jonk are important tributaries of the Mahanadi River and several water resources projects, including Ravishankar Sagar, Tandula, Parry, Jonk, Hasdeo Bango, Kharang, Kodar, Mand, and Maniyari, have been constructed in Chhattisgarh to utilize the water resources potential of the river. These water resource projects are designed and operated based on the current climatic condition where the changed situation may need a prior and proper assessment of future climate extremes for revising operation plans and irrigation releases. Most of these projects are multipurpose projects to supply water for drinking, industrial, and recreational purposes besides irrigation and any significant change due to climate may impact the water supply to crops in the Kharif season, which may be detrimental to crop yield and large-scale rural economy. A location map of the upper Mahanadi basin in the Chhattisgarh

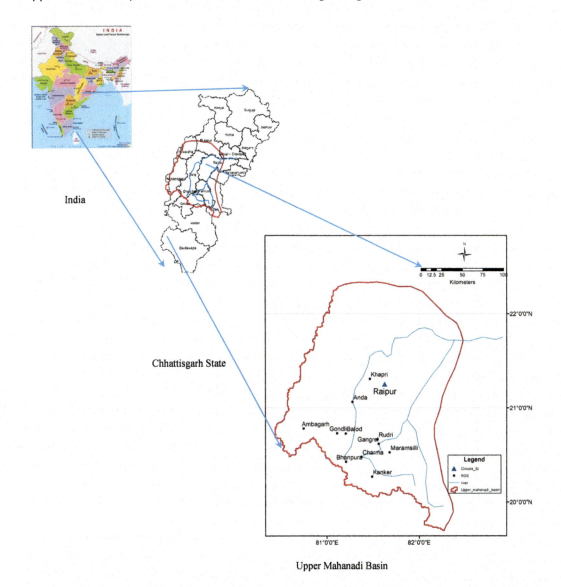

FIGURE 12.1 Location map of the upper Mahanadi basin in India.

state of India is presented in Figure 12.1. The locations of different rain gauge stations selected for the study are shown in Figure 12.2. The observed climate data (min and max temperature) of the Raipur climate station from 1981 to 2014 and rainfall of the Ambagarh, Anda, Bhanpura, Balod, Chamra, Gondli, Khapri, Maramsilli, Rudri, and Gangrel rain gauge stations from 1981 to 2014 have been used for drought analysis and modeling.

12.5 MODEL DEVELOPMENT

12.5.1 Artificial Neural Network Models

ANNs are the computer programs inspired by biological neural networks to simulate a system in way the human brain processes information. Similar to a human brain, ANNs also have neurons that are interconnected to one another and linked to various layers of the networks. The advantage

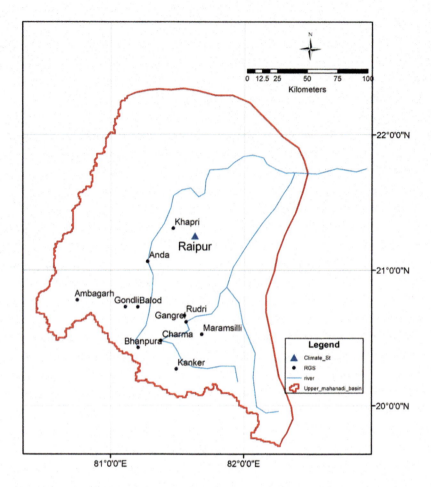

FIGURE 12.2 Location of climate observatory and rain gauge stations in the upper Mahanadi basin.

of using ANNs is to rapidly simulate a nonlinear system and develop models for the problems in which input and output data relation is undefined. In this study, a feedforward multilayer percep-tron (MLP)–based ANN model with the Levenberg–Marquardt (LM) backpropagation algorithm is trained to forecast the SPI. MLP architecture of the ANN has wide application in hydrological modeling and forecasting. The typical architecture of MLPs consists of an input layer, one or more hidden layers, and an output layer.

12.5.2 FUZZY MODEL

Fuzzy logic is a soft computing technique for solving problems by generalizing the standard logic so that the degree of truth anywhere can be processed between 0.0 and 1.0. It is a very powerful technique and is widely used in control systems and the processing of information. According to Lotfi Zadeh (1965), who first developed the concept of fuzzy logic, unlike classical logic, human decision-making generally does not have very crisp answers and has any possibility between Yes and No, e.g., Certainly Yes, Possible, Cannot Say, Not Possibly, and Certainly Not. The fuzzy logic algorithm is very powerful and useful as (i) it allows a higher level of abstraction originating from our knowledge and experience with modeling complex systems, (ii) it can work from approximate data and provide acceptable solutions/answers, and (iii) it helps to deal with the uncertainty. In the beginning, Japan successfully applied the fuzzy logic technique in several applications. The tech-nique shows increasing acceptance in various engineering and medical fields and from control theory to artificial intelligence. Every application demonstrates the benefits of fuzzy logic, such as being a robust algorithm, high performance, simple reasoning process, less development time in comparison to the conventional model, lower cost, and high productivity. Recently, the fuzzy logic algorithm has also been applied by various researchers in water resources modeling and forecasting problems.

12.5.3 Crisp Set and Fuzzy Sets

The crisp set S is a group of objects belonging to a given universe X having identical properties, such as countability and finiteness. A crisp set S can be defined as a group of elements over the universal set X, where a random element belongs to the set A or not; if so, we say $x \in X$; if not, we say $x \in X$. In ordinary (nonfuzzy) set theory, there are only two possible ways: elements either fully belong to a set or are fully excluded from it. The membership $\mu_S(x)$ of a set S as a subset of the universe X is defined as follows (for all $x \in X$):

$$\mu_S(x) = \begin{cases} 1, & \text{iff } x \in S \\ 0, & \text{iff } x \notin S \end{cases}$$

It indicates that an element x is either a member of set $S(\mu_S(x) = 1)$ or not $S(\mu_S(x) = 0)$. This crisp classification is suitable in science/mathematics, however, it cannot represent the human way of thinking and decision-making. Table 12.3 presents the differences between the crisp set and the fuzzy set. Fuzzy logic replaces the set of truth values {0, 1} by the entire unit interval [0, 1], and it maps each element $x \in X$ to a degree of membership in the interval [0, 1]. Figure 12.3 presents the difference between fuzzy logic and Boolean logic.

12.5.4 Membership Function Assignment and Rule Generation

For the development of the membership function, the first step is to classify the input and output data space as low, medium, and high. After the classification of the data space, membership functions are assigned. First, the data points with the highest membership grades are selected. The midpoint of these data points is assigned as membership grade one. Then a membership grade M (0 < M < 1) is assigned.

Figure 12.4 presents a membership function where, a_{li} and b_{li} are the left and right half-width of the function, and x is the average distance of the membership vertex to the right and left edge. It is represented as

$$\frac{x}{b_{ij}} = \frac{1-C}{1} \Rightarrow b_{ij} = \frac{x}{1-C} \tag{12.1}$$

TABLE 12.3
Crisp Set versus Fuzzy Set

Crisp Set	Fuzzy Set
This set has fixed boundaries.	This sets does not have fixed boundaries, and the boundaries of one set overlap another set.
Elements in a set can be a membership or nonmembership of that set.	Some elements have part membership in different sets and are represented by membership degrees.
Crisp sets are based on bi-valued logic.	It is based on the infinite-valued logic.
It works on binary logic and is used in digital and expert systems modeling.	It introduces imprecision and vagueness.

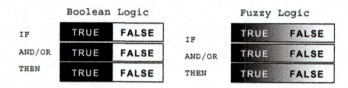

FIGURE 12.3 Boolean logic versus fuzzy logic.

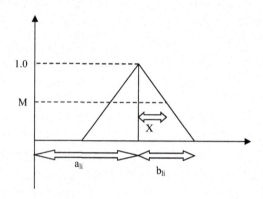

FIGURE 12.4 The triangular membership function.

where C is a parameter to be described by experience or using optimization techniques. Values of C vary typically from 0.5 to 0.8. For developing a fuzzy rule–based model, first, the input and output data are portioned, then appropriate fuzzy membership functions assigned. Next, rules in the form of if–then are developed.

12.5.5 STEPS FOR DEVELOPING FUZZY LOGIC MODEL

A fuzzy rule–based model can be developed as:

- First, objectives of the modeling problem and criteria are to be defined.
- Input and output relationships are to be determined and then selection of a minimum number of input variables.
- Define the input and output data sets by fuzzy membership functions.
- Now divide the modeling problem using the input–output membership function into a series of if–then rules so that the desired output can be obtained.
- Carry out required pre- and post-processing.
- Tune the rules and membership functions for the desired output. In Figure 12.5 a broad classification of major steps used in developing the fuzzy model is presented.

12.5.6 PERFORMANCE MEASURES

The performance of drought forecasting models have been measured using three commonly used performance criteria. The available input and output data is divided into two parts: one part is used for calibration, and another for validation of ANN and fuzzy logic models. Various performance criteria, such as root mean square error (RMSE), model efficiency (Nash and Sutcliffe, 1970), and

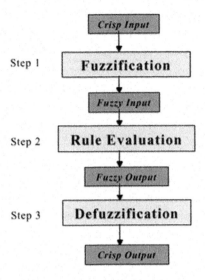

FIGURE 12.5   Steps for developing fuzzy model.

coefficient of correlation (R), are applied to evaluate the performance of the developed models during calibration and validation. The higher values of the coefficient of correlation and efficiency, and lower values of RMSE would indicate better model performance. These performance criteria are defined by the following equations:

$$RMSE = \sqrt{\frac{\sum_{i=1}^{n}\left(SPI_i - ForecastedSPI_i\right)^2}{n}}$$ (12.2)

$$\text{NS Model Efficiency} = 1 - \frac{\sum_{i=1}^{n}\left(SPI_i - ForecastedSPI_i\right)^2}{\sum_{i=1}^{n}\left(SPI_i - Average\,SPI\right)^2}$$ (12.3)

Coefficient of Correlation =

$$1 - \frac{\sum_{i=1}^{n}\left(SPI_i - Average\,SPI\right)\left(ForecastedSPI_i - Average\,ForecastedSPI_i\right)}{\sum_{i=1}^{n}\left(SPI_i - Average\,SPI\right)^2 \sum_{i=1}^{n}\left(ForecastedSPI_i - Average\,ForecastedSPI_i\right)^2}$$ (12.4)

$$\text{Persistence index}\,\left(PERS\right) = 1 - \frac{\sum_{i=1}^{n}\left(SPI_i - ForecastedSPI_i\right)^2}{\sum_{i=1}^{n}\left(SPI_i - SPI_{i-L}\right)^2}$$ (12.5)

where N is the number of observations and L is the forecast lead time. Persistence compares a model under study and the naive model. When PERS $\leq$ 0, then the model under study performs worse or no better than the easy-to-implement naive model. PERS = 1 indicates an exact forecast.

12.6 RESULTS AND DISCUSSION

12.6.1 ANN MODELS

12.6.1.1 Model Inputs

The first step toward developing any data-driven models like ANNs and fuzzy is the identification of input and output variables. In number of hydrological modeling studies it has been demonstrated that the improved forecasting of a hydrological variable can be obtained by considering the same variable values at previous time steps in the input model structure (Lohani and Singh, 2012). Therefore, the forecasting of SPI at six-month lead time has been carried out forecasting the SPI at different time steps by considering the previous time step SPI in the input. Table 12.4 presents the input structure for the six-month lead time SPI forecast.

The SPI values of first through fifth lead periods have been used in the six-month lead period forecast of SPI. Therefore, five steps have been considered in the model structure as presented in Table 12.4. Further, the model is calibrated and validated using the known values of the previous time steps SPI input data vector. In this process known values/forecasted values of the SPI have been used to compute the next SPI values as the case may be (from step 1 to 5; see Table 12.4). These computed SPI values were used as input for forecasting of the SPI after the second or third step onward. It was also inferred that using forecasted SPI as input to these models, as shown in Table 12.4, may carry over the error from first step to second and so on, but these carryover errors may not be significant due to the higher accuracy of soft computing models, i.e., fuzzy and ANN models (Lohani et al., 2007a, 2007b).

12.6.1.2 Training of ANN Models

In this study, the feedforward backpropagation ANN network was applied. This network consists of input neurons consisting of the previous time steps SPI as shown in Table 12.5 and a single output neuron (SPI) with single hidden layer. For developing a fuzzy model, all the input and output data were scaled between 0 and 1. For finalizing the ANN model structure, a number of trial runs have been made by varying the number of neurons between three to nine in the hidden layer. The gradient descent with momentum weight and bias learning function was used for adjusting the weights and biases of the ANN network during model training. The criteria selected to avoid overtraining was through generalization of ANNs for which the improvement in the developed model was checked after each iteration. The default performance function for feedforward networks is the mean square error.

TABLE 12.4
Model Input Structure (Six-Months Lead Time)

Recursive Order	Input Structure	Output
Step 1	SPI(t–4), SPI(t–3), SPI(t–2), (SPI t–1), SPI (t)	SPI(t+1)
Step 2	SPI(t–3), SPI(t–2), SPI (t–1), SPI (t), SPI(t+1)	SPI(t+2)
Step 3	SPI(t–2), SPI(t–1), SPI(t), (SPI t+1), SP(t+2)	SPI(t+3)
Step 4	SPI(t–1), SPI(t), SPI (t+1), SPI (t+2), SPI(t+3)	SPI(t+4)
Step 5	SPI(t), SPI(t+1), SPI(t+2), (SPI t+3), SPI(t+4)	SPI(t+5)
Final step	SPI(t+1), SPI(t+3), SPI (t+4), SPI(t+4), SPI(t+5)	SPI(t+6)

Note: SPI(t–i) is the SPI value at the previous ith interval, and SPI(t+1) is the SPI value at the ith lead period.

TABLE 12.5

ANN Models for 6-Month Forecasts of SPI 3 in Upper Mahanadi Basin

Basin Station	Nodes in hidden layer	Calibration			Validation		
		R^2	RMSE	NS Efficiency	R^2	RMSE	NS Efficiency
Ambagarh	3	0.7732	0.2737	0.7505	0.7637	0.2782	0.7415
Anda	3	0.7819	0.2834	0.7569	0.7724	0.2913	0.7423
Balod	3	0.7914	0.2784	0.7717	0.7851	0.2844	0.7643
Bhanpura	3	0.7791	0.2751	0.7555	0.7677	0.2872	0.7424
Gondli	3	0.7923	0.2765	0.7722	0.7806	0.2921	0.7557
Kanker	3	0.7719	0.2673	0.7519	0.7632	0.2802	0.7482
Khapri	3	0.7802	0.2722	0.7612	0.7721	0.2819	0.7501
Dhamtari	3	0.7778	0.2619	0.7549	0.7662	0.2789	0.7439
Maramsilli	3	0.7954	0.2781	0.7737	0.7819	0.2934	0.7589

Note: Column 3 is the ANN architecture detailing the number of nodes in the input, hidden, and output layers, respectively.

12.6.1.3 ANN Drought Forecast Model

Results of the six-month lead time forecast of SPI 3 using the ANN model for Ambagarh, Anda, Bhanpura, Balod, Chamra, Gondli, Khapri, Maramsilli, Rudri, and Gangrel rain gauge stations are presented in Figure 12.6. It is inferred from Figure 12.6 that the ANN model has good generalization ability for SPI 3, and the extreme values forecasts correspond to the extreme values SPI 3. Scatterplots of the observed and forecast SPI 3 for Ambagarh, Anda, Bhanpura, Balod, Chamra, Gondli, Khapri, Maramsilli, Rudri, and Gangrel rain gauge stations are presented in Figure 12.7. The scatterplot in Figure 12.7 shows several points significantly above or below the 45° line indicating a certain level of overestimation or underestimation in the ANN model results. The developed ANN models show the value of R^2 ranging from 0.7719 to 0.7954, and RMSE from 0.2619 to 0.2834, and NS (Nash–Sutcliffe) efficiency from 0.7505 to 0.7737 during calibration. During the ANN model validation, the value of R^2 ranges from 0.7637 to 0.7851, RMSE from 0.2782 to 0.2934, and NS efficiency from 0.7415 to 0.7643. The R^2 values ranging between 0.7719 to 0.7954 during calibration and 0.7637 to 0.7851 during validation show a good correlation between observed and predicted results at six-months lead time.

12.6.2 Fuzzy Model

12.6.2.1 Identification of Takagi–Sugeno (TS) Fuzzy Model

Input data vectors presented in Table 12.3 were also used for the development of the SPI forecasting model using the Takagi–Sugeno (TS) fuzzy algorithm. Various steps defined in the previous sections were used to develop a TS fuzzy model for forecasting of the SPI. First, fuzzy membership functions and the number of fuzzy rules were determined using the fuzzy clustering approach, and then the TS fuzzy model was optimized using least squares estimation. Using the fuzzy clustering approach, the input and output data were classified in different clusters so that the similarity within a cluster is larger than that among clusters. The input–output data sets were normalized within the unit hypercube, as the similarity matrices are generally highly sensitive to the range of elements in the input vectors. Here in this study, the subtractive clustering method was used for the generation of membership and the fuzzy rule base. Further, a number of trial runs were carried out to obtain an optimum model structure. In trial runs, the cluster radius (r_a) was varied between 0.1 and 1 with

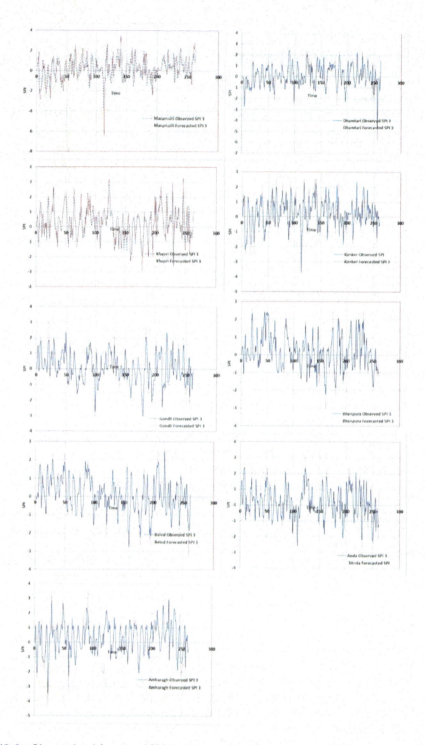

FIGURE 12.6 Observed and forecasted SPI 3 at six-month lead period using ANN.

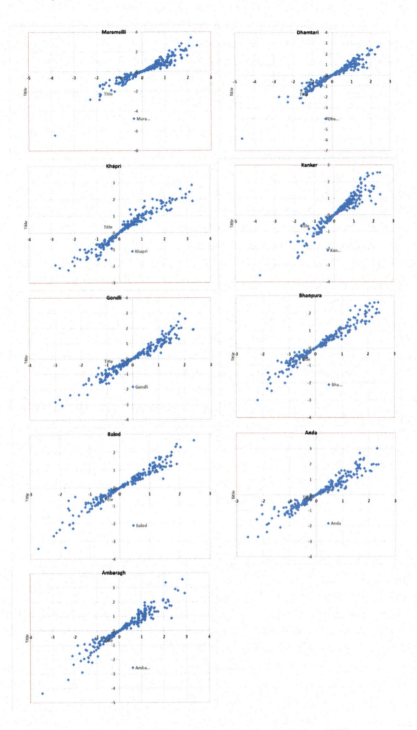

FIGURE 12.7　Observed and forecasted SPI 3 at six-month lead period using ANN.

step size as 0.01. The fuzzy model developed for SPI data sets have different rules ranging from 4 to 7 for discharge–sediment rating curves. For every input vector, a membership degree to each fuzzy set greater than 0 is computed from the Gaussian membership function. Therefore, all the rules fire simultaneously for each combination input and thus provide a crisp output value for a given input data vector using Equation 12.6. Performance indices, such as RMSE, between the computed and observed discharge and the correlation coefficient were used to finalize the optimal parameter combination of the model (Lohani, et al., 2006).

12.6.2.2 Fuzzy Inference System

12.6.2.2.1 Fuzzy Clustering

Fuzzy clustering divides the data points into clusters on the basis of similarity between items, and some of the data points can belong in more than one cluster. Computationally, it is much easier to create fuzzy boundaries in comparison to finalize one cluster for one point. Structure and parameter identification are the two main issues in TS fuzzy modeling. Structure identification involves generation of fuzzy sets, membership functions, and if–then rules. Parameter identification involves identification of the consequent parameter on the basis of an objective function. There are two approaches for rule generation: (i) manual and (ii) data driven. Manual inspection for rule generation has its limitations. The fuzzy clustering–based data-driven approach identifies natural grouping of the data from a large data set and thus produces a concise representation of a system's behavior. The data-driven approach for rule generation has been used in various applications, as it provides promising results. Fuzzy c-means clustering (Bezdek, 1973; Jang et al., 2002), mountain clustering (Yager and Filev, 1994), Gaustafron–Kessel (GK) fuzzy clustering (Gustafson and Kessel, 1979), and subtractive clustering (Chiu, 1994) are used to identify fuzzy clusters. In subtractive clustering, computation is simply proportional to the number of data points and independent of the problem dimension. Each cluster center D_i derived by subtractive clustering is described a as fuzzy rule. The identification of cluster centers is based on the potential value allocated to each data point:

$$P_i = \sum_{j=1}^{N} \exp\left(-4.\frac{\|x_i - x_j\|^2}{r_a^2}\right) \tag{12.6}$$

where P_j is the potential value, x_i is the data point, N is a set of data points, r_a (≥ 0) is the cluster radius, and $\|x_i - x_j\|$ is the Euclidean distance.

The data point showing the maximum value of the potential is assigned as the first cluster center $D1$. After this, the potential values of the remaining data points are revised by subtracting the influence of first cluster center. In the remaining data set, the point showing the maximum modified potential is considered as a second cluster center (D_2). This process is continued to identify other cluster centers. Now, the modified potential after calculation of the jth cluster center is

$$P_i = P_i - P_j^* \exp\left(-4.\frac{\|x_i - D_j\|^2}{r_b^2}\right) \tag{12.7}$$

where r_b ($r_b > r_a \geq 0$) is the radius causing a quantifiable decrease in potential of neighborhood data points and finally avoids closely spaced cluster centers. This procedure is continued so as to generate an adequate number of clusters with a stopping criterion given by Chiu (1994). The identified cluster centers (D_i^*, $i = 1,k$) are the centers of the fuzzy rules' premise of input data vector x. Further, the Gaussian function is used to define membership, and the degree to which rule i is fulfilled is defined by

$$\mu_i(x) = \exp\left(-4.\frac{\left\|x - D_i^*\right\|^2}{r_a^2}\right) \tag{12.8}$$

12.6.2.2.2 Takagi–Sugeno (TS) Fuzzy Model

A linguistic fuzzy model has fuzzy sets in both antecedents and consequents of the rules, whereas in the TS fuzzy model the consequents are expressed as (crisp) functions of the input variables (Takagi and Sugeno, 1985). These data-driven fuzzy models are relatively easy to identify and their structure can be readily analyzed. A TS fuzzy model consists of a set of rules R_i, $i = 1, ..., k$:

R_i: if x_1 is A_{i1} and if x_2 is A_{i2} ... and if x_n is A_{in}

$$\text{then } y = f_i(x_1, x_2, ..., x_n) \tag{12.9}$$

where $x_1, x_2, ..., x_n$ are the antecedents and y is the consequent, A_{ij} are fuzzy sets, and $f_i(x_1, x_2, ..., x_n)$ is linear.

$$f_i(x_1, x_2, ..., x_n) = a_1x_1 + a_2x_2 + ... + a_nx_n + a_{n+1} \tag{12.10}$$

The output of the TS fuzzy model is computed by

$$y = \sum_{i=1}^{M} f_i(x_1, x_2, ...x_n).\Phi_i(x) \tag{12.11}$$

where M is the number of fuzzy rules. $\varphi_i(x)$ are called basis functions, which normalize the degree of rule fulfillment by using the product t-norm:

$$\Phi_i(x) = \frac{\mu_i(x)}{\beta(x)} \tag{12.12}$$

where

$$\mu_i(x) = \prod_{j=1}^{p} \mu_{ij}(x_j) \tag{12.13}$$

$$\beta(x) = \sum_{i=1}^{M} \mu_i(x) \tag{12.14}$$

where x_j is the jth element in the current data vector, p is the dimension of the input data vector, and μ_{ij} is the membership degree of x_j to the fuzzy set describing the jth premise part of the ith rule.

The subtractive clustering allocates a set of rules and antecedent membership functions to model the data behavior. Further, global linear least squares is applied to generate consequent equations (Equation 12.11) for each rule. The subtractive clustering method produces Gaussian membership functions (Equation 12.8) as fuzzy sets that gives a membership value more than 0 for every input.

12.6.3 FUZZY DROUGHT FORECAST MODEL

Results of six-months lead time forecast of the SPI 3 using the TS fuzzy model for the Ambagarh, Anda, Bhanpura, Balod, Chamra, Gondli, Khapri, Maramsilli, Rudri, and Gangrel rain gauge stations are presented in Figure 12.8. The figure depicts that the ANN model has good generalization ability for SPI 3 and the extreme values forecasts correspond to the extreme values SPI 3. Scatterplots of the observed and forecast SPI 3 for the Ambagarh, Anda, Bhanpura, Balod, Chamra, Gondli, Khapri, Maramsilli, Rudri, and Gangrel rain gauge stations are presented in Figure 12.9. The scatterplots in Figure 12.9 show several points significantly above or below the 45° line indicating a certain level of overestimation or underestimation in the TS fuzzy model results. The

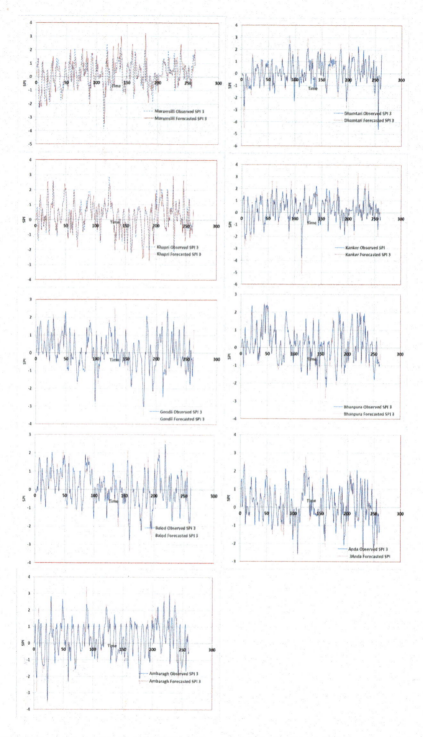

FIGURE 12.8 Observed and forecasted SPI 3 at six-month lead period using TS fuzzy model

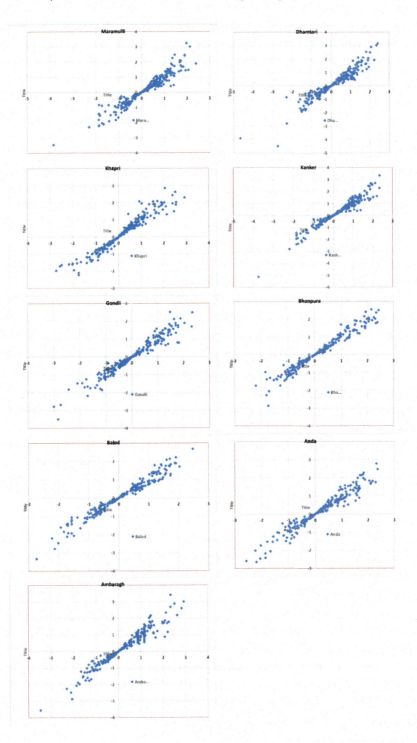

FIGURE 12.9 Observed and forecasted SPI 3 at six-month lead period using TS fuzzy model.

developed fuzzy models show that the value of R^2 ranges from 0.7823 to 0.8109, RMSE from 0.2527 to 0.2742, and NS efficiency from 0.7597 to 0.7829 during calibration. During the ANN models validation the value of R^2 ranges from 0.7701 to 0.8078, RMSE from 0.2692 to 0.2821, and NS efficiency from 0.7511 to 0.7747. The R^2 values ranging between 0.7823 to 0.8109 during calibration and 0.7701 to 0.8078 during validation show a good correlation between observed and predicted results at six-months lead time.

12.7 DISCUSSION

In this study, the proposed forecast models for SPI 3 are presented for forecast lead times of six months. A SPI 3 forecast of six-months lead time signifies a warning period of six months for the drought indicator SPI 3. Table 12.2 shows the input structure of the selected soft computing models at different gauging stations applied in this study for six-months forecast lead time. As shown in Table 12.2, the soft computing models are developed recursively to obtain the desired six-months lead time forecast. Performance of both ANN and fuzzy models were evaluated using different performance measures and are presented in Tables 12.5 and 12.6. The performance of the developed models have been verified using the persistence index.

When the persistence index of a model lies between 0 and 1 its performance is considered to be better than a model showing a persistence index of 0 or below 0. Here, both ANN and fuzzy models have a persistence index greater than 0. ANN models had a PERS ranging from 0.42 to 0.48, and fuzzy models had PERS ranging from 0.46 to 0.54. The correlation coefficient or coefficient of determination (R^2) describes the correlation between the observed and forecasted SPI 3. Both ANN and fuzzy models show consistent results in forecasting of SPI 3 as indicated by the R^2 varying from 0.7719 to 0.7954 for calibration data sets and from 0.7637 to 0.7851 for validation data sets of ANN models, and from 0.7823 to 0.8109 for calibration data sets and 0.7701 to 0.8078 for validation data sets of fuzzy models (Tables 12.5 and 12.6).

As determined from Table 12.6, slightly better correlation is observed with the fuzzy model in comparison to the ANN model for all the rain gauge stations, i.e., Ambagarh (0.7732 to 0.7892 for calibration data and 0.7637 to 0.7814 for validation data), Anda (0.7819 to 0.7934 for calibration data and 0.7724 to 0.7882 for validation data), Balod (0.7914 to 0.8102 for calibration data and 0.7851 to 0.8078 for validation data), Bhanpura (0.7791 to 0.7823 for calibration data and 0.7677 to 0.7701 for validation data), Gondli (0.7923 to 0.8009 for calibration data and 0.7806 to 0.7983 for validation data), Kanker (0.779 to 0.7829 for calibration data and 0.7632 to 0.7767 for validation data), Khapri (0.7802 to 0.7978 for calibration data and 0.7721 to 0.7904 for validation data), Dhamtari (0.7778 to 0.7885 for calibration data and 0.7662 to 0.7794 for validation data), Maramsilli (0.7954 to 0.8109 for calibration data and 0.7819 to 0.8023 for validation data). The root mean square error is the quantitative measure of the model error, and it was found good for all ANN and fuzzy models as is verified by its low values (Tables 12.5 and 12.6). All the fuzzy models show the low value of RMSE in comparison to ANNs. Scatterplots between observed and forecasted SPI 3 data are shown in Figures 12.7 and 12.9 for the Ambagarh, Anda, Balod, Bhanpura, Gondli, Kanker, Khapri, Dhamtari, and Maramsilli sites. These scatterplots indicate that the fuzzy logic and ANN model can forecast SPI 3 more accurately.

This study has shown that the soft computing models ANNs and fuzzy logic can be used as effective means of forecasting drought in the upper Mahanadi basin. This is likely due to the fact that ANNs and fuzzy logic are effective in modeling nonlinear components of time series data. The input structure of the models is same for all the sites and therefore it makes the models convenient for operational purposes.

The results for all the sites from both ANN and fuzzy logic models generally show that SPI 3 forecasts were more accurate when using TS fuzzy models.

The forecast accuracy of both ANN and fuzzy models does not differ significantly at different sites. In general, the performances of ANN and fuzzy models were comparable. There have been

TABLE 12.6

TS Fuzzy Models for Six-Month Forecasts of SPI 3 in Upper Mahanadi Basin

Basin Station	Number of Rules/ Gaussian Membership Function	Calibration				Validation		
		R^2	RMSE	NS Efficiency	R^2	RMSE	NS Efficiency	
Ambagarh	5	0.7892	0.2619	0.7620	0.7814	0.2692	0.7511	
Anda	5	0.7934	0.2742	0.7661	0.7882	0.2821	0.7612	
Balod	5	0.8102	0.2607	0.7789	0.8078	0.2682	0.7703	
Bhanpura	5	0.7823	0.2623	0.7602	0.7701	0.2706	0.7543	
Gondli	5	0.8009	0.2609	0.7829	0.7983	0.2792	0.7741	
Kanker	5	0.7829	0.2538	0.7621	0.7767	0.2621	0.7548	
Khapri	5	0.7978	0.2598	0.7743	0.7904	0.2676	0.7619	
Dhamtari	5	0.7885	0.2527	0.7597	0.7794	0.2612	0.7512	
Maramsilli	5	0.8109	0.2633	0.7819	0.8023	0.2721	0.7747	

Note: Column 3 is the TS fuzzy architecture detailing the number of nodes in the input, hidden, and output layers, respectively.

studies that have shown that the performance of fuzzy logic–based models are in general slightly better than the ANN models (Lohani et al., 2007a, 2007b). It is inferred from Figures 12.6 and 12.8 that there are not ample time step errors in the six-month lead period forecasting of SPI 3.

12.8 CONCLUSIONS

This study investigated the ability of soft computing models to forecast drought. This study also evaluate the performance of the ANNs and fuzzy logic–based methods for drought forecasting. Overall, fuzzy logic–based models were found to provide better results than the ANN model types used to forecast six-month lead time values of SPI 3 for the Ambagarh, Anda, Balod, Bhanpura, Gondli, Kanker, Khapri, Dhamtari, and Maramsilli sites in the upper Mahanadi basin. Fuzzy models showed high values of coefficient of correlation between observed and forecast SPI 3, as well as lower RMSE and MAE (mean absolute error) values compared to ANN models. Further studies need to be done to determine which soft computing model is appropriate for forecasting the SPI in basins having dissimilar climates and physical characteristics. Future studies should also attempt to couple data-driven drought forecasting models with uncertainty analysis, such as bootstrapping or boosting ensembles.

REFERENCES

Adamowski, J. 2008. Development of a short-term river flood forecasting method for snowmelt driven floods based on wavelet and cross-wavelet analysis. Journal of Hydrology 353, 247–266.

Appa Rao, G. 1981. Atmospheric energetics over India during drought and normal monsoon. *Mausam* 32, 67–78.

Appa Rao, G. 1987. *Drought Climatology.* Jal Vigyan Sameeksa, NIH, Roorkee, India, Vol. 1, pp. 43–53.

Bacanli, U.G., Firat, M., Dikbas, F. 2008. Adaptive neuro-fuzzy inference system for drought forecasting. *Stochastic Environmental Research and Risk Assessment* 23(8), 1143–1154.

Barros, A., Bowden, G. 2008. Toward long-lead operational forecasts of drought: An experimental study in the Murray-Darling River Basin. Journal of Hydrology 357(3–4), 349–367.

Beven, K. 2006. A manifesto for the equifinality thesis. Journal of Hydrology 320, 18–36.

Bezdek, J.C. 1973. *Fuzzy Mathematics in Pattern Classification.* PhD thesis, Applied Math. Center Cornell University, Ithaca.

Bordi, I., Sutera, A. 2007. Drought monitoring and forecasting at large scale. In: Rossi, G. et al. (Eds.), *Methods and Tools for Drought Analysis and Management.* Springer.

Chiu, S. 1994. Fuzzy model identification based on cluster estimation. *Journal on Intelligent Fuzzy Systems,* 2, 267–278.

Cutore, P., Di Mauro, G., Cancelliere, A. 2009. Forecasting palmer index using neural networks and climatic indexes. Journal of Hydrologic Engineering 14, 588–595.

Dracup, J.A., Lee, K.S., Paulson, E.N. 1980. On the definition of droughts. *Water Resources Research,* 16(2), 297–302.

Edwards, D.C., McKee, T.B. 1997. *Characteristic of 20th Century Drought in the United States at Multiple Timescales, Climatology Report.* Colorado State University, Fort Collins.

Gustafson, D.E., Kessel, W.C. 1979. Fuzzy clustering with a fuzzy covariance matrix. *Proceedings of IEEE Transactions on Fuzzy Systems* 1(3), 195–204.

Han, P., Wang, P.X., Zhang, S.Y., Zhu, D.H. 2010. Drought forecasting based on the remote sensing data using ARIMA model. *Mathematical Computer Modelling* 51(11), 1398–1403.

Hayes, M.J., Svoboda, M.D., Wilhite, D.A., Vanyarkho, O.V. 1999. Monitoring the 1996 drought using the standardized precipitation index. *Bulletin of American Meteorological Society,* 80, 429–438.

India Meteorological Department. 2002. *Southwest Monsoon Endof-Season Report,* India.

Jang, J.-S.R., Sun, C.-T., Mizutani, E. 2002. *Neuro-Fuzzy and Soft Computing.* Prentice Hall of India Private Limited, New Delhi.

Karamouz, M., Rasouli, K., Nazil, S. 2009. Development of a hybrid index for drought prediction: Case study. Journal of Hydrologic Engineering 14, 617–627.

Lohani, A.K., Goel, N.K., Bhatia, K.K.S. 2005a. Real time flood forecasting using fuzzy logic. *Hydrological Perspectives for Sustainable Development,* Volume I, Eds. M. Perumal, Allied Publishers Pvt. Ltd., New Delhi, 168–176.

Lohani, A.K., Goel, N.K., Bhatia, K.K.S. 2005b. Development of fuzzy logic based real time flood forecasting system for river Narmada in Central India. In International Conference on Innovation Advances and Implementation of Flood Forecasting Technology, ACTIF/Floodman/Flood Relief, October, 2005, Tromso, Norway, www.Actif.cc.net/conference2005/proceedings.

Lohani, A.K., Goel, N.K., Bhatia, K.K.S. 2006. Takagi-Sugeno fuzzy inference system for modeling stage-discharge relationship. *Journal of Hydrology*, 331, 146–160.

Lohani, A.K., Goel, N.K., Bhatia, K.K.S. 2007a. Reply to comments provided by Z. Sen on "Takagi–Sugeno fuzzy system for modeling stage-discharge relationship" by A.K. Lohani, N.K. Goel and K.K.S. Bhatia. *Journal of Hydrology*, 337(1–2), 244–247.

Lohani, A.K. Goel, N.K., Bhatia, K.K.S. 2007b. Deriving stage–discharge–sediment concentration relationships using fuzzy logic. *Hydrological Sciences–Journal*, 52(4), 793–807.

Lohani, A.K., Goel, N.K., Bhatia, K.K.S. 2014. Improving real time flood forecasting using fuzzy inference system. *Journal of Hydrology*, 509, 25–41.

Lohani, A.K., Goel, N.K., Bhatia, K.K.S. 2011. Comparative study of neural network, fuzzy logic and linear transfer function techniques in daily rainfall-runoff modeling under different input domains, *Hydrological Processes*, 25, 175–193.

Lohani, R.K., Singh, R.D. 2012. Hydrological time series modeling: A comparison between adaptive neuro fuzzy, neural network and auto regressive techniques. *Journal of Hydrology*, 442–443(6), 23–35.

Marj Fatehi, A., Meijerink, A.M.J. 2011. Agricultural drought forecasting using satellite images, climate indices and artificial neural network. *Int J Remote Sens* 32(24):9707–9719.

McKee, T.B., Doesken, N.J., Kliest, J. 1993. The relationship of drought frequency and duration to time scales. Proceedings of 8th Conference on Applied Climatology, American Meteorological Society, Boston, pp. 179–184.

Mishra, A.K., Desai, V.R., Singh, V.P. 2007. Drought forecasting using a hybrid stochastic and neural network model, *Journal of Hydrologic Engineering*, 12(6), 626–638.

Mishra, A.K., Desai, V.R. 2005. Drought forecasting using stochastic models. *Stochastic Environmental Research and Risk Assessment* 19 (5), 326–339.

Mishra, A.K., Desai, V.R. 2006. Drought forecasting using feed-forward recursive neural network. Ecological Modelling 198(1–2), 127–138.

Morid, S., Smakhtin, V., Bagherzadeh, K. 2007. Drought forecasting using artificial neural networks and time series of drought indices. *Journal of Climatology*, 27, 2103–2111.

Nash, J., Sutcliffe, J. 1970. River flow forecasting through conceptual models part I: A discussion of principles. *Journal of Hydrology*, 10, 282–290.

Palmer, W.C. 1965. Meteorologic Drought, US Weather Bureau, *Research Paper* No. 45.

Paulo, A.A., Ferreira, E., Coelho, C., Pereira, L.S. 2005. Drought class transition analysis through Markov and Loglinear Models: An approach to early warning. *Agriculture Water Management*, 77, 59–81.

Pongracza, R., Bogardib, I., Duckstein, L. 1999. Application of fuzzy rule-based modeling technique to regional drought. *Journal of Hydrology*, 224, 100–114.

Rouault, M., Richard, Y. 2003, Intensity and spatial extension of drought in South Africa at different time scales. *Water SA*, 29(4), 489–500.

Singh, R.S. 2000. *Drought Assessment in Arid Western Rajasthan*. Central Arid Zone Research Institute, Jodhpur, India.

Steinemann, A. 2003. Drought indicators and triggers: A stochastic approach to evaluation. *Journal of the American Water Resources Association*, 39, 1217–1233.

Subramanyam, V.P. 1964. Climatic water balance of Indian arid zones. Proceedings of Symposium on Indian Arid Zone, Jodhpur, pp. 405–411.

Takagi, T., Sugeno, M. 1985. Fuzzy identification of systems and its application to modeling and control. *IEEE Transactions on Systems, Man and Cybernetics*, 15(1), 116–132.

Vicente-Serrano, S. M., Beguería, S., López-Moreno, J. I. 2010. A multi-scalar drought index sensitive to global warming: The standardized precipitation evapotranspiration index – SPEI. *Journal of Climate* 23, 1696–1718.

Yager, R.R. Filev, D.P. 1994. Generation of fuzzy rules by mountain clustering. *Journal of Intelligent and Fuzzy Systems*, 2, 209–219.

Zadeh, L.A. 1965. Fuzzy sets. *Information and Control*, 8, 338–353.

13 A Copula-Based Joint Deficit Index for the Analysis of Droughts in New Zealand

Tommaso Caloiero and Rasoul Mirabbasi

CONTENTS

13.1 INTRODUCTION

Climate change is arguably one of the most important global challenges the world has been facing during the 21st century, and due to its negative effects on global water supplies, agricultural production, and human health, it poses further issues (Feulner 2017). For example, the increased frequency and impacts of drought are usually considered key characteristics of climate change (Field 2012), and drought is expected to become more frequent in the 21st century in some seasons and areas (IPCC 2013). In fact, drought-prone areas have increased globally, going from 16.2% in the period 1902–1949 to 41.1% in the period 1950–2008 (Wang et al. 2014), and in the next 90 years, the majority of the regions of the world are projected to suffer more severe and widespread droughts (Dai 2013). Drought is one of the extreme climatic phenomena with the greatest impact on the population due to its effects on the availability of water and therefore on economic activities such as agriculture (Lesk et al. 2016) and tourism. Moreover, drought can severely affect human health (Stanke et al. 2013) and ecosystems (Alary et al. 2014). For these reasons, several researchers in recent years have analyzed drought in several parts of the world in order to better understand, identify, document, and monitor such a complex phenomenon (e.g., Hannaford et al. 2011; Fang et al. 2013; Hua et al. 2013; Sirangelo et al. 2015; Sirangelo et al. 2017; Caloiero et al. 2018a, Caloiero and Veltri 2019; Gaitan et al. 2020; Guo et al. 2021). When a drought event occurs, some deficits in several hydrologic variables (precipitation, streamflow, soil moisture, snowpack, groundwater levels, and reservoir storage) are observed; thus, a single definition of drought is not possible, and drought is often classified into meteorological, agricultural, hydrological, and socioeconomic drought according to the types of deficits (Wilhite et al. 2000). For example, meteorological droughts are based on deficits in precipitation, agricultural droughts on deficits in soil moisture, and hydrologic droughts on streamflow deficits (Dracup et al. 1980). In the analysis of a drought, three main characteristics must be considered: duration, severity, and frequency (Tsakiris et al. 2007). Consequently, drought indices are used to determine the status of drought (Buttafuoco et al. 2018). In fact, drought indices

DOI: 10.1201/9781003276548-13

allow summarizing different data (rainfall, snowpack, streamflow, and other water supply indicators) into a comprehensive picture of drought occurrence (Heim 2002). In order to quantify drought severity, several indices derived from hydrometeorological variables have been proposed in the literature. These include the Palmer Drought Severity Index (Palmer 1965), the Crop Moisture Index (Palmer 1968), the Surface Water Supply Index (Shafer and Dezman 1982), the Standardized Precipitation Index (McKee et al. 1993), the Reclamation Drought Index (Weghorst 1996), the Effective Drought Index (Byun and Wilhite 1999), the Streamflow Drought Index (Nalbantis and Tsakiris 2009), the Standardized Precipitation Evapotranspiration Index (Vicente-Serrano et al. 2010), and the Agricultural Reference Index for Drought (Woli et al. 2012). Among these indices, the Standardized Precipitation Index (SPI) has found widespread application worldwide because it can be evaluated for different timescales and allows the analysis of different drought categories (Caloiero et al. 2018b). Moreover, evaluation of the SPI requires only precipitation data, making it easier to calculate than more complex indices, and it allows the comparison of drought conditions in different regions and for different aggregations (Caloiero et al. 2016). For these reasons, the World Meteorological Organization (WMO 1985) recommends that the SPI be used as the primary meteorological drought index.

Although the SPI is regarded as one of the most robust and effective drought indices, sometimes drought monitoring based on a single variable or index could be inadequate because drought-related impacts can be complex. To overcome this issue, a new index named the Joint Deficit Index (JDI), consisting in joint distributions of multiple SPIs using copula functions, has been proposed (Kao and Govindaraju 2010). This index allows for a month-by-month drought assessment and to determine the amount of precipitation required for reaching normal conditions. The JDI has been largely applied, especially in Iran. For example, by means of the JDI, Ramezani et al. (2019) investigated meteorological drought and its trend (using the modified nonparametric Mann–Kendall test) in Iran and its eastern neighboring countries in the period 1971–2014. Bazrafshan et al. (2020) compared the efficiency of the SPI and the JDI in drought monitoring by performing trend analysis and a spatial analysis of drought characteristics in the arid and semi-arid regions of Iran. Mirabbasi et al. (2013) evaluated drought conditions in the northwest of Iran using the JDI approach and copulas, and developed a method for predicting the wetness conditions based on exceedance probability thresholds determined from climatological data. Finally, a comparison between two multi-index drought indices, i.e., the JDI and the MSPI (Multivariate Standardized Precipitation Index), from different statistical aspects in diverse climates of Iran has been performed by Bazrafshan et al. (2015).

Agriculture is obviously one of the main sectors affected when drought hits. In New Zealand agriculture is one of the largest sectors of the tradable economy; therefore, drought episodes can have significant ecological, social, and economic impacts (Palmer et al. 2015). In fact, although rainfall in New Zealand is fairly abundant, isolated drought events at the regional level are not unusual in this country. The widespread drought event that affected New Zealand from late 2007 to the end of autumn 2008 caused damages of about 2.8 billion New Zealand dollars (MAF 2009). The 2013 drought in New Zealand was estimated to have caused the GDP (gross domestic product) to fall by 0.6% (Kamber et al. 2013). In this study, drought events in New Zealand have been evaluated using the JDI. This study aims to identify the most severe drought events that have affected New Zealand and to analyze the evolution of the drought episodes through the identification of the JDI trend in the period 1951–2012.

13.2 STUDY AREA AND DATA

New Zealand is a country located between 34°10′50″ and 46°55′50″ S in the Southern Hemisphere, in the southwestern Pacific Ocean, 2500 km east of the Australian continent. It consists of two main islands and about 600 smaller-sized islands. The two main islands are the North Island and the South Island, separated by the Cook Strait (Figure 13.1). The South Island is the largest land mass

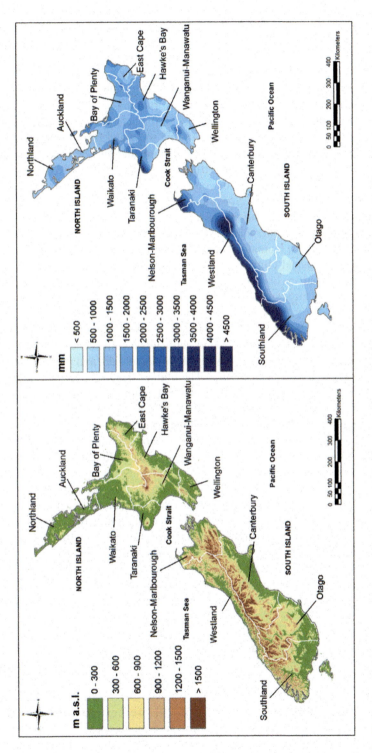

FIGURE 13.1 Digital elevation model of New Zealand (left) and spatial distribution of the mean annual precipitation (right).

of New Zealand and is crossed along its length by the Southern Alps, which, reaching an altitude of 3724 m, divide the island between the Canterbury Plains on the east side and the rough coastlines on the west side. The North Island is less mountainous than the South Island; however, a series of narrow mountain ranges that rises up to 1700 m characterize this part of the country (Figure 13.1).

Three physical factors essentially influence New Zealand's climatic conditions (Oliver 2005). First, New Zealand is situated in the Pacific Ocean, in the interaction zone between the air masses coming from the tropical zones and the oceanic currents from Antarctica. As a result, the first is responsible for the humid weather affecting New Zealand, with temperatures ranging between 18°C and 21°C, while the latter lead to colder temperatures, ranging between 7°C and 13°C, and high rainfall amounts. Therefore, because of the presence of the ocean, New Zealand's climate is characterized by cool summers and mild winters with some snow and frost due to the cold air coming from Antarctica. The second-factor affecting air circulation, and thus New Zealand's climate, is the mountains crossing both the major islands (Griffiths 2011; Jiang et al. 2013). In fact, the Southern Alps crossing the South Island south–north constitute a barrier against airflow coming in from the south–southwest (Tomlinson 1976). Specifically, they block the air that would flow upstream, producing lee troughs on the downwind side as a consequence of their dynamic uplift and vertical mixing on the upwind edge. Considerable differences in rainfall amounts can occur over the short distance between the windward mountain slopes and the rain shadows a few kilometers eastward. Finally, New Zealand's seasonal climate is influenced by the neighboring Australian continental landmass whose eastern land/sea boundary presents a cyclogenesis region, which moves toward the Tasman Sea leading to very large spatial differences in precipitation over short distances (Trenberth 1991; Sinclair 1994).

The National Institute of Water and Atmospheric Research of New Zealand (NIWA 2013) carried out an accurate investigation of New Zealand's climate on both islands, whose results are available for consultation. According to NIWA, warm, humid summers and mild winters characterize the North Island, making it a subtropical area. The highest degree of climatic instability and the rainiest period of the year occurs in winter. Summer and autumn present tropical storms with high winds and severe precipitation. The central region of the North Island typically has stable, dry, and warm summers, but unstable cool winters. The southwestern part of the North Island generally presents more stable weather in summer and at the beginning of autumn, with usually warm summers and not too cold winters. A dry, sunny climate and warm dry settled summer weather, with heavy rainfall coming from the east or the southeast characterize the eastern North Island (NIWA 2013). The northern part of the South Island, instead, is the sunniest region in New Zealand, presenting warm dry and stable weather in summer. The western South Island, although usually characterized by extreme, considerably higher than average, yearly precipitation, may experience occasional dry spells toward the end of the summer and in the winter months, with intense precipitation from the northwest. The Southern Alps heavily influence the climate of the eastern and inland areas of the South Island. In particular, the summer season in this area presents long periods of dryness and low mean annual rainfall. Overall, unsettled weather from the south and southwest across the sea affects the majority of the southern New Zealand area, which also experiences cool coastal breezes (NIWA 2013).

In order to perform a reliable drought analysis, a high-quality database is required. Within this purview, for this study, the New Zealand National Climate database of NIWA has been selected. In fact, this database presents high-quality data with near-complete records and, for this reason, it has been largely used in numerous studies on New Zealand's climate (Salinger and Mullan 1999; Griffiths et al. 2003; Dravitzki and McGregor 2011; Caloiero 2017, 2018, 122020). For the following analyses, 33 monthly rainfall data for the period 1951–2012 were extracted and used, after performing record error checks and metadata analyses for inhomogeneities detection (Figure 13.2 and Table 13.1). Moreover, because of the presence of some missing data in the series, a gap-filling procedure based on the Isohyetal method was applied.

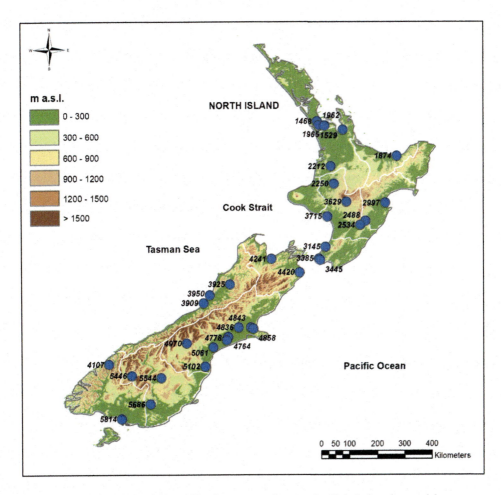

FIGURE 13.2 Location of the selected 33 rain gauge stations on the digital elevation model.

13.3 METHODOLOGY

13.3.1 THE JOINT DEFICIT INDEX (JDI)

In the present study, the JDI was used to assess meteorological droughts. To calculate the JDI, multivariate distribution of precipitation margins (univariate cumulative probability distribution functions) $\{u_1, u_2, \ldots, u_{12}\}$, with different timescales from 1 to 12 months for each month of the year (January, February, etc.) through copula functions, is generated. Then, this cumulative probability is converted into a standard normal variable with mean 0 and variance 1. Precipitation margins with different timescales from 1 to 12 months $\{u_1, u_2, \ldots, u_{12}\}$ are created based on the concept of a modified Standardized Precipitation Index (SPI^{mod}). To calculate the modified SPI, Kao and Govindaraju (2010) suggested that the w-month precipitation X_w be summed with the ending month and represented by X_w^{month}, where the *month* transcript represents one of the months of January, February, ..., December. In other words, the w-months precipitation time series, $X_w(t)$, is divided into 12 subseries:

$$X_w^{month}(g) = X_w\big(12(g-1)+m\big) = X_w(t) \tag{13.1}$$

TABLE 13.1

ID, Name, and Coordinates of the 33 Selected Rain Gauges

ID	Station Name	Latitude	Longitude
1468	Auckland, Owairaka	−36.893	174.726
1529	Thames 2	−37.159	175.551
1874	Opotiki	−38.006	177.287
1962	Auckland Aero	−37.008	174.789
1965	Auckland, Ardmore	−37.034	174.964
2212	Te Kuiti High School	−38.334	175.153
2250	Taumarunui	−38.888	175.260
2488	Kopua	−40.079	176.274
2534	Dannevirke	−40.210	176.109
2997	Napier Nelson Pk	−39.501	176.911
3145	Paraparaumu Aero	−40.907	174.984
3385	Wellington, Kelburn	−41.286	174.767
3445	Wellington Aero	−41.322	174.804
3629	Waiouru Treatment Plant	−39.464	175.663
3715	Wanganui, Spriggens Park Ews	−39.939	175.045
3909	Hokitika Aero	−42.716	170.988
3925	Reefton Ews	−42.117	171.860
3950	Greymouth Aero	−42.462	171.191
4107	Milford Sound	−44.674	167.922
4241	Nelson Aero	−41.299	173.226
4420	Grassmere Salt Works	−41.729	174.145
4764	Winchmore Ews	−43.793	171.795
4778	Ashburton Council	−43.897	171.747
4836	Darfield	−43.493	172.137
4843	Christchurch Aero	−43.493	172.537
4858	Christchurch Gardens	−43.531	172.619
4970	Lake Tekapo, Air Safaris	−44.002	170.441
5061	Orari Estate	−44.127	171.308
5102	Waimate	−44.741	171.036
5446	Queenstown	−45.037	168.663
5544	Ophir 2	−45.106	169.611
5686	Tapanui	−45.944	169.257
5814	Invercargill Aero	−46.417	168.331

where g is the year index; m is the month index and is equal to 1 (January), 2 (February), ..., 12 (December); and t is the time index, $t = 12(g-1)+m$.

For example, X_1^{Jan} represents January precipitation, and X_8^{Oct} represents the eight-month precipitation accumulation from March to October. In this way, samples in each X_w^{month} group are collected annually. It is clear that as long as $w \le 12$, the samples will have no overlap. In other words, the degree of autocorrelation between the data will be greatly reduced. On the other hand, samples within the same X_w^{month} group are subject to similar seasonal effects and, therefore, seasonal variations are considered in an appropriate manner. By fitting the statistical distribution to each group separately (i.e., generating $u_w^{Jan} = F_{X_w^{Jan}}\left(x_w^{Jan}\right)$, $u_w^{Feb} = F_{X_w^{Feb}}\left(x_w^{Feb}\right)$, ..., $u_w^{Dec} = F_{X_w^{Dec}}\left(x_w^{Dec}\right)$), the

TABLE 13.2

Classification of Wetness Condition Based on the SPI Values and Corresponding Event Probability Limits

Wetness Condition	SPI Intervals	Probability Limit
Extremely wet	SPI ≥2	≥97 %
Severely wet	2 > SPI ≥ 1.5	93.3–97.7 %
Moderately wet	1.5 > SP I≥ 1	84.1–93.3 %
Normal	1 > SPI > –1	15.9–84.1 %
Moderately dry	–1 ≥ SPI > –1.5	6.7–15.9 %
Severely dry	–1.5 ≥ SP I> –2	2.3–6.7 %
Extremely dry	SPI ≤–2	≤2.3 %

Source: McKee et al. (1993).

cumulative distribution function for each group is calculated and converted to the normal cumulative distribution function or $\varphi^{-1}\left(u_w^{month}\right)$ to calculate the SPI^{mod}.

$$SPI_w^{mod} = \varphi^{-1}\left(u_w^{month}\right) = \varphi^{-1}\left(F_{X_w^{month}}\left(x_w^{month}\right)\right) \tag{13.2}$$

In other words, instead of showing the cumulative probability, the value of the SPI^{mod} is presented with a standard normal variable (mean 0 and standard deviation of 1). Drought severity classification based on SPI^{mod} values is presented in Table 13.2, proposed by McKee et al. (1993) for the conventional SPI. The results of research by many scientists show that the most suitable possible distribution function for fitting rainfall data is the gamma distribution function (Thom 1966; Mirabbasi et al. 2012). To determine the overall state of the drought, modified SPIs at different timescales should be considered together. One of the comprehensive statistical processes is to create a joint distribution of multiple modified SPIs through the copulas. Kao and Govindaraju (2010) developed the JDI using the copula distribution function to provide a scientific (probability-based) description of the overall drought condition.

In order to create the JDI, Gaussian and empirical copulas can be used to generate the dependency structure of the $\{u_1, u_2, \ldots, u_{12}\}$ set. However, due to the mathematical complexity of 12-dimensional Gaussian copulas, Kao and Govindaraju (2010) used empirical copulas for this purpose. Using empirical copulas instead of theoretical copula functions is reliable when the length of data used is large. The choice of $\{u_1, u_2, \ldots, u_{12}\}$ in forming high-dimensional copulas increases the complexity of the dependency model. However, because the duration of droughts shows wide time variations, droughts can only be well described by considering different periods (from 1 to 12 months). The annual cycle naturally takes into account seasonal effects (Mirabbasi et al. 2013). In addition, this structure allows for a month-by-month assessment of future conditions. Kao and Govindaraju (2010) did not take into account margins longer than 12 months ($j > 12$) because they found that the samples used for $j > 12$ overlapped and led to biased results.

The copula functions have been proposed by Sklar (1959). A copula is actually the cumulative probability $P\left[U_1 \le u_1, \ldots U_{12} \le u_{12}\right] = t$ of sample margins $\{u_1, u_2, \ldots, u_{12}\}$. As each margin indicates the moisture deficit conditions for each given time period, the deficiency conditions are specified with t. Clearly, a smaller cumulative probability of t indicates drier conditions (drought at different timescales), and a larger value of t indicates wetter conditions. Assuming that t reflects the severity of the joint drought, the probability of occurrence of events smaller than or equal to t

(i.e., events drier than a certain threshold) will be very useful. For this purpose, the definition of the copula distribution function (K_C) is used, because the copula distribution function is in fact the same cumulative probability ($K_C(t) = P\left[C_{U_1,U_2,...,U_{12}}(u_1,u_2,...,u_{12}) \leq t\right]$). The special advantage of using K_C is that it allows the probabilistic criterion of joint deficiency conditions to be calculated, which can be interpreted as an indicator of joint drought. In fact, K_C is the same as the joint cumulative distribution function of $C_{U_1,U_2,...,U_{12}}$ (Nelsen 2006). Therefore, the JDI is defined as follows (Kao and Govindaraju 2010):

$$JDI = \varphi^{-1}\left(K_C\right) \tag{13.3}$$

Similar to the SPI^{mod}, the positive values of the JDI ($K_C > 0.5$) indicate the overall wet conditions, negative values of JDI ($K_C < 0.5$) indicate the overall dry condition, and JDI = 0 ($K_C = 0.5$) shows normal conditions. Since the JDI is on an inverse normal scale (similar to SPI and SPI^{mod}), the classification of droughts based on SPI can also be used for JDI (Table 13.2).

The most important feature of the JDI is the evaluation of overall deficiency conditions based on the dependency structure of deficiency indices with different timescales. The JDI calculation steps are shown in Figure 13.3.

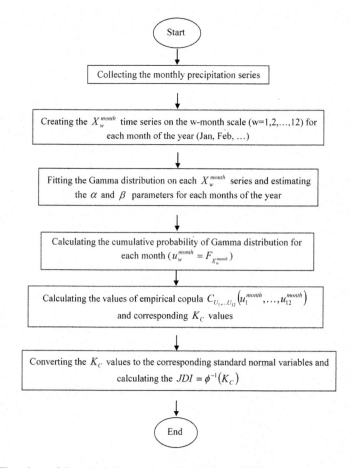

FIGURE 13.3 Flowchart of the calculating the Joint Deficit Index (JDI).

13.3.2 TREND ANALYSIS

In order to analyze possible trends and to assess the trend statistical significance in JDI series, the Mann–Kendall (MK) test (Mann 1945; Kendall 1962) has been used.

In order to evaluate the significance of the trend, first the statistic S must be estimated as

$$S = \sum_{i=1}^{n-1} \sum_{j=i+1}^{n} \text{sgn}\left(x_j - x_i\right) ; \quad \text{with} \quad \text{sgn}\left(x_j - x_i\right) = \begin{cases} 1 & \text{if} \quad \left(x_j - x_i\right) > 0 \\ 0 & \text{if} \quad \left(x_j - x_i\right) = 0 \\ -1 & \text{if} \quad \left(x_j - x_i\right) < 0 \end{cases} \quad (13.4)$$

in which x_j and x_i are the variable values in the years j and i (with $j > i$), respectively, and n is the dimension of the series.

Given independent and randomly ordered values, for $n > 10$, the statistic S is distributed following a normal distribution with 0 mean and variance:

$$Var\left(S\right) = \left[n\ (n-1)\ (2\ n+5) - \sum_{i=1}^{m} t_i\left(t_i - 1\right)\left(2\ t_i + 5\right) \right] \Big/ 18 \quad (13.5)$$

where m is the number of tied groups and ti is the number of observations in the i-th group.

Finally, the standardized statistic Z_{MK}, can be computed as

$$Z_{MK} = \begin{cases} \dfrac{S-1}{\sqrt{Var\left(S\right)}} & \text{for } S > 0 \\ 0 & \text{for } S = 0 \\ \dfrac{S+1}{\sqrt{Var\left(S\right)}} & \text{for } S < 0 \end{cases} \quad (13.6)$$

By applying a two-tailed test, for a specified significance level α, the significance of the trend can be evaluated. In particular, in this work, the rainfall series have been examined for a significance level equal to 95%.

13.4   RESULTS

In order to characterize the JDI values evaluated for the 33 series, according to the threshold defined in Table 13.2, wet and drought classes have been extracted for all stations and displayed in Table 13.3. In particular, the number of occurrences in four classes, i.e., wet conditions (WC), moderate drought (MD), severe drought (SD), and extreme drought (ED), have been estimated. As a result, in only seven stations were extreme drought classes identified. Four of these stations are located in the North Island and three on the South Island. Moreover, among the seven stations, only station ID 5544 is situated in an inland area, while the others are near the coasts.

In order to analyze the intra-annual drought distribution, the results of the mean JDI values on a monthly timescale have been presented in Figure 13.4 as a boxplot. Results evidenced predominant wet conditions in all the months. In fact, only in June, November, and December did the mean JDI values show negative values, even though very close to 0. Conversely, in February, July, August, October, and, especially, September, mean JDI values higher than 0 were detected. Moreover, the mean JDI values never reach drought conditions, while in February and March severe wet values have been identified. It is important to note that for 32 out of 33 stations, the occurrence of wet

TABLE 13.3

Drought Occurrences Based on JDI

ID	MD	SD	ED	WC	ID	MD	SD	ED	WC
1468	67	31	23	118	3950	71	0	0	117
1529	65	29	13	117	4107	65	34	0	118
1874	75	30	5	117	4241	69	30	0	118
1962	74	21	0	118	4420	73	21	0	118
1965	58	38	0	118	4764	59	53	0	118
2212	58	20	0	118	4778	69	29	0	118
2250	62	18	0	118	4836	68	19	0	118
2488	70	32	0	118	4843	55	49	0	118
2534	51	29	0	117	4858	62	18	28	118
2997	66	35	0	118	4970	69	33	0	117
3145	63	23	0	118	5061	65	32	0	118
3385	70	33	0	117	5102	68	32	7	116
3445	74	30	13	118	5446	62	14	0	116
3629	55	19	0	117	5544	65	20	18	117
3715	66	26	0	117	5686	68	23	0	118
3909	81	26	0	118	5814	72	19	0	117
3925	72	28	0	118					

MD, moderate drought; SD, severe drought; ED, extreme drought; WC, wet conditions.

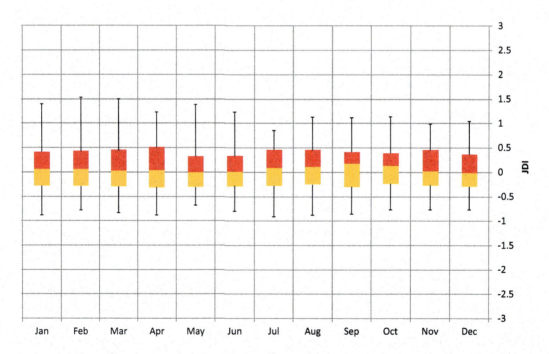

FIGURE 13.4 Boxplot on the mean monthly JDI values. The bottom and top of the box are the second and third quartiles, the band inside the box is the median, and the ends of the whiskers represent the minimum and maximum of all of the data.

conditions is higher than the one obtained for dry conditions; the only exception is represented by station ID 1468, which presented 121 and 118 months with dry and wet conditions, respectively.

Figure 13.5 shows, for the period 1951–2012, the temporal distribution of the drought conditions in the various stations, thus allowing an immediate detection of the worst dry events. As a result, the second half of the 20th century has been characterized by several dry episodes. The first relevant drought events took place in September 1951 with 10 and 15 rain gauges showing ED and SD, respectively. Another significant occurrence dates to October 1961, when ED and SD conditions were detected in 10 and 15 rain gauges, respectively. Dry conditions were also present in the period October–November 1958, with more than 10 rain gauges affected by ED; between July and December 1969, with 19 stations presenting ED values; and across 1972 and 1973, when for 8 consecutive months 10 rain gauges showed JDI values lower than –1.5.

According to the JDI values, one of the worst drought events of the 21st century has been detected between 1982 and 1983, when SD conditions were identified for 9 consecutive months. Moreover, 11 and 8 stations presented ED values in May 1982 and February 1983, respectively. In April 1988 17 stations showed SD values, while ED conditions were identified in 6 stations. The last relevant drought events of the past century took place in January 1998, with 8 and 17 rain gauges showing ED and SD, respectively.

As regards the current century, 3 main drought events have been identified. The first one occurred in 2001 when more than 10 stations evidenced SD values for 3 months. The second one refers to August 2005, when ED and SD conditions were detected in 8 and 16 rain gauges, respectively. The last one took place between 2007 and 2008 when from November to February SD values affected more than 10 stations, with a maximum of 16 stations in January 2008.

Four years, characterized by the main drought events, have been better analyzed by means of boxplots in which the bottom and top of the box are the second and third quartiles, the band inside the box is the median, and the ends of the whiskers represent the minimum and maximum of all of the data (Figure 13.6).

As regards 1969, a clear negative behavior of the JDI values can be easily identified, with several stations presenting average values below 0 and lower than –1 in four stations. In one station (ID 4858) the maximum negative value evidenced ED conditions (Figure 13.6a). The negative behavior of the JDI values is more marked in 1973. In fact, this year, only two stations showed mean values above 0, even though in all the stations the maximum negative values did not exceed the ED threshold (Figure 13.6b). The year 1983 showed a completely different behavior than the other years. In fact, in 1983 almost an equal number of stations evidenced mean JDI values above or below 0, with the first ones mainly concentrated in the South Island. Thus, the distribution of the JDI values seems to indicate that the drought event that occurred in 1983 affected the North Island (Figure 13.6c). Finally, in 1998, although the majority of the stations showed a higher concentration of negative values, ten stations showed maximum positive values higher than 2, thus evidencing extreme wet conditions (Figure 13.6d).

The results of the trend analysis applied to the monthly JDI values are presented in Figure 13.7 and Figure 13.8. In particular, Figure 13.7 shows the percentages of stations with a positive or negative trend. As a result, a clear negative trend was detected in all the months. For example, in May, more than 30% of the stations showed a negative trend, while no stations evidenced significant positive trends. Generally, the number of stations showing a negative trend is higher in the autumn months. In fact, 27.3% of the series showed negative trends in March and April, while only 6.1% and 3.0% evidenced opposite tendencies, respectively. Also in February, July, and December, about 21% of the stations evidenced a negative trend behavior, with a positive trend maximum percentage equal to 12.1% in July. Only in October was the percentage of stations presenting a positive trend (15.2%) higher than the one evidencing a negative trend (12.1%).

Specifically, in January significant trends were identified only in the South Island with negative values in the Canterbury and in the Otago areas, and positive ones in the Westland and Southland regions. In February, the trend behavior is similar to the one obtained in January for the South

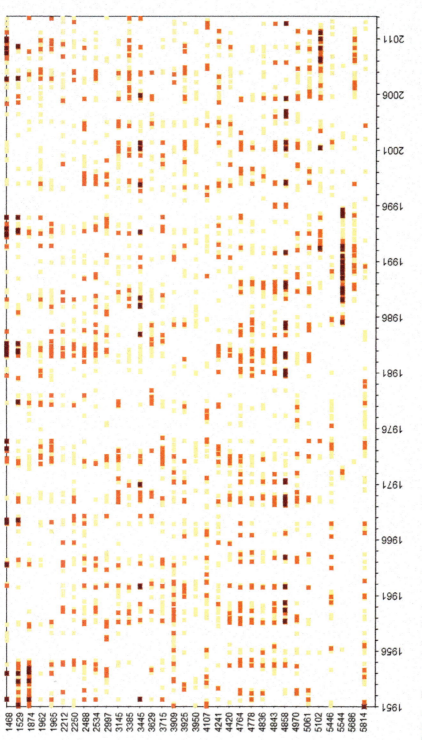

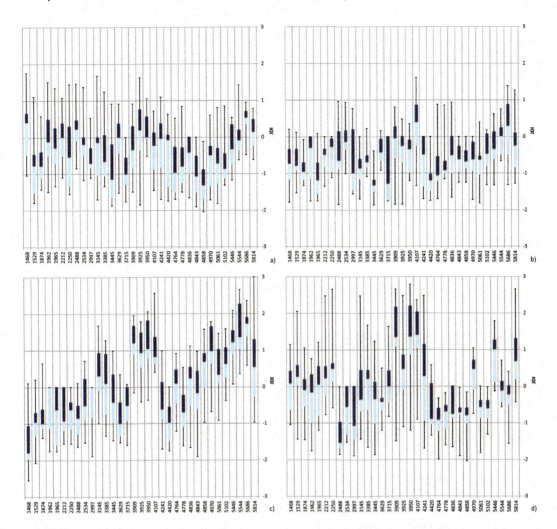

FIGURE 13.6 Boxplot on the JDI values evaluated for (a) 1969, (b) 1973, (c) 1983, and (d) 1998. The bottom and top of the box are the second and third quartiles, the band inside the box is the median, and the ends of the whiskers represent the minimum and maximum of all of the data.

Island, while negative trends have been also detected in the Auckland and Wellington regions of the North Island. In March and April, a similar spatial trend distribution was obtained, with negative trends identified in the Auckland and Wellington regions in the North Island, and in the Otago, Canterbury, and Nelson–Marlborough areas in the South Island. Opposite trend signs have been detected in the Westland region. Differently from April, in March the Southland region showed a positive trend. In May, the negative trends were distributed across both the islands, in particular in the North Island in the Auckland and Wellington regions, and in all the regions of the South Island with the exception of the Southland. In both June and July, a negative trend was detected in the Canterbury region, in the Otago region, and in the eastern side of the Nelson–Marlborough region in the South Island, and in the Auckland and Wellington regions in the North Island. Instead, a positive trend was evaluated in the Bay of Plenty and Westland regions in June, and in Hawke's Bay, in the Southland and Westland regions in July. An equal trend distribution was detected in August and September, with negative trends affecting the Auckland, Wellington, Otago, and Canterbury regions, and an opposite behavior identified in the Wanganui-Manawatu, Southland, and Westland

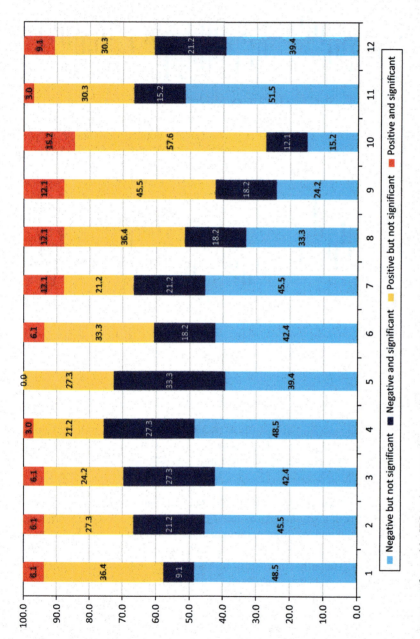

FIGURE 13.7 Percentages of monthly JDI series presenting a positive or negative trend. From a spatial point of view, the negative trends seem to be concentrated in the eastern side of both the islands, and in particular of the South Island (Figure 13.8).

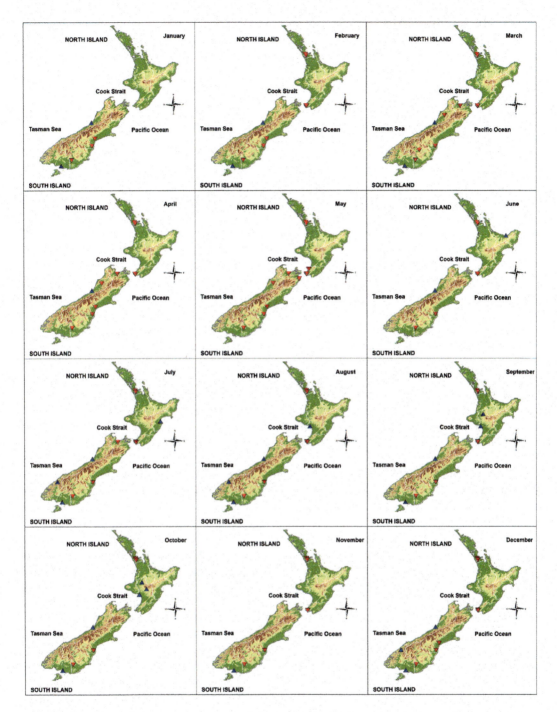

FIGURE 13.8 Spatial results of the monthly trend analysis. Blue and red triangles indicate positive and negative trends, respectively.

regions. In October the spatial distribution of the trends was similar to the previous two months, but with no significant trends identified in the Wellington area. Finally, in November and December, a negative trend was detected in both the islands, in the Canterbury, Otago, Auckland, and Wellington regions, while positive trends were identified only in the South Island, in the Southland region for both November and December, and only in the Westland area only in December.

13.5   CONCLUSIONS AND DISCUSSION

The analysis of drought, in an area where agriculture is one of the largest sectors of the tradable economy, is particularly important for the significant ecological, social, and economic impacts of drought events. Moreover, current global-level assessments suggest that droughts are expected to both increase and decrease, depending upon geographic location, following future climate change (Wang 2005). Within this context, in this study, drought was analyzed in New Zealand, which, based on the latest climate and impact modeling, can expect more droughts in the future in some locations (Clark et al. 2011). In particular, an assessment of meteorological droughts in New Zealand has been conducted based on monthly precipitation data from 33 rain gauge stations in the period 1951 to 2012. First, time series of the modified Standardized Precipitation Index (SPI^{mod}) were constructed for different periods based on monthly mean precipitation values to determine drought categories. Then the analysis of meteorological drought was provided by means of the JDI, which has been demonstrated to produce a more accurate characterization of drought conditions than the SPI (Mirabbasi et al. 2013).

In order to detect the intra-annual drought distribution, the mean JDI values on a monthly timescale were evaluated. Results evidenced predominant wet conditions in all the months and, especially, in July, August, October, and September, and reaching severe wet values in February and March. Although wet conditions prevail over dry conditions, several drought episodes have been identified in the study period, both in the past and in the present century, affecting large areas mainly located on the central-east side of both islands. The oldest drought events occurred in 1951, with about 45% of the station presenting at least severe drought, in 1958 and in 1961, when in more than 30% of the rain gauges extreme drought values were evaluated. Three long drought episodes have been identified in 1969, lasting for six months; across 1972 and 1973, for eight consecutive months; and between 1982 and 1983, when severe drought conditions were detected in nine consecutive months. The last relevant drought event of the past century took place in 1998, with more than 50% of the stations at least affected by severe drought. After the year 2000, three main drought events have been identified: in 2001, involving about one-third of the stations; in 2005; and across 2007 and 2008 when severe drought values were identified in about half of the stations. Finally, a trend analysis was conducted at a monthly scale, evidencing a clear negative trend in all the months, but in particular in the autumn months when the highest number of stations showing negative trends was detected. This negative trend was mainly identified in the Canterbury and Otago regions, in the South Island, and in the Auckland and Wellington regions, in the North Island, thus evidencing an increase in drought trends in all the areas that are presently subject to drought, supporting what past studies have evidenced (Mullan et al. 2005). In fact, the results of this study confirm the geographic pattern of change found by Mullan et al. (2005), who mainly detected a drought increase in the future projections on the east coast and no change in drought projections for the west coast of the South Island. Specifically, as also evidenced by Clark et al. (2011), the results of this study highlight that key agricultural regions on the eastern side such as the Canterbury Plains are the most consistently vulnerable areas, together with other regions in the North Island.

ACKNOWLEDGMENTS

The authors would like to thank the National Institute of Water and Atmospheric Research for providing access to the New Zealand meteorological data from the National Climate Database.

REFERENCES

Alary, V., Messad, S., Aboul-Naga, A., Osman, M.A., Daoud, I., Bonnet, P., Juanes, X. and Tourrand, J.B. 2014. Livelihood strategies and the role of livestock in the processes of adaptation to drought in the Coastal Zone of Western Desert (Egypt). *Agricultural Systems*, 128, pp. 44–54.

Bazrafshan, J., Nadi, M. and Ghorbani, K. 2015. Comparison of empirical copula-based joint deficit index (JDI) and multivariate standardized precipitation index (MSPI) for drought monitoring in Iran. *Water Resources Management*, 29(6), pp. 2027–2044.

Bazrafshan, O., Zamani, H. and Shekari, M. 2020. A copula-based index for drought analysis in arid and semi-arid regions of Iran. *Natural Resource Modeling*, 33(1), p. e12237.

Buttafuoco, G., Caloiero, T., Ricca, N. and Guagliardi, I. 2018. Assessment of drought and its uncertainty in a southern Italy area (Calabria region). *Measurement*, 113, pp. 205–210.

Byun, H.R. and Wilhite, D.A. 1999. Objective quantification of drought severity and duration. *Journal of Climate*, 12(9), pp. 2747–2756.

Caloiero, T. 2017. Drought analysis in New Zealand using the standardized precipitation index. *Environmental Earth Sciences*, 76, p. 569.

Caloiero, T. 2018. SPI trend analysis of New Zealand applying the ITA technique. *Geosciences*, 8(3), p. 101.

Caloiero, T. 2020. Evaluation of rainfall trends in the South Island of New Zealand through the innovative trend analysis (ITA). *Theoretical and Applied Climatology*, 139, pp. 493–504.

Caloiero, T. and Veltri, S. 2019. Drought assessment in the Sardinia Region (Italy) During 1922–2011 using the standardized precipitation index. *Pure and Applied Geophysics*, 176, pp. 925–935.

Caloiero, T., Sirangelo, B., Coscarelli, R. and Ferrari, E. 2016. An analysis of the occurrence probabilities of wet and dry periods through a stochastic monthly rainfall model. *Water*, 8(2), p. 39.

Caloiero, T., Veltri, S., Caloiero, P. and Frustaci, F. 2018a. Drought analysis in Europe and in the Mediterranean basin using the standardized precipitation index. *Water*, 10(8), p. 1043.

Caloiero, T., Coscarelli, R., Ferrari, E. and Sirangelo, B. 2018b. Occurrence probabilities of wet and dry periods in Southern Italy through the SPI evaluated on synthetic monthly precipitation series. *Water*, 10(3), p. 336.

Clark, A., Mullan, A. and Porteous, A. 2011. *Scenarios of Regional Drought Under Climate Change*. NIWA Client Report WLG2012-32. Wellington: National Institute of Water and Atmospheric Research.

Dai, A.G. 2013. Increasing drought under global warming in observations and models. *Nature Climate Change*, 3, pp. 52–58.

Dracup, J.A., Lee, K.S. and Paulson, E.G. 1980. On the definition of droughts. *Water Resources Research*, 16(2), pp. 297–302.

Dravitzki, S. and McGregor, J. 2011. Extreme precipitation of the Waikato region, New Zealand. *International Journal of Climatology*, 31(12), pp. 1803–1812.

Fang, K., Gou, X., Chen, F., Davi, N. and Liu, C. 2013. Spatiotemporal drought variability for central and eastern Asia over the past seven centuries derived from tree-ring based reconstructions. *Quaternary International*, 283, pp. 107–116.

Feulner, G. 2017. Global challenges: Climate change. *Global Challenges*, 1, pp. 5–6.

Field, C.B. 2012. *Managing the Risks of Extreme Events and Disasters to Advance Climate Change Adaptation: Special Report of the Intergovernmental Panel on Climate Change*. Cambridge and New York: Cambridge University Press.

Gaitán, E., Monjo, R., Pórtoles, J. and Pino-Otín, M.R. 2020. Impact of climate change on drought in Aragon (NE Spain). *Science of The Total Environment*, 740, p. 140094.

Griffiths, G.M. 2011. Drivers of extreme daily rainfall in New Zealand. *Weather and Climate*, 31, pp. 24–49.

Griffiths, G.M., Salinger, M.J. and Leleu, I. 2003. Trends in extreme daily rainfall across the South Pacific and relationship to the South Pacific Convergence Zone. *International Journal of Climatology*, 23(8), pp. 847–869.

Guo, H., Wang, R., Garfin, G.M., Zhang, A., Lin, D., Liang, Q. and Wang, J. 2021. Rice drought risk assessment under climate change: Based on physical vulnerability a quantitative assessment method. *Science of the Total Environment*, 751, p. 141481.

Hannaford, J., Lloyd-Hughes, B., Keef, C., Parry, S. and Prudhomme, C. 2011. Examining the large-scale spatial coherence of European drought using regional indicators of precipitation and streamflow deficit. *Hydrological Processes*, 25(7), pp. 1146–1162.

Heim, R.R. 2002. A review of twentieth-century drought indices used in the United States. *Bulletin of the American Meteorological Society*, 83(8), pp. 1149–1165.

Hua, T., Wang, X.M., Zhang, C.X. and Lang, L.L. 2013. Temporal and spatial variations in the Palmer Drought Severity Index over the past four centuries in arid, semiarid, and semihumid East Asia. *Chinese Science Bulletin*, 58, pp. 4143–4152.

IPCC 2013. *Summary for Policymakers. Fifth Assessment Report of the Intergovernmental Panel on Climate Change*. Cambridge and New York: Cambridge University Press.

Jiang, N., Griffiths, G. and Lorrey, A. 2013. Influence of large-scale climate modes on daily synoptic weather types over New Zealand. *International Journal of Climatology*, 33(2), pp. 499–519.

Kamber, G., McDonald, C. and Price, G. 2013. *Drying out: Investigating the Economic Effects of Drought in New Zealand*. Wellington: Reserve Bank of New Zealand.

Kao, S.C. and Govindaraju, R.S. 2010. A copula-based joint deficit index for droughts. *Journal of Hydrology*, 380(1–2), pp. 121–134.

Kendall, M.G. 1962. *Rank Correlation Methods*. New York: Hafner Publishing Company.

Lesk, C., Rowhani, P. and Ramankutty, N. 2016. Influence of extreme weather disasters on global crop production. *Nature*, 529, pp. 84–87.

MAF 2009. *Regional and National Impacts of the 2007–2008 Drought*. Tai Tapu: Butcher Partners Ltd,

Mann, H.B. 1945. Nonparametric tests against trend. *Econometrica*, 13(3), pp. 245–259.

McKee, T.B., Doeskin, N.J. and Kleist, J. 1993. The relationship of drought frequency and duration to time scales. In: *Proceedings of the 8th Conference on Applied Climatology*, pp. 179–184.

Mirabbasi, R., Fakheri-Fard, A. and Dinpashoh, Y. 2012. Bivariate drought frequency analysis using the copula method. *Theoretical and Applied Climatology*, 108, pp. 191–206.

Mirabbasi, R., Anagnostou, E.N., Fakheri-Fard, A., Dinpashoh, Y. and Eslamian, S. 2013. Analysis of meteorological drought in Northwest Iran using the joint deficit index. *Journal of Hydrology*, 492, pp. 35–48 .

Mullan, B., Porteous, A., Wratt, D. and Hollis. M. 2005. *Changes in Drought Risk with Climate Change. Changes in Drought Risk with Climate Change*. NIWA Client Report: WLG2005-23. Wellington: National Institute of Water & Atmospheric Research Ltd.

Nalbantis, I. and Tsakiris, G. 2009. Assessment of hydrological drought revisited. *Water Resources Management*, 23, pp. 881–897.

Nelsen, R.B. 2006. *An Introduction to Copulas*. New York: Springer.

NIWA 2013. Overview of New Zealand Climate. Website: National Institute of Water & Atmospheric Research Ltd. http://www.niwa.co.nz/education-and-training/schools/resources/climate/overview.

Oliver, J.E. 2005. *Encyclopedia of World Climatology*. Amsterdam: Springer.

Palmer, W.C. 1965. *Meteorological Drought, Res.* Paper number 45. Washington DC: Weather Bureau.

Palmer, W.C. 1968. Keeping track of crop moisture conditions, nationwide: The new crop moisture index. *Weatherwise*, 21(4), 156–161.

Palmer, J.G., Cook, E.R., Turney, C.S.M., Allen, K., Fenwick, P., Cook, B., O'Donnell, A.J., Lough, J.M., Grierson, P.F.G. and Baker, P. 2015. Drought variability in the eastern Australia and New Zealand summer drought atlas (ANZDA, CE 1500–2012) modulated by the Interdecadal Pacific Oscillation. *Environmental Research Letters*, 10, p. 124002.

Ramezani, Y., Tahroudi M.N. and Ahmadi, F. 2019. Analyzing the droughts in Iran and its eastern neighboring countries using copula functions. *Quarterly Journal of the Hungarian Meteorological Service*, 123(4), pp. 435–453.

Salinger, M.J. and Mullan, A.B. 1999. New Zealand climate: Temperature and precipitation variations and their links with atmospheric circulation 1930–1994. *International Journal of Climatology*, 19(10), pp. 1049–1071.

Shafer, B.A. and Dezman, L.E. 1982. Development of a surface water supply index (SWSI) to assess the severity of drought conditions in snowpack runoff areas. In: Proceedings of the Western Snow Conference, pp. 164–175.

Sinclair, M.R. 1994. An objective cyclone climatology for the Southern Hemisphere. *Monthly Weather Review*, 122(10), pp. 2239–2256.

Sirangelo, B., Caloiero, T., Coscarelli, R. and Ferrari, E. 2015. A stochastic model for the analysis of the temporal change of dry spells. *Stochastic Environmental Research and Risk Assessment*, 29, pp. 143–155.

Sirangelo, B., Caloiero, T., Coscarelli, R. and Ferrari, E. 2017. Stochastic analysis of long dry spells in Calabria (Southern Italy). *Theoretical and Applied Climatology*, 127, pp. 711–724.

Sklar, A. 1959. Fonctions de répartition à n dimensions et leurs marges. *Publications de l'Institut de Statistique de L'Université de Paris*, 8, pp. 229–231.

Stanke, C., Kerac, M., Prudhomme, C., Medlock, J. and Murray, V. 2013. Health effects of drought: A systematic review of the evidence. *PLOS Currents Disasters*, June 5 edition 1.

Thom, H.C.S. 1966. *Some Methods of Climatological Analysis. WMO Technical Note Number 81.* Geneva: Secretariat of the World Meteorological Organization.

Tomlinson, A.I. 1976. Climate. In: Ward, I. (ed.) *New Zealand Atlas.* Wellington: Government Printer, pp. 82–89.

Trenberth, K.E. 1991. Storm tracks in the Southern Hemisphere. *Journal of the Atmospheric Sciences*, 48(19), pp. 2159–2178.

Tsakiris, G., Pangalou, D. and Vangelis, H. 2007. Regional drought assessment based on the Reconnaissance Drought Index (RDI). *Water Resources Management*, 21, pp. 821–833.

Vicente-Serrano, S.M., Beguería, S. and López-Moreno, J.I. 2010. A multiscalar drought index sensitive to global warming: The standardized precipitation evapotranspiration index. *Journal of Climate*, 23, pp. 1696–1718.

Wang, G. 2005. Agricultural drought in a future climate: Results from 15 global climate models participating in the IPCC 4th assessment. *Climate Dynamics*, 25, pp. 739–753.

Wang, Q., Wu, J., Lei, T., He, B., Wu, Z., Liu, M., Mo, X., Geng, G., Li, X., Zhou, H. and Liu, D. 2014. Temporal-spatial characteristics of severe drought events and their impact on agriculture on a global scale. *Quaternary International*, 349, pp. 10–21.

Weghorst, K.M. 1996. *The Reclamation Drought Index: Guidelines and Practical Applications.* Denver: Bureau of Reclamation.

Wilhite, D.A., Hayes, M.J. and Svodoba, M.D. 2000. Drought monitoring and assessment in the U.S. In: Voght, J.V., Somma, F. (Eds) *Drought and Drought Mitigation in Europe.* Dordrecht: Kluwers.

WMO 1985. *Standardized Precipitation Index – User Guide, WMO No 1090.* Geneva: World Meteorological Organization.

Woli, P., Jones, J.W., Ingram, K.T. and Fraisse, C.W. 2012. Agricultural reference index for drought (ARID). *Agronomy Journal*, 104, pp. 287–300.

14 Comparative Copula-Based Multivariate Meteorological Drought Analysis
A Case Study from Northeast India

P. Kanthavel, Deepak Jhajharia, Ghanashyam
Singh Yurembam, and Rasoul Mirabbasi

CONTENTS

14.1 INTRODUCTION

Natural hazards, such as, floods, droughts, and earthquakes, are caused due to various climatic, geological, and hydrological factors. Drought is one of the most commonly occurring natural hazards, and is mainly associated with a lack of rainfall in a place or a region and to some extent with the rising temperature in that place or region. The occurrence of drought is common over all types of climatic conditions and not endemic to only arid and semiarid regions. However, the frequency and severity of drought may change according to the type of climate, such as, in a region, an arid climate is more likely to have moderate to severe drought-like conditions and occur more often than in a humid-type region. Interestingly, the northeastern region of India that has places, like Cherrapunji (Meghalaya), having a total annual rainfall of 15,000 mm has been reported to experience more frequent unparalleled drought events than the western region of India in the recent decade (Parida and Oinam, 2015). The northeast region observed a 54% probability of drought occurrence, which was

DOI: 10.1201/9781003276548-14

double the probability of drought occurrence (27%) over western India (more drought-prone areas with arid to semiarid climate) in the last 15 years from 2000 to 2014 based on the southwest monsoon rainfall data from Assam and Meghalaya in the northeast region and Gujarat and Rajasthan in western India.

Different types of drought indices are already available in the literature. For example, McKee et al. (1993) developed the Standardized Precipitation Index (SPI) based on rainfall. Keyantash and Dracup (2004) also developed an aggregate drought index (ADI) covering all forms of drought. Tsakiris et al. (2007) used the Reconnaissance Drought Index (RDI) based on rainfall value and potential evapotranspiration as the input parameters. Waseem et al. (2015) developed a composite drought index (CDI) for multivariate drought analysis. Tigkas et al. (2013) developed the DrinC (drought indices calculator) based on the RDI, SDI (Streamflow Drought Index), and SPI. In the present study, the SPI was chosen for studying the meteorological drought over two stations from northeast India having different types of environmental conditions because of its simplicity and flexibility of use and the requirement of rainfall data as the only input parameter for defining the drought event. Besides identifying drought events using the aforementioned indices, further advanced research on drought attributes, i.e., drought duration, severity, and interval, has been done by researchers using advanced statistical tools and soft computing techniques in recent decades. The multivariate analysis of drought variables by copula theory is one such advanced topic explored first by Shiau (2006). Shiau reported that it would be most appropriate to study drought events through probability theory and stochastic process methods. Mirabbasi et al. (2012) also mentioned that drought events are related to stochastic phenomena, such as rainfall or streamflow, and applied two-dimensional copulas to investigate the drought characteristics of a region in northwest Iran. Various researchers used the copula-based multivariate analysis in the recent decade to study and compare the behavior of different types of drought variables in different climatic conditions (Shiau and Modarres, 2009; Chen et al., 2013; Maeng et al., 2017; Azam et al., 2018; Hangshing and Dabral, 2018).

The probabilistic characterization of drought is important in the accurate assessment of water resources and the planning and management of drought in view of the water scarcity due to meteorological drought in almost all kinds of environments, and also in view of the prevalent and well-established global warming in the climate change scenario (Rossi et al., 1992). Therefore, two different copula families (Archimedean and elliptical) were used in the multivariate drought analysis to study the three different types of drought variables in the present study. The inspiration to undertake the drought analysis study in the northeast region was derived from the findings of Parida and Oinam (2015) who reported that monsoon rainfall in the northeast region gradually dropped in recent years (from 2001 to 2014) in comparison to the 1980s to 1990s and found unprecedented drought based on the percentage departure from rainfall in the region. Thus, a meteorological drought study based on rainfall data from two different climatic conditions of northeast India, i.e., the subtropical highland climate in Gangtok (Sikkim) and the humid subtropical climate in Imphal (Manipur), was undertaken through sophisticated copula modeling. Drought events were characterized by three attributes – severity, duration, and interval – through the SPI for one-month timescale for both stations by using rainfall data for the last half-century.

14.2 DATA AND METHODOLOGY

14.2.1 STUDY AREA AND DATA

Northeast India, rich in biodiversity and a monsoon-dominated climatic system, of which nearly 60% of the geographical area is under forest cover and contains two major river basin systems, i.e., Brahmaputra and Barak. A large number of people of the region practice shifting cultivation (jhum) and cultivate various crops, such as paddy, tea, jute, oilseeds, forest produce, and other horticultural crops (e.g., orange, banana, pineapple, and jackfruit). The region contains one of the few remaining rainforests of the Indian subcontinent and the majority of crop area is under rainfed agriculture, which is highly vulnerable to climate change and anthropogenic-induced global warming. Two stations, namely, Imphal and Gangtok were selected in this study from the northeast region situated in the states of Manipur and Sikkim, respectively (Figure 14.1). Gangtok (27.33° N, 88.6° E, and elevation of 1700 m above mean sea level) is the capital of Sikkim with average annual rainfall of 3500 mm and lies in the Eastern Himalayan ranges of a subtropical highland climate (Cwb based on the Köppen–Geiger system). Imphal (24.81° N, 93.95° E, and elevation of 780 m above mean sea level) is the capital of Manipur with an average annual rainfall of 1500 mm. It lies in the valley of the Manipur River and in the bank of the Barak River surrounded by the northeastern hills , which have a humid subtropical climate (Cwa based on the Köppen–Geiger system). The monthly rainfall data of both stations used in the drought analysis were obtained from the India Meteorological Department, Pune, for the last 51 years from 1966 to 2016. Figure 14.2 shows the monthly rainfall time series of both stations.

14.2.2 STANDARDIZED PRECIPITATION INDEX (SPI)

The SPI can be calculated in various timescales, such as SPI-1, SPI-3, and SPI-6, describing different types of drought conditions. The short-range (long-range) timescale reveals important information about soil moisture deficiency (reservoir and groundwater storage). The procedure related to the SPI one-month timescale used in this chapter is as follows. First, the rainfall values being the continuous data are well-fitted with the gamma distribution. Then the fitted gamma distribution is transformed to the standard normal distribution with mean zero and standard deviation one. This final normalized format is known as the SPI. The right-hand (left-hand) side of the standard normal distribution curve indicates wet (dry) conditions. The climate type is classified from extremely wet to extremely dry based on the SPI values, as given by McKee et al. (1993). The SPI values and corresponding climate conditions are summarized in Table 14.1.

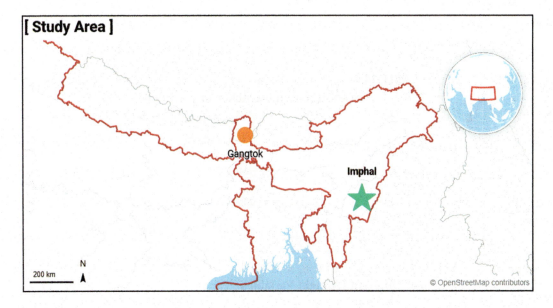

FIGURE 14.1   Location map of the two study areas selected from northeast India.

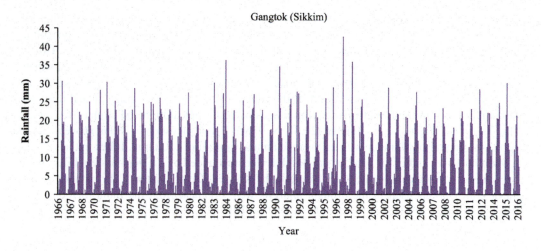

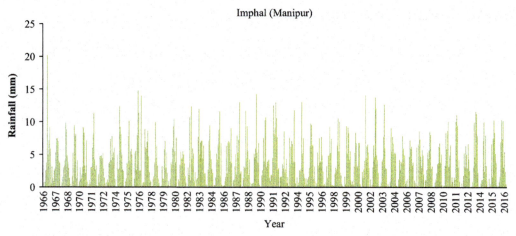

FIGURE 14.2   Monthly rainfall data of Gangtok and Imphal stations (northeast India).

TABLE 14.1

SPI Classifications Based on McKee et al. (1993)

SPI	Classification
2.00>	Extremely wet
1.50 to 1.99	Very wet
1.00 to 1.49	Moderately wet
−0.99 to 0.99	Near normal
−1 to −1.49	Moderate drought
−1.50 to −1.99	Severe drought
−2.00 <	Extreme drought

14.2.2.1   Definition of Different Drought Variables

The drought events using the monthly rainfall data of a place are identified, and the drought variables are defined based on SPI values that are less than zero (Shiau, 2006). The following drought variables are used in this study, which is defined as given next.

- **Drought duration (D)** is defined as the number of consecutive intervals (months) where SPI is below the specific threshold value of zero (Shiau, 2006).
- **Drought severity (S)** is defined as the cumulative SPI value during a drought period $S = \sum_{i}^{D} SPI_i$, where SPI_i is the value of SPI in the ith timescale (Mishra and Singh, 2010).
- **Drought interval time (I)** is defined as the period elapsing from the starting point of a drought to next drought beginning (Song and Singh 2010).

14.2.3 Best Model Fitting

The selection of the best-fitted copula model is done based on the Akaike information criteria (AIC) (Akaike 1974; Bozdogan 2000) and Schwarz information criterion (SIC) (Neath and Cavanaugh 1997). The minimum values of the AIC and the SIC indicate the best-fitted model in comparison to other models used in this study. The expression of the AIC and the SIC are explained as follows:

$$\text{AIC} = -2logL\left(\hat{\theta}\right) + 2k \tag{14.1}$$

$$\text{SIC} = 2logL\left(\hat{\theta}\right) + klogn \tag{14.2}$$

where θ is the set (vector) of model parameters; $L\left(\hat{\theta}\right)$ is the likelihood of the candidate model given the data when evaluated at the maximum likelihood estimate of θ; k is the number of estimated parameters in the candidate model; and n is the number of observations.

14.2.4 Univariate Distribution Analyses

In this study, in the univariate analysis, each drought variable is separately fitted with various marginal distributions, namely, normal, lognormal, Weibull, and gamma. The best-fit distribution is selected based on the lowest value of the AIC. The univariate return period of drought variables is estimated using (Shiau and Shen, 2001)

$$T_D = \frac{EL}{1 - F_D\left(d\right)} \tag{14.3}$$

where EL is the expected drought interval time, and $F_D\left(d\right)$ indicates cumulative distribution functions of drought duration. For the remaining drought variables, the return period is also calculated in the same way.

14.2.5 Bivariate Analyses

The multivariate analysis for various univariate variables was simplified after the introduction of the copula theory, which possesses the key advantage of being applicable in various marginal distributions of univariate variables. The copula runs based on the Sklar theorem, which states that if $FX_1, X_2, ..., Xm(x_1, x_2, ..., x_m)$ is multivariate distributions of m correlated random variables with their marginal distributions $FX1(x_1)$, $FX2(x_2)$, ..., $FXm(x_m)$, with the copula C:

$$FX_1, X_2, ..., X_m(x_1, x_2, ..., x_m) = C(FX_1(x_1), FX_2(x_2), ..., FX_m(x_m)) \tag{14.4}$$

The bivariate return period is expressed as (Shiau, 2003)

$$T_{DS}^{OR} = \frac{EL}{P\left[D > d \cup S > s\right]} = \frac{EL}{1 - C\left\{F_D\left(d\right), F_S\left(s\right)\right\}} \tag{14.5}$$

T_{DS}^{OR} is the joint return period for $D \geq d$ or $S \geq s$

$$T_{DS}^{AND} = \frac{EL}{P\left[D > d \cap S > s\right]} = \frac{EL}{1 - F_D(d) - F_S(s) + C\left\{F_D(d), F_S(s)\right\}} \qquad (14.6)$$

T_{DS}^{AND} is the joint return period for $D \geq d$ and $S \geq s$, and $F_D(d)$ and $F_S(s)$ indicate cumulative distribution functions of drought duration and severity, respectively. $C\left\{F_D(d), F_S(s)\right\}$ indicates the copula-based joint distribution function of the drought duration and severity, respectively.

14.2.6 TRIVARIATE ANALYSIS

In the trivariate analysis, three drought variables were considered simultaneously. The trivariate probability analysis is calculated with the help of two kinds of copula families, i.e., asymmetric Archimedeans and elliptical. The trivariate return period is expressed as (Shiau, 2003)

$$T_{OR} = \frac{EL}{P\left[D > d \cup S > s \cup I > i\right]} = \frac{EL}{1 - C\left(F_D(d), F_S(s), F_I(i)\right)} \qquad (14.7)$$

$$T_{AND} = \frac{EL}{P\left[D > d \cap S > s \cap I > i\right]}$$

$$= \frac{EL}{1 - F_D(d) - F_S(s) - F_I(i) + C\left(F_D(d), F_S(s)\right) + C\left(F_D(d), F_I(i)\right) + C\left(F_S(s), F_I(i)\right) - C\left(F_D(d), F_S(s), F_I(i)\right)}$$

$$(14.8)$$

where, $F_D(d)$, $F_S(s)$, and $F_I(i)$ denotes cumulative distribution functions of drought duration, severity, and interval, respectively; $C\left(F_D(d), F_S(s)\right), C\left(F_D(d), F_I(i)\right)$, and $C\left(F_S(s), F_I(i)\right)$ denotes joint distributions for duration–severity, duration–interval, and severity–interval respectively; and $C\left(F_D(d), F_S(s), F_I(i)\right)$ denotes joint distributions for the duration–severity–interval.

14.3 RESULTS AND DISCUSSIONS

With the help of the aforementioned methods, the results of drought analysis for two sites in northeast India with two different types of climates are mentioned in the following sections.

14.3.1 SPI-1 DROUGHT ANALYSIS

Figure 14.3 show the various values of SPI-1 for both Gangtok and Imphal. The most negative and the second most negative SPI-1 values of –5.14 and –4.26 (–3.47 and –2.95) were witnessed in 1996 and 2016 (1984 and 2004), respectively, indicating the extreme dry condition in Gangtok (Imphal). While, the most positive SPI-1 value of 3.25 (3.35) was witnessed in 1984 (1976) indicating the extreme wetness condition over Gangtok (Imphal). These SPI-1 values were used to produce the overall condition of wetness or dryness over both the stations for the last 51 years. Table 14.2 demonstrates various types of climate conditions based on the SPI-1 for both the stations. According to Table 14.2 indicating details of the SPI-1, out of a total 612 months from 1966 to 2016 the drought months resulted in 89 (14.54%) months in Gangtok and 78 (12.75%) months in Imphal. Out of various types of drought conditions, the moderate dry condition is dominant in both stations. Near-normal conditions were witnessed in more than 70% of the total months over both stations. The

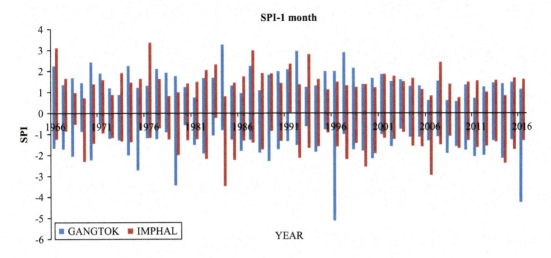

FIGURE 14.3   SPI-1 for Gangtok and Imphal.

occurrences of drought events increased in the second half of the study period compared to first half over both stations, implying either a change in rainfall pattern or maybe the effect of anthropogenic-induced climate change in the northeast region during the second half of the 51-year period.

14.3.1.1   SPI Drought Variables

Three drought attributes, namely, duration, severity, and interval time, as defined in Section 14.2, were obtained by using the SPI-1 values for both the stations. Table 14.3 shows these three drought variables and their parameters for Gangtok and Imphal. The maximum drought severity of magnitude 11.58 (in 1996) and 3.47 (in 1984) was observed over Gangtok and Imphal, respectively. Whereas the lowest values of severity of magnitude 1.01 and 1.02 were observed over Gangtok and Imphal, respectively.

   In the case of drought duration, most of the drought events take one month and a few drought events take two months; the maximum drought duration encountered was four months in the Gangtok station compared to Imphal, which had two months as the maximum drought duration. The interval time regarded maximum as 27 months and 36 months over Gangtok and Imphal, respectively.

14.3.1.2   Dependency Measures Among Drought Variables

While measuring the dependency measures among drought variables, the major problem involved is measuring the amount of association or dependence between these variables. The Pearson linear correlation coefficient considers only linear dependence among variables, which is not so robust in the presence of heavy-tailed variables (Mirabbasi et al., 2012). Instead, nonparametric methods are useful in view of the aforementioned fault in the Pearson linear correlation coefficient method. In this study, two nonparametric methods, namely, Spearman's ρ and Kendall's τ, have been applied. In the case of copula modeling, Kendall's τ value helps in decision-making of the selection of the copula family to be applied (Nelsen 2007). Mirabbasi et al. (2012) may also be referred to check the criterion for selecting the type of copula. The association between different variables were identified by using the p-value procedure, in which the null hypothesis (H_O) is accepted when the observed p-value of the selected variables is less than or equal to the critical value, i.e., at the 5% level of significance ($\alpha = 0.05$) indicating no association between the selected drought variables. If the observed p-value of the selected variables is more than 0.05, then the null hypothesis is rejected and the alternate hypothesis (H_a) is accepted indicating the association between the selected drought variables. Figure 14.4 and Table 14.4 clearly explains the dependence between three drought

TABLE 14.2

Study Months Classified According to SPI-1 for Gangtok (Imphal)

MONTH	Near Normal	Moderately Wet	Very Wet	Extremely Wet	Moderately Dry	Severely Dry	Extremely Dry
January	34(40)	4(9)	2(2)	1(0)	6(0)	4(0)	0(0)
February	36(38)	4(3)	2(0)	1(2)	3(4)	4(4)	1(0)
March	37(40)	6(6)	2(0)	0(1)	4(1)	0(1)	2(2)
April	33(35)	7(4)	1(2)	1(2)	4(5)	4(2)	1(1)
May	40(37)	5(3)	2(4)	0(1)	1(2)	2(2)	1(2)
June	36(35)	4(3)	2(2)	2(1)	4(6)	2(3)	1(1)
July	42(34)	5(3)	0(4)	0(0)	2(8)	1(1)	1(1)
August	38(32)	4(1)	0(6)	1(0)	5(10)	1(1)	2(1)
September	39(37)	3(5)	0(1)	3(2)	2(3)	3(2)	1(1)
October	33(36)	5(7)	4(1)	0(0)	7(3)	0(1)	2(3)
November	34(36)	4(5)	5(2)	0(1)	4(7)	1(0)	3(0)
December	31(41)	6(7)	3(1)	1(2)	6(0)	3(0)	1(0)
Total	433(441)	57(56)	23(25)	10(12)	48(49)	25(17)	16(12)

Note: Values in brackets indicate the values for the Imphal station.

TABLE 14.3

Drought Variables and Parameters for the Selected Stations of Northeast India

Station	Duration						Severity						Interval					
	Mean	S.D.	Min	Max			Mean	S.D.	Min	Max			Mean	S.D.	Min	Max
Gangtok	1.187	0.5118	1	4			1.88	1.425	1.01	11.58			7.78	6.55	2	27
Imphal	1.15	0.3568	1	2			1.76	0.63	1.02	3.47			8.78	7.17	2	36

Note: Min, Max, and S.D. indicate minimum, maximum, and standard deviation, respectively.

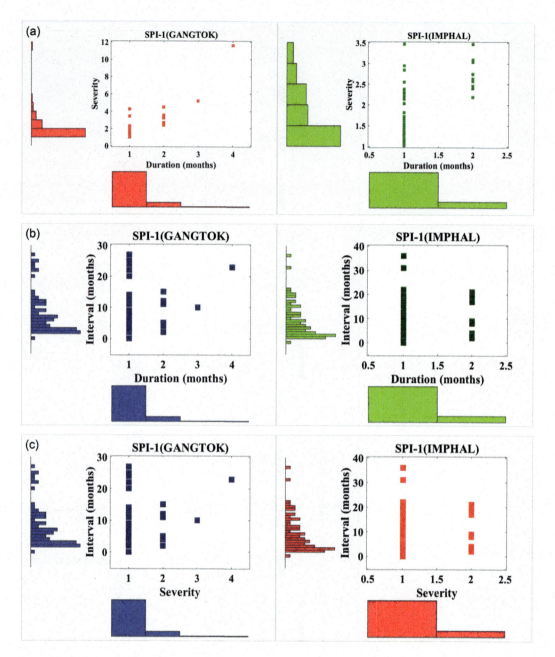

FIGURE 14.4 Dependence between drought variables: (a) duration–severity, (b) duration–interval, (c) severity–interval.

variables for both stations. No association between the drought duration and severity was observed in the case of both selected stations, as the null hypothesis (H_0) was accepted at $\alpha = 0.05$ through both tests (see Table 14.4). However, a strong association was observed between drought duration and interval, and drought severity and interval in the case of both stations, as the null hypothesis of no association was rejected through both the non-parametric tests and the alternate hypothesis, i.e., H_a was accepted at $\alpha = 0.05$.

TABLE 14.4

Dependency Measures between Drought Variables for Both Stations

SPI-1	Spearman's ρ		Kendall's τ	
	Gangtok	Imphal	Gangtok	Imphal
D-S	0.58(**2.8e–08**)	0.54(**1.7e–06**)	0.49(**2.3e–07**)	0.45(**9.2e–06**)
S-I	0.145(*0.21*)	0.108(-*0.38*)	0.1(*0.22*)	-0.092(*0.28*)
D-I	0.12(*0.30*)	0.008(*0.94*)	0.1(*0.31*)	0.007(*0.95*)

Notes: D-S, duration and severity; S-I, severity and interval; D-I, duration and interval. The values given in brackets indicate the p-value, and the bold (italic) values of p indicate the acceptance (rejection) of the null hypothesis indicating no (a strong) association between the drought variables.

14.3.2 UNIVARIATE DROUGHT ANALYSIS

Tables 14.5a and 14.5b show the results obtained in the case of four different marginal distributions (normal, lognormal, Weibull, and gamma) of three different drought variables along with their estimated parameters for the Gangtok and Imphal stations, respectively. The best-fitted marginal distribution is selected based on the lowest value of the AIC and the highest value of the log-likelihood. The results of marginal distributions showed that the lowest (highest) values of the AIC (log-likelihood) were observed entirely in case of lognormal distribution for all three drought variables of Gangtok and Imphal. Thus, the lognormal is the best-fitted marginal distribution for all three drought variables for both stations. Also, Figures 14.5a, 14.5b, and 14.5c show the results of univariate fitting of the best-fitted (lognormal) marginal distributions graphically for drought severity, interval, and duration, respectively, for both selected stations.

TABLE 14.5A

Marginal Distributions of Drought Variables along with Estimated Parameters for Gangtok

Variable	Distribution	Parameters						
		Mean	S.D.	Shape	Scale	MLKH	AIC	
Duration	Normal	1.19	0.51			−55.69	115.38	
	Weibull	1.18	0.54	2.33	1.34	−50.67	105.33	
	Lognormal	**1.17**	**0.35**			**−23.08**	**50.15**	BF
	Gamma	1.19	0.39	9.28	0.13	−32.94	69.88	
Severity	Normal	1.89	1.43			−132.49	268.97	
	Lognormal	**1.83**	**0.89**			**−85.26**	**174.52**	BF
	Weibull	1.91	1.23	1.59	2.13	−108.16	220.32	
	Gamma	1.89	0.96	3.83	0.49	−96.70	197.39	
Interval	Normal	7.89	6.53			−243.36	497.76	
	Lognormal	**7.91**	**7.26**			**−216.72**	**437.43**	BF
	Weibull	7.96	6.06	1.33	8.65	−222.30	448.60	
	Gamma	7.89	5.89	1.80	4.39	−220.48	444.96	

Note: S.D., standard deviation; MLKH, maximum likelihood; AIC, Akaike information criteria; BF, best fit.

TABLE 14.5B

Marginal Distributions of Drought Variables along with Estimated Parameters for Imphal

Variable	Distribution	Parameters						
		Mean	S.D.	Shape	Scale	MLKH	AIC	
Duration	Normal	1.15	0.36			−25.91	55.82	
	Lognormal	**1.14**	**0.29**			**−7.92**	**19.83**	**BF**
	Weibull	1.14	0.40	3.15	1.28	−27.39	58.78	
	Gamma	1.15	0.30	14.34	0.08	−13.66	31.32	
Severity	Normal	1.76	0.63			−64.57	133.14	
	Lognormal	**1.76**	**0.59**			**−54.58**	**113.16**	**BF**
	Weibull	1.77	0.65	2.95	1.98	−63.42	130.84	
	Gamma	1.76	0.58	9.17	0.19	−57.17	118.33	
Interval	Normal	8.91	7.14			−226.26	456.52	
	Lognormal	**9.01**	**8.28**			**−204.93**	**413.85**	**BF**
	Weibull	8.98	6.68	1.36	9.80	−208.74	421.48	
	Gamma	8.91	6.56	1.85	4.83	−207.27	418.54	

Note: S.D., standard deviation; MLKH, maximum likelihood; AIC, Akaike information criteria; BF, best fit.

With the help of the selected univariate marginal distribution, all three drought variables with theoretical return periods of 2, 5, 10, 20, 50, and 100 years have been estimated for Gangtok and Imphal. Table 14.6 demonstrates the computed univariate drought variables with six different theoretical return periods ranging from 2 to 100 years for both stations. Drought severity and duration of 3.4 and 1.8 (2.7 and 1.6) was observed over Gangtok (Imphal) for a theoretical return period of 10 years. Similarly, the higher values of univariate drought duration and severity for all the remaining return periods were observed over Gangtok in comparison to Imphal, which proves that Gangtok is more likely to be drought affected with larger drought duration and more severe in nature.

14.3.3 BIVARIATE DROUGHT ANALYSIS

In the case of the bivariate drought modeling, the symmetric copula belonging to the Archimedean family and the elliptical copula family were applied in this study. The best-fit copula was selected based on the lowest values of AIC and SIC and the highest log-likelihood value. Tables 14.7a and 14.7b show the estimated copula parameters and their performances for five different copulas for Gangtok and Imphal, respectively. Results reveal that the normal copula is found to be the best-fitted copula among all cases except in one case (drought severity at Gangtok). In the case of drought severity at Gangtok, Clayton is found to be the best-fitted copula.

The drought risk depends upon the return period of a drought variable. The return period is calculated in the form of AND and OR type in order to identify the drought risk associated with the selected variable of that station. By considering the occurrences of any one of the variables among the two selected variables in the OR type, the shortest return period is observed in comparison to the theoretical return period indicating moderate risk with high frequency. Table 14.9 shows the bivariate return period of drought variables in case of Gangtok and Imphal. Table 14.9 shows that the values of joint return period in case of the OR type, and it is worth mentioning that all cases but one of the joint return period of each bivariate over the Gangtok station observed higher values in comparison to the Imphal station. This indicates that the Imphal station is prone to drought events of moderate risk with higher frequencies. Table 14.9 also shows the values of the joint return period

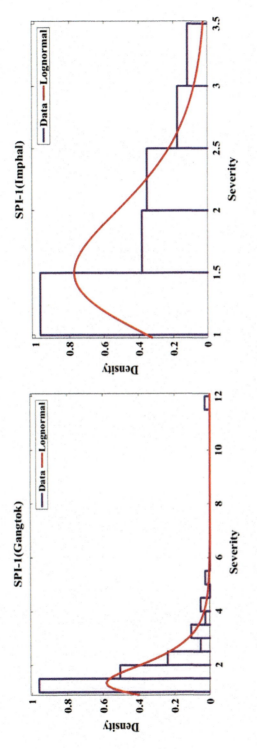

FIGURE 14.5 a Univariate distribution fitting in the case of drought severity (lognormal as best-fitted marginal distribution).

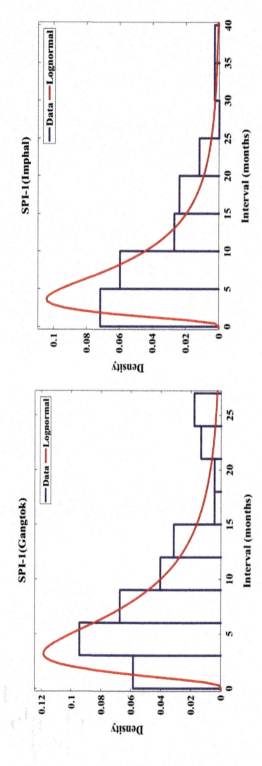

FIGURE 14.5 b Univariate distribution fitting in the case of drought interval (lognormal as best-fitted marginal distribution).

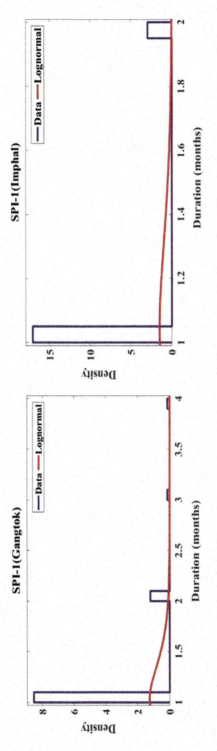

FIGURE 14.5 c Univariate distribution fitting in the case of drought duration (lognormal as best-fitted marginal distribution).

TABLE 14.6

Univariate Drought Variables and Return Period for Both Stations

Year	Gangtok			Imphal		
	Duration	Severity	Interval	Duration	Severity	Interval
2	1.3	2.0	8.4	1.2	1.9	8.9
5	1.6	2.9	14.2	1.4	2.4	15.3
10	1.8	3.4	19.1	1.6	2.7	20.8
20	2.0	4.0	24.7	1.7	3.0	27.1
50	2.2	4.8	33.5	1.9	3.4	36.6
100	2.4	5.3	40.7	2.0	3.7	44.8

TABLE 14.7A

Estimated Copula Parameters and Comparison of Their Performance for Gangtok

Variable	Copula	Parameters	AIC	SIC	MLKH	BF
D/S	Clayton	$\theta = 2.41$	**−26.18**	**−21.72**	**15.18**	**BF**
	Gumbel	$\theta = 1.88$	−24.16	−19.69	14.16	
	Frank	$\theta = 6.85$	−14.41	−9.94	9.29	
	t	$\sigma = 0.77, \vartheta = 7.23$	−6.85	−2.39	5.51	
	Normal	$\rho = 0.77$	2.10	4.36	−0.02	
S/I	Normal	$\rho = 0.18$	**1.33**	**3.60**	**0.36**	**BF**
	Clayton	$\theta = 0.02$	1.39	5.86	1.39	
	Gumbel	$\theta = 1.07$	1.76	6.23	1.20	
	Frank	$\theta = 0.91$	2.64	7.11	0.76	
	t	$\sigma = 0.16, \vartheta = 1894.91$	3.60	8.07	0.28	
D/I	Normal	$\rho = 0.15$	**0.57**	**2.84**	**0.74**	**BF**
	t	$\sigma = 0.14, \vartheta = 4680320.67$	2.27	6.74	0.95	
	Clayton	$\theta = 0.06$	3.08	7.55	0.54	
	Gumbel	$\theta = 1.02$	3.17	7.64	0.50	
	Frank	$\theta = 0.79$	3.49	7.96	0.34	

Note: AIC, Akaike information criteria; SIC, Schwarz information criterion; MLKH, maximum likelihood; BF, best fit; D, duration; S, severity; I, interval.

for the AND type, and it is worth mentioning that all cases but one of the joint return period of each bivariate at the Imphal station observed higher values in comparison to the Gangtok station. This indicates that Gangtok station is more likely to witness drought events of extreme risk with shorter frequencies. Figures 14.6, 14.7, and 14.8 depict contours of the AND and OR type return period for both stations.

TABLE 14.7B

Estimated Copula Parameters and Comparison of Their Performance for Imphal

Variable	Copula	Parameters	AIC	SIC	MLKH	BF
D/S	Normal	$\hat{A} = 0.61$	**−18.38**	**−16.22**	**10.22**	**BF**
	t	$\sigma = 0.61, \vartheta = 18.10$	−16.39	−12.13	10.29	
	Clayton	$\theta = 1.88$	−13.66	−9.41	8.92	
	Gumbel	$\theta = 1.40$	−12.71	−8.46	8.45	
	Frank	$\theta = 5.09$	−9.40	−5.15	6.79	
S/I	Normal	$\rho = -0.08$	**1.94**	**4.10**	**0.06**	**BF**
	t	$\sigma = -0.18, \vartheta = 4.86$	2.29	6.54	0.95	
	Frank	$\theta = -0.93$	3.33	7.58	0.43	
	Gumbel	$\theta = 1.00$	3.55	7.80	0.32	
	Clayton	$\theta = 0.00$	4.46	8.72	−0.14	
D/I	Normal	$\rho = 0.01$	**2.04**	**4.20**	**0.01**	**BF**
	t	$\sigma = 0.01, \vartheta = 4669185.59$	3.58	7.84	0.30	
	Gumbel	$\theta = 1.00$	4.15	8.41	0.01	
	Clayton	$\theta = 0.00$	4.17	8.42	0.01	
	Frank	$\theta = 0.02$	4.18	8.44	0.00	

Note: AIC, Akaike information criteria; SIC, Schwarz information criterion; MLKH, maximum likelihood; BF, best fit; D, duration; S, severity; I, interval.

14.3.4 TRIVARIATE DROUGHT ANALYSIS

All the three drought variables were considered simultaneously in the trivariate analysis. The trivariate probability analysis was calculated through two kinds of copula families, i.e., asymmetric Archimedeans and elliptical. In the trivariate analysis, the asymmetric Archimedean copula family and elliptical copula contain two and three parameters, respectively. The results of the trivariate analysis indicate that the asymmetric Clayton copula and the normal copula are selected as the best-fit copula for Gangtok and Imphal, respectively (Table 14.8). The trivariate return period provides superior assessment for the drought risk management in comparison to the bivariate return period based on only two variables at a time.

Table 14.9 explains about the theoretical return periods of the two stations. In case of the trivariate OR type return period, slightly higher magnitude values are observed for the Gangtok station in comparison to the Imphal station in all the theoretical return periods (see Table 14.9), which implies moderate drought risk with higher frequency at Imphal. The results of AND type return period analysis shows that Gangtok station witnessed comparatively lower values in the first three theoretical return periods of 2, 5, and 10 years, which implies the extreme drought risk with shorter frequencies at Gangtok. In the remaining three theoretical return periods (20, 50, and 100 years), Imphal witnessed comparatively lower values in comparison to Gangtok, which implies extreme drought risk with shorter frequencies at Imphal.

TABLE 14.8A

Estimated Copula Parameters and Their Performance in Trivariate Drought Analysis for Gangtok

Variable	Copula	Parameters				AIC	SIC	MLKH	BF
D/S/I	Clayton	$\theta_1 =2.41$	$\theta_2 =0.09$			**−18.67**	**−16.40**	**10.36**	**BF**
	Gumbel	$\theta_1 =1.88$	$\theta_2 =1.05$			−13.41	−11.15	7.73	
	Frank	$\theta_1 =6.85$	$\theta_2 =0.73$			−8.73	−6.47	5.39	
	T	$\sigma_{12} =0.78$	$\sigma_{23} =0.12$	$\sigma_{13} =0.15$	$\vartheta = 22.75$	0.06	8.76	4.25	
	Normal	$\rho_{12} =0.77$	$\rho_{23} =0.15$	$\rho_{13} =0.18$		2.18	8.80	2.08	

Note: AIC, Akaike information criteria; SIC, Schwarz information criterion; MLKH, maximum likelihood; BF, best fit; D, duration; S, severity; I, interval.

TABLE 14.8B

Estimated Copula Parameters and Their Performance in Trivariate Drought Analysis for Imphal

Variable	Copula	Parameters				AIC	SIC	MLKH	BF
D/S/I	Normal	$\rho_{12} = 0.61$	$\rho_{23} = 0.01$	$\rho_{13} = -0.08$		**−14.21**	**−7.93**	**10.29**	**BF**
	T	$\sigma_{12} = 0.62$	$\sigma_{23} = 0.00$	$\sigma_{13} = -0.12$	$\vartheta =18.62$	−14.30	−6.06	11.47	
	Clayton	$\theta_1 = 1.88$	$\theta_2 = 0.04$			−3.07	−0.91	2.57	
	Gumbel	$\theta_1 = 1.40$	$\theta_2 = 1.01$			−2.42	−0.27	2.24	
	Frank	$\theta_1 = 5.09$	$\theta_2 = -0.12$			0.18	2.34	0.94	

Note: AIC, Akaike information criteria; SIC, Schwarz information criterion; MLKH, maximum likelihood; BF, best fit; D, duration; S, severity; I, interval.

14.4 CONCLUSIONS

In the present study, meteorological drought in terms of three characteristics (severity, duration and interval) was studied by using the SPI-1 based on rainfall data from 1966 to 2016 from two stations in northeast India with different climatic conditions, i.e., a subtropical highland climate in the Eastern Himalayan ranges and a humid subtropical climate. The drought months and severity of drought is higher in case of Gangtok in comparison to Imphal, which is likely due to high rainfall (Shiau and Modarres, 2009). In the univariate drought analysis, lognormal distribution was the best-fitted marginal distribution. The drought variables identified for theoretical return periods imply that risk associated with drought variables is high at Gangtok, i.e., duration and severity at Gangtok is high in comparison to Imphal for the corresponding theoretical return period. The drought interval is comparatively high at Imphal in comparison to Gangtok. Bivariate and trivariate drought analyses based on two different copula families were also carried out. In the joint probability drought analysis, the Gaussian (normal) and Clayton copulas were found to be the best-fitted copulas. The OR type

TABLE 14.9A

Bivariate and Trivariate Return Periods of Drought Variables for Gangtok

Duration	Severity	Interval	Joint Return Period (AND–OR)						TRI Return Period	
			TDS	T'DS	TSI	T'SI	TDI	T'DI	TDSI	T'DSI
1.3	2.0	8.4	2.90	1.54	4.88	1.24	5.46	1.27	6.89	1.10
1.6	2.9	14.2	10.23	4.02	28.94	3.02	30.20	2.94	40.98	2.32
1.8	3.4	19.1	22.56	7.72	90.50	5.63	106.64	5.73	151.14	4.33
2.0	4.0	24.7	53.20	16.16	314.28	11.30	390.28	11.60	496.94	8.27
2.2	4.8	33.5	140.66	38.43	1629.58	28.97	1809.48	27.47	1693.03	19.90
2.4	5.3	40.7	252.42	62.04	3546.66	45.45	6678.91	56.96	3253.46	38.94

Note: TDS denotes AND type and T'DS denotes 'OR' type bivariate return period of duration–severity.

TABLE 14.9B

Bivariate and Trivariate return periods of drought variables in the case of Imphal station

| Duration | Severity | Interval | Joint Return Period (AND–OR) | | | | | | | | TRI Return Period | | |
			TDS	T'DS	TSI	T'SI	TDI	T'DI			TDSI	T'DSI
1.2	1.9	8.9	3.29	1.48	6.80	1.24	5.52	1.23		10.45	1.05
1.4	2.4	15.3	10.74	3.10	51.89	2.80	29.52	2.53		60.28	2.16
1.6	2.7	20.8	30.13	6.42	221.44	5.28	145.70	5.40		160.53	4.02
1.7	3.0	27.1	64.90	11.06	988.39	10.27	477.27	9.62		375.75	7.72
1.9	3.4	36.6	233.15	28.29	7357.10	25.30	3358.84	25.39		1040.14	18.73
2.0	3.7	44.8	519.76	51.23	33657.01	50.19	11438.51	46.70		2148.42	36.87

Note: TDSI denotes AND type and T'DSI denotes OR type trivariate return period of duration–severity–interval.

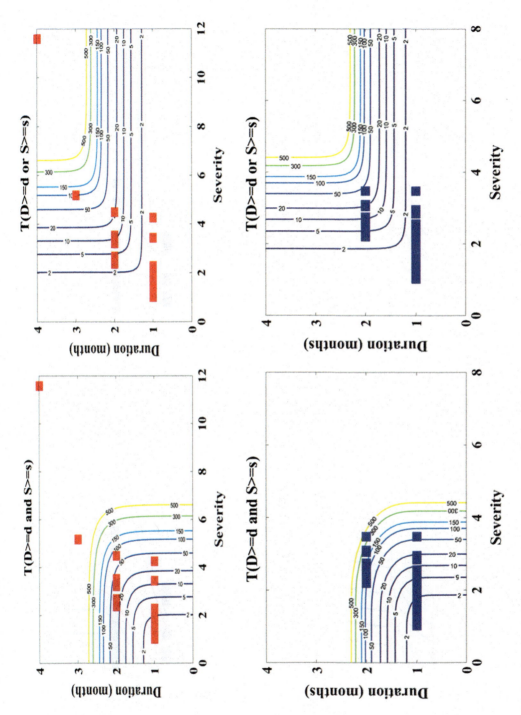

FIGURE 14.6 Bivariate return period analysis in the case of drought duration–severity for Gangtok (top) and Imphal (bottom).

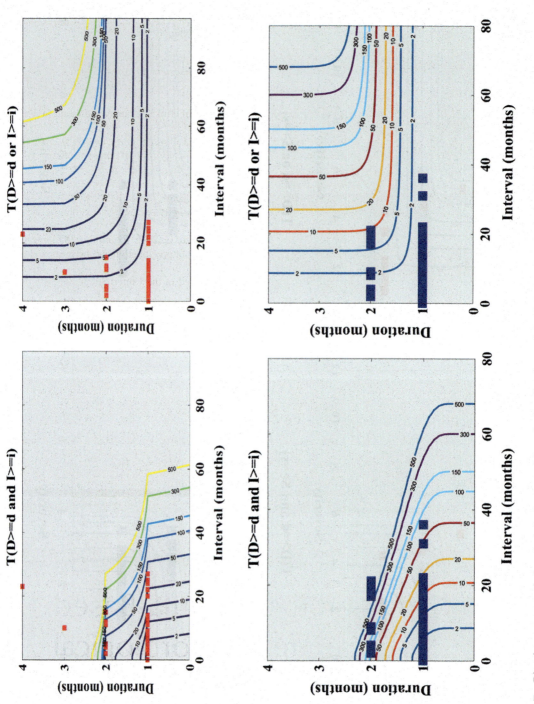

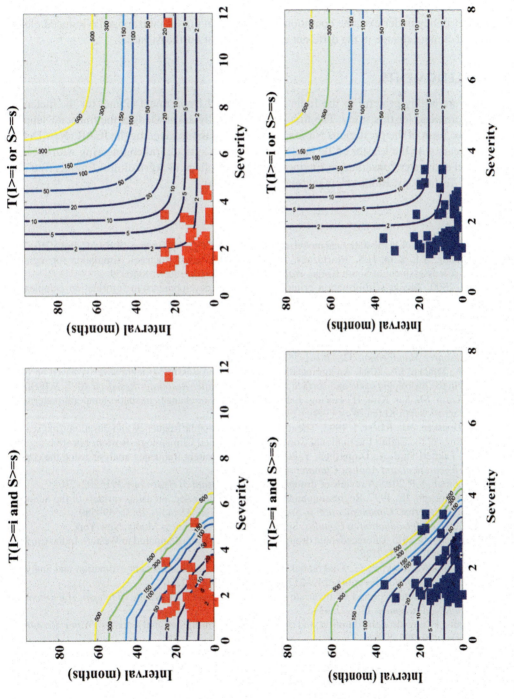

FIGURE 14.8 Bivariate return period analysis in the case of drought severity–interval for Gangtok (top) and Imphal (bottom).

bivariate and trivariate return period analyses showed Imphal is prone to drought events of moderate risk with higher frequencies in comparison to Gangtok. The AND type bivariate and trivariate return period analyses showed Gangtok to be extreme drought-risk prone with shorter frequency in comparison to Imphal. The results of the present study on joint probability multivariate return period analysis of different drought variables will help the water resources planners in devising a suitable water policy under extreme conditions in the present era of anthropogenic-induced global warming and climate change in the different environments of northeast India.

ACKNOWLEDGMENTS

The authors are highly grateful to the India Meteorological Department (Pune) for the meteorological data provided for this study. The first author gratefully acknowledges the financial support received in the form of the prestigious National Talent Scholarship given by the ICAR (New Delhi) during two years with the Master of Technology program in agricultural engineering with specialization in soil and water conservation Engineering at the College of Agricultural Engineering and Post-Harvest Technology, Ranipool, Gangtok, Sikkim (India).

REFERENCES

Akaike, H. 1974 A new look at the statistical model identification. *IEEE Trans Autom Control* AC-19(6):716–723.
Azam, M., Maeng, S.J., Kim, H.S., Murtazaev, A. 2018. Copula-based stochastic simulation for regional drought risk assessment in South Korea. *Water* **10**: 359, DOI: 10.3390/w10040359.
Bozdogan, H. 2000. Akaike's information criterion and recent developments in information complexity. *Journal of Mathematical Psychology* **44**: 62–91.
Chen, L., Singh, V.P., Guo, S., Mishra, A.K., Guo, J. 2013. Drought analysis using copulas. *Journal of Hydrologic Engineering* 18(7): 797–808.
Hangshing, L., Dabral, P.P. 2018. Multivariate frequency analysis of meteorological drought using Copula. *Water Resource Management* **32**: 1741–1758.
Keyantash, J.A., Dracup, J.A. 2004. An aggregate drought index: Assessing drought severity based on fluctuations in the hydrologic cycle and surface water storage. *Water Resources Research* 40(9): W09304.
Maeng, S.J., Azam, M., San Kim, H., Hwang, J.H. 2017. Analysis of changes in spatio-temporal patterns of drought across South Korea. *Water* 9(9): 679.
McKee, TB., Doesken, NJ., Kliest, J. 1993. The relationship of drought frequency and duration time scales. Proceedings of the Eighth International Conference on Applied Climatology. Boston, 179–184.
Mirabbasi, R., Fakheri-Fard, A., Dinpashoh, Y. 2012. Bivariate drought frequency analysis using the copula method. *Theoretical and Applied Climatology* 108: 191–206.
Mishra, A.K., Singh, V.P. 2010. A review of drought concepts. *Journal of Hydrology* **391**: 202–216.
Neath, AA, Cavanaugh, JE. 1997. Regression and time series model selection using variants of the Schwarz information criterion. *Communication in Statistics- Theory and Methods* 26(3):559–580.
Nelsen, R.B. 2007. *An Introduction to Copulas*. Springer Science & Business Media, New York.
Parida, B.R., Oinam, B. 2015. Unprecedented drought in North East India compared to Western India. *Current Science* **109** (110): 2121–2126.
Rossi, G., Benedini, M., Tsakiris, G., Giakoumakis, S. (1992) On regional drought estimation and analysis. *Water Resources Management* 6(4):249–277. doi:10.1007/BF00872280.
Shiau, J.T. 2003. The return period of bivariate distributed hydrological events. *Stochastic Environment Research Risk Assessment* 17(1–2): 42–57.
Shiau, J.T. 2006. Fitting drought duration and severity with two-dimensional copulas. *Water Resources Management* **20**: 795–815.
Shiau, J.T., Modarres, R. 2009. Copula-based drought severity-duration-frequency analysis in Iran. *Meteorological Applications* 16: 481–489.
Shiau, J.T., Shen, H.W. 2001. Recurrence analysis of hydrologic droughts of differing severity. *Journal of Water Resources Planning Management* **127**(1): 30–40.
Song, S., Singh, VP. 2010. Frequency analysis of droughts using the Plackett copula and parameter estimation by genetic algorithm. *Stochastic Environmental Research and Risk Assessment* **24**:783–805.

Tigkas, D., Vangelis, H., Tsakiris, G. 2013. The Drought Indices Calculator (DrinC). Conference Paper, 1333–1342. https://www.researchgate.net/publication/245542402.

Tsakiris, G., Pangalou, D., Vangelis, H. 2007. Regional drought assessment based on the Reconnaissance Drought Index (RDI). *Water Resources Management* 21: 821–833.

Waseem, M., Ajmal, M., Kim, T.W. 2015. Development of a new composite drought index for multivariate drought assessment. *Journal of Hydrology* **527**: 30–37.

15 Multivariate Assessment of Drought Using Composite Drought Index

Rasoul Mirabbasi and Deepak Jhajharia

CONTENTS

15.1 INTRODUCTION

Drought, a natural phenomenon, is the result of a lack of rainfall compared to the expected or normal rainfall in the selected region under investigation, and it should not be confused with the dry period specific to low rainfall areas, which is a permanent feature of a climate of selected region. Drought, an unwanted phenomenon in terms of its occurrence in the long term, is currently one of the major global problems related to water scarcity. It can be reduced with proper planning, but drought in Iran is becoming a severe humanity crisis throughout the country. There is a strong necessity for evaluation studies related to drought, planning, and preparedness to deal with the increase in damages caused by drought at the international, regional, and local levels. Iran, one of the drier regions of the world due to the climate and extensive climatic changes in most parts of country, has experienced many droughts in the past decades causing lots of damage to the economy and to the society as a whole.

Researchers use drought indices to investigate and evaluate drought. Until now, many indices have been developed and used by various researchers to study and evaluate different types of droughts (meteorological, hydrological, groundwater, agricultural, socioeconomic droughts). Precipitation is one of the most important variables used in the definition of drought, and more than 90% of studies related to drought have been conducted based on it (Morid et al., 2006). The Standardized Precipitation Index (SPI) was introduced by McKee et al. (1993). This index is calculated for each location based on the long-term precipitation time series for a given time scale. Morid et al. (2006) compared seven meteorological drought indices for monitoring droughts in Iran, and the results showed that two indices, namely, the SPI and EDI (Effective Drought Index), performed better than the other five indices selected to analyze drought events in Iran. Although the SPI is widely used all over the world today, it also has some limitations. Therefore, despite these limitations, Kao

DOI: 10.1201/9781003276548-15

and Govindaraju (2010) thought of modifying this widely used method and proposed the modified Standard Precipitation Index (SPImod). Mirabbasi Najafabadi et al. (2014) studied the long-term monitoring of droughts in Urmia, Iran, using different indices and concluded that the modified SPI provides more accurate results in drought assessment than the conventional SPI. Also, Mahmoudi and Mirabbasi (2015) compared the SPI and modified SPI for drought assessment in Kohkiloyeh and Boyer Ahmad provinces, Iran, and they concluded that the SPI cannot accurately estimate the seasonal changes of rainfall and that the modified SPI is a suitable substitute for it.

Groundwater is considered one of the most important water resources in any region, and its correct understanding and basic exploitation can play an important role in the sustainable development of the economic and social activities of a region, especially in arid and semiarid regions. Lack of proper knowledge and excessive exploitation of these resources will result in irreparable damages, such as severe and irreversible drops in the groundwater levels decrease in the flow of wells and changes in the groundwater flow pattern. For this purpose, it is necessary to carry out accurate studies of the fluctuations of groundwater levels in order to be aware of the groundwater situation and its optimal management. Akbari et al. (2009) reported that the occurrence of drought, overharvesting, population increase, increase in cultivated area, and a large number of harvesting wells are the main factors of the drop in the groundwater level of the aquifers of the Mashhad Plain, Iran. Mohammadi et al. (2012) studied the spatial and temporal changes in the groundwater level of Kerman Plain, Iran, and observed that the drop of the groundwater level has occurred in most places, especially in the northern and western exits.

Khan et al. (2008) investigated the relationship between drought and groundwater level in one of the plains of Australia and concluded that there is a strong relationship between the SPI and the shallow groundwater level of the region. Mendicino & Senatore (2008) proposed an index called the Groundwater Resource Index (GRI) and investigated the drought in the Calabria region of Italy by using the SPI and the GRI. They observed that the geological features of the basin are effective on the GRI and cause a delay in the change of the GRI compared to the SPI. On the other hand, the larger the timescale of the SPI, the higher its correlation with the GRI in this region. In addition, the GRI is more suitable than the SPI in predicting the condition of groundwater resources.

Abbasi et al. (2016) studied the temporal and spatial changes in the groundwater level of the Qorveh and Dehgolan Plains, Iran, and its relationship with drought. The results showed that the highest values of significant positive correlation coefficients were between the SPI-24 and the GRI-48 with a delay of 48 months, and in 89.8% of the wells, the SPI had a significant positive correlation with the GRI with a delay of 48 months. Therefore, groundwater drought has occurred with a time delay compared to meteorological drought, and during the studied period, the water level of piezometric wells decreased. Nasabpour et al. (2017) investigated the relationship between the SPI and the GRI in the Kerman–Baghein Plain, Iran. The results showed that the most severe meteorological drought in Kerman–Baghein Plain occurred in 2000 with severity value of –1.7, and the worst drought occurred in 2004 with severity value of 1.5. Based on the GRI, in the case of the Kerman–Baghein Plain, the driest period was observed in the year 2009 and the wettest period was observed in the year 2004. Drought is basically defined in relation to the lack of precipitation, but nowadays everyone has accepted that drought is a multivariable phenomenon and the fluctuations of any of the atmospheric and climatic variables can cause an increase or decrease in the intensity of this phenomenon. In the event of drought, the increase in temperature, evapotranspiration, and decrease in runoff cause the effects of precipitation deficiency to intensify (Govindaraju, 2013).

In the last few decades, efforts have been made to describe drought and find the pattern of precipitation of dry seasons in time and place to face drought conditions through drought indices. Single or univariate indices are useful to a large extent for specific locations and specific purposes and applications, but they do not provide a comprehensive picture of drought characteristics as a multivariate climate phenomenon. In addition, evaluating drought with a single-variable index may not be enough for significant and reliable drought assessment and decision-making. Therefore, efforts to develop multivariate indices based on a combination of different drought indices or different drought indicators in order to provide a comprehensive picture of the conditions and characteristics of drought have begun in previous years and are still ongoing. Waseem et al. (2015) developed a Composite Drought Index (CDI) for multivariate evaluation of droughts. The results showed that in comparison with univariate indices, such SPI and Streamflow Drought Index (SDI), the CDI provides a more comprehensive description of hidden variability in individual drought characteristics. In addition, it seems that the developed CDI is a flexible and effective physical index that depends on the weather conditions of the studied area.

The Boroujen Plain is one of the important plains situated in Chaharmahal and Bakhtiari province, Iran, in terms of agriculture and population. Groundwater is one of the most important sources of water supply for agriculture in the Boroujen Plain. In recent decades, the sudden increase in population and the development of agriculture and industry in the Boroujen Plain has caused an increase in the demand and a large extraction of groundwater resources, and the lack of appropriate recharge has caused a severe drop in the groundwater level in the Boroujen Plain. The aim of the present study is to evaluate drought conditions in the Boroujen Plain in the period of 1985–2015 using the CDI, which is formed by combining two drought indices, namely, the GRI and the SPImod plus the potential evapotranspiration. The potential evapotranspiration over the Boroujen Plain was obtained by using the Hargreaves–Samani method.

15.2 MATERIALS AND METHOD

15.2.1 STUDY AREA

Boroujen Plain, situated in the province of Chaharmahal and Bakhtiari, Iran, was chosen as the study area in this research to evaluate drought using the CDI. The Boroujen Plain, located between 31°51′ to 32°5′ north latitude and 51°6′ to 51°26′ east longitude, extends up to an area of 666.35 square kilometers, of which 381.14 square kilometers are alluvial formation and the rest is covered by mountains. Figure 15.1 shows the location of the Boroujen Plain. The elevation of the studied

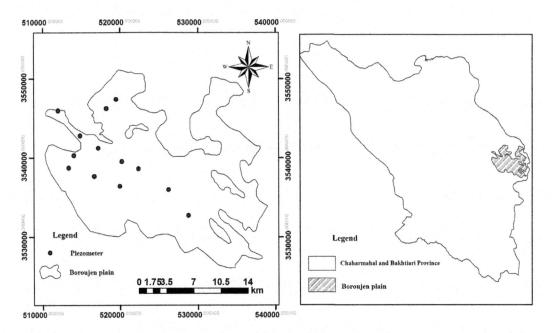

FIGURE 15.1 Location of the Boroujen Plain in Chaharmahal and Bakhtiari province, and piezometric wells in the plain.

area is between a minimum of 2121 meters above mean sea level and a maximum of 2983 meters above mean sea level. There are 286 wells in the Boroujen Plain, which annually empty about 34 million cubic meters of groundwater resources of the plain and to deliver to various consumers. More than 90% of the groundwater of the Boroujen Plain is used in the agricultural sector in the agricultural season. Part of the drinking water in the city of Boroujen is also supplied from these wells. In the Boroujen Plain, groundwater changes are monitored on a monthly basis through 17 observation wells, of which 13 piezometers have data of suitable length (see Figure 15.1).

15.2.2 Used Data

In this research, the monthly rainfall data was used to calculate the SPImod, and the monthly data of the groundwater level of 13 piezometers in the Boroujen Plain to calculate the GRI. The maximum and minimum temperature of the Boroujen station for the ETP calculation for the period of 31 years

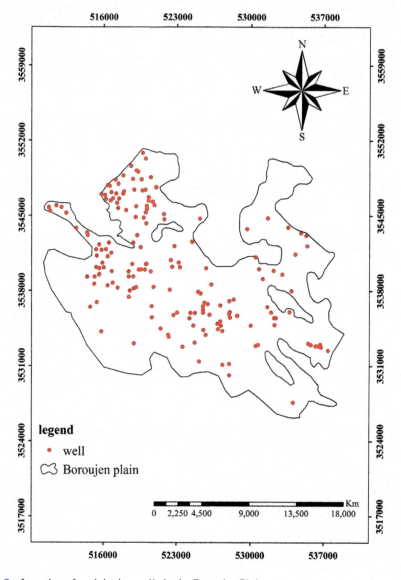

FIGURE 15.2 Location of exploitation wells in the Boroujen Plain.

from 1985 to 2015 were used in the drought analysis in the plain. The data used in this study were obtained from Chaharmahal and Bakhtiari Regional Water Company and the Iran Meteorological Organization (IRIMO).

With the increase in population and the development of agricultural and industrial activities, the demand for water has also increased. This increase in the need for water has caused deep and semideep wells to be dug in the plains of the region. As shown in Figure 15.2, 286 exploitation wells have been dug in the Boroujen Plain, and the concentration of these wells is mostly in the north and center parts of the plain.

After collecting the data and reconstructing the missing data, the monthly rainfall time series during a common period of 31 years (1985–2015) was used to calculate the SPImod drought index in the Boroujen Plain. The groundwater level data of 13 piezometric wells for 31 years (1985–2015) were used to calculate the GRI in the Boroujen Plain. Piezometric wells were selected for the study considering various cases such as having long-term data, low fluctuations of the water table level, and low missing data.

The daily maximum, minimum, and average temperature data in Celsius during a common time period were used to calculate the potential evapotranspiration in the Boroujen Plain. Finally, in this study, a CDI, which is a combination of three variables (GRI, SPI, and ETP), was created and used to investigate drought in the Boroujen Plain. In the following, the method of calculating the SPImod and GRI as well as the ETP will be explained.

15.2.3 Modified Standardized Precipitation Index (SPImod)

The SPImod was presented by Kao and Govindaraju (2010) with the aim of correcting the disadvantages of the SPI method. In this method, first, the monthly rainfall data is extracted and then the X_w^{month} time series is formed on the scale of w months (w = 1, …, 6, …, 12) for each month of the year as follows:

$$X_w^{month}(g) = X_w(12(g-1)+m) = X_w(t) \tag{15.1}$$

where g is the year index; m is the month index and is equal to 1 (January), 2 (February), …, 12 (December); and t is the time index equal to t = 12(g – 1) + m.

Thus, the samples in each X_w^{month} group are collected annually. It is clear that as long as w ≪ 12, the samples will not overlap. In other words, the degree of autocorrelation between the data will be greatly reduced. On the other hand, samples within the same X_w^{month} group are subject to similar seasonal effects. Therefore, seasonal variations are considered in a suitable way.

By fitting the statistical distribution to each group separately (i.e., creating $F_{X_w^{month}}$), SPImod is calculated similar to the conventional SPI from the following equation (Mirabbasi et al., 2013):

$$SPI_w^{month} = \varnothing^{-1}\left(u_w^{month}\right) = \varnothing^{-1}\left(F_{X_w^{month}}\left(x_w^{month}\right)\right) \tag{15.2}$$

The gamma cumulative probability is converted to a standard normal variable. In this method, the values of the standard normal variable (Z) are calculated as follows:

For $0 < H(x) < 0.5$

$$Z = SPI = -\left(t - \frac{C_0 + C_1.t + C_2.t^2}{1 + d_1.t + d_2.t^2 + d_3.t^3}\right) \tag{15.3}$$

For $0.5 < H(x) < 1$

$$Z = SPI = +\left(t - \frac{C_0 + C_1.t + C_2.t^2}{1 + d_1.t + d_2.t^2 + d_3.t^3}\right) \tag{15.4}$$

where

$$t = \sqrt{\ln\left(\frac{1}{\left(H(x)\right)^2}\right)} \qquad 0 < H(x) < 0.5 \qquad (15.5)$$

$$t = \sqrt{\ln\left(\frac{1}{\left(1-H(x)\right)^2}\right)} \qquad 0.5 < H(x) < 1 \qquad (15.6)$$

In the preceding relationships, C_0, C_1, C_2, d_1, d_2, and d_3 are equal to 2.52, 0.80, 0.01, 1.43, 0.12, and 0.001, respectively. Finally, the SPI_w^{month} are arranged in the chronological order.

According to Table 15.1, the wetness condition is divided into seven categories from extremely wet conditions to extremely dry conditions based on the SPI.

15.2.4 GROUNDWATER RESOURCE INDEX (GRI)

Mendicino and Senatore (2008) proposed an index called the Groundwater Resource Index, considered to be a reliable index for modeling, monitoring, and forecasting the drought situation for the Mediterranean region. The different lithological conditions of the Earth have an important effect on the response of the underground waters of the basin in the summer season compared to the winter season rains. In general, GRI values have many spatial changes and are sensitive to the lithology characteristics of the studied area. The most important feature of this index is its high correlation with the average runoff in some rivers of the basin in predicting summer droughts. The value of the GRI is calculated by using the following relationship:

$$GRI_{y.m} = \frac{D_{y.m} - \mu_{D.m}}{\sigma_{D.m}} \qquad (15.7)$$

where $GRI_{y.m}$ is the index value in month m of year y, $D_{y.m}$ is the water level in month m of year y, $\mu_{D.m}$ is the average data of water level in month m for year D, and $\sigma_{D.m}$ is the standard deviation of the data of the groundwater level of month m for D years. The classification of the wetness condition based on the GRI is the same as the SPI and is given in Table 15.1.

TABLE 15.1

Classification of Wetness Conditions According to the GRI and SPI Values

Wetness Condition	SPI Interval
Extremely wet	$SPI \geq 2$
Severely wet	$2 > SPI \geq 1.5$
Moderately wet	$1.5 > SPI \geq 1$
Normal	$1 > SPI > -1$
Moderately dry	$-1 \geq SPI > -1.5$
Severely dry	$-1.5 \geq SPI > -2$
Extremely dry	$SPI \leq -2$

15.2.5 POTENTIAL EVAPOTRANSPIRATION (ETP)

Zarei et al. (2015) examined different methods to calculate the ETP. Based on their study, they proposed the Hargreaves–Samani method to calculate potential evapotranspiration in Chaharmahal and Bakhtiari province. Therefore, in this research, the Hargreaves–Samani (1982) method was used to calculate the ETP at the Boroujen station. The formula for calculating the ETP through the Hargreaves–Samani method is as follows:

$$ETP = 0.0023*Ra*\left(T_{mean} + 17.8\right)*\left(T_{max} - T_{min}\right)^{0.5} \tag{15.8}$$

where T_{max}, T_{min}, and T_{mean} values are the maximum, minimum, and average daily temperature (in °C), respectively. Ra is the maximum possible solar radiation to the Earth's surface (in mm/day).

Ghamarnia et al. (2012) showed that for areas that are large enough, and using reliable data and a sufficient number of stations, the initial coefficient of 0.0023 used in the Hargreaves–Samani equation does not need to be adjusted and can be used with high confidence.

15.2.6 COMPOSITE DROUGHT INDEX (CDI)

The method of calculating the CDI created in this study is derived from the method proposed by Waseem et al. (2015). They created a composite drought index that was composed of precipitation, discharge, Normalized Difference Vegetation Index (NDVI), and land surface temperature (LST).

In this method, two types of variables are considered. The first group of variables are directly proportional to the occurrence of drought, which are represented by Y (its larger value indicates the aggravation of drier conditions and vice versa). The second category is the variables that are inversely proportional to the occurrence of drought and are indicated by X (its larger value indicates wetter conditions and vice versa).

In this study, existing variables, including the GRI and SPImod (variable X), and ETP evapotranspiration (variable Y) were considered. However, the number of X and Y variables can be increased depending on the availability of data. In the following, the steps for calculating the CDI are presented:

1. The historical data set (time series of monthly average corresponding to the year (t_i () of variables X and Y is arranged in a matrix (Equation 15.9), which is column n and row m as follows:

$$DB_k = \begin{matrix} t_1 \\ \vdots \\ t_m \end{matrix} \left[\begin{pmatrix} X_{11} & \cdots & X_{1n} \\ \vdots & \ddots & \vdots \\ X_{m1} & \cdots & X_{mn} \end{pmatrix} \begin{pmatrix} Y_{11} & \cdots & Y_{1n} \\ \vdots & \ddots & \vdots \\ Y_{m1} & \cdots & Y_{mn} \end{pmatrix} \right] \tag{15.9}$$

To evaluate the drought, DB_k is a single matrix (n × m) containing average monthly variables X and Y in the duration of drought K (weekly, monthly, or annually).

In this study, (K is monthly) t_i is the corresponding year and I = 1, 2, ..., m. X_{ij} is the ith average value of the jth variable of X (here SPI and GRI), Y_{ij} is the ith average value of the jth variable of Y (in this study ETP) and j = 1, 2, ..., n. In the DB_k matrix, brackets are used to distinguish between X and Y variables.

2. DB_k includes variables with different dimensions, so by using $N_{ij} = \dfrac{X_{ij}}{\sum_{i=1}^{m} abs\left(X_{ij}\right)}$ and

$N_{ij} = \dfrac{Y_{ij}}{\sum_{i=1}^{m} abs\left(Y_{ij}\right)}$ are converted into dimensionless variables.

The dimensionless variables are stored in a new matrix NDB_k as follows:

$$NDB_k = \begin{matrix} t_1 \\ \vdots \\ t_m \end{matrix} \begin{pmatrix} N_{11} & \cdots & N_{1n} \\ \vdots & \ddots & \vdots \\ N_{m1} & \cdots & N_{mn} \end{pmatrix} \tag{15.10}$$

where N_{ij} is the corresponding standardized value.

3. In the next step, the corresponding weights of the j variables are determined based on the entropy of the variable. Entropy, in the field of information theory, was introduced by Shannon (1948) to measure the degree of uncertainty of information.

Entropy is a more effective measure than variance to describe information and data features (Rajsekhar et al., 2014). Entropy weights were introduced in order to present a balanced relationship between variables and provide unbiased relative weights based on the individual variability of the data. A greater entropy weight indicates a greater change in the investigated variable.

$$Na_{ij} = \frac{abs(X_{ij})}{\sum_{i=1}^{m} abs(X_{ij})} \tag{15.11}$$

$$Na_{ij} = \frac{abs(Y_{ij})}{\sum_{i=1}^{m} abs(Y_{ij})} \tag{15.12}$$

$$EN_j = \frac{1}{\ln m} \sum_{i=1}^{m} N_{ij} \ln(Na_{ij}) \tag{15.13}$$

$$DS_j = 1 - EN_j \tag{15.14}$$

$$Ew_j = \frac{DS_j}{\sum_{j=1}^{n} DS_j} \tag{15.15}$$

where EW_i is the weight assigned to the variables; So that $\sum_{j=1}^{n} EW_i = 1$. DS_j is a measure of entropy between the jth variables.

4. In the next step, the wettest possible conditions of the study area (MWC) (Equation 15.16) were defined by selecting a set of maximum historical data values of the jth variables belonging to X and minimum values of the jth variable belonging to Y. To define the driest possible conditions (MDC), the corresponding inverse values were used (Equation 15.17):

$$MWC = a_1^+, a_2^+, \ldots, a_n^+ \tag{15.16}$$

$$MDC = a_1^-, a_2^-, \ldots, a_n^- \tag{15.17}$$

where

$$a_j^+ = \{\max N_{ij}, j \ X; \min N_{ij}, j \ Y\} \tag{15.18}$$

$$a_j^- = \{\min N_{ij}, j \ X; \max N_{ij}, j \ Y\} \tag{15.19}$$

5. So far, a large number of similarity criteria have been created and used in different fields (Cha, 2007). In the present study, the differences in current conditions (PC) (data row N_{ij} of data related to year t_i, for example for t_1, PC = $(N_{11}, N_{12}, ..., N_{1N})$ and similarly for the rest of t_1) in comparison with MWC and MDC were estimated with and using the weighted Euclidean distance (according to Equations 15.20 and 15.21):

$$S_i^+ = \sqrt{\sum_{j=1}^{n} Ew_j \left(S_{ij}^+\right)^2} \tag{15.20}$$

$$S_i^- = \sqrt{\sum_{j=1}^{n} Ew_j \left(S_{ij}^-\right)^2} \tag{15.21}$$

where S_i^+ is the weighted Euclidean distance between MWC and PC, and S_i^- is the weighted Euclidean distance between MDC and PC:

$$S_{ij}^- = N_{ij} - a_j^+ \tag{15.22}$$

$$S_{ij}^+ = N_{ij} - a_j^- \tag{15.23}$$

Finally, the generated CDI (Equation 15.24) is estimated based on the similarity of each PC to NWC. Since the value of CDI (which varies between 0 and 1) is defined based on relative similarity to NWC, its highest value (close to 1.0) indicates the wettest conditions and vice versa. In addition, the severity of drought is defined and classified based on the calculated value of CDI_i (which varies from 0.0 to 1.0) according to the criteria defined by Waseem et al. (2015). Table 15.2 shows the classification criteria for the drought conditions based on the CDI values.

$$CDI_i = \frac{S_j^+}{S_j^- + S_j^+} \tag{15.24}$$

In the present study, a code was written and used in the MATLAB® software to calculate the indicators used for evaluating the drought conditions in the Boroujen Plain.

TABLE 15.2
Classification of Drought Conditions According to the CDI Values

Drought Condition	CDI Interval
Extremely dry	CDI < 0.1
Severely dry	0.2 > CDI ≥ 0.1
Moderately dry	0.3 > CDI ≥ 0.2
Mild dry	0.4 > CDI ≥ 0.3
Near normal to normal	0.5 > CDI ≥ 0.4
Above normal	CDI ≥ 0.5

Source: Waseem et al. (2015).

15.3 RESULTS AND DISCUSSION

The results of drought evaluation in the Boroujen Plain using the SPImod at Boroujen station and through the GRI at piezometers (No. 6 is used as an example) during a 31-year period from 1985 to 2015 are given in Figures 15.3 and 15.4, respectively (piezometer No. 6 is used as an example).

As seen from Figure 15.3, the number of wet months at the Boroujen station during the studied period was more than the number of dry months based on the SPImod. Therefore, in terms of meteorological drought during the 31-year period, the number of wet periods was more than the number of dry periods, although severe droughts were also recorded.

As shown in Figure 15.4, the one-month GRI time series in the case of piezometer No. 6 show that at the beginning of the study period (1985) until the end of 1999, the longest continuous wet

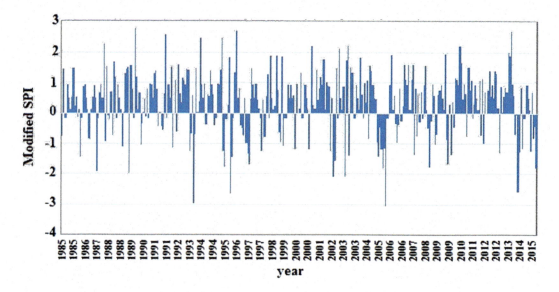

FIGURE 15.3   Time series of the SPImod of the Boroujen station for a 31-year period (1985–2015).

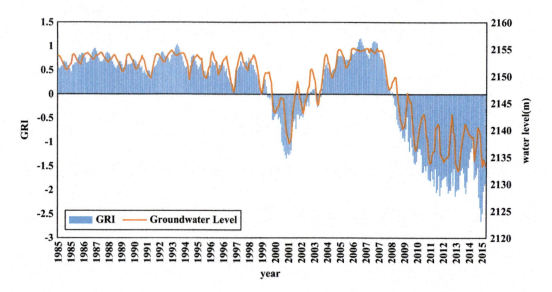

FIGURE 15.4   Time series of the GRI and groundwater level at monthly timescale in piezometer No. 6 during 1985–2015.

period occurred, with recording of the highest amount of 1.24 in November 1993. It shows that the condition of the Boroujen aquifer was good until 2008. The longest period of drought started from December 2008 and continued until the end of the study period in the year 2015.

The most severe one-month GRI value at piezometer No. 6 was obtained as –2.65 in the month of August in 2015. The results of the investigation of groundwater drought with the GRI indicate an increase in groundwater drought in recent years. Therefore, the comparison of Figure 15.2 with Figures 15.3 and 15.4 shows that the main reason for the occurrence of groundwater drought in the Boroujen Plain was not the decrease in precipitation but the increase in the amount of harvesting from the aquifer due to the development of agricultural lands, which is the main cause of the deterioration of the aquifer.

Also, the delineation map of the GRI in monthly timescales in March 2015 is presented in Figure 15.5. Comparison of the spatial changes of the GRI values in March 2015 (see Figure 15.5) with the distribution map of harvesting wells in Boroujen Plain (see Figure 15.2) shows that the groundwater

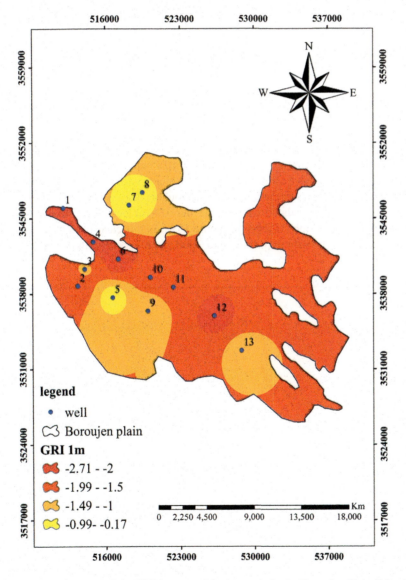

FIGURE 15.5 Delineation map of the GRI in a monthly timescale in the Boroujen Plain (March 2015).

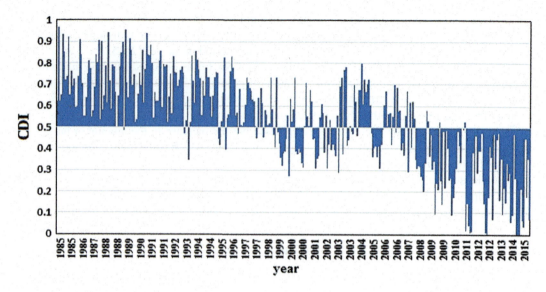

FIGURE 15.6   Time series of the CDI from 1985 to 2015 at piezometer No. 6 in the Boroujen plain.

drought situation in the Boroujen Plain is worse in the areas where the harvesting wells are more concentrated in the selected region.

Hosseinpour et al. (2016) analyzed the meteorological droughts and groundwater drought conditions using the modified SPI and GRI in the Shahrekord Plain aquifer. The results showed that the recent droughts in the plain have affected the groundwater resources with a time delay. Also, as the timescale of the indices increases, the frequency of droughts decreases and their duration increases.

The time series of the calculated CDI during the 31-year period at the Boroujen station and piezometer No. 6 (as an example) is given in Figure 15.6. As it is clear from the figure, at the beginning of the study period (in 1985) until the end of 2000, the longest continuous wet period occurred.

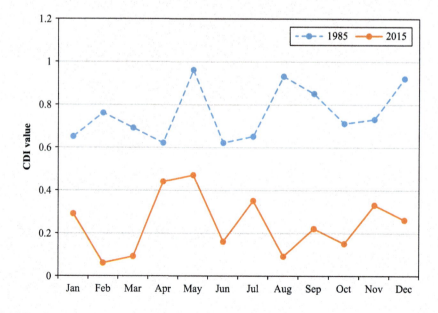

FIGURE 15.7   CDI values for the Boroujen Plain for the years 1985 to 2014.

Also, the highest value associated with the wet event at piezometer No. 6 was 0.97 in the month of November in the year 1989. During this period, the area also faced drought for a few months. After 2000 to 2008, consecutive dry and wet months were recorded in the plain. After that, the longest drought period began, which continued from the middle of 2008 to the end of the study period (2015). The most severe value of the CDI in this period in piezometer No. 6 was recorded as 0 during the months of August to November 2014.

Figure 15.7 shows the values of the CDI from the beginning (1985) to the end (2015) of the study period. As it is clear from these Figures, the value of the index in 2015 in all the twelve months is less than the corresponding values in the last 31 years, and the values of the index in all months are less than 0.5, which indicates the occurrence of drought during the last years of the study period. In 2015, extremely severe drought occurred in the plain in the months of February, March, and August, and in months of June and October, the drought was found to be severe in the plain.

The results of the drought study through the CDI indicate the occurrence of severe droughts in recent years, especially in 2014 and 2015 (the end of the study period), because of the increase in evapotranspiration in recent years due to the increase in average air temperature and the increase in the harvest of groundwater sources by the agriculture and allied sectors. Therefore, the main reason for the droughts in the region can be attributed to the lack of proper management during the past three to four decades.

15.4 CONCLUSIONS

In this study, a composite drought index was developed by combining the modified SPI, GRI, and potential evapotranspiration. This index simultaneously provides a description of the meteorological and groundwater drought conditions, and was applied to study drought conditions in the Boroujen Plain, Iran. The CDI showed that the Boroujen Plain has faced severe drought since 2008, especially at the end of the study period in the year 2015. The zoning map of the GRI over the plain shows the conditions of very severe drought in the whole plain in 2015. Also, due to the high concentration of exploitation wells in the northern and center parts of the plain, these areas have a worse drought situation than other areas. Zare Bidaki & Khoshnoudmotlagh (2013) reported that the most important factor of groundwater loss in the Shahrekord Plain is overexploitation, which is in total agreement with our findings based on the CDI in this study. The downward trend of the CDI in the plain in recent years indicates the effectiveness of factors, such as decreasing precipitation, increasing air temperature and evapotranspiration, and overexploitation of water for drinking and agriculture due to the increase in population. The share of increase in exploitation of the aquifer has been more influential than other factors. Finally, analysis of the results shows that if management strategies are not considered to reduce the amount of exploitation of the aquifer, more severe droughts will follow in the coming years, which will cause irreparable damage, such as land subsidence in the plains.

ACKNOWLEDGMENTS

The authors gratefully acknowledge the agencies that provided the data used in this study, namely, the Chaharmahal and Bakhtiari Regional Water Company, Chaharmahal and Bakhtiari Province, Iran; and the Iran Meteorological Organization,.

REFERENCES

Abbasi, F., Farzadmehr, J., Chapi, K., Bashiri, M., Azarakhshi, M. (2016). Spatial and temporal variations of groundwater quality parameters in Qorveh- Dehgolan plain and its relationship with drought. *Hydrogeology*. 1(2), 11–23. https://doi.org/10.22034/hydro.2016.5002

Akbari, M., Jorgeh, M.R., Madani Sadat, H. (2009). Assessment of decreasing of groundwater-table using Geographic Information System (GIS) (Case study: Mashhad Plain Aquifer). *Journal of Water and Soil Conservation*. 16(4), 63–78.

Cha, S.H. (2007). Comprehensive survey on distance/similarity measures between probability density functions. *International Journal of Mathematical Models and Methods in Applied Sciences.* 1(4), 300–307.

Ghamarnia, H., Rezvani, V., Khodaei, E., Mirzaei, H. (2012). Time and place calibration of the Hargreaves equation for estimating monthly reference evapotranspiration under different climatic conditions. *Journal of Agricultural Science.* 4(3), 111–122.

Govindaraju, R.S. (2013). Special issue on data-driven approaches to droughts. *Journal of Hydrologic Engineering.* 18(7), 735–736.

Hargreaves, G.H., Samani, Z.A. (1982). Estimating potential evapotranspiration. *Journal of Irrigation and Drainage Engineering.* 108(3), 223–230.

Hosseinpor, Z., Radfar, M., Mirabbasi, R. (2016). Analysis of meteorological and groundwater droughts using modified SPI and GRI indices in Shahrekord plain aquifer. 8th International Conference on Integrated Natural Disaster Management, Tehran, Iran.

Kao, S.C., Govindaraju, R.S. (2010). A copula-based joint deficit index for droughts. *Journal of Hydrology.* 380(1), 121–134.

Khan, S., Gabriel, H.F., Rana, T. (2008). Standard precipitation index to track drought and assess impact of rainfall on water tables in irrigation areas. *Irrigation Drainage System.* 22(2), 159–177.

Mahmoudi, A., Mirabbasi, R., (2015). Comparison of the performance of SPI and modified SPI indices in detecting drought events (Case study: Dehdasht city, Kohgilooyeh and Boyer Ahmed Province). Second National Conference on Water Crisis, Shahrekord, Iran.

Mckee, T.B., Doseken, N.J., Kleist, J. (1993).The relationship of drought frequency and duration times scales. Eightieth Conference on Applied Climatology, 17-22 January 1993, Anaheim, Califonia.

Mendicino, G., Senatore, A. (2008). A Ground Water Resource index (GRI) for drought monitoring and forecasting in a Mediterranean climate. *Journal of Hydrology.* 357(3), 282–302.

Mirabbasi, R., Anagnostou, E.N., Fakheri-Fard, A., Dinpashoh, Y., Dinpashoh, S. (2013). Analysis of meteorological drought in northwest Iran using the Joint Deficit Index. *Journal of Hydrology.* 492, 35–48.

Mirabbasi Najafabadi, R., Fakheri-Fard, A., Dinpashoh, Y., Eslamian, S. (2014). Longterm drought monitoring of Urmia using joint deficit index (JDI). *Water and Soil Science.* 23(4), 87–103.

Mohammadi, S., Salajegheh, A., Mahdavi, M., & Bagheri, R. (2012). An investigation on spatial and temporal variations of groundwater level in Kerman plain using suitable geostatistical method (During a 10-year period). 19(1), 60–71. https://doi.org/10.22092/ijrdr.2012.103069

Morid, S., Smakhtin, V., Moghaddasi, M. (2006). Comparison of seven meteorological indices for drought monitoring in Iran. *International Journal of Climatology.* 26(7), 971–985.

Nasabpour, S., Heydari, E., Khosravi, H., (2017). Investigating the relationship between the SPI index and the GRI index (case study: Kerman-Baghein Plain). The 4th Conference on Environmental Planning and Management. Tehran, Iran.

Rajsekhar, D., Singh, V.P., Mishra, A.K. (2014). Multivariate drought index: An information theory based approach for integrated drought assessment. Journal of Hydrology. 526, 164–182.

Shannon, C.E. (1948). A mathematical theory of communication. Reprinted with corrections from the Bell System. *Technical Journal.* 27(3), 379–423.

Waseem, M., Ajmal, M., Kim, T.W. (2015). Development of a new composite drought index for multivariate drought assessment. *Journal of Hydrology.* 527, 30–37.

Zare Bidaki, R., Khoshnoudmotlagh, S. (2013). The effect of overexploitation on groundwater resources of Shahrekord Plain. *Water Management in Dry Areas.* 1, 15–24.

Zarei, M., Tabatabaei, S., Babazadeh, H., Sedghi, H. (2015). Determining the best radiation model for Hargreaves-Samani equation in Shahrekord plain using lysimeter data. *Iranian Water Researches Journal.* 9(3), 47–56.

16 Multimodel Ensemble–Based Drought Characterization over India for 21st Century

Vivek Gupta and Manoj Kumar Jain

CONTENTS

16.1 INTRODUCTION

Drought is a creeping natural phenomenon requiring long-term predictions for policy makers to effectively lay out the plan for water management. Significant spatial and temporal variability in water resources accompanied with population growth, different living standards, and agricultural and industrial practices possess various types of stresses resulting in scarcity of water in different regions at different time intervals. Sea level rise, change in monsoon, short-duration but frequent severe storms and floods, and prolonged and more severe drought can be observed in the near future (Brenkert and Malone, 2005). Since the Indian economy mainly depends on agriculture, which primarily depends on the annual monsoon, the future patterns of precipitation distribution and probable occurrence of drought throughout the country are needed to alleviate losses that occur during disasters and for developing adaption approaches to deal with climate change and its effects on the water resources of the country (Lal et al., 2001).

Although a huge amount of literature is available for assessing the impacts of climate change on droughts globally and regionally, there exists uncertainty in drought prediction partly due to the selection of drought indices and partly due to the selection of the model (Kim et al., 2015). Drought patterns have been found to vary spatially as well as temporally. An increasing trend of drought severity has been found in Iran (Abarghouei et al., 2011), Nebraska (Wu et al., 2015), southern Europe (Lehner et al., 2017), North America (Rind et al., 1990), and also globally (Dai, 2011; Sheffield and Wood, 2007). However, a decreasing trend has been identified in a large region of China (Wang et al., 2014), most of Asia (Kim and Byun, 2009), and some parts of Iran (Golmohammadi, 2012). In India, very few studies have been reported on the assessment of the impacts of climate change on drought. In recent decades, drought intensities and the percentage of drought affected area in India increased due to global warming and mainly due to warming of the equatorial Indian Ocean (Kumar et al., 1994; Kumar et al., 2013). In the monsoon period, wet days have significantly reduced along the east coast pointing to extra intense rainfall events and prolonged dry spells (Roy et al., 2003). The present study has been undertaken mainly to investigate the likely occurrence of mesoscale meteorological drought over India for different climatic predictions on a spatiotemporal scale.

DOI: 10.1201/9781003276548-16

For this purpose, SPI-12 has been utilized as the drought indicator. Future projected precipitation data from various global climate models (GCMs) has been obtained for representative concentration pathway (RCP) 8.5. A divisive clustering algorithm called the K-means clustering algorithm has been exploited to find the homogeneous drought regions for future climatic conditions in India. For finding the optimum number of clusters, the average silhouette width and Dunn index have been used. Trend analysis of drought characteristics has also been performed in this study using the Mann–Kendall test and Sen's slope test. Further, periodicity analysis of droughts was conducted by using the Fourier transform approach.

16.2 K-MEANS CLUSTERING

MacQueen (1967) developed an unsupervised divisive learning algorithm commonly known as K-means clustering for classifying multivariate data. Equation 16.1 shows the objective function which K-means algorithm tries to minimize:

$$J = \sum_{j=1}^{K}\sum_{i=1}^{M} Y_i - C_j^{2} \tag{16.1}$$

where $Y_i - C_j$ is the Euclidean distance from the ith data point to the jth cluster center in multidimensional space of data attributes. M is the total number of data points and K is the number of clusters.

The steps involved in K-means clustering are as follows:

1. Coordinates of K-cluster centers are guessed randomly.
2. Each data point is assigned to a cluster whose center has a minimum distance from that data point.
3. Coordinates of the center are updated by averaging the Euclidean distances of all data points from the center of the cluster to which the data points are assigned, as shown in Equation 16.2:

$$Z_j = \frac{1}{n_j}\left(\sum_{\forall Y_p \in C_j} Y_p\right) \tag{16.2}$$

where n_j is the total number of feature vectors in the jth cluster and Y_p are their coordinates for the jth cluster.

4. The process of assigning the data points to the cluster and updating the cluster is repeated until convergence.

Since the final result is very sensitive to the initial guess of the cluster center, the algorithm is repeated again and again with different cluster centers for the best results (Dalton et al., 2009). The optimum number of clusters needs to be worked out before applying the K-means clustering algorithm for getting well-separated compact clusters. Cluster validity indices such as the Dunn index (Dunn, 1974), average silhouette width index (ASWI) (Rousseeuw, 1987), and the Davies and Bouldin index (DBI) (Davies and Bouldin, 1979) are generally used to obtain the optimum number of clusters. In this study, we have used the average silhouette width index for validation of clusters.

The silhouette width can be calculated as follows:

$$S(i) = \frac{\alpha(i) - \beta(i)}{\max\big(\alpha(i), \beta(i)\big)} \tag{16.3}$$

where $\alpha(i)$ is the mean distance of ith data point with other data points of the same cluster and $\beta(i)$ is the average dissimilarity of a data point with neighboring clusters. Here a neighboring cluster of a data point is considered as the cluster (other than the cluster that contains the data point) whose average dissimilarity is the lowest from a data point.

16.3 PERIODICITY ANALYSIS USING FOURIER ANALYSIS

Fourier analysis comprises the decomposition of a time series as various sinusoidal functions (Bloomfield, 2000; Pollock, 2009). Thus, a cyclic time series can be written in the form

$$y_t = \sum_{j=0}^{m} \left[A_j \sin(\omega_j t) + B_j \sin(\omega_j t) \right] + \varepsilon_t \qquad (16.4)$$

where t is time; A_j and B_j, $j = 0, \ldots, m$ are estimable parameters; and $\in_t$, $t = 1, \ldots, n$ are identically independent random variables with zero mean and unit variance. The value of $m = n/2$ if n is even and $(n-1)/2$ if n is odd, and $\omega_j = (2\pi j / n)$ are frequency included in cyclic fluctuation. Parameters A_j and B_j can be estimated with linear regression using the following expressions (Pollock, 2009):

$$A_j = \frac{2\sum_{t=1}^{n} \sin(\omega_j t) y_t}{n} \qquad (16.5)$$

$$B_0 = \frac{1}{n} \sum_{t=1}^{n} y_t \qquad (16.6)$$

$$B_j = \frac{2\sum_{t=1}^{n} \cos(\omega_j t) y_t}{n}, j = 1, \ldots, m \qquad (16.7)$$

$$A_j = \frac{2\sum_{t=1}^{n} \sin(\omega_j t) y_t}{n} \qquad (16.8)$$

The factor $\theta_j = \sqrt{(A_j^2 + B_j^2)}; j = 0, \ldots, m$ can be defined as the amplitude of the jth periodic component and indicates significance of components within the sum.

16.4 RESULTS AND DISCUSSION

The minimum value of the DBI and maximum value of the ASWI represent the optimum number of clusters. For RCP 8.5, the maximum value of the ASWI is for two clusters and the minimum value of the DBI is for five clusters but second highest value of the ASWI can be seen at seven clusters for which DBI also shows second least value. Therefore, the optimum number can be considered as seven. The cluster generated are shown in Figure 16.1, and it is important to notice that the generated drought homogeneous clusters are in accordance with different climate zones in India and also with homogeneous regions generated by Pattanaik (2007) and Parthasarathy et al. (1987).

The Gumbel distribution is used to generate the SPI at a scale of 12 months for each grid point across India. Further, various drought characteristics were calculated using the theory of runs. A run is considered when the SPI value falls below –1. The Mann–Kendall and Sen's slope tests were applied to test the temporal trends in drought characteristics for each grid points. It can be seen

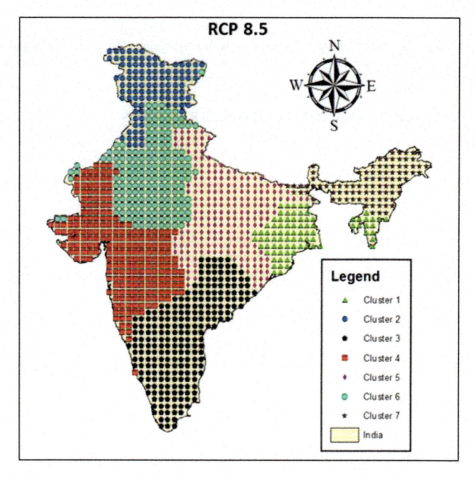

FIGURE 16.1 Homogeneous drought clusters generated using *K*-means clustering.

from Table 16.1 that the Sen's slope for drought magnitude is positive, indicating a likely decrease in drought risk with time since the more negative value of magnitude corresponds to the more severe drought, thus the positive slope represents decreasing drought risk. In addition, drought duration is exhibiting a decreasing trend and the interval between subsequent droughts is showing a positive trend, indicating the likelihood of fewer shorter droughts as time progresses in the 21st century.

TABLE 16.1

Average Value of Sen's Slope for Different Clusters and for All of India

	Magnitude	Severity	Duration	Interval
Cluster 1	0.6621	0.0123	−0.3801	2.0655
Cluster 2	0.1576	0.0015	−0.0939	−0.2484
Cluster 3	1.2934	0.0315	−0.6319	1.6839
Cluster 4	0.3953	0.0025	−0.2143	2.9335
Cluster 5	0.2465	−0.0001	−0.1589	1.7122
Cluster 6	0.6800	0.0109	−0.4066	1.4067
Cluster 7	0.4247	0.0109	−0.1827	0.6163
All India	0.5754	0.0190	−0.2687	1.5688

The Sen's slope value of drought interval is positive for almost all clusters, except Cluster 2, which indicates that the interval between two consecutive drought events is likely to increase in the future for almost the whole country, except in Jammu and Kashmir and some parts of Punjab. It is interesting to observe from Figure 16.2 that reduction of drought risk is more in South India as compared to other parts of the country, since in Southern India, magnitude and duration of drought show significant positive and negative trends, respectively. Parts of Gujarat, Maharashtra, Madhya Pradesh, Uttar Pradesh, and middle and southern Northeastern India are likely to witness positive increasing trends in the drought interval resulting in fewer droughts in these regions in the future.

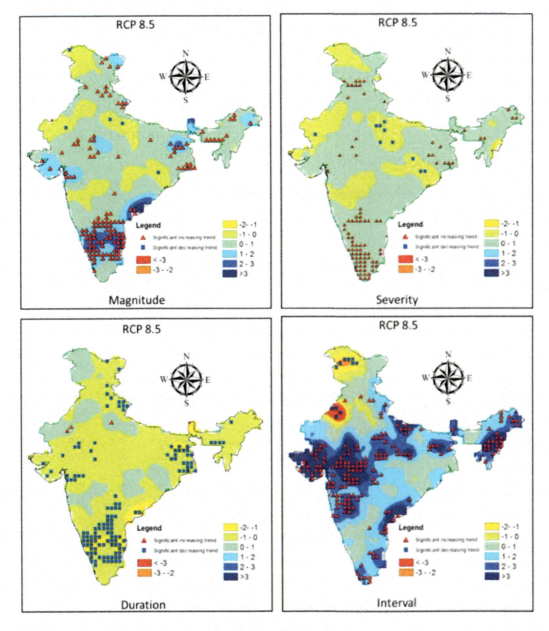

FIGURE 16.2 Spatial distribution of temporal trend of drought magnitude, severity, duration, and drought interval.

16.5 PERIODICITY ANALYSIS USING FOURIER TRANSFORM

Periodicity analysis of drought in monsoon months was carried out using Fourier analysis. Only significant periodicities above the 90% confidence interval have been considered. For examining the hidden cycles in drought series of SPI-12 values of all four monsoon months, periodograms of series were calculated for all 1153 grid points all over the India. Table 16.2 summarizes results of significant peaks for periodograms of different clusters and as well as for all of India. Results from Table 16.2 indicate that mainly periodicities of 2.3 years dominate in the SPI series of all monsoon

TABLE 16.2

Percentage of Area under Significant Peaks with Various Significant Periodicity Values

Periodicity (Years)	39.5	26.3	19.8	15.8	11.3	7.9	5.6	4.4	3	2.3
					June					
All India	10.5	7.9	12.4	11.4	16.7	20	26.5	36.6	37.1	71.5
Cluster 1	11	9.9	8.8	13.2	11	20.9	20.9	34.1	28.6	85.7
Cluster 2	11.5	7.7	10.6	10.6	19.2	19.2	25	36.5	39.4	30.8
Cluster 3	8	6.3	12.2	12.2	15.1	24.4	30.7	37.4	42	94.1
Cluster 4	8.7	6.9	13	13	16	25.5	32	39	43.7	97.4
Cluster 5	10.7	7.1	17	14.3	17.9	18.3	21.9	33.5	34.8	76.8
Cluster 6	4.8	9.6	14.5	14.5	19.3	16.9	37.3	48.2	30.1	72.3
Cluster 7	7.7	9.9	12.6	11	18.1	14.3	30.2	33.5	36.3	51.6
	July									
All India	13.9	10.2	12.3	9.9	14.4	17.6	23.8	38.2	36.9	75.5
Cluster 1	14.3	9.9	15.4	14.3	14.3	19.8	16.5	35.2	47.3	75.8
Cluster 2	13.5	7.7	11.5	7.7	15.4	17.3	26	41.3	37.5	39.4
Cluster 3	13.9	10.9	8	9.2	13	16.8	27.7	39.5	35.3	88.7
Cluster 4	14.7	11.7	8.7	10	13.9	17.7	29	41.1	36.8	91.8
Cluster 5	12.5	8.9	13.8	9.8	15.2	17.4	21	37.9	39.3	81.3
Cluster 6	7.2	6	15.7	7.2	9.6	20.5	22.9	32.5	34.9	57.8
Cluster 7	18.1	11	16.5	14.3	15.9	17.6	25.3	37.9	39	67.6
	August									
All India	8.7	8.8	14.1	9.8	15.6	18.1	25.2	38.6	37.9	72.7
Cluster 1	7.7	5.5	11	8.8	16.5	24.2	25.3	37.4	37.4	89
Cluster 2	8.7	1.9	10.6	4.8	19.2	26	21.2	40.4	35.6	59.6
Cluster 3	10.1	16.4	15.1	9.7	15.5	18.1	23.5	39.1	39.9	90.3
Cluster 4	10.8	17.3	16	10.4	16.5	19	24.7	40.7	41.6	93.5
Cluster 5	6.3	6.7	13.8	8.5	12.5	18.8	30.4	33	40.6	69.6
Cluster 6	1.2	3.6	7.2	7.2	26.5	20.5	28.9	42.2	37.3	54.2
Cluster 7	6	7.7	18.7	13.7	11	14.3	23.6	34.6	33.5	55.5
	September									
All India	8.6	6.9	9.4	8.2	13.8	19.4	26.5	35.7	36.1	63.7
Cluster 1	12.1	8.8	12.1	9.9	17.6	13.2	23.1	36.3	31.9	82.4
Cluster 2	4.8	8.7	14.4	5.8	13.5	19.2	35.6	30.8	40.4	42.3
Cluster 3	11.3	7.1	8	6.3	15.5	18.1	28.6	41.2	36.1	88.2
Cluster 4	12.1	7.8	8.7	6.9	16.5	19	29.9	42.9	37.7	91.3
Cluster 5	6.3	3.6	6.7	7.6	14.7	21	29.9	33.5	37.9	64.7
Cluster 6	7.2	7.2	6	12	8.4	20.5	21.7	28.9	30.1	32.5
Cluster 7	4.4	5.5	10.4	11	11.5	23.1	20.3	39	37.4	42.9

months for all regions of India. Moreover, there are mainly three interannual cycles of 2.3, 3, and 4.4 years, which cover more than 30% percent of the country. Another interesting point that could be observed from Table 16.2 is that the percentage of area under significant peaks has an almost similar value for all clusters, which indicates that the significant periodicities would spread uniformly across all regions of India.

16.6 CONCLUSIONS

Spatiotemporal analysis of occurrence of drought was performed over India using 80 years (2021–2100) of multimodel ensemble averages of projected precipitation data obtained from nine different GCMs. Principal component analysis and the K-means clustering algorithm were applied to form homogeneous drought regions for spatial analysis. Spatial patterns of temporal trends of drought severity, magnitude, duration, and intervals were analyzed using Mann–Kendall and Sen's slope analyses. Further, spatial ranking of severe drought based on severity and duration were also analyzed. Furthermore, temporal variations of the area under drought and minimum annual SPI were analyzed to find the most vulnerable time periods in projected droughts. Periodicity analysis was also carried out using Fourier transform to find hidden periodicities in the SPI-12 series of monsoon months. Projected drought intervals are likely to increase over Gujarat, Maharashtra, Madhya Pradesh, and Uttar Pradesh for RCP 8.5. Interannual significant periodicities of 2.3 to 4.4 years are likely to be distributed uniformly in all clusters of drought regions of India during the 21st century.

REFERENCES

Abarghouei H.B., Zarch M.A.A., Dastorani M.T., Kousari M.R., Zarch M.S. (2011) The survey of climatic drought trend in Iran. *Stochastic Environmental Research and Risk Assessment* 25:851–863.

Bloomfield P. (2000) *Fourier Analysis of Time Series*. Hoboken, NJ: Wiley.

Brenkert A.L., Malone E.L. (2005) Modeling vulnerability and resilience to climate change: A case study of India and Indian states. *Climatic Change* 72:57–102.

Dai A.G. (2011) Characteristics and trends in various forms of the Palmer drought severity index during 1900–2008. *Journal of Geophysical Research-Atmospheres* 116: D12115.

Dalton L., Ballarin V., Brun M. (2009) Clustering algorithms: On learning, validation, performance, and applications to genomics. *Current Genomics* 10:430–445.

Davies D.L., Bouldin D.W. (1979) A cluster separation measure. *IEEE Transactions on Pattern Analysis and Machine Intelligence* ITPIDJ 0162-8828, PAMI-1:224–227.

Dunn J.C. (1974) Well-separated clusters and optimal fuzzy partitions. *Journal of Cybernetics* 4:95–104.

Golmohammadi F. (2012) Investigating knowledge, attitude and skills of elite farmers about of importance and effects of climate changes in agriculture and extension education activities in confronting them (in South Khorasan Province-Iran). *International Journal of Science and Engineering Investigations* 1:46–54.

Kim D.W., Byun H.R. (2009) Future pattern of Asian drought under global warming scenario. *Theoretical and Applied Climatology* 98:137–150.

Kim H., Park J., Yoo J., Kim T.-W. (2015) Assessment of drought hazard, vulnerability, and risk: A case study for administrative districts in South Korea. *Journal of Hydro-environment Research* 9:28–35.

Kumar K.R., Kumar K.K., Pant G. (1994) Diurnal asymmetry of surface temperature trends over India. *Geophysical Research Letters* 21:677–680.

Kumar P., Wiltshire A., Mathison C., Asharaf S., Ahrens B., Lucas-Picher P., Christensen J.H., Gobiet A., Saeed F., Hagemann S. (2013) Downscaled climate change projections with uncertainty assessment over India using a high resolution multi-model approach. *Science of the Total Environment* 468(Supplement):S18–S30.

Lal M., Nozawa T., Emori S., Harasawa H., Takahashi K., Kimoto M., Abe-Ouchi A., Nakajima T., Takemura T., Numaguti A. (2001) Future climate change: Implications for Indian summer monsoon and its variability. *Current Science* 81:1196–1207.

Lehner F., Coats S., Stocker T.F., Pendergrass A.G., Sanderson B.M., Raible C.C., Smerdon J.E. (2017) Projected drought risk in 1.5 degrees C and 2 degrees C warmer climates. *Geophysical Research Letters* 44:7419–7428.

MacQueen J. (1967) Some methods for classification and analysis of multivariate observations. Proceedings of the Fifth Berkeley Symposium on Mathematical Statistics and Probability, Oakland, CA. pp. 281–297.

Parthasarathy B., Sontakke N.A., Monot A.A., Kothawale D.R. (1987) Droughts floods in the summer monsoon season over different meteorological subdivisions of India for the period 1871–1984. *Journal of Climatology* 7:57–70.

Pattanaik D. (2007) Analysis of rainfall over different homogeneous regions of India in relation to variability in westward movement frequency of monsoon depressions. *Natural Hazards* 40:635–646.

Pollock S.D. (2009) Statistical Fourier analysis: Clarifications and interpretations. *Journal of Time Series Econometrics* 1:1–47.

Rind D., Goldberg R., Hansen J., Rosenzweig C., Ruedy R. (1990) Potential evapotranspiration and the likelihood of future drought. *Journal of Geophysical Research-Atmospheres* 95:9983–10004.

Rousseeuw P.J. (1987) Silhouettes: A graphical aid to the interpretation and validation of cluster analysis. *Journal of Computational and Applied Mathematics* 20:53–65.

Roy R., Schmor B.J., Dominique R., Liu B.C., Hernandez-Mateo F., Juarez-Ruiz J.M., Das S.K. (2003) Transition metal-catalyzed cross-coupling reactions toward the synthesis of glycomimetics. *Abstracts of Papers of the American Chemical Society* 225:U252–U252.

Sheffield J., Wood E.F. (2007) Characteristics of global and regional drought, 1950–2000: Analysis of soil moisture data from off-line simulation of the terrestrial hydrologic cycle. *Journal of Geophysical Research-Atmospheres* 112:D17.

Wang W.X., Zuo D.D., Feng G.L. (2014) Analysis of the drought vulnerability characteristics in Northeast China based on the theory of information distribution and diffusion. *Acta Physica Sinica* 63:229201.

Wu D., Qu J.J., Hao X.J. (2015) Agricultural drought monitoring using MODIS-based drought indices over the USA Corn Belt. *International Journal of Remote Sensing* 36:5403–5425.

17 Drought Characteristics and Forecasting Under Climate Change Conditions
A Case Study of Indonesia

Wanny K. Adidarma and Flavia D. Frederick

CONTENTS

17.1 INTRODUCTION

In recent decades, the observed global temperature has increased. This is closely related to changes in the hydrological cycle on a large scale, such as an increase in water vapor content in the atmosphere, changes in the rainfall pattern, the intensity and duration of rainfall, changes in soil moisture, and surface runoff. These changes indicate significant spatial and interdecadal variability.

DOI: 10.1201/9781003276548-17

This is proven by the increase of heavy rainfall frequency in several regions and even globally. In addition, the number of areas that are classified as dry had increased twice since 1970 (IPCC, 2007).

The impact of climate change on drought has begun to be felt in several regions in Indonesia, as well as globally, with the increasing intensity of drought at the end of this decade (Adidarma et al., 2009). Indonesia is one of the countries that is closely dependent on the agriculture sector. The agricultural sector, as the largest user of water resources, needs to be given greater attention because it plays a determining factor in the success of food security. The problem of drought is a crucial matter for Indonesia. Therefore, there is a need to know the drought trends that occur in Indonesia. This study will discuss two existing trends: the impact of climate change on the amount of rain and the impact of meteorological drought trends on agricultural drought.

17.1.1 DECLINING TRENDS IN RAINFALL IN DRY SEASON WILL NOT NECESSARILY LEAD TO RISING SEVERE DROUGHT

17.1.1.1 Preface and Study Area

This study discusses how much the impact of climate change affects the amount of rainfall and meteorological drought on Java Island, one of the main islands in Indonesia. The island of Java is the economic center of Indonesia and almost half of Indonesia's population lives on the island of Java. The area selection is based on the availability of continuous and long rainfall data in Java, in the Cirebon, Pekalongan, and the Kedu regions. The selected sites are shown on the location map of Indonesia (see Figure 17.1).

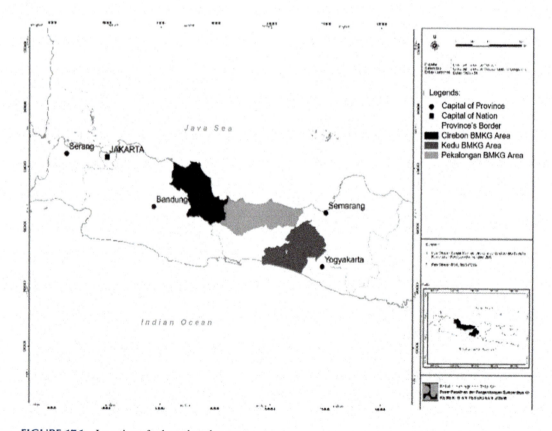

FIGURE 17.1 Location of selected study area.

17.1.1.2   Analysis Results

17.1.1.2.1   Characteristics of Monthly and Annual Rainfall

The existing ground stations are selected through the data screening and data filling stages; the number of ground stations is 29 for Cirebon, 23 for Kedu, and 28 for Pekalongan. The average monthly rainfall in each region can be seen in Figure 17.2. January, February, and March had the highest totals is in the Pekalongan, and the lowest rainfall is in the dry season and the rainy season in Kedu Region.

Annual rainfall in Pekalongan is 3075 mm, Kedu is 3025 mm, and Cirebon is 2315 mm. The 30-year rainfall grouping does not indicate a significant difference in the rainfall, except for the Kedu coefficient of variation (CV) which increases in year groups 2 and 3, as shown in Table 17.1. This indicates that the distribution of annual rainfall is temporally more uneven. The spatially high coefficient of variation occurs in Pekalongan. The smallest monthly and annual rainfall occurs in Cirebon, while the largest coefficient of variation in monthly rainfall is also in Cirebon, as shown in Table 17.2. The coefficient of large variation means that the margin in rainfall from month to month is getting immense.

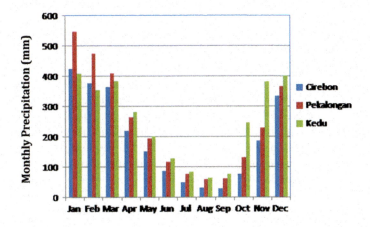

FIGURE 17.2   Monthly rainfall of three selected areas.

TABLE 17.1

Average Annual Rainfall and Coefficient of Variation for Annual Rainfall

		Year Groups		
		1	**2**	**3**
	Annual Rainfall			
RegionSpatial CV	**Parameter**	**1916–1940**	**1941–1970**	**1971–2000**
Pekalongan	Average	3027.4	3031.6	3160.7
CV = 0.3007	CV	0.1185	0.1596	0.1156
Kedu	Average	3065.1	3124.6	2925.9
CV = 0.2357	CV	0.1565	0.2331	0.2208
Cirebon	Average	2399.6	2309.1	2265.8
CV = 0.2732	CV	0.1487	0.1684	0.1234

TABLE 17.2

Average Monthly Rainfall and Coefficient of Variation for Monthly Rainfall

Region	Annual Rainfall Parameter	Year Groups		
		1 1916–1940	**2** 1941–1970	**3** 1971–2000
Pekalongan	Average	253.4	251.5	263.9
	CV	0.7555	0.7657	0.7517
Kedu	Average	258.2	258.6	244.0
	CV	0.6584	0.6752	0.7016
Cirebon	Average	200.4	191.9	189.3
	CV	0.8149	0.8250	0.8140

17.1.1.2.2 Test for Seasonal Rainfall

In order to apply the test, the seasons are divided into the annual (calendar year) rainy season, dry season, and wet season. The tests for the existence of trends and changes in distribution are carried out in the annual rainy season, dry and wet seasons. There are three periods: 1916–1940, 1941–1970, and 1971–2000. Based on the results from the trend testing in the three regions using various periods, the dry season and the rainy season are divided using the period of March–August (MAMJJA) and September–February (SONDJF). The number of ground stations that are evenly distributed throughout the regions manage to illustrate the condition of the entire related area.

Table 17.3 and Figure 17.3 show the rainfall percentage and the result from the Mann–Whitney test for the MAMJJA and SONDJF periods. From the study, at a glance, it can be seen that the Cirebon region is worst affected by a reduction in annual rainfall, where this region has a tendency for the rainfall in the dry season to decrease significantly at 41.4% of the ground stations, while the Pekalongan area has an increasing trend in the annual rainy season and rainy season because it contains a positive trend, i.e., 32.4% of the area is significant or has a tendency to increase. The trend of reduction and change in the rainfall in the dry season implies that the events of drought will be more severe. This temporary conclusion needs more accurate verification using a different series, namely, drought that takes into account the events of rain from month to month and is able to describe the actual drought.

17.1.1.2.3 Test for Drought Intensity and Duration

17.1.1.2.3.1 Whole Year   The drought index for each month is described with the Standardized Precipitation Index (SPI) value. SPI-1 (timescale 1 month), SPI-3 (timescale 3 months), and SPI-6 (timescale 6 months). The consecutive negative SPI drought index will form a certain duration and amount of drought and end with drought intensity, which is the number of droughts divided by the duration of the droughts. The number of droughts is dimensionless and describes the severity of the drought. The duration of drought is monthly, and therefore the intensity of the drought is 1/month. The average drought intensity and duration can be seen in Table 17.4. Changes in the maximum annual drought intensity and duration for SPI of all scales are always less than 2%/year, including the small changes. The average intensity fluctuation of SPI-1 at each ground station was greater than SPI-3 and SPI-6, while the average drought duration of SPI-6 was greater than SPI-1 and SPI-3.

TABLE 17.3

Ground Stations Percentage after the Trend Existence Test

Region	Annual (January–December)			Dry Season (March–August)			Rainy Season (September–February)		
	Positive	Negative	Null	Positive	Negative	Null	Positive	Negative	Null
Cirebon	3.5	37.9	58.6	0.0	41.4	58.6	6.9	10.3	82.8
Kedu	0.0	26.1	73.9	0.0	26.1	73.9	0.0	13.6	86.4
Pekalongan	20.6	5.9	73.5	2.9	17.6	79.5	32.4	2.9	64.7

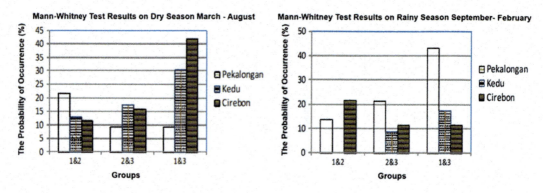

FIGURE 17.3 Ground stations (percentage) that experienced a change in distribution (Group 1: 1916–1940; Group 2: 1941–1970; Group 3: 1971–2004).

TABLE 17.4

Drought Intensity (1/Month) and Drought Duration (Months)

Region	Average Intensity			Average Duration		
	SPI-1	SPI-3	SPI-6	SPI-1	SPI-3	SPI-6
Cirebon	0.93	0.62	0.46	3.62	6.07	7.53
Kedu	0.90	0.57	0.41	3.95	6.77	8.38
Pekalongan	0.96	0.64	0.45	4.13	6.34	8.01

17.1.1.2.3.2 Maximum Frequency of Drought Intensity per Decade for Three-Months Timescale The extreme events of each decade is described by the maximum value of the intensity and duration of drought. This is done as a study of the tendency for the maximum value to increase spatially and temporally on the frequency side. The frequency of certain intensity events in each decade (see Figure 17.4) shows that the tendency of increasing the frequency of extreme events (drought intensity of more than 1.5) is significant in the Kedu region.

From the previous discussion, the average maximum intensity value is between 1.0 and 1.5, therefore extreme values are taken at intensities of more than 1.0 and 1.5. Frequency calculations are carried out for the number of ground stations experiencing a certain intensity and the number of ground station years, or the number of events for one decade in all posts with a certain intensity. The number of ground stations represents the spatial distribution of these intensities and the number of ground station years indicates the frequency of these occurrences.

Figure 17.5 shows the percentage from the number of ground stations and number of events for a certain period, which consists of eight points representing eight decades. The existence of trends from the data series with eight values of percentage can be tested with the Mann–Kendall test. The intensity frequency of more than 1.0 and 1.5 generally increases in the last decades (seventh and eighth).

17.1.1.2.3.3 Conclusion of Drought Trend in Java Island Monthly rainfall data is secondary data used to calculate the presence of annual and seasonal rainfall trends and to test changes in distribution. There are two operations (studies) to indicate the existence of trends in all regions. The first operation is the first derivative of the rain data into an annual maximum drought intensity series, and the third operation or third derivative of the rainfall data is

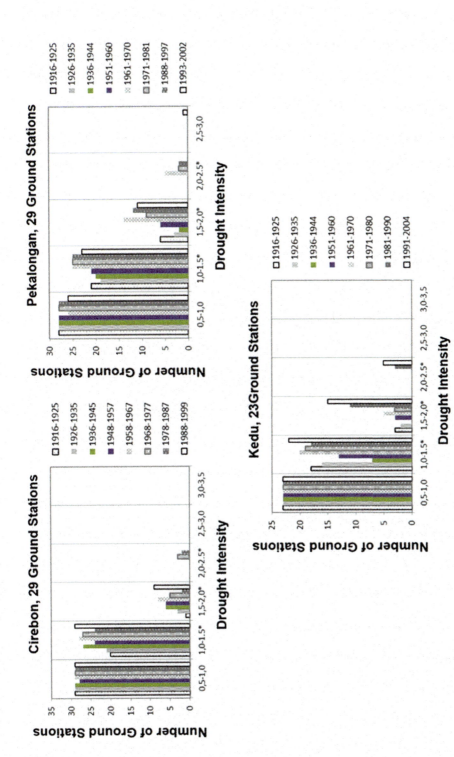

FIGURE 17.4 The frequency of events of certain intensity in each decade for the Cirebon, Pekalongan, and Kedu regions.

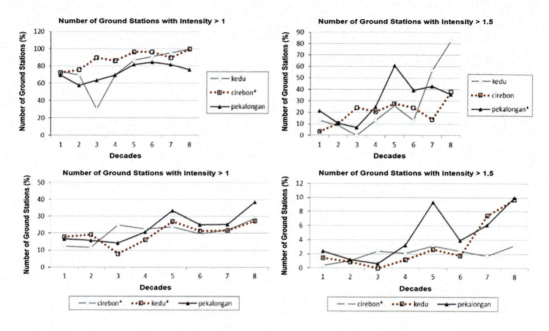

FIGURE 17.5 Spatial distribution and occurrence frequency for drought intensity >1 and >1.5.

the frequency of drought intensity events of more than 1.0 and more than 1.5 in each decade. In this series, there is a new trend, especially those with more than 1.5. Extreme events are represented by intensities of more than 1.0 and 1.5 as has been proven in previous studies.

Spatially, the frequency of extreme drought at the beginning of the decade was less than 20% of the area and gradually increased and reached 80% at the end of the eighth decade. Cirebon only reached 40% of the area at the end of the decade, while Pekalongan fluctuated so that the increasing trend was less significant. Kedu has a significant tendency to increase the frequency of extreme events (drought intensity of more than 1.5).

The Cirebon region with significantly reduced rainfall is not accompanied by an increase in the frequency of extreme drought, which is indicated by a drought intensity of more than 1.5. The increase in the amount of rain in Pekalongan is not followed by a decrease in the frequency of extreme drought events. The Kedu region, which for decades with rainfall that is almost constant in 75% of the area and only decreases in 25% of the area, could have an increasing trend of extreme drought events.

17.2 AGRICULTURE DROUGHT TREND

17.2.1 Preface and Study Area

Changes in drought intensity have created vulnerability in the agricultural sector, especially rice and palawija (secondary crops) production. One of the efforts that can be made to reduce losses due to drought is mitigation measures that can only be carried out if a study on the characteristics of agricultural drought is available. If meteorological drought only considers rainfall data and hydrological drought focuses on rain and flow data only, then agricultural drought pays attention to many aspects that have an impact on rice production. For the needs of planning for irrigation networks and reservoirs, it is often necessary to study agricultural drought, which is able to produce drought characteristics in relation to water stress on agriculture. The characteristics of agricultural drought involve a ranking of water deficit (drought) expressed in terms of years and are referred to as the return period, which is a reflection of supply (rain or river discharge) and crop needs.

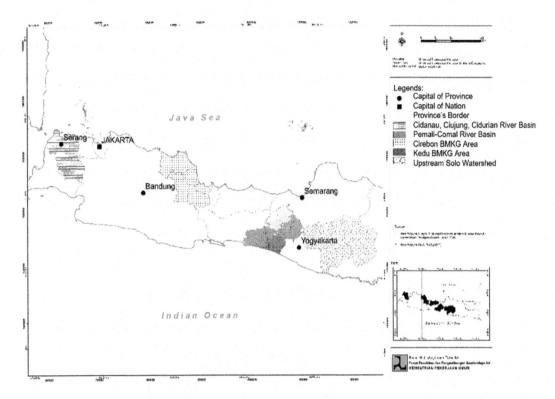

FIGURE 17.6 Location of selected study area.

This study about agriculture drought requires a fairly long series of rain data, therefore the study location is chosen based on the availability of data in that location. The study locations include several areas on Java Island, such as the Pemali–Comal River Basin, Cidanau–Ciujung–Cidurian Watershed, and Upper Solo Watershed, as shown in Figure 17.6. The Pemali–Comal River Basin, covering an area of 4842.84 km², consists of more than 20 watersheds and the largest of which are two – the Pemali Watershed with approximately 1200 km² of area, and the Comal Watershed, covering an area of 850.5 km² – while the Upper Solo Watershed is estimated to be 5987 km².

17.2.2 ANALYSIS RESULTS

17.2.2.1 Data Condition

At the Pemali–Comal River Basin, there were 15 ground stations used from 1951–2010. The Upper Solo Watershed used 11 ground stations from 1975 to 2005. The smallest average annual rainfall is between 2000 and 2500 mm found in the Upper Solo Watershed; the highest rainfall was in the Pemali–Comal River Basin at 3800 mm.

The test of agricultural drought, which is calculated from meteorological drought and compares it with its impact, in this case is with data on the number of hectares of rice fields affected by drought every month from 1989 to 2010, available only in the Pemali–Comal Watershed.

17.2.2.2 Agriculture Drought

The series of average monthly rainfall data in an area is calculated in advance for seven regions/watersheds/river basins. The agricultural drought is calculated which is a meteorological drought against the water requirement threshold, for rice plants of 220 mm/month and for palawija (secondary crops) 120 mm/month (Oldeman, 1975). The output of agricultural drought is in the form of drought intensity in mm/month and duration of drought in months; a zero value of both indicates that rainfall falls above the threshold and is a very wet year. The intensity and duration of this annual drought are used as the input for calculations related to the return period so as to produce outputs in the form of drought intensity and duration with various return periods. The greater the return period, the higher the drought severity. The annual drought intensity and duration series are shifted to the return period series so that the return period distribution or severity ranking can be detected by dividing the time period into two to three groups, with the data length of each group ranging from 15 to 30 years.

To prove that the aforementioned drought deserves to be called an agricultural drought, the series of drought intensity and duration are tested for their relationship with its impact, with the number of hectares of rice fields affected by drought. The test results show a strong relationship on the drought intensity index multiplied by drought duration for the Pemali–Comal River Basin.

From the aforementioned studies, it can be seen that the drought index in the form of drought intensity and duration for rice has a strong relationship with its impact (the rice field cannot produce any rice due to drought), so the drought method can be classified as an agricultural drought.

17.2.2.3 Pemali–Comal River Basin

In 1995–2010, the probability of occurrence for agriculture drought, especially for rice, with the intensity below the average decreased compared with previous years, meanwhile the probability of occurrence for the return period of two to five years increased significantly to more than four times. The drought intensity and duration of palawija (secondary crops) do not significantly change from period 1 to periods 2 and 3; the change in probability in the form of an increase or decrease is still below 50% (Figure 17.7).

17.2.2.4 Cidanau–Ciujung–Cidurian River Basin

In 1976–2007, the decrease in the frequency of rice drought duration events occurred, which was below the average compared to the previous year, especially in 1916–1945, and there was an increase in the frequency of rice drought duration events on the return period of one to five years, as seen in Figure 17.8. The intensity of drought has not changed significantly over time. Likewise, with the palawija (secondary crops) drought, there was no significant change in frequency both for the intensity and duration of the drought.

17.2.2.5 Upper Solo Watershed

In 1993–2010, the decrease in the frequency of rice drought duration occurred. The intensity was below the average and an increase in the frequency of drought with a return period of two to five

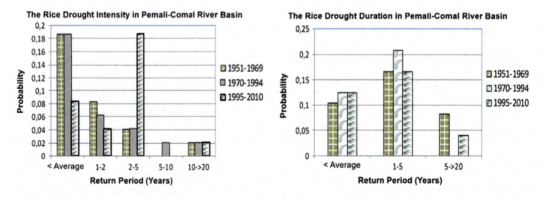

FIGURE 17.7 Probability distribution of drought intensity and duration in three periods in the Pemali–Comal River Basin.

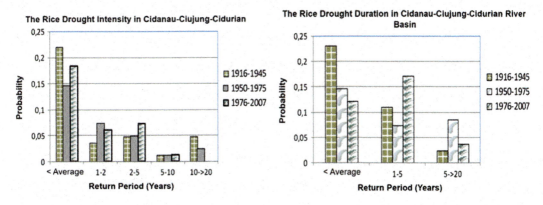

FIGURE 17.8 Probability distribution of drought intensity and duration in three periods in the Cidanau–Ciujung–Cidurian River Basin.

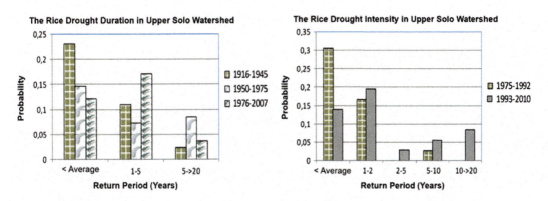

FIGURE 17.9 Probability distribution of rice drought intensity and duration in three periods in the Upper Solo Watershed.

years, five to ten years, and over ten years. For the frequency of the occurrence of drought duration, there is no significant change from time to time, as shown in Figure 17.9. Palawija (secondary crops) drought also changes in frequency, which indicates a change in drought characteristics, namely, the frequency of drought intensity and duration below the average in 1993–2010, which experienced a change in frequency. Meanwhile those with a return period of 1–5 years have increased, as seen in Figure 17.10.

17.2.2.6 Discussion

The agricultural drought method used in this study needs to be tested through the relationship between three data series, i.e., the intensity, duration, and number of hectares of rice fields affected by drought. The complete data availability is only in the Pemali–Comal River Basin, so that testing can only be done in the relevant location. The relationship between the three types of variables can be seen in Figure 17.11. Review of the two graphs suggest that if the drought intensity approaches or is more than 220 mm/month and the drought duration reaches six months or more, the area of rice fields affected by the drought will reach above 10,000 ha. There were several events with intensities exceeding 220 mm/month in 2006 in the Pemali–Comal River Basin, but the duration of the event was five months. After 1998 and from 1999 to 2004, the pattern of drought intensity changed, where events with more than average levels (return period 1–5 years) often occurred.

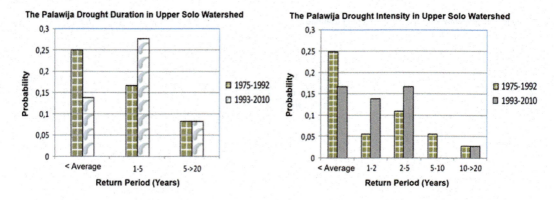

FIGURE 17.10   Probability distribution of palawija drought intensity and duration in three periods in Upper Solo Watershed.

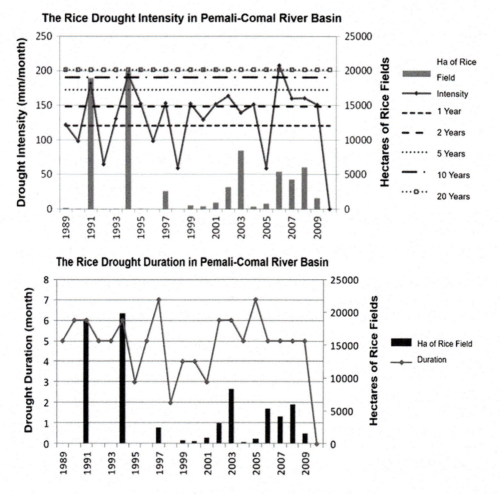

FIGURE 17.11   Series of rice drought intensity and duration and number of hectares of rice fields affected by drought in Pemali–Comal River Basin.

During that period the area of rice fields affected by the drought gradually increased to its peak in 2004. Furthermore, in 2006–2009, the intensity of drought continues to be in a position above the average so that the losses are continuous. Therefore, it can be seen that there is a correlation between the three variables.

17.3 SPATIOTEMPORAL DROUGHT CHARACTERISTICS

17.3.1 PREFACE AND RESULT OF THE STUDY

The spatiotemporal analysis considers the relationship between spatial drought parameters (area affected by the disaster) and temporal drought (related to frequency and return period). The areal–severity–frequency curve is the result of calculations with spatial and temporal dimensions, and provides information on the coverage of areas affected by drought with various ranks, such as the severity on a certain level expressed in terms of return period.

A new data series was developed in the form of a percentage of area affected by drought at a certain level, so that each year the number and average area can be calculated. The data series will follow a certain frequency distribution such as GEV, EVI, gamma, log Pearson III, and lognormal-3 parameters, and EV1 is selected to calculate the area with a certain return period. Overall, the EV1 distribution is fit for calculating the percentage of planned area for various SPIs because other types of distribution often produce values above 1 or more than 100%. In fact, an area that is more than 100% has no meaning. It is commonly used in frequency analysis for drought, such as the areal–severity–frequency curve (Gottschalk and Perzyna, 1993; Kim et al., 2002; Loukas and Vasiliades, 2004).

The drought index with the timescale of one month, SPI-1, is very volatile with a varied duration; the SPI-3 is less volatile with a longer duration. Both still describe meteorological drought conditions rather than hydrological drought (timescale of six months or more) or agricultural drought (WMO, 2012). The larger the timescale, the wider the coverage of the area subject to drought. If the timescale is made to be 9 or 12 months, then the curved shape is more horizontal because the coverage area become almost homogeneous (Loukas and Vasiliades, 2004).

The classification of drought used is less than –2, extremely dry level; between –1.5 and –1.99, severely dry level; between –1 and 1.49, moderately dry level; between –0.5 and 0.99, light level; between 0 and –0.49, the level is near normal. The locations used for this study is in the Pemali–Comal River Basin, with the Pemali Watershed in the west and the Comal Watershed in the center; and the Cirebon Area, Pekalongan Area, and Upper Solo Watershed.

17.3.2 DISCUSSION

Figure 17.12 to Figure 17.17 discuss the study areas. It can be seen that the scope of extreme drought coverage (less than –2) is always less than 20% for SPI-1 and less than 30% for SPI-3, for a return period of 2 years to 100 years. The scope for severe drought extent (less than –1.5) is less than 25% for SPI-1 and less than 40% for SPI-3 for a return period of 2 years to 100 years. At a drought level near to normal (SPI less than 0), the area extent of SPI-1 ranges from 40% to 90%, except for the Comal Watershed (30%–80%) and the Pemali–Comal River Basin approaching 40% to close to 80% for the 2-year return period up to 100 years. For SPI-3, the broad extent of drought is close to normal, ranging from 40% to 100%, except for the Comal Watershed (30%–90%).

Significant dry years were in 1967, 1982, 1997 and 2002. The characteristics of the drought of these years according to the areal–severity–frequency study, as shown in Figures 17.12 to 17.17 are as follows:

- Cirebon region (see Figure 17.12): For SPI-1, 1982 approached the 10-year return period; 1997 combined the 50-year return period from moderate to extreme levels and to 25 years at normal to moderate levels; and 1967 approached the 25-year return period. For SPI-3, it

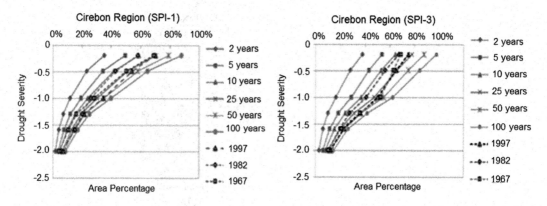

FIGURE 17.12   Areal–severity–frequency of Cirebon Region, severity measured by SPI-1 and SPI-3.

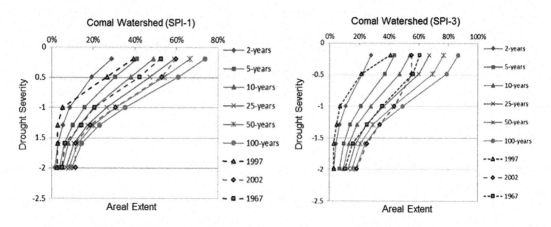

FIGURE 17.13   Areal–severity–frequency of Comal Watershed, severity measured by SPI-1 and SPI-3.

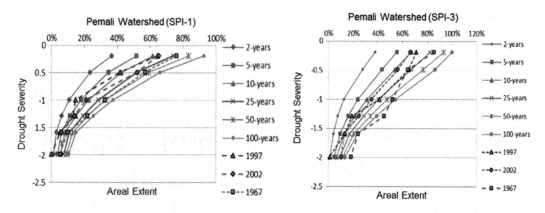

FIGURE 17.14   Areal–severity–frequency of Pemali Watershed, severity measured by SPI-1 and SPI-3.

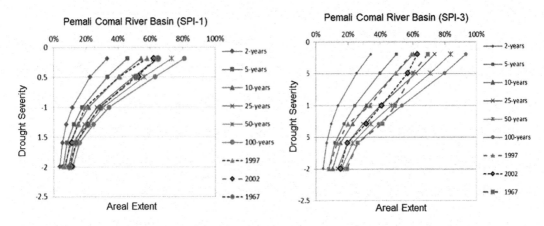

FIGURE 17.15   Areal–severity–frequency of Pemali–Comal River Basin, severity measured by SPI-1 and SPI-3.

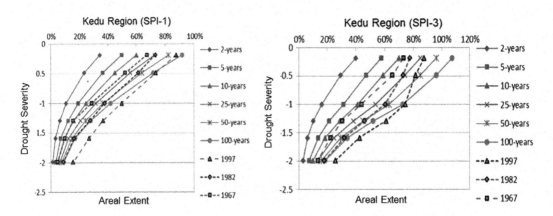

FIGURE 17.16   Areal–severity–frequency of Kedu region, severity measured by SPI-1 and SPI-3.

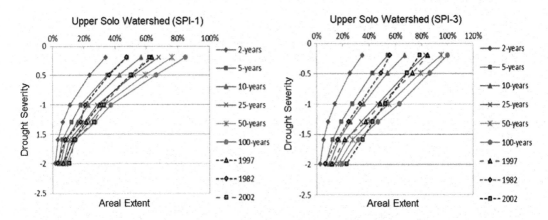

FIGURE 17.17   Areal–severity–frequency of Upper Solo Watershed, severity measured by SPI-1 and SPI-3.

is almost similar to SPI-1, except for 1967 where levels are near normal leading to a 10-year return period.

- Comal Watershed (see Figure 17.13): For SPI-1, 1967 had moderate to extreme levels in the 10-year return period and near normal to moderate levels between the 10- and 25-year return periods; 2002 approached 50 years; and the 1997 level was moderate to extreme with 2-year return period, near normal drought, and moderate level 5-year return period. For SPI-3, in 2002, a 100-year return period from moderate extreme and near to normal moderate level 25-year return period; 1967 is approaching a 25-year return period.

- Pemali Watershed (see Figure 17.14): SPI-1, 1967, 50-year return period; 2002 between 10- and 25-year return period; 1967, between 10- and 25-year return period. SPI-3, 1967, 100-year return period for severe to moderate levels, and the rest of the small rank turns into 10-year return period; and for 1997 between 10- and 25-year return period.

- Pemali–Comal River Basin (see Figure 17.15): For SPI-1, 1997 is close to 10-year return period; 2002 and 1967 are closer to 25-year return period. For SPI-3, 1997, 5- to 10-year return period, severe to extreme levels in 5 years and near normal to severe levels in 10 years; and 2002 is in a 25-year return period at moderate to extreme levels remaining mild in the 10-year return period.

- Kedu region (see Figure 17.16): SPI-1, both 1997 and 1967 are close to 100-year return period, for 1997 with a coverage area of 20% to 100% and 1967 at a level near to normal to moderate approaching a 50-year return period with coverage less than 20% to 80%; years 1967 between a 10- and 25-year return period with wide coverage between 10% and 65%. SPI-3, 1997 at moderate to extreme levels of 100-year return period and dry levels of return period of 25- to 100-year return period (wide coverage 30%–80%). The year 1982 was between 25 and 100 years, the high return period was always for severe extreme, and 1967 between 10- and 25-year return period.

- Upper Solo Watershed (see Figure 17.17): SPI-1, 1997 and 2002 between 25- and 100-year return period, less than 20% to 60% of the area; in 1982 the return period was 5–10 years. SPI-3, for severe and extreme 100-year return period, normal to severe level between 26-100-year return period, 1997 moderate to extreme, 10-year return period, and the rest that is near to normal for a 5-year return period.

From the observed incidence of dry years, only the Kedu region experienced a dryness of up to 90% for SPI-1 in 1997. For other incidents, the dryness varied between 60% and 80%. In the Cirebon, Kedu, and Upper Solo Watershed areas in 1997 there was a high return period drought, but this was not the case with the Pemali–Comal River Basin.

Figures 17.18 to 17.20 show that the incidence of drought in 1997 was not as severe as in 2002, the study below examines the characteristics of drought from different sides. The area of

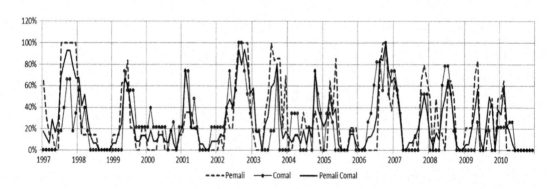

FIGURE 17.18 Percentage of area affected by drought more than normal (SPI <0.7) in the Pemali–Comal River Basin using SPI-3.

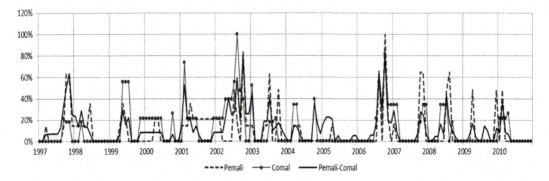

FIGURE 17.19   Percentage of area affected by drought more than moderate (SPI <-1.3) in the Pemali–Comal River Basin using SPI-3.

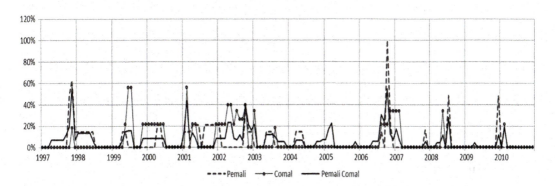

FIGURE 17.20   Percentage of area affected by drought more than severe (SPI <-1.6) in the Pemali–Comal River Basin using SPI-3.

the Pemali–Comal River Basin is approximately 4837.2 km², the westernmost area of the Pemali Watershed is 1248 km² and the Comal Watershed in the center of 825 km², the rest consists of small watersheds with a total of close to 30, with the spatiotemporal distribution conditions of the drought as follows:

1. Deficit or drought that is more than the average level occurs almost every year with a wide range of 0% to 100%, i.e., from no deficit (all surpluses) to all areas affected by a deficit (see Figure 17.18).
2. For drought levels that are more than moderate, the 100% broad coverage only occurred two times, in 2002 and 2006 in the Comal and Pemali Watershed, respectively. The coverage reached 80%, in three years in 1997, 2002, 2003 and 2006, for the Pemali Watershed, and the Pemali–Comal River Basinmali.
3. In the more-than-severe level, 100% wide coverage was in 2006 in the Pemali Watershed, and 80% coverage in 1997 in the Pemali Watershed as well. Both events lasted only one month.
4. Spatially, the incidence of 2002 (see Figures 17.19 and 17.20) was more severe because it occurred for more than two months. The deficit of drought was so that the average area in one year became the largest.

17.4 DROUGHT FORECASTING

When a drought occurs in an area, it is often not realized by the community because of the impact has not been felt yet. This occurs due to the lack of information regarding the onset, the end, and the magnitude of the drought, which should be calculated and used as a basis for estimating the possible impacts so that efforts to deal with drought can be carried out as quickly as possible long before the impact is felt. One of the parameters that can be used as a measure of drought severity is the drought index.

One of the methods to calculate the drought index that is often used is the SPI. The SPI was designed to quantitatively determine the rain deficit in an area with various timescales. In forecasting the drought index, additional parameters are needed to facilitate the prediction process. There is a tendency that drought occurs more frequently and increases in intensity and lengthens in duration, according to studies from Adidarma et al. (2009) and SDA (2012). One of the reasons is the El Niño phenomenon, which occurs more frequently. The El Niño–Southern Oscillation is thought to have a very close relationship with drought events, especially the magnitude and its character, and La Niña is related to thunderstorm events. This parameter is called teleconnection. The weather and climate of a specific location on earth usually has a very strong relationship with other locations, connected through atmospheric order linkages (Burroughs 2003). The teleconnection used in this study uses the Oceanic Niño Index (ONI) value, which is the deviation from the monthly sea surface temperature to its long-term average running for three months, and it covers the region of 5° N–5° S and 170° W–120° W (Adidarma, 2015).

This study will discuss two methods related to the purpose of doing drought prediction: in the context of mitigation in the Limboto–Bolango–Bone and Sumbawa River Basins, and the Annual Reservoir Operation Plan at Batutegi Dam and Nipah Dam.

17.4.1 MITIGATION

17.4.1.1 Study Location and Data

Limboto–Bolango–Bone (LBB) River Basin is a cross-province river basin located in Gorontalo Province, North Sulawesi Province, and Central Sulawesi Province. The LBB River Basin has an area of 5253 km², covering three large watersheds (Limboto Watershed, Bolango Watershed, Bone Watershed) and 62 small watersheds. The Sumbawa River Basin is located in Nusa Tenggara Barat Province. The Sumbawa River Basin has an area of 15,414.5 km² with 555 watersheds. Figure 17.21 has maps of the LBB and Sumbawa River Basins.

The study for both locations use rainfall data from the Tropical Rainfall Measuring Mission (TRMM). There are 22 ground stations in the LBB River Basin. However, the data are not adequate temporally or spatially. Spatially, the location of each ground station is not spatially well distributed; most of the ground stations are located at the center of the river basin. Temporally, there are many missing data, and the rainfall data at every station is of a different time period. There are 16 ground rainfall stations in the Sumbawa River Basin. However, this study does not have ground station data from this river basin. Since the ground station data are not adequate temporally or spatially for the LBB River Basin and the lack of ground station data for the Sumbawa River Basin, the TRMM rainfall data is used as the input for the analysis from January 1, 1998–September 30, 2019. (Adidarma, Frederick, et al. 2019).

For the LBB River Basin, the average rainfall in a year is 157.86 mm. The dry months occur from July to October, and February is also considered a dry month because the rainfall that occurs in that month is below the average (Figure 17.22).

In the Sumbawa River Basin, the average rainfall in a year is 107.25 mm. The dry months occur from May to October (Figure 17.23).

The SPI-3 analysis results for both river basins can be seen in Figures 17.24 and 17.25.

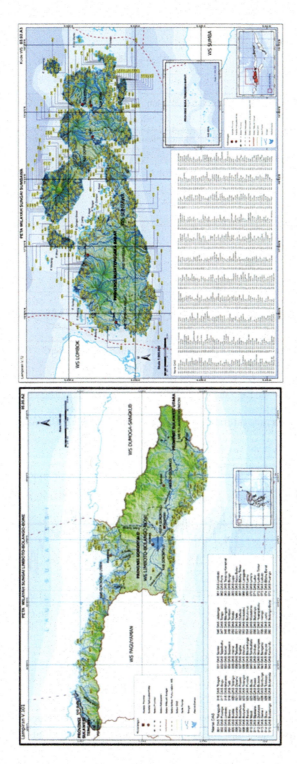

FIGURE 17.21 (Left) LBB and (right) Sumbawa River Basin maps.

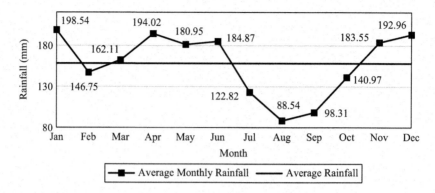

FIGURE 17.22   Monthly average rainfall pattern for the LBB River Basin from 1998 to 2019.

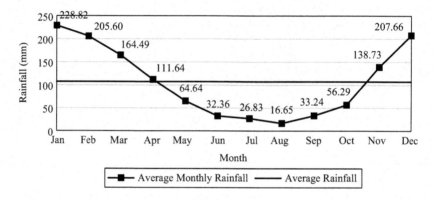

FIGURE 17.23   Monthly average rainfall pattern for the Sumbawa River Basin from 1998 to 2019.

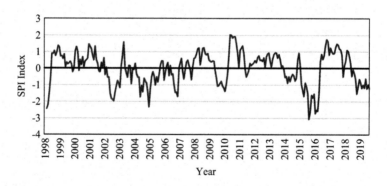

FIGURE 17.24   SPI-3 for the LBB River Basin from 1998 to 2019.

17.4.1.2   Forecast Model and Results

The method used for the forecast model is the function of the teleconnection parameter, the ONI. The SPI-3 and ONI have the same timescale, three months. Historical ONI data is used for the parameters of the resulting equations. The results of this forecast model are the SPI equation for each month using four scenarios. The best equation is chosen from the smallest root mean square error (RMSE) value. The first model uses the second-level polynomial equation (Equation 17.1).

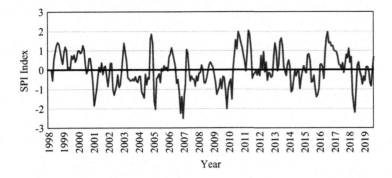

FIGURE 17.25 SPI-3 for the Sumbawa River Basin from 1998 to 2019.

The second model uses multilinear regression (Equation 17.2). The matrix of each model and model scenario can be seen in Table 17.5. This forecast model is only made for the dry season. Although drought also occurs in wet months, the impact is smaller for the water availability than the drought impact in dry months.

$$y_1 = ax^2 + bx + c \qquad\qquad (17.1)$$
$$y_2 = ax_1 + bx_2 + cx_3 + d \qquad\qquad (17.2)$$

where y_2 is the forecasted SPI on the current analyzed month for model 2.4; x_1 is the ONI from one month before the current analyzed month; x_2 is the ONI from two months before the current analyzed month; x_3 is the ONI on the current analyzed month; and a, b, c, d are constants obtained from the trial and error procedure.

The summary of the models can be seen in Table 17.6 for the LBB River Basin and in Table 17.7 for the Sumbawa River Basin. Based on the smallest RMSE occurrence, the best model for the LBB River Basin is model 2.3, and for Sumbawa it appears that 50% of the best model is for model 2.3 and another 50% is for model 2.1. With the consideration of the RMSE value, the model chosen for the Sumbawa River Basin is model 2.1.

Both of the selected models have the same limitation, which is the forecast time is only for one month ahead, because one of the inputs from both models is observed in the SPI from one month

TABLE 17.5
Model Matrices

Model	Input	Forecast Capability
1	Forecasted ONI on the C-month	6 months from the C-month
2.1	Observed SPI from A-Month	1 month from the C-month
	Observed SPI from B-Month	
	Forecasted ONI on the C-month	
2.2	Observed SPI from A-Month	2 months from the C-month
	Forecasted ONI on the B-Month	
	Forecasted ONI on the C-month	
2.3	Historical ONI on the A-Month	1 month from the C-month
	Observed SPI from B-Month	
	Forecasted ONI on the C-month	
2.4	Forecasted ONI from A-Month	6 months from the C-month
	Forecasted ONI on the B-Month	
	Forecasted ONI on the C-month	

TABLE 17.6

Summary Results for LBB River Basin

Model	February (DJF) RMSE	r	July (MJJ) RMSE	r	August (JJA) RMSE	r	September (JAS) RMSE	r	October (ASO) RMSE	r
1	0.656	0.748	0.747	0.581	0.663	0.698	0.617	0.739	0.5	0.83
2.1	0.473	0.868	0.457	0.815	0.26	0.94	**0.41**	0.942	0.35	0.93
2.2	0.65	0.75	0.59	0.718	0.414	0.845	0.417	0.941	0.46	0.85
2.3	**0.444**	0.905	**0.395**	0.892	**0.259**	0.938	0.429	0.91	**0.33**	0.91
2.4	0.657	0.755	0.684	0.593	0.658	0.72	0.643	0.759	0.52	0.83

before. This limitation makes the forecast capability less effective for forecasting. Model 1 and model 2.4 have the longest time for forecast because all of the inputs are in the ONI. With the consideration from the value of the RMSE for each model, it can be concluded that model 2.4 is better than model 1. The list of constants for each month is in Table 17.8 and Table 17.9.

The maximum RMSE of this model for the LBB River Basin is 0.684 and for the Sumbawa River Basin is 0.698. For the Sumbawa River Basin, May and June have a weak correlation, therefore these months are not suited for forecasting the SPI. Considering the classification of the SPI has an increment of 0.5, the RMSE from these models are on acceptable range because the RMSE is around 0.5. The graph of comparison between the SPI observed and SPI forecast from model 2.4 for each month in the two river basins are shown in Figure 17.26 and Figure 17.27.

17.4.2 Annual Reservoir Operation Plan

17.4.2.1 Batutegi Dam

The Batutegi Dam is located in Lampung Province with a watershed area of 422 km² (Figure 17.28), storage volume of 750.3 million m³, and estimated inflow of 841.1 million m³/year. The performance of the reservoir is included as a multiyear dam. This dam is able to irrigate 80% from 55,373 ha rice fields, which is the Way Sekampung System Irrigation Area (consisting of Feeder Canal 1 and Feeder Canal 2), and supplies 1.1 m³/s of raw water and 1.6 m³/s of maintenance flow. The remaining 20% is served by the weir and dams on the downstream of the Batutegi Dam.

In operating a reservoir, it is necessary to determine the wet years and dry years of the future in order to regulate the amount of outflow that will be discharged from the outlet in accordance with the predetermined reservoir operation rules. From Table 17.10 it can be seen that there is a strong association between the reservoir water level elevation with SPI-12 compared to the SPI on a smaller scale than 12 months, because this reservoir is a multiyear dam. Furthermore, the SPI-12 forecast will be implemented, especially in the dry season, based on the historical and forecast ONI. In dry years, there needs to be an adjustment in the cropping pattern and a reduction in the area of irrigation that will be irrigated to only Feeder Canal 1, therefore it is necessary to forecast the dry years for adjustment preparation.

The wet/dry year forecast procedure based on the SPI-12 wet/dry level is shown in Figure 17.29 using a statistical approach, in this case the multilinear regression equation. This section only discusses the SPI-12 forecast for December, May, and April, and then formulates it to the wet/dry year indicator for reservoir operation purposes.

The SPI-12 forecast as a measuring tool for dry/wet level is predicted from independent variables such as historical SPI-12 and historical ONI and forecasted ONI with equations as shown in Table 17.11 through trial and error by minimizing the RMSE and maximizing the correlation coefficient,

TABLE 17.7
Summary Results for Sumbawa River Basin

Model	May (MAM)		June (AMJ)		July (MJJ)		August (JJA)		September (JAS)		October (ASO)	
	RMSE	r	RMSE	r	RMSE	r	RMSE	r	RMSE	r	RMSE	r
1	0.84	0.31	0.92	0.35	0.743	0.483	0.62	0.641	0.47	0.55	0.663	0.74
2.1	**0.62**	0.68	0.59	0.66	**0.408**	0.828	0.517	0.558	0.389	0.79	**0.632**	0.75
2.2	0.85	0.34	0.78	0.49	0.588	0.656	0.484	0.791	0.48	0.56	0.636	0.76
2.3	0.65	0.67	**0.51**	0.75	0.548	0.77	**0.424**	0.784	**0.362**	0.81	0.639	0.76
2.4	0.86	0.34	0.87	0.38	0.635	0.552	0.494	0.813	0.503	0.58	0.698	0.73

TABLE 17.8
Constants for Model 2.4 in LBB River Basin

Month	a	b	C	d
February (DJF)	-0.1818	-0.4762	0.0611	-0.0336
July (MJJ)	-2.0512	1.2673	-0.1499	0.0024
August (JJA)	-1.3843	0.4423	-0.1441	-0.1328
September (JAS)	-1.4114	0.3406	-0.1444	-0.2419
October (ASO)	-1.3619	0.5736	-0.1438	-0.1489

TABLE 17.9
Constants for Model 2.4 in Sumbawa River Basin

Month	a	b	c	d
May (MAM)	-0.0989	0.0197	-0.3324	-0.0306
June (AMJ)	-0.2011	1.3155	-1.4863	-0.0204
July (MJJ)	-0.2092	1.5807	-1.9257	0.0281
August (JJA)	-0.2124	2.1783	-2.2003	-0.0867
September (JAS)	-0.2056	1.3721	-1.4715	0.0307
October (ASO)	-0.1781	-0.5834	-0.0659	-0.1913

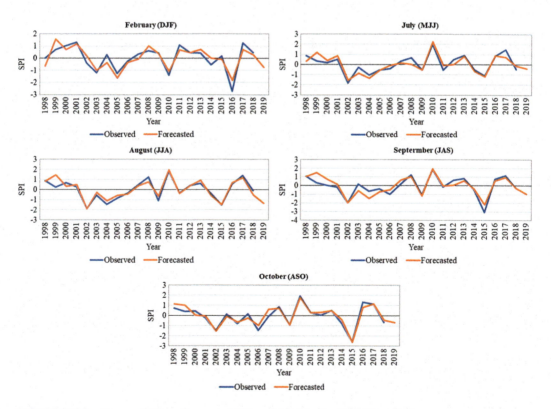

FIGURE 17.26   Observed SPI vs. forecasted SPI model 2.4 for the LBB River Basin.

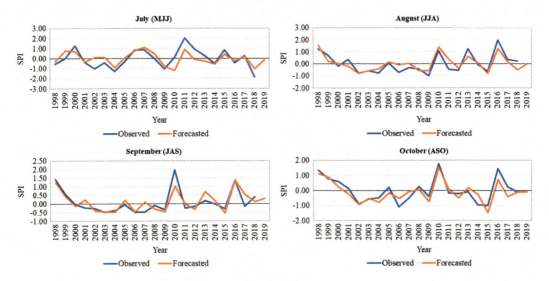

FIGURE 17.27   Observed SPI vs. Forecasted SPI model 2.4 for the Sumbawa River Basin.

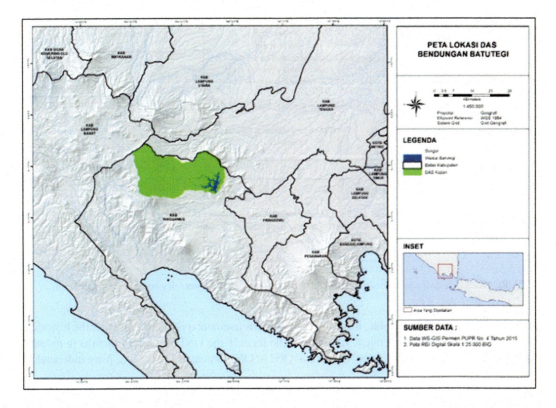

FIGURE 17.28   Batutegi Dam location.

as shown in the last two columns in Table 17.11. The proximity between the SPI-12 forecast and actual results can be seen in Figure 17.30.

The equation in Table 17.11 uses the following regression:

$$Y_n = ax_1 + bx_n + cx_{n-1} + \ldots, p\, x_{n-6} + q \qquad (17.3)$$

TABLE 17.10

Correlation Coefficient for Each Variable for Batutegi Dam

	RWL	SPI-12	SPI-3	ONI
RWL (Reservoir Water Level)	1			
SPI-12	**0.6776**	1		
SPI-3	0.4037	0.4886	1	
ONI (JASON)	−0.4455	−0.4275	**−0.6278**	1

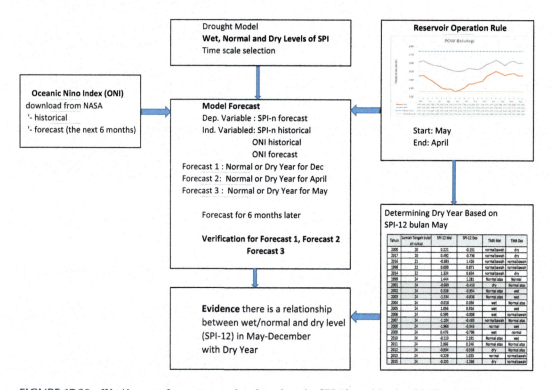

FIGURE 17.29 Wet/dry year forecast procedure based on the SPI-12 wet/ dry level in Batutegi Dam.

where Y_n is a dependent variable; n is the SPI-12 months to come; q is a constant; x_1 is the historical SPI-12; x_n, ..., x_{n-1} are independent variables in the form of the ONI for certain months (n months, n−1 months, n−2 months, and so on) (historical ONI or ONI forecast); and a, b, ..., p are independent variable coefficients.

17.4.2.2 Nipah Dam

The Nipah Dam is located in the Nipah River in Madura Island, East Java Province, with a watershed area of 73.87 km^2 (Figure 17.31). The Nipah Dam has a multipurpose potential, e.g., irrigation water for 1911 ha of rice fields, water resources conservation, and tourism/recreation. This dam is included as a within-year dam, with an inflow of 115.25 million m^3 and an effective storage volume of 2.29 million m^3. The dam's current is greater than the reservoir capacity, because in 2014 (seven years after its establishment and it cannot be operated), the spillway crest was forced to be lowered from an elevation of +46.15 m to the elevation of +43.0 m due to social problems that occurred in

TABLE 17.11

Equation and the SPI-12 Forecast Performance for Batutegi Dam

Time Base May Year n

Forecast-1 — Now: May (Year n)

	SPI 12 (Latest)	Year n (Forecast)				Year n−1 (Historical)		Constant	Performance	
		ONI ASO	ONI SON	ONI JAS	ONI ASO	ONI SON	ONI JAS		RMSE	Correl
Forecast SPI-12 Dec Year n	0.3525 (May (Year n))	3.213	−0.632	−3.85	−1.19	0.87	0.797	−0.081	0.398	0.908

Forecast-2 — Now: Jan (Year n+1)

	SPI 12 (Latest)	Year n+1 (Forecast)				Year n (Historical)					Constant	Performance	
		ONI SON	ONI ASO	ONI JAS	ONI AMJ	ONI SON	ONI ASO	ONI JAS	ONI AMJ	ONI MAM		RMSE	Correl
Forecast SPI-12 Apr (Year n+1)	0.9285 (Dec (Year n))	2.2077	−1.9348	−0.9952	0.0416	1.2487	−1.5255	1.2769	0.7765	−2.3554	1.5284	−0.0227	0.3667	0.93

Forecast-3 — Now: Jan (Year n+1)

	SPI 12 (Latest)	Year n+1 (Forecast)				Year n (Historical)					Constant	Performance	
		ONI SON	ONI ASO	ONI JAS	ONI AMJ	ONI SON	ONI ASO	ONI JAS	ONI AMJ	ONI MAM		RMSE	Correl
Forecast SPI-12 May (Year n+1)	1.2078 (Dec (Year n))	2.5886	−3.5012	1.5622	−0.5454	0.0322	−1.6828	3.4010	−3.2416	1.6238	0.0496	0.3497	0.93

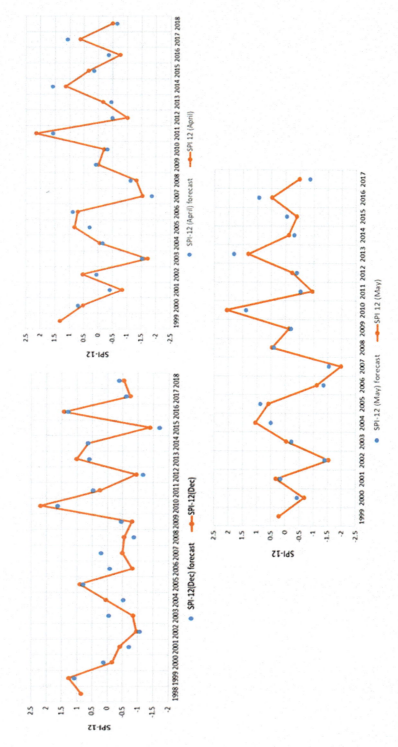

FIGURE 17.30 Observed SPI-12 vs. forecasted SPI-12 in Batutegi Dam.

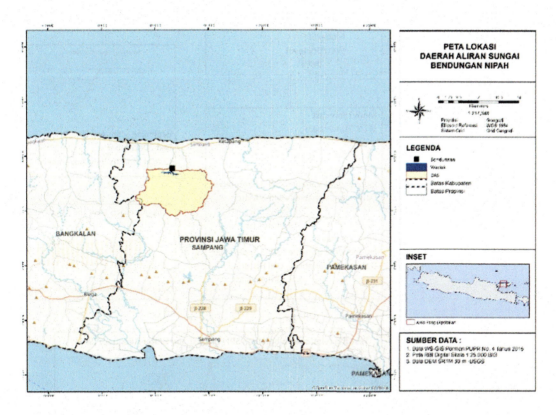

FIGURE 17.31 Nipah Dam location.

TABLE 17.12

Correlation Coefficient for Each Variable in Nipah Dam

	RE	SPI-12	SPI-9	SPI-6	SPI-3
RE (Reservoir Elevation)	1				
SPI-12	0.0934	1			
SPI-9	0.1289	0.8869	1		
SPI-6	0.2076	0.7276	0.8699	1	
SPI-3	0.2342	0.4946	0.6372	0.7655	1
ONI SON	−0.4944	−0.1966	−0.4484	−0.6847	−0.6637
ONI Jan–Dec	−0.1537	−0.2283	−0.3304	−0.3318	−0.2404
ONI JJASON	−0.4692	−0.1725	−0.3683	−0.6179	**−0.6236**

the area caused by the flooding of the lands around the dam reservoir. Thus, the carrying capacity decreased by 50%. The reservoir filling stage was carried out at the end of 2015 and after that it was operating as usual.

From Table 17.12, it appears that there is a strong association between the reservoir water level elevation with SPI-3 and SPI-6 compared to SPI on a scale greater than six months because this reservoir is a within-year dam.

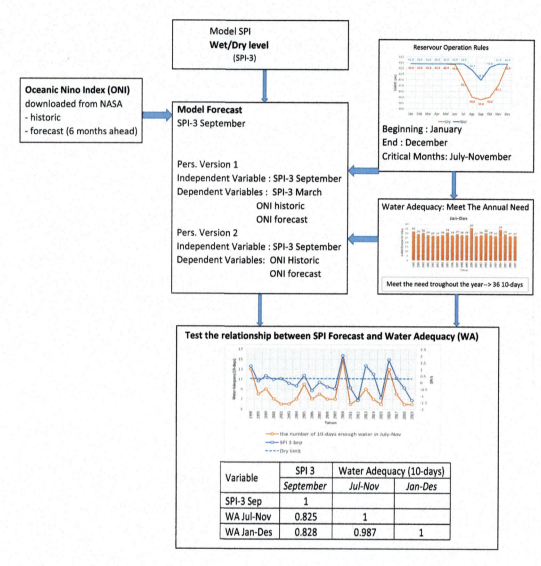

FIGURE 17.32   Wet/dry year forecast procedure based on the SPI-3 wet/dry level in the Nipah Dam.

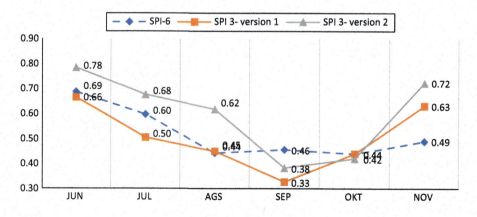

FIGURE 17.33   Error in the SPI forecast for July–November in the Nipah dam.

TABLE 17.13

Equation and the SPI-3 Forecast Performance for Nipah Dam

Equation	SPI-3		ONI						Constant	Objective Function	
	Mar	JAS	JJA	MJJ	AMJ	MAM	FMA	JFM		RMSE	Correl
Version 1	0.2936	−1.0462	−0.0341	−0.0146	−0.2531	0.9187	0.1608	0.0035	−0.0053	0.33	0.81
Version 2	—	−0.5921	−0.6866	−0.0997	0.1141	1.1663	−0.1384	0.0057	0.1070	0.38	0.74

In Figure 17.32, upper right side, the reservoir operation rules for the Nipah Reservoir for seven months created excess water, in other words, the reservoir was able to meet the needs of irrigation water, however, during July to November a water shortage occurred. The dry year prediction is only needed during July–November so that irrigation services can be adjusted or reduced. Based on the conducted trial and error, it is indicated that the best forecast with the smallest error (RMSE) occurred in September, as seen in Figure 17.33. The equation is built using SPI-3 and SPI-6 forecast functions from SPI-3 or SPI-6 in March and ONI in various months. Equation version 1, as mentioned in Table 17.13, uses the March SPI and various ONIs, while version 2 only uses the various month ONIs.

The equation (as mentioned in Table 17.13) uses the regression as follows:

$$Y_n = ax_1 + bx_n + cx_{n-1} + \ldots, p\, x_{n-6} + q \tag{17.4}$$

where Y_n is a dependent variable; SPI-3 for the next September; q is a constant; x_1 is n–6 months SPI-3 or September SPI-3; $x_n, \ldots, x_{n-1}$ are independent variables in the form of the ONI for certain months (n months, n–1 months, n–2 months, and so on) (historical ONI or ONI forecast); and a, b, $\ldots$, p are independent variable coefficients.

17.5 CONCLUSION

In this study, the rainfall trend in Java Island, Indonesia was investigated using Mann–Kendall test. The results indicated a significant decrease in rainfall in this island, which has led to an increase in the number of drought events and their intensity. In the following, drought conditions in the study area were analyzed using the Standardized Precipitation Index (SPI) in different time scales. The trend of drought time series showed that the occurrence of severe droughts had a significant increasing trend. The results of spatial analysis indicated that the frequency of extreme drought at the beginning of the decade was less than 20% of the area and gradually increased and reached 80% at the end of the eighth decade. Cirebon only reached 40% of the area at the end of the decade, while Pekalongan fluctuated so that the increasing trend was less significant. Kedu has a significant tendency to increase the frequency of extreme events.

The investigation of the trend of the occurrence of agricultural droughts, especially for rice crop, showed that the occurrence of severe droughts has increased, so that the probability of occurrence for the return period of 2 to 5 years increased significantly to more than 4 four times.

REFERENCES

Adidarma, W.K. 2015. *Model Pendukung Penanggulangan Kekeringan Berbasis Disaster Risk Management.* Bandung: PT. Dunia Pustaka Jaya.

Adidarma, W.K., M. Lanny, and U. Adelia. 2009. "Apakah Trend Hujan di Musim Kemarau yang Berkurang Akan Menimbulkan Intensitas Kekeringan yang Bertambah Parah?." Lokarkarya "Identifikasi Dampak Perubahan Iklim pada Sektor Sumberdaya Air dari Program: Penguatan IPTEK Adaptasi dan Mitigasi Perubahan Iklim", Kementerian Negara Riset dan Teknologi Balai Irigasi. Bekasi, 21 April.

Adidarma, W.K., F. Frederick, D. Yudianto, R. Mohammad, and O. Subrata. 2019. "Drought Forecasting in Limboto Bulango Bone River Basin in Gorontalo Province." 6th International Seminar of HATHI. Kupang.

Burroughs, W. 2003. *Climate into the 21st Century.* Cambridge: Cambridge University Press.

Gottschalk, L, and G. Perzyna. 1993. "Low Flow Distribution along a River", Extreme Hydrological Events: Precipitation, Floods and Droughts (Proceedings of the Yokohama Symposium, July 1993), IAHS Publication no. 213, 33–41.

IPCC. 2007. "Fourth Assessment Report: 'Climate Change 2007 (AR4)'." *Working Group I Report "The Physical Science Basis".*

Kim, T.-W., J.B. Valdes Valde, and J. Aparicio. 2002. "Frequency and Spatial Characteristics of Droughts in the Conchos River Basin." *International Water Resources Association.* Mexico City: Water International. 420–430.

Loukas, A, and L. Vasiliades. 2004. "Probabilistic Analysis of Drought Spatiotemporal Characteristics in Thessaly Region." *Natural Hazards and Earth System Science* 4: 719–731.

Oldeman, R.L. 1975. *An Agro-climatic Map of Java*. Bogor: Central Research Institute for Agriculture.

SDA, Puslitbang. 2012. "*Laporan Penelitian Prakiraan Dan Pengendalian Kekeringan Serta Pengembangan Peta Resiko Banjir dan Kekeringan*." Bandung.

WMO. 2012. *Standardized Precipitation Index User Guide (WMO - No.1090)*. Geneva: World Meteorological Organization.

18 Stochastic Modeling of Water Deficit over Different Agroclimatic Zones of Karnataka

GVS Reddy, Sita Ram Bhakar, and Rohitashw Kumar

CONTENTS

DOI: 10.1201/9781003276548-18

18.1 INTRODUCTION

Drought is a weather-related creeping natural disaster. It is a phenomenon associated with the scarcity of rain and thus water. Drought is considered to be the most complex but least understood of all natural hazards affecting more people than any other hazards. It results from subnormal, untimely, and uneven distribution of rainfall. Drought has been a matter of serious concern to man since ancient times, and even today, it is an outstanding example of man's helplessness before nature's large-scale and formidable phenomena. Drought is a natural phenomenon that occurs periodically all over the globe. It is a major factor of uncertainty that continues to haunt Indian agriculture and economy. The year 2002 was one of the worst drought spells in recent times (Kalsi et al., 2006). Definitions of drought are almost as numerous as there are publications on the subject. Lack of sufficient water to meet the normal water requirements of the locality is a common and a fairly satisfactory definition of drought. The perception of drought is different for different people. To the farmer, drought is a period during which normal farm operations are hampered. To the hydrologist, with prolonged deficiency of rainfall, drought occurs with marked depletion of surface water and consequent drying up of reservoirs, lakes, streams, and rivers; cessation of spring flows; and a fall in groundwater levels. To the meteorologist, it is a situation when there is a significant decrease of rainfall from the normal values over a given area. On the other hand, to the agricultural scientist, drought occurs when the soil moisture and rainfall are inadequate during the growing season to support healthy crop growth to maturity and cause extreme water stress. Thus, droughts can be classified as hydrological, meteorological, and agricultural.

Study of drought characteristics is one of the important aspects of rainfed farming as well as water resources planning, management, and allocation of irrigation water. Most studies refer to streamflow or meteorological drought analysis for a particular basin or a specific historical event without giving much emphasis to the quantitative and qualitative aspects of the drought. The success or failure of crops particularly under rainfed conditions is closely linked with the rainfall pattern. Coincidence of dry spells with the sensitive phenological stages of crops causes damage to the crop development. Hence, simple criteria related to sequential phenomenon like dry and wet spells and prediction of probability of onset and termination of the wet season could be used to obtain specific information needed for crop planning and for carrying out agricultural operations (Khichar et al., 1991).

Rainfed area predominates the agricultural scenario in Karnataka state in India. Out of a total area of 19.05 million hectares (Mha), 10.7 Mha are suitable for cultivation, in which 70% of the cultivable land is dependent on rainfall. Better management of agricultural practices in such areas is possible by analysis of rainfall and characterization of drought. Various developmental agencies involved in tackling drought problems in Karnataka lack information about the drought characteristics and forecast. As a result, rainfall and water resources are utilized in a very inefficient way leading to crop failure, even in normal rainfall years due to shorter or longer drought spells. This calls for the study of drought characterization and modeling to develop appropriate control measures and consequent management of drought.

18.2   STOCHASTIC MODELING

A mathematical model representing a stochastic process is called stochastic model or time series model. The water deficit from an area is stochastic in nature since it is affected by climatic parameters. The stochastic nature of a water deficit can be represented by simulation or mathematical modeling, which is a basic tool to generate the desired parameter with greater accuracy. Stochastic modeling of water deficit as a time series is important for selection of suitable crop variety, scheduling of irrigation, and drought management planning. A stochastic model explains the extent of dependence of the present observation on the past observation. Therefore, stochastic modeling of water deficit may provide good insight and understanding of the process for useful application in crop planning of the region.

18.3   CROP PLANNING

In dry land areas, crop production depends on vagaries of nature. Characterization of drought during a crop season helps in planning the most effective cropping pattern and developing suitable supplementary irrigation facilities. Recognizing these concerns, it is necessary to analyze, characterize, and model the drought situation under changing climatic patterns over different agroclimatic zones in Karnataka state.

18.4   STUDY AREA AND DATA

The state of Karnataka ($11°30'–18°30'$ N latitude and $74°15'–78°30'$ E longitude) occupies the western part of the Deccan Plateau, India (Figure 18.1). Karnataka covers an area of 191,773 km^2 and is bounded by the Arabian Sea to the west and has common borders with Andhra Pradesh to the east, Tamil Nadu to the south and southeast, Kerala to the southwest, Goa to the northwest, and Maharashtra to the north. The area selected covers all ten agroclimatic zones of Karnataka. Ten meteorological stations located in different agroclimatic zones of the state were selected for the study (see Table 18.1). The soils of Karnataka can broadly be classified into six groups. They are (i) red soils, (ii) black soils, (iii) lateritic soils, (iv) alluvio-colluvial soils, (v) forest soils, and (vi) coastal soils.

The meteorological data and data pertaining to soils and cropping pattern were collected from various sources. Data – namely, maximum and minimum temperatures, relative humidity, sunshine duration, wind speed, and rainfall – for Bidar, Raichur, Chitradurga, Mysore, Belgaum, Chickmagalur, and Mangalore were obtained from the India Meteorological Department, Pune. Meteorological data of Bangalore and Bellary were collected from observatories of the University of Agricultural Sciences, Bangalore; and the Central Soil and Water Conservation Research and Training Institute, Bellary. Information regarding soil characteristic features was collected from the National Bureau of Soil Survey and Land Use Planning, Bangalore. The information pertaining to crops and a map of study area were collected from the Karnataka State Department of Agriculture, Bangalore.

18.5   WATER BALANCE EQUATION

The water balance statement when applied to hydrologic equations states that in a specified period of time, all water entering a specified area must either go into storage within its boundaries, be consumed there in, be exported there from, or flow out. So for its computation, a procedure introduced

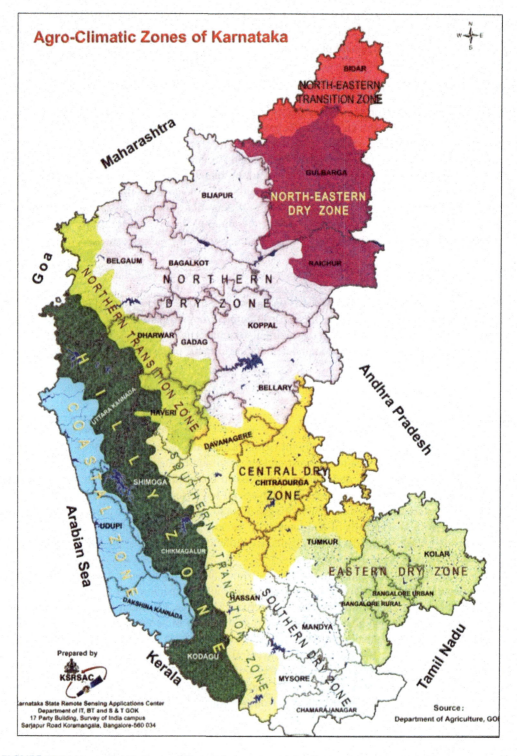

FIGURE 18.1 Agroclimatic zones of Karnataka.

TABLE 18.1

Location of Stations under Different Agroclimatic Zones of Karnataka

S. No.	Agroclimatic Zone	Station	Latitude, (°N)	Longitude (°E)	Elevation (m)
1	North-eastern transition zone	Bidar	17.92	77.53	664
2	North-eastern dry zone	Raichur	16.20	77.35	400
3	Northern dry zone	Bellary	15.32	76.85	445
4	Central dry zone	Chitradurga	14.23	76.43	733
5	Eastern dry zone	Bangalore	12.97	77.58	930
6	Southern dry zone	Mysore	12.30	76.70	767
7	Southern transition zone	Shivamoga	13.93	75.63	571
8	Northern transition zone	Belgaum	15.85	74.62	747
9	Hilly zone	Chickmagalur	13.25	75.75	371
10	Coastal zone	Mangalore	12.92	74.88	102

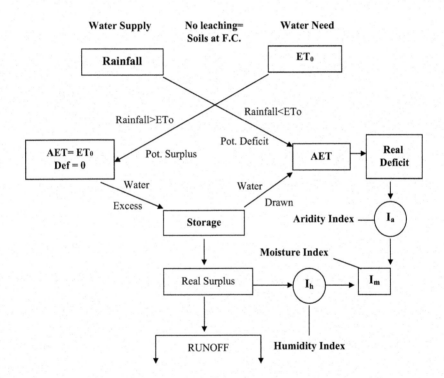

FIGURE 18.2 A generalized flow diagram of the climatic water balance.

by Thornthwaite and Mather (1957) was used. Figure 18.2 shows a generalized flow diagram of the climatic water balance.

The procedure for computation of different water balance elements is given below.

Available water (AW) is the soil moisture between field capacity and permanent wilting point. Field capacity (FC) is the amount of water that the soil holds against drainage by gravity (at 1/3 bar). The permanent wilting point (PWP) is the moisture content in soil when plants permanently wilt and will not recover (at 15 bar).

18.6   REFERENCE EVAPOTRANSPIRATION (ET_0)

Reference evapotranspiration (ET_0) is the rate of evapotranspiration from a hypothetical reference with assumed crop height at 0.12 m, a fixed surface resistance of 77 s/m, and albedo of 0.23, closely resembling the evapotranspiration from an extensive surface of green grass of uniform height, actively growing, completely shading ground, and with adequate water. Reference evapotranspiration was computed using the Penman–Monteith equation as per the procedure given by Allen et al. (1998):

$$ET_0 = \frac{0.408 \, \Delta \, (R_n - G) + \gamma \, \dfrac{900}{T+273} \, U_2 \, (e_a - e_d)}{\Delta + \gamma \, (1 + 0.34 \, U_2)} \tag{18.1}$$

where ET_0 is the reference evapotranspiration (mm/day), Δ is the slope of saturation vapor pressure function (kPa/°C), G is the soil heat flux (MJ/m²day), T is the mean daily temperature (°C), R_n is the net radiation (MJ/m²day) = $R_{ns} - R_{nl}$, R_{ns} is the net short wave radiation (MJ/m²day), R_{nl} is the net long wave radiation (MJ/m²day), γ is the psychrometric constant (kPa/°C), e_a is the saturation vapor pressure at temperature T (kPa), e_d is the saturation vapor pressure at dew point (kPa), and U_2 is average daily wind speed at 2 m height (m/s).

18.7   WEEKLY MOISTURE EXCESS AND DEFICIT (P-ET_0)

The difference between rainfall (P) and reference evapotranspiration gives the weekly moisture excess and deficit. A negative value of this difference indicates a moisture deficit, which means the amount by which the rainfall fails to supply the potential water need of area. A positive difference is moisture excess, which is the amount of excess water available for soil moisture replenishment and also for runoff.

18.7.1   Accumulated Potential Water Loss and Gain (Acc (P-ET_0))

Values of P-ET_0 are summed up week by week as an aid in the computational steps that follow, which are designated as Acc (P-ET_0).

18.7.2   Storage (STOR)

In order to carry out the calculation of soil moisture storage, it is necessary to know from where one should begin. According to Thornthwaite and Mather (1957), the starting value of soil moisture content can only be obtained by assuming the value of soil moisture storage is equal to the moisture-holding capacity after the period of rain ceases during the wet season of the year. As the soil dries, the rate of evapotranspiration decreases. According to Thornthwaite and Mather (1957), the release of moisture is an exponential function. Considering the this principle, the following formula was used for computing the soil moisture storage at the end of each week:

$$STOR = AWC \times e^{\left(\frac{Acc(RF - ET_0)}{AWC} \right)} \tag{18.2}$$

where STOR is the actual storage of soil moisture (mm), AWC is the available water-holding capacity of soil (mm), RF is rainfall (mm), and ET_0 is the reference evapotranspiration (mm).

18.7.3   Change in Storage (ΔSTOR)

The positive changes in soil storage are termed as soil moisture recharge. The negative changes are termed as soil moisture utilization, when the value in storage is above the water-holding capacity; it is assumed that there is no change in soil storage.

18.7.4 ACTUAL EVAPOTRANSPIRATION (AET)

Actual evapotranspiration (AET) is considered to take place at the potential rate, when precipitation exceeds the reference evapotranspiration during a particular week and also when moisture in the soil is near field capacity. However, after soil moisture was depleted to a point where ability of the soil to transmit the moisture was reduced, the actual rate of ET was sharply reduced. Therefore, weekly AET was calculated from the following equations:

1. (a) When $P > ET_0$

$$AET = ET_0 \qquad (18.3)$$

2. (b) When $P < ET_0$

$$AET = P + abs(\Delta STOR) \qquad (18.4)$$

From these equations, it is clear that when precipitation is less than ET_0, then AET is equal to precipitation plus absolute value of change in soil moisture storage than previous week.

18.7.5 WATER DEFICIT (DEF)

The amount by which the actual evapotranspiration (AET) and reference evapotranspiration differ in any week is the water deficit (DEF). Water deficit only exists when $P-ET_0$ is negative and is calculated by the following equation:

$$DEF = ET_0 - AET \qquad (18.5)$$

18.7.6 WATER SURPLUS (SUR)

The water surplus is the amount of positive $P-ET_0$ that remains in excess after recharging the soil to the field capacity by the following equation:

$$SUR = P - AET \qquad (18.6)$$

18.7.7 CLIMATOLOGICAL INDICES

On the basis of the aforementioned parameters, climatological indices, including the humidity index (I_h), aridity index (I_a), moisture index (I_m), and moisture adequacy index (I_{ma}), were computed using the following expressions (Thornthwaite and Mather, 1957):

$$I_h = \frac{SUR}{ET_0} \times 100 \qquad (18.7)$$

$$I_a = \frac{DEF}{ET_0} \times 100 \qquad (18.8)$$

$$I_a = \frac{DEF}{ET_0} \times 100 \quad I_{ma} = \frac{AET}{PET} \times 100 \qquad (18.9)$$

$$I_m = I_h - I_a \qquad (18.10)$$

18.8   STOCHASTIC MODELING OF WATER DEFICIT

In this study, a time series analysis by an autoregressive modeling approach was used to provide weekly forecasts of water deficit (DEF). The weekly water deficits of the selected ten stations, representing different agroclimatic zones of Karnataka, were used for this study. From the water deficit data of 25 years (1982–2006), depending upon the availability of data at respective stations, the data up to 2004 were used for model development, and the remaining two years of data (2005 and 2006) were used in validating the performance of the model.

18.8.1   Time Series Model Development

The time series of water deficit was broken down into a deterministic component in the form of trend and periodic parameter, and a stochastic (random) component consisting of chance and chance-dependent effects. Trend is a long, smooth movement lasting over a span of observations. It may rise or fall. The periodic component is nonstationary in mean and variance, and repetitive over fixed interval of time; whereas the stochastic component can be regarded stationary in mean and variance, and it is constituted by irregular oscillations and random effects that cannot be accounted for physically. A discrete time series is denoted by X_t, t = 1, 2, 3, etc., where X_t is at equidistant time interval and decomposed by an additive type. The additive form provides a reasonable model in most cases and is expressed as

$$X_t = T_t + P_t + S_t \tag{18.11}$$

in which T_t is the trend component, P_t is the periodic component, and S_t is the stochastic component having dependent and independent parts at time t. Modeling of the water deficit series is analyzed and determined following the stepwise procedure.

18.8.2   Trend Component (T_t)

Steady and regular movements in a time series through which the values are on average either increasing or decreasing is termed as trend (Kottegoda, 1980). In order to model the water deficit time series, a null hypothesis of no trend in the series was adopted. For detecting the trend in the time series, the following two statistical tests were performed:

(i) Turning point test: In an observed sequence X_i, i = 1, 2, …, N, where N represents years, and a turning point p occurs at time t = i, if X_i is either greater than or less than the two adjacent values X_{i-1} and X_{i+1}. The expected number of turning points in a random series is $E(p) = \dfrac{2(N-2)}{3}$ and the variance is given by $var(p) = \dfrac{(16N-29)}{90}$. Consequently, p can be expressed as a standard normal deviate, $Z = \dfrac{(p - E(p))}{(var(p))^{1/2}}$. The calculated value of Z was compared with its table value at the p = 0.01 level of significance, viz., +2.58. When the calculated value of Z has been found to exist within limits, then the hypothesis of no trend was not rejected.

(ii) Kendall's rank correlation test: This test, τ test, is based on a proportionate number of subsequent observations that exceed a particular value. For a sequence $X_1, X_2, …, X_N$, the standard procedure is to determine the number of times p occurs in all pairs of observations $(X_i, X_j; j > i)$ that X_j is greater than X_i. For a trend-free series, $E(p) = \dfrac{N(N-1)}{4}$ and the test statistic, $\tau = \left\{ \dfrac{4p}{N(N-1)} \right\} - 1$. $var(\tau) = \dfrac{2(2N+5)}{9N(N-1)}$ and standard normal deviate,

$$Z = \frac{\tau}{\left[\text{var}(\tau)\right]^{\frac{1}{2}}}$$. The calculated value of Z was compared with the table value at p = 0.05 or 0.01 level of significance. When calculated value of Z has been found to exist within the limits, the null hypothesis of no trend in the series was not rejected.

18.8.3 PERIODIC COMPONENT (P$_T$)

In a time series, periodicity can be represented through a system of sine and cosine functions, after the trend component, if present, has been estimated and removed. The procedure used is termed harmonic analysis. The existence of P$_t$ was identified by the serial correlogram, i.e., a graph of auto-correlation coefficients, r$_k$ against lag k.

Autocorrelation also called serial correlation, r$_k$, is a very useful tool for investigating the structure of a time series with regard to how observations separated by a fixed period of time are inter-related, i.e., the degree to which the observations in any one year is dependent upon the magnitude of the observations in the years preceding it. It was assumed for simplification that the trend component was removed. It was also used for identification of the deterministic and stochastic components in a series. The serial correlation coefficient, SCC (r$_k$), for an observed series was computed using the expression

$$r_k = \frac{\sum_{i=1}^{N-k} X_i X_{i+k} - \frac{1}{N-k}\left(\sum_{i=1}^{N-k} X_i\right)\left(\sum_{i=1}^{N-k} X_{i+k}\right)}{\left[\sum_{i=1}^{N-k} X_i^2 - \frac{1}{N-k}\left(\sum_{i=1}^{N-k} X_i\right)^2\right]^{\frac{1}{2}}\left[\sum_{i=1}^{N-k} X_{i+k}^2 - \frac{1}{N-k}\left(\sum_{i=1}^{N-k} X_{i+k}\right)^2\right]^{\frac{1}{2}}} \tag{18.12}$$

The tolerance limits for the correlogram of an independent series was computed as

$$r_k(95\%) = \frac{-1 \pm 1.96\sqrt{N-k-1}}{N-k} \tag{18.13}$$

18.8.4 STOCHASTIC COMPONENT (S$_T$)

The stochastic component is constituted by various random effects, which cannot be estimated exactly. An autoregressive (AR) stochastic model was used for presentation of the time series. In this model, the current value of the process was expressed as a finite, linear aggregate of values of the process and variates that was completely random. But, the stochastic component was obtained by subtracting the periodic component and time-dependent structure. Nevertheless, an autoregressive model of order p, AR(p), can mathematically be expressed as

$$S_t = \varphi_{p,1} S_{t-1} + \varphi_{p,2} S_{t-2} + ... + \varphi_{p,p} S_{t-p} + a_t = \sum_{k=1}^{p} \varphi_{p,k} S_{t-k} + a_t \tag{18.14}$$

where $\varphi_{p,k}$ is the autoregressive model parameters, k = 1, 2, ..., p; and a$_t$ is an independent random number.

18.8.5   SUM OF SQUARES ANALYSIS

In this method, the sum of squares of residuals and deviation of observed series from their mean value were calculated to get the R^2 by ratio and use it as a tool to assess adequacy of model. If Y_t denotes the observed value of water deficit series, $\bar{Y}$ denotes the mean of the observed water deficit series, and $\hat{Y}$ denotes the fitted value of the water deficit, then the sum of squares of residuals is

$$A = \sum \left(Y_t - \hat{Y}_t \right)^2 \tag{18.15}$$

The sum of squares of deviations of observed values from their means is

$$B = \sum \left(Y_t - \bar{Y}_t \right)^2 \tag{18.16}$$

Then R^2 is the measure required and is given as

$$R^2 = \left(B - A \right) / B \tag{18.17}$$

If all the residuals are zero, i.e., $R^2 = 1$, it indicates perfect fitting of the formulated model. The closer the value of R^2 is to unity, the better is the goodness of fit.

18.8.6   VALIDATION OF STOCHASTIC MODEL

A major application of modeling time series is to generate or forecast future values or data. Generating the water deficit time series for the entire sampling period was first checked using the weekly developed water deficit models. Forecasting was made for two years (2005 and 2006) ahead. The generated/forecasted values from each model were compared to the observed data. The variation of the generated/forecasted and observed series was presented graphically with respect to time. The correlation coefficient of the mean generated series and mean observed series was determined.

18.9   CROP PLANNING OVER DIFFERENT AGROCLIMATIC ZONES

Distribution of rainfall, soil characteristics, and other agroclimatic parameters are mainly responsible for adopting various cropping patterns suited to different locations. So, with the integrated approach of drought analysis and available natural resources, the cropping pattern for different locations of the state are suggested. Selection of cropping pattern also depends on the length of growing season, which is decided on the basis of the period of moisture availability for crop growth. The length of growing season was determined by the Thornthwaite and Mather (1957) method and based on onset and cessation of rainfall. After onset of monsoon, the growing season starts when $I_{ma} > 0.5$ and ends when $I_{ma} < 0.25$. However, if I_{ma} (where I_{ma} is the moisture adequacy index, which is the ratio of AET/ET_0) is less than 50% during active growth stages of the crop, then it is considered as drought. The crops and cropping patterns for different agroclimatic zones were selected on the basis of soil characteristics, length of growing season, rainfall distribution, and dry spell analysis. The specific measures to be considered at respective agroclimatic zones of the state of Karnataka have been suggested for taking up sustainable crop production.

18.10   RESULTS AND DISCUSSION

This study has been undertaken with the objective of drought modeling different agroclimatic zones of Karnataka. For this study, weekly rainfall data for a period of 30 to 31 years (1976–2006) and data for other meteorological parameters, like maximum and minimum temperature, relative humidity, wind speed, and sunshine hours, for 25 years (1982–2006) were collected and used in this study. An attempt has been made to analyze rainfall pattern on an annual and seasonal basis. The mean onset and cessation of the rainy season were computed, and initial and conditional probabilities of dry spells have been determined. The water deficit and surplus water periods were determined through the water balance technique and artificial neural networks technique has been used in water deficit estimation. Water deficit data are stochastic in nature and, therefore, stochastic analysis of water deficit has been carried out. Validation of developed time series models has been made by comparing generated and observed values of water deficit. Information pertaining to length of the crop growing season, dry spell periods, and rainfall analysis has been utilized for crop planning over different agroclimatic zones of Karnataka.

18.11 ANALYSIS OF RAINFALL PATTERN AND ESTIMATION OF DRY SPELLS

Rainfall, being the only source of water under dryland agriculture, provides a lot more insight for proper planning with respect to crop production. The weekly rainfall data of ten stations over different agroclimatic zones of Karnataka for a period of 30 to 31 years (1976–2006) were analyzed on an annual and seasonal basis, and the results are presented as follows.

18.11.1 STATISTICAL ANALYSIS OF ANNUAL RAINFALL

Annual rainfall data were analyzed statistically and parameters, including mean, standard deviation (SD), coefficient of variation (CV), and coefficient of skewness (C_S), are shown in Table 18.2. Mangalore is the highest rainfall-receiving zone having an average annual rainfall of 3636.1 mm with SD of 621.1 mm. The coefficient of variation of rainfall is observed to be highest at Bellary with CV of 34.5% and lowest at Chickmagalur with CV of 16.3%. If CV is within the threshold limits of variability, it is considered that the rainfall is highly dependable. The CV of six stations falls within the threshold limit (<25%), which shows highly dependable rainfall conditions, and the threshold limit of CV is exceeded at Bellary, Bangalore, Mysore, and Belgaum. The coefficient of skewness ranged from –0.79 at Chickmagalur to 1.51 at Mysore. The coefficient of skewness is observed to be positive at all stations except Chickmagalur. Results of annual rainfall at different stations revealed that there is a large spatial and temporal variation of rainfall in the state.

TABLE 18.2
Annual Rainfall Parameters over Different Agroclimatic Zones of Karnataka

Station	Mean (mm)	Max (mm)	Min (mm)	SD (mm)	CV (%)	Skewness
Bidar	905.3	1316.3	620.1	193.4	21.4	0.37
Raichur	724.2	1085.9	470.7	174.4	24.1	0.43
Bellary	475.0	914.1	188.2	163.7	34.5	0.68
Chitradurga	633.5	1016.3	354.2	157.3	24.8	0.26
Bangalore	923.9	1363.0	528.2	234.3	25.4	0.27
Mysore	786.1	1577.0	467.4	221.3	28.2	1.51
Shivamoga	1117.8	1564.2	678.3	230.3	20.6	0.40
Belgaum	999.3	1850.7	584.8	286.9	28.7	1.20
Chickmagalur	1404.1	1811.0	850.9	229.4	16.3	–0.79
Mangalore	3636.1	5083.5	2545.0	621.1	17.1	0.28

SD, standard deviation; CV, coefficient of variation.

18.11.2 TREND OF ANNUAL RAINFALL

Computation of trend values provides a tool to ascertain if a locality is getting drier or wetter. If rainfall series show a positive (negative) trend, the area would be considered to be getting wetter (drier). The trend of annual rainfall was found to be nonsignificant for all stations. However,

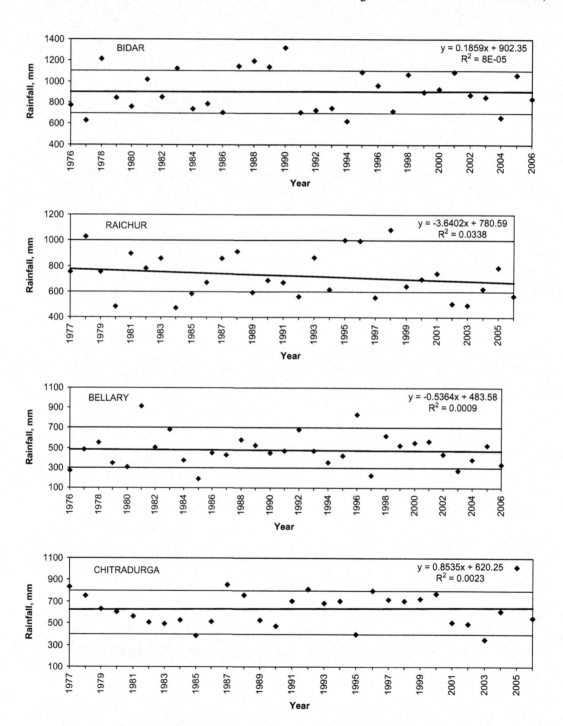

FIGURE 18.3 Trend of annual rainfall at Bidar, Raichur, Bellary, and Chitradurga.

slopes of annual rainfall are positive at five locations (Bidar, Chitradurga, Bangalore, Mysore, and Belgaum) and negative at the remaining five stations (Raichur, Bellary, Shivamoga, Chickmagalur, and Mangalore) (see Figures 18.3–18.5). But the slopes are not significant for all locations, indicating erratic distribution of rainfall over the state. Correlation coefficient values (R^2) of developed regression equations were found to be very low (less than 0.1) and were found to be nonsignificant at all the stations.

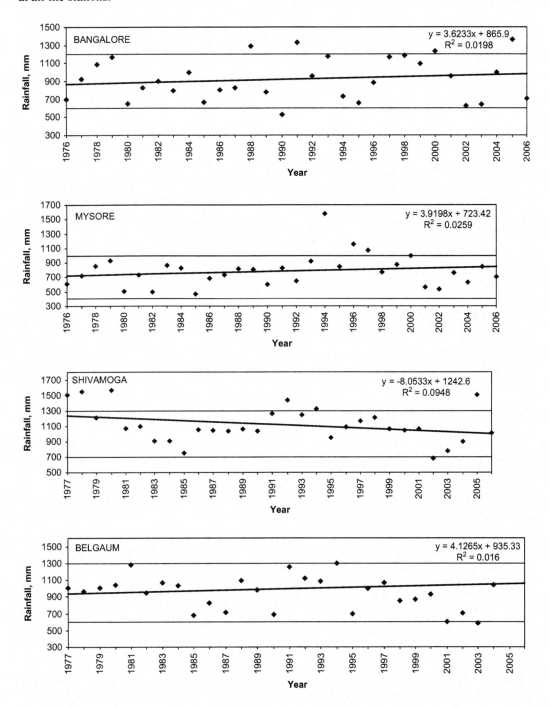

FIGURE 18.4 Trend of annual rainfall at Bangalore, Mysore, Shivamoga, and Belgaum.

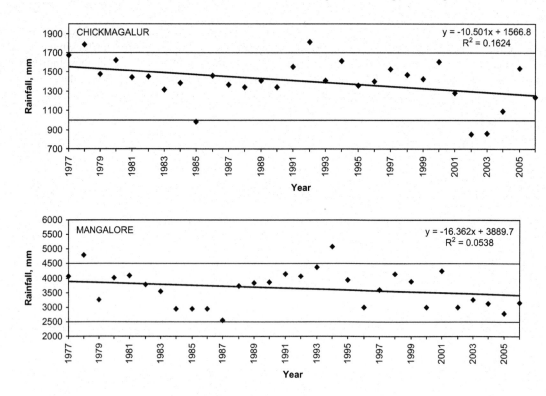

FIGURE 18.5   Trend of annual rainfall at Chickmagalur and Mangalore.

18.11.3   ANALYSIS OF SEASONAL RAINFALL

Different parameters of kharif (23rd–39th standard meteorological week (SMW)), rabi (40th–8th SMW), and summer (9th–2nd SMW) rainfall over different agroclimatic zones of Karnataka have been evaluated. Percentagewise distributions of seasonal rainfall over all stations revealed that Mangalore received the highest percentage (85%) and Mysore the lowest percentage (46%) of kharif rainfall. The highest (lowest) 31% (8%) of rabi season rainfall was observed at Chitradurga (Mangalore). Bellary, Chitradurga, Bangalore, and Mysore received the highest percentage of rabi season rainfall (Table 18.3), which is due to contribution of rainfall from the northeastern monsoon since these stations lie in the eastern and southern parts of Karnataka (Figure 18.4).

18.11.4   REFERENCE EVAPOTRANSPIRATION

Reference evapotranspiration (ET_0) refers to total water losses through evaporation from soil and transpiration from vegetation under unlimited water supply.

Many researchers have used various methods for calculating reference evapotranspiration at different locations. However, several scientists (Hatfield and Allen, 1997; Kashyap and Panda, 2001; Goyal, 2004; Bhakar et al., 2006; Goyal, 2005; Salam and Mazrooe, 2006; Biggs et al., 2008) have evaluated and preferred the Penman–Monteith equation for calculating reference evapotranspiration using meteorological data of particular regions. Weekly ET_0 values for different agroclimatic zones of Karnataka were computed using Equation 18.1. The annual ET_0 values for different stations over different agroclimatic zones of Karnataka are presented in Table 18.4. The annual mean ET_0 values varied from 1594.80 mm with the CV of 4.03% at Mangalore to 2123.54 mm with the CV of 4.82% at Raichur.

TABLE 18.3
Seasonal Rainfall Parameters over Different Agroclimatic Zones of Karnataka

Station	Mean (mm)	Max (mm)	Min (mm)	SD (mm)	CV %	Skewness
Kharif Season						
Bidar	676.8	1079.6	341.5	167.8	24.8	0.42
Raichur	510.5	813.1	230.4	156.0	30.6	0.42
Bellary	251.4	509.9	59.2	107.0	42.6	0.33
Chitradurga	326.8	569.9	152.5	107.1	32.8	0.83
Bangalore	497.1	835.5	216.3	163.8	32.9	0.28
Mysore	362.8	684.5	174.4	128.1	35.3	0.49
Shivamoga	725.9	987.5	323.8	153.8	21.2	−0.83
Belgaum	728.0	1507.2	374.4	257.4	35.4	1.11
Chickmagalur	919.7	1222.2	351.1	205.2	22.3	−1.09
Mangalore	3086.0	4398.9	1993.9	623.7	20.2	−0.23
Rabi Season						
Bidar	146.9	405.2	23.9	104.0	70.8	0.84
Raichur	136.6	372.1	24.8	95.3	69.8	1.12
Bellary	140.2	289.4	8.3	81.7	58.3	0.27
Chitradurga	195.7	376.9	24.2	98.5	50.3	-0.01
Bangalore	246.6	694.1	107.0	136.9	55.5	1.92
Mysore	236.0	796.3	79.3	159.7	67.7	1.98
Shivamoga	258.1	483.5	82.1	117.6	45.5	0.26
Belgaum	141.6	283.5	8.0	65.4	46.2	0.00
Chickmagalur	283.7	631.3	77.8	138.4	48.8	0.82
Mangalore	280.4	752.8	83.7	152.1	54.2	1.29
Summer Season						
Bidar	81.7	271.9	5.1	48.8	59.7	1.88
Raichur	77.1	207.6	6.0	47.2	61.2	0.66
Bellary	83.4	184.1	3.6	47.5	57.0	0.29
Chitradurga	111.0	262.3	16.4	55.4	50.1	0.71
Bangalore	180.1	354.6	59.8	69.4	38.5	0.19
Mysore	187.3	357.6	63.5	84.6	45.2	0.24
Shivamoga	133.8	363.3	47.9	68.3	51.0	1.39
Belgaum	129.7	360.1	38.9	78.0	60.1	1.18
Chickmagalur	200.7	380.6	55.7	84.0	41.9	0.49
Mangalore	269.7	827.0	2.2	218.0	80.8	1.08

SD, standard deviation; CV, coefficient of variation.

18.11.5 WEEKLY WATER DEFICIT

The difference between reference evapotranspiration and actual evapotranspiration will account for the weekly water deficit. The graphs showing the difference between weekly ET_0 and AET values depicting weekly water deficit periods over different agroclimatic zones of Karnataka are presented in Figures 18.6–18.8.

The difference between ET_0 and AET (water deficit) was observed to be more during the 1st–22nd SMW, and the water deficit ranged from 25.97 to 45.01 mm at Bidar, 34.83 to 52.97 mm at

TABLE 18.4

Annual Reference Evapotranspiration (ET_0) over Different Agroclimatic Zones of Karnataka

Station	Mean ET_0 (mm)	Max ET_0 (mm)	Min ET_0 (mm)	SD (mm)	CV (%)
Bidar	1830.38	1916.08	1735.35	46.68	2.55
Raichur	2123.54	2388.68	1957.95	102.37	4.82
Bellary	2013.22	2276.58	1839.99	129.83	6.45
Chitradurga	1930.61	2100.24	1711.84	108.98	5.64
Bangalore	1689.18	1779.96	1556.76	62.16	3.68
Mysore	1791.77	1975.34	1620.13	109.64	6.12
Shivamoga	1758.66	1865.38	1689.59	48.60	2.76
Belgaum	1638.54	1729.10	1555.79	47.00	2.87
Chickmagalur	1630.44	1810.44	1481.91	79.08	4.85
Mangalore	1594.80	1743.27	1495.94	64.19	4.03

SD, standard deviation; CV, coefficient of variation.

Raichur, 24.13 to 47.66 mm at Bellary, and 15.28 to 52.13 mm at Chitradurga (Figure 18.6). During the kharif season (23rd–39th SMW), the water deficit was observed to be less ranging, from 0.37 to 16.76 mm at Bidar; 3.54 to 22.29 mm at Raichur; 8.29 to 24.41 mm at Bellary, and 3.66 to 13.30 mm at Chitradurga. From the 40th SMW onward, the increasing trend of water deficit was observed, and it ranged from 4.58 to 29.34 mm at Bidar, 2.96 to 32.12 mm at Raichur, 4.86 to 25.63 mm at Bellary, and 2.46 to 30.23 mm at Chitradurga.

The water deficit during the 1st–22nd SMW ranged from 6.51 to 40.78 mm at Bangalore, 5.84 to 45.78 mm at Mysore, 10.68 to 45.77 mm at Shivamoga, and 10.22 to 42.03 mm at Belgaum (Figure 18.7). The water deficit during the kharif season was observed to be much less at Shivamoga and Belgaum in comparison to Bangalore and Mysore. During the kharif season, the water deficit ranged from 1.62 to 9.47 mm at Bangalore, 3.54 to 7.67 mm at Mysore, 0.17 to 5.97 mm at Shivamoga, and 0.16 to 6.00 mm at Belgaum, respectively. From the 40th SMW onward, the water deficit showed an increasing trend at all stations, ranging from 1.58 to 24.87 at Bangalore, 3.29 to 29.09 mm at Mysore, 1.54 to 32.88 mm at Shivamoga, and 1.98 to 28.58 mm at Belgaum.

18.11.6 WEEKLY WATER SURPLUS

The quantity of surplus water generated during different weeks was estimated in climatic water balance study at different agroclimatic zones as well as the probabilities of occurrence of surplus water of 20, 40, 60, 80, and 100 mm during the 23rd–43rd SMW of the rainy season when potential surplus of water is expected in different agroclimatic zones. During the 23rd–43rd SMW of the rainy season, the probability of 20 mm surplus water ranged between 0% and 52% at Bidar, 0% and 40% at Raichur, 0% and 32% at Bellary, 0% and 48% at Chitradurga, 4% and 68% at Bangalore, 0% and 40% at Mysore, 16% and 72% at Shivamoga, 4% and 60 % at Belgaum, 28% and 84% at Chickmagalur, and 8% and 100% at Mangalore, respectively.

Similarly, the probability of 40 mm surplus water varied from 0% to 28% at dry zones except Bangalore, 0% to 60% at transition zones and the eastern dry zone (Bangalore), and 8% to 100% at hilly and coastal zones. The probabilities of 60 and 80 mm surplus water were also observed to be high at hilly and transition zones. The results also revealed that the chances of 100 mm surplus water was very low (probability <13%) at dry zones, except Bangalore, followed by transition zones,

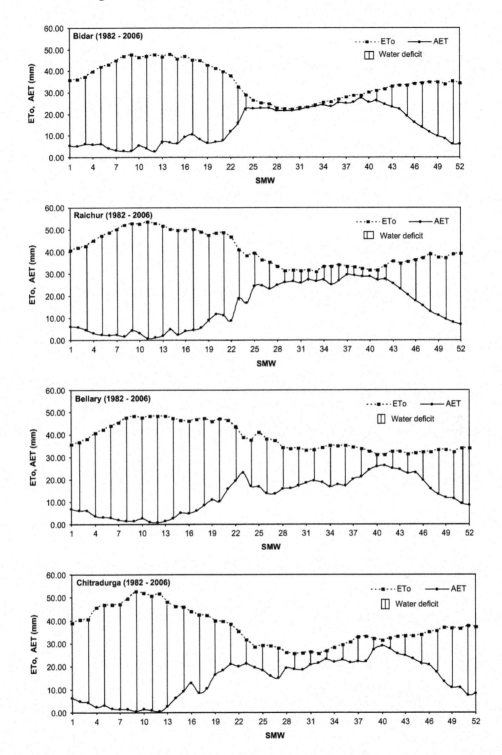

FIGURE 18.6 Difference between ET_0 and AET depicting the weekly water deficit at Bidar, Raichur, Bellary, and Chitradurga.

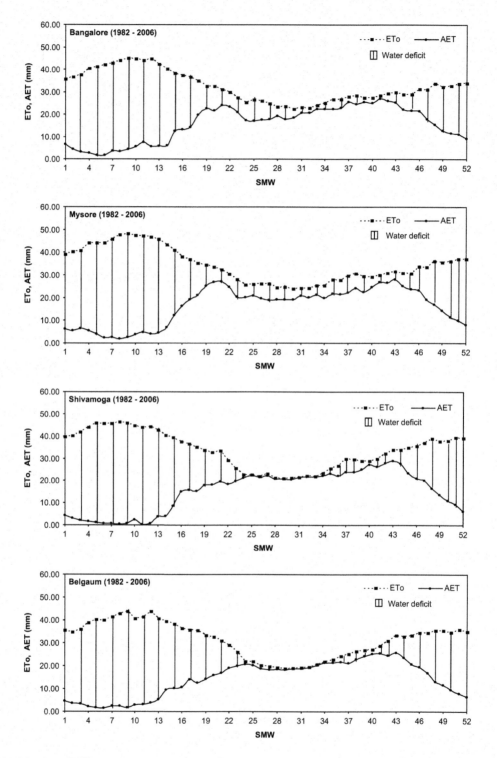

FIGURE 18.7 Difference between ET_0 and AET depicting the weekly water deficit at Bangalore, Mysore, Shivamoga, and Belgaum.

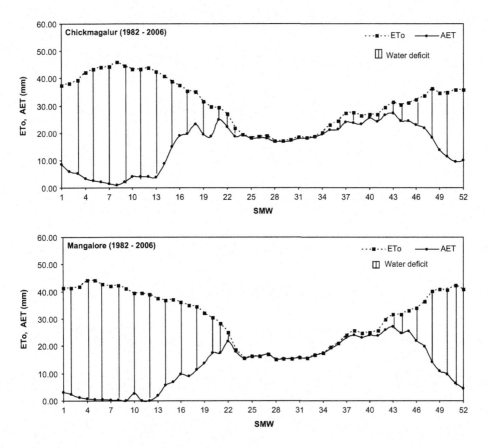

FIGURE 18.8 Difference between ET_0 and AET depicting weekly water deficit at Chickmagalur and Mangalore.

Bangalore, and Chickmagalur (probability up to 32%). However, at the coastal zone, the chance of 100 mm surplus water was high (probability up to 92 %).

18.11.7 AGRICULTURAL DROUGHT

Drought years were identified and their intensities were assessed from departure of annual aridity indices from the normal value. The taxonomy of droughts, on the basis of their intensity, over different agroclimatic zones is depicted in Figures 18.9–18.12.

Drought years and their severities over different zones are given in Table 18.5. Disastrous droughts were observed at six stations (Bellary, Bangalore, Mysore, Shivamoga, Chickmagalur, and Mangalore). Droughts of other severities were observed in all the zones in a random manner. The percentage of drought years of various categories in different zones varied from 44.00% to 56.00% with an overall average of 49.60%. Total drought years varied from 11 to 14 during the study period (1982–2006) in different zones, indicating the frequency of nearly one drought year in every two years. The results pertaining to agricultural drought indicated that drought of the same category does not prevail simultaneously over all the agroclimatic zones, but usually it occurs sporadically all over the state in a number of times ranging from moderate to disastrous.

The results also indicated that droughts are not only common in dry and transition zones but equally occur sporadically over hilly and coastal zones (heavy rainfall areas).

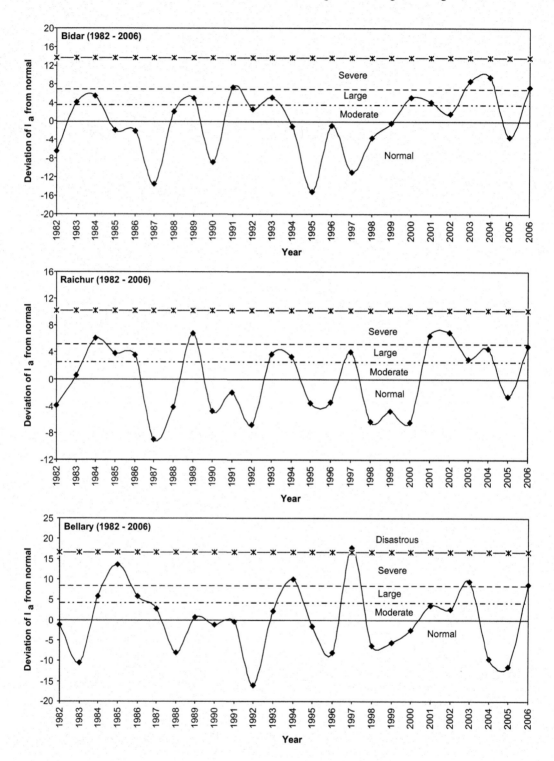

FIGURE 18.9 Aridity index at Bidar, Raichur, and Bellary.

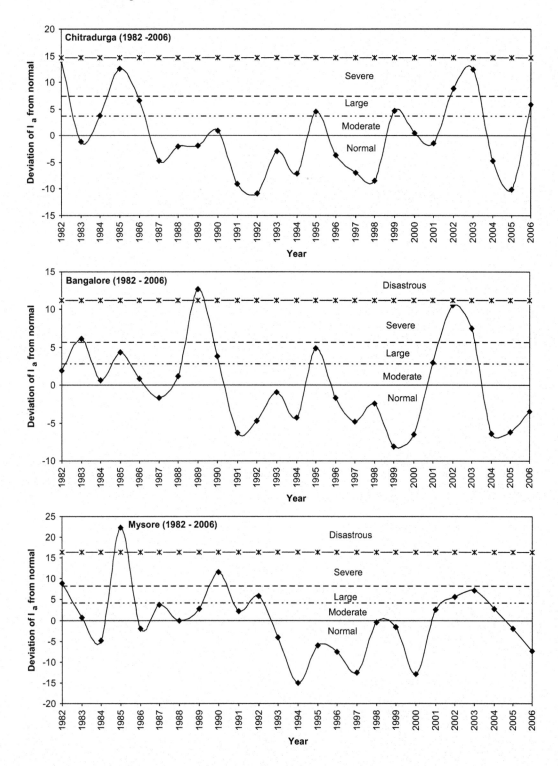

FIGURE 18.10   Aridity index at Chitradurga, Bangalore, and Mysore.

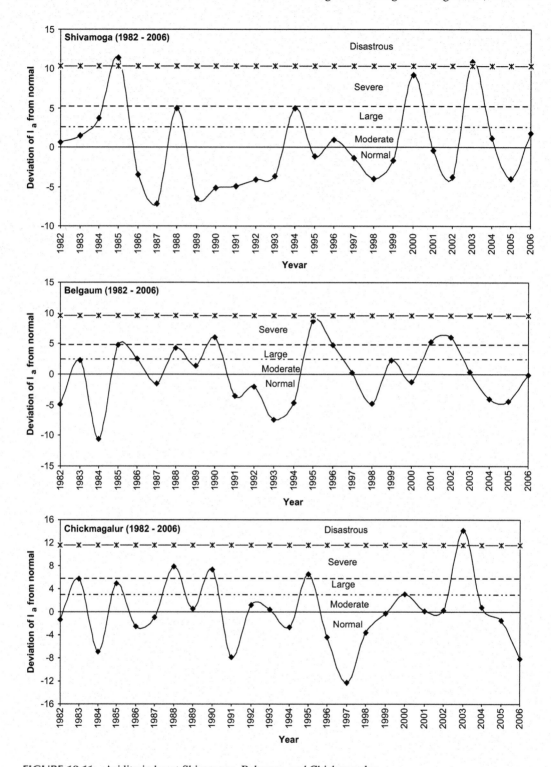

FIGURE 18.11   Aridity index at Shivamoga, Belgaum, and Chickmagalur.

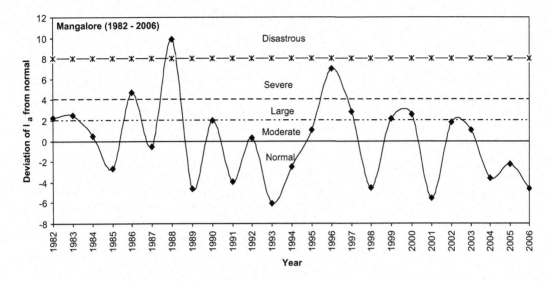

FIGURE 18.12   Aridity index at Mangalore.

TABLE 18.5
Agricultural Drought Years and Their Severities over Different Agroclimatic Zones of Karnataka

Station	Severity of Drought				Total Drought Years
	Moderate	Large	Severe	Disastrous	
Bidar	1988, 1992, 2002	1983, 1984, 1989, 1993, 2000, 2001	1991, 2003, 2004, 2006	—	13 (52.0)
Raichur	1983	1985, 1986, 1993, 1994, 1997, 2003, 2004, 2006	1984, 1989, 2001, 2002	—	13 (52.0)
Bellary	1987, 1989, 1993, 2001, 2002	1984, 1986	1985, 1994, 2003, 2006	1997	12 (48.0)
Chitradurga	1990, 2000	1984, 1986, 1995, 1999, 2006	1982, 1985, 2002, 2003	—	11 (44.0)
Bangalore	1982, 1984, 1986, 1988	1985, 1990, 1995, 2001	1983, 2002, 2003	1989	12 (48.0)
Mysore	1983, 1987, 1989, 1991, 2001, 2004	1992, 2002, 2003	1982, 1990	1985	12 (48.0)
Shivamoga	1982, 1983, 1996, 2004, 2006	1984, 1988, 1994	2000	1985, 2003	11 (44.0)
Belgaum	1983, 1989, 1997, 1999, 2003	1985, 1986, 1988, 1996	1990, 1995, 2001, 2002	—	13 (52.0)
Chickmagalur	1989, 1992, 1993, 2001, 2002, 2004	1983, 1985, 2000	1988, 1990, 1995	2003	13 (52.0)
Mangalore	1984, 1990, 1992, 1995, 2002, 2003	1982, 1983, 1997, 1999, 2000	1986, 1996	1988	14 (56.0)
Total years of different severity	43 (34.7)	43 (34.7)	31 (25)	7 (5.6)	124 (49.6)

Note: Figures in parenthesis indicate percentage of drought years.

18.12   STOCHASTIC MODELING OF WATER DEFICIT

Weekly water deficit data were analyzed using time series analysis techniques, and a stochastic weekly water deficit generator was developed over all selected stations of the state. The time series of weekly water deficit was assumed to be composed of a deterministic component and stochastic component. The mathematical procedure of stochastic modeling has been used in investigating the structure of time series of weekly water deficit for the different agroclimatic zones of Karnataka. The results are presented in the following sections.

18.12.1   TREND COMPONENT

For identification of the trend component, annual water deficit series was used to suppress periodic component (Kottegoda, 1980). The annual water deficit series was obtained from 23 years (1982–2004) of water deficit data and used for identification and detection of the trend component in the series. The hypothesis of no trend in the series was made and checked using the test statistic (Z) procedure for turning point and Kendall's rank correlation test.

   The results for turning point for ten stations reveal that all the absolute test statistics values varied from 0.52 to 1.03. On the other hand, the absolute values of the test statistics obtained from Kendall's rank correlation test for all but one station were found to vary from 0.013 to 1.94. However, test statistic value obtained through the Kendall's rank correlation test for the Chitradurga station was found to be –2.10. The estimated values of the test statistics (Z) for both tests were within the acceptable range at 0.01 level of significance for all agroclimatic zones. Hence, the hypothesis of no trend was accepted. Therefore, the trend component in the water deficit series of different agroclimatic zones is absent and the observed series were found to be trend free.

18.12.2   PERIODIC COMPONENT

The periodic component is easily identified by its cyclic phenomenon imposed on the series, provided the series is trend free. Presence of a periodic component in a series is detected through the construction of a correlogram, which is a graph showing the relationship between the autocorrelation functions on the ordinate and lag K on the abscissa. For proper identification and interpretation of the autocorrelogram, the estimate of the autocorrelation function for weekly water deficit was made for all the agroclimatic zones of Karnataka.

18.12.2.1   Autocorrelation Analysis

The autocorrelation functions for the weekly time series up to lag 104 were determined from Equation 18.12, which has been used to identify the periodic component in the time series. The correlogram for the weekly observed series along with the tolerance limit for an independent series are shown in Figures 18.13 to 18.15. The resultant oscillating shape of the autocorrelogram shows the presence of periodicity in the time series. Moreover, in all stations, the weekly serial correlogram has peaks at about 52 and 104, and troughs at about 26 and 78, with a base period of 52. It is also seen that some of the autocorrelograms of the time series fall out of the confidence limits for all the stations, indicating the presence of time-dependent series, i.e., $X_{(t+1)}$ is dependent on $X_{(t)}$, and $X_{(t+2)}$ is dependent on $X_{(t+1)}$, and so on. Values of lag 1 autocorrelation coefficients for the series lies outside the range of confidence limit and is significantly different from 0, which confirms the mutual dependence in water deficit observed series.

18.12.2.2   Stochastic Component

The stochastic component that was obtained by the removal of the periodic component from the observed series is treated as a random variable. The autoregressive stochastic model of order p, AR (p), was applied to the stochastic component of the weekly water deficit series to generate a synthetic

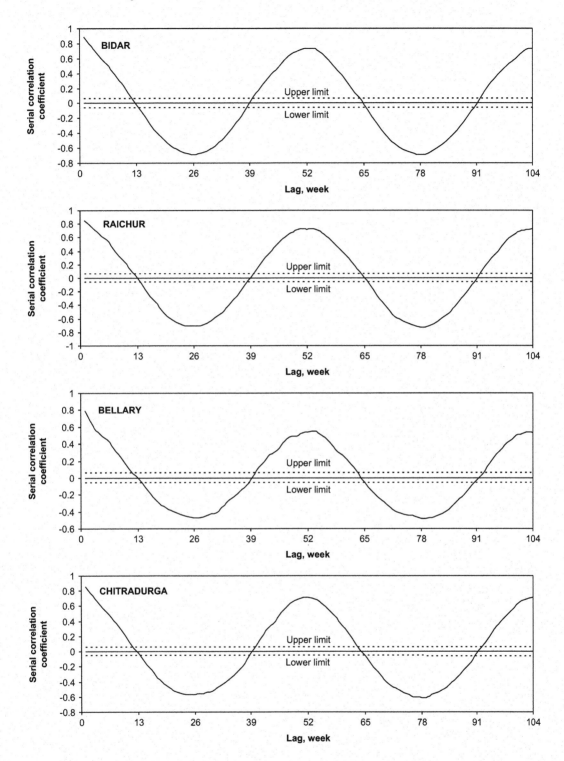

FIGURE 18.13 Correlogram of weekly water deficit for Bidar, Raichur, Bellary, and Chitradurga.

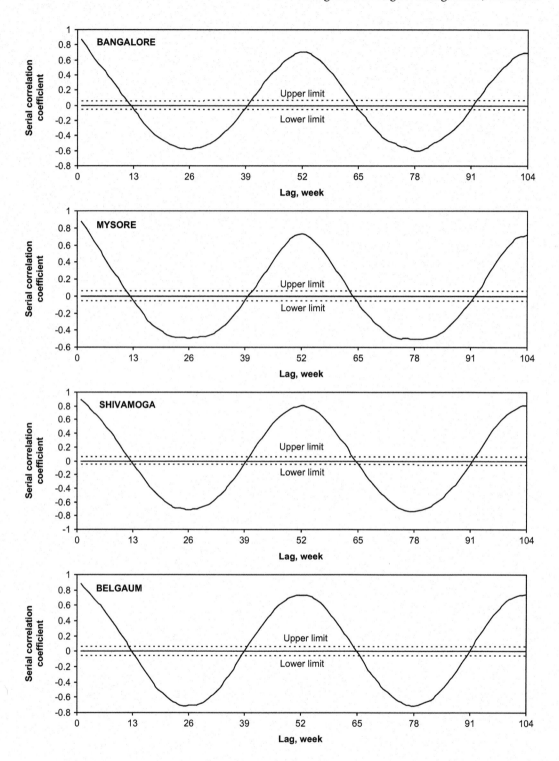

FIGURE 18.14   Correlogram of weekly water deficit for Bangalore, Mysore, Shivamoga, and Belgaum.

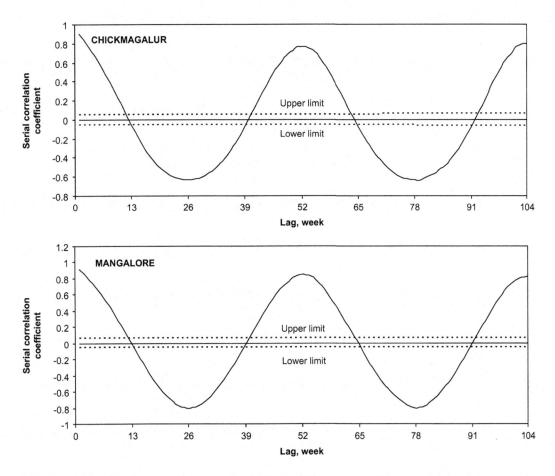

FIGURE 18.15   Correlogram of weekly water deficit for Chickmagalur and Mangalore.

data sequence of the weekly series. In the autoregressive model, the current value of a variable was equated to the weighted sum of a preassigned number of past values and variates that is completely random.

18.12.2.3   Model Parameter Estimation

For modeling the stochastic component, the autocorrelation function estimated from Equation 18.14 is required. Selection and estimation of autoregressive parameters is based upon the determined values of the autocorrelation function of different lags (Table 18.6).

The final forms of expression for estimating the stochastic component in the weekly water deficit series for the different regions are as follows:

$$\text{Bidar } S_t = 0.825S_{t-1} + 0.072S_{t-2} + a_t \tag{18.18}$$

$$\text{Raichur } S_t = 0.670S_{t-1} + 0.217S_{t-2} + a_t \tag{18.19}$$

$$\text{Bellary } S_t = 0.680S_{t-1} + 0.142S_{t-2} + a_t \tag{18.20}$$

$$\text{Chitradurga } S_t = 0.747S_{t-1} + 0.129S_{t-2} + a_t \tag{18.21}$$

$$\text{Bangalore } S_t = 0.799S_{t-1} + 0.084S_{t-2} + a_t \tag{18.22}$$

$$\text{Mysore } S_t = 0.791S_{t-1} + 0.099S_{t-2} + a_t \tag{18.23}$$

TABLE 18.6

Model Order and Autoregressive Parameters of Weekly Water Deficit Series

Region	Model Order	$\Phi_{(p,k)}$	Values
Bidar	2	$\Phi_{(2,1)}$	0.825
		$\Phi_{(2,2)}$	0.072
Raichur	2	$\Phi_{(2,1)}$	0.670
		$\Phi_{(2,2)}$	0.217
Bellary	2	$\Phi_{(2,1)}$	0.680
		$\Phi_{(2,2)}$	0.142
Chitradurga	2	$\Phi_{(2,1)}$	0.747
		$\Phi_{(2,2)}$	0.129
Bangalore	2	$\Phi_{(2,1)}$	0.799
		$\Phi_{(2,2)}$	0.084
Mysore	2	$\Phi_{(2,1)}$	0.791
		$\Phi_{(2,2)}$	0.099
Shivamoga	2	$\Phi_{(2,1)}$	0.745
		$\Phi_{(2,2)}$	0.164
Belgaum	2	$\Phi_{(2,1)}$	0.745
		$\Phi_{(2,2)}$	0.153
Chickmagalur	2	$\Phi_{(2,1)}$	0.778
		$\Phi_{(2,2)}$	0.136
Mangalore	2	$\Phi_{(2,1)}$	0.826
		$\Phi_{(2,2)}$	0.101

$$\text{Shivamoga } S_t = 0.745S_{t-1} + 0.164S_{t-2} + a_t \tag{18.24}$$

$$\text{Belgaum } S_t = 0.745S_{t-1} + 0.153S_{t-2} + a_t \tag{18.25}$$

$$\text{Chickmagalur } S_t = 0.778S_{t-1} + 0.136S_{t-2} + a_t \tag{18.26}$$

$$\text{Mangalore } S_t = 0.826S_{t-1} + 0.101S_{t-2} + a_t \tag{18.27}$$

18.12.2.4 Model Identification

Residual variance has been used as a criterion to determine the order of the autoregressive model. Residual variance at different lags was computed for the weekly water deficit series of different zones. A suitable model order is the order for which the value of residual variance is minimum. Here, the minimum residual variance has been observed for the first-order model in the weekly series of all the agroclimatic zones. However, the serial autocorrelation of residual series from the first-, second-, and third-order models were compared in each weekly series, and the residual series that gave well within the minimum range of values was selected to represent the stochastic component of the autoregressive model. According to the comparison made, for all the regions of Karnataka, the second-order autoregressive model was selected for the weekly water deficit series. The stochastic component was estimated by fitting the autoregressive parameters for the selected model order.

18.12.2.5 Residual Series

The S_t series of the stochastic component, which was obtained after the removal of the deterministic components from the time series, consists of two components. These are the dependent part represented by an autoregressive model, and the independent or residual part that is completely random,

TABLE 18.7

Model Structure of Weekly Water Deficit Series

Region	Mean (mm week^{-1})	Periodic Component (mm week^{-1})	Stochastic Component (mm week^{-1})
Bidar	20.80	$10.79\text{Cos}(2\pi t/p) + 19.09\text{Sin}(2\pi t/p)$ $-3.20\text{Cos}(4\pi t/p) - 1.78\text{Sin}(4\pi t/p)$	$0.825S_{t-1} + 0.072S_{t-2} + a_t$
Raichur	26.04	$9.00\text{Cos}(2\pi t/p) + 22.48\text{Sin}(2\pi t/p)$ $-1.94\text{Cos}(4\pi t/p) - 0.99\text{Sin}(4\pi t/p)$	$0.670S_{t-1} + 0.217S_{t-2} + a_t$
Bellary	26.09	$2.81\text{Cos}(2\pi t/p) + 17.41\text{Sin}(2\pi t/p)$ $-2.13\text{Cos}(4\pi t/p) + 3.71\text{Sin}(4\pi t/p)$	$0.680S_{t-1} + 0.142S_{t-2} + a_t$
Chitradurga	22.78	$11.52\text{Cos}(2\pi t/p) + 17.82\text{Sin}(2\pi t/p)$ $-2.90\text{Cos}(4\pi t/p) + 4.43\text{Sin}(4\pi t/p)$	$0.747S_{t-1} + 0.129S_{t-2} + a_t$
Bangalore	17.01	$11.60\text{Cos}(2\pi t/p) + 14.87\text{Sin}(2\pi t/p)$ $-0.61\text{Cos}(4\pi t/p) + 5.51\text{Sin}(4\pi t/p)$	$0.799S_{t-1} + 0.084S_{t-2} + a_t$
Mysore	18.44	$13.34\text{Cos}(2\pi t/p) + 14.38\text{Sin}(2\pi t/p)$ $-1.43\text{Cos}(4\pi t/p) + 7.54\text{Sin}(4\pi t/p)$	$0.791S_{t-1} + 0.099S_{t-2} + a_t$
Shivamoga	18.65	$15.71\text{Cos}(2\pi t/p) + 16.33\text{Sin}(2\pi t/p)$ $-1.57\text{Cos}(4\pi t/p) + 4.99\text{Sin}(4\pi t/p)$	$0.745S_{t-1} + 0.164S_{t-2} + a_t$
Belgaum	17.37	$13.65\text{Cos}(2\pi t/p) + 14.69\text{Sin}(2\pi t/p)$ $-1.56\text{Cos}(4\pi t/p) + 2.41\text{Sin}(4\pi t/p)$	$0.745S_{t-1} + 0.153S_{t-2} + a_t$
Chickmagalur	15.94	$14.61\text{Cos}(2\pi t/p) + 14.92\text{Sin}(2\pi t/p)$ $-1.83\text{Cos}(4\pi t/p) + 6.39\text{Sin}(4\pi t/p)$	$0.778S_{t-1} + 0.136S_{t-2} + a_t$
Mangalore	17.85	$16.84\text{Cos}(2\pi t/p) + 15.93\text{Sin}(2\pi t/p)$ $0.27\text{Cos}(4\pi t/p) + 3.66\text{Sin}(4\pi t/p)$	$0.826S_{t-1} + 0.101S_{t-2} + a_t$

also called white noise. The residual series, a_t, is obtained by deducting the generated series, which is the sum of the periodic and stochastic components, from the observed time series.

18.12.2.6 Model Structure

The model structure of a time series constitutes the sum of the trend, periodic, and stochastic components. Since the observed weekly series in each agroclimatic region are trend free, the submodels of the periodic and stochastic components are added together to form the newly developed model structure of the water deficit series. The mathematical structure of the additive model can now be presented for weekly water deficit series as given in Table 18.7 for different agroclimatic zones. The formulated model structure was used to generate a similar sequenced series of weekly water deficit. The formulated models were subjected to diagnostic checking for either rejection or acceptance. Goodness-of-fit tests for observed weekly water deficit models were carried out for all the zones. Two different tests of goodness of fit – sum of squares analysis and autocorrelation analysis – were used for diagnostic checking.

In the sum of squares analysis, the sum of squares of residuals and deviation of the observed series from their mean value were estimated. The value of R^2 obtained for each agroclimatic zone for the weekly series model was found to range from 0.9667 at Bellary to 0.9964 at Mangalore. The developed model has the best goodness of fit to generate weekly water deficit series.

18.13 AUTOCORRELATION ANALYSIS

The autocorrelation function of weekly residual series was estimated for lag k for each region. The value of autocorrelation function was plotted against the lag to obtain a correlogram. The resulting correlogram of weekly series, along with the confidence limit at 0.01 level of significance, for each agroclimatic zone are shown in Figures 18.16–18.18. The results show that for all lags, the

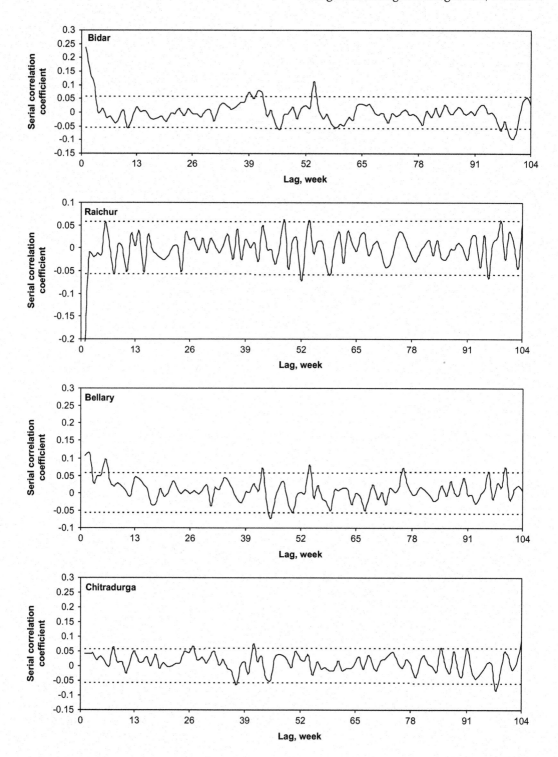

FIGURE 18.16 Correlogram of the residual series of weekly water deficit for Bidar, Raichur, Bellary, and Chitradurga.

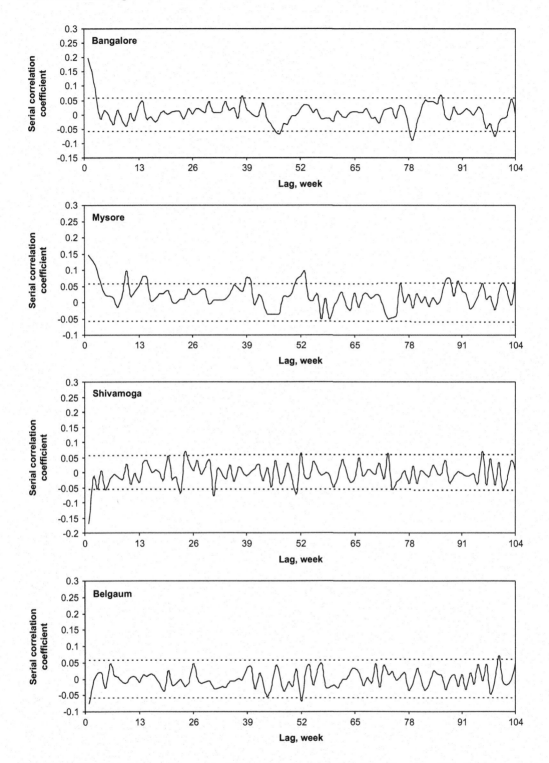

FIGURE 18.17 Correlogram of the residual series of weekly water deficit for Bangalore, Mysore, Shivamoga, and Belgaum.

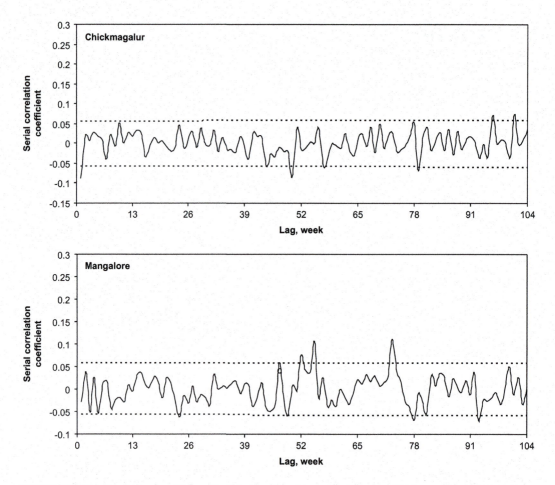

FIGURE 18.18   Correlogram of the residual series of weekly water deficit for Chickmagalur and Mangalore.

autocorrelation function falls fairly within the confidence limits. So we can use these models and generate values of weekly water deficit and compare the statistical characteristics of observed and generated series.

18.13.1   VALIDATION OF STOCHASTIC MODEL

Comparison of the generated with the observed water deficit series was made to validate the model. Figures 18.19–18.22 depict the variation of observed and generated weekly water deficit series for the modeling period (1982–2004) for all the stations. The figures indicate that there is a close agreement between the observed and generated weekly water deficit during different meteorological weeks for the modeling period. Values of mean, standard deviation, coefficient of skewness, kurtosis, and variance show that for the observed and generated series statistical characteristics are not significantly different and that the formulated models are significantly adequate for generating the weekly water deficit.

The values of the correlation coefficient (r) between the observed and generated weekly water deficit series at different stations for the modeled period (1982–2004) were worked out, and the correlation coefficient ranged from 0.9743 at Bellary to 0.9967 at Mangalore in weekly series. The model structures formulated can be used for long-term prediction of the weekly water deficit for their respective agroclimatic stations. The preceding results for predicted and observed weekly

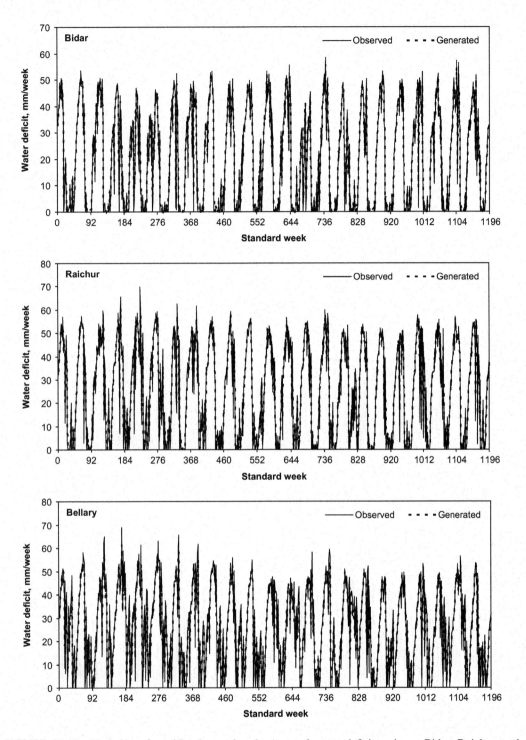

FIGURE 18.19 Variation of weekly observed and generated water deficit series at Bidar, Raichur, and Bellary (1982–2004).

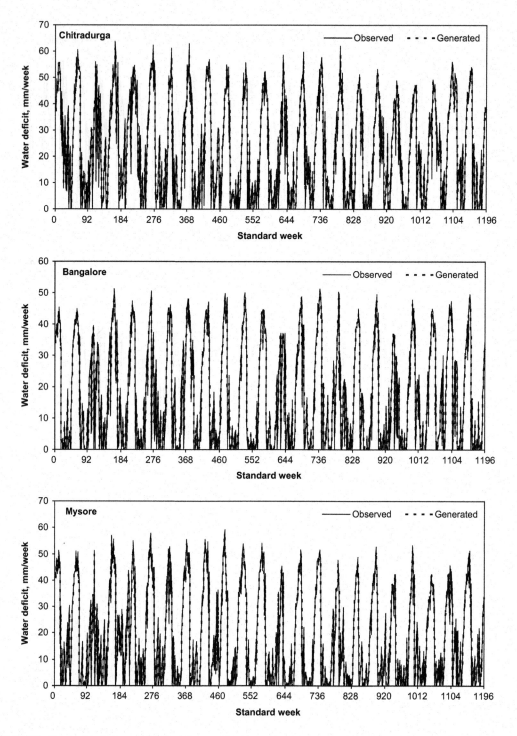

FIGURE 18.20 Variation of weekly observed and generated water deficit series at Chitradurga, Bangalore, and Mysore (1982–2004).

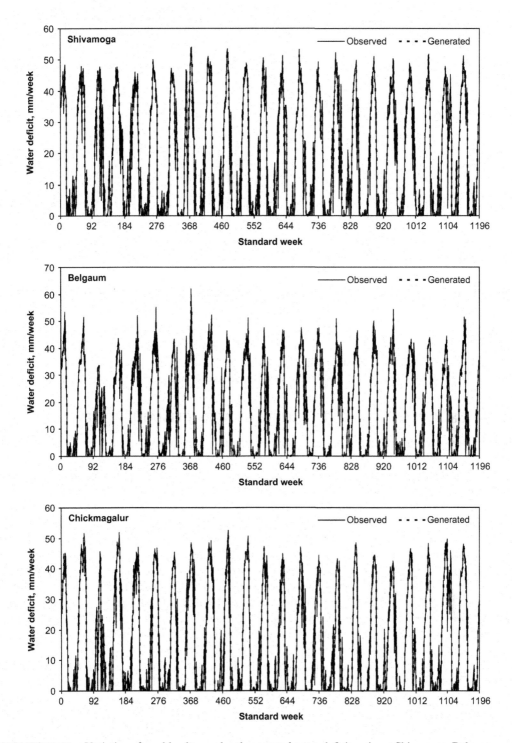

FIGURE 18.21 Variation of weekly observed and generated water deficit series at Shivamoga, Belgaum, and Chickmagalur (1982–2004).

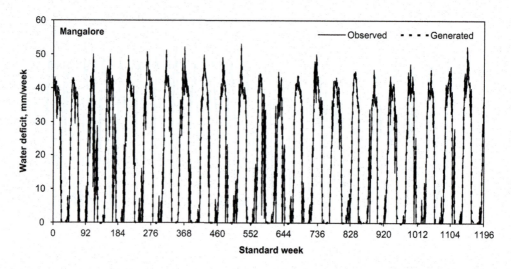

FIGURE 18.22 Variation of weekly observed and generated water deficit series at Mangalore.

water deficit series of two years (2005 and 2006) indicate adequacy of the model for predicting water deficit series for respective periods in different agroclimatic zones of Karnataka. Values of the correlation coefficient and statistical characteristics for the observed and predicted series confirm the reliability of the weekly water deficit model for generation of data.

18.14   CROP PLANNING OVER DIFFERENT AGROCLIMATIC ZONES

The selection of crops and cropping pattern for a rainfed region mainly depends on the quantum and distribution pattern of rainfall during the crop growing season and soil characteristics. It becomes essential to plan cropping sequences such that critical crop stages escape continuous severe dry spells safely or, if not possible, then to select suitable drought-resistant crops or adopt *in situ* moisture conservation practices to maintain soil moisture conditions. In this direction, an integrated resources–based crop planning was suggested for Karnataka based on rainfall parameters, soil characteristics, length of growing season, and dry spells. The detailed resources-based agricultural crop planning is presented in Table 18.8 over selected stations. The expected annual rainfall at 30% probability ranges from 550.7 mm at Bellary to 4063.7 mm at Mangalore, whereas at the 70% probability level, the expected annual rainfall ranges from 383.7 mm at Bellary to 3149.2 mm at Mangalore. At 50% probability, it was noticed that at all stations it is closer to the average annual rainfall, which hints for certainty of getting the average annual rainfall amount every other year or once in two years. The probability of two or three consecutive dry weeks over all stations indicated that there are chances of getting intermediate dry spells at all stations.

However, more dry spells were observed at the Bidar and Chitradurga stations, followed by Raichur, Mysore, Bangalore, transition zones, and heavy rainfall zones (Chickmagalur and Mangalore). The length of growing season (LGS) was determined from two criteria, viz., the difference between onset and withdrawal of rainy season and based on moisture adequacy index (I_{ma}) values. Based on former criteria, the LGS season varied from 16 weeks at Bellary to 23 weeks at Chickmagalur and Mangalore. Based on I_{ma} values, due to availability of moisture in the soil profiles, the LGS was found to be three to four weeks more compared with criteria of onset and ending on rainy season. At normal conditions, the date of sowing was decided based on the regular onset of the rainy season and it varied from the 22nd SMW (Chickmagalur and Mangalore) to the 27th SMW at Bellary. Based on soil, LGS, and rainfall characteristics, the suitable cropping pattern is suggested for all stations (Table 18.8). It is revealed that double cropping is possible at all stations, except Bellary and Chitradurga. However, *in situ* moisture conservation practices are very much

TABLE 18.8

Integrated Resources–Based Agricultural Crop Planning at Bidar

Parameter	Description	Remarks
Rainfall	Average annual rainfall (AAR): 905.3 mm	AAR at P(30%) = 1057.7 mm AAR at P(50%) = 851.9 mm AAR at P(70%) = 764.3 mm
Soil type	Major area: Shallow to medium black clay Remaining area: Red lateritic	—
2/3 consecutive dry weeks	2D > 50% = Nil, 2D > 25% = 26, 27, 34, 39th SMW during kharif	Intermittent dry spells are likely to occur
	2D > 50% = After 40th SMW onward during rabi	Moisture conservation practices provide more benefit
	3D > 25% = Nil, during kharif	
	3D > 25% = After 40th SMW onward during rabi	
Length of growing season (LGS)	LGS: 19 weeks based on onset and end of monsoon LGS: 23 weeks based on I_{ma} values	—
Date of sowing at normal conditions	24th SMW (June 11–17)	Late sowing up to 28th SMW (July 9–15)
Cropping pattern	**Monocropping/intercropping:** Hybrid sorghum + pigeonpea Green gram/black gram + pigeonpea Sesamum + pigeonpea Groundnut + pigeonpea **Double cropping:** Black gram and rabi sorghum Black gram and sunflower Sunflower and chickpea/rabi sorghum Green gram and chickpea/rabi sorghum Sesamum and rabi sorghum/sunflower Chickpea/coriander and rabi sorghum	• Opting for monocropping or double cropping depends on onset of rainy season. • In case of monocropping, farmers should always prefer intercropping for sustainable crop production. • Choice of crop selection depends on farmer's preference (based on economic, food habit, and land and labor factors). • Short-duration varieties are most preferred.

necessary at all stations for taking up sustainable crop production, and, wherever possible, it is necessary to harvest excess runoff from fields to be used for providing supplementary irrigations at critical crop growth stages. It is also suggested to incorporate perennial components like tree species for fuel, timber, and fodder; and horticultural plants and perennial grass species at all zones for improving the sustainability of the rainfed ecosystem.

18.15 CONCLUSIONS

This study was undertaken with the objectives to develop and validate an appropriate stochastic model of water deficit, and to plan suitable cropping patterns on the basis of drought analysis for different agroclimatic zones of Karnataka by using rainfall data from 1976 to 2006. Weekly ET_0 and water balance component values were estimated by the Penman–Monteith equation and the Thornthwaite and Mather bookkeeping technique, respectively. Water deficit data were also analyzed to develop and validate appropriate stochastic models. Suitable crops and cropping systems were also suggested. The results of the study are summarized as follows:

(i) Annual rainfall analysis of ten stations representing different agroclimatic zones indicated that the mean rainfall at Bellary was lowest with a higher value of the coefficient of variation compared to other stations. At four stations (Bellary, Bangalore, Mysore, and

Belgaum), the CV of annual rainfall exceeded the threshold limit (>25%), which highlights the extremely erratic distribution of rainfall, and the trend of annual rainfall was found to be nonsignificant. The CV of kharif season rainfall was found to be within the threshold limit (50%) at all stations, which indicates that occurrence of mean kharif rainfall is highly dependable, whereas the CV of the rabi season was within the threshold limit (<50%) at only three stations. Higher values of CV (>150%) of weekly rainfall indicated erratic distribution of rainfall over all stations during rabi and summer. During kharif, a few weeks were observed to be above the threshold limit at Raichur, Bellary, Chitradurga, Bangalore, Mysore, and Belgaum.

(ii) The mean weekly ET_0 among all stations was observed to be highest in dry agroclimatic zones (Raichur followed by Bellary, Chitradurga, Mysore, and Bangalore). However, in kharif season, higher values of weekly ET_0 were observed at Bellary. The mean weekly water deficit values revealed that even during kharif season, every station experienced water deficits of different intensities. However, the water deficit was observed to be more during the 1st–22nd SMW and after the 40th SMW. Distribution of weekly rainfall, which is highly erratic and uncertain, results in a water deficit of various categories in different agroclimatic zones, producing intermittent droughty conditions. Surplus water availability in different agroclimatic zones clearly indicated that there are chances of getting excess water in the rainy season in all the zones, which can be effectively harvested and used in water deficit periods. A climatic shift study at stations under consideration indicated that Raichur, Bellary, and Chitradurga were found to be predominantly in the semiarid climate; Bidar, Bangalore, Mysore, and Belgaum in the dry subhumid condition; Shivamoga between a dry and moist subhumid climate; and Chickmagalur in humid and Mangalore in per humid climates.

(iii) Meteorological drought evaluation indicated a frequency of one drought year in every twelve to five years. Whereas in the case of agricultural drought, a frequency of one drought year of varying intensity was observed for every two to two and half years. Developed autoregressive stochastic models were found to be adequate for forecasting the weekly water deficit as seen from the highly significant values of correlation coefficient between the observed and generated, and observed and predicted water deficit series.

ACKNOWLEDGMENTS

The authors gratefully acknowledge the Karnataka State Department of Agriculture, the National Data Centre (Pune), ICAR-NBSS & LUP (Bangalore), ICAR-IISWC Research Centre (Bellary), and University of Agricultural Sciences (Bangalore) for providing the map and the data used in this study.

REFERENCES

Allen, R.G., I.S. Pereira, D. Daes and M. Smith. 1998. Crop evapotranspiration, guideline for computing crop water requirements. *FAO Irrigation and Drainage*, Paper 56. Rome, Italy.
Bhakar, S.R., R.V. Singh and H. Ram. 2006. Stochastic modeling of wind speed at Udaipur region. *Journal of Agricultural Engineering.* **43**: 1–7.
Biggs, T.W., P.K. Mishra and H. Turral. 2008. Evapotranspiration and regional probabilities of soil moisture stress in rainfed crops, Southern India. *Agricultural and Forest Meteorology.* **148**: 1585–1597.
Goyal, R.K. 2004. A reference evapotranspiration model for arid region of Rajasthan. *Indian Journal of Soil Conservation.* **32**: 10–15.
Goyal, R.K. 2005. Determination of pan coefficient for estimation of reference evapotranspiration for Jodhpur (Rajasthan). *Journal of Agrometeorology.* **7**: 307–310.
Hatfield, J.L. and R.G. Allen. 1997. Evapotranspiration estimate under deficient water supplies. *Journal of Irrigation and Drainage Engineering. ASCE.* **123**: 301–308.

Kalsi, S.R., R.K. Jenamani and H.R. Hatwar. 2006. Meteorological features associated with Indian drought in 2002. *Mausam*. **57**: 459–474.

Kashyap, P.S. and R.K. Panda. 2001. Evaluation of crop evapotranspiration estimation methods and development of crop-coefficient for potato crop in sub humid region. *Agricultural Water Management*. **50**: 9–25.

Khichar, M.L., R. Singh and V.U.M. Rao. 1991. Water availability periods for crop planning in Haryana. *International Journal of Tropical Agriculture*. **1**: 301–305.

Kottegoda, N. T. 1980. *Stochastic Water Resource Technology*. The MacMillan Press ltd., London. pp. 384.

Salam, M.A. and S.A. Mazrooe. 2006. Evapotranspiration estimates and water balance of Kuwait. *Journal of Agrometeorology*. **8**: 243–247.

Thornthwaite, C.W. and J.R. Mather. 1957. *Instructions and Tables for Computing Potential Evapotranspiration and the Water Balance*. Publication No. 10 Laboratory of Climatology, Centerton, NJ.

19 Tree Ring–Based Drought and Flood Analyses from the Himalayan Region
Limitations, Challenges, and Future Perspectives

Mayank Shekhar, Ayushi Singh, Bency David, Nidhi Tomar, Ipsita Roy, Parminder S. Ranhotra, and A. Bhattacharyya

CONTENTS

19.1 INTRODUCTION

In the context of unprecedented global warming, extreme climatic events like droughts and floods are the greatest challenges to socioeconomic values. The temperature increase might accelerate the severity of drought conditions due to high evapotranspiration and drying of the land surface (IPCC, 2013). Moreover, warmer climates could alter precipitation, such as its amount, intensity, and frequency in many parts of the world (Goswami et al., 2006; Kumar et al., 2006; Rajeevan et al., 2008; IPCC, 2013). There are extensive studies and methodologies adopted for the management of droughts and floods so that such adverse situations can be judiciously mitigated.

During the last millennia, the Earth's climate system experienced two significant anomalies: the warm Medieval Warm Period (MWP) about 900 to 1300[*] and the cool Little Ice Age (LIA) about 1500 to 1850 (Grove, 2001; Graham et al., 2011). However, there remains a lack of spatial homogeneity in the duration and dynamics of these two anomalies (Matthes, 1939; Lamb, 1965; Bradley and Jonest, 1993; Hughes and Diaz, 1994; Grove, 2001; Bradley et al., 2003; Mann et al., 2009) and thus ensuing discussions regarding their regional hydroclimatic exodus. On the Indian subcontinent, drought events are generally related to the weak Indian summer monsoon (ISM). The weakening of the ISM by 1300 (Gupta et al., 2003; Sinha et al., 2011a, 2011b) was responsible for the high-intensity monsoon megadrought (MMD) events in the core monsoon region (Gupta et al., 2003; Sinha et al., 2007, 2011b; Tiwari et al., 2006) and Andaman Islands (Laskar et al., 2013).

---

[*] All years are CE.

DOI: 10.1201/9781003276548-19

These recorded MMD events during the LIA phase (1350–1850) caused by the decrease in monsoon precipitation also correspond with the widespread MMD episodes in China (Zhang et al., 2011) and other parts of Asia (Fleitmann et al., 2003, 2007; Wang et al., 2005; Hu et al., 2008). The post-LIA timeframe experienced a relative increase in precipitation but with intermittent extreme events of drought and floods, as recorded meteorologically during the past century (IPCC, 2013; Pathak et al., 2019). It is, therefore, crucial to have a precise understanding of such past hydroclimatic anomalies and their driving mechanisms in continuation to present climate on a finer temporal scale so as to remodel the future trends precisely.

Reconstructed high-resolution climatic records of the past few centuries by the tree-ring analysis are of great significance for a better understanding of recent past climate and related events (Mann et al., 1999). However, the Himalayan region is limited with the availability of only a few long tree-ring records (Bhattacharyya and Shah, 2009; Shekhar, 2014; Shekhar et al., 2017). This lack of a long tree-ring data network from this climatically sensitive highland is a major hindrance to understanding the hydroclimatic events on the regional scale and their global teleconnections.

The Himalayan region is one of the principle areas for tree-ring research because of its huge geography, great floral diversity, climate, and physiognomy, thus providing many potential tree species and sites for such studies. However, systematic dendrochronological reviews with the accurate dating of tree rings in terms of a calendar year of their formation through cross-dating started only during the 1980s (Bhattacharyya et al., 1988). Among the various applications of tree-ring analysis, climate reconstruction gained much significance, especially in addressing the global warming trend by the reconstruction of past temperature anomalies worldwide (Cook et al., 1995; Hughes et al., 1999; Mann et al., 1999; Jacoby et al., 2000; D'Arrigo et al., 2001; Mann et al., 2012; Anchukaitis et al., 2012; Babst et al., 2017) including the Asian region (Esper et al., 2002; Cook et al., 2003; Zhu et al., 2009; Shi et al., 2010; Chen et al., 2013; Zhang et al., 2014; Liu et al., 2019). In last couple of decades, the Himalayan region has also in focus because of its crucial role in monsoon dynamics. Moisture is an essential factor influencing the growth of trees, and at any area, the precipitation, coupled with other climatic factors, affects the tree-growth response. In recent times, past moisture variabilities have been assessed from different regions of the Himalayas by the reconstruction of various moisture regimes, such as precipitation, streamflow, and floods, based on tree-ring width (Cook et al., 2003; Bhattacharyya and Shah, 2009; Yadav et al., 2011; Shah et al., 2014; Krusic et al., 2015; Thapa et al., 2015; Panthi et al., 2017; Mangave et al., 2020).

The analysis of past drought indices is an essential means to understand the long-term moisture variability. These indices include the Palmer Drought Severity Index (PDSI), Standardized Precipitation Index (SPI), and Standardized Precipitation Evapotranspiration Index (SPEI). Cook et al. (2010) well demonstrated the Asian monsoon failure and its regional significance using the PDSI, which was linked to monsoon failure in relation to sea surface temperature and sea-level pressure anomalies on a global perspective. Moreover, the monsoon floods are also the most frequent and widespread climatic event in the Himalayan region (Gardner, 2002). Tree rings have also been used to reconstruct past disasters, especially floods (Stoffel et al., 2010; Ballesteros-Canovas et al., 2015) based on the concept that trees affected by the hydrogeomorphic processes will record the information of events in their growth rings (Shroder, 1978; Stoffel and Corona, 2014). Tree-ring series analysis of such affected trees can be used to infer the frequency and magnitude of floods (Ballesteros-Cánovas et al., 2017).

Here we provide an advanced review of the tree-ring research and its applications pursued from the Himalayan region, emphasizing the past climate, drought, and flood analyses, together with discussions on the limitations, challenges, and future perspective for this region.

19.2 HIMALAYAN FORESTS AND THEIR POTENTIAL FOR TREE-RING ANALYSIS

A large number of tree species, especially conifers, growing in the Himalayas have been successfully analyzed by dendrochronology. Conifers are best suited for tree-ring analysis because of their clear annual growth-ring boundary, longevity, and ease in collecting samples, in comparison to the broad-leaved tree species. Moreover, wide lateral (northwest to northeast) and altitudinal (subtropical to subalpine) distribution of conifers in the Himalayan region, either making a pure conifer forest or mixed with broad-leaved taxa (Champion and Seth, 1968), make them suitable for studying the captured climatic signals from different climatic regimes of the Himalaya. It is estimated that conifers are represented by about 17 genera with 60 species (Sahni, 1990), some listed in Table 19.1,

TABLE 19.1
Distribution and Probable Age of Various Himalayan Conifers and Their Utility in Environment Reconstruction

Species	Distribution	Probable Age (Years)	Applicability of Tree-Ring Data to Different Science Domains
Taxus wallichiana Zucc.	All along the Himalaya from Kashmir Himalaya through Khasi Hills, Naga Hills close to Himalaya to Arunachal Pradesh	Probably over 1000	Ecology (Chaudhary et al., 1999; Yadav and Singh, 2002)
Pinus merkusii Jungh. & de Vriese	Lohit district of Arunachal Pradesh at around 2500 m	350	Climate (Buckley et al., 2005) Ecology (Shah and Bhattacharyya, 2012)
Pinus wallichiana A.B. Jacks.	All along the Himalayas from Kashmir to Arunachal Pradesh, 1800–3700 m; rare in Sikkim and in a considerable portion of Kumaon	400–500	Ecology (Yadava et al., 2017) Climate Buckley et al., 2005; Shah and Bhattacharyya, 2009; Yadav, 2009; Chaudhary et al., 2013) Natural hazards (Bhattacharyya et al., 2017) Flood (Ballesteros Cánovas et al., 2017) Glacier (Singh and Yadav, 2000; Shekhar et al., 2017)
Pinus gerardiana Wallich ex D. Don	Bashar westward to Kashmir, Chitral, North Baluchistan and Afghanistan, Kinnaur; 1800–3000 m	1000	Climate (Yadav, 1991; Singh and Yadav, 2007; Singh et al., 2009)
Pinus roxburghii Sargent	Entire Himalayas from Pakistan to Arunachal Pradesh at 450–2300 m, except Kashmir Valley proper	327	Climate (Pant and Borgaonkar, 1984; Singh and Yadav, 2014; Shekhar et al., 2018) Natural hazards (Ballesteros Cánovas et al., 2017; Bhattacharyya et al., 2017) Ecology (Shah et al., 2013) Forest fire
Pinus kesiya Royle ex Gordon	Khasi and Naga Hills close to Himalaya	350	Ecology Chaudhary and Bhattacharyya, 2002; Shah and Bhattacharyya, 2012)
Tsuga dumosa (D. Don) Eichler [syn. *T. brunoniana*]	Eastern Kumaon to Arunachal Pradesh	600	Climate (Borgaonkar et al., 2018)
Picea smithiana (Wall.) Boiss.	In the western Himalayas, from Afghanistan to Kumaon at 2150–3300 m	331	Climate (Borgaonkar et al., 2011; Thapa et al., 2015) Ecology (Thapa et al., 2017)
Abies pindrow (Royle ex D. Don) Royle	Kashmir Himalaya to Central Himalaya Nepal, 2300–3350 m	360	Climate (Chaudhary et al., 2013; Ram and Borgaonkar, 2014) Glacier history (Bhattacharyya et al., 2001; Shekhar et al., 2017)

(Continued)

TABLE 19.1 (CONTINUED)

Distribution and Probable Age of Various Himalayan Conifers and Their Utility in Environment Reconstruction

Species	Distribution	Probable Age (Years)	Applicability of Tree-Ring Data to Different Science Domains
Abies spectabilis (D. Don) Mirbel	All along the Himalayas from Pakistan to Arunachal Pradesh, Tibet; 2800–4300 m; common between 3300–4100 m	370	Glacier history (Bhattacharyya et al., 2001; Shekhar et al., 2017) Ecology (Yadav et al., 2004; Shrestha et al., 2017) Natural hazards (Bhattacharyya et al., 2017)
Abies densa Griffith	Toward east of Nepal, Darjeeling to Arunachal Pradesh, at 2750–3950 m	400	Climate (Bhattacharyya and Chaudhary, 2003) Glacier history (Shekhar, 2014) Ecology (Shah and Bhattacharyya, 2012) Streamflow (Shekhar and Bhattacharyya, 2015) Paleoseismic dating (Bhattacharyya et al., 2008)
Cedrus deodara (Roxburgh ex D. Don) G. Don	Kashmir Himalaya, Garhwal, Kurnauli Valley (West Nepal); 1200–3300 m	1200	Climate Bhattacharyya and Yadav, 1989; Singh and Yadav, 2005; Yadav, 2009; Yadav et al., 2015, 2017; Shah et al., 2018) Snowfall (Yadav and Bhutiyani, 2013) Glacier history Ecology and streamflow (Shah et al., 2013; Misra et al., 2015) Natural hazards (Laxton and Smith, 2009)
Larix griffithii Hooker f.	East Nepal to Arunachal Pradesh, Chumbi Valley (Tibet), northeast Upper Myanmar, 2400–3700 m; common from 2900 to 3300 m	250	Climate (Chaudhary and Bhattacharyya, 2000; Shekhar, 2014; Yadava et al., 2015) Streamflow (Shah et al., 2014)
Juniperus polycarpos K. Koch	Afghanistan, Baluchistan, Cagan Valley, Kashmir Lahul to over Kumaon, west Tibet, 2500–4300 m.	Probably over 2000	Ecology (Yadav et al., 2006)
Podocarpus neriifolius D. Don	Sikkim, Bhutan, Kameng District	Not dated	—

in which *Juniperus* (7 spp), *Pinus* (7 spp), *Abies* (4 spp), *Picea* (4 spp), and *Cedrus* (1 sp) are common and widely distributed. In the western Himalayas, *Cedrus deodara, Juniperus polycarpas, Pinus roxburghii, P. wallichiana, P. gerardiana, Abies spectabilis*, and *A. pindrow*, along with the broad-leaved taxa, namely, *Betula utilis, B. alnoides*, and species of *Quercus, Alnus*, and *Rhododendron* are the main forest subjects. In the eastern Himalayas, the main constituents are *Abies densa, Larix* sp., and *Juniperus wallichiana*. Most of the conifers are found promising in terms of their annual growth development, but so far only some of them have been used for climate reconstruction. Two taxa, viz., *Cedrus deodara* and *Pinus gerardiana*, growing in the western Himalayas are found excellent for this purpose (Bhattacharyya et al., 1988; Yadav, 2009). Several conifers growing at the upper temperate belt extend to the subalpine zone and the prominent trees *Juniperus, Abies spectabilis, Abies densa, Picea smithiana, Pinus wallichiana*, and *Larix*, and the associated broadleaf taxa *Betula utilis* are found suitable for tree-ring analysis because of fewer problems with missing and false rings (Fritts, 1976).

19.3   APPLICATIONS OF TREE-RING DATA IN THE HIMALAYAN PERSPECTIVE

19.3.1   Past Climate Reconstructions

Climate in the Himalayan region varies over short distances with distinct changes in the precipitation regime due to strong topographic forcing. However, the temperature shows synchronization on a regional scale (Yadav et al., 1997, 1999; Shekhar, 2014; Shah et al., 2018, 2019), indicating that temperature-sensitive high-resolution proxy data could provide temperature information for the Himalayas on the regional scale. Previous climatic studies based on tree rings have been carried out in the Indian Himalayas, Nepal, Pakistan, and Bhutan (Cook et al., 2003; Bhattacharyya and Shah, 2009; Shah et al., 2014; Krusic et al., 2015; Thapa et al., 2015; Zafar et al., 2016; Asad et al., 2017; Gaire et al., 2017; Panthi et al., 2017). Many studies have been carried out concerning the climate reconstruction in the Himalayan region (Figures 19.1 and 19.2; Tables 19.2 and 19.3), especially the reconstruction of pre-monsoon temperature (Bhattacharyya and Yadav, 1992; Yadav et al., 1997, 1999). A 598-year (1390 to 1987) reconstruction of spring (March–May) temperatures from the western Himalayan region, using a ring-width chronology of Himalayan cedar (*Cedrus deodara* (D. Don) G. Don) (Yadav et al., 1999), is one of the longest reconstructions available so far for this region.

There are some analyses on the reconstruction of precipitation from several sites of arid to cold-arid regions of the western Himalayas (Yadav and Park, 2000; Yadav and Singh 2002; Singh and Yadav, 2005; Singh et al., 2006, Singh et al., 2009; Yadav et al., 2011, 2014, 2015; Yadava et al., 2015). Yadav et al. (2014) developed an annually resolved February–May precipitation record extending back to 1730 using tree-ring width data of Himalayan *Cedrus deodara* (cedar) for the first time from the Kumaon region and recorded the pre-instrumental droughts of the 1740s, 1780s, and 1840s in large part of the western Himalayas. Twentieth-century decadal-scale droughts occurred in the 1920s, and 1960s to early 1970s, and pluvial episodes in the 1910s and 1980s.

Based on tree-ring data of *Cedrus deodara* from the moisture-stressed sites of western Himalaya, Yadav et al. (2013) reconstructed a November to April snow water equivalent (SWE), which extends back to 1460. A significant relationship observed between SWE reconstruction and January–March Chenab River flow revealed its potential utility in understanding the water resource availability in the long-term perspective. The reconstructed interannual to decadal-scale variations in precipitation anomalies are of great significance as variability in the pre-monsoon precipitation has a direct impact on rabi crop productivity in the Himalayan region. Recently, Yadav et al. (2017) used annually resolved tree ring–width chronologies of Neoza pine (*Pinus gerardiana*) and Himalayan cedar from the semiarid region of Kishtwar, Jammu and Kashmir, India. The reconstructed April–May SPI extends back to 1439 (~576 years). The reconstruction revealed long-term droughts during the 15th to early 17th centuries followed by a general wetting, with 1984–2014 being the wettest interval in the past 576 years. In another study by Shah et al. (2018) from the Lidder (Liddar) Valley, Kashmir Himalaya, using the ring-width of *Cedrus deodara*, the reconstructed April–June precipitation showed a prolonged decadal dry period between 1822 and 1887, while the wettest period occurred during the early 19th century and the latter part of 20th century.

Climatic reconstructions from the eastern Himalayas are limited (Bhattacharyya and Chaudhary, 2003; Shah, 2007; Shah et al., 2014; Shekhar, 2014). Initially Chaudhary et al. (1999) evaluated the dendroclimatic potentiality of several conifer species, viz., *Abies densa, Juniperus indica, Larix griffithiana, Pinus roxburghii, P. wallichiana, P. kesiya, P. merkusii, Taxus baccata,* and *Tsuga dumosa,* growing at diverse ecological sites in the eastern Himalaya. All these taxa were found potentially datable and chronologies were developed. Ring-width data of *Larix griffthiana,* a subalpine deciduous conifer growing in the Sange area, Arunachal Pradesh, has been reported suitable for the May temperature reconstruction (Chaudhary and Bhattacharyya, 2000). In another study, *Larix griffithiana* from high-elevation North Sikkim was used for the reconstruction of mean temperature of late-summer months (July, August, and September) which extended back to

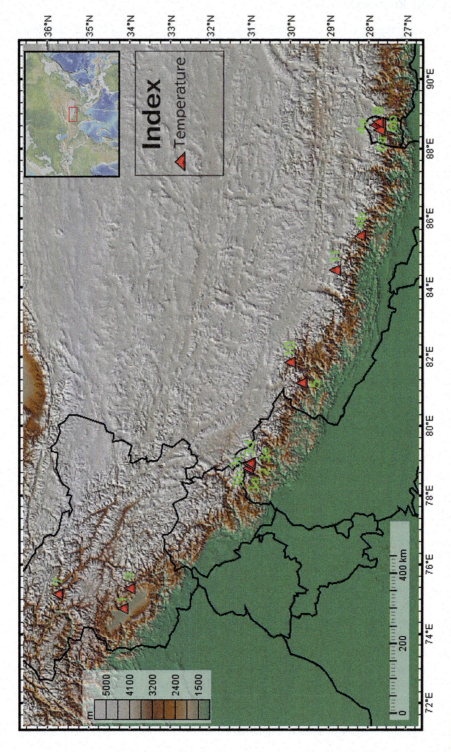

FIGURE 19.1 Locations of the temperature reconstruction sites in the Himalayan region.

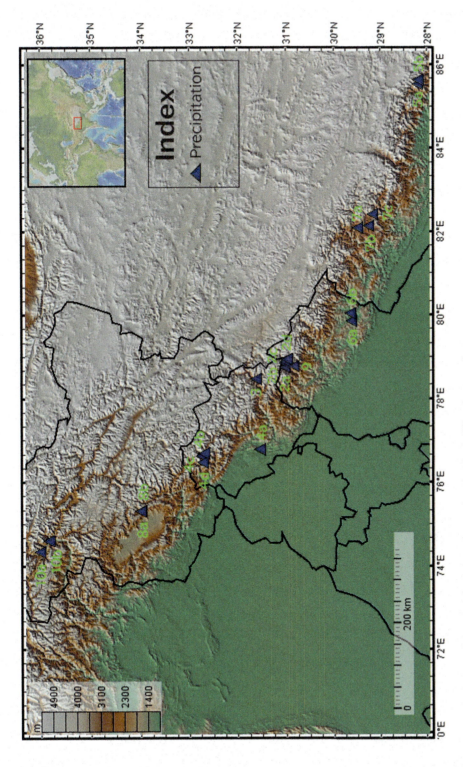

FIGURE 19.2 Locations of the precipitation reconstruction sites in the Himalayan region.

TABLE 19.2

Tree Ring–Based Temperature Reconstruction (Warm and Cool Temperature Years) in and around Adjoining Areas of the Himalayas

Site	Species	Time Span of Reconstruction	Warm Years	Cool Years	Reference
Timang Village	*Picea smithiana* (Boiss), *Pinus wallichiana* (A.B. Jacks), *Abies spectabilis* (D. Don spach), *Taxus* sp., *Tsuga Dumosa*	1399–2017	1600–1625 1633–1657 1682–1704 1740–1752 1779–1795 1936–1945 1956–1972	1658–1681 1705–1722 1753–1773 1796–1874 1900–1936 1937	
Simikot, Humla	*Abies spectabilis* (D. Don, Spach)	1718–2011	1742–1760 1780–1810 1892–1905 1985–2011	1815–1825 1833–1849 1864–1882 1907–1925 1932–1945	Gaire et al., 2020
Lidder Valley	*Pinus wallichiana*	1840–2014	1866–1892 1982–2012 1870–1875 1879–1880 2001–2012	1902–1939 1949–1967 1864, 1894, 1933, 1957, 1967 1977–1978	Shah et al.,2019
Dambung, Sikhim	*Tsuga dumosa* (D. Don)	1705–2008	1713–1735 1823–1827 Extremes: 1724, 1823, 1824, 1825	1816–1819 1831–1837 1856–1859 1884–1887 Extremes: 1816, 1817, 1832, 1833, 1857, 1877,1884, 1885	Borgaonkar et al., 2018
Giligit Valley Hunza Valley	*Picea smithiana, Pinus gerardiana*	1523–2013	1571, 1601, 1916, 1622, 1625 Decadal: 1911–1920 1599–1608 1698–1707 1860–1869 1642–1651	1574, 1688, 1660, 1588, 1579 Decadal: 1683–1692 1584–1593 1949–1950 1552–1563 1901–1910	Zafar et al., 2016
Khaptad National Park Lokhanda	*Picea smithiana* (wall) Boiss	1591–2012		1537–1622 1731–1780 1817–1846	Thapa et al., 2015
Lachen Lachung	*Larix griffithiana* (Lindl. et Gord.) Hort ex Carr.	1852–2005	Decadal: 1996–2005 1915–1919 1991–1995 1996–2000 2001–2005	Decadal: 1927–1936 1852–1856 1883–1887 1928–1932 1933–1937	Yadava et al., 2015
Gangotri	*Cedrus deodara* (Roxb.) (G. Don)	1960–2000	1998, 2002	1573–1622 1731–1780 1817–1846	Yadav et al., 2004

(Continued)

TABLE 19.2 (CONTINUED)

Tree Ring–Based Temperature Reconstruction (Warm and Cool Temperature Years) in and around Adjoining Areas of the Himalayas

Site	Species	Time Span of Reconstruction	Warm Years	Cool Years	Reference
Yumthang T. Gompa	*Abies densa*	1504–1995	1777–1779	1782–1786	Bhattacharyya and Chaudhary, 2003
			1843–1843	1830–1831	
			1904–1906	1899	
			1926–1927	1933–1975	
			1980–1982	Decadal:	
			Decadal:	1801–1810	
			1978–1987	1893–1902	
			1813–1822	1781–1790	
			1905–1914	1827–1836	
			1837–1846	1850–1859	
			1960–1969	1929–1938	
				1969–1978	
Harshil Dharali Mukhaba Kopang Bhaironghati Gangotri Juma Kos	*Cedrus deodara* (D. Don) (G. Don)	1480–1987	1662–1691	1631–1650	
			1713–1722	1964–1984	
			1797–1806		
			1945–1974		
Kashmir Valley	*Cedrus deodara* (D. Don) (G. Don)	1698–1988	1713–1722	1730–1750	Yadav et al., 1997
	Pinus wallichiana		1748–1757	1730s, 1780s,	
	Picea smithiana		1850–1870	1810s, 1830s	
			1856–1875		
Gangotri	*Cedrus deodara* (D. Don) (G. Don)	1390–1987	1474–1493	1435–1454	Yadav et al., 1999
			1617–1636	1952–1971	
			1649–1668	1515–1534	
			1784–1803	1819–1838	
			1541–1560	1916–1935	

1852 and showed warming since the 1930s, with 1996–2005 being the warmest phase in the past ~150 years (Yadav et al., 2015). *Pinus kesiya* growing in and around the Shillong plateau has also been analyzed to evaluate its suitability for tree-ring analysis and the tree-ring data reported to be linked with sea surface temperature, teleconnected among different climate phenomena. Buckley et al. (2005) showed a strong correlation of *Pinus merkusii* ring-width chronology with the tropical Indian and Pacific Ocean bands in seasons preceding the summer monsoon. Recently, Borgaonkar et al. (2018) used *Tsuga dumosa* (D. Don) to reconstruct the late-summer mean temperatures (July, August, September) of Sikkim, eastern Himalayas, for the past 302 years (1705–2008). An overall steady decreasing trend in the regional surface temperature has been observed since 1705; cold conditions during 1816–1819, 1831–1837, 1856–1859, and 1884–1887; with extreme cold years of 1816, 1817, 1832, 1833, 1857, 1877, 1884, and 1885; and warm conditions prevailed during 1713–1735 and 1823–1827. Earlier, Bhattacharyya and Chaudhary (2003) also reconstructed July–September temperatures that extended back to 1757 based on ring-width data of *Abies densa* growing at T Gompa in Arunachal Pradesh, and Yumthang in Sikkim. The reconstructions showed that the warmest and

TABLE 19.3

Tree Ring–Based Precipitation Reconstructions (High and Low Precipitation Years) in and around Adjoining Area of the Himalayas

Site	Species	Time Span of Reconstruction	High Precipitation Years	Low Precipitation Years	Reference
Jutial and Bagroat Valley	*Picea smithiana*	1540–2016	1540–1568	1569–1577	Khan et. al., 2020
			1578–1597	1598–1612	
			1613–1620	1621–1629	
			1632–1637	1638–1654	
			1655–1672	1673–1680	
			1681–1696	1697–1720	
			1721–1727	1728–1739	
			1740–1752	1753–1761	
			1762–1776	1777–1793	
			1794–1800	1801–1840	
			1841–1859	1860–1874	
			1875–1913	1914–1932	
			1933–1959	1960–1985	
			1986–1997	1998–2011	
Dingad Valley	*Abies pindrow* (Royle ex D. Don) *Picea smithiana* (Wall.) Boiss) *Aesculus indica* (Wall. Ex Camb) Hook)	1743–2015	Decadal: 1752–1762 1766–1776 1799–1809 1817–1827 1885–1895 Climate: 1746–1776 1796–1826 1870–1900	Decadal: 1777–1787 1837–1847 1898–1908 1965–1975 1983–1993 2000–2010 Climate: 1837–1867 1899–1929 1946–1976 1982–2012	Singh et.al., 2019
Sarbal Khadnaad	*Cedrus deodara*	1688–2014	1723–1725 1731–1736 1748–1773 1810–1821 1888–1905 1950–1957 1973–1999 2007–2010	1726–1730 1737–1747 1774–1788 1791–1804 1806–1809 1822–1887 1906–1924 1927–1941 1943–1949 1958–1972 2000–2006	Shah et.al., 2018
Mugu Jumla Guthichaur	*Abies spectabilis*	1763–2013	1844–1848 1850–1862 1864–1866 1878–1886 1894–1897 1902–1907 1909–1917 1937–1943 1945–1950 1971–1984 1986–1988 2000–2008	1873–1877 1921–1923 1925–1929 1951–1956 1958–1962 1994–1996	Gaire et al., 2017

(Continued)

TABLE 19.3 (CONTINUED)

Tree Ring–Based Precipitation Reconstructions (High and Low Precipitation Years) in and around Adjoining Area of the Himalayas

Site	Species	Time Span of Reconstruction	High Precipitation Years	Low Precipitation Years	Reference
			Extreme highs: 1855, 1886, 1912, 1914, 1920, 1934, 1947, 2001, 1888, 1889, 1942, 1943, 1976, 1977, 1881, 1882, 1883, 1979–1983	Extreme lows: 1892, 1921, 1935, 1954, 1967, 1985, 1999, 1994–1995, 1873–1875	
Gangolihat Jageshwar	*Cedrus deodara* [(Roxb.) G. Don]	1730–2012	1733–1737 1751–1755 1758–1762 1772–1776 1788–1792 1862–1866 1911–1915	1744–1748 1782–1786 1812–1816 1846–1850 1920–1924 1964–1968 1972–1976 1995–1999	Yadav et al., 2014
Kyanjing Gompa Langtang Village	*Betula utilis* (D. Don)	1552–2009	1620, 1634, 1735	1608, 1787, 1999	Dawadi et al., 2013
Udaipur Madgram Ratoli Tindi Kukumseri	*Cedrus deodara* [(Roxb.) G. Don]	1330–2003			Yadav et al., 2011
Kinnaur	*Cedrus deodara* [(Roxb.) G. Don]	1310–2005 AD	1359–1408 1723–1772 1949–1998	1471–1520 1584–1633 1773–1822	Singh et al., 2009
Harshil Dharali Mukhaba Jangla Bhaironghati Gangotri	*Cedrus deodara* [(Roxb.) G. Don]	1560–1999	1573–1582 1602–1611 1646–1655 1666–1675 1727–1736 1731–1740 1751–1760 1818–1827 1956–1965 1982–1991	1622–1631 1658–1667 1677–1686 1705–1714 1781–1790 1793–1802 1806–1815 1856–1865 1903–1912 1940–1949	Singh et al., 2006
Gangotri	*Cedrus deodara* (D. Don, G. Don)	1171–1988	1355–1374 1434–1453 1948–1967 1516–1535 1564–1583	1233–1252 1210–1229 1856–1875 1472–1491 1781–1800	Yadav and Park, 2000

coldest ten-year periods within the entire time span, respectively, occurred in 1978–1987 (+0.25°C) and 1801–1810 (–0.31°C).

19.3.2 DROUGHT RECONSTRUCTIONS FROM THE HIMALAYAS

Conifers, especially growing in the semiarid area of the western Himalayan region, are useful in successful drought reconstructions. The tree ring–width response for the moisture-stressed sites in different parts of the world had been developed to provide an understanding of temporal and spatial patterns of droughts (Woodhouse and Overpeck, 1998; Cook et al., 2004; Touchan et al., 2005, 2007, 2011; Esper et al., 2007; Stahle et al., 2007; Woodhouse et al., 2010; Burnette and Stahle, 2013; Griffin and Anchukaitis, 2014; Stahle et al., 2016). During the past two decades, considerable progress has also been made to develop annually resolved tree ring–based drought and hydrological records from Asia (Sheppard et al., 2004; Davi et al., 2006; Li et al., 2006, 2007; Liang et al., 2006; Treydte et al., 2006; Yin et al., 2008; Cook et al., 2010; Shao et al., 2010; Zhang et al., 2011; Ram, 2012; Sano et al., 2012; Yang et al., 2012, 2014; Yadav, 2013). The logic behind this physiology is that tree growth in the moisture-stressed regions are usually favored by the available moist regimes, such as streams, wet springs, or increased precipitation (Yadav and Park, 2000; Singh and Yadav, 2005; Singh et al., 2009; Yadav, 2011; Yadav et al., 2011).

For the Himalayan region, the application of tree-ring data in drought reconstruction has accelerated in recent years, especially from 2010 onward (Figure 19.3 and Table 19.4). The PDSI has been established for different regions of the western Himalayas (Cook et al., 2010; Ram, 2012; Sano et al., 2012; Yadav, 2013; Shekhar et al., 2018; Singh et al., 2019). The first longer-time drought index reconstruction was the June–July–August PDSI extending back to 1300 using the tree-ring chronology network from the monsoon region of Asia including the western Himalaya, available as the "Asian Drought Atlas" (MADA) by Cook et al. (2010). Ram (2012) developed summer (April–September) PDSI extending back to 1820 using the ring-width chronologies of *Abies pindrow* and *Picea smithiana* from the Jammu and Kashmir region of Himalaya. The October–May PDSI reconstruction by Yadav in 2013 revealed the 20th-century features with dominant decadal-scale pluvial phases (1981–1995, 1952–1968, and 1918–1934) as compared to the severe droughts in the early seventeenth century (1617–1640) as well as in the late 15th to early 16th (1491–1526) centuries. Long-term persistent droughts were found around 1491–1526 and 1617–1640, with the later period being the most severe in the context of the last seven centuries. A drying trend observed in the western Himalayas in recent decades begun after 1995. It is discussed that the droughts during the past two decades occurred because of hydrological demand due to intense agricultural activities on irrigated lands. Another drought reconstruction based on the SPI from Kumaon Himalaya (Yadav et al., 2015) established the SPI relationship with crop productivity (wheat and barley) for the Almora area, western Himalaya, from 1974 to 2010. The crop failure in the Kumaon and Garhwal region was found consistent with negative SPI-May values in observational as well as tree ring–based reconstruction.

Sano et al. (2012) additionally used the $\delta^{18}O$ chronology of *Abies spectabilis* from Nepal to understand the relationship with the mean PDSI series prepared using the data available on a single-grid point close to tree-ring sites as well as three other grid points from distant locations in the Indian plains. They revealed a reduction in monsoon and increasing dryness for the past two centuries in the Nepal Himalayas, also consistent with other proxy records across the Himalayas and Tibet. Subsequent reconstruction of spring self-calibrating Palmer Drought Severity Index (scPDSI) from Nepal, central Himalaya (Panthi et al., 2017), revealed a continuous shift toward drier conditions since the early 1980s. This may be the response of increasing spring temperatures since the 1980s in the central Himalaya (Thapa et al., 2015). The early 19th century (during 1801–1804 and 1807–1828) also experienced severe-to-extreme spring drought episodes in the central Himalayas followed by frequent spring drought conditions during the 20th century as well. Singh

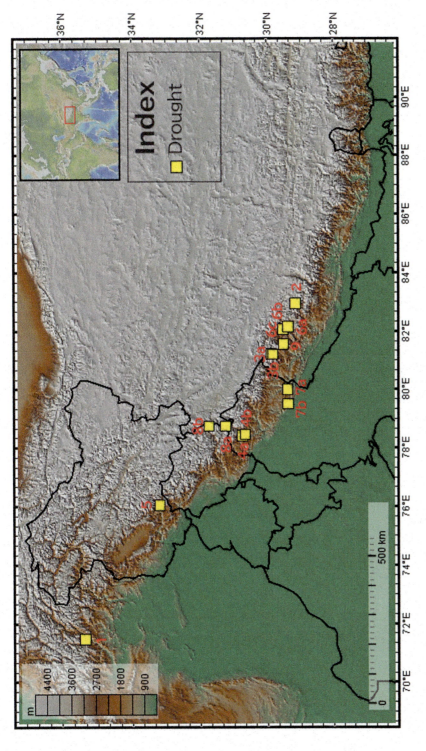

FIGURE 19.3 Locations of the drought reconstruction sites in the Himalayan region.

TABLE 19.4

Tree Ring–Based Drought Reconstructions (Drought and Pluvial Years) in and around the Adjoining Area of the Himalayas

Site	Species	Time Span of reconstruction	Drought Years	Pluvial Years	Reference and Index
Hindu Kush range of Pakistan	*Cedrus deodara* (Roxb.) G. Don	1550–2017	1593–1598, 1602–1608, 1631–1645, 1647–1660, 1756–1765, 1785–1800, 1870–1878, 1917–1923	1663–1675, 1687–1708, 1771–1773, 1806–1814, 1844–1852, 1932–1935, 1965–1969, 1990–1999	Ahmad et al., 2020 PDSI
Chhetti and Ranghadi of Api Nampa Conservation Area	*Tsuga dumosa*	1707–2013	1741–1750, 1781–1789, 1950–1959, 2009–2013 1921–1960 (long dry phase) 1975–1993 (longest dry phase) Driest phase: 1715, 1741, 1789	1733–1739, 1819–1823, 1850–1853, 1912–1914, 1737, 1738, 1903, 2001 (wettest phase)	SPEI
Uttarkashi	*Cedrus deodara Pinus geradiana*	1767–2014	1777–1786, 1787–1796, 1797–1806, 1827–1836, 1877–1886, 1897–1906, 1907–1916, 1957–1966, 1987–1996, 1997–2006, 2007–2014	1767–1776, 1807–1816, 1817–1826, 1837–1846, 1847–1856, 1857–1866, 1867–1876, 1887–1896, 1907–1916, 1928–1936, 1937–1946, 1947–1956, 1967–1976, 1977–1986	Shekhar et al., 2018SPEI
Dolpo area of Trans–Himalayan region	*Pinus wallichiana*	1611–2015	1707 (extreme) 1705 (severe), 1706, 1784, 1786, 1809, 1810, 1813, 1821, 1849, 1858, 1861, 1909, 1967, 2006, 2009	1702–1706, 1783–1789, 1769–1798, 1812–1814, 1816–1827, 1846–1848, 1856–1864, 1873–1877, 1879–1882, 1899–1901, 1908–1911, 1967–1968	Gaire et.al., 2017

(Continued)

TABLE 19.4 (CONTINUED)

Tree Ring–Based Drought Reconstructions (Drought and Pluvial Years) in and around the Adjoining Area of the Himalayas

Site	Species	Time Span of reconstruction	Drought Years	Pluvial Years	Reference and Index
Kishtwar, Jammu & Kashmir	*Cedrus deodara* (Roxb.) G. Don	1509–2014	1809–1808 1819–1828, 1837–1846, 1950–1959	1762–1771 1782–1791 1856–1865 1970–1979 2000–2009	Singh et al., 2017 SPI
Rara National Park (RNP)	*Picea smithiana*	1498–2013	1747, 1755, 1757–1758, 1763, 1811–1813, 1819–1820, 1873, 1892, 1908, 1916, 1935, 1959, 1967, 2010	1883–1891 (very wet) 1725–1727 (moderately wet) 1731, 1783–1798, 1829–1832, 1977–1981, 2002–2003	Panthi et al., 2017 scPDSI
Kumaon	*Cedrus deodara* (Roxb.) G. Don		1920–1924, 1782–1786, 1812–1816, 1744–1748, 1964–1968	1911–1915, 1723–1727, 1788–1792, 1758–1762, 1733–1737	Yadav et al., 2015 SPI
Humla district, western Nepal	*Abies spectabilis*	1778–2000	1986–1995	1794–1803	Sano et al., 2012
Pahalgam, Srinagar	*Abies pindrow*, *Picea smithiana*	1677–1980	Low soil moisture: 1830–1834, 1848–1849, 1859–1876, 1887–1901, 1933–1953, 1964–1973	High soil moisture: 1836–1839, 1857, 1927–1930, 1979–1981	Ram 2012 PDSI
Kerala Tekkedy, Narangathara, Nellikooth	*Tectona grandis* L. f.	1481–2003	1484–1496, 1508–1515, 1538–1542, 1564–1570, 1641–1703, 1758–1766, 1810–1838	1497–1507, 1516–1537, 1544–1563, 1571–1640	Borgaonkar et al., 2010

et al. (2017) reconstructed the SPI of May based on the ring-width chronology of *Cedrus deodara* from the Kishtwar region of the western Himalayas, and discussed the cause of drought as a failure of monsoon precipitation. They showed pluvial phases during the 1950s and 1990s, and dry phases during the 1970s. The wettest phase of the 1990s was followed by a distinct drying phase since 2000s. The reconstruction further revealed that the long-term droughts observed in the late 18th (1760s–1790s) and 19th (1850s–1870s) centuries were also consistent with the low precipitation in spring season (March–May) in the cold semiarid region of Kinnaur, western Himalaya (Yadava et al., 2016). In recent decades since the 2000s, the increasing drought noted in Kashmir, has also been observed in large parts of the western Himalayas largely due to weakening of the westerlies (Yadav and Bhutiyani, 2013; Yadava et al., 2016). The observational records for the Pir Panjal region also revealed a decreasing trend in snowfall since the 1990s due to weakening of westerlies.

The reconstruction of winter (November–March) drought using the SPEI from the western Himalayas (Shekhar et al., 2018) revealed the 9 extreme droughts and 11 wet events since the late 17th century (1773, 1795, 1877, 1878, 1883, 1899, 1904, 1966, and 1984). Similarly, extreme wet years were reconstructed during 1772, 1802, 1813, 1815, 1820, 1837, 1848, 1854, 1873, 1895, and 1972. Many of the reconstructed droughts show synchronization with the major known years of drought events of the Indian subcontinent, viz., East India Drought (1790–1796) during the end of 18th century; Late Victorian Great Drought during 1876–1878 and the Indian famine (1899–1900) over west and central India during the end of the 19th century; and the Bihar famine of the 20th century during 1965–1967. The time span of winter droughts in this reconstruction are found to correspond to most of the drought periods of the South Asian region, which are mostly connected to the strength of southwestern monsoon circulation. Gaire et al. (2019) based on February–August scP-DSI reconstruction from Nepal, central Himalayas, also revealed the captured signatures of similar historic megadrought events of Asia: Strange Parallels drought (1756–1768), East India Drought (1790–1796), and late Victorian Great Drought (1876–1878), which also shows that these famines were widespread. Also, reconstructed scPDSI showed many episodes with prolonged drought during the 19th and 20th century showing the regional extent.

19.3.3 FLOOD AND STREAMFLOW RECONSTRUCTIONS FROM THE HIMALAYAS

Reconstruction of flood events is a challenging aspect of tree-ring studies that has been attempted recently from the Himalayas (Figure 19.4 and Table 19.5). Preliminary analysis was carried out to determine the suitability of trees for the dating of flood events for the Kullu area of Himachal, western Himalayas (Bhattacharyya et al., 2017; Bhattacharyya and Shekhar, 2018). Further, based on tree-ring dating of flood scars in both conifers and broad-leaved taxa, at least 66 flood incidents were recorded to have taken place since 1965 (Ballesteros-Cánovas et al., 2017). The reconstructed data on flash flood occurrences in the Kullu district suggests that in the last five decades, there were five distinct phases of flood activity: (i) high flood activities between 1967–1981, 1988–1995, and 2003–2014; and (ii) low flood activities between 1981–1987 and 1996–2001 (Ballesteros-Cánovas et al., 2017).

The tree-ring data of Himalayan conifers and broad-leaved taxa have also been used to reconstruct streamflow (Figure 19.4 and Table 19.5) with the hypothesis that the streamflow discharge model is based on a strong relationship between tree-ring records and its contemporary instrumental records of river discharge, enabling the development of river discharge for a more extended period of tree-ring data, i.e., beyond the available instrumental discharge data. Using tree-ring data of *Abies densa* growing in the vicinity of the river gauge station, mean discharge of Zemu Chuu in the upper reaches of River Teesta at Lachen, North Sikkim, was reconstructed (Shekhar and Bhattacharyya, 2015). In the reconstructed discharge data the 23 high-discharge events and 21 low-discharge years were recorded for the January to April over the past 222 years (1775–1996). This January–April streamflow reconstruction data set would be of great significance in river water management when water scarcity is acute during those months in the northeast Himalaya. Earlier, Shah et al. (2014), based on the ring width of *Larix griffithina*, analyzed the river discharge of the same river for the period of the previous year March to the current year of February extending back to 1790. The streamflow is also found to be lower than the average during the years of monsoon failure, the East India Drought of 1792–1796, and past significant droughts of 1876–1878, also providing validity to the reconstruction.

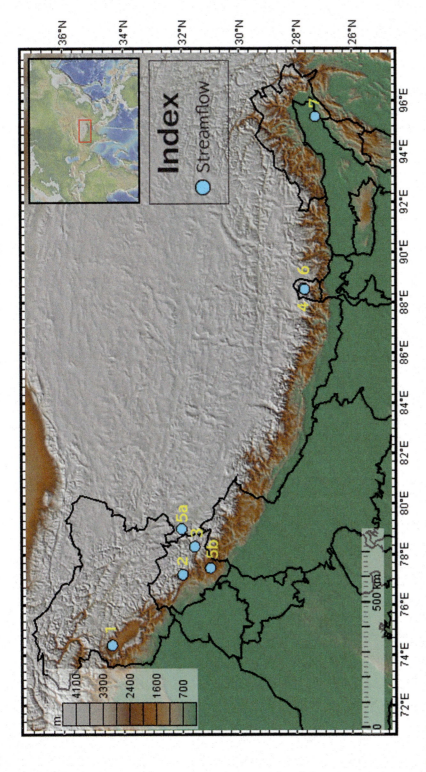

FIGURE 19.4 Locations of the streamflow reconstruction sites in the Himalayan region.

TABLE 19.5

Tree-Ring–Based Streamflow Reconstructions (High and Low Flow Years) in and around Adjoining Area of the Himalayas

Site	Species	Time Span of the Reconstruction	High Flow Years	Low Flow Years	Reference
Kinnaur, Himachal Pradesh (western Himalayas)	*Cedrus deodara* (Roxburgh ex D. Don) G. Don	1660 –2004	1751–1760, 1922–1931, 1730–1739, 1822–1831	1779–1788, 1812–1821, 1740–1749, 1907–1916	Misra et al., 2015
Sikkim (eastern Himalayas)	*Abies densa* Griffith	1775–1996	1778, 1786, 1791,1792, 1802, 1810, 1821, 1823, 1830, 1833, 1835, 1842, 1851, 1852, 1853, 1870, 1873, 1900, 1915, 1930, 1931, 1946, 1975	1782, 1783, 1799, 1814, 1815, 1825, 1837,1839, 1840, 1844, 1848, 1860, 1864, 1866, 1905, 1921, 1932, 1940, 1952, 1967, 1975	Shekhar and Bhattacharyya, 2015
Sikkim (eastern Himalayas)	*Larix griffithii* Hooker f.		1823–1835, 1879–1890, 1926–1946, 1980–1989	1791–1805, 1813–1822, 1914–1925	Shah et al., 2014
Beas River Kullu (western Himalayas)	*Cedrus deodara* (Roxburgh ex D. Don) G. Don	1834–1984	1840–1843, 1893–1897, 1922–1926, 1935–1940, 1942–1945	1847–1851, 1863–1866, 1927–1929, 1960–1962, 1972–1974	Shah et al., 2013
Kinnaur, Himachal Pradesh (western Himalayas)	*Pinus gerardiana* Wallich ex D. Don and *Cedrus deodara* (Roxburgh ex D. Don) G. Don	1295–2005	1817–1846, 1954–1983, 1953–2002	1673–1722, 1450s–1510s, 1540s–1560s, 1610s–1710s, 1770s–1820s	Singh and Yadav, 2013
Upper Indus Basin (Hunza Valley region of northern Pakistan)	*Cedrus deodara* (Roxburgh ex D. Don) G. Don *Pinus gerardiana* Wallich ex D. Don *Pinus wallichiana* A.B. Jacks *Picea smithiana* (Wall.) Boiss	1452–2008	1988–2008	1572 to 1683 1637 to 1663	Cook et al., 2013

19.4 CONCLUSIONS

The review of tree-ring study or dendrochronological analysis of the Himalayan region reveals that most of this region's conifers are suitable for the different aspects of tree-ring research. Besides a large number of studies on climate reconstruction mainly from the western Himalayan region, some exploratory tree ring analyses have been performed on its other diversified applications, e.g., flash flood, drought, and streamflow reconstruction. These successful reconstructions reveal tree-ring data and their suitability in the various diversified application in the extension of climate, drought, etc. data for the Himalayan region. Most of the tree-ring analyses here are based on ring width. Other tree-ring parameters, such as tree-ring isotope and vessel size, are less often used to understand past climatic change and other aspects of environmental, ecological, and geomorphological studies. Extensive efforts are now required to collect tree-ring samples from new geographical areas and ancient wood, leftover stumps of logging in forest sites, and old timbers. Subfossil wood may provide materials for more extended tree ring chronology and its application in various climate and related sciences.

ACKNOWLEDGMENTS

The authors would like to express their gratitude to Director, BSIP, for encouragement and permission to publish this review paper (see BSIP/RDCC/Publication No.). Authors PSR and MS also extend their sincere thanks to the in-house project funding and BSRA fellowship, respectively. IR would like to thank INSPIRE fellowship (IF150998) for providing fellowship during the tenure of this work.

REFERENCES

Ahmad, S., Zhu, L., Yasmeen, S., Zhang, Y., Li, Z., Ullah, S., Han, S. and Wang, X., 2020. A 424-year tree-ring-based Palmer Drought Severity Index reconstruction of Cedrus deodara D. Don from the Hindu Kush range of Pakistan: Linkages to ocean oscillations. *Climate of the Past*, 16(2), 783–798.

Anchukaitis, K.J., Breitenmoser, P., Briffa, K.R., Buchwal, A., Büntgen, U., Cook, E.R., D'arrigo, R.D., Esper, J., Evans, M.N., Frank, D., Grudd, H., 2012. Tree rings and volcanic cooling. *Nature Geoscience*. 5(12), 836–837.

Asad, F., Zhu, H., Liang, E., Ali, M., Hamayun, M., Sigdel, S.R., Khalid, M., Hussain, I., 2017. Climate signal in tree-ring width chronologies of *Pinus wallichiana* from the Karakoram mountains in Northern Pakistan. *Pakistan Journal of Botany*. 49(6), 2466–2473.

Babst, F., Poulter, B., Bodesheim, P., Mahecha, M.D., Frank, D.C., 2017. Improved tree-ring archives will support earth-system science. *Nature Ecology and Evolution*. 1(2), 1–2.

Ballesteros-Cánovas, J.A., Stoffel, M., St George, S., Hirschboeck, K., 2015. A review of flood records from tree rings. *Progress in Physical Geography*. 39(6), 794–816.

Ballesteros-Cánovas, J.A., Trappmann, D., Shekhar, M., Bhattacharyya, A. and Stoffel, M., 2017. Regional flood-frequency reconstruction for Kullu district, Western Indian Himalayas. *Journal of Hydrology* 546, 140–149.

Bhattacharyya, A., Chaudhary, V., 2003. Late-summer temperature reconstruction of the eastern Himalayan region based on tree-ring data of *Abies densa*. *Arctic, Antarctic, and Alpine Research*. 35(2), 196–202. https://doi.org/10.1657/1523-0430 (2003)035[0196:LTROTE]2.0.CO;2

Bhattacharyya, A., Shah, S.K., 2009. Tree-ring studies in India past appraisal, present status and future prospects. *IAWA Journal*. 30(4), 361–370.

Bhattacharyya, A., Shekhar, M., 2018. Scenario of tree ring study of the Himalayas: Present status and future outlook. In: Das, A.P., Bera, S. (Eds.), *Plant Diversity in the Himalaya Hotspot Region*. Bishen Singh Mahendra Pal Singh, Dehra Dun (India). pp. 668–676.

Bhattacharyya, A., Yadav, R.R., 1989. Growth and climate relationship in Cedrus deodara from Joshimath, Uttar Pradesh. *Palaeobotanist*. 38, 411–414.

Bhattacharyya, A., Yadav, R.R., 1992. Tree growth and recent climatic changes in the western Himalaya. *Geophytology*. 22, 255–260.

Bhattacharyya, A., Chaudhary, V., Gergan, J.T., 2001. Tree ring analysis of *Abies pindrow* around DokrianiBamak (Glacier), western Himalayas, in relation to climate and glacial behaviour: Preliminary results. *Palaeobotanist*. 50, 71–75.

Bhattacharyya, A., LaMarche, V.C., Telewski, F.W., 1988. Dendrochronological reconnaissance of the conifers of northwest India. *Tree-Ring Bulletin*. 48, 21–30.

Bhattacharyya, A., Shah, S.K., Chaudhary, V., 2008. Feasibility of tree-ring data in palaeoseismic dating in Northeast Himalaya. *Journal of Geological Society of India*. 71(3), 419–424.

Bhattacharyya, A., Stoffel, M., Shekhar, M., Cánovas, J.A.B., Trappmann, D., 2017. Dendrogeomorphic potential of the Himalaya–case studies of process dating of natural hazards in Kullu valley, Himachal Pradesh. *Current Science*. 113(12), 2317.

Borgaonkar, H.P., Gandhi, N., Ram, S., Krishnan, R., 2018. Tree-ring reconstruction of late summer temperatures in northern Sikkim (eastern Himalayas). *Palaeogeography Palaeoclimatology Palaeoecology*. 504, 125–135.

Borgaonkar, H.P., Sikder, A.B., Ram, S., 2011. High altitude forest sensitivity to the recent warming: A tree-ring analysis of conifers from western Himalaya, India. *Quaternary International*. 236, 158–166. https://doi.org/10.1016/j.quaint.2010.01.016

Borgaonkar, H.P., Sikder, A.B., Ram, S., Pant, G.B., 2010. El Niño and related monsoon drought signals in 523-year-long ring width records of teak (Tectona grandis LF) trees from south India. *Palaeogeography, Palaeoclimatology, Palaeoecology*. 285(1–2), 74–84.

Bradley, R.S., Jonest, P.D., 1993. 'Little Ice Age' summer temperature variations: Their nature and relevance to recent global warming trends. *The Holocene*. 3(4), 367–376.

Bradley, R.S., Hughes, M.K., Diaz, H.F., 2003. Climate in medieval time. *Science*. 302(5644), 404–405.

Buckley, B.M., Cook, B.I., Bhattacharyya, A., Dukpa, A.D., Chaudhary, V., 2005. Global surface temperature signals in pine ring width chronologies from southern monsoon Asia. *Geophysical Research Letters*. 32(20), L20704.

Burnette, D.J., Stahle, D.W., 2013. Historical perspective on the dust bowl drought in the central United States. *Climatic Change*. 116(3–4), 479–494.

Cánovas, J.B., Trappmann, D., Shekhar, M., Bhattacharyya, A., Stoffel, M., 2017. Regional flood-frequency reconstruction for Kullu district, Western Indian Himalayas. *Journal of Hydrology*. 546, 140–149.

Champion, H.G., Seth, S.K., 1968. *A Revised Survey of the Forest Types of India*. Government of India Press, Delhi. pp.404.

Chaudhary, V., Bhattacharyya, A., 2000. Tree ring analysis of *Larix griffthiana* from the Eastern Himalayas in the reconstruction of past temperature. *Current Science*. 79, 1712–1716.

Chaudhary, V., Bhattacharyya, A., 2002. Suitability of *Pinus kesiya* in Shillong, Meghalaya for tree-ring analyses. *Current Science*. 83(8), 1010–1015.

Chaudhary, V., Bhattacharyya, A., Yadav, R.R., 1999. Tree-ring studies in the Eastern Himalayan region: Prospects and problems. *IAWA Journal*. 20, 317–324.

Chaudhary, V., et al., 2013. Reconstruction of August–September temperature, in North-Western Himalaya since AD 1773, based on tree-ring data of *Pinus wallichiana* and *Abies pindrow*. In: Kotlia, B.S. (Ed.), *Holocene: Perspectives, Environmental Dynamics and Impact Events*, Nova Science Publishers, 145–156.

Chen, Y., Zong, Y., Li, B., Li, S., Aitchison, J.C., 2013. Shrinking lakes in Tibet linked to the weakening Asian monsoon in the past 8.2 ka. *Quaternary Research*. 80(2), 189–198.

Cook, E.R., Anchukaitis, K.J., Buckley, B.M., D'Arrigo, R.D., Jacoby, G.C., Wright, W.E., 2010. Asian monsoon failure and megadrought during the last millennium. *Science*. 328(5977), 486–489.

Cook, E.R., Briffa, K.R., Meko, D.M., Graybill, D.A., Funkhouser, G., 1995. The 'segment length curse' in long tree-ring chronology development for palaeoclimatic studies. *The Holocene*. 5(2), 229–237.

Cook, E.R., Esper, J., D'Arrigo, R., 2004. Extra-tropical Northern Hemisphere temperature variability over the past 1000 years. *Quaternary Science Reviews*. 23, 2063–2074.

Cook, E.R., et al., 2013. Five centuries of Upper Indus River flow from tree rings. *Journal of Hydrology*. 486, 365–375. https://doi.org/10.1016/j.jhydrol.2013.02.004

Cook, E.R., Krusic, P.J., Jones, P.D., 2003. Dendroclimatic signals in long tree-ring chronologies from the Himalayas of Nepal. *International Journal of Climatology: A Journal of the Royal Meteorological Society*. 23(7), 707–732.

D'Arrigo, R., Jacoby, G., Frank, D., Pederson, N., Cook, E., Buckley, B., Nachin, B., Mijiddorj, R., Dugarjav, C., 2001. 1738 years of Mongolian temperature variability inferred from a tree-ring width chronology of Siberian pine. *Geophysical Research Letters*. 28(3), 543–546.

Davi, N.K., Jacoby, G.C., Curtis, A.E., Baatarbileg, N., 2006. Extension of drought records for central Asia using tree rings: West-central Mongolia. *Journal of Climate*. 19(2), 288–299.

Dawadi, B., Liang, E., Tian, L., Devkota, L.P. and Yao, T., 2013. Pre-monsoon precipitation signal in tree rings of timberline Betula utilis in the central Himalayas. *Quaternary International*. 283, 72–77.

Esper, J., Cook, E.R., Schweingruber, F.H., 2002. Low-frequency signals in long tree-ring chronologies for reconstructing past temperature variability. *Science*. 295(5563), 2250–2253.

Esper, J., Frank, D., Büntgen, U., Verstege, A., Luterbacher, J., Xoplaki, E., 2007. Long-term drought severity variations in Morocco. *Geophysical Research Letters*. 34(17), 1–5.

Fleitmann, D., Burns, S.J., Mangini, A., Mudelsee, M., Kramers, J., Villa, I., Neff, U., Al-Subbary, A.A., Buettner, A., Hippler, D., Matter, A., 2007. Holocene ITCZ and Indian monsoon dynamics recorded in stalagmites from Oman and Yemen (Socotra). *Quaternary Science Reviews*. 26(1–2), 170–188.

Fleitmann, D., Burns, S.J., Mudelsee, M., Neff, U., Kramers, J., Mangini, A., Matter, A., 2003. Holocene forcing of the Indian monsoon recorded in a stalagmite from southern Oman. *Science*. 300(5626), 1737–1739.

Fritts, H.C., 1976. *Tree Rings and Climate*. Press, London.

Gaire, N.P., Bhuju, D.R., Koirala, M., Shah, S.K., Carrer, M., Timilsena, R., 2017. Tree-ring based spring precipitation reconstruction in western Nepal Himalaya since AD 1840. *Dendrochronologia*. 42, 21–30.

Gaire, N.P., Dhakal, Y.R., Shah, S.K., Fan, Z.X., Bräuning, A., Thapa, U.K., Bhandari, S., Aryal, S., Bhuju, D.R., 2019. Drought (scPDSI) reconstruction of trans-Himalayan region of central Himalaya using *Pinus wallichiana* tree-rings. *Palaeogeography Palaeoclimatology Palaeoecology*. 514, 251–264.

Gaire, N.P., Fan, Z.X., Shah, S.K., Thapa, U.K., and Rokaya, M.B., 2020. Tree-ring record of winter temperature from Humla, Karnali in central Himalaya: A 229 years-long perspective for recent warming trend. *Geografiska Annaler: Series A, Physical Geography*. 102(3), 297–316.

Gardner, J.S., 2002. Natural hazards risk in the Kullu district, Himachal Pradesh, India. *Geographical Review*. 92(2), 282–306.

Goswami, B.N., Venugopal, V., Sengupta, D., Madhusoodanan, M.S., Xavier, P.K., 2006. Increasing trend of extreme rain events over India in a warming environment. *Science*. 314(5804), 1442–1445.

Graham, N.E., Ammann, C.M., Fleitmann, D., Cobb, K.M., Luterbacher, J., 2011. Support for global climate reorganization during the "Medieval Climate Anomaly". *Climate Dynamics*. 37(5–6), 1217–1245.

Griffin, D., Anchukaitis, K.J., 2014. How unusual is the 2012–2014 California drought? *Geophysical Research Letters*. 41(24), 9017–9023.

Grove, J.M., 2001. The Initiation of the "Little Ice Age" in regions round the North Atlantic. *Climatic Change*. 48(1), 53–82.

Gupta, A.K., Anderson, D.M., Overpeck, J.T., 2003. Abrupt changes in the Asian southwest monsoon during the Holocene and their links to the North Atlantic Ocean. *Nature*. 421, 354–357.

Hu, C., Henderson, G.M., Huang, J., Xie, S., Sun, Y., Johnson, K.R., 2008. Quantification of Holocene Asian monsoon rainfall from spatially separated cave records. *Earth and Planetary Science Letters*. 266(3–4), 221–232.

Hughes, M.K., Diaz, H.F., 1994. Was there a 'Medieval Warm Period', and if so, where and when? *Climatic Change*. 26(2–3), 109–142.

Hughes, M.K., Vaganov, E.A., Shiyatov, S., Touchan, R., Funkhouser, G., 1999. Twentieth-century summer warmth in northern Yakutia in a 600-year context. *The Holocene*. 9(5), 629–634.

IPCC, 2013. Summary for policymakers. In: Stocker, T.F. et al. (Eds.), *Climate Change 2013. The Physical Science Basis. Contribution of Working Group III to the Fifth Assessment Report of Intergovernmental Panel on Climate Change*. Cambridge University Press, Cambridge and New York.

Jacoby, G.C., Lovelius, N.V., Shumilov, O.I., Raspopov, O.M., Karbainov, J.M., Frank, D.C., 2000. Long-term temperature trends and tree growth in the Taymir region of northern Siberia. *Quaternary Research*. 53(3), 312–318.

Khan, A., Chen, F., Ahmed, M., Zafar, M.U., 2020. Rainfall reconstruction for the Karakoram region in Pakistan since 1540 CE reveals out-of-phase relationship in rainfall between the southern and northern slopes of the Hindukush-Karakorum-Western Himalaya region. *International Journal of Climatology*. 40(1), 52–62.

Krusic, P.J., Cook, E.R., Dukpa, D., Putnam, A.E., Rupper, S., Schaefer, J., 2015. Six hundred thirty-eight years of summer temperature variability over the Bhutanese Himalaya. *Geophysical Research Letters*. 42(8), 2988–2994.

Kumar, K.R., Sahai, A.K., Kumar, K.K., Patwardhan, S.K., Mishra, P.K., Revadekar, J.V., Kamala, K., Pant, G.B., 2006. High-resolution climate change scenarios for India for the 21st century. *Current Science*. 90(3), 334–345.

Lamb, H.H., 1965. The early medieval warm epoch and its sequel. *Palaeogeography, Palaeoclimatology, Palaeoecology.* 1, 13–37.

Laskar, A.H., Yadava, M.G., Ramesh, R., Polyak, V.J., Asmerom, Y., 2013. A 4 kyr stalagmite oxygen isotopic record of the past Indian Summer Monsoon in the Andaman Islands. *Geochemistry Geophysics Geosystems.* 14(9), 3555–3566.

Laxton, S.C., Smith, D.J., 2009. Dendrochronological reconstruction of snow avalanche activity in the Lahul Himalaya, Northern India. *Natural Hazards.* 49(3), 459–467. https://doi.org/10.1007/s11069-008-9288-5

Li, J., Chen, F., Cook, E.R., Gou, X., Zhang, Y., 2007. Drought reconstruction for north central China from tree rings: The value of the Palmer drought severity index. *International Journal of Climatology: A Journal of the Royal Meteorological Society.* 27(7), 903–909.

Li, J., Gou, X., Cook, E.R., Chen, F., 2006. Tree-ring based drought reconstruction for the central Tien Shan area in northwest China. *Geophysical Research Letters.* 33(7), 1–5.

Liang, E., Liu, X., Yuan, Y., Qin, N., Fang, X., Huang, L., Zhu, H., Wang, L., Shao, X., 2006. The 1920s drought recorded by tree rings and historical documents in the semi-arid and arid areas of northern China. *Climatic Change.* 79(3–4), 403–432.

Liu, Y., Wang, L., Li, Q., Cai, Q., Song, H., Sun, C., Liu, R., Mei, R., 2019. Asian summer monsoon-related relative humidity recorded by tree ring δ ^{18}O during last 205 years. *Journal of Geophysical Research: Atmospheres.* 124(17–18), 9824–9838.

Managave, S., Shimla, P., Yadav, R.R., Ramesh, R., Balakrishnan, S., 2020. Contrasting centennial-scale climate variability in high mountain Asia revealed by a tree-ring oxygen isotope record from Lahaul-Spiti. *Geophysical Research Letters.* 47(4), e2019GL086170.

Mann, M.E., Bradley, R.S., Hughes, M.K., 1999. Northern hemisphere temperatures during the past millennium: Inferences, uncertainties, and limitations. *Geophysical Research Letters.* 26(6), 759–762.

Mann, M.E., Fuentes, J.D., Rutherford, S., 2012. Underestimation of volcanic cooling in tree-ring-based reconstructions of hemispheric temperatures. *Nature Geoscience.* 5(3), 202–205.

Mann, M.E., Zhang, Z., Rutherford, S., Bradley, R.S., Hughes, M.K., Shindell, D., Ammann, C., Faluvegi, G., Ni, F., 2009. Global signatures and dynamical origins of the Little Ice Age and Medieval Climate Anomaly. *Science.* 326(5957), 1256–1260.

Matthes, F.E., 1939. Report of committee on glaciers, April 1939. *Eos, Transactions American Geophysical Union.* 20(4), 518–523.

Misra, K.G., Yadav, R..R., Misra, S., 2015. Satluj river flow variations since AD 1660 based on tree-ring network of Himalayan cedar from western Himalaya, India. *Quaternary International.* 371, 135–143. https://doi.org/10.1016/j.quaint.2015.01.015

Pant, G.B., Borgaonkar, H.P., 1984. Growth rate of Chir pine (*Pinus roxburghii*) trees in Kumaun area in relation to regional climatology. *Himalayan Research & Development.* 3(2), 1–5.

Panthi, S., Bräuning, A., Zhou, Z.K., Fan, Z.X., 2017. Tree rings reveal recent intensified spring drought in the central Himalaya, Nepal. *Global and Planetary Change.* 157, 26–34.

Pathak, R., Sahany, S., Mishra, S.K., Dash, S.K., 2019. Precipitation Biases in CMIP5 Models over the South Asian Region. *Scientific Reports.* 9(1), 1–13.

Rajeevan, M., Bhate, J., Jaswal, A.K., 2008. Analysis of variability and trends of extreme rainfall events over India using 104 years of gridded daily rainfall data. *Geophysical Research Letters.* 35(18), 1–6.

Ram, S., 2012. Tree growth–climate relationships of conifer trees and reconstruction of summer season Palmer Drought Severity Index (PDSI) at Pahalgam in Srinagar, India. *Quaternary International.* 254, 152–158.

Ram, S., Borgaonkar, H.P., 2014. Climatic response of various tree ring parameters of fir (*Abies pindrow*) from Chandanwadi in Jammu and Kashmir, western Himalaya, India. *Current Science.* 106(11), 1568–1576.

Sahni, K.C., 1990. *Gymnosperms of India and Adjacent Countries.* Bishen Singh Mahendral Pal Singh.

Sano, M., Ramesh, R., Sheshshayee, M.S., Sukumar, R., 2012. Increasing aridity over the past 223 years in the Nepal Himalaya inferred from a tree-ring δ ^{18}O chronology. *The Holocene.* 22(7), 809–817.

Shah, S.K., 2007. *Analysis of Climatic Changes in NORTH-EAST HIMALAYA and Its Comparison with Western Himalaya During Late Quaternary.* University of Lucknow, Lucknow, India. (PhD thesis).

Shah, S.K., Bhattacharyya, A., 2009. Tree-ring analysis of sub-fossil woods of *Pinus wallichiana* from Ziro Valley, Arunachal Pradesh, Northeast Himalaya. *Journal of the Geological Society of India.* 74, 503–508.

Shah, S.K., Bhattacharyya, A., 2012. Spatio-temporal growth variability of three Pinus species of Northeast Himalaya with relation to climate. *Dendrochronologia.* 30(4), 266–278.

Shah, S.K., Bhattacharyya, A., Shekhar, M., 2013. Reconstructing discharge of Beas river basin, Kullu valley, western Himalaya, based on tree-ring data. *Quaternary International.* 286, 138–147.

Shah, S.K., Bhattacharyya, A., Chaudhary, V., 2014. Streamflow reconstruction of Eastern Himalaya River, Lachen 'Chhu', North Sikkim, based on tree-ring data of *Larix griffithiana* from Zemu Glacier basin. *Dendrochronologia.* 32(2), 97–106.

Shah, S.K., Pandey, U., Mehrotra, N., 2018. Precipitation reconstruction for the Lidder Valley, Kashmir Himalaya using tree-rings of *Cedrus deodara. International Journal of Climatology.* 38, 758–773.

Shah, S.K., Pandey, U., Mehrotra, N., Wiles, G.C., Chandra, R., 2019. A winter temperature reconstruction for the Lidder Valley, Kashmir, Northwest Himalaya based on tree-rings of *Pinus wallichiana. Climate Dynamics.* 53(7–8), 4059–4075.

Shao, X., Xu, Y., Yin, Z.Y., Liang, E., Zhu, H., Wang, S., 2010. Climatic implications of a 3585-year tree-ring width chronology from the northeastern Qinghai-Tibetan Plateau. *Quaternary Science Reviews.* 29(17–18), 2111–2122.

Shekhar, M., 2014. *Application of Multi-proxy Tree-ring Parameters in the Reconstruction of Climate vis-à-vis Glacial Fluctuation from the Eastern Himalaya.* University of Lucknow, Lucknow, India. (PhD thesis).

Shekhar, M., Bhattacharyya, A., 2015. Reconstruction of January–April discharge of Zemu Chuu–A first stage of Teesta River North Sikkim Eastern Himalaya based on tree-ring data of fir. *Journal of Hydrology: Regional Studies.* 4, 776–786.

Shekhar, M., et al., 2017. Himalayan glaciers experienced significant mass loss during later phases of little ice age. *Scientific Reports.* 7, 10305. https://doi.org/10.1038/s41598-017-09212-2

Shekhar, M., Pal, A.K., Bhattacharyya, A., Ranhotra, P.S., Roy, I., 2018. Tree-ring based reconstruction of winter drought since 1767 C.E. from Uttarkashi, Western Himalaya. *Quaternary International.* 479, 58–69.

Sheppard, P.R., Tarasov, P.E., Graumlic, L.J., Heussner, K.U., Wagner, M., Österle, H., Thompson, L.G., 2004. Annual precipitation since 515 BC reconstructed from living and fossil juniper growth of northeastern Qinghai Province China. *Climate Dynamics.* 23(7–8), 869–881.

Shi, J., Cook, E.R., Lu, H., Li, J., Wright, W.E., Li, S., 2010. Tree-ring based winter temperature reconstruction for the lower reaches of the Yangtze River in southeast China. *Climate Research.* 41(2), 169–175.

Shrestha, K.B., Chhetri, P.K., Bista, R., 2017. Growth responses of Abies spectabilis to climate variations along an elevational gradient in Langtang National Park in the central Himalaya, Nepal. *Journal of Forest Research.* 22(5), 274–281. https://doi.org/10.1080/13416979.2017.1351508

Shroder, J.F., 1978. Dendrogeomorphological analysis of mass movement on Table Cliffs Plateau, Utah. *Quaternary Research.* 9, 168–185. https://doi.org/10.1016/0033-5894 (78)90065-0

Singh, J., Yadav, R.R., 2000. Tree-ring indications of recent glacier fluctuations in Gangotri, Western Himalaya, India. *Current Science.* 79(11), 1598–1601.

Singh, J., Yadav, R.R., 2005. Spring precipitation variations over the western Himalaya, India, since AD 1731 as deduced from tree rings. *Journal of Geophysical Research: Atmospheres.* 110(D1), 1–8.

Singh, J., Yadav, R.R.., 2007. Dendroclimatic potential of millennium-long ring-width chronology of *Pinus gerardiana* from Himachal Pradesh, India. *Current Science.* 93(6), 833–836.

Singh, J., Yadav, R.R., 2013. Tree-ring-based seven century long flow records of Satluj River, western Himalaya, India. *Quaternary International.* 304, 156–162. https://doi.org/10.1016/j.quaint.2013.03.024

Singh, J., Yadav, R.R., 2014. Chir pine ring-width thermometry in western Himalaya, India. *Current Science.* 735–738. https://doi.org/10.18520/cs%2Fv106%2Fi5%2F735-738

Singh, J., Park, W.K., Yadav, R.R., 2006. Tree-ring-based hydrological records for western Himalaya, India, since AD 1560. *Climate Dynamics.* 26, 295–303.

Singh, J., Singh, N., Chauhan, P., Yadav, R.R., Bräuning, A., Mayr, C., Rastogi, T., 2019. Tree-ring δ ^{18}O records of abating June–July monsoon rainfall over the Himalayan region in the last 273 years. *Quaternary International.* 532, 48–56.

Singh, J., Yadav, R.R., Wilmking, M., 2009. A 694-year tree-ring based rainfall reconstruction from Himachal Pradesh, India. *Climate Dynamics.* 33, 1149–1158.

Singh, V., Yadav, R.R., Gupta, A.K., Kotlia, B.S., Singh, J., Yadava, A.K., Singh, A.K., Misra, K.G., 2017. Tree ring drought records from Kishtwar, Jammu and Kashmir, northwest Himalaya, India. *Quaternary International.* 444, 53–64.

Sinha, A., Berkelhammer, M., Stott, L., Mudelsee, M., Cheng, H., Biswas, J., 2011b. The leading mode of Indian Summer Monsoon precipitation variability during the last millennium. *Geophysical Research Letters.* 38(15), 1–5.

Sinha, A., Cannariato, K.G., Stott, L.D., Cheng, H., Edwards, R.L., Yadava, M.G., Ramesh, R., Singh, I.B., 2007. A 900-year (600 to 1500 AD) record of the Indian summer monsoon precipitation from the core monsoon zone of India. *Geophysical Research Letters.* 34(16).

Sinha, A., Stott, L., Berkelhammer, M., Cheng, H., Edwards, R.L., Buckley, B., Aldenderfer, M., Mudelsee, M., 2011a. A global context for megadroughts in monsoon Asia during the past millennium. *Quaternary Science Reviews*. 30(1–2), 47–62.

Stahle, D.W., Cook, E.R., Burnette, D.J., Villanueva, J., Cerano, J., Burns, J.N., Griffin, D., Cook, B.I., Acuna, R., Torbenson, M.C., Szejner, P., 2016. The Mexican Drought Atlas: Tree-ring reconstructions of the soil moisture balance during the late pre-Hispanic, colonial, and modern eras. *Quaternary Science Reviews*. 149, 34–60.

Stahle, D.W., Fye, F.K., Cook, E.R., Griffin, R.D., 2007. Tree-ring reconstructed megadroughts over North America since AD 1300. *Climatic Change*. 83(1–2), 133.

Stoffel, M., Corona, C., 2014. Dendroecological dating of geomorphic disturbance in trees. *601 Tree-Ring Research*. 70, 3–20.

Stoffel, M., Bollschweiler, M., Butler, D.R., Luckman, B.H., 2010. *Tree Rings and Natural Hazards: A State-of-art* (Vol. 41). Springer Science & Business Media.

Thapa, U.K., et al., 2017. Tree growth across the Nepal Himalaya during the last four centuries. *Progress in Physical Geography*. 41(4), 478–495. https://doi.org/10.1177/0309133317714247

Thapa, U.K., Shah, S.K., Gaire, N.P., Bhuju, D.R., 2015. Spring temperatures in the far-western Nepal Himalaya since AD 1640 reconstructed from *Picea smithiana* tree-ring widths. *Climate Dynamics*. 45(7–8), 2069–2081.

Tiwari, M., Ramesh, R., Somayajulu, B.L.K., Jull, A.J.T., Burr, G.S., 2006. Paleomonsoon precipitation deduced from a sediment core from the equatorial Indian Ocean. *Geo-Marine Letters*. 26(1), 23–30.

Touchan, R., Akkemik, Ü., Hughes, M.K., Erkan, N., 2007. May–June precipitation reconstruction of southwestern Anatolia, Turkey during the last 900 years from tree rings. *Quaternary Research*. 68(2), 196–202.

Touchan, R., Anchukaitis, K.J., Meko, D.M., Sabir, M., Attalah, S., Aloui, A., 2011. Spatiotemporal drought variability in northwestern Africa over the last nine centuries. *Climate Dynamics*. 37(1–2), 237–252.

Touchan, R., Funkhouser, G., Hughes, M.K., Erkan, N., 2005. Standardized precipitation index reconstructed from Turkish tree-ring widths. *Climatic Change*. 72(3), 339–353.

Treydte, K.S., Schleser, G.H., Helle, G., Frank, D.C., Winiger, M., Haug, G.H., Esper, J., 2006. The twentieth century was the wettest period in northern Pakistan over the past millennium. *Nature*. 440(7088), 1179–1182.

Wang, Y., Cheng, H., Edwards, R.L., He, Y., Kong, X., An, Z., Wu, J., Kelly, M.J., Dykoski, C.A., Li, X., 2005. The Holocene Asian monsoon: Links to solar changes and North Atlantic climate. *Science*. 308(5723), 854–857.

Woodhouse, C.A., Overpeck, J.T., 1998. 2000 Years of drought variability in the central United States. *Bulletin of the American Meteorological Society*. 79(12), 2693–2714.

Woodhouse, C.A., Meko, D.M., MacDonald, G.M., Stahle, D.W., Cook, E.R., 2010. A 1,200-year perspective of 21st century drought in southwestern North America. *Proceedings of the National Academy of Sciences*. 107(50), 21283–21288.

Yadav, R.R.., 1991.Tree ring research in India: An overview. *Palaeobotanist*. 40, 394–398.

Yadav, R.R., 2009. Tree ring imprints of long-term changes in climate in western Himalaya, India. *Journal of Biosciences*. 34(5), 699–707.

Yadav, R.R., 2011. Tree ring evidence of a 20th century precipitation surge in the monsoon shadow zone of the western Himalaya, India. *Journal of Geophysical Research: Atmospheres*. 116(D2), 1–10.

Yadav, R.R., 2013. Tree ring–based seven-century drought records for the Western Himalaya, India. *Journal of Geophysical Research: Atmospheres*. 118(10), 4318–4325.

Yadav, R.R., Bhutiyani, M.R., 2013. Tree-ring-based snowfall record for cold arid western Himalaya, India since AD 1460. *Journal of Geophysical Research: Atmospheres*. 118(14), 7516–7522.

Yadav, R.R., Braeuning, A., Singh, J., 2011. Tree ring inferred summer temperature variations over the last millennium in western Himalaya, India. *Climate Dynamics*. 36(7–8), 1545–1554.

Yadav, R.R., Park, W.K., 2000. Precipitation reconstruction using ring-width chronology of Himalayan cedar from western Himalaya: Preliminary results. *Journal of Earth System Science*. 109(3), 339–345.

Yadav, R.R., Singh, J., 2002. Tree-ring-based spring temperature patterns over the past four centuries in western Himalaya. *Quaternary Research*. 57(3), 299–305.

Yadav, R.R., Gupta, A.K., Kotlia, B.S., Singh, V., Misra, K.G., Yadava, A.K., Singh, A.K., 2017. Recent wetting and glacier expansion in the northwest Himalaya and Karakoram. *Scientific Reports*. 7(1), 1–8.

Yadav, R.R., Misra, K.G., Kotlia, B.S., Upreti, N., 2014. Premonsoon precipitation variability in Kumaon Himalaya, India over a perspective of ~300 years. *Quaternary International*. 325, 213–219.

Yadav, R.R., Misra, K.G., Yadava, A.K., Kotlia, B.S., Misra, S., 2015. Tree-ring footprints of drought variability in last~ 300 years over Kumaun Himalaya, India and its relationship with crop productivity. *Quaternary Science Reviews.* 117, 113–123.

Yadav, R.R., Park, W.K., Bhattacharyya, A., 1997. Dendroclimatic reconstruction of April–May temperature fluctuations in the western Himalaya of India since A.D. 1698. *Quaternary Research.* 48, 187–191.

Yadav, R.R., Park, W.K., Bhattacharyya, A., 1999. Spring-temperature variations in western Himalaya, India, as reconstructed from tree-rings: AD 1390–1987. *The Holocene.* 9, 85–90.

Yadav, R.R., Singh, J., Dubey, B., Chaturvedi, R., 2004. Varying strength of relationship between temperature and growth of high-level fir at marginal ecosystems in western Himalaya, India. *Current Science.* 86, 1152–1156.

Yadav, R.R., Singh, J., Dubey, B., Misra, K.G., 2006. A 1584-year ring width chronology of juniper from Lahul, Himachal Pradesh: Prospects of developing millennia-long climate records. *Current Science.* 90(8), 1122–1126.

Yadava, A.K., Braeuning, A., Singh, J., Yadav, R.R., 2016. Boreal spring precipitation variability in the cold arid western Himalaya during the last millennium, regional linkages, and socio-economic implications. *Quaternary Science Reviews.* 144, 28–43.

Yadava, A.K., Yadav, R.R., Misra, K.G., Singh, J., Singh, D., 2015b. Tree ring evidence of late summer warming in Sikkim, northeast India. *Quaternary International.* 371, 175–180.

Yang, B., Kang, S., Ljungqvist, F.C., He, M., Zhao, Y., Qin, C., 2014. Drought variability at the northern fringe of the Asian summer monsoon region over the past millennia. *Climate Dynamics.* 43(3–4), 845–859.

Yang, B., Qin, C., Shi, F., Sonechkin, D.M., 2012. Tree ring-based annual streamflow reconstruction for the Heihe River in arid northwestern China from AD 575 and its implications for water resource management. *The Holocene.* 22(7), 773–784.

Yin, Z.Y., Shao, X., Qin, N., Liang, E., 2008. Reconstruction of a 1436-year soil moisture and vegetation water use history based on tree-ring widths from Qilian junipers in northeastern Qaidam Basin, northwestern China. *International Journal of Climatology: A Journal of the Royal Meteorological Society.* 28(1), 37–53.

Zafar, M.U., Ahmed, M., Rao, M.P., Buckley, B.M., Khan, N., Wahab, M., Palmer, J., 2016. Karakorum temperature out of phase with hemispheric trends for the past five centuries. *Climate Dynamics.* 46(5–6), 1943–1952.

Zhang, J., Chen, F., Holmes, J.A., Li, H., Guo, X., Wang, J., Li, S., Lü, Y., Zhao, Y., Qiang, M., 2011. Holocene monsoon climate documented by oxygen and carbon isotopes from lake sediments and peat bogs in China: A review and synthesis. *Quaternary Science Reviews.* 30(15–16), 1973–1987.

Zhang, Y., Shao, X.M., Yin, Z.Y., Wang, Y., 2014. Millennial minimum temperature variations in the Qilian Mountains, China: Evidence from tree rings. *Climate of the Past.* 10(5), 1763–1778.

Zhu, H.F., Fang, X.Q., Shao, X.M., Yin, Z.Y., 2009. Tree ring-based February–April temperature reconstruction for Changbai Mountain in Northeast China and its implication for East Asian winter monsoon. *Climate of the Past.* 5(4), 661–666.

20 Remote Sensing Capabilities for Observational Drought Assessment

Khodayar Abdollahi and Zahra Eslami

CONTENTS

20.1 INTRODUCTION

Environmental and meteorological added values of remote sensing (RS) data are among the most important perspectives of the application of remote sensing in drought assessment. RS products provide mature data sources for advanced land change analysis, hence, recently they have paved a valuable way to monitor droughts, especially through identifying and detecting the spatial-temporal landscape changes remotely. The information is collected through various platforms including ground-based (such as surface radars or ground-based research instruments), aerospace (such as balloons, helicopters, and aircraft), and space (satellites and shuttles) bases that use sensors to collect electromagnetic spectrum for getting information. Thanks to recent progress, a wide range of sensors are available for better decision-making and acquiring enhanced information on a range of different scales (West et al., 2019).

Depending on the electromagnetic energy source, these systems may be active or passive remote sensing. Both remote sensing techniques are based on the fact that the ratio of the reflected electromagnetic radiation versus transmitted or absorbed energy on a surface is varying. These changes in the characteristics of surface features may be characterized by tone, size, texture, shape, or contrast in a specific wavelength. The amount of absorbed or radiated energy by different phenomena under drought conditions depends on a number of factors, including the distance between the source and surface, the wavelength of the radiation, passage environment, and surface characteristics. For example, the spectral reflections of wet and dry soils are clearly different.

DOI: 10.1201/9781003276548-20

Drought is a complex hazard in natural ecosystems that affects a variety of sectors in natural, economic, and social systems (Wilhite, 2000; Morgan; Vicente-Serrano 2016). Due to the complexity of drought characteristics, it demonstrates a spatiotemporal variability; for this reason, remote sensing can be a useful tool for drought monitoring. Over time, our capability to deal with the complexities of early warning and impressive monitoring of drought conditions has improved.

20.2 USEFULNESS OF REMOTE SENSING FOR DROUGHT MONITORING

Recent technological progress in collecting satellite and RS data explicitly has given an opportunity to use RS data for drought assessment purposes (Thenkabail and Gamage, 2004; Senay et al., 2015; West et al., 2019). A close look at recent advances shows that remotely sensed data have provided a better prospect for monitoring of drought-related spatiotemporal changes and hydrological components, such as rainfall, soil moisture, groundwater, and evapotranspiration (AghaKouchak et al., 2015). On the other hand, it provides an opportunity to quantify the effects of drought, vegetation health, and observational changes in land cover.

The use of meteorological measurements from ground stations for drought assessment may be challenging, as these types of data sets commonly are comprised of different record lengths and variable qualities (AghaKouchak and Nakhjiri, 2012; AghaKouchak et al., 2015). Compared to in-ground observations, some key advantages of RS products are the reliability of collected data, their spatial coverage ,and regular daily to weekly observations (Barrett et al., 1990; Barrett and Herschy, 1989; Morgan, 1989; Kogan, 1997; Ahmadalipour et al., 2017). Due to high spatial and temporal coverage, RS-based products can provide real-time global information. This may provide a unique opportunity to improve water resources management (Sheffield et al., 2018; AghaKouchak et al., 2015). From the perspective of the drought management cycle, RS could provide a variety of data that could be found useful for preparedness, mitigation plans, risk reduction, and crisis management (Figure 20.1). A key advantage of RS data over ground station recordings is that RS data reflect the dynamics of phenomena, such as plant phenology, biomass burning, canopy stress, and other hydrological dynamics in generally acceptable time and space scales (Asner and Alencar, 2010). However, in practice, in many instances, both meteorological and RS data assimilation systems are used to get

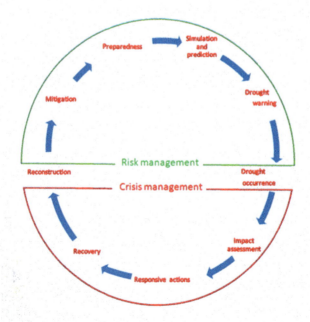

FIGURE 20.1   The series of drought-related stages in relation to risk mitigation and crisis management.

TABLE 20.1

RS Methods Compared to Traditional Methods

Remote Sensing Data	Ground-Based Data
Data collection covers a wide spatial range	Meteorological stations cannot be built anywhere
Sufficientcoverage	Insufficient station data
Low maintenance cost	High maintenance cost
Often collected automatically	Needs a large personnel network
Periodic data collection	Frequent data collection
Data is regional or surface	Data is point-based and does not show regional changes
Most data can be analyzed visually	Data is not primarily visual

the different indices. Such a uniform framework is especially useful for studying integrated drought monitoring (Su et al., 2017).

Drought indices are simplifying procedures that provide a quantitative map of the hazard by means of reducing the complexity of the drought phenomenon. Although from spatial interpolation of *in situ* data it is possible to gain a variety of valuable information relating to the severity, duration, and frequency of the droughts, in practice, many factors affect the quality of interpolated output maps (Rhee et al., 2010). This uncertainty highlights the value of RS in providing distributed hydro-meteorological characteristics. Over time, spatial resolution of RS data has been improved. Global observation, near real-time recording, fixed data recording, and improved spatial resolution are the main benefits of satellite-based sensors to traditional gauged observations (Heumann, 2011; Barrett et al., 1990; Barrett and Herschy, 1989; Morgan, 1989; AghaKouchak et al., 2015). RS data can be an appropriate source for drought monitoring, especially in areas without measurements or regions with limited *in situ* measurements. Whereas station data exists, combining RS data with station data may yield better results (Table 20.1).

20.3   DROUGHT OBJECT MODELING

Drought object modeling (DOM) is an ongoing scientific concept in drought assessment that converts RS images into geographic phenomena with unique attributes and known behaviors (Woryboys et al., 1990; Budd, 2000; Stein et al., 2009; Rulinda et al., 2010). Object identification in DOM was found to be one of the closest ideas to the "knowledge-base actual identification" concept that takes advantage of geographical information systems for automatic interpretation of RS images (Batty et al., 1998). Recorded digital numbers by the satellite sensors is an attribute for satellite images; and behavior refers to the condition that droughts takes place or the vegetation cover dies due to drought occurrence (Tucker, 1979; Budd, 2000; UNISDR, 2009). DOM utilizes image processing techniques to partition the digital values image into multiple classified segments with approximately homogeneous states (Stein et al., 2009).

Various classifier procedures, such as ridges detection, mathematical morphology, and homogeneity identification, in single band or multiple bands can be employed in DOM to identify and classify raster pixels (Stein et al., 2009; Berhan et al., 2011). Classification may be accomplished through a variety of statistical algorithms including k-nearest neighbor, naive Bayes, and fuzzy classifiers (Berhan et al., 2011). With the advances in computer hardware, RS images could be used as data. Analysis of large sets of RS images to obtain summarized understandable and practical information is called "image mining" (Stein et al., 2009). Recently intelligent systems or neural network-based image classifiers have been applied for image object identification (An et al., 2019). Novel approaches like deep learning RS image classifications are also on the horizon(Li et al., 2018).

20.4 SATELLITE DATA FOR DROUGHT ANALYSIS

Temporal and spatial variations of drought patterns play an important role in the vulnerability of drought effects. Local observations are generally based on a time-limited sampling of a few point sources that do not have a good spatial distribution. However, RS solves this problem because it is suitable for wide geographical spatial coverage. RS data, such as digital elevation models, land cover, satellite precipitation data, evapotranspiration products, and soil moisture products, could have different applications in drought assessment. This chapter briefly introduces six mentioned main RS datasets that are primarily needed for drought analysis.

20.4.1 SATELLITE-BASED PRECIPITATION DATA

Reduction in precipitation amount is a controlling factor in creating persistent drought events (Damberg and AghaKouchak, 2014; Van Loon and Laaha, 2015; Yan et al., 2018). Hence, accurate long-term precipitation datasets are essential for drought monitoring and analysis, as rainfall is characterized by major temporal and spatial variability (Zeng et al., 2012; Zhu et al., 2019). Both passive and active (radar) RS methods are used to derive satellite-based precipitation data (Kidd and Levizzani, 2011). At present, the development of RS techniques has provided data in remote regions with large-scale coverage. Some available satellite-based precipitation products for drought assessment are presented in Table 20.2.

20.4.2 REMOTELY SENSED DATA FOR SNOW COVER

Knowledge of snow cover dynamics along with its depth are two fundamental requirements for the characterization of hydrological droughts (Shaban, 2009, Kumar et al., 2014, West et al., 2019). However, from the hydrological viewpoint, we have to include the following drought-relevant snow parameters to our list: snow water equivalent, snow albedo, and snow melt (Kongoli et al., 2012;

TABLE 20.2

Common Satellite-Based Precipitation Products

Data Set	Resolution	Timescale	Coverage	Period	Reference
GPCP	2.5°	Monthly	Global	1979–present	Adler et al.(2003)
CMAP	2.5°	Monthly	Global	1979–present	Xie et al.(2003); Xie and Arkin (1997)
CMORPH	0.25°/8 km	30 min/3 h/Daily	60°S–60°N	2002–present	Joyce et al. (2004)
CPC-Global	0.5°	Daily	Global land	2006–present	Xie et al. (2010)
GPCP 1dd	1.0°	Daily	Global	1996–present	Huffman and Bolvin (2013)
GPCP_PEN_v2.2	2.5°	5daily	Global	1979–2014	Xie et al. (2003)
GPM	0.1°	30 min/3 h/daily	60°S–60°N	2015–present	(Hou et al., 2008, 2014)
GSMaP	0.1°	1 h/daily	60°S–60°N	2002–2012	Ushio et al. (2009)
MSWEP	0.1°/0.5°	3 h/daily	Global	1979–present	Beck et al. (2017)
PERSIANN-CCS	0.04°	30 min/3, 6 h	60°S–60°N	2003–present	Sorooshian et al. (2000)
PERSIANN-CDR	0.25°	3, 6 h/Daily	60°S–60°N	1983–present	Ashouri et al. (2015)
TRMM 3B42	0.25°	3 h/Daily	50°S–50°N	1998–present	Huffman et al. (2007)
TRMM 3B43	0.25°	Monthly	50°S–50°N	1998–present	Huffman et al. (2007)

Source: Adopted from Sun et al. (2018).

TABLE 20.3

Relative Responses of Sensors against Snow Properties

Property	Gamma Rays	Visible/Near Infrared	Thermal Infrared	Microwaves
Albedo	—	H	—	—
Depth	M	M(Shallow)	L	M
Liquid water content	—	L	L	H
Snow cover	L	H	M	H
Snow soil interface	L	—	N	H
Snow water content	H	(Shallow)	L	H
Snowmelt	—	L	L	M
Stratigraphy	—	—	—	H
Temperature	—	—	M	L

Source: Rango (1993).
Notes: H, high; M, medium; L, low.

Painter et al., 2013; Molotch and Margulis, 2008; Aghakouchak et al., 2015). Table 20.3 shows the suitability of snow properties versus RS band responses. Considering the fact that many snow-covered areas are located in remote regions, RS provides a unique opportunity for data acquisition in term of parameters such as snow extent, snow water equivalent, and snow volume (Koike andSuhama, 1993). Table 20.4 shows a list of common sensors used for snow detection.

TABLE 20.4

Selected RS Products in Monitoring Snow

Data Set	Resolution	Timescale	Coverage	Period	Reference
GOES	Bands 1: 1000, Bands 2–4: 4000, band 5: 8000	3 hours	Global	1975–present	Dietz et al. (2012)
Landsat 1–3/MSS	79	18 days	185 km swath width	1972–1983	Dietz et al. (2012)
Landsat 4–5/TM	30 (band 6:20)	16 days	185 km swath width	1982–present	Dietz et al. (2012)
Landsat 7/ETM+	30 (band 6: 60; band 7: 15)	16 days	185 km swath width	1999–present	Dietz et al. (2012)
Quick SCAT/Seawinds	25 km	Daily	1800 km swath width	1999–2009	Dietz et al. (2012)
NOAA/AVHRR	1090	Daily	2399 km swath width	1978–present	Dietz et al. (2012)
SPOT	1150	1–2 days	2200 km swath width	1998–present	Dietz et al. (2012)
Terra & Aqua/MODIS	Bands 1–2: 250; Bands 3–7: 500; Bands 8–36: 1000	2 per day	2330 km swath width	2000–present	Dietz et al. (2012)
VIIRS	400–750	16 days	3040 km swath width	Since 2012	Dumont and Gascoin (2016)

20.4.3 Leaf Area Index Products

Leaf area index (LAI) is a dimensionless quantity that can be defined as the area of foliage in the vegetal cover per ground area (Chen andCihlar, 1996). LAI plays as a controlling factor in the exchange of water and energy among the soil, the atmosphere, and the plant (Hueteet al., 2006; Wright et al., 1996). This ecological characteristic commonly is used for water balance estimation (Hueteet al., 2006; Wright et al., 1996), estimation of deforestation rate (Costa and Floi, 2000; Fensholtet al., 2004), and spatial and temporal changes in interception (Vargas et al., 2002). Satellites themselves are not capable to capture LAI values directly. Existing RS approaches for generating LAI maps rely on empirical models such as vegetation indices–LAI relationships, canopy reflectance modeling, or hybrid models (Barbu et al., 2014). Statistical correlation between LAI and the Standardized Precipitation Index for some regions of the world shows LAI derived from satellite (e.g., MODIS, GLASS, etc.) products are useful indicators for assessment of vegetation response to drought (Kim et al., 2017). In terms of practicality, *in situ* techniques are less favorable, as RS methods are capable of providing large-scale spatial converge without damaging the vegetation canopy (Gowdaet al., 2015). Ground truth measurements are used as validation generally (Ariza-Carricondo et al., 2019; Mourad et al., 2020). Interested readers are directed to a recent review by Fang et al.(2019) who have provided an extensive overview on LAI products, applications, and validation (see Table 20.5).

20.4.4 Land Surface Temperature

Estimation of land surface temperature (LST) is a useful information for analysis of global warming, risk of desertification, and drought severity. Only a few sensors are capable of detecting all the required parameters for LST at high accuracy (Sun and Pinker, 2003). Most RS products make use of the thermal band for derived LST. Upward land radiation is controlled by LST, and for this reason it has a great value in term of climate studies. Hu et al. (2019) suggest that a combined usage of LST and NDVI (Normalized Difference Vegetation Index) products may be useful in the assessment of agricultural drought. To assess drought severity, drought indicators are generally used to identify the degree of environmental anomaly over the longterm. Table 20.6 shows a number of commonly

TABLE 20.5
Selected Common LAI Products

Dataset	Resolution	Timescale	Period	Reference
CYCLOPES	1/112°	10 days	1997–2007	Baret et al. (2007)
EUMETSAT	1.1 km	10 days	Since 2015	García Haroet al. (2018)
GA-TIP	1 km	8 days	2002–2011	Disney et al. (2016)
GEOV2	1 km	8 days	Since 2000	Xiao et al. (2014)
GLASS	1 km	Monthly	1998–2006	Deng et al.(2006)
GLOBMAP	500 m	8 days	Since 2000	Liu, et al. (2012)
JRC-TIP	0.01°	16 days	Since 2000	Pinty et al. (2011)
MERIS	300 m	10 days	2003–2011	Tum et al. (2016)
MISR	1.1 km	daily	Since 2000	Diner et al. (2008)
MODIS	500 m	4 days	Since 2000	Huang et al. (2008)
PROBA-V	300 m	10 days	Since 2014	Baret et al. (2016)
University of Toronto (UofT)	250 m	10 days	2003	Gonsamo and Chen (2014)
VIIRS	500 m	8 days	Since 2012	K. Yan et al. (2018)

Source: Fang et al. (2019).

TABLE 20.6

Remote Sensing-Based Drought Monitoring Indices and Their Thresholds

Index	Formula	State			Reference
		Severe Drought Condition	Normal Condition	Wet Condition	
Normalized Difference Vegetation Index (NDVI)	$NDVI = \dfrac{NIR - RED}{NIR + RED}$	−1	Depends on the study area	+1	Tucker (1979)
Drought severity	$NDVI_{dev} = NDVI_i - NDVI_{mean}$	−1	0	+1	Thenkabail and Gamage (2004)
Temperature condition	$TCI = 100 \times (BT_{max} - BT_i)/ (BT_{max} - BT_{min})$	0 %	50 %	100 %	Kogan (1995, 1997)
Vegetation condition	$VCI = 100 \times (NDVI_{max} - NDVI_i)/ (NDVI_{max} - NDVI_{min})$	0 %	50 %	100 %	Kogan (1995, 1997)

Notes: RED is red band reflectance and NIR is near-infrared band reflectance, NDVI deviation. Long term mean, BT_{max}: max brightness temperature, BT_i: brightness temperature at *i*thmonth, BT_{min}: min brightness temperature.

used indices. Table 20.7is a summary of the characteristics of the selected satellite-based LST products (Sun et al., 2018).

20.4.5 SATELLITE-BASED EVAPOTRANSPIRATION PRODUCTS

Evapotranspiration is the second largest component in the global water budget (Krajewski et al., 2006). Under drought conditions, evapotranspiration turns into a more significant factor as its intensification imposes water stress to the hydrological system. Most likely the RS-based scheme is the only approach that has the potential to assess evapotranspiration for large scales. The methods for estimation of evapotranspiration with RS and the surface energy balance approaches are reviewed by Liou and Kar (2014). Different forms of energy balance are generally used for simulation of thermal-based evapotranspiration. A short list of the global RS evapotranspiration products is presented in Table 20.8.

20.4.6 SOIL MOISTURE REMOTE SENSING AND DATA ASSIMILATION

Among the various hydrological variables, soil moisture is a crucial factor in drought assessment, especially for the case of agricultural drought (Paredes-Trejo& Barbosa, 2017; Zhu et al., 2019). Soil moisture is a key indicator of agricultural drought, as it significantly influences both plant growth and productivity (Boken et al., 2005; Wilhite, 2005). In view of water balance modeling, the state of the hydrological system can be seen as the portioning of precipitation in form of variables, like snowpack storage, interception, surface runoff, soil moisture, and groundwater (McNamara et al., 2011). Soil moisture is the result of other operating hydrologic components, thus it is greatly variable in spatial and temporal scales. Spatial distribution of soil moisture can be monitored for assessing drought stress using RS (Jupp et.al., 1998). Despite RS data values for large-scale modeling, *in situ* soil moisture measurements, like the International Soil Moisture Network (ISMN) database, are

TABLE 20.7

Characteristics of Selected Satellite-Based LST Products

Dataset	Resolution	Timescale	Coverage	Period	Reference
AVHRR	1 km	Daily	Global	1978–present	Sun and Pinker(2003)
Landsat ETM+	60m*	16 days	Global	1982–present	Tomlinson et al.(2011)
MODIS	1 km	Daily	Global	2000–present	Tomlinson et al.(2011)
ASTER	90m	Daily		Since 1999	Tomlinson et al.(2011)

Source: Adopted from Sun et al. (2018).
* Resampled to 30 m.

TABLE 20.8

Major Global RS-Based Evapotranspiration

Data Set	Resolution	Timescale	Coverage	Period	Reference
AVHRR-E	8 km	Daily	global	1983–2013	Zhang et al. (2010)
GLASS-E	1 km	8 days	Global	2000–now	Yao et al. (2014)
GLEAM	0.25°	Daily	Global	1980–2016	Miralles et al. (2011)
LST- E	0.05°	Daily	Global	2000–2013	Raoufi and Beighley (2017)
MOD16	1 km	8 days	Global	2000–now	Mu et al. (2011)
MTE-E	0.5°	Monthly	Global	1982–2011	Jung et al. (2011)
PML-E	0.5°	Monthly	Global	1982–2012	Zhang et al. (2016)
PT-JPL	1°	Monthly	Global	1986–1993	Fisher et al. (2008)
WB-E	0.5°	Annual	Global	1982–2009	Zeng et al. (2012)

Source: Yang (2019).

crucial controlling points for the calibration and validation (Liang and Wang, 2020). Large-scale soil moisture monitoring commonly is considered in drought studies; for this reason, RS data takes precedence over the collected data from ground-based networks (Kerr, 2007; Njoku and Entekhabi, 1996; Peng et al., 2016; Abbaszadeh et al., 2018). The satellite revisit period is one of the major limitations for drought monitoring. The temporal problem of the daily coverage has been resolved by the cost of coarser spatial resolution (~20 to 50 km) (Brocca et al., 2017; Abbaszadeh et al., 2018). Satellite soil moisture spatial data only considers the top layer of soil (0–5 cm) (Xu et al., 2020). Surface moisture can be estimated using a variety of RS instruments, including microwave, visible spectrum, and thermal infrared sensors.

20.4.6.1 Microwave Wavelengths

Microwave wavelength ranges between 0.5 and 100 cm. Since the sensitivity of microwave wave sensors to soil moisture is high, these sensors are commonly used in both active and inactive water-related applications (Njoku andEntekhabi, 1996; Engman, 1998). Due to the high resolution of the recorded images of soil moisture, the synthetic-aperture radar (SAR) sensor is one of the most widely used active sensors (Ahmad et al., 2011) (Table 20.9). Information obtained from inactive RS sensors has also led to the presentation of global soil moisture maps (Walker and Houser, 2001; Ahmad et al., 2011). High spatial resolution is a key characteristic of the derived

TABLE 20.9
Microwave Soil Moisture Products

Data Set	Resolution	Timescale	Channel	Period
Scanning Multichannel Microwave Radiometer (SMMR)	140	Daily	Multiple from 6.6 GHz	1978–1987
Special Sensor MicrowaveImager (SSM/I)	25	Daily	Multiple from 19.4 GHz	1987–present
Microwave Imager TRMM	25	Daily	Multiple from 10.7 GHz	1997–2015
Scatterometer ERS	50	3–4 days	5.3 GHz	1992–present
Advanced Microwave Scanning Radiometer (AMSR-E)	25	Daily	Multiple from 6.9 GHz	2002–2011
Advanced Scatterometer (ASCAT)	25	1–2 days	5.3 GHz	2006–present
Soil Moisture Ocean Salinity Satellite (SMOS)	35	2–3 days	1.4 GHz	2009–present
Advanced Microwave Scanning Radiometer 2 (AMSR2)	25	Daily	Multiple from 6.9 GHz	2012–present
Soil Moisture Active Passive (SMAP)	3, 9, 36	2–3 days	1.4 GHz	2015–present

Source: Ahmadalipour et al. (2017).

soil moisture RS data from active RS, while a high temporal resolution is expected from the inactive sensors.

20.4.6.2　Visible Band Wavelengths

The visible spectrum band is wavelengths from about 380 to 740 nanometers andis less sensitive to changes in soil moisture than the short-wave infrared band (Yuan et al., 2019). Although there are many factors that affect the out-coming reflectance from the soil surface, in the visible band of the electromagnetic spectrum, the reflection is mainly affected by soil moisture and texture (Thomasson et al., 2001; Kaleita et al., 2005). Previous studies have made known there exists an inverse relationship between soil moisture and surface spectral reflectance (Figure 20.2) (Post et al., 2000; Galvao et al., 2001).

20.4.6.3　Near-Infrared Spectrum Wavelengths

The infrared band (thermal radiation) of the electromagnetic spectrum starts from the upper boundary of the red visible light with wavelengths of 780 nanometers to 1 millimeter. A combination of thermal infrared and interpretive techniques is commonly used to estimate changes in land surface temperature that is indirectly related to soil moisture. Combining the optical spectrum with thermal infrared can provide better results than soil moisture investigation (Zhang and Zhou, 2016). Meanwhile, the usefulness of the 1450 nanometer waveband for predicting reliable soil moisture content was reported by some researchers (Zhang and Zhou, 2016). For passive RS, cloudiness issues area major problem. Microwave measurements are generally used as an auxiliary or alternative data source in areas with cloud cover interception.

20.4.6.4　Soil Moisture and Data Assimilation

Combining observational data with governing rules of the hydrological system improves the drought modeling results, particularly in the cases of dealing with problems such as avoiding overparameterization, obtaining an optimal measurement network, estimating the parameters, filling the missing data, performing reanalyses, and reducing systematic errors.

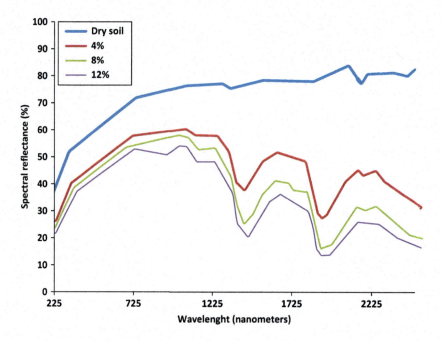

FIGURE 20.2   Relation between spectral reflectance and soil moisture (Bogrekci and Lee, 2006).

Data assimilation methods can be classified into two main categories: statistical methods and variational methods. Various data assimilation techniques, such as successive correction (Cressman, 1959; Bratseth, 1986), optimal interpolation method (Lorenzo et al., 2017), 3D/4D variational analysis, and the Kalman filter (Kalman, 1960), are available in the literature. These approaches are mostly based on a statistical procedure to minimize the sum of the residuals of estimated values (Zhangand Moore, 2014).This mathematical procedure turns out to a better estimation of the state of the hydrological system than what could be achieved by applying the model itself or using the data alone (Rodell, 2012). For instance, in order to generate a global scale time series of the groundwater storage for drought monitoring, GRACE satellite observations were combined with the Catchment Land Surface Model by Li et al. (2019) that resulted in estimating unobserved quantities.

Because of the nature of soil moisture, a combination of information with different sources is needed for its estimation (Moradkhani, 2008). The limitation of satellite observations for soil moisture estimation, inconsistency in temporal and spatial resolution, coverage gaps, and the inaccessibility to deeper soil moisture are among the major reasons that explain the appropriateness of data assimilation for soil moisture estimation. Improvements in the estimation accuracy of soil moisture data is a common report after such applications. Kumar et al. (2014) reported that finer timescale drought assessment can be improved by soil moisture data assimilation. Xu et al. (2020) reported a better simulation from combining Soil Moisture Active Passive (SMAP) satellite soil moisture observations with *in situ* measurements. Wagner et al. (1999) applied an exponential filter to describe soil moisture dynamics at the root zone using satellites. Baldwin et al. (2019) estimated root zone soil moisture across a catchment in the eastern United States using passive microwave satellite data.

Kalman filtering is a common method that is used for data assimilation, and produces soil moisture estimates of unknown values that tend to show a better accuracy than independent RS data (Li et al., 2019). A theoretical framework based on the Kalman filter method was proposed for soil moisture data assimilation by Xu et al. (2020).

20.6 CONCLUSIONS

RS is an innovative technology in a wide range of applications/monitoring such as collecting physical characteristics of land surface, vegetation health, acquiring drought information, detecting the level of drought severity, and analyzing the extent. RS information retrieved by various sensors is vital for environmental management during droughts. The present chapter attempts to review the use of innovative RS datasets, such as satellite-based land cover, soil surface moisture, precipitation, land surface temperature, snow cover, and evaporation, in drought monitoring. Meanwhile from the perspective of drought assessment, soil moisture plays a key role. RS data can be used directly, or it may also be used for simulating water and energy balance quantitatively. By taking advantage of distributed hydrological models, RS products may be used to stimulate hydrological components. Some of the most important applications of RS data are as follows:

1. Use of topographic data (such as digital elevation models).
2. Classify land surface, such as open water, leaf index data, and NDVI maps.
3. Use of satellite-derived and radar precipitation data for hydrological modeling.
4. Use of surface electromagnetic reflection data and microwave products such as the Scanning Multichannel Microwave Radiometer(SMMR) and Advanced Scatterometer (ASCAT) for evaluating current and past soil moisture.
5. Determine land surface temperature in the field of energy balance.
6. Remotelysensed data for cover snow area and snow water equivalent.
7. Use ofRS products to configure, calibrate, and validate hydrological models.

In areas that have access to long-term high-quality gauged data, interpolation/extrapolation approaches can be used. In sites with poor data availability, a combination of the RS data and hydrological models may be applied to produce acceptable results. Otherwise, for incomplete, inconsistent datasets a combination of RS along with *in situ* measurements can be used. The integration of terrestrial and satellite data has many benefits for drought monitoring, assessment, and forecasting. However, in practice, integrating remote sensing data with one another or integrating them with station data is a complex task. The process may end up with some issues, such as uncertainty in RS data, scale problems, and combining different datasets/sensor types. As errors such as shift, rescaling, skew, perspective, and rotation due to ground and satellite movements are taking place, corrections on satellite data are often needed. For similar reasons, in drought studies calibration and validation of RS-feed hydrological models are essential to help a modeler select the appropriate parameters and minimize errors.

BIBLIOGRAPHY

Adler, R. F., Huffman, G. J., Chang, A., Ferraro, R., & Ping-Ping, X. (2003). The version-2 global precipitation climatology project (GPCP) monthly precipitation analysis (1979–present). *Journal of Hydrometeorology*, 4, 1147–1167.

Abbaszadeh, P., Moradkhani, H., & Zhan, X. (2018). Downscaling SMAP radiometer soil moisture over the CONUS using an ensemble learning method. *Water Resources Research*, 55(1), 324–344.

Aghakouchak, A., & Nakhjiri, N. (2012). A near real-time satellite-based global drought climate data record. *Environmental Research Letters*, 7(4).

Aghakouchak, A., Farahmand, A., Melton, F. S., Teixeira, J., Anderson, M. C., Wardlow, B. D., and Hain, C. R. (2015). Reviews of geophysics remote sensing of drought: Progress, challenges. *Reviews of Geophysics*, 53, 1–29. https://doi.org/10.1002/2014RG000456.

Ahmadalipour, A., Moradkhani, H., Yan, H., & Zarekarizi, M. (2017). Remote sensing of drought: Vegetation, soil moisture, and data assimilation. In: *Remote sensing of hydrological extremes*. Springer, Cham, pp. 121–149.

Ahmed, A., Zhang, Y., & Nichols, S. (2011). Review and evaluation of remote sensing methods for soil-moisture estimation. *SPIE Reviews*, 2(1), 028001.

An, J., Li, W., Li, M., Cui, S., and Yue, H. (2019). Identification and classification of maize drought stress using deep convolutional neural network. *Symmetry*, 11(2), 256.

Ariza-Carricondo, C., Di Mauro, F., De Beeck, M. O., Roland, M., Gielen, B., Vitale, D., Ceulemans, R., & Papale, D. (2019). A comparison of different methods for assessing leaf area index in four canopy types. *Central European Forestry Journal*, 65(2), 67–80. https://doi.org/10.2478/forj-2019-0011.

Ashouri, H., Hsu, K.-L., Sorooshian, S., Braithwaite, D. K., Knapp, K. R., Cecil, L. D., … Prat, O. P. (2015). PERSIANN-CDR: Daily precipitation climate data record from multi-satellite observations for hydrological and climate studies. *Bulletin of the American Meteorological Society*, 96(1), 69–83.

Asner, G. P., & Alencar, A. (2010). Drought impacts on the Amazon forest: The remote sensing perspective. *New Phytologist*, 187(3), 569–578.

Baldwin, D., Manfreda, S., Lin, H., & Smithwick, E. A. (2019). Estimating root zone soil moisture across the Eastern United States with passive microwave satellite data and a simple hydrologic model. *Remote Sensing*, 11(17), 2013.

Barbu, A. L., Calvet, J. C., Mahfouf, J. F., & Lafont, S. (2014). Integrating ASCAT surface soil moisture and GEOV1 leaf area index into the SURFEX modelling platform: A land data assimilation application over France. *Hydrology and Earth System Sciences*, 18(1), 173.

Baret, F., Hagolle, O., Geiger, B., Bicheron, P., Miras, B., Huc, M., et al. (2007). CYCLOPES global products derived from VEGETATION part 1: Principles of the algorithm. *Remote Sensing of Environment*, 110(3), 275–286.

Baret, F., Weiss, M., Verger, A., & Smets, B. (2016). ATBD for LAI, FAPAR and FCover from PROBA-V products at 300m Resolution (Geov3). Imagines_rp2. 1_atbd-lai, 300.

Barrett, E. C., & Herschy, R. W. (1989). Opportunities for satellite remote sensing in hydrology and water management. *Geocarto International*, 4(2), 11–18.

Barrett, E. C., Beaumont, M. J., & Herschy, R. W. (1990). Satellite remote sensing for operational hydrology: Present needs and future opportunities. *Remote Sensing Reviews*, 4(2), 451–466.

Batty, M., Dodge, M., Doyle, S., & Smith, A. (1998). Modelling virtual environments. In *Geocomputation: A Primer*. Chichester: John Wiley, pp. 139–161.

Beck, H. E., van Dijk, A. I. J. M., Levizzani, V., Schellekens, J., Miralles, D. G., Martens, B., & de Roo, A. (2017). MSWEP: 3-hourly 0.25 degrees global gridded precipitation (1979–2015) by merging gauge, satellite, and reanalysis data. *Hydrology and Earth System Sciences*, 21(1), 589–615.

Berhan, G., Hill, S., Tadesse, T., & Atnafu, S. (2011). Using satellite images for drought monitoring: A knowledge discovery approach. *Journal of Strategic Innovation and Sustainability*, 7(1), 135.

Bessis, J. L., Béquignon, J., & Mahmood, A. (2004). The international charter "space and major disasters" initiative. *Acta Astronautica*, 54(3), 183–190. https://doi.org/10.1016/S0094-5765(02)00297-7.

Bogrekci, I., & Lee, W. S. (2006). Effects of soil moisture content on absorbance spectra of sandy soils in sensing phosphorus concentrations using UV-VIS-NIR spectroscopy. *Transactions of the ASABE*, 49(4), 1175–1180.

Boken, V. K., Cracknell, A. P., & Heathcote, R. L. (2005). *Monitoring and Predicting Agricultural Drought: A Global Study*. Oxford University Press, Oxford.

Bratseth, A. M. (1986). Statistical interpolation by means of successive corrections. *Tellus A*, 38(5), 439–447.

Brocca, L., Crow, W. T., Ciabatta, L., Massari, C., De Rosnay, P., Enenkel, M., … Wagner, W. (2017). A review of the applications of ASCAT soil moisture products. *IEEE Journal of Selected Topics in Applied Earth Observations and Remote Sensing*, 10(5), 2285–2306.

Budd, T. (2000). *Understanding Object-Oriented Programming with Java*. Addison Wesley Longman, Reading, MA.

Chen, J. M., & Cihlar, J. (1996). Retrieving leaf area index of boreal conifer forests using landsat TM images. *Remote Sensing of Environment*, 55(2), 153–162. https://doi.org/10.1016/0034-4257(95)00195-6.

Costa, M. H., & Foley, J.2000. Combined effects of deforestation and doubled atmospheric CO2 on the climate of Amazonia. *Journal of Climate*, 13, 18–34.

Cressman, G. P. (1959). An operational objective analysis system. *Monthly Weather Review*, 87(10), 367–374.

Damberg, L., & A. AghaKouchak (2014). Global trends and patterns of drought from space. *Theoretical and Applied Climatology*, 117(3), 441–448.

Deng, F., Chen, J. M., Plummer, S., Chen, M., & Pisek, J. (2006). Algorithm for global leaf area index retrieval using satellite imagery. *IEEE Transactions on Geoscience and Remote Sensing*, 44(8), 2219–2229.

Dietz, A. J., Kuenzer, C., Gessner, U., & Dech, S. (2012). Remote sensing of snow–a review of available methods. *International Journal of Remote Sensing*, 33(13), 4094–4134.

Diner, D. J., Martonchik, J. V., Borel, C., Gerstl, S. A. W., Gordon, H. R., Knyazikhin, Y., & Verstraete, M. M. (2008). Multi-angle imaging spectro radiometer (MISR) level 2 surface retrieval algorithm theoretical basis (JPL D-11401, Revision E). Jet propulsion laboratory (JPL/NASA), Pasadena, 78.

Disney, M., Muller, J.-P.,Kharbouche, S., Kaminski, T., Voßbeck, M., Lewis, P., & Pinty, B. (2016). A new global fAPAR and LAI dataset derived from optimal albedo estimates: Comparison with MODIS products. *Remote Sensing*, 8(4), 275.

Dumont, M., & Gascoin, S. (2016). Optical remote sensing of snow cover. In *Land Surface Remote Sensing in Continental Hydrology*, pp. 115–137.

Engman, E. T. (1998). *Remote sensing in hydrology*. Geophysical Monograph Series, 108, 165–177. https://doi .org/10.1029/GM108p0165.

Fang, H., Baret, F., Plummer, S., & Schaepman-Strub, G. (2019). An overview of global leaf area index (LAI): Methods, products, validation, and applications. *Reviews of Geophysics*, 57(3), 739–799.

Fensholt, R., Sandholt, I., & Rasmussen, M. S. (2004). Evaluation of MODIS LAI, fAPAR and the relation between fAPAR and NDVI in a semi-arid environment using in situ measurements. *Remote Sensing of Environment*, 91(3–4), 490–507.

Fisher, J. B., Tu, K. P., & Baldocchi, D. D. (2008). Global estimates of the land–atmosphere water flux based on monthly AVHRR and ISLSCP-II data, validated at 16 FLUXNET sites. *Remote Sensing of Environment*, 112(3), 901–919.

Galvão, L. S., Pizarro, M. A., & Epiphanio, J. C. N. (2001). Variations in reflectance of tropical soils: Spectral-chemical composition relationships from AVIRIS data. *Remote Sensing of Environment*, 75(2), 245–255.

García Haro, F. J., Campos-Taberner, M., Muñoz-Marí, J., Laparra, V., Camacho, F., Sánchez-Zapero, J., & Camps-Valls, G. (2018). Derivation of global vegetation biophysical parameters from EUMETSAT Polar System. *ISPRS Journal of Photogrammetry and Remote Sensing*, 139, 57–74. https://doi.org/10 .1016/j.isprsjprs.2018.03.005

Gavahi, K., Abbaszadeh, P., Moradkhani, H., Zhan, X., & Hain, C. (n.d.). Multivariate assimilation of remotely sensed soil moisture and evapotranspiration for drought monitoring. 1–39. https://doi.org/10.1175/JHM -D-20-0057.1.

Gonsamo, A., & Chen, J. M. (2014). Improved LAI algorithm implementation to MODIS data by incorporating background, topography, and foliage clumping information. *IEEE Transactions on Geoscience and Remote Sensing*, 52(2), 1076–1088.

Gowda, P., Oommen, T., Misra, D., Schwartz, R., Howell, T., & Wagle, P. (2015). Retrieving leaf area index from remotely sensed data using advanced statistical approaches. *Journal of Remote Sensing & GIS*, 5, 156.

Heumann, B. W. (2011). Satellite remote sensing of mangrove forests: Recent advances and future opportunities. *Progress in Physical Geography*, 35(1), 87–108.

Hou, A., Jackson, G. S., Kummerow, C., & Shepherd, J. M. (2008). Global precipitation measurement. In: S. Michaelides (Ed.), *Precipitation: Advances in Measurement, Estimation, and Prediction*. Springer, Berlin, pp. 131–170.

Hou, A. Y., Kakar, R. K., Neeck, S., Azarbarzin, A. A., Kummerow, C. D., Kojima, M., ... & Iguchi, T. (2014). The global precipitation measurement mission. *Bulletin of the American Meteorological Society*, 95(5), 701–722.

Hu, X., Ren, H., Tansey, K., Zheng, Y., Ghent, D., Liu, X., & Yan, L. (2019). Agricultural drought monitoring using European Space Agency Sentinel 3A land surface temperature and normalized difference vegetation index imageries. *Agricultural and Forest Meteorology*, 279, 107707.

Huang, D., Knyazikhin, Y., Wang, W., Deering, D. W., Stenberg, P., Shabanov, N., et al. (2008). Stochastic transport theory for investigating the three dimensional canopy structure from space measurements. *Remote Sensing of Environment*, 112(1), 35–50.

Huete, A., Didan, K., Shimabukuro, Y., Ratana, P., Saleska, S., Hutyra, L., Yang, W., Nemani, R., Myneni, R., 2006. Amazon rainforests green-up with sunlight in the dry season, *Geophysical Research Letters*, 33(6), 4.

Huffman, G. J., & Bolvin, D. T. (2013). *TRMM and Other Data Precipitation Data Set Documentation*. NASA, Greenbelt, 28(2.3), 1.

Huffman, G. J., Bolvin, D. T., Nelkin, E. J., Wolff, D. B., Adler, R. F., Gu, G.,... & Stocker, E. F. (2007). The TRMM Multisatellite Precipitation Analysis (TMPA): Quasi-global, multiyear, combined-sensor precipitation estimates at fine scales. *Journal of Hydrometeorology*, 8(1), 38–55.

Huyck, C., Verrucci, E., & Bevington, J. (2014). Remote sensing for disaster response: A rapid, image-based perspective. In *Earthquake Hazard, Risk and Disasters*. Elsevier Inc.https://doi.org/10.1016/B978-0-12 -394848-9.00001-8.

Joyce, R. J., Janowiak, J. E., Arkin, P. A., & Xie, P. (2004). CMORPH: A method that produces global precipitation estimates from passive microwave and infrared data at high spatial and temporal resolution. *Journal of Hydrometeorology*, 5, 487–503.

Jung, M., Reichstein, M., Margolis, H. A., Cescatti, A., Richardson, A. D., Arain, M. A., … Gianelle, D. (2011). Global patterns of land-atmosphere fluxes of carbon dioxide, latent heat, and sensible heat derived from eddy covariance, satellite, and meteorological observations. *Journal of Geophysical Research: Biogeosciences*, 116(G3).

Jupp, D. L., Tian, G., McVicar, T. R., Qin, Y., & Fuqin, L. (1998). Soil moisture and drought monitoring using remote sensing I: Theoretical background and methods. *EOC Report*, 1, 16–21.

Kaleita, A. L., Tian, L. F., & Hirschi, M. C. (2005). Relationship between soil moisture content and soil surface reflectance. *Transactions of the ASAE*, 48(5), 1979–1986.

Kalman, R. E. (1960). A new approach to linear filtering and prediction problems. *Journal of Basic Engineering*, 82(1), 35.

Kerr, Y. H. (2007). Soil moisture from space: Where are we? *Hydrogeology Journal*, 15(1), 117–120.

Kidd, C., & Levizzani, V. (2011). Status of satellite precipitation retrievals. *Hydrology & Earth System Sciences*, 15(4).

Kim, K., Wang, M. C., Ranjitkar, S., Liu, S. H., Xu, J. C., & Zomer, R. J. (2017). Using leaf area index (LAI) to assess vegetation response to drought in Yunnan province of China. *Journal of Mountain Science*, 14(9), 1863–1872.

Kogan, F. N.1995. Droughts of the late 1980s in the United States as derived from NOAA polar orbiting satellite data. Weather in the United States. *Bulletin of American Meteorological Society* 76: 655–668.

Kogan, F. N.1997. Global drought watch from space. *Bulletin of American Meteorological Society* 78:621–636.

Koike, T., & Suhama, T. (1993). Passive-microwave remote sensing of snow. *Annals of Glaciology*, 18, 305–308. https://doi.org/10.1017/s0260305500011691

Kongoli, C., Romanov, P., & Ferraro, R. (2012). Snow cover monitoring from remote-sensing satellites: Possibilities for drought assessment. *Remote Sensing of Drought: Innovative Monitoring Approaches*, March, 359–386. https://doi.org/10.1201/b11863.

Krajewski, W. F., Anderson, M. C., Eichinger, W. E., Entekhabi, D., Hornbuckle, B. K., Houser, P. R., … & Wood, E. F. (2006). A remote sensing observatory for hydrologic sciences: A genesis for scaling to continental hydrology. *Water Resources Research*, 42(7), 1–13.

Kumar, S. V., Peters-Lidard, C. D., Mocko, D., Reichle, R., Liu, Y., Arsenault, K. R., … Cosh, M. (2014). Assimilation of remotely sensed soil moisture and snow depth retrievals for drought estimation. *Journal of Hydrometeorology*, 15(6), 2446–2469.

Li, B., Rodell, M., Kumar, S., Beaudoing, H. K., Getirana, A., Zaitchik, B. F., … & Tian, S. (2019). Global GRACE data assimilation for groundwater and drought monitoring: Advances and challenges. *Water Resources Research*, 55(9), 7564–7586.

Li, Y., Zhang, H., Xue, X., Jiang, Y., & Shen, Q. (2018). Deep learning for remote sensing image classification: A survey. *Wiley Interdisciplinary Reviews: Data Mining and Knowledge Discovery*, 8(6), e1264.

LiangS.& WangJ.(2020).Soil moisture contents. In *Advanced Remote Sensing*. Academic Press. Elsevier, pp. 685–711. https://doi.org/10.1016/b978-0-12-815826-5.00018-0

Liou, Y. A., & Kar, S. K. (2014). Evapotranspiration estimation with remote sensing and various surface energy balance algorithms: A review. *Energies*, 7(5), 2821–2849.

Liu, J., Pattey, E., & Jégo, G. (2012). Assessment of vegetation indices for regional crop green LAI estimation from Landsat images over multiple growing seasons. *Remote Sensing of Environment*, 123, 347–358.

Lorenzo, A. T., Morzfeld, M., Holmgren, W. F., & Cronin, A. D. (2017). Optimal interpolation of satellite and ground data for irradiance nowcasting at city scales. *Solar Energy*, 144, 466–474.

McNamara, J. P., Tetzlaff, D., Bishop, K., Soulsby, C., Seyfried, M., Peters, N. E., … & Hooper, R. (2011). Storage as a metric of catchment comparison. *Hydrological Processes*, 25(21), 3364–3371.

Miralles, D. G., De Jeu, R. A. M., Gash, J. H., Holmes, T. R. H., & Dolman, A. J. (2011). Magnitude and variability of land evaporation and its components at the global scale. *Hydrology and Earth System Sciences*, 15(3), 967–981.

Molotch, N. P., & Margulis, S. A. (2008). Estimating the distribution of snow water equivalent using remotely sensed snow cover data and a spatially distributed snowmelt model: A multi-resolution, multi-sensor comparison. *Advances in Water Resources*, 31(11), 1503–1514. https://doi.org/10.1016/j.advwatres.2008.07.017.

Moradkhani, H. (2008). Hydrologic remote sensing and land surface data assimilation. *Sensors*, 8(5), 2986–3004.

Morgan, J. (1989). Satellite remote sensing in meteorology and climatology- status, perspectives and challenges, Deutsche Meteorologen-TagungueberAtmosphaere, Ozeane, Kontinente, Kiel, Federal Republic of Germany, May 16–19, 1989. *Ann. Meteorol.*, 26, 39–43.

Mourad, R., Jaafar, H., Anderson, M., & Gao, F. (2020). Assessment of leaf area index models using harmonized Landsat and Sentinel-2 surface reflectance data over a semi-arid irrigated landscape. *Remote Sensing*, 12(19). https://doi.org/10.3390/RS12193121.

Mu, Q., Zhao, M., & Running, S. W. (2011). Improvements to a MODIS global terrestrial evapotranspiration algorithm. *Remote Sensing of Environment*, 115(8), 1781–1800.

Njoku, E. G., & Entekhabi, D. (1996). Passive microwave remote sensing of soil moisture. *Journal of Hydrology*, 184(1–2), 101–129. https://doi.org/10.1016/0022-1694(95)02970-2.

Obre, C. A., Silva Dias, M. A. F., Culf, A., Polcher, J., Gash, J. H., Marengo, J., Avissar, R.2004. The Amazonian climate. In: Kabat, P., et al. (Eds.), *Vegetation, Water, Humans and the Climate*. Springer Verlag, New York, pp. 79–92.

Painter, T. H., Seidel, F. C., Bryant, A. C., McKenzie Skiles, S., & Rittger, K. (2013). Imaging spectroscopy of albedo and radiative forcing by light-absorbing impurities in mountain snow. *Journal of Geophysical Research Atmospheres*, 118(17), 9511–9523. https://doi.org/10.1002/jgrd.50520.

Paredes-Trejo, F., & Barbosa, H. (2017). Evaluation of the SMOS-derived soil water deficit index as agricultural drought index in Northeast of Brazil. *Water*, 9(6), 377.

Peng, J., Loew, A., Zhang, S., & Wang, J. (2016). Spatial downscaling of global satellite soil moisture data using temperature vegetation dryness index. *IEEE Transactions on Geoscience and Remote Sensing*, 1(54), 558–566.

Pinty, B., Andredakis, I., Clerici, M., Kaminski, T., Taberner, M., Verstraete, M. M., et al. (2011). Exploiting the MODIS albedos with the two stream inversion package (JRC-TIP): 1. Effective leaf area index, vegetation, and soil properties. *Journal of Geophysical Research*, 116, D09105.

Post, D. F., Fimbres, A., Matthias, A. D., Sano, E. E., Accioly, L., Batchily, A. K., & Ferreira, L. G. (2000). Predicting soil albedo from soil color and spectral reflectance data. *Soil Science Society of America Journal*, 64(3), 1027–1034.

Rango, A. (1993). Snow hydrology processes and remote sensing. *Hydrological Processes*, 7(2), 121–138.

Raoufi, R., & Beighley, E. (2017). Estimating daily global evapotranspiration using penman–monteith equation and remotely sensed land surface temperature. *Remote Sensing*, 9(11), 1138.

Rhee, J., Im, J., & Carbone, G. J. (2010). Monitoring agricultural drought for arid and humid regions using multi-sensor remote sensing data. *Remote Sensing of Environment*, 114(12), 2875–2887.

Rodell, M. (2012). 11 satellite gravimetry applied to drought monitoring. In *Remote Sensing of Drought: Innovative Monitoring Approaches* (Vol. 261).

Rulinda, C. M., Bijker, W., & Stein, A. (2010). Image mining for drought monitoring in eastern Africa using MeteosatSEVIRI data. *International Journal of Applied Earth Observation and Geoinformation*, 12, S63–S68.

Senay, G. B., Velpuri, N. M., Bohms, S., Budde, M., Young, C., Rowland, J., & Verdin, J. P. (2015). Drought monitoring and assessment: Remote sensing and modeling approaches for the famine early warning systems network. In *Hydro-Meteorological Hazards, Risks and Disasters*. Elsevier, pp. 233–262.

Shaban, A. (2009). Indicators and aspects of hydrological drought in Lebanon. *Water Resources Management*, 23(10), 1875–1891.

Sheffield, J., Wood, E. F., Pan, M., Beck, H., Coccia, G., Serrat-Capdevila, A., & Verbist, K. (2018). Satellite remote sensing for water resources management: Potential for supporting sustainable development in data-poor regions. *Water Resources Research*, 54(12), 9724–9758.

Sheykhmousa, M., Kerle, N., Kuffer, M., & Ghaffarian, S. (2019). Post-disaster recovery assessment with machine learning-derived land cover and land use information. *Remote Sensing*, 11(10). https://doi.org/10.3390/rs11101174.

Shunlin Liang, J. W. (2019). *Advanced Remote Sensing:Terrestrial Information Extraction and Applications*. Academic Press. libgen.lc.pdf

Sorooshian, S., Hsu, K. L., Gao, X., Gupta, H. V., Imam, B., & Braithwaite, D. (2000). Evaluation of PERSIANN system satellite-based estimates of tropical rainfall. *Bulletin of the American Meteorological Society*, 81(9), 2035–2046.

Stein, A., Hamm, N. A. S., & Ye, Q. (2009). Handling uncertainties in image mining for remote sensing studies. *International Journal of Remote Sensing*, 30(20), 5365–5382.

Su, Z., He, Y., Dong, X., & Wang, L. (2017). Drought monitoring and assessment using remote sensing. *Remote Sensing of Hydrological Extremes*, February 2019, 151–172.

Sun, D., & Pinker, R. T. (2003). Estimation of land surface temperature from a geostationary operational environmental satellite (GOES-8). *Journal of Geophysical Research: Atmospheres*, 108(D11).

Sun, Q., Miao, C., Duan, Q., Ashouri, H., Sorooshian, S., & Hsu, K. L. (2018). A review of global precipitation data sets: Data sources, estimation, and intercomparisons. *Reviews of Geophysics*, 56(1), 79–107.

Thenkabail, P. S., & Gamage, M. S. D. N.2004. The use of remote sensing data for drought assessment and monitoring in Southwest Asia (Vol. 85). *Iwmi*.

Thomasson, J. A., Sui, R., Cox, M. S., & Al–Rajehy, A. (2001). Soil reflectance sensing for determining soil properties in precision agriculture. *Transactions of the ASAE*, 44(6), 1445.

Tomlinson, C. J., Chapman, L., Thornes, J. E., & Baker, C. (2011). Remote sensing land surface temperature for meteorology and climatology: A review. *Meteorological Applications*, 18(3), 296–306.

Tucker, C. J.1979. Red and photographic infrared linear combinations for monitoring vegetation. *Remote Sensing of Environment*, 8, 127–150.

Tum, M., Günther, K., Böttcher, M., Baret, F., Bittner, M., Brockmann, C., & Weiss, M. (2016). Global gap-free MERIS LAI time series (2002–2012). *Remote Sensing*, 8(1), 69.

UNISDR United Nations Secretariat of the International Strategy for Disaster Reduction. (2009). *Drought Risk Reduction Framework and Practices: Contributing to the Implementation of the Hyogo Framework for Action*. Geneva, Switzerland.

Ushio, T., Sasashige, K., Kubota, T., Shige, S., Okamoto, K., Aonashi, K., & Kawasaki, Z. I. (2009). A Kalman filter approach to the global satellite mapping of precipitation (GSMaP) from combined passive microwave and infrared radiometric data. *Journal of the Meteorological Society of Japan*, 87A, 137–151.

Van Loon, A. F., & Laaha, G. (2015). Hydrological drought severity explained by climate and catchment characteristics. *Journal of Hydrology*, 526, 3–14.

Vargas, L. A., Andersen, M. N., Jensen, C. R., & Jørgensen, U. (2002). Estimation of leaf area index, light interception and biomass accumulation of *Miscanthus sinensis* 'Goliath'from radiation measurements. *Biomass and Bioenergy*, 22(1), 1–14.

Vicente-Serrano, S. M. (2016). Foreword: Drought complexity and assessment under climate change conditions. *Cuadernos de InvestigacionGeografica*, 42(1), 7–11.

Wagner, W., Lemoine, G., & Rott, H. (1999). A method for estimating soil moisture from ERS scatterometer and soil data. *Remote Sensing of Environment*, 70(2), 191.

Walker, J. P., & Houser, P. R. (2001). A methodology for initializing soil moisture in a global climate model: Assimilation of near-surface soil moisture observations. *Journal of Geophysical Research Atmospheres*, 106(D11), 11761–11774. https://doi.org/10.1029/2001JD900149.

Wang, L., & Qu, J. J. (2009). Satellite remote sensing applications for surface soil moisture monitoring: A review. *Frontiers of Earth Science in China*, 3(2), 237–247. https://doi.org/10.1007/s11707-009-0023-7.

West, H., Quinn, N., & Horswell, M. (2019). Remote sensing for drought monitoring & impact assessment: Progress, past challenges and future opportunities. *Remote Sensing of Environment*, 232(June), 111291. https://doi.org/10.1016/j.rse.2019.111291.

Wilhite, D. A. (2000). Chapter 1 Drought as a natural hazard. *Drought: A Global Assessment*, 1, 3–18.

Wilhite, D. A. (Ed.). (2005). *Drought and Water Crises: Science, Technology, and Management Issues*. CRC Press.

Worboys, M. F., Hearnshaw, H. M., & Maguire, D. J. (1990). Object-oriented data modelling for spatial databases. *International Journal of Geographical Information Systems*, 4(4), 369–383.

Wright, I., Nobre, C. A., Tomasella, J., Da Rocha, H. R., Roberts, J., Vertamatti, E., Culf, A., Alvala , R., Hodnett, M., Ubarana, V., 1996. Towards a GCM surface parameterisation for Amazonia. In: Gash, J., Nobre, C., Roberts, J., Victoria, R. (Eds.), *Amazon Deforestation and Climate*. John Wiley & Sons, Chichester, UK, pp. 473–504.

Xiao, Z., Liang, S., Wang, J., Chen, P., Yin, X., Zhang, L., & Song, J. (2014). Use of general regression neural networks for generating the GLASS leaf area index product from time series MODIS surface reflectance. *IEEE Transactions on Geoscience and Remote Sensing*, 52(1), 209–223.

Xie, P., & Arkin, P. A. (1997). Global precipitation: A 17-year monthly analysis based on gauge observations, satellite estimates, and numerical model outputs. *Bulletin of the American Meteorological Society*, 78, 2539–2558.

Xie, P., Chen, M., & Shi, W. (2010). CPC global unified gauge-based analysis of daily precipitation, preprints, 24th conference on hydrology, Atlanta, GA, American Meteorological Society (Vol. 2).

Xie, P. P., Janowiak, J. E., Arkin, P. A., Adler, R., Gruber, A., Ferraro, R., … Curtis, S. (2003). GPCP pentad precipitation analyses: An experimental dataset based on gauge observations and satellite estimates. *Journal of Climate*, 16(13), 2197–2214.

Xu, L., Abbaszadeh, P., Moradkhani, H., Chen, N., & Zhang, X. (2020). Continental drought monitoring using satellite soil moisture, data assimilation and an integrated drought index. *Remote Sensing of Environment*, 250, 112028.

Yan, G., Liu, Y., & Chen, X. (2018). Evaluating satellite-based precipitation products in monitoring drought events in southwest China. *International Journal of Remote Sensing*, 39(10), 3186–3214. https://doi.org/10.1080/01431161.2018.1433892.

Yang, Y. (2019). Remotely sensed evapotranspiration. *Observation and Measurement of Ecohydrological Processes*, 155.

Yao, Y., Liang, S., Li, X., Hong, Y., Fisher, J. B., Zhang, N., … Jiang, B. (2014). Bayesian multimodel estimation of global terrestrial latent heat flux from eddy covariance, meteorological, and satellite observations. *Journal of Geophysical Research: Atmospheres*, 119(8), 4521–4545.

Yuan, J., Wang, X., Yan, C. X., Wang, S. R., Ju, X. P., & Li, Y. (2019). Soil moisture retrieval model for remote sensing using reflected hyperspectral information. *Remote Sensing*, 11(3), 366.

Zeng, H., Li, L., & Li, J. (2012). The evaluation of TRMM multisatellite precipitation analysis (TMPA) in drought monitoring in the Lancang River Basin. *Journal of Geographical Sciences*, 22(2), 273–282. https://doi.org/10.1007/s11442-012-0926-1.

Zhang, K., Kimball, J. S., Nemani, R. R., & Running, S. W. (2010). A continuous satellite derived global record of land surface evapotranspiration from 1983 to 2006. *Water Resources Research*, 46(9).

Zhang, D., & Zhou, G. (2016). Estimation of soil moisture from optical and thermal remote sensing: A review. *Sensors*, 16(8). https://doi.org/10.3390/s16081308.

Zhang, Z., & Moore, J. C. (2014). *Mathematical and Physical Fundamentals of Climate Change*, Chapter 9. Elsevier.

Zhang, Y., Peña-Arancibia, J. L., McVicar, T. R., Chiew, F. H., Vaze, J., Liu, C., … Miralles, D. G. (2016). Multi-decadal trends in global terrestrial evapotranspiration and its components. *Scientific Reports*, 6, 19124.

Zheng, G., & Moskal, L. M. (2009). Retrieving leaf area index (LAI) using remote sensing: Theories, methods and sensors. *Sensors*, 9(4), 2719–2745. https://doi.org/10.3390/s90402719.

Zhu, Q., Luo, Y., Zhou, D., Xu, Y. P., Wang, G., & Gao, H. (2019). Drought monitoring utility using satellite-based precipitation products over the Xiang River Basin in China. *Remote Sensing*, 11(12), 1–17. https://doi.org/10.3390/rs11121483.

21 Four Decades of Satellite Data for Agricultural Drought Monitoring throughout the Growing Season in Central Chile

Francisco Zambrano Bigiarini

CONTENTS

DOI: 10.1201/9781003276548-21

21.1 CONTEXT

21.1.1 CLIMATE-RELATED DISASTERS AND IMPACTS ON AGRICULTURE

In the current climate change scenario, food security is threatened by diminished crop yields and livestock and fisheries (Wollenberg et al. 2016). For 2004–2014, the estimation of the global economic losses due to climate-related disasters is $100 billion, and the agriculture sector absorbed 31% (FAO 2016). Within almost the same period, global warming impacted China's economy by about $820 million in the corn and soybean sectors (Chen, Chen, and Xu 2016). Moreover, Stevanović et al. (2016) indicate that climate change may impact agricultural welfare globally, and the annual loss could reach 0.3% of the future total gross domestic product (GDP) at the end of the century. Moore et al. (2017) reveal that the total damage net cost of agriculture is about $8.5 per ton of CO_2. The three most critical climate-related disasters affecting agriculture are flood, drought, and storms.

Damages and losses by drought in crops are 14.9%, livestock is 87.6%, and fisheries and aquaculture represent 9.8% (FAO 2016). The economic impact of drought is difficult to measure due to the complex nature of the disaster. However, several estimations have been made. For example, the 2001–2002 drought in Canada would have impacted $3.5 billion; meanwhile, in Australia, the 2002–2003 drought economic losses would have reached 1% of the GDP (Ding et al. 2011). The estimated cost of agriculture during the 2012 drought in the United States was $30 billion (Smith and Matthews 2015), and the impact of the 2014 drought on California's agriculture was about $2.2 billion (Howitt et al. 2014). This makes drought one of the most dangerous disasters impacting agriculture. It is expected that future droughts will be more frequent, longer lasting, and more intensive (Cook et al. 2015), but the risk could be reduced by cutting CO_2 emissions (Ault 2020).

Therefore, countries should advance quickly toward the implementation of adaptation (Rickards and Howden 2012) and mitigation practices. The Food and Agriculture Organization (FAO) promotes climate-smart agriculture (Lipper et al. 2014) which aims "to enhance the capacity of the agricultural systems to support food security, incorporating the need for adaptation and the potential for mitigation into sustainable agriculture development strategies". A transformational adaptation in some cases might be needed (Rickards and Howden 2012). However, to succeed in implementing adaptation and mitigation practices, we need a comprehensive understanding of drought. Earth observation satellite-based technologies have become of transcendental help in the last decade (West et al. 2019) to achieve this goal.

21.1.2 DROUGHT IN CHILE

Drought is classified into four types: meteorological, hydrological, agricultural, and socioeconomic (Wilhite and Glantz 1985). A prolonged deficit of precipitation corresponds to meteorological drought and is the first to appear. When the deficit is persistent over time and impacts the hydrological system it is a called hydrological drought. When the lack of water impacts vegetation development and diminishes its productivity, this is known as agricultural drought, and the impact triggered on society and economy is called socioeconomic drought. The first definition of drought types (Wilhite and Glantz 1985) attributed the occurrence to climate variability being solely affected by precipitation and temperature, but Van Loon et al. (2016) redefines it by considering the anthropogenic influence of drought, reinforcing the need to account for the understanding of how human beings impact it. Aridity and the drought area has increased since 1970 over the Amazon, northeastern Brazil, southern Africa, and Central Europe (Dai 2011), but Chile's case has been less studied (Garreaud et al. 2017, 2020).

The change in rainfall patterns due to climate change has impacted Chile with a prolonged annual rainfall deficit that has lasted for more than ten years. Because of its unprecedented duration, it has been named a megadrought (Garreaud et al. 2017). The start of this megadrought can be dated to 2007–2008, when much of the Chilean territory was declared under agricultural emergency by drought by the Ministry of Agriculture (Zambrano et al. 2016, 2017, 2018). The teleconnection of the El Niño–Southern Oscillation (ENSO) and the rainfall patterns over Central Chile used to be very strong for decades (1960–2000) (Montecinos and Aceituno 2003) but has been weak during the megadrought, rather it has demonstrated to be connected to positive pressure anomalies over the central-eastern subtropical Pacific and negative pressure anomalies over the Amundsen–Bellingshausen Sea (Garreaud et al. 2020). This prolonged decrease in precipitation has affected the entire hydrological system within Central Chile, depleting the water reserves in the Andes Mountains, aquifers, reservoirs, lakes, and dams (Boisier et al. 2018). The recent analysis of the vegetation condition for season 2019–2020 (until December 2019) showed that Central Chile is facing an extreme anomaly regarding its normal state (2000–2019); this is particularly critical over crops, grassland, and forest (Zambrano et al. 2020). As a result, crops will be diminished in their production and quality at harvest. Some farmers have decided to leave cultivated surfaces abandoned because they do not have sufficient water available for irrigation. Earth observation data has become crucial to provide information for the understanding of drought around the globe (AghaKouchak et al. 2015; West, Quinn, and Horswell 2019), and for Chile, I have been working with the use of remote sensing data for agricultural drought monitoring, analysis, and prediction (Zambrano et al. 2016, 2017, 2018, 2020).

This chapter will be focused on agricultural drought in Chile and how satellite-based Earth observation has been helping with its monitoring and understanding. The following sections address satellite data and will describe commonly used remote sensing data sets; and agricultural drought monitoring, which will describe some of the indices that have been used. The chapter will also discuss the main findings of the last 40 years, and Final remarks address the potentially unexplored issues that could help advance the study of agricultural drought in Chile.

21.1.3 Agricultural Drought Definition

Mishra and Singh (2010) define agricultural drought as *a period with declining soil moisture and consequent crop failure without any reference to surface water resources*. Information regarding crops is needed to monitor agricultural drought. The availability of data, such as, type, surface, yield, and phenology are not easily accessible, especially in developing countries like Chile. The principal source of information about agriculture in the country is the agricultural census made for the Ministry of Agriculture, but the last one was completed in 2007 and it has not been updated since. To overcome the lack of information, satellite data have been used as proxies for vegetation conditions (Rouse et al. 1974), development (Vrieling et al. 2018), and biomass production (Rigge et al. 2013; Meroni et al. 2014). Thus, these proxies are used to study agricultural drought as indicators of vegetation status (Rhee et al. 2010). Indirectly, a common practice is the analysis of vegetative response to environmental variables such as precipitation, evapotranspiration, and soil moisture (Quiring and Ganesh 2010; Zambrano et al. 2016, 2017). This section describes the used satellite data sets for agricultural drought in Chile and the derived indices.

21.2 CHARACTERIZATION OF CHILE

21.2.1 Climate, Topography, and Land Cover

The Andes Mountains are one of the primary geological forms that determine the landscape, climate, and agriculture conditions in Chile. The prevailing climate is temperate Mediterranean (Csb), but north 31° S is cold desert (Bwk) according to the Köppen climate classification system (Kottek et al. 2006). There is an east–west and north–south pattern for elevations, precipitation, temperature, and land cover. North 34° S dominates an elevation above 1400 m asl, south of this latitude the mountain holds to the east (Figure 21.1a). Mean annual precipitation varies from less than 300 mm in the north to 1800 mm in the south (Figure 21.1b). The temperature pattern is not as strong

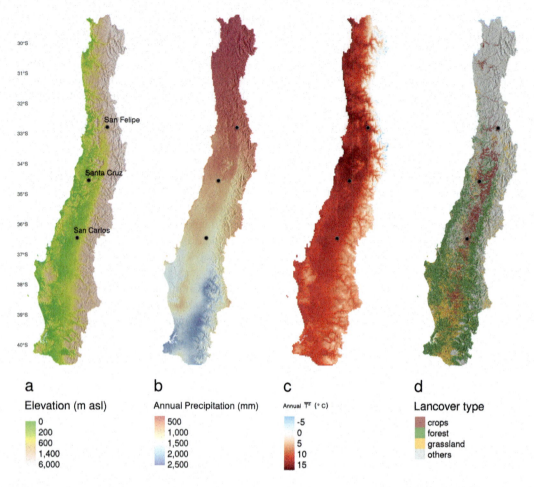

FIGURE 21.1   Central Chile (a) elevation, (b) landcover, (c) annual precipitation (mm), and (d) mean annual temperature (°C).

as precipitation, but north 34°, there is a slight raise of mean annual temperature (Figure 21.1c). Regarding the land use, the total surface of vegetation of the area is primarily 59,665 km² of forest, 28,109 km² of grassland, crops correspond to 15,848 km², and the rest are other types (e.g., barren, shrub, and urban) (Figure 21.1d) (Zhao et al. 2016).

21.2.2   AGRICULTURE

Chile's central part has 90% of the land used for agriculture (29°–41° S), which is severely affected by the megadrought (INE 2007). The main proportion of cultivated land in Chile is in the Central Valley, which corresponds to the depression between the Chilean Coastal Range and the Andes Mountains, with elevations ranging from 200 m to 400 m (Figure 21.1a). The small availability of water north of 32°40′ S makes it difficult for the development of agriculture, and only irrigated crops are cultivated, including grapes, avocado, and clementine. Also, vineyards, mainly for pisco production and a little for wine, and horticulture such as lettuce, green beans, artichoke, corn, and carrots in the transversal valleys that run east–west from the Andes to the Pacific. Fruits like walnut, grapes, avocado, plum, cherry, apple, hazelnut, blueberry, and raspberry; and vineyards, industrial crops (e.g., corn, rice, wheat, oats), and horticulture (e.g., corn, lettuce, onion, pumpkin,

Type	Product Name	Short Name	Version	Coverage	Spatial Resolution	Time span	Frequency	Reference
Vegetation	Normalized Difference Vegetation Index-3rd generation (NDVI) using the Global Inventory Monitoring and Modeling System (GIMMS)	NDVI3g	3	Global	1/12 °	1981-2015	By-monthly	Tuker et al. (2005)
	NOAA Climate Data Record (CDR) of AVHRR Surface Reflectance	NOAA CDR AVHRR	4	Global	0.05°	1981-NRT	daily	Vermote (2019)
	Vegetation Indices 16-Day L3 Global 250m	MOD13Q1.006	6	Global	250m	2000-NRT	16-day	Didan, K. (2015)
	Vegetation Indices 16-Day L3 Global 500m	MOD13A1.006	6	Global	500m	2000-NRT	16-day	Didan K. (2015)
	Vegetation Indices 16-Day L3 Global 1km	MOD13A2.006	6	Global	1km	2000-NRT	16-day	Didan K. (2015)
	Vegetation Indices Monthly L3 Global 1km	MOD13A3.006	6	Global	1km	2000-NRT	monthly	Didan K. (2015)
Land Surface Temperature and Emissivity	Land Surface Temperature/Emissivity Daily L3 Global 1km	MOD13.A1.006	6	Global	1km	2000-NRT	8-day	Wan et al. (2015)
	Land Surface Temperature/Emissivity 8-Day L3 Global 1km	MOD13A2.006	6	Global	1km	2000-NRT	monthly	Wan et al. (2015)
Land Cover Type/Dynamics	Land Cover Type Yearly L3 Global 500m	MCD12Q1.006	6	Global	500m	2001-2018	yearly	Friedl and Sulla-Menashe (2019)
	Land Cover Dynamics Yearly L3 Global 1km	MCD12Q2.006	6	Global	1km	2001-2017	yearly	Friedl and Sulla-Menashe (2019)
Evapotranspiration/Latent Heat Flux	Net Evapotranspiration 8-Day L4 Global 500m	MOD16A2.006	6	Global	500m	2000-NRT	8-day	Running, Mu. And Zhao (2017)
Precipitation	Climate Hazards Group InfraRed Precipitation with Station data (CHIRPS)	CHIRPS v2	2	50°S - 50°N	0.05°	1981-NRT	daily,pental, Dekadal, monthly	Funk et al. (2014)

FIGURE 21.2 Description of Earth observation data set for drought monitoring.

melon, carrot, watermelon, asparagus) dominate between 32°40′ and 37°42′ S. Farther south, the land is mainly used for raising cattle for beef and dairy production, and includes croplands with cereals (e.g., wheat, oat, barley), and to a lesser extent, fruit (e.g., hazelnut, apples, blueberry, cherry) (ODEPA 2015).

21.3 SATELLITE-BASED DATA SETS FOR MONITORING AGRICULTURAL DROUGHT

21.3.1 LAND

Since the satellite era, the increase in research findings has been significant for many matters, including climate change and how it affects vegetation. The launch on July 23, 1972, of Landsat-1 corresponds to the first satellite that can monitor land cover globally. The Landsat 1 (formerly ERTS-1) results in the derivation and early uses of the Normalized Difference Vegetation Index (NDVI); Rouse et al. 1974), which by using the absorbed and reflected radiation from vegetation gives a measure of vegetation development and quality. Since the launch, there has been an increased number of missions and countries taking part. Belward and Skøien (2015) makes a careful study regarding satellite mission and highlights the existing global land cover data sets having a medium resolution that benefits global and regional studies. Two of them are the AVHRR and MODIS.

21.3.1.1 Advanced Very High Resolution Radiometer (AVHRR)

Provided by the National Oceanic and Atmospheric Administration (NOAA), the Advanced Very High Resolution Radiometer (AVHRR) is a sensor that measures the reflectance of the Earth in five spectral bands of 1 km spatial resolution. The AVHRR sensors have been aboard multiple satellites: the first was TIROS-N launched on October 13, 1978, and the last was the Metop-C launched on November 7, 2018. Two of the most used data sets derived from AVHRR are as follows:

1. Normalized Difference Vegetation Index, third generation (NDVI3g) using the Global Inventory Monitoring and Modeling System (GIMMS). This NDVI time series was developed within the framework of the GIMMS project (Tucker et al. 2005; Pinzon and Tucker 2014). It is carefully assembled from different AVHRR sensors and accounts for various deleterious effects.
2. NOAA Climate Data Record (CDR) of AVHRR Surface Reflectance. This data set contains gridded daily surface reflectance and brightness temperatures derived from the AVHRR sensors onboard eight NOAA polar-orbiting satellites: NOAA-7, -9, -11, -14, -16, -17, -18, and -19. Surface reflectance from AVHRR channels 1 and 2 (at 640 and 860 nm) is the NOAA CDR.

21.3.1.2 Moderate Resolution Imaging Spectroradiometer (MODIS)

The Moderate Resolution Imaging Spectroradiometer (MODIS) is a sensor instrument aboard the Terra and Aqua satellites. Aqua's orbit around the Earth passes from south to north over the equator in the afternoon, while Terra passes from north to south across the equator in the morning. Terra and Aqua MODIS view the entire Earth's surface every one to two days, acquiring data in 36 spectral bands (https://modis.gsfc.nasa.gov/about/). The MODIS Science Team is divided into four discipline groups: atmosphere, calibration, land, and ocean. Each of the science teams work independently with different products, data formats/structures, and webpages. The products developed for the land team are useful for agricultural drought (https://modis.gsfc.nasa.gov/data/dataprod/).

Each one of the MODIS products has different types, which is reflected in its short name. If the product was gathered by the Terra satellite, the three first letters would be MOD; if it was gathered by Aqua, MYD; or combined satellites, MCD. Following these letters come two numbers that reflect the product. For example, vegetation indices are number 13, then the short name is MOD13. After that, the short name has a combination of letters and numbers that identify the spatial resolution and the temporal frequency. Finally, the product's short name ends with a dot followed by the number of the version, and the last version is 006. For example, the sixth version of the product vegetation indices (13) obtained by the Terra satellite (MOD), with a frequency of 16 days and 250 m of spatial resolution (Q1) has the short name MOD13Q1.006.

For more information on naming convention, visit https://lpdaac.usgs.gov/data/get-started-data/collection-overview/missions/modis-overview/.

A description of the products used (or being used) to study agricultural drought in Chile is next. All the products also have multiple layers.

1. Vegetation Indices (VI), MOD13. The two first layers gather the indices NDVI and the EVI (Enhanced Vegetation Index), and the rest are quality flag, reflectance (red, blue, nir, mir), zenith angle (view and sun), relative azimuth angle, composite day of the year, and pixel reliability. The spatial resolution could be 250 m, 500 m, 1 km, or 5.6 km (CMG, Climate Modeling Grid). The composition could be 16 days or 1 month.
2. Emissivity, Land Surface Temperature (LST), MOD11. Provides per-pixel land surface temperature and emissivity. The first layer is daytime LST; other layers correspond to quality, nighttime LST, local time (day/night), view zenith angle (day/night), emissivity, and clear sky coverage (day/night). This product could have a spatial resolution of 1 km or 5.6 km (CMG), and the composition could be daily, eight days, or monthly.
3. Land Cover Types/Dynamics, MCD12. These products are gathered from the Terra and Aqua satellites; these are at 500 m and 5.6 km (CMG) of spatial resolution and are generated yearly. The land cover product (MCD12Q1) provides five classification schemes for land cover type in addition to layers for the land cover property, land cover property assessment, and land cover quality control. The land cover dynamics product (MCD12Q2) provides estimates of the vegetation phenology timing at global scales. Furthermore, MCD12Q2 includes information related to the range and summation of the EVI computed

from MODIS surface reflectance data at each pixel. It identifies the vegetation growth, maturity, and senescence that mark seasonal cycles. It has 500 m spatial resolution and yearly frequencies.

4. Evapotranspiration (ET), MOD16. Derived from data obtained from the Terra satellite, the algorithm is based on the Penman–Monteith equation, which includes inputs of daily meteorological reanalysis data along with MODIS remotely sensed data products such as vegetation property dynamics, albedo, and land cover. It has five layers: total evapotranspiration (ET), average latent heat flux (LE), total potential evapotranspiration (PET), average potential latent heat flux (PLE), and evapotranspiration quality control flags (ET QC). It is developed at a spatial resolution of 500 m, and frequencies of eight days and monthly (Mu et al. 2007).

21.3.2 Precipitation

TIROS-1 was the first Television and Infrared Observation Satellite for weather monitoring and was launched on 1st April 1960 (Kidd 2001). Since then, weather satellites have become invaluable global climate measurement tools at regular intervals (Sun et al. 2018). One of the principal climatic variables that can be retrieved from satellite quasi-globally is precipitation. The sensors onboard that help to derive precipitation can be classified into three categories: visible/infrared (VIS/IR), passive microwave (PMW), and active microwave sensors (Michaelides et al. 2009). Several products have been developed using the data gathered from multiple satellites; this has allowed improving the coverage, quality, and accuracy (Huffman et al. 2007). There are more than 20 satellite-derived precipitation products that has been evaluated in different part of the globe (Beck et al. 2017; Bharti and Singh 2015; Salio et al. 2015; Sun et al. 2018) and one relevant satellite-based precipitation product that has shown higher accuracies compared with *in situ* weather stations and can be used for agricultural drought monitoring in Chile (Zambrano et al. 2017) is CHIRPS v2.

21.3.2.1 Climate Hazards Group InfraRed Precipitation with Station Data (CHIRPS)

Climate Hazards Group InfraRed Precipitation with Station data (CHIRPS) is a 30-plus year quasi-global rainfall data set. Spanning 50° S–50° N (and all longitudes), starting in 1981 to the near present, CHIRPS incorporates 0.05° resolution satellite imagery with *in situ* station data to create gridded rainfall time series appropriate for trend analysis and seasonal drought monitoring (Funk et al. 2014).

21.4 DROUGHT INDICES

The two most relevant climate variables related to drought are precipitation and temperature. The deficit of rainfall is the primary driver that triggers agricultural drought, and high temperatures contribute to intensifying it (Diffenbaugh et al. 2015; Yang et al. 2020). When we look at the local scale, the drought condition varies between crops because of two main factors: water demand (climate factors, crop type, growth type, total growing period, crop water needs, and drought resistance), and crop water supply (soil water), which in turn depend on precipitation availability (Mishra et al. 2015). Thus, many indices have been developed in the past decades for the understanding of drought at a regional and global scale (Heim 2002; Zargar et al. 2011). Precipitation deficit has been evaluated as an indirect factor for agricultural drought (Tian, Yuan, and Quiring 2018) through the Standardized Precipitation Index (SPI; Iwata et al. 2012) and the Standardized Precipitation Evapotranspiration Index (SPEI; Vicente-Serrano, Beguería, and López-Moreno 2010). A more direct measure of agricultural drought has been made through the use of spatiotemporal series of vegetation indices, which are related to vegetation conditions (Rouse et al. 1974), development (Vrieling et al. 2018), and biomass production (Rigge et al. 2013; Meroni et al. 2014).

21.4.1 STANDARDIZED PRECIPITATION INDEX (SPI)

The World Meteorological Organization (WMO) recommends the SPI as an index to characterize droughts (Hayes et al. 2011). This index uses historical cumulative rainfall data (>30 years) to derive a statistic index that gives a measure of rainfall deficit probability. This index's main feature is its multiscale calculation capability, which could be used considering different accumulated precipitation ranges. This characteristic allows it to be related to different types of droughts. Short-term SPI (<9 months) has shown a strong relationship with soil moisture and vegetation response, which is useful for analyzing agricultural drought. I will provide an example of the SPI's derivation for a timescale of 12 months (SPI-12) for the weather station Quinta Normal located in Santiago, Chile. The Quinta Normal is the station with the longest historical record.

First, to be more explicit about SPI, I will calculate SPI-12 for April; the procedure for the remaining months and timescales follow the same routine. It is needed to calculate the 12-month accumulated precipitation (cumPP-12) for April. Then, we sum the rainfall from March 2019 to April 2020 and do the same for 1951 to 2019. We will then have the accumulated precipitation of the past 12 months for April from 1951 to 2020. Figures 21.3a and 23.1b show the density and empirical cumulative distribution function (ECDF), respectively. The probability distribution of cumPP does not follow a Gaussian shape, and it is more similar to gamma or Weibull distribution. Thus, it presents two difficulties for drought monitoring: (1) it is skewed, and the mean and median do not fit; and (2) its properties depend on the amount of rainfall, which in turn depend on the location and climate condition. Therefore, it is not comparable between weather stations. Further, we need a measure that allows comparison. Thereby, Iwata et al. (2012) develop the SPI, which transforms the distribution of the cPP to a normal standard distribution, with a mean of 0 and a standard deviation of 1.

$$z = \frac{x - \mu}{\sigma} \tag{21.1}$$

For the transformation, it must fit a distribution to the cumPP, commonly gamma or Pearson III. It is used in the SPEI package (Beguería and Vicente-Serrano 2017) in the R environment (R Core Team 2020). For the fitting, the default parameters on SPEI are based on unbiased probability weighted moments (ub-pwm), but two other methods could be chosen (i.e., maximum likelihood and plotting position) (Vicente-Serrano et al. 2010; Beguería et al. 2014). Figures 21.3c and 21.3d show the corresponding normal standardized distribution for April and a timescale of 12 months.

SPI-12 for April values can be compared among different locations. Values above 0 indicate that cumPP-12 is higher (wetter) than average (1951–2020), and below 0 reflects drier conditions. Thereby, we could classify drought intensity according to the probability of occurrence. In Figure 21.4 the SPI-12 for April is plotted, together with the classes of drought intensity. It could be seen from SPI-12 for April that in the last 70 years, the weather station Quinta Normal has had three Aprils in extreme drought conditions: 1969, 1999, and 2020. Also, from the previous 13 years (since 2008), eight of them have been below –0.5 standard deviation (mild drought).

The SPI time series can be obtained by repeating the procedure for all months. The monthly time series of SPI-12 are shown in Figure 21.5. From 2009 forward, there is a tendency to drier conditions that have not been revealed before, and this period has been called a megadrought (Garreaud et al. 2017).

A relatively new index, the SPEI (Vicente-Serrano et al., 2010) was developed following SPI's methodology incorporating the effect of temperature as a simplified water balance of precipitation less evapotranspiration (D = P – ET). The SPI and SPEI have been used to study agricultural drought in many parts of the world by analyzing different timescales and their relation to crop yield and soil moisture (Rhee et al. 2010; Potopová et al. 2015; Zhang et al. 2017; Tian et al. 2018).

Perhaps one of the main restrictions to using these indices is their dependency on weather station density for appropriate spatial coverage. For undeveloped and developing countries, generally,

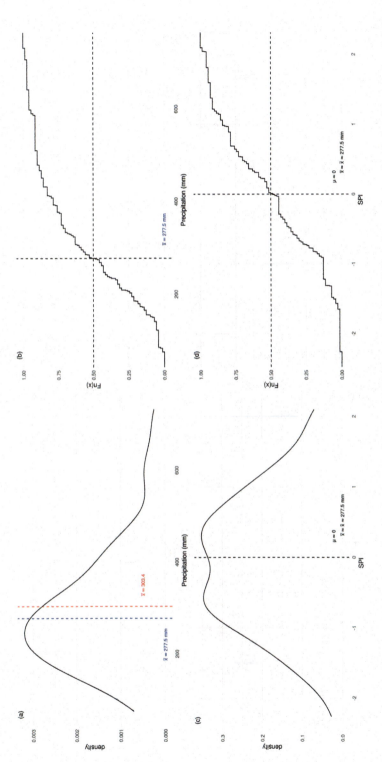

FIGURE 21.3 Distribution of cumulative 12-month precipitation (cumPP-12) and SPI-12 for April 1951–2020 at the weather station Quinta Normal. (a and d) Density. (b and d) Empirical cumulative distribution function (ECDF).

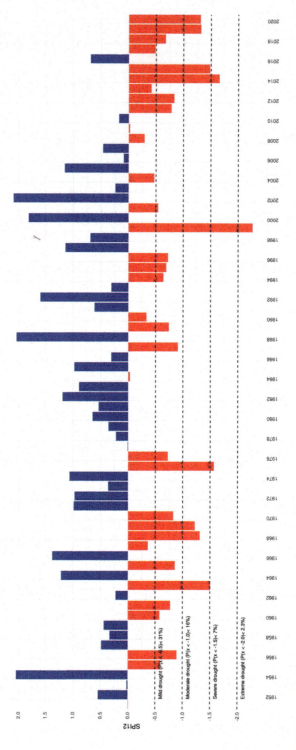

FIGURE 21.4　SPI-12 for April at weather station Quinta Normal, Santiago, Chile.

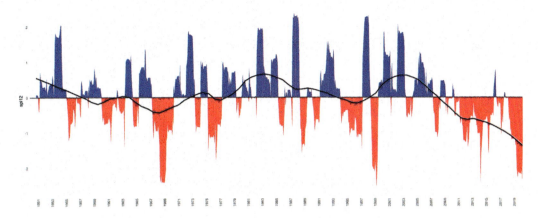

FIGURE 21.5 SPI-12 obtained by repeating the same procedure for all months at Quinta Normal, Santiago, Chile.

there is a lack of weather stations and historical records of precipitation and temperature. However, recently, satellite-derived products with a long-term history of rain (Funk et al. 2015; Ashouri et al. 2015) and spatially distributed at different spatial resolutions can be used to overcome this issue. One example is the CHIRPS v2 data set as we describe further in this chapter.

21.4.2 VEGETATION-BASED DROUGHT INDICES

The NDVI has been widely used for vegetation status monitoring and biomass productivity, thus filling the gap due to the lack of crop yield data for the spatial analysis of vegetation. The NDVI is a simple index derived from spectral reflectance in the red and infrared wavelengths:

$$NDVI = \frac{\rho_{nir} - \rho_{red}}{\rho_{nir} + \rho_{red}} \tag{21.2}$$

Green leaves absorb energy through chlorophyll on the wavelength of red and blue, which is used for photosynthesis, and in the green spectrum, chlorophyll reflect most of the energy. The leaves' internal structure reflects most of the light in the near-infrared wavelength (NIR). Thus, the NDVI can monitor vegetation greenness as an indicator of status, quantity, and development.

21.4.2.1 Gap Filling of NDVI Time Series

A known issue about passive remote sensing systems is that they are affected by atmospheric conditions because of cloudiness and difficulty to cross solar radiation through the sky (Julien and Sobrino 2018; Cai et al. 2017; Atzberger and Eilers 2011). To remove the noise due to atmospheric conditions, different gap-filling methods have been used, such as harmonic analysis of time series (Roerink et al. 2000), iterative interpolation for data reconstruction (Julien and Sobrino 2010), Savitzky–Golay filter (Jönsson and Eklundh 2004), and locally weighted polynomial regression (lowess) (Zambrano et al. 2016; Cleveland 1981). The method's parameters should be carefully calibrated; the quality of the smoothed NDVI obtained depends on these values. However, there is still no consensus on a unique optimal method for NDVI reconstruction (Julien and Sobrino 2018). For the study of drought using vegetation-based indices, this is a first-order issue that could lead to identifying misleading drought events. Thus, the first step in using the NDVI time series for drought analysis must be focused on diminishing these effects by removing the noise.

Figure 21.6 shows the bimonthly time series of NDVI3g from 1981 to 2015 for three important agriculture zones in Central Chile: San Felipe, Santa Cruz, and San Carlos (see Figure 21.1). The

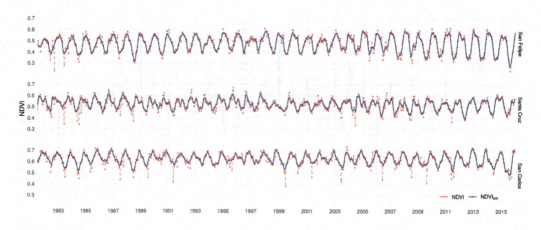

FIGURE 21.6 Time series of NDVI3g and smoothed NDVI $\left(NDVI_{sm}\right)$ based on lowess filter for 1981–2015 for three locations over Central Chile.

red line is the pure NDVI time series, and the blue line is the smoothed NDVI (NDVI $_{sm}$) in which values affected by atmospheric conditions were attempted to be removed. Once the NDVI is reconstructed, the NDVI $_{sm}$ time series can be used to derive vegetation-based indices. In the following sections, I will refer to the NDVI $_{sm}$ as NDVI.

21.4.2.2 NDVI z-Score (zNDVI)

In Figure 21.6 each NDVI line corresponds to the NDVI value over a pixel of ~8 × 8 km ^{2} (pixel size for NDVI3g), which could be associated with greenness, quantity, and/or development of vegetation. Thus, the shape of the NDVI through the season reflects vegetation's phenological behavior: lower values for winter/autumn when vegetation is less developed and/or have less canopy cover, and higher values for opposite spring/summer results. The pure NDVI does not allow identification of drought because the values are not comparable in time or space due to vegetation type and on the phenological stage. Thus, Burgan and Hartford (1993) introduced the relative greenness that measures the distance of the NDVI for a given date regarding its historical mean, also known as an anomaly of NDVI $\left(aNDVI = NDVI_t - \overline{NDVI}\right)$. The aNDVI addresses the issue about the comparison in time due to different phenological stages. In turn, it could be known when the NDVI is below its historical mean, which could be used as a measure of drought. Nevertheless, the aNDVI should not be compared in space with other vegetation types because its value depends on the magnitude that NDVI reaches, which depends on the vegetation type. Thereby, Peters et al. (2002) derived the NDVI-based Standardized Vegetation Index (SVI), which corresponds to a z-score of NDVI (zNDVI).

The zNDVI is based on the assumption that the NDVI follows a normal probability distribution; hence, it is affected by skewed distributions. Figure 21.7 shows the density distribution per each period within the year of the NDVI time series, considering 35 years of NDVI data. The magnitude and the timing of the NDVI through the season change for San Carlos, San Felipe and Santa Cruz.

The shape of the distributions shows to be Gaussian around the middle of the season (max NDVI), and moving toward the start/end of the season, the shape, in some cases, turns skewed, which affects the zNDVI derived. The zNDVI is calculated per each period, which depends on the data source used. In this case, the periods are bimonthly. Then, the zNDVI considers the distribution of values of NDVI per period, as shown in Figure 21.7, and calculates the z-score over it. Thus, the values of the zNDVI represent values from a normal standard distribution, which have a mean of 0 and a standard deviation of 1. Because the zNDVI is calculated for the NDVI regarding each period, its values do not depend on the vegetation's stage (i.e., phenology) and neither on the type.

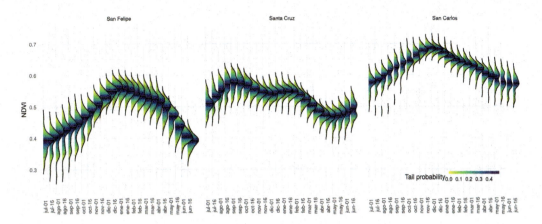

FIGURE 21.7   Vertical density probability function for 1981–2015 for each period within the year (bimonthly) of the NDVI data. Colors regard tail probability that is 0.5 at the median.

Hence, it allows being compared in time and space, analogously to the SPI. Figure 21.8 shows the zNDVI values for 1981–2015 for the three locations. In this figure, it is possible to compare the drought for the three locations for 1981–2015. If we focus on 2014–2015, we can see that the duration of the drought was longer in San Carlos, followed by San Felipe and Santa Cruz; but if we want to find where the intensity was higher, it is clear that San Carlos reaches a zNDVI value near to −4 for late 2015 in comparison with higher (>−3) values for the other locations. Thus, the zNDVI eases the drought analysis based on a z-score per period.

21.4.2.3   Vegetation Condition Index (VCI)

One of the first drought indices based on NDVI time series is the Vegetation Condition Index (VCI) proposed by Kogan (1995). The VCI corresponds to a scaled index, which estimates drought by measuring the distance of the actual NDVI to its historical min ($NDVI_{min}$) for the specific date regarding the historical amplitude of NDVI ($NDVI_{max} - NDVI_{min}$).

$$VCI^t = \frac{\left(NDVI^t - NDVI^t_{min} \right)}{\left(NDVI^t_{max} - NDVI^t_{min} \right)} \tag{21.3}$$

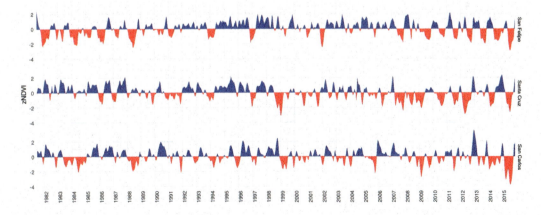

FIGURE 21.8   zNDVI for 1981–2015 for three locations within Central Chile.

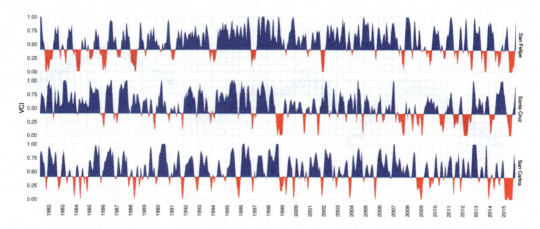

FIGURE 21.9   VCI for 1981–2015 for three locations within Central Chile.

where t is each NDVI period, and $NDVI^t_{max}$ and $NDVI^t_{min}$ are the long-term maximum and minimum, respectively, per period t.

Like the zNDVI (and SPI), the VCI removes seasonal behavior due to climate condition variation through the year, allowing a "fair" comparison in time and space. Like the zNDVI, the VCI is biased if the probability distribution of the NDVI is skewed. Moreover, the scaled indices such as the VCI are more sensitive to outliers than the standardized ones. Figure 21.9 shows overestimations of values above 0 (blue shaded area), this is because most of the period through the winter months and near the start/end of the season present right-skewed distributions that exaggerate the positive values of VCI. In the probability distribution of a VCI, the tails occur more frequently than a normal distribution; therefore, they tend to overestimate drought events, in this case as high as three times more for values below VCI = 0.1. In Figure 21.10 a–c it can be appreciated that values below the median are more likely to occur with VCI rather than zNDVI.

21.4.2.4   Cumulative NDVI over the Season z-Score (zcNDVI)

The seasonal biomass productivity of vegetation has been shown to have a strong relationship with the temporal integration of vegetation indices over the season (Jung et al. 2008; Rigge et al. 2013),

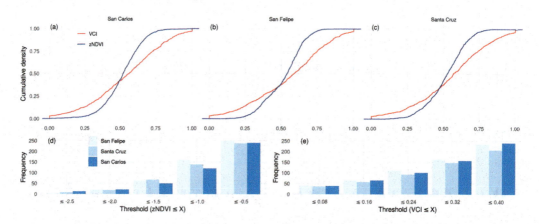

FIGURE 21.10   (a–c) Comparison of the empirical cumulative distribution functions (ECDFs) between the VCI and zNDVI (rescaled [0,1]) for three agricultural zones within central Chile; and frequency of drought identified per five thresholds of (d) zNDVI and (e) VCI.

and as such can be used to estimate crop yields (e.g., Funk and Budde 2009). To obtain a proxy for the primary productivity of croplands (Zambrano et al. 2018), calculate the cumulative NDVI over the growing season per unit ($cNDVI^t$) and per season as

$$cNDVI^t = \sum_{t=SOS}^{EOS} NDVI^t \tag{21.4}$$

where the *SOS* (start of season) and *EOS* (end of season) vary per unit, and t refers to each 16-day NDVI composite. For the calculation, the pixel or other zone could be used as a unit (e.g., administrative unit) representing the agriculture condition.

The $cNDVI^t$ is transformed into a standardized anomaly to reflect how much the vegetation's primary productivity is above or below normal. This can be written as

$$zcNDVI^t = \frac{cNDVI^t - \overline{cNDVI^t}}{\sigma\left(cNDVI^t\right)} \tag{21.5}$$

where $\overline{cNDVI}$ corresponds to the unit-level multiannual average and σ ($cNDVI^S$) to the standard deviation of $cNDVI^S$ between the 2000–2001 and 2016–2017 seasons.

Following is the example for the pixels located into San Felipe, Santa Cruz, and San Carlos. The time series of NDVI3g since 1981 were analyzed to retrieve the season's stages SOS and EOS over each pixel (Figure 21.11). The estimation was made based on the methodology proposed by Vrieling et al. (2018). The first step for phenology retrieval is adjusting a double hyperbolic tangent function to the vegetation index's time series. The function can be written as (Meroni et al. 2014; Vrieling et al. 2018)

$$VI\left(t\right) = a_0 + a_1 \cdot tanh + 1\frac{t}{2} + a_4 \cdot tanh \cdot a_6 + 1\frac{t}{2} - a_4 \tag{21.6}$$

where t is time (days). The function was fitted with the Levenberg–Marquardt least squares method implemented in the package minpack.lm (Elzhov et al. 2016) in R (R Core Team 2020). As

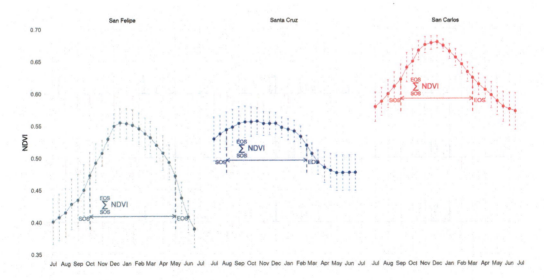

FIGURE 21.11 Median values and the standard error for 1981–2015 of the NDVI, showing the estimated start (SOS) and end of season (EOS).

mentioned by Vrieling et al. (2018), this routine requires an initial estimate of the parameter values. In this case, the same criteria was followed as in Vrieling et al. (2018) to define the initial parameter estimates:

- a_0 : the minimum VI value. The minimum VI in the first half of the time series.
- $a_1 \left[a_4 \right]$: the VI amplitude of the green-up [senescence] phase. The difference between the maximum VI and the minimum VI in the first [second] half of the time series.
- $a_2 \left[a_5 \right]$: the inflection point (days) for the green-up [senescence phase]. The midpoint between the start [end] of time series and the time of maximum VI.
- $a_3 \left[a_6 \right]$: control of the slope at the inflection point for both phases (day 1). Initialized at 0.02 [−0.02].

Based on the fitted model and the temporal range, the SOS and EOS parameters will be determined using a threshold approach (White et al. 1997) as follows:

- SOS: start of the season, defined as the first moment when the fitted function reaches 50% of the amplitude between the maximum and the preceding minimum VI. This value and the EOS will be adjusted iteratively according to the plant's observational images, which will be the ground-truth of the phenological stages.
- EOS: end of the season, defined as the first moment of the senescence phase when the fitted function reaches 50% of the amplitude between the maximum and minimum VI.

Figure 21.11 reveals the shape of the NDVI curve through the season, having lower magnitudes for the pixel located north (San Felipe), and as moves to the south, the NDVI reach higher values (San Carlos). For this example, the arbitrary start date was July 8, which works well for the pixels in San Felipe and San Carlos, but not for the one located at Santa Cruz. The start date for the analysis must be chosen carefully and depend on each pixel, and not be a fixed date. Nevertheless, I will continue with the example.

In Figure 21.12, the zcNDVI was calculated per each time step (16 days) as the season moves on. The use of the cNDVI allows us to focus just on the season and worry about how it develops. The zcNDVI allows having the accumulated anomaly per period as the season moves forward (depending on the time frequency of the data set used). The zcNDVI$_i$ corresponds to the ith period. For the rest of the chapter, I will refer to zcNDVI at EOS as $zcNDVI_{eos}$.

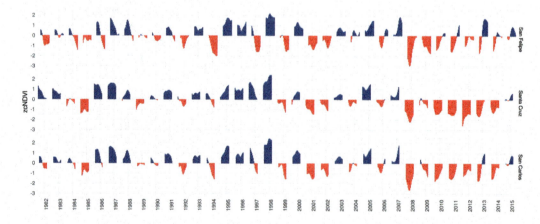

FIGURE 21.12   Time series of the standardized anomaly of the cumulative NDVI (zcNDVI) over the season between 1981 and 2015.

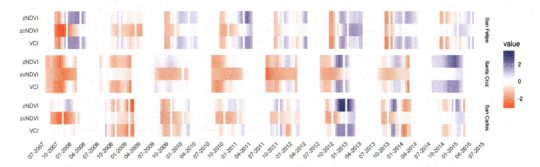

FIGURE 21.13 Comparison of time series of the drought indices zNDVI, VCI, and zcNDVI for 2007–2015.

21.4.2.5 Comparison of zNDVI, VCI, and zcNDVI

Figure 21.13 compared the three vegetation difference–based indices analyzed, it can be seen that the VCI and zNDVI have similar behavior, although the VCI, as described earlier, overestimates extreme drought events. But, when VCI and zNDVI are compared with the zcNDVI, there are more significant differences. For example, in San Carlos for season 2007–2008, the zcNDVI shows extreme values when the zNDVI and VCI indicate mild or no drought.

Further, the variance explained between indices by the statistical measure r^2 was calculated. The VCI and zNDVI have higher r^2, but when compared with zcNDVI the values decrease significantly, reaching values ranging between 0.04 and 0.26 for VCI and 0.06 and 0.26 for zNDVI. The VCI and zNDVI explain just a lower proportion of the variance in the zcNDVI. The zcNDVI has shown to have a high r^2 with production (tonnes) at regional scale over Chile (Zambrano et al. 2018), and also the use of cumulative VI at local scale showed improvement over standard VI to estimate grain yield (Venancio et al. 2020). Nonetheless, more research is needed to fully comprehend the power of the zcNDVI to retrieve vegetation productivity and thus for drought monitoring.

21.5 DROUGHT INSIGHTS IN CHILE SINCE 1981 USING ZCNDVI

For this section, I used the Landcover 2014 for Chile developed by Zhao et al. (2016), the AVHRR 5 km (1981–2019), and MODIS (2000–2020) data sets to spatially explore drought impact on vegetation over Central Chile by use of the zcNDVI.

The land cover was resampled from 30 m to ~500 m (MODIS) and ~4 km (AVHRR) to fit with the zcNDVI time series and the growing season rasters SOS and EOS. The analysis was focused on land cover types of crops, forest, and grassland (Figure 21.14a). Besides the three land covers, Central Chile was longitudinally divided into three zones: north (29°–33° S), center (33°–37° S), and south (37°–41° S).

21.5.1 Growing Season

First, the characteristic growing season was estimated with the MCD12Q2 collection 6 product. This product provides yearly data from 2001 to 2017. I used the median value per pixel over the range of years, considering a single season. All the pixels of a defined growing season were considered, as shown in Figures 21.14b and 21.14c. Central Chile has a longitudinal pattern for the growing season from north to south due mainly to the different climatic conditions (see Figure 21.1), which in turn define the type of vegetation (forest/crops/grassland types).

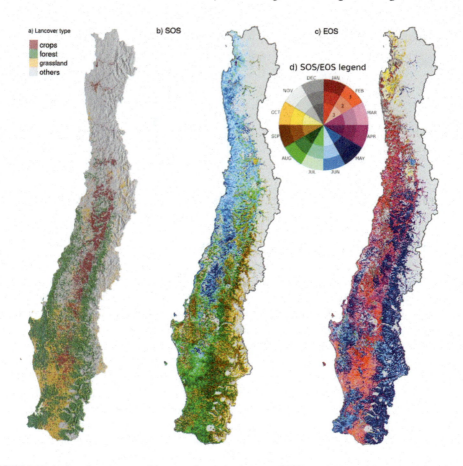

FIGURE 21.14 (a) Land cover type, (b) start of growing season (SOS), and (c) end of growing season (SOS) over Central Chile derived from the product MCD12Q2.006 (MODIS). (d) Legend for SOS and EOS following Vrieling et al. (2018); the outer circle represents the first 10-day period of each month.

21.5.2 Droughts for 2000–2020 (MODIS)

Figure 21.15 shows the values of the $zcNDVI_{eos}$, when the zcNDVI reaches the time of EOS per each pixel, derived from MOD13A1 data set at ~500 m of spatial resolution. Each pixel has its own EOS according to Figure 21.14c.

The year range 2000–2020 highlights 2001–2002, 2007–2008, and 2019–2020 as the three seasons of lower values of $zcNDVI_{eos}$. For the season 2001–2002, the drought is located to the center and south of Central Chile, 2007–2008 was kind homogeneous over Central Chile, and season 2019–2020 shows the lower values concentrated in the north part.

Analyzing by land cover type and by zone (Figure 21.16), in the north, season 2019–2020 reached lower values having a zcNDVI of –2.74 in the forest, –2.05 in grasslands, and –0.845 in crops. For the center, crops, forest, and grassland had around –0.5 for season 2019–2020, showing lower values in season 2007–2008 and 2000–2001. In the south, since 2016, the values of $zcNDVI_{eos}$ have been positive, showing an increase in vegetation development.

21.5.3 Droughts for 1981–2020 (AVHRR)

In the following, I will analyze the drought in Central Chile since 1981 derived from the zcNDVI from AVHRR data at 4 km and derived from the SPI at different timescales by using estimates of precipitation from the CHIRPS v2 data set.

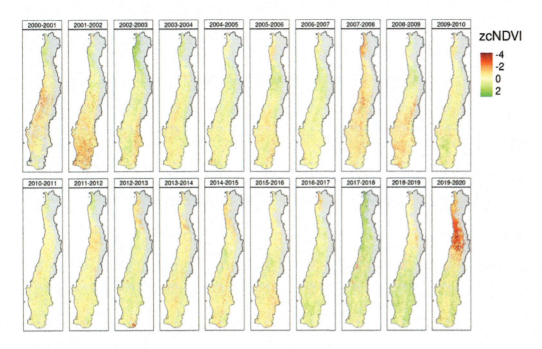

FIGURE 21.15 Maps of zcNDVI at EOS within Central Chile for 2001–2020.

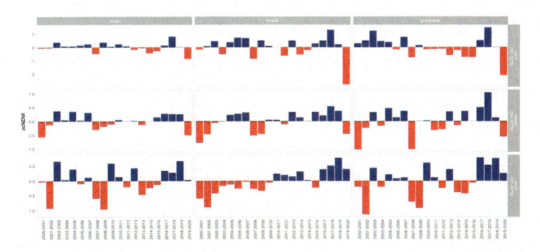

FIGURE 21.16 Average value of zcNDVI reached at EOS by land cover type (crops, forest, and grassland) and per zone (north, central, and south) within Central Chile for 2001–2020.

21.5.4 ZCNDVI Derived from AVHRR NDVI 5 KM

Figure 21.17 shows the values of the $zcNDVI_{eos}$, highlights the seasons 1988–1989 and 2019–2020 as the two more extreme droughts. When the zcNDVI is averaged by land cover type and zone (Figure 21.18), a particular pattern of positive and negative values arises. From 1981 to 2000, zcNDVI reveals a low productivity pattern for the forest, grassland, and crops in Central Chile; and the values are lower for the center and south part of Central Chile. After that, for 2000–2017, the $zcNDVI_{eos}$ shows positive values all around Central Chile, except for grassland north from 37° S.

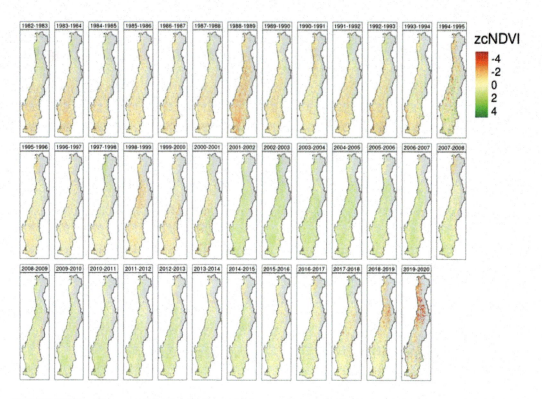

FIGURE 21.17   Maps of zcNDVI at EOS within Central Chile for 1981–2020.

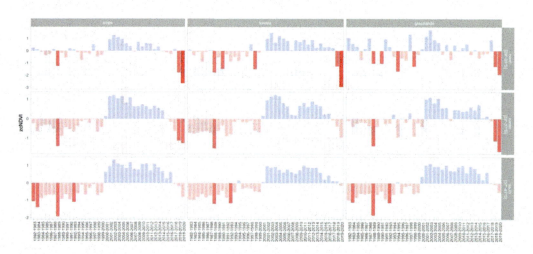

FIGURE 21.18   Average value of zcNDVI reached at EOS by land cover type (crops, forest, and grassland) and per zone (north, central, and south) within Central Chile for 1981–2020.

Finally, since 2017 there arises another low productivity pattern. Surprisingly, the inflection point time occurs each 17–18 years.

The most extreme negative anomaly was found for 2019–2020 in the north zone for the forest and crops type reaching values of –2.91 and –2.62. This represents an occurrence of 1.8 and 4.4 times every 1000 years. In the center, grasslands were –1.68 for 2019–2020; forest and crops reach –1.52 and –1.41 for 1988–1989. Also, crops have values under –1 since 2018–2019.

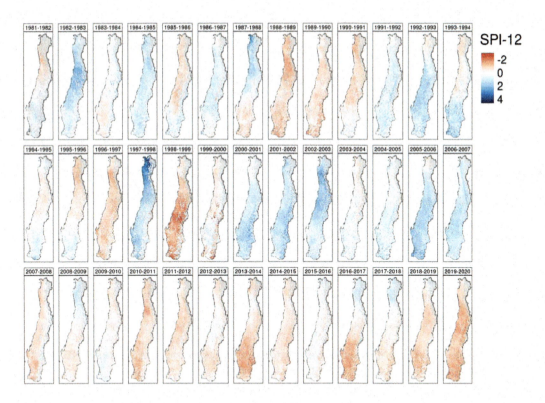

FIGURE 21.19 *SPI$_{eos}$*-12 for 1981–2020 over Central Chile.

It is essential to indicate that this is a quick exercise for evaluating *zcNDVI$_{eos}$* over Chile. The aim is to remark the importance of considering the growing season for vegetation impact analysis of drought. In this case, the land cover for 2014 was used and the characteristic growing season derived from 2001–2017.

21.5.5 SPI DERIVED FROM CHIRSP V2

At timescales of 3, 6, 12, 24, and 36 months, the SPI was calculated from the CHIRPS v2 decadal data set (see section 21.3.2.1). Then to compare it against the zcNDVI, we need the SPIs at the time of EOS. Figure 21.19 shows the SPI$_{eos}$-12 since 1981 for Central Chile. Chile has suffered a persistent drought since 2007–2008, as shown in the third row of Figure 21.19.

21.5.5.1 Correlation Analysis between zcNDVI and SPI

To understand how much of the variability on zcNDVI is explained by precipitation deficit, in this section I calculate the coefficient of determination (r^2) between *zcNDVI$_{eos}$* and the *SPI$_{eos}$* at timescales of 3, 6, 12, 24, and 36 months.

The deficit of precipitation of 12 months at EOS (SPI$_{eos}$-12) shows a higher impact on vegetation (i.e., crops, forest, and grassland) north from 34° S, followed by the 24-months deficit (Figures 21.20 and 21.21). As shown in Figure 21.1c, northern 34° S annual precipitation is less than 500 mm (arid), and moving to the south from 34° S, annual precipitation increases until more than 2500 mm (humid). Thereby, for the arid part of Central Chile, the precipitation deficit is a key variable determining vegetation development. Nevertheless, for the humid part, the precipitation deficit has a lower or no impact on the vegetation condition. This does not mean that precipitation does not play a key role, but the deficit has a minor impact due that the amount of precipitation in the humid part.

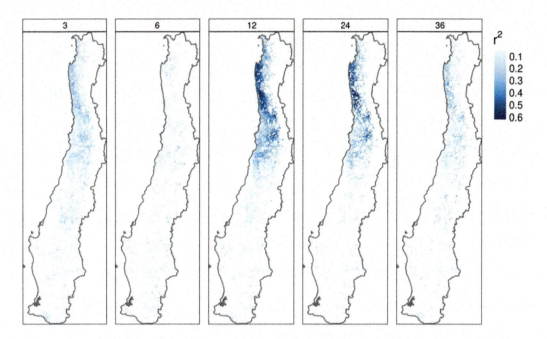

FIGURE 21.20 Map of the determination coefficient (r^2) for the correlation between zcNDVI$_{eos}$ and SPI$_{eos}$ for 1982–2020 at timescales of 3, 6, 12, 24, and 36 months.

21.6 FINAL REMARKS

In this chapter, I have made a comprehensive analysis of agricultural drought in Central Chile. I describe how drought has been monitored, the leading indices used, and the primarily available Earth observation data sets. I have highlighted the importance of fine-tuning agricultural drought analysis considering the period more relevant for agriculture development, the growing season. I analyzed the last 40 years of drought using the standardized anomaly of cumulative NDVI within the growing season (zcNDVI) and its relation with precipitation deficit measured by the SPI derived from the CHIRPS v2 data set.

Some key points that have come out from the chapter are:

- There is a plethora of Earth observation data sets that can be used for drought monitoring and analysis. This chapter describes some of the most used data sets for regional and long-term analysis, such as the NDVI derived from the AVHRR (~4 km) sensor and the precipitation satellite estimates from CHIRPS (~5 km), both having data since 1981. The medium resolution MODIS data set for land monitoring includes the vegetation indices (MOD13), land surface temperature (MOD11), land cover dynamics (MCD12), and the evapotranspiration product (MOD16). All MODIS products have varying spatial resolutions of 250 m, 500 m, and 1 km, with data available since 2000.
- Drought indices should be of simple interpretation for allowing comparison in time and space. For example, the SPI is a drought index recommended by the WMO for drought monitoring. It allows evaluating different timescales of precipitation deficit (from 1 to >48 months), and can be compared over space because it is measured in terms of anomaly with respect to its normal condition over a period of time (usually more than 30 years). Similarly, for vegetation drought, standardized indices have been proposed such as the zNDVI or scaled indices as the VCI. These indices depend on the probability distribution

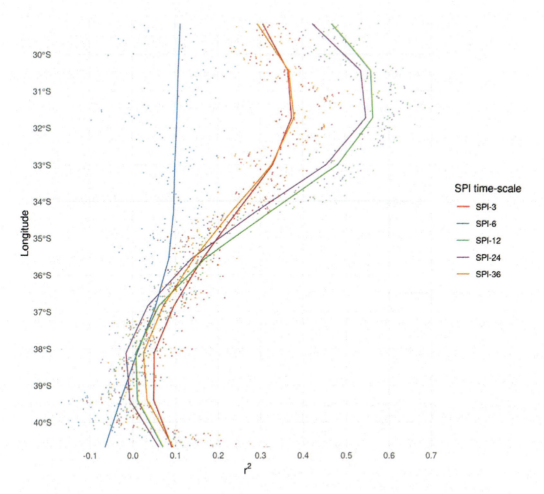

FIGURE 21.21 Longitudinal pattern of the determination coefficient (r^2) for the correlation between zcNDVI$_{eos}$ and SPI$_{eos}$ for 1982–2020 at timescales of 3, 6, 12, 24, and 36 months.

of the NDVI, assuming a normal distribution. Both could be affected by outliers, but the VCI is more sensitive to them and tends to overestimate drought.

- I presented the zcNDVI, an index that considers the standardized anomaly (similar to the SPI and zNDVI) for the cumulative NDVI within the growing season. For this index, previously, seasonal phenology metrics such as the SOS and EOS should be derived. This could be derived from the analysis of NDVI time series or be extracted from the MCD12Q2. This index's comparison with zNDVI and VCI is very distinct ($r^2 = 0.06$ to 0.26). The VCI and zNDVI have a similar meaning ($r^2 = 0.75$ to 0.94). The zcNDVI could be derived from the SOS to EOS, with the frequency of the data set used.

- From the analysis of $zcNDVI_{eos}$ by the MODIS (2000–2020) and AVHRR (1981–2020) data sets, it seems that for the period covered by MODIS the vegetation shows a higher productivity pattern. The long-term AVHRR data set has an inflection point in the year 2000. Then, since 2018–2019 the $zcNDVI_{eos}$ turns around to negative values, showing the most extreme values over the vegetation of the north part of Central Chile during season 2019–2020.

- The analysis of SPI$_{eos}$ shows a persistent pattern of the deficit since 2007 for 12 and 24 months (SPI-12 and SPI-24) over Central Chile. The correlation of SPI$_{eos}$ and $zcNDVI_{eos}$

is the strongest for 12 and 24 months north of 34° S. For the arid part of Central Chile, over 40% of the variability is explained for the precipitation deficits of 12 and 24 months. But south of that point, the correlation with the shortage of precipitation decreases dramatically, indicating that the rainfall deficit has a lower impact on vegetation in the humid part of Central Chile.

- This chapter did not consider the impact of temperature or soil moisture on vegetation. Future research must include evapotranspiration (e.g., MOD16) and global soil moisture data sets (e.g., ESACCISM). It could also be used as another long-term data set of climate variables, for example, the TerraClimate data set (Abatzoglou et al. 2018), which has data since 1958.

This chapter uses the land cover for 2014, but also needed is long-term land cover to consider the variation over time. The growing season used considered the characteristic season derived for 2000–2020 (MOD12Q2), which, similar to land cover, should be improved by deriving the season by, e.g., AVHRR, and evaluating the variation over time.

ACKNOWLEDGMENTS

This chapter was financed thanks to the contribution of the FONDECYT project 11190360 of the National Research and Development Agency (ANID), Chile.

REFERENCES

Abatzoglou, John T., Solomon Z. Dobrowski, Sean A. Parks, and Katherine C. Hegewisch. 2018. "TerraClimate, a high-resolution global dataset of monthly climate and climatic water balance from 1958–2015." *Scientific Data* 5 (January): 170191. https://doi.org/10.1038/sdata.2017.191.

AghaKouchak, A., A. Farahmand, F. S. Melton, J. Teixeira, M. C. Anderson, B. D. Wardlow, and C. R. Hain. 2015. "Remote sensing of drought: Progress, challenges and opportunities." *Reviews of Geophysics* 53 (2): 452–80. https://doi.org/10.1002/2014RG000456.

Ashouri, Hamed, Kuo Lin Hsu, Soroosh Sorooshian, Dan K. Braithwaite, Kenneth R. Knapp, L. Dewayne Cecil, Brian R. Nelson, and Olivier P. Prat. 2015. "PERSIANN-CDR: Daily precipitation climate data record from multisatellite observations for hydrological and climate studies." *Bulletin of the American Meteorological Society* 96 (1): 69–83. https://doi.org/10.1175/BAMS-D-13-00068.1.

Atzberger, Clement, and Paul H. C. Eilers. 2011. "Evaluating the effectiveness of smoothing algorithms in the absence of ground reference measurements." *International Journal of Remote Sensing* 32 (13): 3689–3709. https://doi.org/10.1080/01431161003762405.

Ault, Toby R. 2020. "On the essentials of drought in a changing climate." *Science* 368 (6488): 256–60. https://doi.org/10.1126/science.aaz5492.

Beck, Hylke E., Noemi Vergopolan, Ming Pan, Vincenzo Levizzani, Albert I. J. M. Van Dijk, Graham P. Weedon, Luca Brocca, Florian Pappenberger, George J. Huffman, and Eric F. Wood. 2017. "Global-scale evaluation of 22 precipitation datasets using gauge observations and hydrological modeling." *Hydrology and Earth System Sciences* 21 (12): 6201–17. https://doi.org/10.5194/hess-21-6201-2017.

Beguería, Santiago, and Sergio M Vicente-Serrano. 2017. "SPEI: Calculation of the standardised precipitation-evapotranspiration index." https://cran.r-project.org/package=SPEI.

Beguería, Santiago, Sergio M. Vicente-Serrano, Fergus Reig, and Borja Latorre. 2014. "Standardized precipitation evapotranspiration index (SPEI) revisited: Parameter fitting, evapotranspiration models, tools, datasets and drought monitoring." *International Journal of Climatology* 34 (10): 3001–23. https://doi.org/10.1002/joc.3887.

Belward, Alan S., and Jon O. Skøien. 2015. "Who launched what, when and why; trends in global land-cover observation capacity from civilian earth observation satellites." *ISPRS Journal of Photogrammetry and Remote Sensing* 103: 115–28. https://doi.org/10.1016/j.isprsjprs.2014.03.009.

Bharti, Vidhi, and Charu Singh. 2015. "Evaluation of error in TRMM 3B42V7 precipitation estimates over the Himalayan region." *Journal of Geophysical Research* 120 (24): 12, 458–12, 473. https://doi.org/10.1002/2015JD023779.

Boisier, Juan P., Camila Alvarez-Garreton, Raúl R. Cordero, Alessandro Damiani, Laura Gallardo, René D. Garreaud, Fabrice Lambert, Cinthya Ramallo, Maisa Rojas, and Roberto Rondanelli. 2018. "Anthropogenic drying in central-southern Chile evidenced by long-term observations and climate model simulations." *Elementa* 6 (1): 74. https://doi.org/10.1525/elementa.328.

Burgan, R. E., and R. A. Hartford. 1993. "Monitoring vegetation greenness with satellite data." General Technical Report – US Department of Agriculture, Forest Service 297 (INT-297). https://doi.org/10.2737/INT-GTR-297.

Cai, Zhanzhang, Per Jönsson, Hongxiao Jin, and Lars Eklundh. 2017. "Performance of smoothing methods for reconstructing NDVI time-series and estimating vegetation phenology from MODIS data." *Remote Sensing* 9 (12): 1271. https://doi.org/10.3390/rs9121271.

Chen, Shuai, Xiaoguang Chen, and Jintao Xu. 2016. "Impacts of climate change on agriculture: Evidence from China." *Journal of Environmental Economics and Management* 76: 105–24. https://doi.org/10.1016/j.jeem.2015.01.005.

Cleveland, William S. 1981. "LOWESS: A Program for Smoothing Scatterplots by Robust Locally Weighted Regression." *American Statistician* 35 (1). https://doi.org/10.2307/2683591.

Cook, Benjamin I., Toby R. Ault, and Jason E. Smerdon. 2015. "Unprecedented 21st century drought risk in the American Southwest and Central Plains." https://doi.org/10.1126/sciadv.1400082.

Dai, Aiguo. 2011. "Drought under global warming: A review." https://doi.org/10.1002/wcc.81.

Diffenbaugh, Noah S., Daniel L. Swain, Danielle Touma, and Jane Lubchenco. 2015. "Anthropogenic warming has increased drought risk in California." *Proceedings of the National Academy of Sciences of the United States of America* 112 (13): 3931–6. https://doi.org/10.1073/pnas.1422385112.

Ding, Ya, Michael J. Hayes, and Melissa Widhalm. 2011. "Measuring economic impacts of drought: A review and discussion." https://doi.org/10.1108/09653561111161752.

Elzhov, T. V., K. M. A. Mullen, N. Spiess, and B. Bolker. 2016. "minpack.lm: R Interface to the Levenberg Marquardt nonlinear least-squares algorithm found in MINPACK, plus support for bounds." URL https://CRAN.R-project.org/package=minpack.lm. R package version 1.2-1.

FAO. 2016. *Damages and Losses from Climate-related Disasters in Agricultural Sectors.* Food and Agriculture Organization of the United States. http://www.fao.org/3/a-i6486e.pdf.

Funk, C. C., P. J. Peterson, M. F. Landsfeld, D. H. Pedreros, J. P. Verdin, J. D. Rowland, B. E. Romero, G. J. Husak, J. C. Michaelsen, and A. P. Verdin. 2014. "A quasi-global precipitation time series for drought monitoring." *Vol. 832.* U.S. Geological Survey Data Series 832. https://doi.org/110.3133/ds832.

Funk, Chris, and Michael E Budde. 2009. "Phenologically-tuned MODIS NDVI-based production anomaly estimates for Zimbabwe." *Remote Sensing of Environment* 113 (1): 115–25. https://doi.org/10.1016/j.rse.2008.08.015.

Funk, Chris, Pete Peterson, Martin Landsfeld, Diego Pedreros, James Verdin, Shraddhanand Shukla, Gregory Husak, et al. 2015. "The climate hazards infrared precipitation with stations - A new environmental record for monitoring extremes." *Scientific Data* 2 (150066): 21. https://doi.org/10.1038/sdata.2015.66.

Garreaud, René D., Camila Alvarez-Garreton, Jonathan Barichivich, Juan Pablo Boisier, Duncan Christie, Mauricio Galleguillos, Carlos LeQuesne, James McPhee, and Mauricio Zambrano-Bigiarini. 2017. "The 2010–2015 mega drought in Central Chile: Impacts on regional hydroclimate and vegetation." *Hydrology and Earth System Sciences Discussions* 2017: 1–37. https://doi.org/10.5194/hess-2017-191.

Garreaud, René D., Juan P. Boisier, Roberto Rondanelli, Aldo Montecinos, Hector H. Sepúlveda, and Daniel Veloso-Aguila. 2020. "The Central Chile Mega Drought (2010–2018): A climate dynamics perspective." *International Journal of Climatology* 40 (1): 421–39. https://doi.org/10.1002/joc.6219.

Hayes, Michael, Mark Svoboda, Nicole Wall, and Melissa Widhalm. 2011. "The Lincoln declaration on drought indices: Universal meteorological drought index recommended." *Bulletin of the American Meteorological Society* 92 (4): 485–88. https://doi.org/10.1175/2010BAMS3103.1.

Heim, Richard R. 2002. "A review of twentieth-century drought indices used in the United States." *Bulletin of the American Meteorological Society* 83 (8): 1149–65. https://doi.org/10.1175/1520-0477(2002)083<1149:AROTDI>2.3.CO;2.

Howitt, Richard, Josué Medellín-Azuara, and Duncan MacEwan. 2014. *Economic Analysis of the 2014 Drought for California Agriculture.* Center for Watershed Sciences. University of California, Davis, California, 20p. http://watershed.ucdavis.edu.

Huffman, George J., Robert F. Adler, David T. Bolvin, Guojun Gu, Eric J. Nelkin, Kenneth P. Bowman, Yang Hong, Erich F. Stocker, and David B. Wolff. 2007. "The TRMM Multisatellite Precipitation Analysis (TMPA): Quasi-global, multiyear, combined-sensor precipitation estimates at fine scales." *Journal of Hydrometeorology* 8 (1): 38–55. https://doi.org/10.1175/JHM560.1.

Instituto Nacional de Estadísticas (INE). 2007. "VII Censo nacional agropecuario y forestal. Instituto Nacional de Estadística. Informe Agropecuarias 2007." *Respuesta a Author (29-03-2023, 21:27)*.

Iwata, Takashi, Noritoshi Nishiyama, Koshi Nagano, Nobuhiro Izumi, Takuma Tsukioka, Kyukwang Chung, Shoji Hanada, Kiyotoshi Inoue, Masahide Kaji, and Shigefumi Suehiro. 2012. "Preoperative serum value of sialyl Lewis X predicts pathological nodal extension and survival in patients with surgically treated small cell lung cancer." In *Journal of Surgical Oncology* 105:818–24. 8. https://doi.org/10.1002/jso.23002.

Jönsson, Per, and Lars Eklundh. 2004. "TIMESAT: A program for analyzing time-series of satellite sensor data." *Computers and Geosciences* 30 (8): 833–45. https://doi.org/10.1016/j.cageo.2004.05.006.

Julien, Yves, and José A. Sobrino. 2010. "Comparison of cloud-reconstruction methods for time series of composite NDVI data." *Remote Sensing of Environment* 114 (3): 618–25. https://doi.org/10.1016/j.rse.2009.11.001.

Julien, Yves, and José A Sobrino. 2018. "Optimizing and comparing gap-filling techniques using simulated NDVI time series from remotely sensed global data." *International Journal of Applied Earth Observation and Geoinformation* 76: 93–111. https://doi.org/10.1016/j.jag.2018.11.008.

Jung, Martin, Michel Verstraete, Nadine Gobron, Markus Reichstein, Dario Papale, Alberte Bondeau, Monica Robustelli, and Bernard Pinty. 2008. "Diagnostic assessment of European gross primary production." *Global Change Biology* 14 (10): 2349–64. https://doi.org/10.1111/j.1365-2486.2008.01647.x.

Kidd, C. 2001. "Satellite rainfall climatology: A review." *International Journal of Climatology* 21 (9): 1041–66. https://doi.org/10.1002/joc.635.

Kogan, F. N. 1995. "Application of vegetation index and brightness temperature for drought detection." *Advances in Space Research* 15 (11): 91–100. https://doi.org/10.1016/0273-1177(95)00079-T.

Kottek, Markus, Jürgen Grieser, Christoph Beck, Bruno Rudolf, and Franz Rubel. 2006. "World map of the Köppen-Geiger climate classification updated." *Meteorologische Zeitschrift* 15 (3): 259–63. https://doi.org/10.1127/0941-2948/2006/0130.

Lipper, Leslie, Philip Thornton, Bruce M. Campbell, Tobias Baedeker, Ademola Braimoh, Martin Bwalya, Patrick Caron, et al. 2014. "Climate-smart agriculture for food security." *Nature Climate Change* 4 (12): 1068–72. https://doi.org/10.1038/nclimate2437.

Meroni, M, D Fasbender, F Kayitakire, G Pini, F Rembold, F Urbano, and M M Verstraete. 2014. "1-s2.0-S0034425713004239-main.pdf." *Remote Sensing of Environment* 142: 57–68.

Michaelides, S., V. Levizzani, E. Anagnostou, P. Bauer, T. Kasparis, and J. E. Lane. 2009. "Precipitation: Measurement, remote sensing, climatology and modeling." *Atmospheric Research* 94 (4): 512–33. https://doi.org/10.1016/j.atmosres.2009.08.017.

Mishra, Ashok K., Amor V. M. Ines, Narendra N. Das, C. Prakash Khedun, Vijay P. Singh, Bellie Sivakumar, and James W. Hansen. 2015. "Anatomy of a local-scale drought: Application of assimilated remote sensing products, crop model, and statistical methods to an agricultural drought study." *Journal of Hydrology* 526: 15–29. https://doi.org/10.1016/j.jhydrol.2014.10.038.

Mishra, Ashok K., and Vijay P. Singh. 2010. "A review of drought concepts." *Journal of Hydrology* 391 (1–2): 202–16. https://doi.org/10.1016/j.jhydrol.2010.07.012.

Montecinos, Aldo, and Patricio Aceituno. 2003. "Seasonality of the ENSO-related rainfall variability in central Chile and associated circulation anomalies." *Journal of Climate* 16 (2): 281–96. https://doi.org/10.1175/1520-0442(2003)016<0281:SOTERR>2.0.CO;2.

Moore, Frances C., Uris Baldos, Thomas Hertel, and Delavane Diaz. 2017. "New science of climate change impacts on agriculture implies higher social cost of carbon." *Nature Communications* 8 (1): 1607. https://doi.org/10.1038/s41467-017-01792-x.

Mu, Qiaozhen, Faith Ann Heinsch, Maosheng Zhao, and Steven W. Running. 2007. "Development of a global evapotranspiration algorithm based on MODIS and global meteorology data." *Remote Sensing of Environment* 111 (4): 519–36. https://doi.org/10.1016/j.rse.2007.04.015.

ODEPA. 2015. *Chilean Agriculture Overview 2015*. Edited by of the Chilean Ministry of Agriculture. July 2019 Office of Agricultural Studies and Policies (ODEPA). Office of Agricultural Policies, Trade; Information. www.odepa.gob.cl.

Peroni Venancio, Luan, Everardo Chartuni Mantovani, Cibele Hummel do Amaral, Christopher Michael Usher Neale, Ivo Zution Gonçalves, Roberto Filgueiras, and Fernando Coelho Eugenio. 2020. "Potential of using spectral vegetation indices for corn green biomass estimation based on their relationship with the photosynthetic vegetation sub-pixel fraction." *Agricultural Water Management*. https://doi.org/10.1016/j.agwat.2020.106155.

Peters, Albert J., Elizabeth A. Walter-Shea, Lei Ji, Andrés Viña, Michael Hayes, and Mark D. Svoboda. 2002. "Drought monitoring with NDVI-based Standardized Vegetation Index." *Photogrammetric Engineering and Remote Sensing* 68 (1): 71–75.

Pinzon, Jorge E., and Compton J. Tucker. 2014. "A non-stationary 1981–2012 AVHRR NDVI3g time series." *Remote Sensing* 6 (8): 6929–60. https://doi.org/10.3390/rs6086929.

Potopová, Vera, Petr Stepánek, Martin Mozný, Lubos Türkott, and Josef Soukup. 2015. "Performance of the standarised precipitation evapotranspiration index at various lags for agricultural drought risk assessment in the Czech Republic." *Agricultural and Forest Meteorology* 202: 26–38.

Quiring, Steven M., and Srinivasan Ganesh. 2010. "Evaluating the utility of the Vegetation Condition Index (VCI) for monitoring meteorological drought in Texas." *Agricultural and Forest Meteorology* 150 (3): 330–39. https://doi.org/10.1016/j.agrformet.2009.11.015.

R Core Team. 2020. *R: A Language and Environment for Statistical Computing.* Vienna, Austria: R Foundation for Statistical Computing. https://www.r-project.org/.

Rhee, Jinyoung, Jungho Im, and Gregory J. Carbone. 2010. "Monitoring agricultural drought for arid and humid regions using multi-sensor remote sensing data." *Remote Sensing of Environment* 114 (12): 2875–87. https://doi.org/10.1016/j.rse.2010.07.005.

Rickards, L., and S. M. Howden. 2012. "Transformational adaptation: Agriculture and climate change." In *Crop and Pasture Science* 63:240–50. 3. https://doi.org/10.1071/CP11172.

Rigge, Matthew, Alexander Smart, Bruce Wylie, Tagir Gilmanov, and Patricia Johnson. 2013. "Linking phenology and biomass productivity in south dakota mixed-grass prairie." *Rangeland Ecology and Management* 66 (5): 579–87. https://doi.org/10.2111/REM-D-12-00083.1.

Roerink, G. J., M. Menenti, and W. Verhoef. 2000. "Reconstructing cloudfree NDVI composites using Fourier analysis of time series." *International Journal of Remote Sensing* 21 (9): 1911–7. https://doi.org/10.1080/014311600209814.

Rouse, J. W., R. H. Hass, J. A. Schell, D. W. Deering, and J. C. Harlan. 1974. *"Monitoring the vernal advancement and retrogradation (green wave effect) of natural vegetation."* Final Report, RSC 1978–4, Texas A & M University, College Station, Texas.

Salio, Paola, María Paula Hobouchian, Yanina García Skabar, and Daniel Vila. 2015. "Evaluation of high-resolution satellite precipitation estimates over southern South America using a dense rain gauge network." *Atmospheric Research* 163: 146–61. https://doi.org/10.1016/j.atmosres.2014.11.017.

Smith, Adam B., and Jessica L. Matthews. 2015. "Quantifying uncertainty and variable sensitivity within the US billion-dollar weather and climate disaster cost estimates." *Natural Hazards* 77 (3): 1829–51. https://doi.org/10.1007/s11069-015-1678-x.

Stevanović, Miodrag, Alexander Popp, Hermann Lotze-Campen, Jan Philipp Dietrich, Christoph Müller, Markus Bonsch, Christoph Schmitz, Benjamin Leon Bodirsky, Florian Humpenöder, and Isabelle Weindl. 2016. "The impact of high-end climate change on agricultural welfare." *Science Advances* 2 (8): e1501452. https://doi.org/10.1126/sciadv.1501452.

Sun, Qiaohong, Chiyuan Miao, Qingyun Duan, Hamed Ashouri, Soroosh Sorooshian, and Kuo Lin Hsu. 2018. "A review of global precipitation data sets: Data sources, estimation, and intercomparisons." *Reviews of Geophysics* 56 (1): 79–107. https://doi.org/10.1002/2017RG000574.

Tian, Liyan, Shanshui Yuan, and Steven M. Quiring. 2018. "Evaluation of six indices for monitoring agricultural drought in the south-central United States." *Agricultural and Forest Meteorology* 249: 107–19. https://doi.org/10.1016/j.agrformet.2017.11.024.

Tucker, Compton J., Jorge E. Pinzon, Molly E. Brown, Daniel A. Slayback, Edwin W. Pak, Robert Mahoney, Eric F. Vermote, and Nazmi El Saleous. 2005. "An extended AVHRR 8-km NDVI dataset compatible with MODIS and SPOT vegetation NDVI data." 20. 26. https://doi.org/10.1080/01431160500168686.

Van Loon, Anne F., Tom Gleeson, Julian Clark, Albert I. J. M. Van Dijk, Kerstin Stahl, Jamie Hannaford, Giuliano Di Baldassarre, et al. 2016. "Drought in the Anthropocene." *Nature Geoscience* 9 (2): 89–91. https://doi.org/10.1038/ngeo2646.

Vicente-Serrano, Sergio M., Santiago Beguería, and Juan I. López-Moreno. 2010. "A multiscalar drought index sensitive to global warming: The standardized precipitation evapotranspiration index." *Journal of Climate* 23 (7): 1696–1718. https://doi.org/10.1175/2009JCLI2909.1.

Vrieling, Anton, Michele Meroni, Roshanak Darvishzadeh, Andrew K. Skidmore, Tiejun Wang, Raul Zurita-Milla, Kees Oosterbeek, Brian O'Connor, and Marc Paganini. 2018. "Vegetation phenology from Sentinel-2 and field cameras for a Dutch barrier island." *Remote Sensing of Environment* 215 (September): 517–29. https://doi.org/10.1016/j.rse.2018.03.014.

West, Harry, Nevil Quinn, and Michael Horswell. 2019. "Remote sensing for drought monitoring & impact assessment: Progress, past challenges and future opportunities." *Remote Sensing of Environment* 232 (October). https://doi.org/10.1016/j.rse.2019.111291.

White, Michael A, Peter E Thomton, and Steven W Running. 1997. "A continental phenology model for monitoring vegetation responses to interannual climatic variability." *Global Biogeochemical Cycles* 11 (2): 217–34.

Wilhite, Donald A., and Michael H. Glantz. 1985. "Understanding: The drought phenomenon: The role of definitions." *Water International* 10 (3): 111–20. https://doi.org/10.1080/02508068508686328.

Wollenberg, E., S. J. Vermeulen, E. Girvetz, A. M. Loboguerrero, and J. Ramirez-Villegas. 2016. "Reducing risks to food security from climate change." *Global Food Security* 11: 34–43. https://doi.org/10.1016/j.gfs.2016.06.002.

Yang, Mingxia, Yuling Mou, Yanrong Meng, Shan Liu, Changhui Peng, and Xiaolu Zhou. 2020. "Modeling the effects of precipitation and temperature patterns on agricultural drought in China from 1949 to 2015." *Science of the Total Environment* 711. https://doi.org/10.1016/j.scitotenv.2019.135139.

Zambrano, Francisco, Mario Lillo-Saavedra, Koen Verbist, and Octavio Lagos. 2016. "Sixteen years of agricultural drought assessment of the Biobío region in Chile using a 250 m resolution vegetation condition index (VCI)." *Remote Sensing* 8 (6): 1–20. https://doi.org/10.3390/rs8060530.

Zambrano, Francisco, Brian Wardlow, Tsegaye Tadesse, Mario Lillo-Saavedra, and Octavio Lagos. 2017. "Evaluating satellite-derived long-term historical precipitation datasets for drought monitoring in Chile." *Atmospheric Research* 186 (April): 26–42. https://doi.org/10.1016/j.atmosres.2016.11.006.

Zambrano, Francisco, Anton Vrieling, Andy Nelson, Michele Meroni, and Tsegaye Tadesse. 2018. "Prediction of drought-induced reduction of agricultural productivity in Chile from MODIS, rainfall estimates, and climate oscillation indices." *Remote Sensing of Environment* 219 (December): 15–30. https://doi.org/10.1016/j.rse.2018.10.006.

Zambrano, Francisco, Mauricio Molina, Alejandro Venegas, Julio Molina, and Paulina Vidal. 2020. "Impact of megadrought on vegetation productivity in Chile: Forest lesser resistant than crops and grassland." In: 2020 IEEE International Geoscience and Remote Sensing Symposium, September 26 - October 2, 2020 • Virtual Symposium, p. 1-5.

Zargar, Amin, Rehan Sadiq, Bahman Naser, and Faisal I. Khan. 2011. "A review of drought indices." *Environmental Reviews* 19 (1): 333–49. https://doi.org/10.1139/a11-013.

Zhang, Xiang, Nengcheng Chen, Jizhen Li, Zhihong Chen, and Dev Niyogi. 2017. "Multi-sensor integrated framework and index for agricultural drought monitoring." *Remote Sensing of Environment* 188 (January): 141–63. https://doi.org/10.1016/J.RSE.2016.10.045.

Zhao, Yuanyuan, Duole Feng, Le Yu, Xiaoyi Wang, Yanlei Chen, Yuqi Bai, H. Jaime Hernández, et al. 2016. "Detailed dynamic land cover mapping of Chile: Accuracy improvement by integrating multi-temporal data." *Remote Sensing of Environment* 183 (September): 170–85. https://doi.org/10.1016/j.rse.2016.05.016.

22 Application of Multisource Data for Drought Monitoring and Assessment over the Yellow River Basin, China

Yi Liu, Shanhu Jiang, and Liliang Ren

CONTENTS

22.1 INTRODUCTION

Droughts are long-lasting extreme events that have widespread impacts on a variety of sectors accompanied by serious economic losses (Wilhite and Glantz, 1985). From its physical causative mechanism, the formation and development process of a drought is closely related to water cycling, which involves complex transformations among different hydrologic fluxes and state variables (e.g., precipitation, evapotranspiration, soil moisture, and runoff). According to the usable water concerned, droughts can be classified into four types, namely, meteorological, hydrological, agricultural, and socioeconomic (Mishra and Singh, 2010). This multifaceted nature of drought highlights the necessity of considering multiple hydrometeorological data for reliable drought monitoring and drought assessment.

Hydrometeorological observations from ground meteorological stations, *in situ* agricultural sites, and streamflow gauges are commonly recognized as "ground truth" values that ideally can provide accurate materials for estimating the moisture status. However, due to the limitations of measuring instruments and observation technology, hydrological observations are sometimes found of poor quality, short temporal coverage, and discontinuous, which largely impedes their applications for drought analysis (Ford et al., 2015; Zhu et al., 2016; Peng et al., 2017). Especially for soil moisture and runoff, the lack of dense observation networks in data-sparse regions and large spatial heterogeneity of such variables make the site-based observations generally less representative for indicating the moisture condition over the spatial range.

In recent years, the development of remote sensing techniques and land surface models has promoted the popularization of satellite observations and hydrological simulations for drought monitoring. Different from site-based observations, satellite observations are data series with wide spatial coverage and high horizontal resolution, making them more suitable for tracking the moisture dynamics over the areal extent. A variety of satellite products derived from different spaceborne microwave instruments have been released. For instance, the Tropical Rainfall Measuring Mission (TRMM), Precipitation Estimation from Remotely Sensed Information using Artificial Neural Networks – Climate Data Record (PERSIANN-CDR), Climate Hazards Group Infrared Precipitation with Stations (CHIRPS), Climate Precipitation Center Morphing Method (CMORPH), and TRMM Multisatellite Precipitation Analysis (TMPA) are popular satellite precipitation products widely used for hydrological applications. Based on different systems and mission designs, satellite soil moisture products also experienced continuous updates. The Soil Moisture Ocean Salinity (SMOS), launched in November 2009, is the first satellite devoted to soil moisture measurements (Kerr et al., 2010). In January 2015, the Soil Moisture Active Passive (SMAP) continued to provide combined information on global surface soil moisture with the sensors of an active L-band radar and a passive L-band radiometer (Entekhabi et al., 2010). The soil moisture retrieval data set released by the European Space Agency Climate Change Initiative program (denoted as ESA CCI SM) is representative of the integrated products that utilize the synergy between active and passive products for estimating soil moisture (Dorigo et al., 2011; Albergel et al., 2012; Taylor et al., 2012; Dorigo et al., 2017). In spite of this progress, satellite observations are sometimes intermittent due to the effects of clouds and sensor errors, which limits their applications. The hydrological simulations provide another alternative for obtaining continuous data series by simulating the hydrological process at fine spatiotemporal resolutions. The Global Land Data Assimilation System (GLDAS) and North America Land Data Assimilation System (NLDAS) are two representatives that provide long-term simulations of multiple hydrological variables at different scales. They also serve as valuable references for intermittent data gap filling of satellite observations and data assimilation (Xia et al., 2014). At the same time, uncertainties that originate from model parameterization, model structure, and external forcing should not be neglected when using these products (Raoult et al., 2018).

This chapter aims at evaluating the capabilities of satellite observations and hydrological model simulations in drought monitoring. Three precipitation products –PERSIANN-CDR, CHIRPS, and the Global Precipitation Climatology Centre (GPCC) gridded gauge analysis precipitation data set – were selected to evaluate their performances in meteorological drought monitoring. A weekly standardized soil moisture index was proposed by using the ESA CCI SM product, and the applicability of the new drought index for drought characterization was analyzed by comparing its performance with other reference indices. Based on the hydrological simulations of the variable infiltration capacity (VIC) model constructed in the Yellow River Basin in China, we compared two methods of flash drought identification and explored the role of hydrometeorological variables in formulating flash droughts.

22.2 METHODOLOGY

In this section, detailed formulas for constructing standardized drought indices and flash drought identification are provided.

22.2.1 STANDARDIZED DROUGHT INDEX

The Standardized Precipitation Index (SPI) is representative of standardized drought indices (SI), which use a probabilistic approach to recognize the multiscalar nature of droughts with flexible timescales available (Mckee et al., 1993). This mathematical normalization algorithm was employed by other standardized drought indices, such as the Standardized Precipitation Evapotranspiration Index (SPEI), Standardized Soil Moisture Index (SSI), and Standardized Runoff Index (SRI), and they share the same drought classification criterion as the SPI. The difference among these standardized drought indices mainly lies in the variables considered for calculating moisture deficiency (d). The SPI considers the variation of precipitation, whereas for the SPEI, both precipitation and potential evapotranspiration are incorporated for measuring d; likewise, soil moisture content for the SSI and runoff for the SRI (McKee et al., 1993; Vicente-Serrano et al., 2010; Sheffield et al., 2004; Shukla and Wood, 2008). The timescale of a standardized drought index is related to the accumulation (or average) month of d. To remove the impact of seasonality, the original time series of d were firstly subdivided into 12 monthly series, and for each month, the accumulated d can be derived as

$$d_{i,j}^{k} = \sum_{l=13-k+j}^{12} d_{i-1,l} + \sum_{l=1}^{j} d_{i,l} \qquad j < k$$

or (22.1)

$$d_{i,j}^{k} = \sum_{l=j-k+1}^{j} d_{i,l} \qquad j \geq k$$

where k is the predefined timescale, which commonly ranges from 1 to 48 months; and $d_{i,j}$ denotes the moisture departure in the month j of year i.

Then a theoretical probability distribution, such as the three-parameter log-logistic, generalized extreme value (GEV), Pearson type III, or lognormal probability, was chosen to fit the derived d series. The probability density function and its cumulative form are

$$f(x; \mu, \sigma, \xi)$$

(22.2)

$$F(x; \mu, \sigma, \xi)$$

where μ, σ, and ξ are the corresponding locations, scale, and shape parameters of a probability distribution, respectively; and $f(x; \mu, \sigma, \xi)$ and $t(x)$ are the probability density function (PDF) and cumulative density function (CDF), respectively. The performance of the selected theoretical probability distribution was evaluated based on the Kolmogorov–Smirnov test. Following the classical approximation standardization method, a standardized drought index can easily be obtained as the standardized values of a cumulative density function $F(x)$:

$$SI = W - \frac{a_0 + a_1 W + a_2 W^2}{1 + b_1 W + b_2 W^2 + b_3 W^3}$$

(22.3)

where $p = 1 - F(x) \leq 0.5$ and $W = \sqrt{-2\ln(p)}$. If $p > 0.5$, then $p = 1 - p$ and $SI = 1 - SI$. Values of constants are set as $a_0 = 2.515517$, $a_1 = 0.802853$, $a_2 = 0.010328$, $b_1 = 1.432788$, $b_2 = 0.189269$, and $b_3 = 0.001308$.

The preceding steps can also be extended to the weekly times series of hydrological variables (i.e., the original time series are processed into 52 weekly series) to generate a weekly standardized drought index. In this chapter, the monthly SPI, and weekly SPI, SPEI, and SSI were used to indicate the process of droughts. In addition, a weekly self-calibrating Palmer Drought Severity Index, which couples the variable infiltration capacity model in its hydrologic accounting section (denoted as VIC-scPDSI) (Zhu et al., 2018), was employed to evaluate the performance of weekly standardized drought indices in drought monitoring.

22.2.2 INTENSIFICATION RATE–BASED FLASH DROUGHTS IDENTIFICATION

Soil moisture has a close relationship with vegetation conditions and is commonly selected for flash drought monitoring and assessments. In this chapter, we propose a quantitative method to identify flash droughts by focusing on the rate of intensification (RI) (Liu et al., 2020). Unlike a traditional, slowly evolving drought, a flash drought is characterized by the rapid depletion of soil moisture resulting from a period of abnormally warm and dry weather conditions. In this sense, flash drought is most likely to occur in the onset-development phase of the drought event. As shown in Figure 22.1a, suppose t_1 is the onset time that the soil layer is experiencing the abnormally dry conditions and has the potential to precede a drought, and the time node t_5 represents the stationary point that moisture deficits suffer abrupt changes from rapid decline to smooth fluctuations (e.g., $t_6 \sim t_{15}$ in Figure 22.1a) or even present an increased pattern (e.g., $t_{15} \sim t_{18}$ in Figure 22.1a). In other words, the stationary point can be viewed as the termination of rapid soil moisture reduction, which may emerge at or before the peak of drought intensity. With this in mind, the identification of flash droughts is further generalized as two questions: How do you extract the onset-development phase (i.e., the time period from $t_1 \sim t_5$) of droughts? How fast should the intensification be to be recognized as a flash drought?

The weekly soil moisture percentiles were employed to depict the drought process (Figure 22.1b). Specifically, drought events are extracted when the soil moisture falls below a predetermined value. Similar to previous studies, the threshold adopted here is twofold: (a) soil moisture is less than the 40th percentile, and (b) the peak drought intensity must fall below the 20th percentile (Otkin et al., 2018). The onset time (i.e., t_1 in Figure 22.1a) of a flash drought event, therefore, is defined as the first week when the soil moisture falls below the 40% percentile. As for the stationary point (i.e., t_5 in Figure 22.1a), a univariate polynomial function is employed to determine its location along the horizontal axis (namely, the date in Figure 22.1a). When a linear regression function is employed, the peak of drought intensity is chosen as the stationary point. With respect to the nonlinear case, we increase the order of the polynomial in sequence (e.g., linear polynomials, quadratic, cubic, …, nth-order polynomial) until a minimum value of 0.95 for the deterministic coefficient R^2 of a fitting polynomial is attained. In calculus, the stationary point (i.e., t_5 in Figure 22.1a) can be located when the first derivative of the constructed polynomial equals 0 (i.e., $\dfrac{\partial Y}{\partial X} = 0$). In other words, the polynomial function is only used to judge when the flash drought event terminates but would not participate in estimating the magnitude of the flash drought intensification. With the extracted stationary point, the mean intensification rate (RI_{mean}) of a drought event during its onset-development phase can be calculated as

$$RI_{mean} = \frac{1}{n} \sum_{i=1}^{n} \left[\frac{SM(t_{i+1}) - SM(t_i)}{t_{i+1} - t_i} \right], \quad t_1 \leq t_i \leq t_n \tag{22.4}$$

where t_1 denotes the onset time, and t_n represents the stationary point. A similar concept of RI_{mean} is also employed in several previous studies. For instance, Ford and Labosier (2017) recommended that the soil moisture content dropping from the 40th percentile to below the 20th percentile in no less than 4 pentads (equivalent to a RI_{mean} value of 6.5 percentile/week) could be recognized as flash

(a) Development process of flash drought

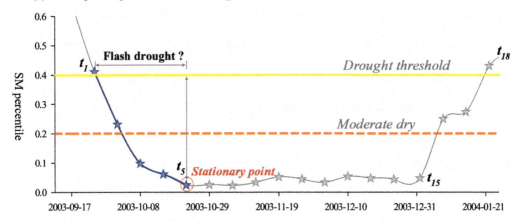

Problem 1: how to extract the onset-development phase (i.e., the time period of $t_1 \sim t_5$) of droughts?

Problem 2: how fast the intensification is could be recognized as a flash drought?

(b) Framework of flash drought identification

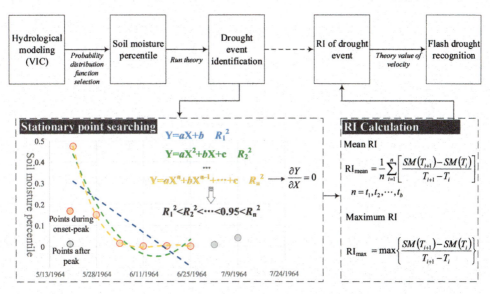

FIGURE 22.1 (a) Schematic overview of the evolution process of flash drought indicated by soil moisture percentile. t_1 represents the onset time, t_5 denotes the stationary point where rapid depletion of soil moisture ends, and the period from t_6 to t_{18} represents the recovery stage where soil moisture increases gradually then returns to the normal condition. (b) Generalized flowchart of identifying flash drought based on intensification rate. The left dashed rectangle shows how to find the stationary point with the polynomial fitting method. The left rectangle gives the formulas for calculating the rate of intensification.

droughts. However, in reality, there may exist some exceptional cases where the RI_{mean} is lower than the predetermined value (i.e., 6.5 percentile/week) but with rather high instantaneous RI_i values at a certain moment. We determined that such cases should also be considered flash droughts. On these grounds, the instantaneous maximum intensification rate (RI_{max}) during the onset-development phase is introduced:

$$RI_{\max} = \max \left\{ \frac{SM(t_{i+1}) - SM(t_i)}{t_{i+1} - t_i} \right\}, \quad t_1 \leq t_i \leq t_n \tag{22.5}$$

Flash drought is recognized when either the condition of RI_{mean} or RI_{max} is met over a sufficiently long enough period of time.

In addition, Mo and Lettenmaier (2015) defined two types of flash droughts, i.e., heat wave and precipitation deficit flash droughts (denoted as HWFD and PDFD, respectively), and they were compared with flash droughts identified by RI (denoted as RIFD):

$$HWFD : T_{anomaly} > \sigma; AET_{anomaly} > 0; P_{anomaly} < 0; SM < 40\% \, percentile \tag{22.6}$$

$$PDFD : P < 40\% \, percentile; AET_{anomaly} < 0; T_{anomaly} > \sigma; SM < 40\% \, percentile \tag{22.7}$$

where $T_{anomaly}$, $P_{anomaly}$, and $AET_{anomaly}$ represent the anomaly of weekly air temperature, precipitation, and AET, respectively; and σ represents the standard deviation of $T_{anomaly}$ series.

22.3   CASE STUDY

22.3.1   STUDY AREA

The Yellow River Basin (YRB) is located in northern China, extending from 96° E–119° E longitude and 32° N–42° N latitude. The river originates from the Tibetan Plateau and flows into the Bohai Sea through nine provinces. With a total length of 5464 km, the Yellow River controls a drainage area of 752,440 km², ranking as the second largest river in China. The elevation of the YRB ranges from 0 to 6253 m above the mean sea level, with topography generally decreasing from northwest to southeast with an average slope of 2.4‰. The Tibetan Plateau, Loess Plateau, and flood plains are the three main geomorphic types. The climate of the YRB is complex and varies greatly. Seasonally, it is wet and rainy in summer, and cold and dry in winter. The spatial distribution of precipitation in the area decreases from the southeast toward the west, and the average annual precipitation is 466 mm. The average annual temperature is between −4°C and 14°C.

22.3.2   DATA SETS

Three precipitation products, i.e., PERSIANN-CDR, CHIRPS V2.0, and GPCC V8.0 – were used to explore and compare their drought monitoring utilities in this region during 1983–2016. The PERSIANN-CDR, with a resolution of 0.25° and spatial coverage of 60° N/S, was developed by the Center for Hydrometeorology and Remote Sensing at the University of California, Irvine, and advised by the 2.5° monthly Global Precipitation Climatology Project data. CHIRPS V2.0 was designed by the Climate Hazards Group and obtained by merging the IR-derived precipitation and the ground station observations (over 20,000 gauges). The temporal and spatial resolution of the CHIRPS v2.0 product is 0.25° and 50° N/S coverage land from 1981 to the present. The latest GPCC v8.0 monthly precipitation product is derived from approximately 80,000 meteorological stations around the world, and it provides precipitation data with multiple spatial resolutions and a global scale.

Two sources of soil moisture products, the ESA CCI soil moisture (ESA CCI SM) product and the VIC simulations, were employed for drought analysis. The ESA CCI SM was developed by the European Space Agency's Climate Change Initiative program. One appealing advantage of this product is that it uses a sophisticated processing chain to synergistically combine the strengths of records from multiple active and passive microwave sensors by matching the cumulative distribution function. The ESA CCI SM has a rather wide spatiotemporal coverage and is updated at regular intervals. It releases daily surface soil moisture records with $0.25° \times 0.25°$ spatial resolution. The ESA CCI SM COMBINED data sets (v03.3) during 2000–2016 were employed. In addition, the VIC model was run at a spatial resolution of $0.25°$ at a daily time step, calibrated during 1961–1990, and validated during 1991–2012. Generally, the model can capture most of the streamflow variability with the Nash–Sutcliffe coefficients of efficiency varying between 0.65 and 0.94, and the absolute values of relative bias ranging from 0.3% to 10.8%. As for soil moisture verification, the VIC-simulated daily values are aggregated to ten-day resolution and compared with agricultural observations. The correlation coefficients (CCs) are mostly above 0.5, with higher CC values (up to 0.75) concentrated in the middle parts of the YRB. We also compared the distribution tails of soil moisture percentiles derived from VIC simulations against those of *in situ* observations. The probability of detection for four intervals on average is around 0.5~0.6.

In addition, daily meteorological forcing data, including precipitation, maximum and minimum temperature, and average wind speed from the national meteorological stations, were provided by the China Meteorological Data Sharing Service System (http://cdc.cmc.gov.cn/home.do). The third-generation Normalized Difference Vegetation Index (NDVI) data set is derived from NOAA's Advanced Very High-Resolution Radiometer (AVHRR) sensor and released by the Global Inventory Modeling and Mapping Studies (GIMMS) group. The data covers the period 1982–2016 with a spatial resolution of $0.083°$ and a temporal resolution of 16 days, and it can be obtained from the NASA Ames Ecological Forecasting Lab.

Four evaluation coefficients, i.e., the correlation coefficient (CC), root-mean-squared error (RMSE), mean error (ME), and relative bias (BIAS), were employed to evaluate the deviations between different products.

22.3.3 SATELLITE-BASED PRECIPITATION PRODUCTS IN DROUGHT MONITORING

22.3.3.1 Evaluation of the Accuracy of Satellite Precipitation and Reanalysis Precipitation

Figure 22.2 shows the spatial distributions of CC, BIAS, RMSE (mm/month), and ME (mm/month) of the PERSIANN-CDR, CHIRPS, GPCC 8.0 versus China gauge-based monthly precipitation data (CGDPA) over the YRB. It can be seen that the CC values of GPCC 8.0 mostly vary between 0.96 and 1, indicating a good agreement with the China gauge-based monthly precipitation data. For CHIRPS, the CC values mostly range from 0.9 to 0.96 (with several grids reaching up to 0.96–1), and the poor performance was mainly distributed in the northwestern boundary area of the YRB with the lowest CC values. With respect to the PERSIANN-CDR, the CC values in most regions are above 0.86, but almost no grid presents a high value (CC >0.96). Most BIASs range from –0.4 to 0.4, and the values in most areas are in the range –0.1 to 0.1. In the northwest of the Yellow River Basin, the RMSEs of the satellite-derived products mainly range between 10 and 20 mm/month, and vary from 0 to 10 mm/month for GPCC 8.0. The MEs of the three satellite precipitation products are mainly in the range of –5 to 5 mm/month, and the CHIRPS product performs best. The above analysis suggest that the GPCC 8.0 precipitation product has a better consistency with the China gauge-based monthly precipitation data than CHIRPS and PERSIANN-CDR.

22.3.3.2 Satellite Precipitation and Reanalysis Precipitation for Drought Monitoring

Given the general good agreements between the PERSIANN-CDR, CHIRPS, and GPCC 8.0 and the CGDPA product (Figure 22.3), we further analyzed the performances of the three precipitation products in drought monitoring. The 1-, 3-, and 12-month SPI (denoted as SPI-1, SPI-3, and SPI-12, respectively) were calculated using the PERSIANN-CDR, CHIRPS, GPCC 8.0, and CGDPA products. As shown in Figure 22.3, the CC values of the PERSIANN-CDR, CHIRPS, and GPCC 8.0 in most areas of the YRB vary between 0.65–0.85, 0.70–0.85 and 0.90–0.99, respectively, implying that there is a good agreement between the SPI series derived from the satellite-based and reanalysis precipitation products and the CGDPA. In addition, the three precipitation products present

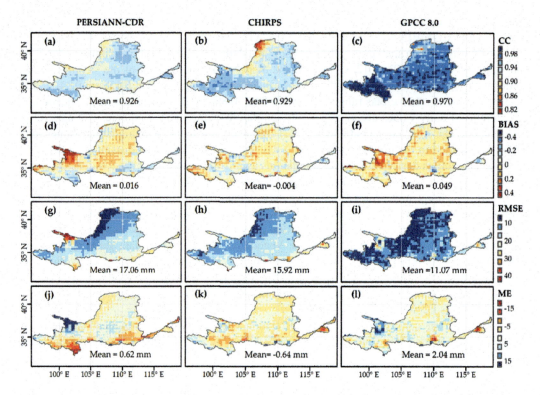

FIGURE 22.2 Spatial distributions of CC, BIAS, RMSE (mm/month), ME (mm/month) of PERSIANN-CDR, CHIRPS, and GPCC 8.0 versus China gauge-based monthly precipitation data (CGDPA) over the YRB.

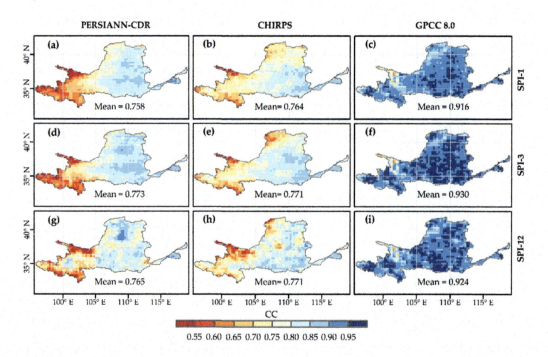

FIGURE 22.3 Spatial distributions of the CC value of SPI-1, SPI-3, and SPI-12 derived from the PERSIANN-CDR, CHIRPS, and GPCC 8.0 versus those from the CGDPA over the YRB.

TABLE 22.1

Drought Characteristics of the Typical 2007 Drought Event over the YRB

Timescale of SPI (Month)	Product	Time Range (Year, Month)	Duration (Month)	Minimum of SPI
1 month	CGDPA	August 1997	1	−1.530
	PERSIANN-CDR	August 1997	1	−1.665
	CHIRPS	August 1997	1	−1.691
	GPCC 8.0	August 1997	1	−1.574
3 months	CGDPA	June 1997 to October 1997	5	−1.390
	PERSIANN-CDR	June 1997 to October 1997	5	−1.481
	CHIRPS	May 1997 to October 1997	6	−1.497
	GPCC 8.0	May 1997 to December 1997	8	−1.452
12 months	CGDPA	August 1997 to May 1998	10	−1.212
	PERSIANN-CDR	August 1997 to May 1998	10	−1.325
	CHIRPS	July 1997 to May 1998	11	−1.410
	GPCC 8.0	August 1997 to May 1998	10	−1.251

similar performances in terms of the timescales of the SPI, where the CC values of SPI-3 are generally higher than those of SPI-1 and SPI-12 across the entire basin. Poor performances of the satellite-derived precipitation products in drought monitoring are mainly distributed in the areas with high elevations, such as the western YRB, and this phenomenon is particularly significant for the PERSIANN-CDR.

We further selected the 1997 drought event to evaluate the performances of the three precipitation products in drought monitoring. Table 22.1 exhibits the onset time, termination time, drought duration, and minimum values of the SPI identified by the four precipitation products. It can be seen that at the 1- and 12-month scales, the GPCC 8.0 is able to capture the onset and termination time of this drought event, and performs best among the three precipitation products. As for the PERSIANN-CDR, except for the considerable differences in the minimum SPI value, the time range and duration of the drought identified by the PERSIANN-CDR generally match with those from the CGDPA at three different timescales. With respect to the CHIRPS product, except for the 1-month timescale, the drought characteristics identified by the CHIRPS product do not agree with those from the CGDPA product. Overall, the GPCC 8.0 product has the best agreement at multiple timescales, followed by PERSIANN-CDR, and these two products are capable alternatives for drought monitoring.

22.3.4 ESACCI Soil Moisture–Based Drought Index

This section focuses on the performance of the ESACCI soil moisture product for drought monitoring. A weekly standardized soil moisture index (denoted as ESA CCI SSI) was derived from the ESA CCI soil moisture data sets, and the time series of the drought index, drought area, and the spatial evolution pattern of the 2002 drought event indicated by the ESA CCI SSI were compared with other three drought indices.

22.3.4.1 Temporal Variability of Drought Indices

Figure 22.4 compares the time series of the standardized soil moisture index derived from the ESA CCI soil moisture data sets (denoted as ESA CCI SSI) and three reference indices (i.e., SPI, SPEI, and VIC-scPDSI) at a weekly timescale. It can be seen that the ESA CCI SSI was significantly ($p < 0.05$) correlated with the other three drought indices, and the correlation with VIC-scPDSI was better than

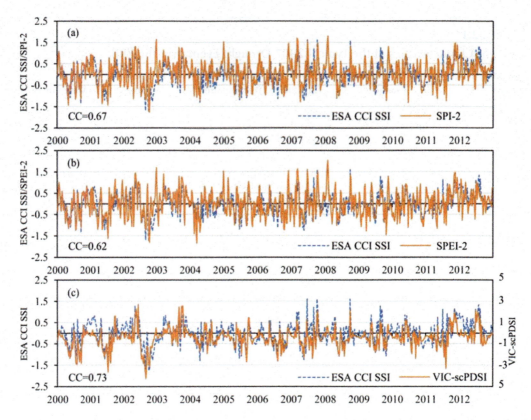

FIGURE 22.4   Time series of ESA CCI SSI against (a) SPI, (b) SPEI, and (c) VIC-scPDSI values averaged over the whole Yellow River Basin during 2000–2012.

the SPI and SPEI, with the CC values of 0.73, higher than that of 0.67 and 0.62, respectively. Comparing the time series of the SPI and SPEI, there was an obvious time lag for the ESA CCI SSI, implying that the agricultural drought detected by the ESA CCI SSI might occur one week later than the meteorological drought reflected by the SPI or SPEI. Additionally, the ESA CCI SSI and VIC-scPDSI had a good temporal consistency during 2000–2012, however, higher values were generally identified by the former, while lower values were usually captured by the latter. This indicated that the drought severity measured by the VIC-scPDSI was slightly more severe than that indicated by ESA CCI SSI.

22.3.4.2   Comparison on Drought Area

The weekly drought area ratio (DAR) for mild to extreme droughts indicated by the ESA CCI SSI, SPI, SPEI, and VIC-scPDSI are displayed in Figure 22.5. Overall, the four indices present the similar pattern that the mild and moderate drought areas were much larger than that at severe and extreme drought. Taking the case of the ESA CCI SSI as an example, the DAR at mild and moderate drought on average was 15.0% and 9.5%, respectively, while that at severe and extreme drought was 5.5% and 1.7%, respectively. This suggested that the YRB was more vulnerable to mild and moderate droughts during 2000–2012.

22.3.4.3   Spatial Evolution of the 2002 Drought

Figure 22.6 compares the spatial evolution of the 2002 drought event indicated by the ESA CCI SSI, SPI, SPEI, VIC-scPDSI, NDVI anomaly, and ESA CCI SM anomaly. Overall, the ESA CCI SSI and ESA CCI SM anomaly presented a highly spatial consistency. The spatial migration pattern of the ESA CCI SSI was generally consistent with that of the SPI and SPEI for June 9–15 and June 23–29,

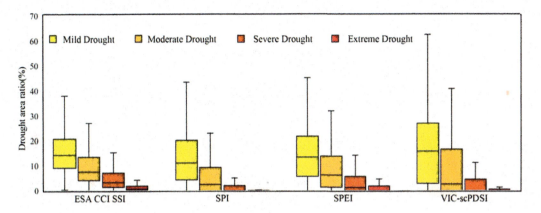

FIGURE 22.5 Box plots of drought area ratio (DAR) for mild to extreme droughts indicated by the ESA CCI SSI, SPI, SPEI, and VIC-scPDSI over the Yellow River Basin during 2000–2012.

however, there was a significant difference during June 16–22. Specifically, the mild and moderate drought indicated by the SPI and SPEI were mainly observed at limited areas (i.e., the central and western regions), while that detected by the ESA CCI SM was distributed in a wide range of the YRB. There was a similar spatial distribution of ESA CCI SSI and VIC-scPDSI, however, the drought captured by the latter was severer than that measured by the former form the perspective of both drought area and drought intensity. In addition, a discrepancy of spatial pattern for the ESA CCI SSI and NDVI anomaly was found in the study period. This might be attributed to the limited information obtained by the SM signal from microwave remote sensing on the deep root zone, where SM is available for the plants (Nicolai-Shaw et al., 2017).

22.3.5 FLASH DROUGHT IDENTIFICATION AND CHARACTERIZATION

According to the method mentioned in Section 22.2.2, flash drought events indicated by two methods were extracted using the VIC model simulated soil moisture series over the YRB during 1961–2012. Figure 22.7 shows the spatial distribution of the frequency of occurrence (FOC; the ratio of weeks under flash drought) for flash drought indicated by the two different methods. For RIFD, the FOC on average varied between 3% and 10%, with high occurrence of flash drought concentrated in the northwestern, southeastern, and southern regions (Figure 22.7a). As for the HWFD and PDFD, different patterns of FOC were observed both in the magnitude and spatial distributions. Specifically, the FOC of HWFD on average was less than 3%, with high values located in the southern region. With respect to the PDFD type, the magnitude of FOC increased to 4%–6%, and spatially the source region experienced a lower flash drought frequency. Taking HWFD and PDFD as a whole (i.e., events of HWFD and PDFD were merged into a new sample for calculating the FOC), the FOC ranged between 5% and 10%, and presented an increased pattern from northwest to southeast (Figure 22.7d). This suggests that different ways for identifying flash droughts would affect drought frequency to some extent. In addition, Wang and Yuan (2018) also analyzed flash drought frequency in China based on the method of Mo and Lettenmaier (2015), and suggested a similar spatial distribution of HWFD and PDFD with our results, but the values of FOC were generally higher. The different source of soil moisture data set (global reanalysis products were employed in their study) might be one possible reason. The time period of interest may be another possible reason (Liu et al., 2016) because their study mainly reflects the overall dry status during 1979–2010, which has a different climate condition than 1961–2012.

To explore how the flash drought identification methods behave in monitoring the spatial evolution of soil moisture, the 1991 flash drought event is selected for further analysis. As shown in

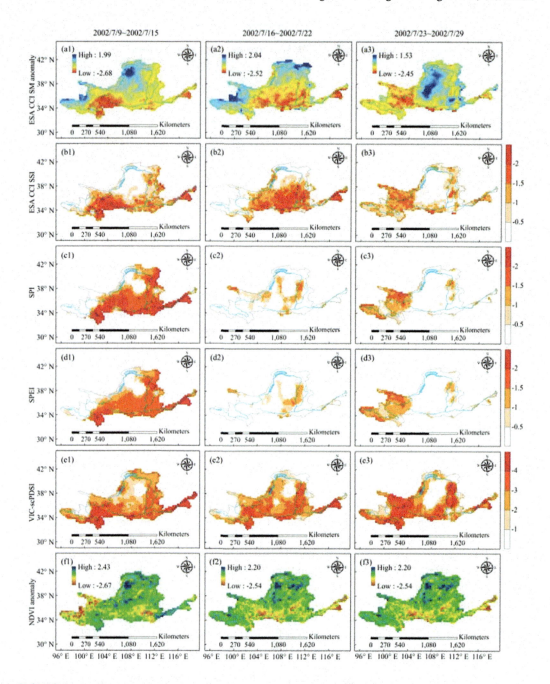

FIGURE 22.6 Spatial evolution of the 2002 drought indicated by (a1–a3) ESA CCI SM anomaly, (b1–b3) ESA CCI SSI, (c1–c3) SPI, (d1–d3) SPEI, (e1–e3) VIC-scPDSI, and (f1–f3) NDVI anomaly.

Figure 22.8a, the drought event indicated by soil moisture percentile experienced continuous reduction starting in June 1991, and sustains a rather low value until the next year. During this process, the period from June to July exhibits a rapid rate of intensification during which the soil moisture percentile sharply decreases from above the 40th percentile to below the 20th percentile over four weeks. Meanwhile, a spatial migration pattern is apparent for the dry condition. As shown in Figure 22.8b, the drought center generally experiences a western–eastern–northern shifting mode, with the

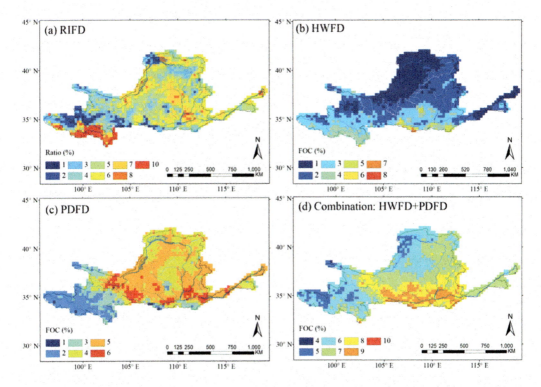

FIGURE 22.7   Spatial distribution of the frequency of occurrence (FOC) for flash droughts indicated by (a) RIFD, (b) HWFD, (c) PDFD, and (d) sum of HWFD and PDFD (i.e., events of HWFD and PDFD were merged into a new sample for calculating FOC).

vegetation types of grassland, cropland, and shrubland suffering in sequence. This spatial variation of vegetation condition is also reflected by the 15-day NDVI anomalies (Figure 22.8c). In the following section, we pay special attention to the abilities of each method to depict the drought migration pattern, as well as in capturing the instantaneous variation of soil moisture.

The initial drought patches, as indicated by the soil moisture percentile, emerged in the source region of the YRB on June 11 (Figure 22.8d). This dry signal is also captured by RI_{mean} and RI_{max}, but missed by HWFD and PDFD. In the following week (June 18), the latter method again fails to display dry conditions. According to the NDVI anomalies (Figure 22.8c), some parts of the source region exhibited negative values on June 21, implying that the vegetation health starts to decline, which also signifies the onset of this flash drought event. In such cases, we consider that the ways of defining HWFD and PDFD are less competent to provide early warning for flash droughts.

22.4   CONCLUSIONS

In this chapter, the Yellow River Basin in China was selected as the research domain and a case study on the applicability of satellite observations and hydrological model simulations in drought characterization was conducted.

The long-term PERSIANN-CDR, CHIRPS, and GPCC 8.0 precipitation products were employed to derive the SPI time series, and their performances in meteorological drought monitoring were compared with the SPI series calculated from the CGDPA gauge measurements during 1983–2016. In most parts of the YRB, the GPCC 8.0 product, which incorporates precipitation data from a greater number of meteorological stations, presents a slightly higher consistency with the CGDPA than the PERSIANN-CDR and CHIRPS products. The two satellite-derived precipitation

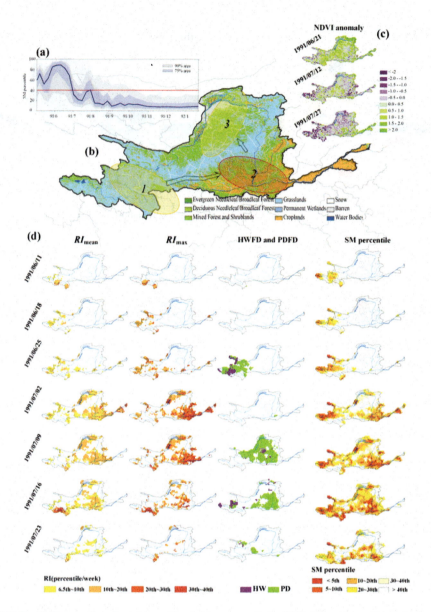

FIGURE 22.8 (a) Time series of weekly soil moisture percentile in 1991. The gray and blue shadows denote the 75th–90th percentile range of soil moisture values over YRB. (b) The three circles with numbers in the middle indicate the moving path of the drought center during June–July in 1991. (c) Spatial evolution of NDVI anomalies (dimensionless, minus the mean then divided by the standard deviation) in June and July 1991. (d) Spatial evolution of the mean and maximum intensification rate of soil moisture, two types of flash droughts defined by Mo and Lettenmaier (2015), and soil moisture percentile in June and July 1991.

products do not perform well in areas of high elevation, such as the western YRB, especially for the PERSIANN-CDR product. Given the high accuracy of precipitation detection, the GPCC 8.0 product could be considered a valuable long-term precipitation data set for analyzing the historical droughts in the YRB. The GPCC 8.0 product could also be used for data correction of the inversion of satellite-derived remote sensing precipitation.

A weekly standardized soil moisture index was derived from the active-passive combined ESA CCI SM product, and its performance for characterizing droughts was compared to the ESA CCI

SM anomalies, the SPI, the SPEI, VIC-scPDSI, and NDVI anomalies. Among the reference indices, ESA CCI SSI presented a relatively higher consistency with VIC-scPDSI and a significantly delayed (one week) response with the SPI and SPEI. Severe and extreme drought areas indicated by the ESA CCI SSI are generally larger than are reflected by the SPI, SPEI, and VIC-scPDSI. An investigation of the 2002 drought event suggests that the ESA CCI SSI could capture the onset and development process of the drought event, and displays a similar spatial pattern with NDVI anomalies in some areas of the YRB. Overall, the ESA CCI SSI is able to reveal the temporal and spatial patterns of agricultural drought during a historical period (2000–2012) over the YRB.

Based on the hydrological simulations of the VIC model constructed in the YRB during 1961–2012, a new quantitative approach concerning the rate of soil moisture depletion during the onset-development phase is proposed, and its performance in terms of monitoring flash drought events, their onset time, and responses to the changes of vegetation are compared with those of PDFD and HWFD. Overall, the rapid intensification–based approach can effectively track the sudden change in moisture status and is recommended for use. Since the approach of PDFD and HWFD neglects the change of soil moisture with time, it cannot ensure that the identified flash droughts all have rapidly evolving characteristics. Considering our research is based on a case study, an evaluation of these flash drought identification methods for different regions and events is further needed. Findings from this study improve our understanding of flash drought, which also have some implications for promoting drought early warning techniques.

REFERENCES

Albergel, C., P. De Rosnay, C. Gruhier, J. Muñoz-Sabater, S. Hasenauer, L. Isaksen, Y. Kerr, and W. Wagner (2012), Evaluation of remotely sensed and modelled soil moisture products using global ground-based in situ observations, *Remote Sens. Environ.*, 118, 215–226.

Dorigo, W.A., et al. (2017), ESA CCI soil moisture for improved Earth system understanding: State-of-the art and future directions. *Remote Sens. Environ.*, 203, 185–215.

Dorigo, W. A., et al. (2011), The International Soil Moisture Network: A data hosting facility for global in situ soil moisture measurements, *Hydrol. Earth Syst. Sci.*, 15(5), 1675–1698.

Entekhabi, D., E. G. Njoku, P. E. Neill, K. H. Kellogg, W. T. Crow, W. N. Edelstein, J. K. Entin, S. D. Goodman, T. J. Jackson, and J. Johnson (2010), The Soil Moisture Active Passive (SMAP) mission, *Proc. IEEE*, 98(5), 704–716.

Ford, T. W., and C. F. Labosier (2017), Meteorological conditions associated with the onset of flash drought in the eastern United States. *Agric. For. Meteor.*, 247, 414–423, https://doi.org/10.1016/j.agrformet.2017.08.031.

Ford, T. W., D. B. McRoberts, S. M. Quiring, and R. E. Hall (2015), On the utility of in situ soil moisture observations for flash drought early warning in Oklahoma, USA. *Geophys. Res. Lett.*, 42, 9790–9798, https://doi.org/10.1002/2015GL066600.

Kerr, Y. H., P. Waldteufel, J.-P. Wigneron, S. Delwart, F. O. Cabot, J. Boutin, M.-J. Escorihuela, J. Font, N. Reul, and C. Gruhier (2010), The SMOS mission: New tool for monitoring key elements of the global water cycle, *Proc. IEEE*, 98(5), 666–687.

Liu, Y., L. Ren, Y. Hong, Y. Zhu, X. Yang, F. Yuan, and S. Jiang (2016), Sensitivity analysis of standardization procedures in drought indices to varied input data selections. *J. Hydrol.*, 538, 817–830.

Liu, Y., Y. Zhu, L. Ren, J. Otkin, E. D. Hunt, X. Yang, F. Yuan, and S. Jiang (2020), Two different methods for flash drought identification: comparison of their strengths and limitations. *J. Hydrometeor.*, 21, 691–704, https://doi.org/10.1175/JHM-D-19-0088.1.

McKee, T.B., N.J. Doesken, J. Kleist (1993), The relationship of drought frequency and duration to time scales. In: Paper Presented at Eighth Conference on Applied Climatology, American Meteorological Society, Anaheim, CA.

Mishra, A.K., V.P. Singh (2010), A review of drought concepts. *J. Hydrol.*, 391 (1–2), 202–216.

Mo, K. C., and D. P. Lettenmaier (2015), Heat wave flash droughts in decline. *Geophys. Res. Lett.*, 42, 2823–2829, https://doi.org/10.1002/2015GL064018.

Nicolai-Shaw N., J. Zscheischler, M. Hirschi, L. Gudmundsson, and S. I. Seneviratne (Dec 2017), "A drought event composite analysis using satellite remote sensing based soil moisture," *Remote Sens. Environ.*, 203, 216–225, https://doi.org/10.1016/j.rse.2017.06.014.

Otkin, J. A., M. Svoboda, E. D. Hunt, T. W. Ford, M. C. Anderson, C. Hain, and J. B. Basara (2018), Flash droughts: A review and assessment of the challenges imposed by rapid onset droughts in the United States. *Bull. Amer. Meteor. Soc.*, 99, 911–919, https://doi.org/10.1175/BAMS-D-17-0149.1.

Peng, J., A. Loew, O. Merlin, and N. E. C. Verhoest (2017), A review of spatial downscaling of satellite remotely sensed soil moisture, *Rev. Geophys.*, 55, 341–366, https://doi.org/10.1002/2016RG000543.

Raoult, N., Delorme, B., Ottlé, C., Peylin, P., Bastrikov, V., Maugis, P., Polcher, J. (2018), Confronting soil moisture dynamics from the ORCHIDEE land surface model with the ESA-CCI product: Perspectives for data assimilation. *Remote Sens.*, 10, 1786.

Sheffield J, Goteti G, Wen F, Wood EF (2004), A simulated soil moisture based drought analysis for the United States. *J. Geophys. Res.*, 109, D24108.

Shukla, S., Wood, A.W. (2008), Use of a standardized runoff index for characterizing hydrologic drought. *Geophys. Res. Lett.*, 35, L02405.

Taylor, C.M., De Jeu, R.A.M., Guichard, F., Harris, P.P., Dorigo, W.A. (2012), Afternoon rain more likely over drier soils. *Nature*, 489, 282–286.

Vicente-Serrano, S.M., Begueria, S., Lopez-Moreno, J.I. (2010), A multiscalar droughti ndex sensitive to global warming: The standardized precipitation evapotranspiration index. *J. Clim.*, 23(7), 1696–1718.

Wang, L., and X. Yuan (2018), Two types of flash drought and their connections with seasonal drought. *Adv. Atmos. Sci.*, 35, 1478–1490.

Wilhite, D.A., Glantz, M.H. (1985), Understanding the drought phenomenon: The role of definitions. *Water Int.*, 10, 111–120.

Xia, Y., Ek, M.B., Mocko, D., Peters-Lidard, C.D., Sheffield, J., Dong, J., Wood, E.F. (2014), Uncertainties, correlations, and optimal blends of drought indices from the NLDAS multiple land surface model ensemble. *J. Hydrometeorol.*, 15, 1636–1650.

Zhu, Y., et al. (2018), Drought analysis in the Yellow River Basin based on a short-scalar Palmer Drought Severity Index. *Water*, 10(11), 1526.

Zhu, Y., et al. (2016), Combined use of meteorological drought indices at multi-time scales for improving hydrological drought detection. *Sci. Total Environ.*, 571, 1058–1068.

23 Analysis of Agricultural Drought in Southwest Iran Using Remote Sensing Indices

Mahshid Karimi, Kaka Shahedi, Tayeb Raziei,
Mirhassan Miryaghoubzadeh, and Ehsan Moradi

CONTENTS

23.1 INTRODUCTION

Drought is a worldwide phenomenon that has important effects on agriculture, economics, ecosystems services, energy, human health, recreation, and water resources (Wilhite, 2000; Wilhelmi and Wilhite, 2002; Keyantash and Dracup, 2004). So far, many definitions of drought have been proposed. However, the lack of a comprehensive and accurate definition of drought and the different meanings of it from different perspectives have prevented a full understanding of the concept of drought. Droughts are classified into four categories: meteorological, socioeconomic, agricultural, and hydrological (Dracup et al., 1980; Orville, 1990). Meteorological drought results from a decrease in precipitation, followed by agricultural drought that occurs when the soil moisture is less than the plant really needs. Hydrological drought leads to a decrease in surface and groundwater. Socioeconomic drought means the lack of water for human needs that causes social and economic anomalies (Zhang et al., 2017). Many drought indicators have been developed and proposed by researchers over the past decades, and researchers have been using these indicators to study the duration of drought, its severity, and the extent of it (Mishra and Singh, 2010). Among the most popular indicators are the Palmer Drought Severity Index (PDSI) (Palmer, 1965), the moisture anomaly index (Z-index) (McKee et al., 1995), the Standardized Precipitation Index (SPI) (McKee et al., 1993), and the Standardized Precipitation Evapotranspiration Index (SPEI) (Vicente-Serrano et al.,

2010). A limited number of measuring stations and their improper distribution have caused remote sensing data to be used for drought monitoring and drought index development (Vicente-Serrano et al., 2015; Bento et al., 2018). The remote sensing data provide substantial and beneficial information about precipitation, soil moisture, and the condition of vegetation, at a large scale (Rhee et al., 2010; Wardlow et al., 2012; Wu et al., 2013; Zhang et al., 2017; Bento et al., 2018). Diversity drought indices have been established, based on remote sensing data, such as the Vegetation Condition Index (VCI), generalized based on the Normalized Difference Vegetation Index (NDVI) (Kogan, 1995). Since drought is always accompanied by temperature anomalies, it led to the development of indicators such as the Temperature Condition Index (TCI), which are based on temperature (Kogan, 1995). Some researchers have also investigated the relationship between ground surface temperature (soil surface and canopy) and air temperature (Moran et al., 1994), thus introducing indicators based on vegetation and temperature such as the Vegetation Health Index (VHI) (Kogan, 2002), Vegetation Temperature Condition Index (VTCI) (Wang et al., 2004; Wan et al., 2004), and Temperature Vegetation Dryness Index (TDVI) (Sandholt et al., 2002). Many studies have been done on drought monitoring using indicators and remote sensing data. Mallick et al. (2009) estimated surface moisture content for cropped soils using a soil wetness index based on land surface temperature (LST), NDVI, and MODIS data sets in India. Du et al. (2013) examined meteorological and agricultural drought using MODIS data sets in Shandong province, China. They used these data to calculate the VCI, TCI, and Precipitation Condition Index (PCI). Finally, the Synthesized Drought Index (SDI) was derived from the aforementioned indicators.

Ghaleb et al. (2015) used remote sensing techniques and Landsat images in Lebanon to investigate drought from 1982 to 2014. They calculated the VHI based on the VCI and TCI, and classified it into five categories: extreme, severe, moderate, mild, and no drought. Amalo et al. (2017) estimated drought using MODIS data sets and remote sensing–based drought indices, namely the TCI, VCI, and VHI in East Java.

Baniya et al. (2019) studied spatial and temporal variations of drought using satellite data–derived VCI values on annual (January–December), seasonal monsoon (June–November), and pre-monsoon (March–May) scales from 1982 to 2015 in Nepal.

Zou et al. (2020) evaluated the performance of three remote sensing–based drought indices the VCI, TCI, and VHI – for drought monitoring in tropical dry forests (TDFs) using MODIS images. Also, the performance of all three indicators was assessed using the SPI (1, 3, 6, 9, 12, 15, 18, 21, and 24 months) for each month (January to December) and each season (dry season, dry-to-wet season, wet season, and wet-to-dry season). The results showed the TCI, VCI and VHI performed well in monitoring meteorological drought in the dry season, poorly in the dry-to-wet season, and moderately in the wet season, respectively.

Since Iran is located in an arid and semiarid region, with an average rainfall of about one-quarter of the world average (250 mm), drought is one of the major problems in the country (Arbabi Sabzevari, 2010). Reducing precipitation and its inappropriate spatial distribution has caused many parts of Iran to be in the dry climatic zone. In addition, the droughts that have occurred in recent years have made water supply in the country one of the most important challenges. The Karkheh watershed is also one of the places affected by the drought problem (Byzedi, 2011; Byzedi et al., 2012). The droughts that have occurred in this basin can cause many challenges in agricultural and economic sectors, and, eventually, environmental, economic, and social crises will occur in a large part of the country that is dependent on this basin. Therefore, the aim of this study is to investigate and evaluate agricultural drought in the Karkheh River Basin using the VCI, TCI, and SWI (Soil Water Index).

23.2 MATERIALS AND METHODS

23.2.1 STUDY AREA

The Karkheh basin, located in the middle and southwestern areas of the Zagros Mountains in Iran, lies between 30°08′ N–35°04′ N and 46°06′ E–49°10′ E (Figure 23.1). This basin stretches over

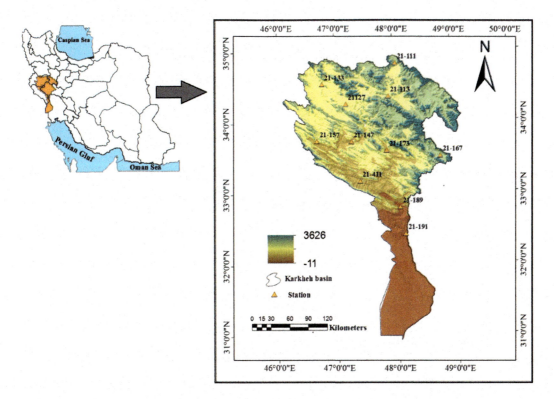

FIGURE 23.1   Location of the study area and rain gauge stations.

50,768 km^2 and approximately 55.5% of the basin is situated in the mountainous areas and the rest in the plains and foothills. The mean annual temperature varies from less than 5°C in the high mountains to 25°C in the southern regions. Figure 23.1 shows the location of the study area and rain gauge stations.

23.2.2   DATA

Monthly precipitation data during the period 2000–2018 were obtained from 11 stations distributed well across the basin. Also, in order to calculate the VCI, TCI, and SWI, and MODIS sensor images with 1 km spatial resolution considering the growing season from 2001 to 2018 downloaded from the US Geological Survey (https://earthexplorer.usgs.gov/) were applied. Due to the fact that the dominant period of vegetation growth and peak vegetation growth in Karkheh River Basin provinces is from the beginning of March to the end of May, the images of March, April, and May from 2001 to 2018 were selected to calculate the VCI, TCI and SWI . Table 23.1 and Table 23.2 illustrate the characteristics of the rain gauge stations and MODIS sensor products in the Karkheh River Basin.

23.2.3   METHODS

23.2.3.1   Vegetation Condition Index (VCI)

The MOD13A2 product provides NDVI values at a per pixel basis at 1 km spatial resolution. Eventually, the VCI is calculated according to Equation 23.1 for each pixel and month of a certain year (Kogan, 1997):

$$VCI_i = \frac{NDVI_i - NDVI_{max}}{NDVI_{max} - NDVI_{min}} \times 100 \qquad (23.1)$$

TABLE 23.1

Characteristics of Selected Rain Gauge Stations in Karkheh River Basin

Station	Longitude	Latitude	Height (m)	Year Established
Aghajan-Bolaghi	48°04'	34°50'	1803	1967
Aran	47°33'	34°24'	1443	1966
Polechehr	47°26'	34°21'	1280	1970
Doabmerk	46°47'	34°33'	1310	1966
Heilian-Seymareh	46°39'	33°44'	703	1966
Dartoot	49°39'	33°44'	907	1981
Dehno	48°47'	33°31'	1770	1969
Doab-visian	48°47'	32°48'	960	1967
Zal Bridge	48°05'	32°48'	300	1969
Bridge Pay	48°09'	32°25'	90	1956
Nazarabad	47°26'	33°11'	571	1977

TABLE 23.2

Characteristics of MODIS Sensor Products

Satellite	Sensor	Spatial Resolution	Product	Index
Terra	MODIS	1 km	MOD13A$_2$	VCI
Terra	MODIS	1 km	MOD11A$_2$	TCI
Terra	MODIS	1 km	MOD13A$_2$ and MOD11A$_2$	SWI

where $NDVI_i$ is the value for the pixel and month, and $NDVI_{max}$ and $NDVI_{min}$ are the maximum and minimum values of NDVI during a specific time period. The VCI value is measured ranging from 0 to 1, while low and high values of the VCI displaying unfavorable condition and favorable conditions, respectively (Amalo et al., 2017).

23.2.3.2   Temperature Condition Index (TCI)

The MOD11A2 product provides LST values at a per pixel basis at 1 km spatial resolution. The TCI is computed by the following equation:

$$TCI_i = \frac{LST_{max} - LST_i}{LST_{max} - LST_{min}} \times 100 \qquad (23.2)$$

where LST_i is the value for the pixel and month, and LST_{max} and LST_{min} are the maximum and minimum values of LST during a specific time period. The TCI value is measured ranging from 0 to 1, while low and high values of the TCI displaying unfavorable and favorable conditions, respectively (Roswintiarti et al., 2011; Wang et al., 2012; Cong et al., 2017).

23.2.3.3   Soil Water Index (SWI)

The SWI shows the relationship between land surface, vegetation, and soil moisture. To extract the SWI, a triangular space concept will be formed between LST and NDVI data (Figure 23.2). This condition occurs if we have a large range of vegetation and areas with different surface moisture in the image. In this figure, both wet and dry edges are obtained by fitting the linear equation to the minimum and maximum values of the Earth's surface temperature in the scatterplot between NDVI and LST, respectively. After forming a triangular space, the SWI for each point can be achieved

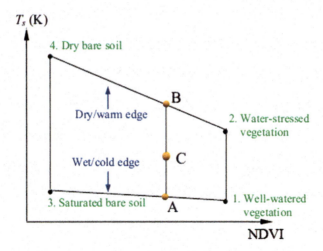

FIGURE 23.2 The hypothetical trapezoidal shape based on the relation between T_s and NDVI (Wang et al., 2012).

using Equation 23.3. The SWI in the dry edge (low water areas) will be 0 and in the wet edge 1 (Mallick et al., 2009; Keshavarz et al., 2014):

$$SWI = \frac{LST_{max(i)} - LST_{(i)}}{LST_{max(i)} - LST_{min(i)}} \qquad (23.3)$$

where i is the pixel number, $LST_{(i)}$ is the LST for ith pixel, and $LST_{min(i)}$ and $LST_{max(i)}$ are the maximum and minimum observed temperature corresponding to the intended pixel:

$$LST_{max} = a_1 + b_1 \times NDVI \qquad (23.4)$$

$$LST_{min} = a_2 + b_2 \times NDVI$$

where a_1 and a_2 are the width from the origin of the fitted lines to the maximum and minimum surface temperatures Also, b_1 and b_2 are the slope of the lines fitted to these values is in order to make the sides of dry and wet.

23.2.3.4 Standardized Precipitation Index (SPI)

The SPI is a very popular meteorological drought index that is computed from long-term monthly precipitation data (Dutta et al., 2015). To calculate the SPI, the station precipitation must first be fitted with different distributions to select the best distribution (usually gamma or Pearson III), which is then transformed into a normal distribution so that the mean SPI for the location and favorable period is 0. According to McKee et al. (1993), among the various series of the SPI, 1-, 3-, and 6-months time periods have been designated as short-term periods, and 12-, 24-, and 48-months periods as long-term periods. Short-term series are used to study agricultural droughts and long-term series are used to identify and study hydrological droughts. Positive SPI values indicate that precipitation is higher than the mean, while negative values show precipitation is lower than the mean. The gamma distribution is normally defined as

$$g(x) = \frac{1}{\beta^{\alpha \, "}(\pm)} x^{\alpha-1} e^{-x/\beta} \qquad (23.5)$$

TABLE 23.3

Classification of SPI, VCI, TCI, and SWI Indicators

SPI	VCI, TCI, and SWI	Category
SPI < –2	0 ≤ VCI, TCI, and SWI < 10	Extremely dry
–2 ≤ SPI < –1.5	10 ≤ VCI, TCI, and SWI < 20	Very dry
–1.5 ≤ SPI <–1	20 ≤ VCI, TCI, and SWI ≤ 30	Moderately dry
–1≤ SPI < 0	30 ≤ VCI, TCI, and SWI ≤ 40	Mild dry
0 < SPI	40 < VCI, TCI, and SWI ≤ 100	No drought

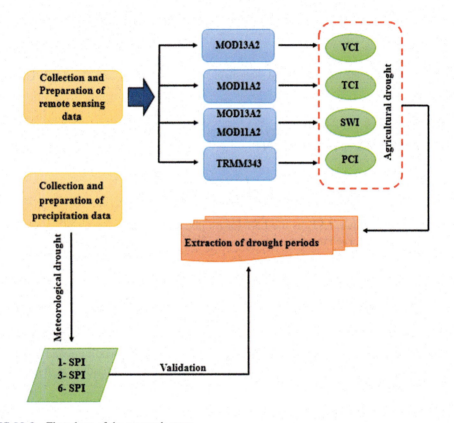

FIGURE 23.3   Flowchart of the research steps.

where $\pm > 0$ is a shape parameter, $\beta > 0$ is a scale parameter, x is the precipitation amount and $\Gamma(\pm)$ is the gamma function.

   Table 23.3 shows the drought status classification based on the SPI, VCI, TCI, and SWI (Du et al., 2013). Figure 23.3 shows the flowchart of the research steps.

23.3   RESULTS AND DISCUSSION

23.3.1   DROUGHT MONITORING BASED ON VCI

The results of the VCI showed that the drought phenomenon expanded for March in the time period of 2001 to 2018, in the years of 2003 (76.2%), 2004 (50.3%), 2005 (62.7%), 2006 (65%), 2008

(83.6%), 2009 (56.4%), 2010 (50%), 2011 (67%), 2012 (91.1%) and 2017 (54.7%). According to the results, the maximum occurrence of this phenomenon was observed in 2012, so that very severe, severe, moderate, and mild drought classes cover 52.9%, 19.1%, 12.2%, and 6.9% of the total basin area, respectively (Table 23.4 and Figure 23.4).

The results of the VCI showed that the drought phenomenon for April in the statistical period of 2001 to 2018, in the years 2001, 2008, 2009, 2011, 2012, and 2015 occurred with more expansion (Table 23.5). According to the results, in 2001, 45.3% of the basin area had no drought and 54.7% had very severe to mild drought conditions. In 2008, drought intensified so that 90.8% of the basin area was in very severe to mild drought conditions and the very severe drought class increased significantly (62.1%). In 2009, the severity of drought decreased (the area of very severe drought class 16.2%) and the area of no drought increased (43.6%). From 2011 to the end of the period, the results of this index showed that the drought phenomenon in 2011 (67%), 2012 (56.6%), 2013 (51.4%), and 2015 (50.4%) occurred with more expansion, and in 2010 (23.2%), 2014 (17.4%), 2016 (24.2%), 2017 (33%), and 2018 (22.9%) showed a lower expansion (Table 23.5 and Figure 23.5).

According to the results of the VCI, there has been a drought phenomenon for May in all years, but in 2001 (75.9%), 2003 (56.2%), 2004 (50.5%), 2005 (56.1%), 2006 (52.5%), 2007 (84.2%), 2008 (88.2%), 2009 (63.8%), 2010 (52.1%), 2011 (50/8 %), 2012 (76.8%), 2013 (66.2%), and 2015 (78.8%) it occurred with more intensity. According to the results, the maximum occurrence of this phenomenon was observed in 2008, so that very severe, severe, moderate, and mild drought classes cover 50.2%, 18.2%, 12.2%, and 7.4% of the total basin area, respectively (Table 23.6 and Figure 23.6).

According to the results of the VCI, a severe drought occurred in parts of the west, east, center, and south in March (Figure 23.4). In 2008 and 2012, the drought was more severe and widespread, covering almost the whole basin. In April, the drought had decreased compared to March. In 2009, 2012, and 2015, most of the central and southern parts faced very severe drought and in 2008 it has spread throughout the basin (Figure 23.5). In May, the severity of drought increased and the most severe drought (severe to very severe) occurred in 2008, 2012, and 2015 (Figure 23.6). In

TABLE 23.4
Drought Category Area According to VCI for March

Category No Drought	Mild Dry	Moderately Dry	Very Dry	Extremely Dry	Year
58.9	15.1	12.5	7.8	5.7	2001
64.2	14.7	11.2	6.5	3.4	2002
23.8	12.2	14.3	15.5	34.1	2003
49.7	21.4	16.5	8.6	3.8	2004
37.3	14.9	18	16.7	13.1	2005
35	17.5	19.2	16	12.3	2006
75.8	10.4	6.5	3.5	3.8	2007
16.4	8.7	14	19.5	41.4	2008
43.6	16.5	14	9.6	16.2	2009
50	17.7	14.1	10.8	7.4	2010
33	20	22.6	15.7	8.7	2011
8.9	6.9	12.2	19.1	52.9	2012
75.3	12.4	7.6	3.4	1.4	2013
72.3	13.5	8.8	3.9	1.5	2014
76	8.9	8	4.6	2.4	2015
88.8	4.3	2.9	2.2	1.8	2016
45.3	16.4	15.3	11.7	11.3	2017
72.9	10.6	7.5	4.9	4	2018

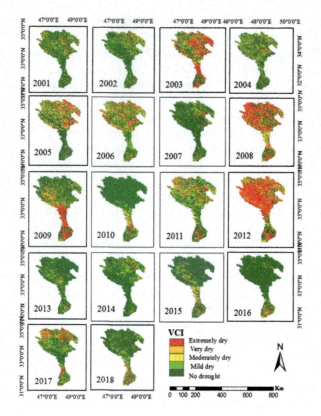

FIGURE 23.4   Spatial distribution of drought intensity in March based on the VCI.

TABLE 23.5
Drought Category Area According to VCI for April

Category No Drought	Mild Dry	Moderately Dry	Very Dry	Extremely Dry	Year
45.3	16.7	16	11.7	10.4	2001
64.1	10.2	9.1	7	9.6	2002
56.2	13	11.1	8.4	11.3	2003
51.5	17.4	15	9.8	6.3	2004
52.6	14.1	12.6	9.9	10.8	2005
67	15.1	10.2	5	2.7	2006
74	9.9	7.2	4.6	4.3	2007
9.2	5.8	9.7	13.1	62.1	2008
43.6	16.5	14	9.6	16.2	2009
76.8	6.5	6	5.5	5.2	2010
33	20	22.6	15.7	8.7	2011
43.4	15.7	14.9	12.1	13.9	2012
48.6	14.9	15.1	11.7	9.7	2013
82.6	8.8	5.1	2.2	1.2	2014
49.6	11.3	13.1	13	13	2015
75.8	9.5	6.9	4.3	3.5	2016
67	12.7	9.2	5.7	5.3	2017
77.1	9.9	6.1	3.6	3.3	2018

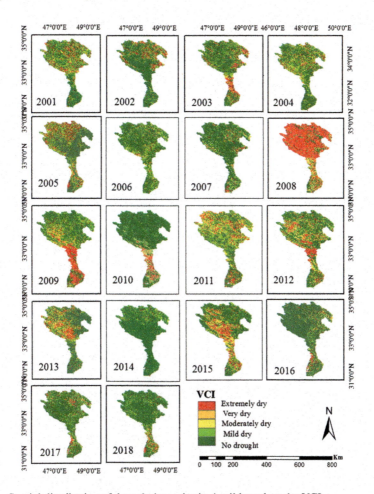

FIGURE 23.5 Spatial distribution of drought intensity in April based on the VCI.

general, the results of the VCI showed that the most severe and widespread drought occurred from March to May in 2008. This finding is consistent with the results of Rezaei Moghadam et al. (2012), Mirmousavi and Karimi (2013), and Rezaei-Banafsheh et al. (2015).

23.3.2 DROUGHT MONITORING BASED ON TCI

According to the results of the TCI, the phenomenon of drought for March in 2003, 2007, 2012, and 2014 was 66.4%, 70.4%, 57.2%, and 94.8%, respectively, of the basin area in very severe to mild drought conditions. According to the results, the maximum occurrence of this phenomenon was observed in 2014, so that very severe, severe, moderate, and mild drought classes cover 77.1%, 8.5%, 5.6%, and 3.5% of the total basin area, respectively (Table 23.7 and Figure 23.7).

In April, the results of the TCI showed that the phenomenon of drought occurred in 2002, 2003, 2004, 2006, 2007, 2007, 2008, 2010, 2016, and 2018 with a greater extent. During 2002, 2003, and 2004, 62.4%, 51.7%, and 52.3%, respectively, of the basin area were exposed to very severe to mild drought conditions. However, in 2006, 2007, 2008, 2010, 2016, and 2018 drought intensified, and 74.8%, 75.4%, 94.4%, 68.8%, 85.5%, and 81% of the basin area had very severe to mild drought conditions, respectively. The maximum occurrence of this phenomenon was observed in 2008, so that very severe, severe, moderate, and mild drought cases covered 64.1%, 14.5%, 9.9%, and 5.8% of the total basin area, respectively (Table 23.8 and Figure 23.8).

TABLE 23.6

Drought Category Area According to VCI for May

No Drought	Mild Dry	Moderately Dry	Very Dry	Extremely Dry	Category Year
24.1	11.2	19	22.2	23.6	2001
64.8	14.7	10.4	6	4.2	2002
43.8	16.5	17.3	12.5	10	2003
49.5	16.3	15.7	11.8	6.8	2004
43.9	18.4	17.9	12.2	7.4	2005
47.5	20	18.9	10.1	3.5	2006
15.8	21.9	26.2	23.7	12.4	2007
11.8	7.4	12.2	18.2	50.5	2008
36.2	14.9	16.6	16.7	15.6	2009
47.9	16.7	18	11.5	5.9	2010
49.2	15.6	15.8	12.2	7.2	2011
23.2	11.1	15.2	19.4	31.2	2012
33.8	13	16.4	18.3	18.6	2013
68.34	13.78	9.98	5.11	2.7	2014
21.2	10.5	12.7	15.7	40	2015
93.45	2.78	1.82	0.99	0.95	2016
67.7	14.3	9.9	4.9	3.3	2017
70.3	12.4	9.2	4.4	3.7	2018

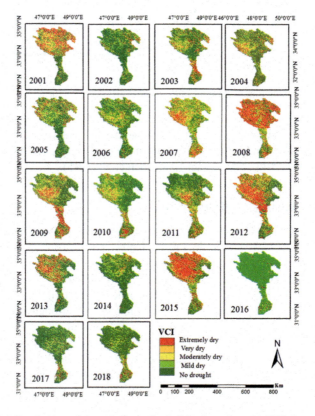

FIGURE 23.6   Spatial distribution of drought intensity in May based on the VCI.

TABLE 23.7

Drought Categories Area According to TCI for March

No Drought	Mild Dry	Moderately Dry	Very Dry	Extremely Dry	Category Year
86.3	10.4	2.7	0.5	0.1	2001
73.4	5.1	6.4	5.6	6.5	2002
33.6	15.7	16.6	14.9	19.2	2003
99.5	0.4	0.1	0	0	2004
51.6	23.4	16	6.1	2.9	2005
93.1	4.5	1.9	0.5	0.1	2006
29.6	27.6	23.1	12.3	7.4	2007
72.7	11.3	7.9	4.4	3.7	2008
76.7	10.7	6.5	3.6	2.6	2009
73	11	8	4.4	3.6	2010
54.9	17.9	12.2	6.7	8.3	2011
42.8	15.5	14	10.5	17.2	2012
95.2	3.4	1.1	0.2	0.1	2013
5.2	3.5	5.6	8.5	77.1	2014
96.9	2	0.8	0.2	0	2015
70.7	17.9	8.3	2.5	0.6	2016
64.7	17	10.3	4.6	3.4	2017
93.1	4.5	1.9	0.5	0.1	2018

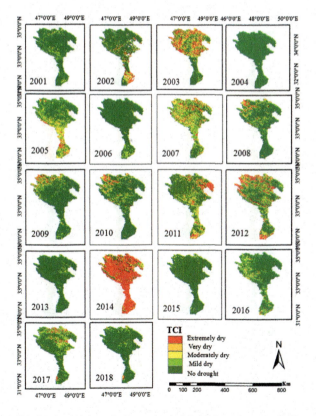

FIGURE 23.7 Spatial distribution of drought intensity in March based on the TCI.

TABLE 23.8

Drought Category Area According to TCI for April

No Drought	Mild Dry	Moderately Dry	Very Dry	Extremely Dry	Category Year
94.1	4.8	1	0.1	0	2001
37.6	20	19.3	13.1	10	2002
48.3	19.5	16.1	4	6.7	2003
47.7	18.3	14.7	10	9.3	2004
66.4	15.8	11.7	4.9	1.2	2005
25.2	22.4	25.8	18.5	8.1	2006
24.6	23	27.8	17.6	7	2007
5.6	5.8	9.9	14.5	64.1	2008
92.7	4.4	2.4	0.5	0.1	2009
31.2	15.1	18.2	16.8	18.6	2010
77.7	14.9	5.6	1.2	0.6	2011
98.7	0.8	0.3	0.1	0.1	2012
88.6	6	3.3	1.3	0.7	2013
59.2	19.8	12.8	5.7	2.5	2014
61.7	20.1	11.7	4.8	1.6	2015
19.5	12.4	21.3	24.9	21.9	2016
84.5	11.3	3.1	0.8	0.3	2017
19	17.5	23.1	21.3	19.1	2018

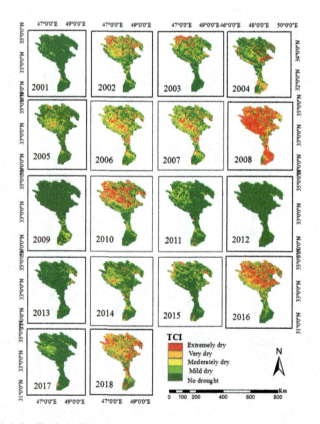

FIGURE 23.8   Spatial distribution of drought intensity in April based on the TCI.

TABLE 23.9

Drought Category Area According to TCI for May

No Drought	Mild Dry	Moderately Dry	Very Dry	Extremely Dry	Category Year
98	1.4	0.4	0.1	0	2001
91.4	6.1	2.1	0.4	0	2002
84	10.5	4.1	1.2	0.2	2003
20.6	20.2	26.8	21.1	11.3	2004
47.3	27.3	15.7	6.5	3.2	2005
78.2	14.1	5.8	1.6	0.3	2006
7.9	6.8	11.6	18.9	54.8	2007
92.6	4.2	2.1	0.6	0.4	2008
86.8	8.7	3.6	0.8	0.1	2009
13.9	16.3	24.9	21.9	23	2010
41.3	19.1	19.6	13.4	6.6	2011
13.4	17.5	22.4	19.9	26.9	2012
79.3	10.2	5.8	2.9	1.8	2013
78	13.8	4.8	2.6	0.8	2014
98.7	0.6	0.2	0.2	0.2	2015
81.1	12.5	5	21.1	0.3	2016
94.7	3	1.1	0.5	0.8	2017
20.6	2.7	4.9	12.1	59.7	2018

In May, the results of the TCI showed the drought phenomenon in the 2004 (79.4%), 2005 (52.7%), 2007 (92.1%), 2010 (86.1%), 2011 (58.7%), 2012 (86.6%), and 2018 (79.4%) occurred with a greater extent during the statistical period. The maximum occurrence of this phenomenon was observed in 2007, so that very severe, severe, moderate, and mild drought cases covered 54.8%, 18.9%, 11.6%, and 6.8% of the total basin area, respectively (Table 23.9 and Figure 23.9).

A survey of the results of the TCI showed that in March, the basin conditions were normal in most years, and only a small part of the east, northwest, and, to some extent, the south was affected by severe drought. In 2014, the drought was more severe and widespread, and covered almost all the basin (Figure 23.7). So that in April, the drought increased compared to March. In 2008, 2010, 2016, and 2018, the central, northern, western, and eastern parts faced very severe drought, and in 2008 it had spread throughout the basin (Figure 23.8). In May, the severity of drought increased, and the most severe drought (severe to very severe) occurred in 2004, 2007, 2010, 2011, 2012, and 2018 (Figure 23.9). So that in 2007 and 2018, except for a small area, almost all of the basin was affected by severe and very severe drought. In general, according to the results of the TCI, the value of this index in most years showed no drought or mild drought conditions. Rezaei Moghadam et al. (2012) and Ebrahimzadeh et al. (2013) achieved the same results in their research.

23.3.3 Drought Monitoring Based on SWI

Based on the SWI results in March, in 2002 (72.3%), 2004 (35.8%), 2005 (57.6%), 2006 (75.1%), 2007 (54%), 2008 (70.7%), 2011 (67.1%), 2012 (61.9%), 2013 (54.2%), 2014 (54.2%), and 2017 (55.9%) more area of the basin had been affected by drought. The maximum occurrence of this phenomenon was observed in 2006, so that very severe, severe, moderate, and mild drought covered 2.1%, 8.6%, 10.7%, and 53.7 % of the total basin area, respectively (Table 23.10 and Figure 23.10).

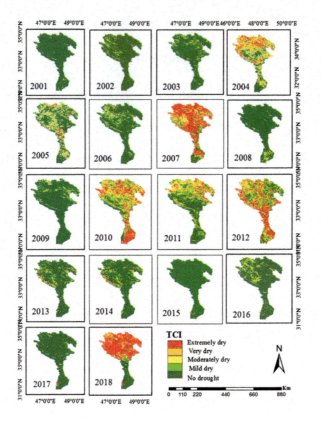

FIGURE 23.9 Spatial distribution of drought intensity in May based on the TCI.

TABLE 23.10

Drought Category Area According to SWI for March

No Drought	Mild Dry	Moderately Dry	Very Dry	Extremely Dry	Category Year
63.6	22	5.5	7.6	1.3	2001
27.7	41	21.2	7.6	2.5	2002
74.3	9.2	6	9.4	1.2	2003
35.8	45.9	7.7	9.2	1.3	2004
42.4	42.7	7.3	6.8	0.7	2005
24.9	53.7	10.7	8.6	2.1	2006
46	34.1	8.4	5.9	5.6	2007
29.3	45.1	13.7	9	2.9	2008
46	29.6	8.5	10.6	5.2	2009
77.9	8	7.9	5.4	0.7	2010
32.9	41	14.7	10.2	1.2	2011
38.1	34.8	15.8	10.4	0.8	2012
45.8	34.1	9	7	4.2	2013
45.8	38.8	10.3	4.7	0.3	2014
75.8	8.7	9.1	4.9	1.6	2015
53.5	28.5	8	9.2	0.8	2016
44.1	32.9	10	7.6	5.4	2017
70.5	13	7.5	7.7	1.3	2018

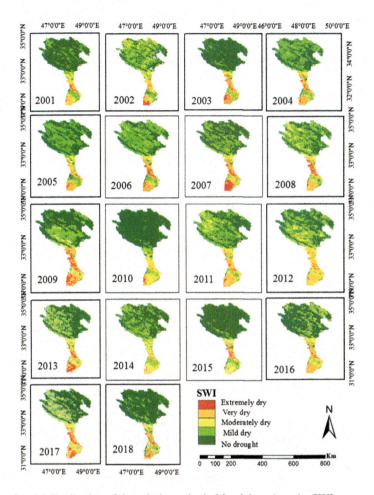

FIGURE 23.10 Spatial distribution of drought intensity in March based on the SWI.

SWI results showed that in April, the phenomenon of drought occurred in all years studied, which, according to the percentage covered by each of the drought severity classes, had a lower incidence than in March. The maximum occurrence of this phenomenon was observed in 2009, so that very severe, severe, moderate, and mild drought covered 0.7%, 9.3%, 15.5%, and 34.4% of the total basin area, respectively (Table 23.11 and Figure 23.11).

According to the SWI results of May, almost the entire basin area was affected by drought, which was not very severe. The maximum occurrence of this phenomenon was observed in 2008, so that very severe, severe, moderate, and mild drought classes covered 2.7%, 7.3%, 20.4%, and 28.6% of the total basin area, respectively (Table 23.12 and Figure 23.12).

According to the results obtained from the SWI, the amount of drought in the southern parts of the basin during the months of March to May is higher in all years (Figures 23.10, 23.11, and 23.12).

23.3.4 CORRELATION BETWEEN REMOTE SENSING INDICES AND SPI

We investigated the correlation between remote sensing indices and the SPI at different timescales. The results showed in that March the SPI3 and SPI6 indices had the highest correlation with the VCI equal to 0.51 and 0.42, respectively, and SPI1 had the highest correlation with the SWI equal to 0.51. Also, SPI1, SPI3, and SPI6 indices showed the lowest correlation equal to 0.25, 0.28, and 0.2 with the TCI, respectively (Figure 23.13). In April, SPI3 and SPI6 had the highest correlation

TABLE 23.11

Drought Category Area According to SWI for April

No Drought	Mild Dry	Moderately Dry	Very Dry	Extremely Dry	Category Year
76.1	7.5	6.1	8.1	2.3	2001
76.8	7	7.2	7.2	1.9	2002
78.9	6	7	8	7.3	2003
69.5	14.3	7.4	8.1	0.6	2004
72	9.7	7.1	8.8	2.4	2005
79	6.4	5.5	7.5	1.6	2006
78.9	6.4	5.6	7.4	1.6	2007
75.2	6.5	7.2	8.7	2.5	2008
40.1	34.4	15.5	9.3	0.7	2009
79.4	6.6	7.7	5.6	0.8	2010
73.5	9.2	8.2	8.6	0.5	2011
73.6	8.7	7.7	8.1	1.9	2012
66.3	13.9	7.8	8.5	3.6	2013
68.9	12.3	7.5	8.8	2.5	2014
72.2	8.8	8.8	8.2	2.1	2015
78.9	3.6	7.5	7	2.9	2016
79.1	3.7	7.4	7.2	2.5	2017
74.8	6.2	7.9	7.8	3.2	2018

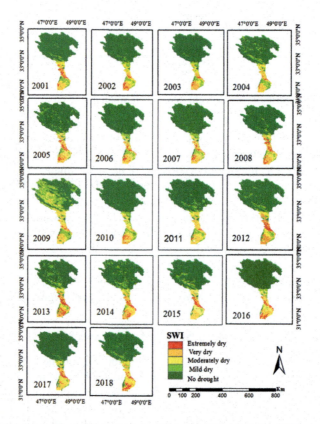

FIGURE 23.11   Spatial distribution of drought intensity in April based on the SWI.

TABLE 23.12

Drought Category Area According to SWI for May

No Drought	Mild Dry	Moderately Dry	Very Dry	Extremely Dry	Category Year
68.9	14.7	9.5	5.1	1.8	2001
72.8	8.7	8.7	8.2	1.5	2002
59.5	17.2	10.6	11.1	1.6	2003
64.7	12.5	9.8	9	4	2004
59.5	19	10.3	8.1	3.1	2005
63.3	13.7	9.8	10.6	2.5	2006
76.3	8.1	8.4	6.7	0.4	2007
41	28.6	20.4	7.3	2.7	2008
63.3	14.2	10.2	11	1.2	2009
68.8	11.6	13.1	5.9	0.6	2010
64.6	12.6	10.8	10.8	1.2	2011
72	13.5	10.7	3.8	0	2012
66.8	12.4	9.8	7.9	3.1	2013
66.3	12.6	10	8	3.1	2014
55.9	23.4	14.5	5	1.2	2015
72.5	8.6	8.8	9.1	0.9	2016
58.6	18.6	8.9	11.6	2.4	2017
78.1	3.8	9.2	7.9	1	2018

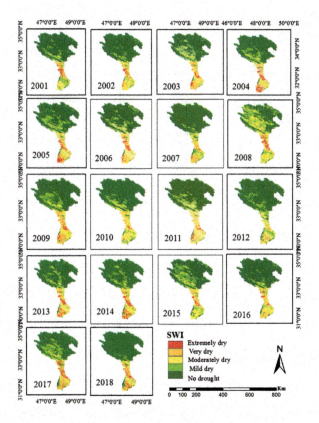

FIGURE 23.12 Spatial distribution of drought intensity in May based on the SWI.

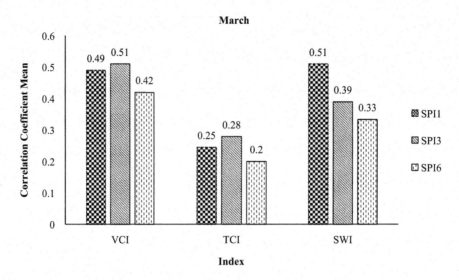

FIGURE 23.13 Correlation coefficients between the SPI and remote sensing indices in March.

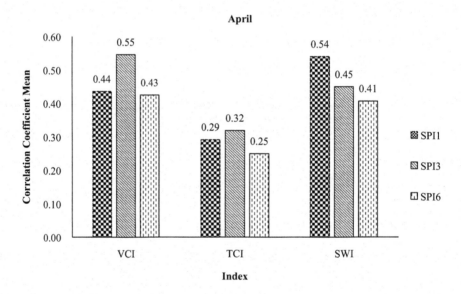

FIGURE 23.14 Correlation coefficients between the SPI and remote sensing indices in April.

with the VCI equal to 0.55 and 0.43, respectively, and also SPI1 had the highest correlation with the SWI equal to 0.54. SPI1, SPI3, and SPI6 showed the lowest correlation equal to 0.29, 0.32, and 0.25, respectively, with the TCI (Figure 23.14). In May, SPI3 and SPI6 had the highest correlation with the VCI equal to 0.43 and 0.49, respectively, and also, SPI1 had the highest correlation with SWI equal to 0.41. SPI1, SPI3, and SPI6 showed the lowest correlation equal to 0.26, 0.34, and 0.23 with the TCI, respectively (Figure 23.15).

According to the results of the VCI, the maximum correlation coefficient was observed with SPI3 in March and April and with SPI6 in May equal to 0.51, 0.55, and 0.49, respectively. This result could indicate a temporal delay between effective precipitation and maximum vegetation

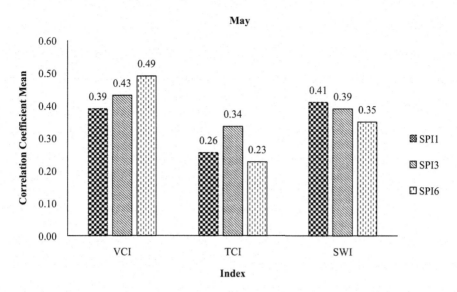

FIGURE 23.15 Correlation coefficients between the SPI and remote sensing indices in May.

growth (Winkler et al., 2017). Given that effective precipitation in the study area usually begins in early winter, this temporal delay seems reasonable. On the other hand, the amount of this temporal delay depends on vegetation type and characteristics, soil situation, and potential evapotranspiration (Winkler et al., 2017). The VCI showed generally better performance with long-term drought indices (SPI3 and SPI6). Zambrano et al. (2016), Winkler et al. (2017), and Jiao et al. (2019) also achieved similar results in their research.

Also, due to the weak correlation between the VCI and SPI1, there is no simultaneous occurrence of meteorological drought and agricultural drought in all years. This finding is consistent the results of Moazzenzadeh et al. (2012), and Fatehi Maraj and Heidarian (2013). Although vegetation is affected by precipitation, the timing of precipitation deficits is decisive for the differing drought patterns based on the VCI (Winkler et al., 2017). Also, due to the fact that vegetation has always been affected by climate and its parameters, it is very complicated to study the relationship and time interval between meteorological drought and agricultural drought, and it seems that in addition to precipitation, other factors such as temperature, evapotranspiration, height, land cover, land use, type of vegetation, type of soil texture and soil moisture, water supply for irrigation of the basin, amount and times of blockage, and opening of dams in the region affect the health and growth of vegetation in the region (Quiring and Papakryiakou, 2003; Bayarjargal et al., 2006; Mirahsani et al., 2017). On the other hand, sensitivity of vegetation to precipitation is dependent on region climate. So that areas with arid and semiarid climates, where water is a limiting parameter, show higher sensitivity than more humid regions (Winkler et al., 2017). Investigation of the correlation between the TCI and SPI in different timescales showed that this index has less correlation than the VCI with SPI1, SPI3, and SPI6, and the highest correlation was observed with SPI3 equal to 0.27. Zhang et al. (2017) and Wang et al. (2019) also achieved similar results in their research. According to the results, the SWI showed a higher correlation with SPI1 equal to 0.51, 0.54, and 0.41 for March, April, and May, respectively, and also had less correlation with the SPI3 and SPI6 indices. Since this index is a criterion for soil moisture in the region, if the delay time between precipitation and this index is considered at shorter intervals, better and more acceptable results will be obtained. Therefore, the SWI performs better to control short-term drought conditions, which is consistent with the findings of Zhang et al. (2017), Wang et al. (2019), and Jiao et al. (2019).

23.4 CONCLUSIONS

In this study, first, the VCI, TCI, and SWI were calculated, and the drought situation in the whole Karkheh River Basin was monitored through these three indices for the time period 2001–2018 in the months of March, April, and May. Then, the performance of each of the indices was analyzed using the SPI and Pearson correlation test. From the comparison of drought indices in the Karkheh basin, the VCI is a suitable index to monitor agricultural drought conditions because of the stronger correlation with SPI3 and SPI6. But none of the indices is considered superior to the other situations. Finally, it can be stated that studies using different indices show different results depending on the location and environmental conditions, therefore, a general comparison between the indices in terms of data accuracy and drought study is not possible. Due to the specific climatic conditions of Iran and the climatic conditions of arid regions, it is not possible to use the same methods and indicators to study drought in different regions and expect exactly the same results. Therefore, the relationship between each of the remote sensing indices with drought in each region should be studied separately, and for each region, depending on the specific climatic conditions and vegetation, an appropriate index to monitor agricultural drought should be considered.

REFERENCES

Amalo, L.F., Hidayat, R. and Haris. 2017. Comparison between remote-sensing-based drought indices in east Java. *IOP Conference Series: Earth and Environmental Science*, 54: 012009. https://doi.org/10.1088/1755-1315/54/1/012009.

Arbabi Sabzevari, A. 2010. The analyze effect drought whit the index of in ternal research precipitation the normal Z in information system geographic in Kashan region. *Journal of Physical Geography*, 3 (7): 105–124. [In Persian]

Baniya, B., Tang, Q., Xu, X., Haile, G.G. and Chhipi-Shrestha, G. 2019. Spatial and temporal variation of drought based on satellite derived vegetation condition index in Nepal from 1982–2015. *Sensors*, 19(2): 430.

Bayarjargal, Y., Karnieli, A., Bayasgalan, M., Khudulmur, S., Gandush, C. and Tucker, C.J. 2006. A comparative study of NOAA–AVHRR derived drought indices using change vector analysis. *Remote Sensing of Environment*, 105(1): 9–22.

Bento, V.A., Gouveia, C.M., DaCamara, C.C. and Trigo, I.F. 2018. A climatological assessment of drought impact on vegetation health index. *Agricultural and Forest Meteorology*, 259: 286–295.

Byzedi, M. 2011. Analysis of hydrological drought based on daily flow series. *World Academy of Science, Engineering and Technology*, 50: 491–496.

Byzedi, M., Siosemardeh, M. Rahimi, A. and Mohammadi, K. 2012. Analysis of hydrological drought on Kurdistan province. *Australian Journal of Basic and Applied Sciences*, 6(7): 255–259.

Cong, D., Zhao, S., Chen, C. and Duan, Z.. 2017. Characterization of droughts during 2001–2014 based on remote sensing: A case study of Northeast China. *Ecological Informatics*, 39: 56–67.

Dracup, J.A., Lee, K.S. and Paulson, E.G. 1980. On the statistical characteristics of drought events. *Water Resources Research*, 16(2): 289–296.

Du, L., Tian, Q., Yu, T., Meng, Q., Jancso, T., Udvardy, P. and Huang, Y. 2013. A comprehensive drought monitoring method integrating MODIS and TRMM data. *International Journal of Applied Earth Observation and Geoinformation*, 23: 245–253.

Dutta, D., Kundu, A., Patel, N.R., Saha, S.K. and Siddiqui, A.R. 2015. Assessment of agricultural drought in Rajasthan (India) using remote sensing derived Vegetation Condition Index (VCI) and Standardized Precipitation Index (SPI). *The Egyptian Journal of Remote Sensing and Space Science*, 18(1): 53–63.

Ebrahimzadeh, S., Bazrafshan, J. and Ghorbani, Kh. 2013. Comparative study between satellite and ground-based drought indices using change vector analysis technique (Case Study of Kermanshah Province). *Journal of Water and Soil*, 27(5): 1034–1045. (in Persian).

Fatehi Maraj, A. and Heydarian, S. A. 2013. Investigation of meteorological, agricultural and hydrological drought using GIS in Khuzestan province. *Iran-Watershed Management Science & Engineering*, 7(23): 19–32. [In Persian]

Ghaleb, F., Mario, M. and Sandra, A.N. 2015. Regional Landsat-based drought monitoring from 1982 to 2014. *Climate*, 3(3): 563–577.

Jiao, W., Tian, C., Chang, Q., Novick, K.A. and Wang, L. 2019. A new multi-sensor integrated index for drought monitoring. *Agricultural and Forest Meteorology*, 268: 74–85.

Keshavarz, M.R., Vazifedoust, M. and Alizadeh, A. 2014. Drought monitoring using a Soil Wetness Deficit Index (SWDI) derived from MODIS satellite data. *Agricultural Water Management*, 132: 37–45. (in Persian).

Keyantash, J.A. and Dracup, J.A. 2004. An aggregate drought index: Assessing drought severity based on fluctuations in the hydrologic cycle and surface water storage. *Water Resources Research*, 40(9): 1–13.

Kogan, F. 2002. World droughts in the new millennium from AVHRR-based vegetation health indices. *Eos, Transactions American Geophysical Union*, 83(48): 557–563.

Kogan, F.N. 1995. Droughts of the late 1980s in the United States as derived from NOAA polar-orbiting satellite data. *Bulletin of the American Meteorological Society*, 76(5): 655–668.

Kogan, F.N. 1997. Global drought watches from space. *Bulletin of the American Meteorological Society*, 78(4): 621–636.

Mallick, K., Bhattacharya, B.K. and Patel, N.K. 2009. Estimating volumetric surface moisture content for cropped soils using a Soil Wetness Index based on surface temperature and NDVI. *Agricultural and Forest Meteorology*, 149: 1327–1342.

McKee, T.B., Doesken, N.J. and Kleist, J. 1993. The Relationship of Drought Frequency and Duration to Time Scales. In Eighth Conference on Applied Climatology, 17–22 *January*, Anaheim, CA, 179–184.

McKee, T.B., Doesken, N.J. and Kleist, J. 1995. Drought monitoring with multiple time scales. In *Proceedings of the Ninth Conference on Applied Climatology*. American Meteorological Society, Dallas, TX, pp. 233–236.

Mirahsani, M.S., Mahini, A. R., Safanian, A. R., Modares, R., Jafari, R. and Mohamadi, J. 2017. Regional drought monitoring of Zayandeh-Rud basin based on index time series changes VCI index MODIS Sensors and SPI index. *Journal of Geography and Environmental Hazards*, 24(1): 1–22. (in Persian)

Mirmousavi, S. H. and Karimi, H. 2013. Effect of drought on vegetation cover using MODIS sensing images case: Kurdistan Province. *Geography and Development*, 11(31): 57–76. [In Persian]

Mishra, A. and Singh, V. P. 2010, A review of drought concepts. *Journal of Hydrology*. 391(1–2): 202–216.

Moazzenzadeh, R., Arshad, S., Ghahraman, B. and Davari, K. 2012. Drought monitoring in unirrigated lands based on the remote sensing technique. *Water and Irrigation Management*, 2(2): 39–52. (in Persian).

Moran, M., Clarke, T., Inoue, Y. and Vidal. A. 1994. Estimating crop water deficit using the relation between surface-air temperature and spectral vegetation index. *Remote Sensing of Environment*. 49(3): 246–263.

Orville, H.D. 1990. AMS statement on meteorological drought. *Bulletin of the American Meteorological Society*, 71(7): 1021–1025.

Palmer, W. C. 1965. *Meteorological Drought*. research paper, US Department of Commerce, Weather Bureau, Washington, DC, 45: 58.

Quiring, S.M. and Papakryiakou, T.N. 2003. An evaluation of agricultural drought indices for the Canadian prairies. *Agricultural and Forest Meteorology*, 118(1–2): 49–62.

Rezaei Banafsheh, M., Rezaei, A. and Faridpor, M. 2015. Analyzing agricultural drought in east Azarbaijan province emphasizing remote sensing technique and vegetation condition index. *Water and Soil Science (Agricultural Science)*, 25(1): 113–123. [In Persian]

Rezaei Moghadam, M.H., Valizadeh Kamran, K.H., Rostamzadeh, H. and Rezaei, A. 2012. Evaluating the adequacy of MODIS in the assessment of drought (case study: Urmia lake basin). Geography and Environmental Sustainability, 2(5): 37–52. [In Persian]

Rhee, J., Im, J. and Carbone, G.J. 2010. Monitoring agricultural drought for arid and humid regions using multi-sensor remote sensing data. *Remote Sensing of Environment*, 114(12): 2875–2887.

Roswintiarti, O., Sofan, P. and Anggraini, N. 2011. Monitoring of drought-vulnerable area in Java Island, Indonesia using satellite remote-sensing data. *Jurnal Penginderaan Jauh dan Pengolahan Data Citra Digital*, 8: 21–34.

Sandholt, I., Rasmussen, K. and Andersen, J. 2002. A simple interpretation of the surface temperature/vegetation index space for assessment of surface moisture status. *Remote Sensing of Environment*, 79(2–3): 213–224.

Vicente-Serrano, S.M., Beguería, S. and López-Moreno, J.I. 2010. A multiscalar drought index sensitive to global warming: The standardized precipitation evapotranspiration index. *Journal of climate*, 23(7):1696–1718.

Vicente-Serrano, S.M., Van der Schrier, G., Beguería, S., Azorin-Molina, C. and Lopez-Moreno, J.I. 2015. Contribution of precipitation and reference evapotranspiration to drought indices under different climates. *Journal of Hydrology*, 526:42–54.

Wan, Z., Wang, P. and Li, X. 2004. Using MODIS land surface temperature and normalized difference vegetation index products for monitoring drought in the southern Great Plains, USA. *International Jour. Remote Sensing*, 25:61–72.

Wang, K., Li, T. and Wei, J. 2019. Exploring drought conditions in the Three River Headwaters Region from 2002 to 2011 using multiple drought indices. *Water*, 11(2): 1–20.

Wang, C., Qi, S., Niu, Z. and Wang, J. 2004. Evaluating soil moisture status in China using the temperature-vegetation dryness index (TVDI). *Canadian Jour. Remote Sensing*, 30:671–679.

Wang, W., Zhang, Z.Z., Wang, X.G. and Wang, H.M. 2012. Evaluation of using the modified water deficit index derived from MODIS vegetation index and land surface temperature products for monitoring drought. In 2012 IEEE International Geoscience and Remote Sensing Symposium, July. 5951–5954.

Wardlow, B.D., Anderson, M.C. and Verdin, J.P. 2012. *Remote Sensing of Drought: Innovative Monitoring Approaches*. CRC Press, 1–411.

Wilhelmi, O.V. and Wilhite, D.A. 2002. Assessing vulnerability to agricultural drought: A Nebraska case study. *Natural Hazards*, 25: 37–58.

Wilhite, D.A. 2000. *Drought as a natural hazard: Concepts and definitions*, 1–18.

Winkler, K., Gessner, U. and Hochschild, V. 2017. Identifying droughts affecting agriculture in Africa based on remote sensing time series between 2000–2016: Rainfall anomalies and vegetation condition in the context of ENSO. *Remote Sensing*, 9(8): 831.

Wu, J., Zhou, L., Liu, M., Zhang, J., Leng, S. and Diao, C. 2013. Establishing and assessing the Integrated Surface Drought Index (ISDI) for agricultural drought monitoring in mid-eastern China. *International Journal of Applied Earth Observation and Geoinformation*, 23: 397–410.

Zambrano, F., Lillo-Saavedra, M., Verbist, K. and Lagos, O. 2016. Sixteen years of agricultural drought assessment of the BioBío region in Chile using a 250 m resolution Vegetation Condition Index (VCI). *Remote Sensing*, 8(6): 530.

Zhang, L., Jiao, W., Zhang, H., Huang, C. and Tong, Q. 2017. Studying drought phenomena in the Continental United States in 2011 and 2012 using various drought indices. *Remote Sensing of Environment*, 190:96–106.

Zou, L., Cao, S. and Sanchez-Azofeifa, A. 2020. Evaluating the utility of various drought indices to monitor meteorological drought in Tropical Dry Forests. *International Journal of Biometeorology*, 64: 701–711.

24 Soil Moisture–Vegetation Stress–Based Agricultural Drought Index Integrating Remote Sensing–Derived Soil Moisture and Vegetation Indices

Gurjeet Singh and Deepak Singh Bisht

CONTENTS

24.1 INTRODUCTION

Droughts, depending upon their characteristics and ability to influence the natural setup, are classified as meteorological, hydrological, agricultural, and socioeconomic droughts. Each of the categories has a specific definition and is known to affect the system differently. Meteorological droughts are the manifestation of rainfall deficit and the length of the dry period. In contrast, hydrological

DOI: 10.1201/9781003276548-24

droughts are the subsequent detrimental impact of meteorological droughts on water availability in terms of river flow, reservoir level, groundwater level, etc. The reduced water availability due to hydrological droughts eventually affects crop production by creating a soil moisture deficit or shortage in irrigation supply, thereby hampering the agricultural output and creating the agricultural drought. The resultant impact of water deficit caused by various drought conditions on the supply side to meet the demand for sustaining people's livelihood is the situation termed as socioeconomic drought.

Since agricultural drought directly impacts society and governance by affecting food production and revenue, it should be assessed and monitored with high priority. Drought monitoring plays a vital role in providing an early warning system for socioeconomic ripples in the event of a prolonged moisture deficit, and therefore accurate assessment of drought is of paramount importance. Although the most common method of drought classification (meteorological drought) provides a potential signature for water availability or water deficit, it alone cannot adequately characterize the agricultural drought. Therefore, more efficient calculation and assessment methods that account for cropping patterns, soil characteristics, and soil moisture patterns to reflect actual crop water stress are needed. Meteorological drought provides information on water deficit and potential drought occurrence based on the deviation of meteorological parameters from the average condition of an area (Bisht et al., 2019). However, agricultural drought is considered to initiate when the soil moisture availability to plants drops to such a level where it starts adversely affecting the crop yield (Pannu and Sharma, 2002). With respect to the agricultural system, it is widely accepted that soil moisture deficits and crop stresses are precursors to agricultural droughts (Wilhite and Glantz, 1985; Ramadas and Govindaraju, 2015).

Consequently, agricultural drought has adverse effects on plants' growth, which affect vegetation health and crop yield. Therefore, vegetation stress can potentially be used as a drought indicator (Chen and Yang, 2011; Belle and Hlalele, 2015; Potopová et al., 2016). However, vegetation stress does not always signify an agricultural drought. Various other factors, such as wildfire, extreme temperature, flood, hail storms, plant diseases, pest attacks, locust swarm, and lack of nutrient supply, also cause vegetation stress.

Several drought monitoring indices have been developed based on the information obtained from soil moisture and vegetation status. Lately, relating crop water stresses with soil moisture deficits has gained momentum as a popular mechanism for agricultural drought monitoring (Maity et al., 2012; Mishra et al., 2015; Ramadas and Govindaraju, 2015, Samantaray et al., 2018). Agricultural drought indices based on vegetation conditions as a drought indicator have also been extensively studied, evaluated, and used in the last few decades (Huete et al., 2002; Deng et al., 2013; Liu and Kogan, 1996). Based on soil moisture and vegetation conditions, some of the noteworthy indices proposed for agricultural drought studies are the Soil Moisture Drought Index (Hollinger et al., 1993), Crop Moisture Index (CMI; Palmer, 1968), Vegetation Condition Index (VCI; Liu and Kogan, 1996), Vegetation Health Index (VHI; Kogan, 2002), Soil Moisture Anomaly Index (SMAI; Bergman et al., 1988), and Soil Moisture Deficit Index (SMDI; Narasimhan and Srinivasan, 2005). Each of these standard indices used for drought monitoring has its advantages and disadvantages. Drought assessment using the aforementioned indices either relies on simulated soil moisture through modeling exercises or rainfall and temperature-derived soil moisture conditions. In addition, all of them are the expressions of either the key hydrologic variable (i.e., soil moisture) or vegetation condition variable (i.e., Normalized Difference Vegetation Index).

A single indicator of drought might not be adequate to identify agricultural drought. It is worth considering a multivariable approach, which would allow integrating multiple agrohydrologic signatures to produce a reliable drought product that expresses the combined effect of various variables (such as soil moisture and vegetation health) on drought assessment. Thus, the present study hypothesizes that crop responses to agricultural drought can be better defined if vegetation and soil moisture stress are accounted for simultaneously. The integration of local-scale soil moisture and vegetation stress as a drought indicator is expected to derive a reliable and robust estimate of agricultural drought compared to the single indices–based droughts indicators, which solely depend on soil moisture or vegetation stress.

24.2 REMOTE SENSING FOR AGRICULTURAL DROUGHT ASSESSMENT

The remote sensing technique provides information on key environmental variables about drought conditions to a global extent, which is useful and essential for agricultural drought monitoring (Rembold et al., 2015). Remote sensing in agricultural drought monitoring relies on unique spectral signatures of the soil surface and vegetation characteristics such as soil moisture, organic matter, vegetation biomass, chlorophyll content, canopy, and soil temperature (Anjum et al., 2011). Remote sensing products are widely used for agricultural drought monitoring due to their advantage over traditional point scale *in situ* observations that cannot provide spatially continuous information about drought conditions. Although these satellite-based observations have adequate spatiotemporal coverage, the trade-off between their spatial and temporal resolutions restricts their utility at the agriculture fields level and during plant growing seasons (Roy et al., 2014). The development of various multisatellite data fusion techniques can overcome these limitations of remote sensing observations (Walker et al., 2012; Hazaymeh and Hassan, 2015; Das et al., 2019). Spatially continuous remote sensing data sets meet the observational requirement in agricultural drought assessment for a broad range of decision-support activities.

24.2.1 AGRICULTURAL DROUGHT ASSESSMENT THROUGH SATELLITE OBSERVATIONS OF SOIL MOISTURE

Although the availability of soil moisture data sets at the point scale has now relatively increased worldwide due to the growing number of ground-based observational networks, the density of observations does not provide a comprehensive understanding of change in soil moisture conditions on the regional scale (Singh et al., 2021). To capture the evolution of soil moisture over time and across space, soil moisture is usually simulated over large areas using meteorological forcing (i.e., rainfall and temperature) or through a root-zone water balance modeling approach. This soil moisture is then utilized in the development of various agricultural drought indices such as the CMI (Palmer, 1968) and SMDI (Narasimhan and Srinivasan, 2005).

The impressive progress of microwave remote sensing in the recent decade (Fernandez-Prieto et al., 2012) has allowed the scientific community to obtain soil moisture estimates at a global extent with different spatial resolutions and frequent revisit intervals (Chan et al., 2018; Das et al., 2019). Microwave remote sensing provides information on volumetric soil moisture based on the dielectric properties of the soil and water (Wang and Qu, 2009). Microwave remote sensing–based products have the potential to provide an accurate measurement of surface soil moisture (~5 cm) to a global extent (Entekhabi et al., 2010; Kerr et al., 2010). The significant increase in global satellite missions–based soil moisture observations (Rebel et al., 2012) has further increased the number of potential applications of satellite soil moisture estimates (Mecklenburg et al., 2016).

The European Space Agency (ESA) has developed a long-term (1978–2016) global soil moisture data set at a spatial resolution of 25 km by merging the observations of various active and passive microwave spaceborne instruments (Dorigo et al., 2015). The current L-band microwave observation–based global satellite missions, namely, the Soil Moisture and Ocean Salinity (SMOS) of the ESA since November 2009 (Kerr et al., 2010), and the Soil Moisture Active Passive (SMAP) of the National Aeronautics and Space Administration (NASA) since January 2015 (Entekhabi et al., 2010), provide the surface soil moisture (~5 cm) globally at a temporal resolution of ~1–2 days within an accuracy range of ~± 0.04 $m^3 m^{-3}$ and multiple spatial scales (36, 25, 9, and 3 km). The use of these microwave remote sensing–based soil moisture estimates for drought monitoring has several advantages, such as (i) global coverage enables drought monitoring over large areas;

(ii) high temporal resolution improves the ability to monitor the onset of drought-related events; (iii) lower frequency of microwave (e.g., L-band) enables all-weather drought monitoring; and (iv) soil moisture measurement is made even when sparse and moderate vegetation is present on the soil surface. However, these soil moisture observations for drought monitoring also have the following limitations: (i) soil moisture measurements have higher uncertainties or unavailability instances over regions with dense vegetation, (ii) coarse resolution (~36 km) products; and (iii) satellite soil moisture validation requires *in situ* observations for different land cover and hydroclimatic regions.

To overcome the limitations of coarser satellite soil moisture estimates, various studies are carried out to improve the spatial resolution (i.e., 1 km, 3 km, and 9 km) of SMAP satellite missions–based products through different multisatellite data fusion techniques (Chan et al., 2018; Das et al., 2018, 2019). Soil moisture estimates based on these satellite missions are validated using point scale (*in situ*) soil moisture measurements over various land covers (i.e., grasslands, woody savannas, and croplands dominated by corn and soybean) and hydroclimatic domains (Colliander et al., 2018; Chan et al., 2018) to support their scientific applications. In one of the studies, Singh et al. (2019a) validated the SMAP-based enhanced soil moisture estimates, i.e., passive-only (9 km gridded) and active-passive SMAP–Sentinel (3 km) products for the Indian conditions under paddy-dominated land cover. The study found that the overall accuracy of the high-resolution SMAP products is satisfactory and encourages the soil moisture user community to use these products for scientific studies and applications. These satellite remote sensing–derived soil moisture products are available continuously and can be utilized for agricultural drought assessment over a large area with reasonable accuracy. In one such application, Mishra et al. (2017) utilized the SMAP soil moisture retrievals (at 36 km spatial resolution) to quantify agricultural drought indices over the United States using the Soil Water Deficit Index (SWDI) proposed by Martínez-Fernández et al. (2015).

24.2.2 AGRICULTURAL DROUGHT ASSESSMENT THROUGH REMOTE VEGETATION OBSERVATIONS

Vegetation biomass, chlorophyll, and plant canopy have a specific response to the incoming electromagnetic spectrum, which is reflected in their spectral signature. The spectral signature of a healthy plant rich in chlorophyll and dense in the canopy will differ from a plant that is under stress due to water shortage or nutrient deficiency. These unique spectral signatures of vegetation biomass, chlorophyll, and canopy characteristics under different conditions aid hugely in agricultural drought monitoring. The optical remote sensing data in the spectral range of 0.4 and 2.5 µm, which accurately classify the vegetation characteristics, have been widely used as inputs to agricultural drought indices (Dalezios et al., 2012). The red, near-infrared (NIR), and shortwave infrared (SWIR) are the commonly used bands under this spectral range to classify vegetation greenness and wetness conditions, and provide distinct inputs for agricultural drought conditions. Healthy vegetation usually contains more chlorophyll and often appears greener, and therefore, has the tendency to absorb most of the red band (visible spectrum) of incident energy while having more reflection in the NIR band. However, the unhealthy and sparse vegetation absorbs most of the NIR spectrum and has more reflection in the visible spectrum (Dalezios et al., 2012). In addition to the red band of the visible spectrum and NIR band, which help assess plant health in growing seasons, the SWIR band has significant capability to respond to vegetation water content that can be utilized for drought monitoring (Dalezios et al., 2012). Vegetation status obtained through remote sensing–based observations has been widely used for drought assessment using various drought indices, viz., Normalized Difference Vegetation Index (NDVI; Tucker and Choudhury, 1987), Leaf Water Content Index (LWCI; Hunt and Rock, 1989), Normalized Difference Water Index (NDWI; Gao, 1996), NDVI anomaly (NDVIA; Anyamba et al., 2001), Vegetation Condition Index (VCI; Kogan, 2002), Standardized Vegetation Index (SVI; Peters et al., 2002), and Vegetation Water Stress Index (VWSI; Ghulam et al., 2008).

24.3 METHODOLOGY

24.3.1 SCOPE OF THE PRESENT METHODOLOGY

Considering various drought indices have different capabilities in agricultural drought detection and a single indicator of drought might not be a robust estimator of agricultural drought, researchers have investigated the unified drought indices (Hao and AghaKouchak, 2013). Various drought indices, which showed different sensitivity to drought conditions even when applied to the same location, have been combined to derive one composite index. For example, the unified drought indices by combining vegetation greenness and wetness conditions are used as the Normalized Difference Drought Index (NDDI; Gu et al., 2007) and Normalized Moisture Index (NMI; Jang et al., 2006). The Vegetation Health Index (VHI; Kogan, 2002) was the first mathematical approach–based unified drought index, wherein the Vegetation Condition Index and Temperature Condition Index were integrated directly through mathematical operations to assess the vegetation health condition for identifying drought-affected areas in agricultural dominated regions. Further, the mathematical operations were also applied to develop the Microwave Integrated Drought Index (MIDI; Zhang and Jia, 2013) for semiarid regions, which integrates the Precipitation Condition Index, Soil Moisture Condition Index, and Temperature Condition Index using a microwave remote sensing data set of different satellite missions. However, predefined weights are needed for combining different drought indices into a unified or composite drought index through mathematical operations. In reality, the weights assigned in mathematical operations may not always be true for all the locations due to spatial heterogeneity existing over a large area. Therefore, it is imperative to combine different drought indices by assigning reasonable weights based on the degree of the importance of each considered indices.

This chapter uses an integrated approach to develop a Soil Moisture–Vegetation Stress–Based Agricultural Drought Index (SVADI) by combining remote sensing observations of soil moisture and vegetation observations for agricultural drought monitoring at a local scale. In this approach, the SMAP surface soil moisture and soil water parameter were used over eastern India to derive a soil moisture stress–based drought index, whereas the Advanced Very High-Resolution Radiometer (AVHRR)–derived NDVI was used to a derive vegetation stress–based drought index over the same area. A spatially dynamic way of combining the soil moisture stress– and vegetation stress–based drought indices was used to estimate the SVADI. Shannon's entropy formula (Soleimani-damaneh and Zarepisheh, 2009) was employed to determine the relative importance of each of the considered drought indices spatially. Shannon's entropy helps achieve the appropriate weight assignments by utilizing the information contained in both indices' time series data sets. The derived SVADI was assessed at a spatial resolution of 9 km on a weekly timescale using SMAP-enhanced passive-only soil moisture products (gridded at 9 km) with a temporal resolution of ~1–2 days. The SMAP–Sentinel active-passive soil moisture product has a higher spatial resolution (3 km) than the SMAP-enhanced passive-only product, however, it was not selected for the study due to its poor temporal resolution (~12 days). The integration of local-scale soil moisture–vegetation stress as a drought indicator can incorporate the location-specific critical factors affecting crop water stress, viz., vegetation health, soil characteristics, and soil moisture condition. The mathematical expressions of the drought indices used in this study are presented in the following sections.

24.3.2 DROUGHT INDICES

24.3.2.1 Vegetation Condition Index (VCI)

The VCI, a widely applied vegetation stress–based drought indicator, was derived using AVHRR multiyear NDVI observations (Kogan and Sullivan, 1993; Kogan, 1995, 1997, 2002). The VCI describes the NDVI-based current vegetation condition of a location (grid cell) concerning the long-term record of the NDVI for the same grid cell. It can be estimated by the normalization of the NDVI with its maximum and minimum range for each grid cell. The value of the VCI ranges from 0 to 1, corresponding to changes in vegetation conditions from extremely unfavorable (vegetation stress) to optimal, respectively. In this study, the VCI is computed on a weekly timescale as follows:

$$VCI = \frac{NDVI_t - NDVI_{\min}}{NDVI_{\max} - NDVI_{\min}} \tag{24.1}$$

where $NDVI_t$ is the NDVI of the current week, and $NDVI_{max}$ and $NDVI_{min}$ are the maximum and minimum values of NDVI in the multiyear time series of NDVI corresponding to the current week.

24.3.2.2   Soil Moisture Stress–Based Index (SMI)

In the present study, the Soil Moisture Stress–Based Index (SMI) is defined based on the concept of the crop water stress–based drought indicator (Rodriguez-Iturbe et al., 1999a, b) and SWDI (Martínez-Fernández et al., 2015). Rodriguez-Iturbe et al. (1999a, 1999b) proposed the concept of "static" crop stress, which is a function of soil moisture stress. Here, the soil moisture stress takes a value of 0 when soil moisture is above the level of incipient stomatal closure (S^*) and reaches a maximum value of 1 when the instant wilting starts (S_W; permanent wilting point). The stress increases nonlinearly between these two limits. A detailed description of the merit of using the concept of a crop water stress–based drought indicator is provided in Ramadas and Govindaraju (2015). Similarly, the SWDI (Martínez-Fernández et al., 2015) has shown the capability to capture agricultural drought conditions by quantifying associated soil moisture stress based on soil moisture and basic soil water parameters, i.e., soil field capacity and wilting point. The SWDI yields positive values when the soil moisture content is higher than the field capacity (excessive water for crop growth); it becomes 0 when the soil moisture is at the field capacity of water content, i.e., no water deficit. Only negative SWDI, which takes place when soil moisture goes below field capacity, signifies the drought condition. The SWDI is considered to be an effective agricultural drought index, as it represents the soil moisture suction capacity (Martínez-Fernández et al., 2015, 2016).

The purpose of incorporating both the concepts of the crop water stress–based drought indicator and the SWDI was to define the SMI similarly to the VCI in the range of 0 (very dry condition) to 1 (wet, favorable condition). The SMI was computed on a weekly timescale using the SMAP surface soil moisture data set as follows:

$$SMI = \begin{cases} 0 & \text{if } SM_t \le WP \\ \dfrac{SM_t - WP}{FC - WP} \\ 1 & \text{if } SM_t \ge FC \end{cases} \tag{24.2}$$

where SM_t is the soil moisture status of the current week (m³/m³), FC is the field capacity of the soil (m³/m³), and WP is the wilting point of the soil (m³/m³).

When the soil moisture status of a week in a specific grid cell of the SMAP product is less than or equal to the wilting point of that grid cell, the SMI has a value equal to 0, whereas the maximum value of 1 when the weekly soil moisture status becomes greater than or equal to the field capacity of the soil. As mentioned earlier, the SMI ranges between 0 and 1, where the values toward 0 signify the drought condition (water deficit), and the values approaching 1 represent adequate moisture availability for crop growth, making it comparable to the VCI. The soil moisture stress decreases nonlinearly between these two limits (Ramadas and Govindaraju, 2015). The primary soil water parameters (i.e., field capacity and wilting point) play a critical role in defining the range over which soil water is available to crops. It is challenging to have *in situ* measurements of these soil water parameters that cover a large swathe of area. A few pedotransfer functions (PTFs) were utilized to

TABLE 24.1

Different Pedotransfer Functions (PTFs) Used for Estimation of Field Capacity (FC) and Wilting Point (WP) Values Using Soil Properties

Source and Details	PTFs for –33 kPa (sfc) and –1500 kPa (sw) Matric Potential
Rawls et al. (1982)(in cm³ cm⁻³)	sfc = 0.2576 + (–0.002) × S + 0.0036 × C + 0.0299 × O sw = 0.0260 + 0.005 × C + 0.0158 × Os
Rawls et al. (1983)(in cm³ cm⁻³)	sfc = 0.3486 + (–0.0018) × S + 0.0039 × C + 0.0228 × O + (–0.0738) × BD sw = 0.0854 +(–0.0004) × S + 0.0044 × C + 0.0122 × O + (–0.0182) × BD where BD is bulk density (in M/gm³)
Rao et al. (1988)(percent by volume)	sfc = 32.1793 + (–0.3184) × S + 0.4174 × C sw = 8.6870 + (–0.068) × S + 0.257 × C
Tomasella and Hodnett (1998)(in cm³ cm⁻³)	sfc = 4.046 + 0.426 × Si + 0.404 × C sw = 0.91 + 0.15 × Si + 0.396 × C
Adhikary et al. (2008)(percent by volume)	sfc = 56.37 + (–0.51) × S + (–0.27) × Si sw = 0.71 + 0.44 × C
Shwetha and Varija (2013)(in cm³ cm⁻³)	sfc = –4.263 + 0.00194 × S + 0.02839 × Si + 5.568 × BD–0.00005 × S × S–0.00011 × S × Si + 0.00106 × S × BD-0.00005 × Si × Si–0.01158 × Si × BD–1.78 × BD × BD sw = –1.076-0.00234 × S-0.00334 × Si+1.920 × BD–0.00003 × S × S+0.00003 × S × Si+0.00101 × S × BD+0.00006 × Si × Si–0.00077 × Si × BD–0.666 × BD × BD where BD is bulk density (in g cm⁻³)

Notes: % sand (S), % silt (Si), % clay (C), % organic matter content (O), bulk density (BD).

define the limits of the soil water parameters for each grid cell in the study region. At matric potentials –33 kPa and –1500 kPa, using soil texture information (percent of sand, silt, and clay, organic matter) and bulk density values, the field capacity (FC) and wilting point (WP), respectively, are estimated using PTFs given in Table 24.1 as described by Samantaray et al. (2018).

24.3.2.3 Soil Moisture–Vegetation Stress–Based Agricultural Drought Index (SVADI)

The SVADI is a synergistic integration of the SMAP-based SMI and AVHRR-derived VCI for agricultural drought monitoring. Both indices range from 0 to 1, corresponding to changes in soil moisture and vegetation conditions from extremely unfavorable to optimal, respectively. The SVADI is computed at a weekly timescale to capture short-term agricultural droughts. The SMI and VCI were combined to define the SVADI by assigning different weights based on the degree of the importance of each of the considered indices in a specific grid cell. This integration of these two drought indices not only allows us to consider the spatial heterogeneity existing over a large area but is also advantageous over the previously investigated drought indices, which were based on predefined weights for combining different drought indices based on mathematical operations. A Shannon's entropy formula (Soleimani-damaneh and Zarepisheh, 2009) was effectively utilized to determine the relative importance of the SMI and VCI. The following steps are used to assign appropriate weights to the SMI and VCI based on their degree of importance for defining the SVADI.

Step 1: Normalization of the time series of each index:

$$I_{ij} = \frac{I_{ij}}{\sum_{i=1}^{m} I_{ij}}, i = 1, 2, \ldots m; j = 1, 2, \ldots n \qquad (24.3)$$

where i represents the input vectors of each time series j.

Step 2: Estimation of entropy E_j as

$$E_j = -E_0 \sum_{i=1}^{m} I_{ij} . \ln I_{ij}, j = 1, 2, \ldots n \qquad (24.4)$$

where entropy constant $E_0 = (\ln m)^{-1}$ is based on the length of the input vector represented as m in Equation 24.3.

Step 3: Degree of diversification:

$$d_j = 1 - E_j, j = 1, 2, \ldots n \qquad (24.5)$$

where n represents the number of time series (j) is used for the analysis.

Step 4: Weight based on the degree of importance of the jth index:

$$w_j = \frac{d_j}{\sum_{j=1}^{n} d_j}, j = 1, 2, \ldots n \qquad (24.6)$$

Step 5: Then the combined index SVADI is given by

$$SVADI_i = \sum_{j=1}^{n} w_j . I_{ij}, i = 1, 2, \ldots m \; j = 1, 2, \ldots n \qquad (24.7)$$

24.3.3 Study Area

Eastern India, consisting of the five Indian states (Figure 24.1) of Bihar, Jharkhand, Chhattisgarh, Odisha, and West Bengal, was considered for this study. The total geographical area of eastern India is about 553,643 km². Most of east India lies on the east coast of India between the Bay of Bengal and the Indo-Gangetic Plain. The overall climate of this region is mainly warm-humid, and a composite climate exists with a cold climate in the extreme north of West Bengal. However, the interior states have a drier and slightly more extreme climate during the winter and summer seasons. The majority of Odisha and West Bengal is a highly warm-humid climate with medium to high rainfall, tropical and short winter with a mild temperature, and experiences very high temperatures during the summer season. The states of Bihar, Jharkhand, and Chhattisgarh fall under a composite climate having high temperatures in summer and cold in winter. Odisha and West Bengal have high humidity as compared to the other states of eastern India. Agriculture is the main occupation for the livelihood of the population in this region. However, the agriculture of this region is mostly rainfed and entirely dependent on the southwest monsoon. Although eastern India has a high mean annual rainfall (1200–2000 mm), the rainfed agriculture of this region frequently suffers from drought. The varying intensities of cyclones, droughts, and floods are reported almost every year in the various parts of the region. The major crop of the study area is paddy, which is usually cultivated during the southwest monsoon season. The SMAP-based soil moisture estimates at 9 km used in this study are validated by Singh et al. (2019a) using intensive *in situ* soil moisture measurements for the paddy-dominated land cover of the eastern Indian state of Odisha. The detailed sampling plan of the *in situ* measurement of soil moisture in the paddy fields is described in Singh et al. (2019b). The selected grids for the validation of SMAP-based 9 km soil moisture products covering *in situ* observations in the paddy-dominated region are presented in Figure 24.1.

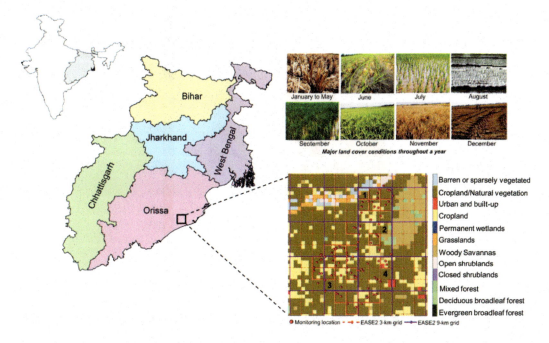

FIGURE 24.1 Map of eastern India region and the state boundaries. Experimental area land use/land coverage (from MODIS IGBP) map showing monitoring locations where soil moisture was measured for validating the SMAP products (red circles). The purple grid represents the SMAP-based soil moisture at a resolution of 9 km, and the thick grids are selected for validation. A view of the major field conditions of paddy crops throughout the year is presented through photographs.

24.3.4 DATA SETS USED

24.3.4.1 Soil Moisture Data Set

The NASA SMAP satellite coarse resolution (~36 km) L-band radiometer (1.41 GHz) observation-based enhanced passive-only soil moisture product (L3_SM_P_E; O'Neill et al., 2016) was used in this study. The SMAP radiometer instrument is designed to provide estimates of volumetric soil moisture in the top 5 cm of soil with an accuracy of 0.04 m^3 m^{-3} (1 standard deviation) over the global land area, excluding regions of snow and ice, mountainous topography, open water, and vegetation with total water content greater than 5 kg m^{-2} (Entekhabi et al., 2010). The L3_SM_P_E product is based on the reconstructed SMAP radiometer measurements at EASE 9 km grid resolution using the Backus–Gilbert optimal interpolation methodology (Chan et al., 2018). In this study, the L3_SM_P_E daily soil moisture product (version 1) is used, which was downloaded from the National Snow and Ice Data Centre (NSIDC; http://nsidc.org/data/SMAP/SMAP-data.html). The daily soil moisture time series was converted to a weekly timescale by averaging the soil moisture products for every calendar week. A weekly time series comprising a total of 165 weeks was developed using L3_SM_P_E daily soil moisture for the duration of April 2015 to May 2018. The spatial map of eastern India's averaged weekly soil moisture status from April 2015 to May 2018 is shown in Figure 24.2. The developed weekly soil moisture time series was used for estimation of the SMI for the study period.

24.3.4.2 Vegetation Data Set

A reprocessed NDVI data set derived from Visible Infrared Imaging Radiometer Suite (VIIRS) and AVHRR satellite observations by the Center for Satellite Applications and Research (STAR) was used for the estimation of the vegetation stress–based index. The reprocessed NDVI data

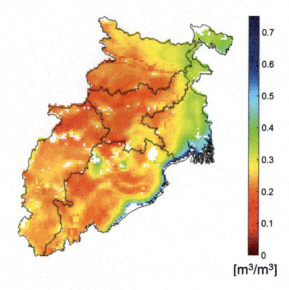

FIGURE 24.2 Spatial map at EASE2 9 km grid resolution of SMAP-enhanced product-based weekly averaged soil moisture from April 2015 to May 2018.

set is noise-free with a pre- and post-launch calibrated process and well-validated (Carter et al., 2020). The noise-free weekly (7 days) composite NDVI data sets at a spatial resolution of 4 km was acquired from the NOAA Office of Satellite and Product Operations (OSPO) website (https://www .star.nesdis.noaa.gov). A total of 165 weeks of the data set from the first week of April 2015 to the last week of May 2018 was acquired from the OSPO website for weekly agricultural drought monitoring. Owing to the different spatial resolutions of the NDVI and soil moisture, the NDVI at 4 km was area gridded at EASE 9 km grid resolution. A weekly averaged NDVI status of eastern India from April 2015 to May 2018 is presented in Figure 24.3.

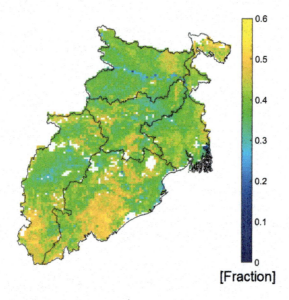

FIGURE 24.3 Spatial map at EASE2 9 km grid resolution of weekly averaged NDVI status (April 2015 to May 2018) using NOAA STAR–provided NDVI product.

24.3.4.3   Soil Properties Data Set

A new global gridded soil information, the SoilGrids250m (Hengl et al., 2017), released by the International Soil Reference and Information Centre (ISRIC), is used in this study. The soil properties, i.e., percent sand, percent silt, percent clay, percent organic matter content, and bulk density, were extracted over eastern India from the SoilGrids250m data set to estimate the field capacity and wilting point values utilizing the PTFs as described in Table 24.1. The available SoilGrids250m data set is aggregated within EASE 9 km grid cells to make a consistent spatial resolution for all the variables used in this study. The SoilGrids250m data set is available under the Open Database License and can be downloaded from https://soilgrids.org/.

24.4   RESULTS AND DISCUSSION

24.4.1   VALIDATION OF SMAP-ENHANCED PASSIVE SOIL MOISTURE PRODUCT

The enhanced passive soil moisture products (L3_SM_P_E) were validated over the paddy-dominated region of eastern India by Singh et al. (2019a). Singh et al. (2019a) carried out the performance analysis of the SMAP L3_SM_P_E product in four selected grids – Grid-1, Grid-2, Grid-3, and Grid-4 – as shown in Figure 24.1 using the *in situ* soil moisture measurements. A linear average of *in situ* observations within each selected EASE2 9 km grid was used to evaluate the performance of the SMAP_L3_SM_P_E product, which mitigates the spatial scale discrepancy between the upscaled *in situ* point observations and the SMAP-enhanced product of the 9 km grid cell. The comparison between SMAP_L3_SM_P_E and upscaled *in situ* soil moisture for two selected grids (Grid-1 and Grid-3) are shown in Figure 24.4, and the quantitative comparison for all four grids in terms of error metrics are presented in Table 24.3. It can be seen in Figure 24.4 that the SMAP-enhanced product has seasonal variability and it follows the trend of rainfall events (e.g., occurrence and dry downs). The quantitative comparison of the whole year's soil moisture dynamics shows that the SMAP-enhanced soil moisture product has a good correlation against *in situ* observations throughout a year with a reasonably acceptable correlation coefficient (R values) of 0.765 to 0.892. The overall performance of the SMAP_L2_SM_P_E product is found to be satisfactory with ubRMSE of 0.060–0.085 $m^3 \ m^{-3}$. The SMAP-enhanced soil moisture product was highly correlated (R value 0.924 to 0.952) against the *in situ* observations during the non-growing season and found to have a very good performance within the SMAP mission requirement, meeting the accuracy range of $\pm 0.04 \ m^3 \ m^{-3}$. However, during the growing season, the performance of these products was challenged due to erroneous ancillary information and found to have dry bias. The overall accuracy of the enhanced SMAP soil moisture products is found to be satisfactory and has

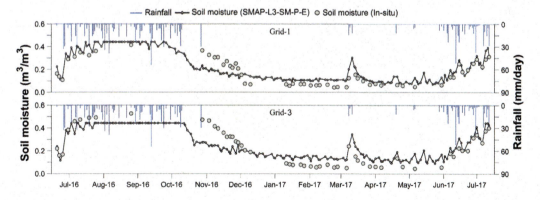

FIGURE 24.4   Time series comparison of SMAP-based soil moisture estimates with respect to upscaled *in situ* soil moisture observations within selected EASE2 9 km grids.

TABLE 24.2

Classification of SVADI for Different Drought Categories

SVADI Value	Drought Category
≥ 0.50	Normal condition (no drought)
0.30 to 0.50	Drought condition
0.10 to 0.30	Severe drought
≤ 0.10	Extreme drought

valuable information that can be utilized for geophysical applications. A detailed validation and methodology are described in Singh et al. (2019a).

24.4.2 COMPUTATION OF SVADI

The SVADI is computed for the entire eastern India region by combining the SMI and VCI, as explained in Section 24.3.2.3. The performance of the SMAP-enhanced product at 9 km grid resolution is found to be satisfactory over the paddy-dominated region of eastern India (as discussed earlier and reported in Singh et al., 2019a) and contains valuable information that can be used for agricultural drought studies. Thus, SMAP-enhanced product-based weekly soil moisture status at 9 km grid resolution was utilized for assessment of the SMI at a weekly timescale based on Equation 24.2 and soil water characteristics (i.e., field capacity and wilting point derived using PTFs given in Table 24.1) over eastern India. The estimated average field capacity and wilting point values over eastern India at the spatial resolution of 9 km are illustrated in Figure 24.5. After computing the SMI, the VCI was computed at a weekly timescale using NOAA STAR–provided NDVI product following Equation 24.1. Both the drought indices SMI and VCI vary between 0 and 1, where the values toward 0 signify the drought condition and the values approaching 1 represent the favorable condition.

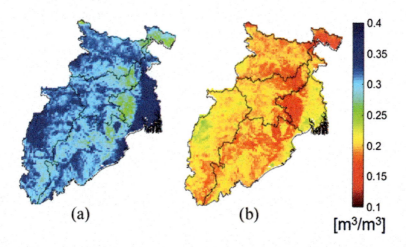

FIGURE 24.5 Spatial map of (a) field capacity and (b) wilting point of the soil over eastern India at EASE2 9 km grid resolution. The spatial maps represent the average value of field capacity and wilting point derived through various pedotransfer functions (PTFs) and soil properties described in Table 24.1.

TABLE 24.3

Performance Assessment of the SMAP-Based Soil Moisture Estimates with *in situ* Observations during Whole Year (WY), Growing Season (GS), and Non-Growing Season (NG)

Grid	ubRMSE (m⁻³ m⁻³)			Bias (m⁻³ m⁻³)			RMSE (m³/m³)			R		
Location	WY	GS	NGS	WY	GS	NGS	WY	GS	NGS	WY	GS	NGS
Grid-1 (17)	0.060	0.078	0.025	0.015	−0.034	0.039	0.062	0.086	0.046	0.830	0.697	0.952
Grid-2 (10)	0.077	0.097	0.034	0.027	−0.037	0.059	0.081	0.104	0.068	0.765	0.416	0.924
Grid-3 (20)	0.085	0.064	0.036	0.005	−0.096	0.054	0.085	0.115	0.065	0.843	0.795	0.943
Grid-4 (12)	0.061	0.063	0.036	0.032	−0.026	0.060	0.069	0.068	0.070	0.892	0.749	0.942

Source: Singh et al. (2019).

Notes: The number of *in situ* point observations within each selected EASE2 9 km grid is shown in parentheses. RMSE: root-mean-square error, ubRMSE: unbiased RMSE, *R*: correlation coefficient.

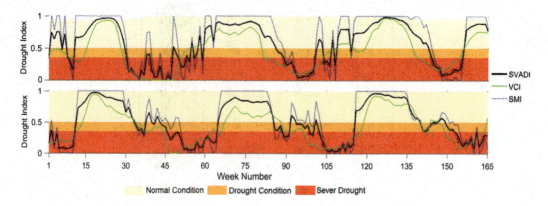

FIGURE 24.6 Comparison of drought monitoring using the SVADI, VCI, and SMI in the selected grid locations of Odisha state having land use/land cover class of cropland/natural vegetation (upper panel) and cropland (lower panel). The values between 0.50 and 1.00 represent the normal condition, 0.35–0.50 represent the drought condition, and values <0.35 represent severe drought. The weights allocated to VCI and SMI at these two locations were 0.474 and 0.526 (upper panel), and 0.426 and 0.574 (lower panel), respectively.

Once the SMI and VCI are estimated, the unified drought index SVADI was defined by assigning appropriate weights to the SMI and VCI based on their degree of importance utilizing Shannon's entropy formula, as described in Section 24.3.2.3. It was found that Shannon's entropy helps in accounting for the spatial heterogeneity existing over a large area and assigns different weights for each grid cell based on the characteristics of the SMI and VCI time series for that particular grid cell. To show the capability of Shannon's entropy formula for defining the SVADI, the comparison of agricultural drought assessment using the SVADI, VCI, and SMI for the selected grid cells of Odisha state across different land use is presented in Figure 24.6. It was found that Shannon's entropy assigns different weights in both of the selected grid cells and helps to classify the appropriate drought class based on the degree of importance of the VCI and SMI. The weekly time series of drought indices show a seasonal pattern of drought evolution in the selected grids (Figure 24.6). The proposed index, SVADI, is site-specific and considers vegetation stress as well as soil water stress. Because the SVADI has a combined signature of both the SMI (which reflects soil moisture stress)

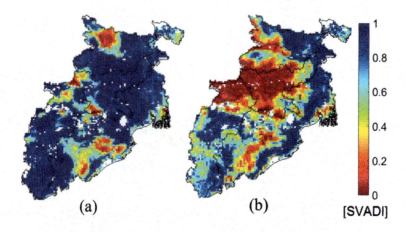

FIGURE 24.7 Area showing agricultural drought situations in eastern India using SVADI during (a) June 22–28, 2015, and (b) September 28–October 4, 2015.

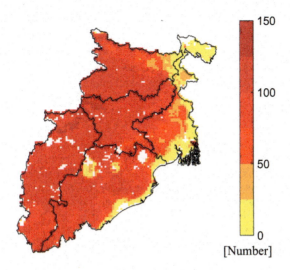

FIGURE 24.8 Agricultural drought hotspot map for eastern India based on well-known VCI and developed SVADI drought indices based on the number of drought events identified from April 2015 to May 2018. The lighter shades indicate fewer drought events, while dark shades correspond to increased drought susceptibility. The white color corresponds to the region with no data.

and VCI (which reflects vegetation health), it can monitor agricultural drought better than that of a stand-alone single-drought index, viz., the SMI or VCI in this case.

The spatial distribution of the prevalent situation of agricultural drought identified using SVADI over eastern India for the weeks of June 22–28, 2015 (wet period) and September 28–October 4, 2015 (dry-down period) are presented in Figure 24.7. The spread of drought conditions is found more after the monsoon season (Figure 24.7b; dry-down period) as compared to the initial period of the monsoon (Figure 24.7a; wet period).

The estimated SVADI at a weekly timescale over eastern India at a 9 km grid location for the period of April 2015 to May 2018 (165 weeks) were utilized for hotspot analysis of agricultural drought. In this regard, the occurrences of severe agricultural drought were estimated across all the grids by counting the number of events whose magnitude exceeded the severe category (Table 24.2) during the 165 weeks. The higher number of severe and extreme drought events indicates drought propensity. These drought hotspots can be identified based on the reasonable consensus achieved from the SVADI-based agricultural drought severity maps shown in Figure 24.8.

The hotspot analysis suggested that other than West Bengal, most eastern Indian states have greater drought susceptibility (Figure 24.8). However, in Odisha, the coastal regions are found to have no drought conditions compared to other parts of the state. South Bihar has greater drought susceptibility than north Bihar. The agriculture in the eastern region of India is predominantly rainfed and vulnerable to hydrological extremes. The identified hotspot region from these drought severity maps needs drought adaptation and mitigation support in the form of additional irrigation facilities and sustainable cultivation practices to minimize crop losses.

24.5 CONCLUSIONS

The proposed research adopted a soil moisture–vegetation stress–based drought index for agricultural drought monitoring over eastern India. The soil moisture stress–based index is a function of the soil moisture and soil characteristics, whereas vegetation stress is derived using NDVI climatology. The site-specific analysis with the integration of soil moisture and vegetation condition is a strength of the proposed framework. The study reveals that Shannon's entropy can effectively assign different weights across the space to classify the appropriate drought class based on the degree of importance of soil moisture stress and vegetation stress conditions. The proposed framework can be helpful to provide a better representation of agricultural drought status compared to previously used single-variable indices. Further, based on drought occurrences, the critically stressed pockets in the region could be identified for drought management and planning.

REFERENCES

Adhikary PP, Chakraborty D, Kalra N, Sachdev CB, Patra AK, Kumar S, et al. (2008). Pedotransfer functions for predicting the hydraulic properties of Indian soils. *Soil Res*, 46(5), 476–484.

Anjum SA, Xie XY, Wang LC, Saleem MF, Man C, Lei W (2011). Morphological, physiological, and biochemical responses of plants to drought stress. *Afr J Agric Res*, 6, 2026–2032.

Anyamba A, Tucker C, Eastman J (2001). NDVI anomaly patterns over Africa during the 1997/98 ENSO warm event. *Int J Remote Sens* 22, 1847–1859.

Belle JA, Hlalele, MB (2015). Vulnerability assessment of agricultural drought hazard: A case of Koti-Se-Phola community council, Thabana Morena, Mafeteng district in Lesotho. *J Geogr Nat Disast*, 5(143), 2167–0587.

Bergman, KH, Sabol, P, Miskus D (1988). Experimental indices for monitoring global drought conditions. *Proceedings 13th Annual Climate Diagnostics Workshop*, US Dept. of Commerce, Cambridge, MA, pp. 190–197.

Bisht DS, Sridhar V, Mishra A, Chatterjee C, Raghuwanshi NS (2019). Drought characterization over India under projected climate scenario. *Int J Climatol*, 39, 1889–1911.

Carter C, Chen M, Jiang Z, Yu Y (2020). Vegetation index product algorithm theoretical basis document (version 2.0), NOAA NESDIS Center for Satellite Applications and Research (STAR). Available at: https://www.star.nesdis.noaa.gov/jpss/documents/ATBD/ATBD_VIIRS_VI_v2.0.pdf

Chan SK, Bindlish R, O'Neill PE, Jackson T, Njoku EG, et al. (2018). Development and assessment of the SMAP enhanced passive soil moisture product, *Remote Sens Environ*, 204, 931–941.

Chen J, Yang Y (2011). A fuzzy ANP-based approach to evaluate region agricultural drought risk. *Procedia Eng*, 23, 822–827.

Colliander A, Jackson TJ, Chan SK, O'Neill P, Bindlish, R, et al. (2018). An assessment of the differences between spatial resolution and grid size for the SMAP enhanced soil moisture product over homogeneous sites. *Remote Sens Environ*, 207, 65–70.

Dalezios NR, Blanta A, Spyropoulos NV (2012). Assessment of remotely sensed drought features in vulnerable agriculture. *Nat Hazards Earth Syst Sci*, 12: 3139–3150.

Das NN, Entekhabi D, Dunbar RS, Chaubell MJ, Colliander A, et al. (2019). The SMAP and Copernicus Sentinel 1A/B microwave active-passive high resolution surface soil moisture product. *Remote Sens Environ*, 233, 111380.

Das, NN, Entekhabi D, Dunbar, RS, Colliander A, Chen, F, Crow W, et al. (2018). The SMAP mission combined Active-Passive soil moisture product at 9 km and 3 km spatial resolutions. *Remote Sens Environ*, 211, 204–217.

Deng M, Di L, Han W, Yagci A, Peng C, Heo G (2013). Web-service-based monitoring and analysis of global agricultural drought. *Photogramm Eng Remote Sens*, 79 (10), 929–943.

Dorigo WA, Gruber A, De Jeu RA, Wagner W, Stacke T, Loew A, Albergel C, Brocca L, Chung D, Parinussa RM, Kidd R (2015). Evaluation of the ESA CCI soil moisture product using ground-based observations. *Remote Sens. Environ*, 162, 380–395.

Entekhabi D, Njoku EG, O'Neill PE, Kellogg KH, et al. (2010). The soil moisture active passive (SMAP) mission. *Proc. IEEE*, 98, 704–716.

Fernández-Prieto D, Van Oevelen P, Su Z, Wagner W (2012). Advances in Earth observation for water cycle science. *Hydrol. Earth Syst Sci*, 16, 543–549.

Gao B (1996). NDWI-A normalized difference water index for remote sensing of vegetation liquid water from space. *Remote Sens Environ*, 58, 257–266.

Ghaleb F, Mario M, Sandra A N (2015). Regional Landsat-based drought monitoring from 1982 to 2004, *Climate*, 3, 563–577.

Ghulam A, Li ZL, Qin Q, et al. (2008). Estimating crop water stress with ETM$^+$ NIR and SWIR data. *Agric For Meteorol*, 148, 1679–1695.

Gu, Y, Brown, JF, Verdin, JP, et al. (2007). A five-year analysis of MODIS NDVI and NDWI for grassland drought assessment over the central Great Plains of the United States. *Geophys Res Lett*, 34, L06407.

Hao Z, AghaKouchak A (2013). Multivariate Standardized Drought Index: A parametric multi-index model. *Adv Water Resour*, 57, 12–18.

Hazaymeh K, Hassan QK (2015). Spatiotemporal image-fusion model for enhancing the temporal resolution of Landsat-8 surface reflectance images using MODIS images. *J Appl Remote Sens*, 9, 096095.

Hengl T, Mendes de Jesus J, Heuvelink GBM, Ruiperez Gonzalez M, Kilibarda M, Blagotić A, et al. (2017). SoilGrids250m: Global gridded soil information based on machine learning. *PLoS ONE*, 12, e0169748.

Hollinger SE, Isard SA, Welford MR (1993). A new soil moisture drought index for predicting crop yields. In: Eighth conference on applied climatology, American Meteorological Society, Anaheim (CA), 17–22 January 1993, AMS, pp 187–190.

Huete A, Didan K, Miura T, Rodriguez EP, Gao X, Ferreira, LG (2002). Overview of the radiometric and biophysical performance of the MODIS vegetation indices. *Remote Sens Environ*, 83, 195–213.

Hunt ER, Rock BN (1989). Detection of changes in leaf water content using near-and middle-infrared reflectance. *Remote Sens Environ*, 30, 43–54.

Jang J, Viau A, Anctil F (2006). Thermal-water stress index from satellite images. *Int J Remote Sens* 27, 1619–1639.

Kerr YH, Waldteufel P, Wigneron, JP, Delwart S, Cabot, F, et al. (2010). The SMOS Mission: New Tool for Monitoring Key Elements of the Global Water Cycle. *Proc IEEE 98*, 666–687.

Kogan FN (1995). Application of vegetation index and brightness temperature for drought detection. *Adv Space Res*, 15, 91–100.

Kogan FN (2002). World droughts in the new millennium from AVHRR-based vegetation health indices. *Eos Trans Am Geophys Union*, 83, 557.

Kogan FN (1997). Global drought watch from space. *Bull Am Meteorol Soc*, 78, 621–636.

Kogan FN, Sullivan J (1993). Development of global drought-watch system using NOAA/AVHRR data. *Adv Space Res*, 13, 219–222.

Liu WT, Kogan F (1996). Monitoring regional drought using the Vegetation Condition Index, *Int J Remote Sens*, 17, 2761–2782.

Maity R, Sharma A, Nagesh Kumar D, Chanda K (2012). Characterizing drought using the reliability-resilience-vulnerability concept. *J Hydrol Eng*, 18, 859–869.

Martínez-Fernández J, Gonzalez-Zamora A, Sanchez N, Gumuzzio A (2015). A soil water based index as a suitable agricultural drought indicator. *J Hydrol* 522, 265–273.

Martínez-Fernández J, Gonzalez-Zamora A, Sanchez N, Gumuzzio A, Herrero-Jimenez C.M. (2016). Satellite soil moisture for agricultural drought monitoring: Assessment of the SMOS derived Soil Water Deficit Index. *Remote Sens Environ*, 177, 277–286.

Mecklenburg S, Drusch M, Kaleschke L, Rodriguez-Fernandez N, Reul N, Kerr Y, Font J, et al., (2016) ESA's soil moisture and ocean salinity mission: From science to operational applications. *Remote Sens Environ*, 180(1), 3–18.

Mishra A, Vu T, Veettil AV, Entekhabi D (2017). Drought monitoring with soil moisture active passive (SMAP) measurements. *J Hydrol*, 552, 620–632.

Mishra AK, Ines AV, Das NN, Khedun CP, Singh VP, Sivakumar B, Hansen JW (2015). Anatomy of a local-scale drought: Application of assimilated remote sensing products, crop model, and statistical methods to an agricultural drought study. *J Hydrol*, 526, 15–29.

Narasimhan B, Srinivasan R (2005). Development and evaluation of Soil Moisture Deficit Index (SMDI) and Evapotranspiration Deficit Index (ETDI) for agricultural drought monitoring. *Agric For Meteorol*, 133, 69–88.

O'Neill PE, Chan SK, Njoku EG, Jackson T, Bindlish R (2016). *SMAP Enhanced L3 Radiometer Global Daily 9 km EASE-grid Soil Moisture*, version 1, Distrib. Act. Arch. Center, NASA Nat. Snow Ice Data Center, Boulder, CO, USA, doi:10.5067/ZRO7EXJ8O3XI.

Palmer WC (1968). Keeping track of crop moisture conditions, nationwide: The new crop moisture index. *Weatherwise*, 21, 156–161.

Pannu US, Sharma TC (2002). Challenges in drought research: Some perspectives and future directions, *Hydrol Sci J*, 47(S), S19–S30, https://doi.org/10.1080/02626660209493019

Peters A, Walter-Shea E, Ji L (2002). Drought monitoring with NDVI-based standardized vegetation index. *Photogarmm Eng Remote Sens*, 68, 71–75.

Potopová V, Boroneanţ C, Boincean B, Soukup J (2016). Impact of agricultural drought on main crop yields in the Republic of Moldova. *Int J Climatol*, 36, 2063–2082.

Ramadas M, Govindaraju RS (2015). Probabilistic assessment of agricultural droughts using graphical models. *J Hydrol*, 526, 151–163.

Rao KS, Narasimha Rao PV, Mohan BK, Venketachalam P (1988). Relation between water retention characteristics of soil and their physical properties. *J Soil Water Conserv, India*, 32, 52–69.

Rawls WJ, Brakensiek DL, Saxtonn, KE (1982). Estimation of soil water properties. *Trans ASABE*, 25, 1316–1320.

Rawls WJ, Brakensiek DL, Soni B (1983). Agricultural management effects on soil water processes part I: Soil water retention and Green and Ampt infiltration parameters. *Trans ASABE*, 26, 1747–1752.

Rebel KT, de Jeu RAM, Ciais P, Viovy N, Piao SL, Kiely G, Dolman AJ (2012). A global analysis of soil moisture derived from satellite observations and a land surface model, *Hydro Earth Syst Sci*, 16, 833–847.

Rembold F, Meroni M, Rojas O, Atzberger C, Ham F, Fillol E (2015). Agricultural drought monitoring using space-derived vegetation and biophysical products: A global perspective. *Remote Sens Water Resour Disasters Urban Stud, Remote Sensing Handbook*, 349–365. CRC Press, Taylor and Francis Group.

Rodriguez-Iturbe I, D'odorico P, Porporato A, Ridolfi L (1999a). On the spatial and temporal links between vegetation, climate, and soil moisture. *Water Resour Res*, 35, 3709–3722.

Rodríguez-Iturbe I, D'Odorico P, Porporato A, Ridolfi L (1999b). Tree-grass coexistence in Savannas: The role of spatial dynamics and climate fluctuations. *Geophys Res Lett*, 26, 247–250.

Roy DP, Wulder MA, Loveland TR, Woodcock CE, Allen RG, Anderson MC, et al. (2014). Landsat-8: Science and product vision for terrestrial global change research. *Remote Sens Environ*, 145, 154–172.

Samantaray AK, Singh G, Ramadas M, Panda RK. (2018). Drought hotspot analysis and risk assessment using probabilistic drought monitoring and severity–duration–frequency analysis. *Hydrol Process*, 33, 432–449.

Shwetha P, Varija K (2013). Soil water-retention prediction from pedotransfer functions for some Indian soils. *Arch Agron Soil Sci*, 59, 1529–1543.

Singh G, Das NN, Panda RK, Colliander A, Jackson T, Mohanty BP, Entekhabi D, Yueh S (2019a). Validation of SMAP soil moisture products using ground-based observations for the paddy dominated tropical region of India. *IEEE Trans Geosci Remote Sens*, 57, 8479–8491.

Singh G, Panda RK, Bisht DS (2021). Improved generalized calibration of an impedance probe for soil moisture measurement at regional scale using Bayesian neural network and soil physical properties. *J Hydrol Eng*, 26, 04020068

Singh G, Panda RK, Mohanty BP (2019b). Spatiotemporal analysis of soil moisture and optimal sampling design for regional scale soil moisture estimation in a tropical watershed of India. *Water Resour Res*, 55, 2057–2078.

Soleimani-damaneh M, Zarepisheh M (2009). Shannon's entropy for combining the efficiency results of different DEA models: Method and application. *Expert Syst Appl*, 36, 5146–5150.

Tomasella J, Hodnett MG (1998). Estimating soil water retention characteristics from limited data in Brazilian Amazonia. *Soil Sci*, 163, 190–202.

Tucker CJ, Choudhury BJ (1987). Satellite remote sensing of drought conditions. *Remote Sens Environ*, 23, 243–251.

Walker JJ, de Beurs KM, Wynne RH, Gao F (2012). Evaluation of Landsat and MODIS data fusion products for analysis of dryland forest phenology. *Remote Sens Environ*, 117: 381–393.

Wang L, Qu JJ (2009). Satellite remote sensing applications for surface soil moisture monitoring: A review. *Front Earth Sci China*, 3, 237–247.

Wilhite DA, Glantz MH (1985). Understanding the drought phenomenon: The role of definitions. *Water Int*, 10, 111–120.

Zhang A, Jia G (2013). Monitoring meteorological drought in semiarid regions using multi-sensor microwave remote sensing data. *Remote Sens Environ*, 134, 12–23.

25 Application of Drought Monitoring Tools for Wildfire Hazard Assessment in Forests of India

N. Kodandapani

CONTENTS

25.1 INTRODUCTION

Recent studies indicate that climatic controls of fuel availability and fuel moisture have a dominant influence on fire regimes in ecosystems across the globe (Littell et al. 2016; Andela et al. 2017). Drought is particularly problematic due to its impacts on both natural and human systems, and is the most commonly occurring climatic extreme with consequences for forest fire activity (Siegert et al. 2001; Aragao et al. 2007; Mann and Gleick 2015). Numerous studies have documented the relationships between interannual variability in drought and the occurrence of fires in different parts of the globe (Archibald et al. 2010; Jolly et al. 2015; Bowman et al. 2017; Kodandapani and Parks 2019). There are several drivers of wildfires, and these include changes in land management and climate change (Bowman et al. 2014; McLauchlan et al. 2020). Approximately, about 3% of the Earth's surface burns each year and landscape fires account for emissions of about 2.0 Gt C year^{-1} (Van der Werf et al. 2017). Anthropogenic-caused climate change has led to a longer fire-weather season (the number of days with fire danger metrics above the median) across the globe, by approximately 20%, between 1970 and 2015 (Jolly et al. 2015).

The impacts of droughts on ecosystem processes such as tree die-off and occurrences of wildfires have been investigated in several parts of the globe (Allen et al. 2010; Dimitrakopoulous et al. 2011; Jolly et al. 2015; Westerling 2016; Parks et al. 2018). Drought effects fire activity directly in ecosystems with changes in fuel moisture, and several metrics have been used in the literature to model fire spread, burned area, and forecast fire activity (Quiring 2009; Littell et al. 2016). Various drought metrics are commonly used in forest fire research, including the Palmer Drought Severity Index (PDSI), which uses a water balance algorithm, wherein precipitation contributes to soil moisture, and temperature-driven evapotranspiration removes soil moisture. Similarly, precipitation anomalies have also been used over variable time intervals, such as monthly or seasonal, to link drought and fire activity in several ecosystems (Aragao et al. 2007; Riley et al. 2013; Abatzoglou and Kolden 2013). A commonly used drought metric to forecast fire occurrences is the Standardized Precipitation Index (SPI). The SPI is a measure of meteorological drought, estimated by the difference of precipitation from the mean, divided by the standard deviation for a specified time period (Riley et al. 2013).

Here, analyses of drought and fire occurrences conducted in two different landscapes in the Western Ghats mountain range are described. The first analysis was conducted at the Nilgiris landscape and the second analysis was conducted at the Uttara Kannada landscape. Although there are several different forest types in the Western Ghats, they can be grouped into two categories (dry versus moist), which depend on the amount of annual rainfall in the Western Ghats. At the landscape scale, climatic conditions are relatively more homogeneous, whereas differences persist in terms of vegetation type and topography, which determines the spatial pattern of fire occurrences at this scale. Accordingly, two landscapes were chosen – one predominantly moist and the other dry – to control for differences in vegetation at this scale, while simultaneously exhibiting important variations in the annual rainfall. Nevertheless, the landscape scale using detailed fire history spatial data spanning from 1996 to 2015 would provide important insights into fire activity and drought. Specifically, this study aimed to (1) assess the temporal pattern of drought at different timescales using the SPI in the two landscapes, and (2) evaluate the relationships between the SPI and wildfire occurrences at the landscape scale.

25.2 METHODS

25.2.1 STUDY AREAS

The Western Ghats is a mountain range in southwest India and is one of the 34 global hotspots of biodiversity (Mittermeier et al. 2005), simultaneously it is the hotspot with the highest human densities (Cincotta et al. 2000). The two landscape-scale study sites were located in the central Western Ghats (see Figure 25.1): the Nilgiris landscape (latitude 11.7° N, longitude 76.5° E) and the Uttara Kannada landscape (latitude 15.2° N, longitude 74.7° E). The elevation of these landscapes varies from 0 to 1450 m and from 39 to 995 m asl, respectively. The mean annual precipitation ranges from about 600 mm to 2000 mm and from 800 to 2500 mm, respectively. The Nilgiris landscape is comprised of tropical dry deciduous forests (65%), tropical moist deciduous forests (10%), tropical dry thorn forests (20%), and settlements (5%) (Kodandapani et al. 2004; Kodandapani et al. 2008). In the Nilgiris landscape, the dry season extends between January and March (Kodandapani et al. 2004). The Nilgiris landscape is comprised of three protected areas: the Bandipur Tiger Reserve, the Mudumalai Tiger Reserve, and Wayanad Wildlife Sanctuary. Protected areas, in general, have various management objectives ranging from strict biodiversity conservation to permitting human activities in certain zones (Jones et al. 2018). In the Mudumalai Tiger Reserve, extraction of non-timber forest products (NTFPs) is banned and so are fodder extractions and grazing, with the exception of the eastern part of the reserve. Similarly, in the Bandipur Tiger Reserve, the extraction of NTFPs is legally banned. However, in the Wayanad Wildlife Sanctuary, the extraction of NTFPs is permitted through the issue of permits (Narendran et al. 2001). The Nilgiris landscape is especially critical from a biodiversity conservation standpoint; it has the highest densities and the largest population of two endangered species in the world, the Asian elephant (*Elephas maximus*) and Bengal tiger (*Panthera tigris*) (Jhala et al. 2015; Mehta et al. 2008).

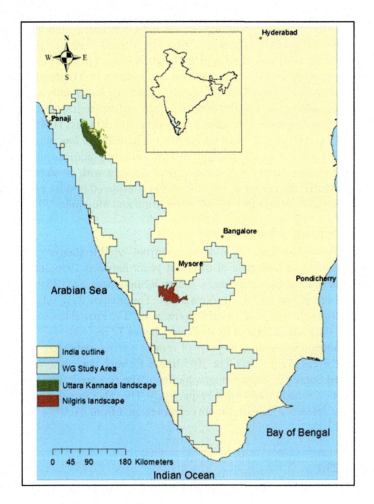

FIGURE 25.1 Location of the Western Ghats study area in India, and the Nilgiris and Uttara Kannada landscapes in the Western Ghats.

The Uttara Kannada landscape is comprised of tropical moist deciduous forests (51%), tropical dry deciduous forests (20%), tropical dry thorn forests (9%), non-forest/agriculture (12%), forest plantations (6%), and water bodies (2%) (Pascal 1986). In the Uttara Kannada landscape, the dry season extends from October to May (Puyravaud et al. 1994). The Uttara Kannada landscape is an intricate socioecological landscape, wherein communities living there have usufruct rights to collect certain forest products. Forests in this landscape are earmarked by the state government for the collection of non-timber forest products by local communities (Puyravaud et al. 1994). Farmers living in the landscape collect leaf litter, native grass species, fuelwood, and graze livestock in forests. These forests are subjected to annual fires, and almost all fires in these two landscapes are mainly due to human causes (Kodandapani et al. 2004; Kodandapani et al. 2008; Mehta et al. 2008; Mondal and Sukumar 2013).

25.2.2 Landscape Fire Analysis and Drought Interactions

25.2.2.1 Rainfall Data Sets and Drought Indicator

Drought and precipitation variability was characterized by applying the SPI (Hayes et al. 1999). This index provides temporal flexibility to assess drought across multiseason and multiyear time intervals. The SPI scales precipitation in units of standard deviations from mean precipitation for each location and time period (Albright et al. 2010). For SPI calculation, the long-term precipitation record over a specific time period for a given location is used to fit a gamma distribution that places the mean long-term SPI for that time interval and location to 0 (Albright et al. 2010). Indian Meteorological Department gridded (0.25° latitude × 0.25° longitude) daily rainfall data sets have been used in this study (Pai et al. 2014). One pixel each in the geographic center of the Nilgiris landscape and the Uttara Kannada landscape, over a period of 30 years (1986–2015), was chosen. Meteorological droughts are defined based on the SPI to characterize low rainfall events. Since the primary interest of this study is in accumulated rainfall deficits to address meteorological drought conditions, the SPI-based definition of meteorological drought with an averaging period of one month is used. Statistically, the period of 1–12 months is considered best for estimating the SPI and the index behaves erratically at shorter timescales (Sharma and Mujumdar 2017).

25.2.2.2 Remote Sensing of Burned Areas

Fire maps for the Nilgiris landscape were developed from satellite imagery (Table 25.1). For the Nilgiris landscape, remote sensing data of different years between 1996 and 2015 were used to delineate annually burned and unburned areas. Indian Remote Sensing (IRS) satellite imagery was used to classify the burned and unburned forest areas. The images were subjected to preliminary processing. such as atmospheric and geometric corrections. The spatial resolution for these images was 23 m, except for the image acquired in 1996, which was 72 m.

For the Uttara Kannada landscape, 16 years of remote sensing data were used to delineate the burned and unburned areas (Table 25.1). The 2015 image for each landscape was georectified using one 1:250000 scanned Survey of India topographic map. The rms error was ±9.2 m and ±11 m for the Nilgiris and Uttara Kannada landscapes, respectively. The other imagery in each landscape was georegistered to these reference images, the rms errors ranged from ±0.07 to 0.1 pixels.

TABLE 25.1

Details of Satellite Data Acquired for the Landscape Scale Analysis

Satellite/Sensor	Date of Acquisition (Nilgiris)	Date of Acquisition (Uttara Kannada)
IRS-1B	06-Mar-1996	
IRS-1C	22-Mar-1997	
IRS-ID	12-Mar-1999	05-Mar-1999
IRS-ID		14-Mar-2000
IRS-ID	01-Mar-2001	24-Mar-2001
IRS-ID	24-Feb-2002	09-Mar-2002
IRS-P6	09-Mar-2004	04-Mar-2004
IRS-P6	04-Mar-2005	11-Apr-2005
IRS-P6	27-Feb-2006	13-Mar-2006
IRS-P6	18-Mar-2007	01-Apr-2007
IRS-P6		02-Mar-2008
IRS-P6	07-Mar-2009	21-Mar-2009
IRS-P6	02-Mar-2010	20-Feb-2010
IRS-P6	01-Feb-2011	04-Apr-2011
IRS-P6	03-Mar-2012	05-Mar-2012
IRS-P6	26-Feb-2013	24-Mar-2013
IRS-P6	29-Mar-2014	12-Apr-2014
IRS-P6	16-Feb-2015	18-Feb-2015

The cloud cover for each of the scenes was less than 10%. The images were subjected to atmospheric corrections by applying the dark-object subtraction method. A methodology specific to the study area was developed by performing supervised classification by generating training sites from burned areas identified on the images (Kodandapani et al. 2008). Through an interactive process, spectral signatures of burned area in each of the three broad forest types were identified. The advantage of this method over using only a single forest type has been the ability to capture the variability in the spectral signature of burned areas due to differences in the structure, phenology, and exposure to soil fractions in these ecosystems. The fire maps were delineated by combining burned areas from each of the three forest types. The fire maps were assessed for their accuracy from field survey fire maps of the Mudumalai Tiger Reserve. The Mudumalai Tiger Reserve is a long-term ecological research (LTER) site and field-surveyed annual fire maps have been maintained since 1989. Fire maps were validated with this data set (Sukumar et al. 1992; Kodandapani et al. 2004). The overall accuracy of the fire maps ranged from 85% to 95% of the burned areas for the 16-year time period.

Statistical significance was assessed at $\alpha = 0.05$. The statistical analyses were conducted in the software R version 3.4.2 (R Development Core Team, 2011).

25.3 RESULTS

25.3.1 RAINFALL PATTERN AND DROUGHT INDICATORS

Mean annual rainfall in the Nilgiris landscape was 865 ± 260 mm (Figure 25.2a), whereas in the Uttara Kannada landscape it was 1047 ± 502 mm (Figure 25.2b). It should be noted that there is considerable spatial variation in annual rainfall in the two landscapes. Rainfall in the Nilgiris landscape is distributed from April to November, and is bimodal, with the first peak in July and the second peak in October (Figure 25.3a). Whereas in the Uttara Kannada landscape, the majority of the rainfall is received from June to September, with a peak in July (Figure 25.3b). Although seasonal droughts are a feature of tropical dry forests, their magnitude and severity depend on climatic water deficits over variable time periods. Figure 25.4 shows the temporal pattern of the SPI at different timescales in the Nilgiris landscape. At shorter timescales (3 months), both dry (SPI < 0) and moist (SPI > 0) periods are frequent. At longer timescales (12 months), the duration of dry periods increases, and there were two droughts between 1995 and 2005 in the Nilgiris landscape – a moderate drought in 1996, and a persistent and prolonged drought of higher magnitude (SPI < –2) between 2002 and 2005 (Figure 25.4). Similarly in the Uttara Kannada landscape, three droughts were observed between 1996 and 2005: two moderate droughts in 1996–1997 and 1999, and a third persistent and prolonged drought in 2002–2004 (Figure 25.5).

25.3.2 FIRE-ROTATIONAL INTERVALS: LANDSCAPE ANALYSES

The mean area burned in the Nilgiris landscape was 118 ± 118 km^2 and in Uttara Kannada landscape the mean burned area was 266 ± 202 km^2. The annual burned area in 2004, 2009, and 2012 were approximately twofold higher compared to the mean annual burned area in the Nilgiris landscape. Similarly, the annual burned area in 2004 and 2005 were twofold higher compared to the mean annual burned area in the Uttara Kannada landscape. Decadal averages revealed a significant difference (P < 0.05) between the burned area for the decade 1996–2005 compared to the decade 2006–2015 in the Nilgiris landscape. Similarly, decadal averages revealed significant differences (P = 0.01) between the burned area for the decade 1999–2005 compared to the decade 2006–2015 in the Uttara Kannada landscape. Both landscapes have experienced substantial fire activity during the last two decades, Figure 25.6 shows the fire history of the two landscapes. Both landscapes were characterized by lengthening fire-rotation intervals; mean fire-return intervals were 237% and 169% shorter in the Nilgiris and Uttara landscapes, respectively, during drought.

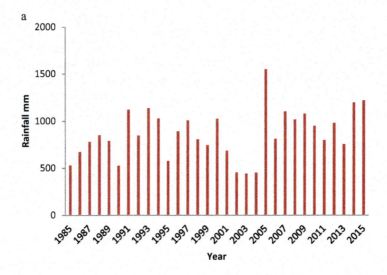

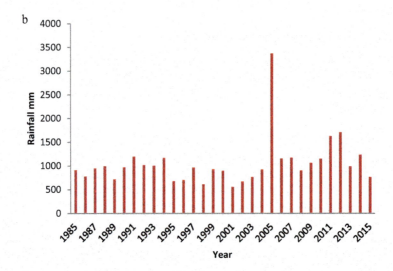

FIGURE 25.2 Annual rainfall from 1985 to 2015 in the two landscapes: (a) Nilgiris, (b) Uttara Kannada.

25.3.3 CORRELATION ANALYSES

In the Nilgiris landscape, coincident drought conditions as revealed by interannual variations in the SPI12 (March) were significantly correlated with variation in the annual log-transformed burned area ($\rho = -0.52$, $P < 0.05$). However, interannual variations in SPI3 (March) and SPI6 (March) were not significantly correlated with annual log-transformed burned area ($P > 0.1$). In the Uttara Kannada landscape, coincident drought conditions as revealed by interannual variations in the SPI12 (March) were significantly correlated with variation in the annual log-transformed burned area ($\rho = -0.64$, $P < 0.01$). Similarly, interannual variations in SPI3 (March) and SPI6 (March) were significantly correlated with variation in the annual log-transformed burned area ($\rho = -0.59$, $P < 0.01$ and $\rho = -0.73$, $P < 0.01$).

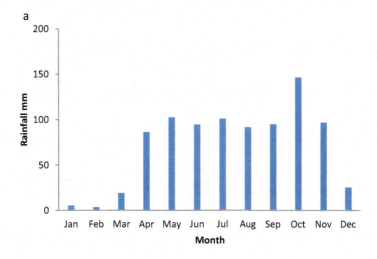

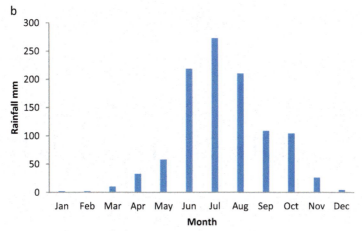

FIGURE 25.3 Mean monthly rainfall in the two landscapes: (a) Nilgiris, (b) Uttara Kannada.

25.4 DISCUSSION

Burned areas showed a decreasing trend at landscape scales in the Western Ghats. During the period 2006–2015 the burned area decreased in comparison to the period 1996–2005: the mean burned area was approximately threefold higher in the Nilgiris and approximately twofold higher in the Uttara Kannada landscapes. Comparing both landscapes, there is a modest increasing trend in annual (SPI12) moist periods especially during the decade 2006–2015. This is in contrast to the dry and drought conditions during the decade 1996–2005. In the Nilgiris and Uttara Kannada landscapes, at least three drought episodes were observed during the period 1996–2005, with the last drought episode being prolonged and intense. Thus, the drought conditions during the 1990s and early 2000s in both landscapes have contributed to increased fire activity. India has witnessed several monsoonal droughts in recent times, especially in 1982, 1987, 2002, and 2009 (Sharma and Mujumdar 2017). Drought conditions have subsequently subsided leading to a decline in burned areas in both landscapes. Similar mitigation in wildland fire potential in response to changing drought conditions has been observed in West African subtropical ecosystems (Jolly et al. 2015).

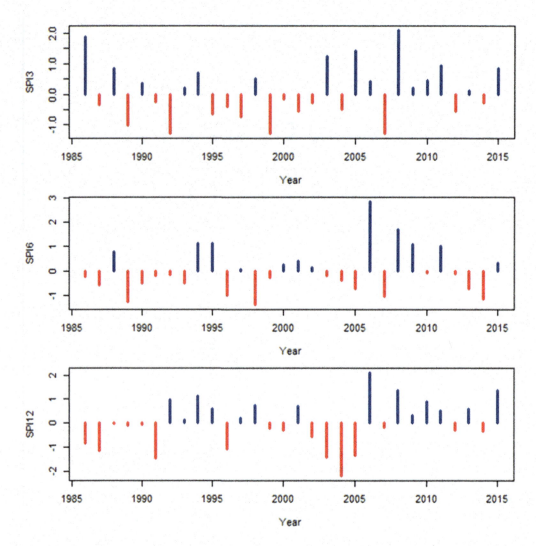

FIGURE 25.4 Changes in the SPI at different timescales in the Nilgiris landscape.

There was a significant negative correlation between coincident annual droughts and annual burned area in both the Nilgiris and Uttara Kannada landscapes. Extremely low rainfall and prolonged dry conditions (20%–40% below average) have been reported in the forested areas of the Western Ghats, for example, in the Nilgiris, from 2000 to 2003 (Suresh et al. 2010). The below-average annual rainfall and the severe monsoonal drought in pan-India in 2002 (Sharma and Mujumdar 2017) have left their fingerprints on the fire activity in both landscapes. Unsurprisingly, extreme drought (SPI < –2) is a significant factor in contributing to the large burned area in 2004 in the Nilgiris landscape (Figure 25.4). Surprisingly, during the same year, the drought episode in the Uttara Kannada landscape was moderate (SPI < –1.3). Nevertheless, the prolonged nature of the drought and poor recovery of ecosystems in the Uttara Kannada landscape have resulted in a heightened state of fire in both 2004 and 2005 (Figure 25.5). This reflects a general trend in drought recovery, with recovery times increasing as precipitation increases (Schwalm et al. 2017). Thus, in the moister Uttara Kannada landscape, the antecedent drought becomes a determining factor for the fuel load. Similar responses of forests to fire occurrences in moist and dry landscapes are reported in other parts of the world (Clarke 1989; Carvalho et al. 2010; Veblen et al. 2002). Recent studies

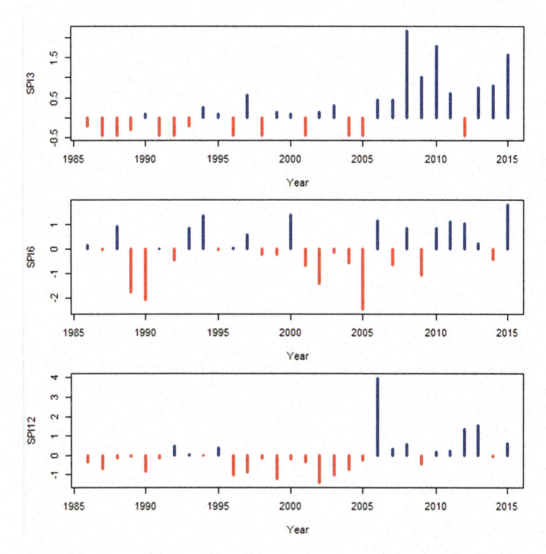

FIGURE 25.5 Changes in the SPI at different timescales in the Uttara Kannada landscape.

indicate drought recovery times were longer in moister regions compared with dry ecosystems. In addition, both multiseasonal and multiyear recovery times have increased from one–two years in 1901–1910 to three–five years in 2001–2010 (Schwalm et al. 2017).

In the early 2000s, the Western Ghats entered a period of prolonged drought, possibly as a result of global climatic change (Kale et al. 2017). The analyses captured a rare drought that was long and prolonged in the Western Ghats in the 2000s, leading to a dramatic increase in the Western Ghats fire activity. From the analyses, it is clear that the wildfires in the Western Ghats are dominated by periods of benign weather followed by extreme drought conditions. Such interannual variability could reflect the impacts of large-scale climate modes, such as the El Niño–Southern Oscillation and the Indian Ocean Dipole, on fire behavior in the region (Kodandapani et al. 2009; Jolly et al. 2015; Kale et al. 2017; Siegert et al. 2001; Chen et al. 2017). Also, the Indian subcontinent and the Western Ghats receive a significant proportion of their annual rainfall from the southwest monsoon. Rainfall data from long-term (1951–2003) observations suggest decreasing trends in both early monsoon and late monsoon rainfall and the number of rainy days (Ramesh and Goswami 2007).

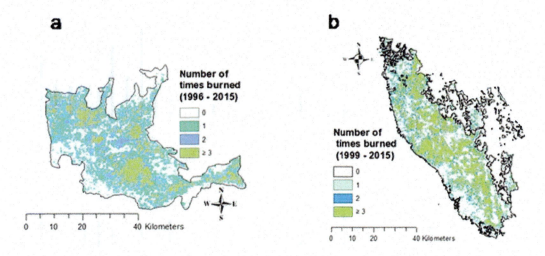

FIGURE 25.6   Spatial pattern of burning in the two landscapes: (a) Nilgiris, (b) Uttara Kannada.

Drought episodes, although not the only factor for increasing burned areas, could be exerting an increasingly significant impact on wildfire activity in the Western Ghats.

The response to drought in both landscapes is similar, although they enjoy different levels of protection. The Nilgiris landscape is a protected area; about one-fourth of all forests in India enjoy similar protection status. These protected areas are critical for the conservation of wildlife and biodiversity (Pringle 2017). The frequent and prolonged drought patterns in similar protected areas and the ensuing fires could have deleterious impacts for the conservation of biodiversity in forested landscapes across the globe (Cochrane 2003). On the contrary, the Uttara Kannada landscape represents an agroforestry landscape, where fire is a pre-eminent tool used for managing ecosystems for human use. Human influences in this landscape have resulted in a matrix of human and natural elements, including shifting cultivation, fires, grazing by cattle, collection of leaf litter, lopping of green leaves for manure, and extraction of timber and firewood (Hegde et al. 1998). Fires are used in similar forests across the globe to manage pastures and woodlands, prepare crop fields, control pests, and manage wildfires (Kull 2004). There exists a strong interaction between landscapes and people, and their use of wildfire as a strategy for conservation and natural resource management in the tropics (Douglas 2017; Laris et al. 2015). Fire in the Uttara Kannada landscape represents a socioecological phenomenon connecting cultural and resource management practices with vegetation types (Khatun et al. 2017). The frequent and prolonged drought events in this landscape, with poor recovery, and the enhanced fire dynamic could have implications for the management of natural resources, biodiversity, and livelihoods in the region (Daniels et al. 1995).

25.5   CONCLUSIONS

The analyses captured a rare and prolonged drought in the Western Ghats in the early 2000s that generated enhanced fire activity across the Western Ghats and especially in the forested landscapes. Increasing intensity, frequency, and area under droughts will have a large positive impact on fire occurrences in the Western Ghats, which is an important driver of global fire regimes. Given that several Global Circulation Models (GCMs) are projecting an increasing probability of droughts in the tropics, the long-term effects of recurrent fires in the Western Ghats could have implications for the management of natural resources and the conservation of biodiversity in the 21st century.

ACKNOWLEDGMENTS

I gratefully acknowledge support for this research by the Council of Scientific and Industrial Research, Government of India. I thank the Indian Meteorological Department (IMD) for providing high-resolution gridded rainfall data. I the thank Indian Space Research Organization (ISRO) for providing the remote sensing data sets for the burned area analyses.

REFERENCES

Abatzoglou JT, Kolden CA (2013) Relationships between climate and macroscale area burned in the western United States. *International Journal of Wildland Fire*, 22, 1003–1020.

Albright TP, Pidgeon AM, Rittenhouse CD, Clayton MK, Flather CH, Culbert PD, Wardlow BD, Radeloff VC (2010) Effects of drought on avian community structure. *Global Change Biology*, 16, 2158–2170.

Allen CD, Macalady AK, Chenchouni H, Bachelet D, McDowell N, Vennetier M, Kitzberger T, Rigling A, Breshears DD, (Ted) Hogg EH, Gonzalez P, Fensham R, Zhang Z, Castro J, Demidova N, Lim JH, Allard G, Running SW, Semerci A, Cobb N (2010) A global overview of drought and heat-induced tree mortality reveals emerging climate change risks for forests. *Forest Ecology and Management*, 259, 660–684.

Andela N, Morton DC, Giglio L, Chen Y, Van Der Werf GR, Kasibhatla PS, DeFries RS, Collatz GJ, Hantson S, Kloster S, Bachelet D, Forrest M, Lasslop G, Li F, Mangeon S, Melton JR, Yue C, Randerson JT (2017) A human-driven decline in global burned area. *Science*, 80, 1356–1362.

Aragao LEOC, Malhi Y, Roman-Cuesta RM, Saatchi S, Anderson LO, Shimabukuro YE (2007) Spatial patterns and fire response of recent Amazonian droughts. *Geophysical Research Letters*, 34, L07701, https://doi.org/10.1029/2006GL028946.

Archibald S, Nickless A, Govender N, Scholes RJ, Lehsten V (2010) Climate and the inter-annual variability of fire in southern Africa: A meta-analysis using long-term field data and satellite derived burned area data. *Global Ecology and Biogeography*, 19, 794–809.

Bowman D, Murphy BP, Williamson GJ, Cochrane MA (2014) Pyrogeographic models, feedbacks and the future of global fire regimes. *Global Ecology and Biogeography*, 23, 821–824. https://doi.org/10.1111/geb.12180

Bowman DMJS, Williamson GJ, Abatzoglou JT, Kolden CA, Cochrane MA, Smith AMS (2017). Human exposure and sensitivity to globally extreme wildfire events. *Nature Ecology and Evolution*. https://doi.org/10.1038/s41559-016-0058.

Carvalho A, Flannigan MD, Logan KA, Gowman LM, Miranda AI, Borrego C (2010). The impact of spatial resolution on area burned and fire occurrence projections in Portugal under climate change. *Climatic Change*, 98, 177–197.

Chen Y, Morton DC, Andela N, van der Werf GR, Giglio L, Randerson JT (2017). A pan-tropical cascade of fire driven by El Nino/Southern Oscillation. *Nature Climate Change*. https://doi.org/10.1038/s41558-017-0014-8

Cincotta RP, Wisnewski J, Engelman R (2000) Human population in the biodiversity hotspots. *Nature*, 404, 990–992.

Clark JS (1989) Effects of long-term water balances on fire regime, North-Western Minnesota. *Journal of Ecology*, 77(4), 989–1004.

Cochrane MA (2003) Fire science for rainforests. *Nature*, 421, 913–919.

Daniels RJR, Gadgil M, Joshi NV (1995) Impact of human extraction on tropical humid forests in the Western Ghats Uttara Kannada, South India. *Journal of Applied Ecology*, 32, 866–874.

Douglas K (2017) Flame resistance: Aboriginal Australian knowledge could be an effective weapon against wildfires. *New Scientist*, 8th July, 37–39.

Dimitrakopoulous AP, Vlahou M, Anagnostopoulou CG, Mitsopoulous ID (2011) Impact of drought on wildland fires in Greece: Implications of climatic change. *Climatic Change*. https://doi.org/10.1007/s10584-011-0026-8.

Hayes MJ, Svoboda MD, Wilhite DA, Vanyarkho OV (1999) Monitoring the 1996 drought using the standardized precipitation index. *Bulletin of the American Meteorological Society*, 80, 429–438.

Hegde V, Chandran MDS, Gadgil M (1998) Variation in bark thickness in tropical forest community of Western Ghats in India. *Functional Ecology*, 12(2), 313–318.

Jhala YV, Qureshi Q, Gopal R (Eds) (2015) *The Status of Tigers in India 2014*. National Tiger Conservation Authority, New Delhi & The Wildlife Institute of India, Dehradun.

Jolly WM, Cochrane MA, Freeborn PH, Holden ZA, Brown TJ, Williamson GJ, Bowman DMJS (2015) Climate induced variations in global wildfire danger from 1979–2013. *Nature Communication*, 6, 7537. https://doi.org/10.1038/ncomms8537.

Jones KR, Venter O, Fuller RA, Allan JR, Maxwell SL, Negret PJ, Watson JEM (2018) One-third of global protected area is under intense human pressure. *Science*, 360, 788–791.

Kale MP, Ramachandran RM, Pardeshi SN, Chavan M, Joshi PK, Pai DS, Bhavani P, Ashok K, Roy PS (2017) Are climate extremities changing fire regimes in India? An Analysis using MODIS fire locations during 2003–2013 and gridded climate data of Indian Meteorological Department. *Proceedings of the National Academy of Sciences, India Section A: Physical Science*, 87(4), 827–843.

Khatun K, Corbera E, Ball S (2017) Fire is REDD+: Offsetting carbon through early burning activities in south-eastern Tanzania. *Oryx*, 51(1), 43–52.

Kodandapani N, Cochrane MA, Sukumar R (2004) Conservation threat of increasing fire frequencies in the Western Ghats, India. *Conservation Biology*, 18, 1553–1561.

Kodandapani N, Cochrane MA, Sukumar R (2008) A comparative analysis of spatial, temporal, and ecological characteristics of forest fires in a seasonally dry tropical ecosystem in the Western Ghats, India. *Forest Ecology and Management*, 256, 607–617.

Kodandapani N, Cochrane MA, Sukumar R (2009) Forest fire regimes and their ecological effects in seasonally dry tropical ecosystems in the Western Ghats, India. In Cochrane, M.A. (Editor), *Tropical Fire Ecology: Climate Change, Land Use and Ecosystem Dynamics*. Springer-Praxis, Heidelberg, Germany.

Kodandapani N, Parks SA (2019) Effects of drought on wildfires in forest landscapes of the Western Ghats, India. *International Journal of Wildland Fire*, 28(6):431–444.

Kull CA (2004) *The Political Ecology of Landscape Burning in Madagascar*. University of Chicago, Chicago.

Littell JS, Peterson DL, Riley KL, Liu Y, Luce CH (2016) A review of the relationships between drought and forest fire in the United States. *Global Change Biology*. https://doi.org/10.1111/gcb.13275.

Laris P, Caillault S, Dadashi S, Jo A (2015) The human ecology and geography of burning in an unstable savanna environment. *Journal of Ethnobiology*, 35, 111–139.

Mann ME, Gleick PH (2015) Climate change and California drought in the 21st century. *PNAS*, 112(13), 3858–3859.

McLauchlan et al. (2020) Fire as a fundamental ecological process: Research Advances and frontiers. *Journal of Ecology*. https://doi.org/10.1111/1365-2745.13403.

Mehta VK, Sullivan PJ, Walter MT, Krishnaswamy J, DeGloria SD (2008) Ecosystem impacts of disturbance in a dry tropical forest in southern India. *Ecohydrology*, 1, 149–160.

Mittermeier R A, Gil PR, Hoffman M, Pilgrim J, Brooks T, Mittermeier CG, Lamoreux J, Da Fonseca GAB (2005) *Hotspots Revisited: Earth's Biologically Richest and most Endangered Terrestrial Ecoregions*, Cemex Mexico.

Mondal N, Sukumar R (2013) Characterising weather patterns associated with fire in a seasonally dry tropical forest in southern India. *International Journal of Wildland Fire*, 23, 196–201.

Narendran K, Murthy IK, Suresh HS, Dattaraja HS, Ravindranath NH, Sukumar R (2001) Non timber forest product extraction, utilisation and valuation: A case study from the Nilgiri Biosphere Reserve, Southern India. *Econ. Bot*, 55(4), 528–538.

Pai DS, Sridhar L, Rajeevan M, Sreejith OP, Sathbai NS, Mukhopadhyay B (2014) Development of a new high spatial resolution (0.25° × 0.25°) Long Period (1901–2010) daily gridded rainfall data set over India and its comparison with existing data sets over the region. *Mausam*, 65, 1–18.

Parks SA, Parisien M-A, Miller C, Holsinger LM, Baggett LS (2018) Fine scale spatial climate variation and drought mediate the likelihood of burning. *Ecological Applications*, 28, 573–586.

Pascal J-P (1986) *Explanatory Booklet on the Forest Map of South India*. French Institute, Pondicherry.

Pringle RM (2017) Upgrading protected areas to conserve wild biodiversity. *Nature*, 546, 91–99.

Puyravaud JP, Pascal JP, Dufour C (1994) Ecotone structure as an indicator of changing forest-savanna boundary (Linganamakki Region, southern India). *Journal of Biogeography*, 21, 581–593.

Quiring SM (2009) Monitoring drought: An evaluation of meteorological drought indices. *Geography Compass*, 3/1:64–88.

R Development Core Team (2011) *R: A Language and Environment for Statistical Computing*. R Foundation for Statistical Computing, Vienna, Austria. http://www.R-project.org.

Ramesh KV, Goswami P (2007) Reduction in temporal and spatial extent of the Indian summer monsoon. *Geophysical Letters*, 34: L23704. https://doi.org/10.1029/2007GL031613.

Riley KL, Abatzoglou JT, Grenfell IC, Klene AE, Heinsch FA (2013) The relationship of large fire occurrence with drought and fire danger indices in the western USA, 1984–2008: The role of temporal scale. *International Journal of Wildland Fire*, 22, 894–909.

Schwalm CR, Anderegg WRL, Michalak AM, Fisher JB, Biondi F, Koch G, Litvak M, Ogle K, Shaw JD, Wolf A, Huntzinger DN, Schaefer K, Cook R, Wei Y, Fang Y, Hayes D, Huang M, Jain A, Tian H (2017) Global patterns of drought recovery. *Nature*, 548, 202–205.

Siegert F, Ruecker F, Hinrichs A, Hoffmann, A A (2001) Increased damage from fires in logged forest during droughts caused by El Nino. *Nature*, 414(22), 437–440.

Sharma S, Mujumdar P (2017) Increasing frequency and spatial extent of concurrent meteorological droughts and heat waves in India. *Scientific Reports*, 7, 15582. https://doi.org/10.1038/s41598-017-15896-3.

Suresh HS, Dattaraja HS, Sukumar R (2010) Relationship between annual rainfall and tree mortality in a tropical dry forest: Results of a 19-year study at Mudumalai, southern India. *Forest Ecology and Management*, 259, 762–769.

Sukumar R, Dattaraja HS, Suresh HS, Radhakrishnan J, Vasudeva R, Nirmala S, Joshi NV (1992) Long-term monitoring of vegetation in a tropical deciduous forest in Mudumalai, southern India. *Current Science*, 62, 608–616.

Van der Werf GR, Randerson JT, Giglio L, van Leeuwen TT, Chen Y, Rogers BM, Kasibhatla PS (2017) Global fire emissions estimates during 1997–2016. *Earth System Science Data*, 9, 697–720. https://doi.org/10.5194/essd-9-697-2017

Veblen TT, Baker WL, Montenegro G, Swetnam TW (2002) *Fire and Climatic Change in Temperate Ecosystems of the Western Americas*. Springer, New York, 464.

Westerling AL (2016) Increasing western US forest wildlife activity: Sensitivity to changes in the timing of spring. *Philosophical Transactions of the Royal Society B*, 371, 20150178. https://doi.org/10.1098/rstb.2015.0178.

26 Hydrological Drought Impacts on River Water Quality of Peninsular River System, Tunga-Bhadra River, India

M. Rajesh, G. Krishna Mohan, and S. Rehana

CONTENTS

26.1 INTRODUCTION

Drought is a natural hazard due to the occurrence of below-normal natural water availability (Tallaksen et al., 1997). Almost 70% of the population lives in drought-prone regions, which account for almost 38% of the land on the planet. According to the American Meteorological Society (AMS, 2013), drought is a natural temporary feature of the climate cycle that quickly wreaks havoc on most regions of the globe. McMahon and Arenas (1982) defined drought as "a period of abnormally dry weather sufficiently prolonged for the lack of precipitation to come to a serious hydrological imbalance and carries connotations of a moisture deficiency with respect to man's usage of water". Takeuchi (1974) defined drought as the condition where the available water is less than the amount expected to satisfy human demand. McGuire and Palmer (1957) defined drought as "monthly or annual precipitation less than some particular percentage of normal". Based on the several definitions of drought by various authors, it can be articulated that drought as a shortage of water availability in terms of precipitation has consequent shortages over streamflow and soil moisture affecting the socioeconomics of a country. Drought can be either a sudden event or a long-term

DOI: 10.1201/9781003276548-26

event, since the precipitation deficit may build up suddenly or it may take months before the stream-flow, reservoir levels, and underground water table begin to reduce in levels and depths (Badripour, 2007; Bond et al., 2008; Yeh et al., 2015). Droughts were observed to be slow to develop, but long lasting (Bond et al., 2008; Yeh et al., 2015).

Historically, any natural hazard was observed to have negative impacts, and, specifically, drought is observed to have its impact in the form of economic, environmental, and social losses (Wilhite, 2000). The shortage of water during drought may incur economic losses to the agricultural sector, power generation, and fisheries and livestock production. The decrease in precipitation causes lack of water and soil moisture and consequent impact on crops, thereby agricultural reducing yields. The decrease in productivity of rangeland due to lack of water results in losses to livestock pro-duction. Low streamflow damages the habitat for fish and other aquatic animals, causing losses to fisheries production, and losses to power generation can be due to the unavailability of the required amount of water for the power generation because of the low streamflow. Drought causes environ-mental and ecological imbalances, resulting in the endangering of hydrology of the area, and ani-mal and plant communities, leading to the environmental losses. The effects of drought on health, increased conflicts, and famine are generally considered as social losses. Drought over an extended period may result in famine, which can have a high negative impact on social losses (Goyal et al., 2017; Wilhite and Glantz, 1985).

Climate is one of the factors responsible for drought propagation (Sheffield et al., 2012). Stahl and Hisdal (2004) provide an overview of hydroclimatological regimes and potential for drought development in different climates around the world. In monsoon climates like India, where the dry and wet seasons alternate due to large-scale atmospheric processes, a drought occurs when the onset of the monsoon is delayed or a complete or partial failure of the monsoon takes place (Flatau et al., 2003; Mishra et al., 2012; Schewe and Levermann, 2012). About 18% of Indian landmass is drought prone, affecting more than 50 million people (Dutta et al., 2013). In this context, drought assessment is crucial for water resources management in India, given that more than 68% of the population is dependent on agriculture.

26.1.1 DROUGHT CLASSIFICATION

According to Wilhite and Glantz (1985), droughts are classified into four approaches, namely, meteorological, hydrological, agricultural, and socioeconomic. The first three approaches deal with prominent hydrological variables such as rainfall, evapotranspiration, streamflow, and soil mois-ture. The last approach deals with the aspects that depend on socioeconomic conditions of a region affected due to drought. The sequence of various forms of drought corresponds to the failure of pre-cipitation (meteorological drought), low streamflow (hydrological drought), decrease in soil mois-ture and crop yields (agricultural drought), and therefore consequent impacts on the socioeconomic activities of humans (socioeconomic drought) (Wilhite and Glantz, 1985), as shown in Figure 26.1.

26.1.1.1 Meteorological Drought

Meteorological drought is defined in terms of the degree of dryness and quantity of precipitation deficit (compared to normal or average precipitation) (Wilhite and Glantz, 1985). Since precipitation patterns and atmospheric conditions are highly region specific, meteorological drought must be con-sidered as region specific. Different regions prefer different approaches or scales to define drought, for example, regions such as Manaus (Brazil) and London identify periods of drought based on the number of days with precipitation less than some specific threshold (Goyal et al., 2017). Other regions, like the Central United States, Northeast Brazil, West Africa, and Northern Australia, characterize their climatic regimes by a seasonal rainfall pattern where the definition based on the number of days with precipitation less than some specified threshold is nonviable. Other regions may identify periods of drought based on monthly, seasonal, or annual timescales (Goyal et al., 2017). Various meteorological drought indices have been developed to monitor and predict drought

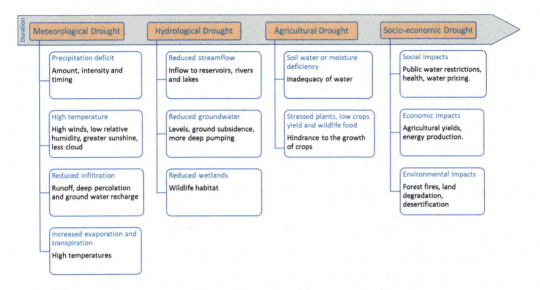

FIGURE 26.1 Various forms of droughts, and the sequence of occurrence and impacts.

events (Table 26.1). Drought indices are calculated by a combination of climatic and meteorological variables, among which precipitation and evapotranspiration are the most essential variables in defining the magnitude and intensity of a drought (Alley, 1984; Chang and Kleopa, 1991).

26.1.1.2 Hydrological Drought

Hydrological drought is defined in terms of the period of precipitation shortfall on surface and subsurface levels (streamflow drought and groundwater drought), which in turn cause the deficit in water in rivers, streams, lakes, and wells. Hydrological drought is observed to commence after a meteorological drought since a deficit in precipitation can cause a deficit in streamflow and reservoirs. Drought is observed to follow the this sequence unless there is a human intervention that may alter it. For example, though precipitation is average in a region, construction of a reservoir/ dam upstream of the river and a less/no release of water may cause hydrological drought directly. Catchment control is an essential aspect for hydrological drought as climate control (Van Lanen et

TABLE 26.1

Meteorological Drought Indices

Index	Input Parameters	Reference
Standardized Precipitation Index (SPI)	P	McKee et al. (1993)
Percent of Normal Index (PNI)	P	Willeke et al. (1994)
Deciles Index (DI)	P	Salehnia et al. (2017)
Effective Drought Index (EDI)	P	Byun and Wilhite (1999)
China Z Index (CZI)	P	Ju et al. (1997), Wu et al. (2001)
Rainfall Anomaly Index (RAI)	P	Salehnia et al. (2017)
Standardized Precipitation Evapotranspiration Index (SPEI)	P, PET	Vicente-Serrano et al. (2010)
Standardized Precipitation Anomaly Index	P	Kironmala and Maity Rajib 2015)

Source: Svoboda and Fuchs, 2016.

Notes: P, precipitation; PET, potential evapotranspiration.

TABLE 26.2

Hydrological Drought Indices

Index	Input Parameters	Reference
Palmer Hydrological Drought Severity Index (PHDI)	P, T, AWC	Palmer (1965)
Standardized Reservoir Supply Index (SRSI)	RD	Gusyev et al. (2015)
Standardized Streamflow Index (SSFI)	SF	Modarres (2007)
Standardized Water-Level Index (SWI)	GW	Bhuiyan (2004)
Streamflow Drought Index (SDI)	SF	Nalbantis and Tsakiris (2009)
Surface Water Supply Index (SWSI)	P, RD, SF, S	Shafer and Dezman (1982)

Source: Svoboda and Fuchs, 2016.

Notes: AWC, available water content; P, precipitation; T, temperature; RD, reservoir; SF, streamflow; S, snowpack; GW, groundwater.

al., 2004); van Vliet et al., 2013). Hydrological droughts are mostly based on below-average streamflow that may be due to human-influenced regulated flows due to diversions, water transfers, and abstractions (Lanen et al., 2013). Hydrological drought can be severe in terms of extremely low reservoir and groundwater levels, which can cause a restriction in water use for irrigation and domestic use (Aghakouchak et al., 2014; Dettinger and Cayan, 2014). Hydrological drought is an extended period of infrequent low streamflow and groundwater flow, unlike a low-flow period, which is the annual cycle of the streamflow. A low flow may occur once or twice a year depending on the climatic conditions, unlike the hydrological drought. Hydrological drought is generally associated with the concept of low flows in rivers, whereas a single hydrological drought can have multiple low-flow events (Zelenhasić and Salvai, 1987). Conventionally, hydrological drought assessment is carried out based on hydrological drought indices to capture the occurrence of natural water availability below average (Van Lanen et al., 2004). Geng and Shen (1992) defined hydrological drought as an eventual and extreme drought, and is the continuity and development of meteorological and agricultural droughts. Similar to meteorological drought indices, which were developed based on precipitation and evapotranspiration, hydrological droughts were majorly formulated based on streamflow, reservoir flows, snowpack, etc. (Table 26.2).

26.1.2 IMPACT OF HYDROLOGICAL DROUGHT ON WATER QUALITY

The water quality of freshwater systems is controlled by climatic variability, and hydrological, biogeochemical, and anthropogenic influences that operate at various temporal and spatial (e.g., global, river basin, local catchment) scales (Mosley, 2015). The potential changes in water quality due to climate change has been assessed through several approaches, such as empirical relations between water quality and climatic trends (Fukushima et al., 2000; Schindler et al., 1996; Williams et al., 1996), and black-box or deterministic modeling approaches to assess potential effects of climate change on surface water quality at the continental or regional scale (Clair and Ehrman, 1996; Krysanova et al., 2005; Mimikou et al., 2000; Wolford and Bales, 1996).

Geng and Shen (1992) defined hydrological drought as an eventual and extreme drought, and is the continuity and development of meteorological and agricultural droughts. During a hydrological drought, river flow lower than its normal water level (or waterline) because of the imbalance between rainfall and surface or underground water (Dracup et al., 1980). Although drought impacts on water quantity are well known (Safavi and Malek Ahmadi, 2015; van Vliet and Zwolsman, 2008), the consequences on water quality are not fully recognized yet (Zwolsman and van Bokhoven, 2007). The low flow and low water levels during hydrological droughts can affect water quality of freshwater systems by, for instance, increasing its residence time; reducing the flushing rate of water bodies with a limited dilution of point source emissions; and the interruption of sediment, organic matter, and nutrient transport (Mishra and Singh, 2010; Mosley, 2015; Palmer and Montagna, 2015; van Vliet and Zwolsman, 2008). Water quality characteristics such as the increase in total dissolved solids and their constituent ions, and the decrease of dissolved oxygen could cause agronomic problems to irrigated agriculture (Maestre-Valero and Martínez-Alvarez, 2010; Sagasta and Burke, 2005). Therefore, water quality is an important variable to consider in the development of irrigation (Assouline et al., 2015).

There have been several studies on the impacts of drought on water quality at local and regional scales. Delpla et al. (2009) studied the impacts of climate change on surface water quality in relation to drinking water production. The impacts of drought were included in the study, but the impacts on non-drinking water were not addressed. Murdoch et al. (2000) and Whitehead et al. (2009) addressed the drought impacts on water quality, but a focus has been given to the broader potential effects of climate change on surface water quality and ecology. Cloke and Hannah (2011), Mishra and Singh (2010), and Pozzi et al. (2013) argue that hydrological drought deserves more attention due to its crucial link with drought impacts. Parmar and Bhardwaj (2013) used dimensional fractal analysis to evaluate the water quality index at Harike Lake, India, and found that the water was severely contaminated and unsuitable for domestic or industrial use. Bhat et al. (2014) examined the impact of all sources of pollution in the Sukhnag stream (Kashmir Himalayas), India, to identify the parameters responsible for spatiotemporal variability in water quality using cluster analysis and factor analysis, and showed that stream water quality is primarily influenced by hydrological runoff and wastewater discharge. They also found that most of the variations in water quality are explained by the natural soluble salts, nonpoint source nutrients, and anthropogenic organic pollutants. Results of regression analysis clearly showed that, in peak flow season, runoff raises the concentration of most inorganic and organic parameters. The increasing level of pollution from the head of the stream to the tail indicated progressive anthropogenic pressure in the downstream areas (Bhat et al., 2014). Udmale et al. (2016) examined rural drinking water availability issues during a recent drought (2012) through 22 focus group discussions in a drought-prone catchment, the Upper Bhima Catchment in Southern Maharashtra State, India. Also, a small chemical water quality study was undertaken to evaluate the suitability of water for drinking purposes based on the Bureau of Indian Standards. Results of the chemical water quality study showed that severe contamination of the drinking water with nitrate–nitrogen, ammonium–nitrogen, and chlorides was found in the analyzed drinking water samples. The results of this study point to an immediate need to investigate the problem of contaminated drinking water sources while designing relief measures for drought-prone areas of India. Dhawale and Paul (2019) examined meteorological and hydrological droughts using the Standardized Precipitation Index (SPI) and Standardized Ground Water Level Index (SWI) in India, especially the Marathwada region of Maharashtra. Results of this study helped in forming mitigating strategies and identifying future drought events and their characteristics. Puczko and Jekatierynczuk-Rudczyk (2020) examined the effects of hydrometeorological extreme events and urban catchment on water quality in small rivers in Białystok (Poland). Results of this study was that the most substantial change in water quality occurs on the second day after rainfall and changed the concentration of some chemical parameters for a long time. Since river water quality may deteriorate to critical values during periods of prolonged low-flow conditions in combination with high water temperatures (Somville and De Pauw, 1982), insight and understanding of the impact of droughts on water quality is essential, especially for rivers that are highly sensitive to drought conditions.

It has been evident from these studies that the water quality can be affected by climate-related mechanisms such as air temperature increase, changes in hydrological factors (e.g., limited dilution of point source emissions during low river flows), terrestrial factors (e.g., changes in vegetation and soil structure), and resource-use factors (e.g., increased water use, increased demand for cooling water) (Murdoch et al., 2000). Since river water quality may deteriorate during periods of prolonged low-flow conditions in combination with high water temperatures (e.g., Somville and De Pauw, 1982), insight and understanding of the impact of drought on water quality is essential, especially for rivers that are highly sensitive to drought conditions (van Vliet and Zwolsman, 2008). This study aims to determine the spatiotemporal distribution of meteorological and hydrological drought of a river basin and assess the impacts of hydrological drought on basic surface water quality parameters at Balehonnur, Hosaritti, and Rattihalli stations, Tunga-Bhadra River, India.

26.2 DATA AND METHODOLOGY

26.2.1 STUDY AREA AND DATA

The Tunga-Bhadra River is a tropical river system that is a major tributary of the Krishna River, the fifth largest river system in central India. The catchment area of the basin is 47,827 km^2 and extended from latitudes 13°06′ to 16°16′ N and longitudes of 74°48′ to 77°31′ E. The Tunga-Bhadra River is a major source for drinking, bathing, crop irrigation, fishing, and livestock water. The river flows for about 531 km in a northeasterly direction through Mysore, Andhra Pradesh, and Telangana, and joins the Krishna beyond Kurnool. The length of the river is 786 km with a total drainage area of the Tunga-Bhadra as 71,417 km^2. The mean annual rainfall in the Tunga-Bhadra basin is 884 mm (NIH, 1992).

The catchment area of the Tunga-Bhadra sub-basin can be divided into three zones depending on the vegetative growth, viz., (i) the Western Ghat belt from Agumbe to Honnali with thick forest and heavy rainfall, (ii) thin vegetative cover from Honnali up to Harihar with moderate rainfall, and (iii) very thin vegetative growth with bare topped hills beyond Harihar and up to Mallapuram with scanty rainfall. The land use in the catchment consists of forest (14.5%), cultivation (59%), pastures (9%), and wasteland (12%); the rest (5.5%) is fallow land (KERS, 1985).

The Tunga-Bhadra River is one of the most polluted river stretches of India, particularly the Bhadra River stretch with primary industrial and municipal effluents (CPCB, 2020). The river location considered for the calculation of hydrological drought are the Balehonnur, Hosaritti, and Rattihalli stations along the Bhadra River, which confluences with the Tunga River to form the Tunga-Bhadra River, a major tributary of Krishna River Basin (Figure 26.2).

The daily discharge and daily precipitation data from 2005 to 2017 recorded at the Balehonnur, Hosaritti, and Rattihalli stations were obtained from the Advanced Centre for Integrated Water Resources Management (ACIWRM), Bengaluru, Karnataka, India. The water quality data were obtained for three stations, namely, downstream of Bhadravathi city, New Bridge (Kodiyal), and Haralahalli Bridge, which are near to the Balehonnur, Rattihalli, and Hosaritti discharge stations, respectively, for 2005–2017 from the ACIWRM (Table 26.3).

26.2.2 METHODOLOGY

26.2.2.1 Standardized Precipitation Index (SPI)

The SPI (Doesken et al., 1991; Hayes, 2006), an internationally recognized index, was developed by McKee et al. (1993) to quantify the precipitation deficit. It was recommended as a standard index worldwide by the World Meteorological Organization (WMO) and the Lincoln Declaration on Drought (Hayes, 2006; Loon, 2015; Stagge et al., 2015). It requires a single input, which is monthly precipitation data with no gaps and for a minimum of 20 years (Teegavarapu, 2017). The flexibility to calculate for multiple timescales (1, 3, 6, 12, 24, and 48 months) or moving averaging windows is an advantage. The impacts of drought on different water resources are indicated through these timescales (Bijaber et al., 2018). Since it standardizes the data, all users of the index will have

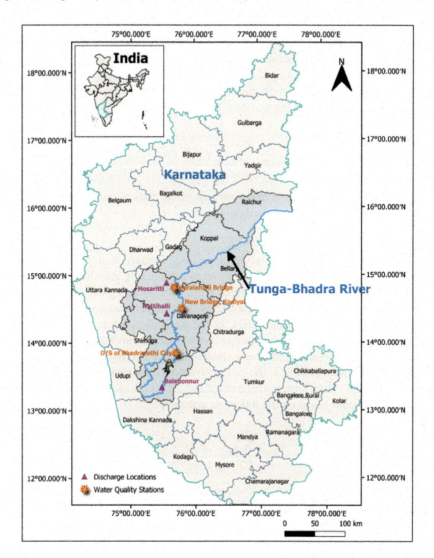

FIGURE 26.2 Location map of the Tunga-Bhadra River and monitoring stations, Karnataka, India.

TABLE 26.3

Details of Selected Monitoring Stations, Karnataka, India

Station Type	Station Name	River	Time Period
Discharge station	Balehonnur	Bhadra River	2005–2017
Discharge station	Hosaritti	Tunga–Bhadra River	2005–2017
Discharge station	Rattihalli	Tunga-Bhadra River	2005–2017
Water quality station	Downstream of Bhadravathi City	Bhadra River	2005–2017
Water quality station	New Bridge, Kodiyal	Tunga-Bhadra River	2005–2017
Water quality station	Haralahalli Bridge	Tunga-Bhadra River	2005–2017

a common basis for both temporal and spatial comparison of index values. The SPI is widely used to study various characteristics of droughts, for example, forecasting, frequency analysis, spatio-temporal analysis, and climate impact studies (Mishra and Desai, 2005a, 2005b; Mishra and Singh, 2010, 2011). To calculate the SPI, a long-term precipitation record at the desired station is first fitted to a probability distribution (e.g., gamma distribution) (Doesken et al., 1991; Hayes, 2006), which is then transformed into a standard normal distribution so that the mean SPI is zero (McKee et al., 1993):

$$SPI = \frac{X_{i,j} - X_{im}}{\sigma} \tag{26.1}$$

where $X_{i,j}$ is the seasonal precipitation at the ith rain gauge station and jth observation, X_{im} is the long-term seasonal mean, and σ is standard deviation.

26.2.2.2 Streamflow Drought Index (SDI)

The Streamflow Drought Index (SDI) was developed by Nalbantis and Tsakiris (2009) using the methodology and calculation of SPI as the basis and by considering monthly streamflow value $Q_{i,j}$, where i is hydrological year and j is the months within the hydrological year. With the methods of normalization associated with the SPI, the SDI was developed using streamflow data. Monthly streamflow values and a historical time series for the streamflow gauge were considered as input parameters. More accurate results can be obtained with longer streamflow data. But, a disadvantage is its consideration of a single-input parameter (Nalbantis and Tsakiris, 2009).

Like the SPI, for estimating the SDI, the cumulative flow values were estimated individually for each month, then the SDI values were calculated for various timescales. The cumulative streamflow volume, $V_{i,k}$, was calculated based on the following equation:

$$V_{i,k} = \sum_{j=1}^{3k} Q_{i,j} \quad k = 1,2,3,4 \tag{26.2}$$

Equation 26.2 includes calculated $V_{i,k}$ values for 3-, 6-, and 12-month periods. The SDI is described with cumulative streamflow volumes $V_{i,k}$ for each reference period k of the ith hydrological year as follows:

$$SDI_{i,k} = \frac{V_{i,k} - \overline{V_k}}{S_k} k = 1,2,3,4 \tag{26.3}$$

where $\overline{V_k}$ and S_k are the mean and the standard deviation of cumulative streamflow volumes for reference period k, respectively. According to the SDI criterion, wet conditions are defined with values greater than zero, whereas drought conditions are defined with values lower than zero.

Descriptions of SDI and SPI states are provided with the criteria in Table 26.4.

26.2.3 Hydrological Drought and Water Quality

Five water quality parameters – water temperature, pH, dissolved oxygen (DO), biochemical oxygen demand (BOD), and nitrates – were considered to study the effect of drought on river water quality. The monthly values of the water quality parameters were taken to study the significance between them during a drought year and reference periods. The mean, minimum, and maximum values were computed, and the p-values obtained from the two-sample t-test were used to test whether differences in water quality were significant at the 95% confidence level. A t-test is a type of inferential statistic used to determine if there is a significant difference between the means of two groups, which may be related in certain features. A paired two-sample t-test determines whether

TABLE 26.4

Description of Hydrological Drought Based on the Streamflow Drought Index (SDI), Standardized Precipitation Index (SPI), and Criterion

Description	Criterion
Non-drought	SDI, SPI $\geq$ 0.0
Mild drought	$-1.0 \leq$ SDI; SPI $<$ 0.0
Moderate drought	$-1.5 \leq$ SDI; SPI < -1.0
Severe drought	$-2.0 \leq$ SDI; SPI < -1.5
Extreme drought	SDI, SPI < -2.0

Source: Nalbantis and Tsakiris (2009).

the difference between the sample means is statistically distinct from a hypothesized difference (Cressie and Whitford, 1986).

26.3   RESULTS AND DISCUSSION

26.3.1   HYDROCLIMATOLOGY

The hydrology of the area was studied by plotting the monthly mean precipitation and monthly mean discharge for the three stations, namely, Balehonnur, Hosaritti, and Rattihalli, for the period from June 1, 2005, to May 31, 2017. From Figure 26.3a, it can be observed that the discharge and precipitation values for Balehonnur station decreased from 2010 onward. The same pattern was observed at Rattihalli station (Figure 26.3c). According to the Karnataka State Natural Disaster Monitoring Centre, the state has been experiencing a decline in rainfall over the past few years along with high fluctuations in the rain patterns. Karnataka receives an annual rainfall of 1135 mm in which the southwest monsoon accounts for about 73% of rainfall. However, since 2011, the state has received an average rainfall of 1033 mm, which is 10% less than the average rainfall (The Times of India, 2019).

Among three stations, the discharge of Balehonnur station was observed to be high, followed by Hosaritti and Rattihalli. The range of discharge at Balehonnur was very high compared to the other stations. This could be due to the distribution of rainfall over these stations (Figure 26.3). The average rainfall of the Balehonnur station was observed to be much higher than the other two stations ,which incur in the variations in the discharge patterns.

26.3.2   HYDROMETEOROLOGICAL DROUGHT ANALYSIS

To study drought over the river basin, the meteorological drought index of the SPI and the hydrological drought index of the SDI were studied. The monthly values of the SPI and SDI were calculated on a 12-month timescale from June 1, 2005, to May 31, 2017, and comparison graphs were plotted between the two indices for each station (Figure 26.4). The intensity of the drought is explained with reference to the thresholds considered in Table 26.4. In this study, we defined −1.5 as a threshold to identify the meteorological and hydrological drought events, represented in Figure 26.4 as rectangular boxes with a, b, c, d as drought years. From the plotted comparison graphs between the SPI and SDI, it was observed that for the Balehonnur station, a moderate drought occurred in June 2012 and June 2013 (box a) and a mild drought was observed between September 2015 and March 2016

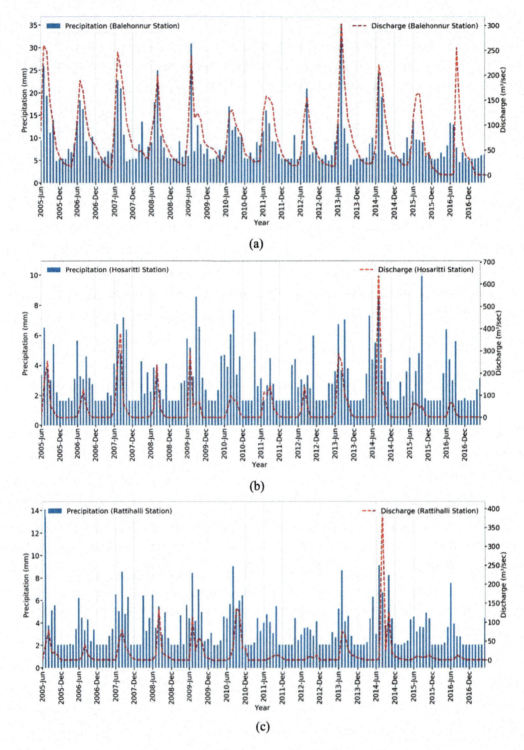

FIGURE 26.3 Mean monthly precipitation and discharge of the Tunga-Bhadra River at the (a) Balehonnur, (b) Hosaritti, and (c) Rattihalli stations, India (2005–2017).

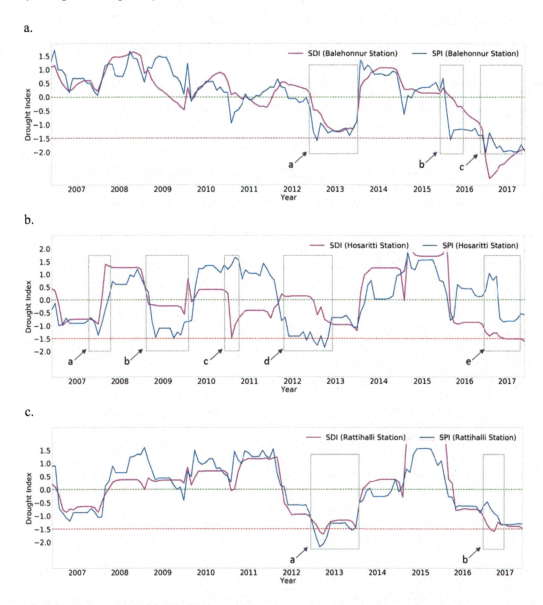

FIGURE 26.4 Trend of the SDI and SPI on a 12-month scale at the Tunga-Bhadra River at the (a) Balehonnur, (b) Hosaritti, and (c) Rattihalli stations, Tunga-Bhadra River, India (2005–2017).

(box b) as shown in Figure 26.4a. While for station Hosaritti, a moderate drought was observed during the period between August 2006 and June 2007 (box a, Figure 26.4b) and September 2012 and June 2013 (box d, Figure 26.4b). For the Rattihalli station, a mild drought was observed during the period between October 2012 and July 2013 (box a) and a moderate drought between October 2015 and November 2016 (box b), as shown in Figure 26.4c. Both meteorological and hydrological droughts have occurred during the same periods for three stations, as shown in Figure 26.4. The meteorological drought response on hydrological drought was noted for three stations with a threshold of –1.5 drought intensity. The years when both meteorological and hydrological drought events occurred were emphasized. More specifically, the hydrological drought events were picked for the possible analysis on river water quality variables. It was observed that based on the SDI values, at the Balehonnur station, a mild hydrological drought occurred from June 2012 to August 2012 followed by a moderate drought until November 2012, which later worsened to severe drought from December 2012 to May 2013. Another event of severe hydrological drought occurred from October 2015 to May 2017 at Balehonnur station.

For the Hosaritti station, a mild hydrological drought was observed during the periods of August 2006 to June 2007 (box a), June 2010 to May 2011 (box c), and September 2012 to June 2013 (box d), as shown in Figure 26.4b. A moderate hydrological drought event was observed from December 2015 to May 2017 (box e), as shown in Figure 26.4b. At the Rattihalli station, a mild drought was observed from August 2006 to July 2017 and October 2011 to July 2013 (box a in Figure 26.4c). A moderate hydrological drought event occurred July–September 2012 and a moderate drought from October 2015 to November 2016 followed by a mild drought from December 2016 to May 2017 (box b), as shown in Figure 26.4c for the Rattihalli station. Table 26.5 describes various meteorological and hydrological drought events at the Balehonnur, Hosaritti and Rattihalli locations.

From Table 26.5, a common drought period from June 2012 to May 2013 was considered for all three stations to demonstrate the impact of hydrological droughts on river water quality. Furthermore, two reference periods – the previous water year (June 2011 to May 2012) and the succeeding water year of drought (June 2013 to May 2014) – were considered in the hydrological drought impact assessment study. A comparison study for various parameters like precipitation, discharge, air and surface water temperatures, and water quality during the drought periods was conducted and compared with the reference wet years (van Vliet and Zwolsman, 2008). Therefore, the study compared the drought year of 2013 with reference wet years of 2012 and 2014 as before and after the occurrence of drought year, respectively.

The climatology of a drought year and two reference years was studied by comparing the precipitation and temperature for the three stations. The annual precipitation and temperatures were compared for the drought year 2013 with reference years 2012 and 2014, as given in Table 26.6. From Table 26.6, it is observed that the annual precipitations were low during the drought period (2013) compared to the other two reference periods (2012 and 2014), except for at the Hosaritti station that received a rainfall of 2.6% more than the previous year. While the annual rainfall decreased, the maximum air temperatures were observed to increase during the drought year. The drought in the current study was observed to be a combined effect of both precipitation and temperature. Comparison of annual precipitation sums and mean and maximum air temperatures at the meteorological station of the Tunga-Bhadra River for the drought period, with the same parameters for the reference periods, also demonstrate the drier and warmer conditions during droughts. The annual

---

TABLE 26.5

Drought Events at Balehonnur, Hosaritti, and Rattihalli Locations over Tunga-Bhadra River Basin

Location	Meteorological Drought	Hydrological Drought
Balehonnur	• June 2010 to March 2011 • November 2011 to June 2013 • June 2015 to May 2017	• June 2012 to May 2013 • October 2015 to May 2017
Hosaritti	• June 2006 to June 2007 • September 2008 to September 2009 • September 2011 to June 2013 • October 2016 to May 2017	• August 2006 to June 2007 • June 2010 to May 2011 • September 2012 to June 2013 • August 2015 to May 2017
Rattihalli	• June 2006 to June 2007 • October 2011 to April 2014 • August 2015 to May 2017	• August 2006 to July 2007 • October 2011 to July 2013 • October 2015 to November 2016 • December 2016 to May 2017

TABLE 26.6

Annual Total Precipitation and Maximum Air Temperature at the Discharge Stations for the Drought Year (2013), Reference Years, and Averaged over the Entire Period 2005–2017

Discharge Location	Parameter	Drought 2013	Reference 2012	2014	Average 2005–2017
Balehonnur	Annual total precipitation (mm)	2826	3102	3590	3257
	Max air temperature (°C)	30.9	30.7	26.7	28.7
Hosaritti	Annual total precipitation (mm)	972	947	1333	1168
	Max air temperature (°C)	32.00	31.40	31.3	31.64
Rattihalli	Annual total precipitation (mm)	1037	1168	1414	1344
	Max air temperature (°C)	31.79	31.23	31.28	31.28

precipitations were substantially lower (2826 mm) during the drought period when compared to the reference periods, while the temperatures were much higher (30.9°C) compared to the post-drought period.

The study compared the hydrological aspects for the drought year 2013 with reference years of 2012 and 2014 by comparing the discharge values, as shown in Figure 26.5. The discharge values at the three stations were observed to be low during the drought period of 2013 compared to the reference periods of 2012 and 2014 (Figure 26.5). As described earlier, even during drought, discharges were observed to be incredibility low at the Rattihalli station (Figure 26.5c). The decrease in the discharges were due to the decrease in the annual precipitation and an increase in the maximum air temperatures. Overall, the study compared the hydroclimatology of the three stations for the drought year and reference years. The lower precipitation values, higher temperatures, and therefore consequent decrease in streamflows were observed during the drought year of 2013 compared to reference years of 2012 and 2014 over the Tunga-Bhadra River Basin.

26.3.3 RIVER WATER QUALITY ANALYSIS

The impact of hydrological drought on river water quality was studied by observing the variation in the water quality parameters during the drought period and the reference wet periods. The variations within each water quality parameter were studied using statistical tests, such as the two-sample t-test (Tables 26.7, 26.8, 26.9). The t-test was conducted on each water quality parameter for a drought year and wet year. According to the two-sample t-test, at the Bhadravathi station, the surface water temperatures, DO, and BOD are observed to differ significantly ($p < 0.05$) during the first reference period of 2012 (before the year of drought) and the drought period of 2013. An increase in river water temperature of about 3.4°C was observed from the reference period of 2012 to the drought year of 2013, whereas the DO was observed to be less during the drought year of 2013 compared to the 2012 reference period. While, the pH was observed to observe to differ significantly after the drought period, with a decrease in mean of 0.3 from the drought period to the second reference period of 2014. At the New Bridge station and Haralahalli station, only pH was observed to differ significantly after drought with a decrease in mean of 0.85 and 0.46, respectively, while all other parameters remain insignificant during the three study years.

Water temperature is one of the prominent variables affecting the physical, biological, and chemical characteristics of a river system and aquatic organisms (Edinger et al., 1968). Ambient air temperature was considered to be the most important variable in defining the water temperature (Webb et al., 2003). Water temperature is also one of the vital physical properties affecting reaction rates. An increase in water temperature can result in a decrease of DO levels, which may lead to anaerobic

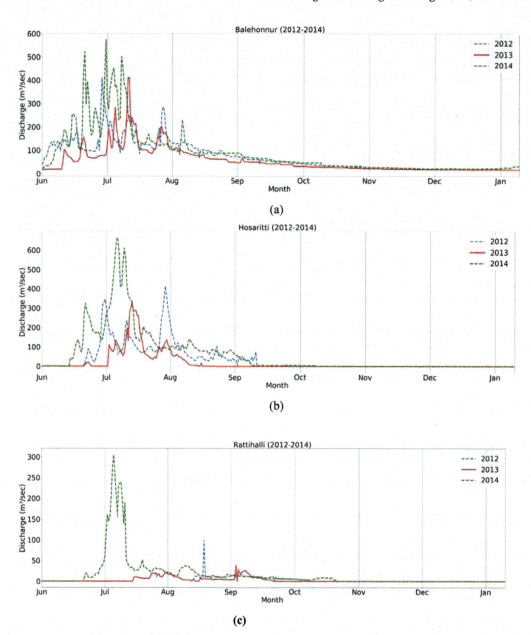

FIGURE 26.5 River flow of the Tunga-Bhadra at the (a) Balehonnur, (b) Hosaritti, and (c) Rattihalli locations during the drought of 2013 and reference periods.

conditions affecting the aquatic life and self-purification capacity of the river system (Rajesh and Rehana, 2022; Rehana et al., 2019). Linear regression models relating air and water temperatures are popular in the prediction of water temperature (Neumann et al., 2003; Rajesh and Rehana, 2021; Rehana and Mujumdar, 2011). The air temperatures and water temperatures were observed (Figures 26.6a, 26.7a, and 26.8a) to follow a linear trend at all the three stations. The water temperatures were observed to increase with the increase in the air temperatures, while the degree of variation was different at each station. The research findings of the present study is in agreement with the results (Rehana and Mujumdar, 2011) for the same case study at the Shimoga station along the Tunga River

TABLE 26.7

Mean and Significant Values of Surface Water Quality Variables Downstream of Bhadravathi for 2013 Drought and Reference Periods

Water Quality Station	Water Quality Variable		Drought	Reference		p-Value	
			2013	2012	2014	2013–2012	2013–2014
Downstream of Bhadravathi	Water temperature	Mean	29	25.6	28.3	**0.02**	
		Minimum	24	24	26		
		Maximum	35	28	30		
	pH	Mean	7.9	7.78	7.6		**0.02**
		Minimum	7.4	7.1	7.2		
		Maximum	8.3	8.2	8.1		
	DO	Mean	6	6.3	6.1		
		Minimum	5	5.1	5.6		
		Maximum	6.6	6.9	7.8		
	BOD	Mean	3.7	4.5	4.1	**0.05**	
		Minimum	3	3	3		
		Maximum	6	6	5		
	Nitrate	Mean	0.3	0.33	0.34		
		Minimum	0.12	0.1	0.02		
		Maximum	0.8	0.9	1.16		

Note: Bold values indicate significant relationships ($p < 0.05$) with the drought year.

TABLE 26.8

Mean and Significant Values of Surface Water Quality Variables at New Bridge NH-4 for 2013 Drought and Reference Periods

Water Quality Station	Water Quality Variable		Drought	Reference		p-Value	
			2013	2012	2014	2013–2012	2013–2014
New Bridge NH-4	Water temperature	Mean	26.1	26.3	26.7		
		Minimum	24	23	24		
		Maximum	30	28	30		
	pH	Mean	8.25	8.37	7.6		**0.001**
		Minimum	7.5	6	7.2		
		Maximum	8.6	9.2	8.2		
	DO	Mean	7.11	6.8	7.3		
		Minimum	6.3	5.2	6.1		
		Maximum	7.6	7.4	7.9		
	BOD	Mean	2.3	2.08	2.59		
		Minimum	1.8	1	2		
		Maximum	3	3	3		
	Nitrate	Mean	0.123	0.15	0.15		
		Minimum	0.08	0.8	0.09		
		Maximum	0.22	0.22	0.58		

Note: Bold values indicate significant relationships ($p < 0.05$) with the drought year.

TABLE 26.9

Mean and Significant Values of Surface Water Quality Variables at Haralahalli Bridge for 2013 Drought and Reference Periods

Water Quality Station	Water Quality Variable		Drought	Reference		p-Value	
			2013	2012	2014	2013–2012	2013–2014
Tungabhadra at Haralahalli Bridge	Water temperature	Mean	25.9	26.16	26.8		
		Minimum	22	23	25		
		Maximum	30	29	32		
	pH	Mean	8.21	8.5	7.75		**0.002**
		Minimum	7.4	7.2	7.4		
		Maximum	8.7	9.2	8.2		
	DO	Mean	7.1	6.9	7.2		
		Minimum	6	5.2	6.1		
		Maximum	7.8	7.4	7.7		
	BOD	Mean	2.5	2.55	2.65		
		Minimum	2	1	2		
		Maximum	3	4	3		
	Nitrate	Mean	0.18	0.16	0.28		
		Minimum	0.1	0.03	0.04		
		Maximum	0.33	0.3	0.7		

Note: Bold values indicate significant relationships ($p < 0.05$) with the drought year.

and the Honnalli station along the Tunga-Bhadra River. The study can conclude that air temperature has a positive response to water temperature for the Tunga-Bhadra River system. The increase of air temperature under climate change has impacted a pronounced increase in water temperatures and resulting DO levels based on general circulation model (GCM) outputs based on statistical down-scaling models over the Tunga-Bhadra River (Rehana and Dhanya, 2018). Furthermore, river water temperatures were observed to differ significantly at the Bhadravathi station ($p < 0.05$), while it fluctuates at the New Bridge station and Haralahalli Bridge with a decrease in the mean value during the drought period. The increased water temperatures during a drought can be correlated with the increase in the air temperature (Mosley et al., 2012).

But unlike the direct relation between air and water temperatures, discharge was observed to follow a quadratic relationship with the water temperature (Figures 26.6b, 26.7b, 26.8b) with a mixed response. The water temperature is observed to decrease with the increase in discharge at the Balehonnur station, while it is observed to decrease to a point and then increase at the Hosaritti and Rattihalli stations. Such an inverse relationship between water temperature and discharge was also noted for other river systems globally (Rehana, 2019; Webb et al., 2003). With such relations between water temperature and streamflows, it can be concluded that the decrease in discharge during drought events can increase the water temperature resulting in reduced water quality in terms of a decrease in DO levels affecting the self-purification capacity of the river system (van Vliet and Zwolsman, 2008). This can also be noted by comparing the water temperature during drought and non-drought years to analyze the river water quality impact during droughts for three stations along the Tunga-Bhadra River. The water temperature for the Bhadravathi station during the drought year of 2013 was noted as 29°C, whereas for the years 2012 and 2014 it was noted as 25.6°C and 28.3°C, respectively (Table 26.7). While the discharge for the Bhadravathi station for the drought year of 2013 was noted as 54 m³/s, for the years 2012 and 2014 it was noted as 72 m³/s and 87 m³/s respectively. This is not consistent with the other stations of New Bridge NH-4 and Haralahalli.

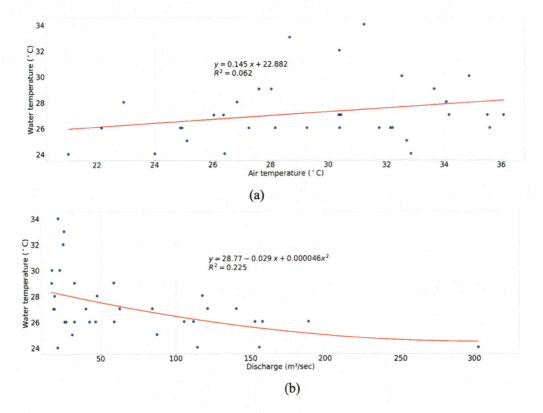

FIGURE 26.6 Relation between monthly (a) water temperature and max air temperature, and (b) water temperature and discharge at Balehonnur for 2012–2014.

It can be noted that river water temperature is also defined by various factors such as excess heat from industries and municipal effluents (Thomann and Mueller, 1987), which may lead to such inconsistency in the temperatures and river flows. The fluctuations in the river water temperatures may also be due to changes in the large portion of flow being derived from the deeper groundwater sources (Wilbers et al., 2009).

According to IS:2296-1982, the tolerance limit of pH is 6.5 to 8.5. A statistically significant difference has been observed in pH values at the New Bridge station and Haralahalli Bridge station, while a mixed response was observed at the Bhadravathi station. The mean pH values at the three water quality stations were observed to be in the tolerance limits, except for a few events at New Bridge and Haralahalli where pH values less than 6.5 and more than 8.5 were observed. Concentration responses of major parameters are mainly determined by their behavior with the fluctuations in the discharge (van Vliet and Zwolsman, 2008). Relatively higher pH values were observed during drought years with lower discharges (Figure 26.9) at the three stations, which is due to the favoring stagnant conditions for algae growth, which uses all the dissolved carbon dioxide among the plants and can cause an increase in pH at the lower water volumes (Gołdyn et al., 2015; van Vliet and Zwolsman, 2008).

DO is an essential river water quality variable and generally considered as the pollution level indicator in water quality monitoring. According to IS:2296-1982, the minimum tolerance limit of DO is 4 mg/l. The DO values at all the stations were observed to be above the tolerance limit. The dissolved oxygen was observed to decrease at the New Bridge station and Haralahalli Bridge station, while it showed a mixed response at the Bhadravathi station. For example, the DO level for the Bhadravathi station in the drought year of 2013 was noted as 6 mg/l, whereas the DO levels for the

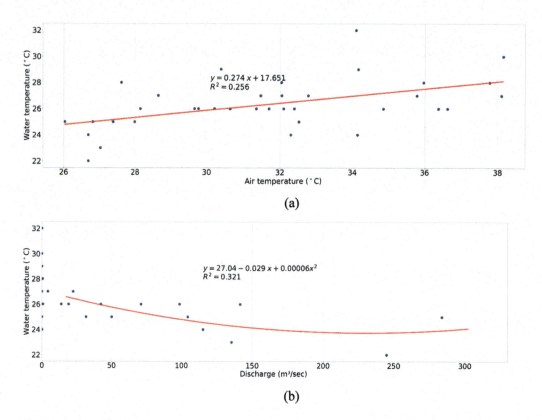

FIGURE 26.7 Relation between monthly (a) water temperature and max air temperature, and (b) water temperature and discharge at Hosaritti for 2012–2014.

reference years of 2012 and 2014 were noted as 6.3 and 6.1 mg/l, respectively. Due to the sufficient water availability for the dilution of pollutants, in non-drought years the DO levels were noted as higher compared to drought years. The decline in DO levels, especially at the lower discharges at the Bhadravathi station (Figure 26.10a, is due to the point source pollutant present (Anderson and Faust, 1972; Chessman and Robinson, 1987) or due to the lower reaeration rates and higher rates of organic matter decomposition by microorganisms (Palmer and Montagna, 2015; Prathumratana et al., 2008; van Vliet and Zwolsman, 2008). The increase in the DO at the New Bridge station and the Haralahalli Bridge station during the lower discharges (Figure 26.10b and c) can be because of an increase in primary production in the river water (Zwolsman and van Bokhoven, 2007). It can be due to the improved effluent quality from the wastewater treatment or fewer untreated sewage inputs due to the reduction in the number of storm events (Attrill and Power, 2000).

The next important river water quality indicator analyzed for the drought and non-drought years is BOD. According to IS:2296-1982, the maximum tolerance limit of BOD is 3 mg/l. The BOD values at the Bhadravathi station were observed to be more than the tolerance limit, while it was within the limit at the other two stations. The minimum and maximum BOD values were observed to be 3 and 6, respectively. A high BOD value indicates that the river is heavily polluted. It can be noted that at Bhadravathi, along the Bhadra River, is one of the most highly polluted river stretches, with industrial pollutants from paper, pulp, rayon, and steel industries, such as from the Mysore paper mill, along with municipal effluents from Bhadravathi city (Rehana and Mujumdar, 2012). Bhadravathi has high BOD values at the lower discharge values (Figure 26.11a). A decrease in BOD values was observed at the other stations at the lower discharge values (Figure 26.11b and c), which are within the tolerance limits.

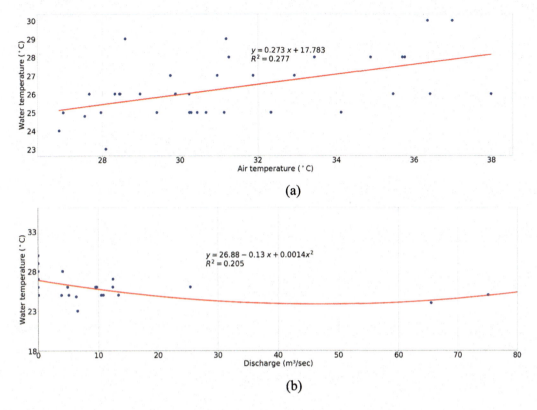

FIGURE 26.8 Relation between monthly (a) water temperature and max air temperature, and (b) water temperature and discharge at Rattihalli for 2012–2014.

According to the Indian Standards for Drinking Water – Specifications (BIS10500:1991), the maximum desirable limit and permissible limits of nitrate content are 45 mg/l and 100 mg/l. Though the nitrates differ significantly ($p < 0.05$) at Haralahalli Bridge, they remain almost insignificant at the other two stations, where the values were observed to be below the desirable limit. The increase in the nitrogen content during the drought can be due to the reducing dilution capacity of the water body (Prathumratana et al., 2008; van Vliet and Zwolsman, 2008; Zwolsman and van Bokhoven, 2007), whereas the decrease in nitrate concentration is due to the reduced agricultural runoff and drainage (Muchmore and Dziegielewski, 1983). The lower nitrate content at the lower discharges (Figure 26.12) can be explained by a reduced supply from soil leaching and overland flow during drought conditions. The negative linear relations found between nitrate and water temperature, except at Haralahalli Bridge (Figure 26.13), may also reflect an intensified uptake of nitrates by algae and increased denitrification (van Vliet and Zwolsman, 2008).

26.4 CONCLUSIONS AND FUTURE DIRECTIONS

The study analyzed hydrological drought impacts on river water quality for a peninsular river system, the Tunga-Bhadra River in India. The drought period of 2013 resulted in low-flow conditions in the Tunga-Bhadra River, particularly at the Bhadravathi River location. The study analyzed annual scale meteorological and hydrological drought indices with the SPI and SDI, respectively, for three stations – Balehonnur, Hosaritti, and Rattihalli – along the Tunga-Bhadra basin. The present framework is useful for temporal and spatial dispersion analysis of drought using the SPI and SDI algorithm, respectively, which performed well in monitoring meteorological and hydrological

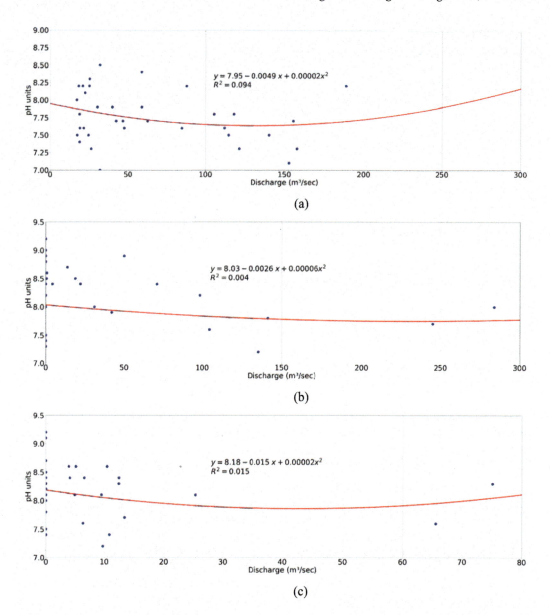

FIGURE 26.9 Relation between monthly discharge and pH at (a) Bhadravathi, (b) Haralahalli Bridge, and (c) New Bridge for 2012–2014.

drought. The lower precipitation values and higher temperatures and, therefore, consequent decrease in streamflows were observed during the drought year of 2013 compared to the reference years of 2012 and 2014 over the Tunga-Bhadra River Basin. The impact of hydrological drought on river water quality was studied by observing the variation in the water quality parameters during the drought period and the reference wet periods. The river water quality monitoring stations were at Bhadravathi, Haralahalli Bridge, and New Bridge with using the water quality indicators of water temperature, DO, BOD, pH, and nitrates. An increase in river water temperature (3.4°C) with a decrease in discharge (25%) have resulted in lower DO (0.3 mg/l) values during the drought year of 2013 compared to the 2012 reference period along at the Bhadravathi station along the Bhadra River. The increased water temperatures during the drought have been correlated with the increase in the

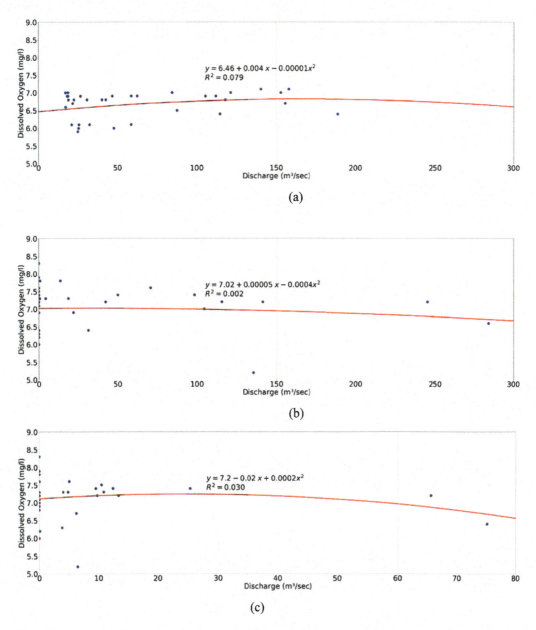

FIGURE 26.10 Relation between monthly discharge and dissolved oxygen at (a) Bhadravathi, (b) Haralahalli Bridge, and (c) New Bridge for 2012–2014.

air temperature. Higher river water temperatures during drought periods can lower oxygen saturation concentration with lower reaeration rates. However, the decrease in discharge during drought events has increased the river water temperature resulting in reduced water quality in terms of a decrease in DO levels affected by the self-purification capacity of the river system. Relatively higher pH values were observed during drought years compared to wet years due to lower water volumes, which favor conditions for algae growth consuming all the dissolved carbon dioxide. The nitrate concentration was observed to be in the desirable limits. Almost all stations experienced at least one severe drought during the study period. As per the standards given by the Central Pollution Control Board (CPCB,

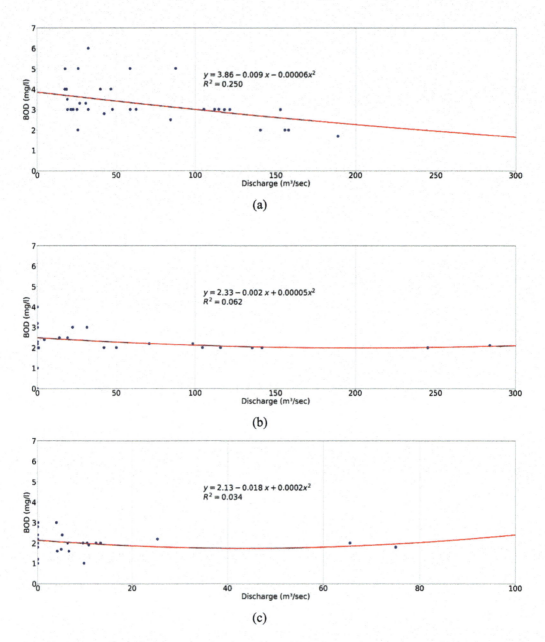

FIGURE 26.11 Relation between monthly discharge and BOD at (a) Bhadravathi, (b) Haralahalli Bridge, and (c) New Bridge for 2012–2014.

2020) for the designated best use of water, the water quality at the Bhadravathi station does not meet any of the criteria due to the high amounts of BOD present. However, at the Haralahalli Bridge station, though the individual parameters were affected due to drought, the overall standards remained the same and the water quality can be used as a drinking water source without conventional treatment but after disinfection, before, during, and after the drought.

This case study demonstrated how the SPI and SDI algorithm could be scaled up and used for assessing drought years. Due to the limitation of the data availability, the present study demonstrated

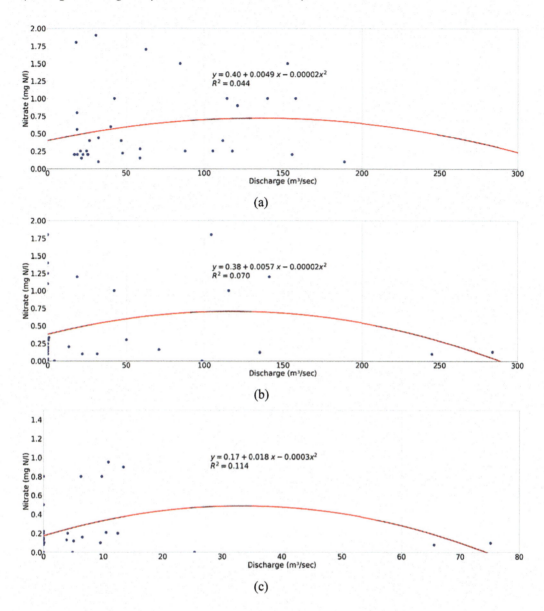

FIGURE 26.12 Relation between monthly discharge and nitrates at (a) Bhadravathi, (b) Haralahalli Bridge, and (c) New Bridge for 2012–2014.

the hydrological drought impacts on water quality with few river locations, which can be extended to other stations based on data availability. Since drought is a multifaceted phenomenon, a coupling of other drought indices may generate enhanced results for analysts and policymakers. This will help in forming mitigating strategies and policy-based solutions, which can be implemented to reduce the impact of drought affecting various sectors. To conclude, this study gives a valuable overview of hydrological drought-driven water quality deterioration and its potential impacts, indicating that further water management strategies should consider monitoring water quality during drought periods.

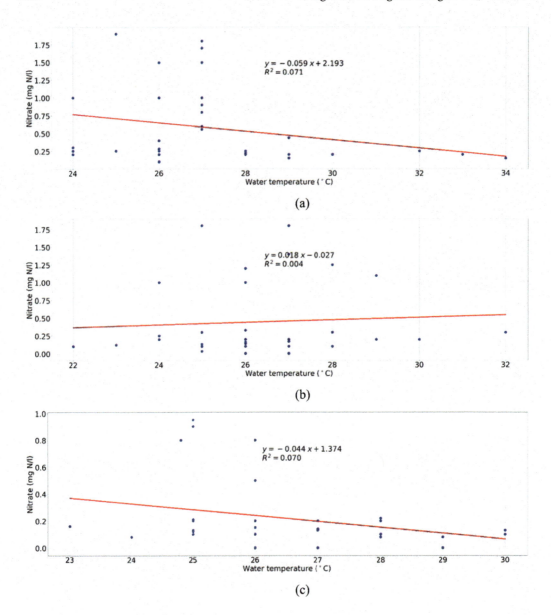

FIGURE 26.13 Relation between monthly water temperature and nitrates at (a) Bhadravathi, (b) Haralahalli Bridge, and (c) New Bridge for 2012–2014.

ACKNOWLEDGMENTS

The authors sincerely thank the Advanced Centre for Integrated Water Resources Management (ACIWRM), Bengaluru, Karnataka, India, for providing Tunga-Bhadra River data. The research work is supported by the Science and Engineering Research Board (SERB), Department of Science and Technology, Government of India, through the Core Research Grant Project No. CRG/2020/002028 to Dr. S. Rehana.

REFERENCES

Aghakouchak, A., Feldman, D., Stewardson, M.J., Saphores, J.-D., Grant, S., Sanders, B., 2014. Australia's drought: Lessons for California. *Science* 343, 1430–1431. https://doi.org/10.1126/science.343.6178.1430

Alley, W.M., 1984. The Palmer Drought severity index: Limitations and assumptions. *Journal of Climate and Applied Meteorology* 23, 1100–1109. https://doi.org/10.1175/1520-0450(1984)023<1100:TPDSIL> 2.0.CO;2

AMS, 2013. American Meteorological Society [WWW Document]. American Meteorological Society. https://www.ametsoc.org/index.cfm/ams/about-ams/ams-statements/statements-of-the-ams-in-force/drought/

Anderson, P.W., Faust, S.D., 1972. Impact of drought on quality in a New Jersey water supply system1. *JAWRA Journal of the American Water Resources Association* 8, 750–760. https://doi.org/10.1111/j.1752-1688 .1972.tb05217.x

Assouline, S., Russo, D., Silber, A., Or, D., 2015. Balancing water scarcity and quality for sustainable irrigated agriculture. *Water Resources Research* 51, 3419–3436. https://doi.org/10.1002/2015WR017071

Attrill, M.J., Power, M., 2000. Modelling the effect of drought on estuarine water quality. *Water Research* 34, 1584–1594. https://doi.org/10.1016/S0043-1354(99)00305-X

Badripour, H., 2007. Role of drought monitoring and management in NAP implementation, in: Sivakumar, M.V.K., Ndiang'ui, N. (Eds.), *Climate and Land Degradation, Environmental Science and Engineering.* Springer, Berlin, Heidelberg, pp. 565–582. https://doi.org/10.1007/978-3-540-72438-4_32

Bhat, S.A., Meraj, G., Yaseen, S., Pandit, A.K., 2014. Statistical assessment of water quality parameters for pollution source identification in Sukhnag stream: An inflow stream of Lake Wular (Ramsar Site), Kashmir Himalaya. *Journal of Ecosystems* 2014, e898054. https://doi.org/10.1155/2014/898054

Bhuiyan, C., 2004. Various drought indices for monitoring drought condition in Aravalli terrain of India. *ISPRS International Journal of Geo-Information* 6, 12–23.

Bijaber, N., El Hadani, D., Saidi, M., Svoboda, M.D., Wardlow, B.D., Hain, C.R., Poulsen, C.C., Yessef, M., Rochdi, A., 2018. Developing a remotely sensed drought monitoring indicator for Morocco. *Geosciences* 8, 55. https://doi.org/10.3390/geosciences8020055

Bond, N.R., Lake, P.S., Arthington, A.H., 2008. The impacts of drought on freshwater ecosystems: An Australian perspective. *Hydrobiologia* 600, 3–16. https://doi.org/10.1007/s10750-008-9326-z

Byun, H.-R., Wilhite, D.A., 1999. Objective quantification of drought severity and duration. *Journal of Climate* 12, 2747–2756. https://doi.org/10.1175/1520-0442(1999)012<2747:OQODSA>2.0.CO;2

Chanda Kironmala, M.R., 2015. Meteorological drought quantification with standardized precipitation anomaly index for the regions with strongly seasonal and periodic precipitation. *Journal of Hydrologic Engineering* 20, 06015007. https://doi.org/10.1061/(ASCE)HE.1943-5584.0001236

Chang, T.J., Kleopa, X.A., 1991. A proposed method for drought monitoring. *JAWRA Journal of the American Water Resources Association* 27, 275–281. https://doi.org/10.1111/j.1752-1688.1991.tb03132.x

Chessman, B.C., Robinson, D.P., 1987. Some effects of the 1982–83 drought on water quality and macroin-vertebrate fauna in the lower La Trobe River, Victoria. *Australian Journal of Marine and Freshwater Research* 38, 289–299. https://doi.org/10.1071/mf9870289

Clair, T.A., Ehrman, J.M., 1996. Variations in discharge and dissolved organic carbon and nitrogen export from terrestrial basins with changes in climate: A neural network approach. *Limnology and Oceanography* 41, 921–927. https://doi.org/10.4319/lo.1996.41.5.0921

Cloke, H.L., Hannah, D.M., 2011. Large-scale hydrology: Advances in understanding processes, dynamics and models from beyond river basin to global scale. *Hydrological Processes* 25, 991–995. https://doi .org/10.1002/hyp.8059

CPCB, 2020. *Central Pollution Control Board.* Wikipedia, The Free Encyclopedia.

Cressie, N.a.C., Whitford, H.J., 1986. How to use the two sample t-test. *Biometrical Journal* 28, 131–148. https://doi.org/10.1002/bimj.4710280202

Delpla, I., Jung, A.-V., Baures, E., Clement, M., Thomas, O., 2009. Impacts of climate change on surface water quality in relation to drinking water production. *Environment International* 35, 1225–1233. https://doi .org/10.1016/j.envint.2009.07.001

Dettinger, M., Cayan, D.R., 2014. Drought and the California delta: A matter of extremes. *San Francisco Estuary and Watershed Science* 12. https://doi.org/10.15447/sfews.2014v12iss2art4

Dhawale, R.K., Paul, S., 2019. *Moving from Crisis Management to Risk Assessment for Drought Planning Using Standardized Precipitation Index (SPI) and Standardized Groundwater Level Index (SWI): Case Study of Marathwada, India.* Global Assessment Report on Disaster Risk Reduction (GAR 2019).

Doesken, N.J., McKee, T., Kleist, J., 1991. *Development of a Surface Water Supply Index for the Western United States. Climatology Report Colorado Climate Center,* Colorado State University, 84.

Dracup, J.A., Lee, K.S., Paulson Jr., E.G., 1980. On the definition of droughts. *Water Resources Research* 16, 297–302. https://doi.org/10.1029/WR016i002p00297

Dutta, D., Kundu, A., Patel, N.R., 2013. Predicting agricultural drought in eastern Rajasthan of India using NDVI and standardized precipitation index. *Geocarto International* 28, 192–209. https://doi.org/10 .1080/10106049.2012.679975

Edinger, J.E., Duttweiler, D.W., Geyer, J.C., 1968. The response of water temperatures to meteorological conditions. *Water Resources Research* 4, 1137–1143. https://doi.org/10.1029/WR004i005p01137

Flatau, M.K., Flatau, P.J., Schmidt, J., Kiladis, G.N., 2003. Delayed onset of the 2002 Indian monsoon. *Geophysical Research Letters* 30. https://doi.org/10.1029/2003GL017434

Fukushima, T., Ozaki, N., Kaminishi, H., Harasawa, H., Matsushige, K., 2000. Forecasting the changes in lake water quality in response to climate changes, using past relationships between meteorological conditions and water quality. *Hydrological Processes* 14, 593–604. https://doi.org/10.1002/(SICI)1099-1085(20000228)14:3<593::AID-HYP956>3.0.CO;2-O

Geng, H., Shen, B., 1992. Definition and significance of hydrological droughts. *Agricultural Research in the Arid Areas* 4, 91–94.

Gołdyn, B., Kowalczewska-Madura, K., Celewicz-Gołdyn, S., 2015. Drought and deluge: Influence of environmental factors on water quality of kettle holes in two subsequent years with different precipitation. *Limnologica* 54, 14–22. https://doi.org/10.1016/j.limno.2015.07.002

Goyal, M., Gupta, V., Eslamian, S., 2017. Hydrological drought: Water surface and duration curve indices, in: S. Eslamian (Ed.), *Handbook of drought and water scarcity: Principles of drought and water scarcity*. Taylor & Francis, London, pp. 45–70.

Gusyev, M., Hasegawa, A., Magome, J., Kuribayashi, D., Sawano, H., Lee, S., 2015. *Drought Assessment in the Pampanga River Basin, the Philippines – Part 1: Characterizing a Role of Dams in Historical Droughts with Standardized Indices*. https://doi.org/10.13140/RG.2.1.4931.4321

Hayes, M.J., 2006. Drought indices, in: *Van Nostrand's Scientific Encyclopedia*. American Cancer Society. https://doi.org/10.1002/0471743984.vse8593

Ju, X.S., Yang, X.W., Chen, L.J., 1997. Research on determination of station indexes and division of regional flood/ drought grades in China. *Journal of Applied Meteorlogy*, 8(1), 26–33.

KERS, 1985. *Reservoir Sedimentation Study Report*. Karnataka Engineering Research Station, Mysore, India.

Krysanova, V., Hattermann, F., Habeck, A., 2005. Expected changes in water resources availability and water quality with respect to climate change in the Elbe River basin (Germany). *Hydrology Research* 36, 321–333. https://doi.org/10.2166/nh.2005.0025

Lanen, H.A.J.V., Wanders, N., Tallaksen, L.M., Loon, A.F.V., 2013. Hydrological drought across the world: Impact of climate and physical catchment structure. *Hydrology and Earth System Sciences* 17, 1715–1732. https://doi.org/10.5194/hess-17-1715-2013

Loon, A.F.V., 2015. Hydrological drought explained. *WIREs Water* 2, 359–392. https://doi.org/10.1002/wat2.1085

Maestre-Valero, J.F., Martínez-Alvarez, V., 2010. Effects of drip irrigation systems on the recovery of dissolved oxygen from hypoxic water. *Agricultural Water Management* 97, 1806–1812. https://doi.org/10.1016/j.agwat.2010.06.018

McGuire, J.K., Palmer, W.C., 1957. The 1957 drought in the eastern united states. *Monthly Weather Review* 85, 305–314. https://doi.org/10.1175/1520-0493(1957)085<0305:TDITEU>2.0.CO;2

McKee, T.B., Doesken, N.J., Kleist, J., 1993. The relationship of drought frequency and duration to time scales. *Environmental Science* 6, 179–184.

McMahon, E.T.A., Arenas, A.D., 1982. *Methods of Computation of Low Streamflow, Studies and Reports in Hydrology*; 36. UNESCO, Paris, 1982.

Mimikou, M.A., Baltas, E., Varanou, E., Pantazis, K., 2000. Regional impacts of climate change on water resources quantity and quality indicators. *Journal of Hydrology* 234, 95–109. https://doi.org/10.1016/S0022-1694(00)00244-4

Mishra, A.K., Desai, V.R., 2005a. Drought forecasting using stochastic models. *Stochastic Environmental Research and Risk Assessment volume* 19, 326–339. https://doi.org/10.1007/s00477-005-0238-4

Mishra, A.K., Desai, V.R., 2005b. Spatial and temporal drought analysis in the Kansabati river basin, India. *International Journal of River Basin Management* 3, 31–41. https://doi.org/10.1080/15715124.2005.9635243

Mishra, A.K., Singh, V.P., 2010. A review of drought concepts. *Journal of Hydrology* 391, 202–216. https://doi.org/10.1016/j.jhydrol.2010.07.012

Mishra, A.K., Singh, V.P., 2011. Drought modeling: A review. *Journal of Hydrology* 403, 157–175. https://doi.org/10.1016/j.jhydrol.2011.03.049

Mishra, V., Smoliak, B.V., Lettenmaier, D.P., Wallace, J.M., 2012. A prominent pattern of year-to-year variability in Indian Summer Monsoon Rainfall. *PNAS* 109, 7213–7217. https://doi.org/10.1073/pnas.1119150109

Modarres, R., 2007. Streamflow drought time series forecasting. *Stochastic Environmental Research and Risk Assessment volume* 21, 223–233. https://doi.org/10.1007/s00477-006-0058-1

Mosley, L.M., 2015. Drought impacts on the water quality of freshwater systems; review and integration. *Earth-Science Reviews* 140, 203–214. https://doi.org/10.1016/j.earscirev.2014.11.010

Mosley, L.M., Zammit, B., Leyden, E., Heneker, T., Hipsey, M., Skinner, D., Aldridge, K., 2012. The impact of extreme low flows on the water quality of the lower Murray River and Lakes (South Australia). *Water Resources Management* 26, 3923–3946. https://doi.org/10.1007/s11269-012-0113-2

Muchmore, C.B., Dziegielewski, B., 1983. Impact of drought on quality of potential water supply sources in the Sangamon River Basin1. *JAWRA Journal of the American Water Resources Association* 19, 37–46. https://doi.org/10.1111/j.1752-1688.1983.tb04554.x

Murdoch, P.S., Baron, J.S., Miller, T.L., 2000. Potential effects of climate change on surface-water quality in North America. *JAWRA Journal of the American Water Resources Association* 36, 347–366. https://doi.org/10.1111/j.1752-1688.2000.tb04273.x

Nalbantis, I., Tsakiris, G., 2009. Assessment of hydrological drought revisited. *Water Resources Management* 23, 881–897. https://doi.org/10.1007/s11269-008-9305-1

Neumann, David W., Balaji, R., Zagona, Edith A., 2003. Regression model for daily maximum stream temperature. *Journal of Environmental Engineering* 129, 667–674. https://doi.org/10.1061/(ASCE)0733-9372(2003)129:7(667)

NIH, 1992. *Quantitative Assessment of Sediment Distribution in the Tungabhadra Reservoir Using Satellite Imagery.* Report No. CS-84/1991-92. National Institute of Hydrology, Roorkee.

Palmer, T.A., Montagna, P.A., 2015. Impacts of droughts and low flows on estuarine water quality and benthic fauna. *Hydrobiologia* 753, 111–129. https://doi.org/10.1007/s10750-015-2200-x

Palmer, W.C., 1965. *Meteorological Drought.* European Environment Agency, US Department of Commerce, Weather Bureau, Research Paper No. 45; 58 p.

Parmar, K.S., Bhardwaj, R., 2013. Water quality index and fractal dimension analysis of water parameters. *International Journal of Environmental Science and Technology* 10, 151–164. https://doi.org/10.1007/s13762-012-0086-y

Pozzi, W., Sheffield, J., Stefanski, R., Cripe, D., Pulwarty, R., Vogt, J.V., Heim, R.R., Brewer, M.J., Svoboda, M., Westerhoff, R., van Dijk, A.I.J.M., Lloyd-Hughes, B., Pappenberger, F., Werner, M., Dutra, E., Wetterhall, F., Wagner, W., Schubert, S., Mo, K., Nicholson, M., Bettio, L., Nunez, L., van Beek, R., Bierkens, M., de Goncalves, L.G.G., de Mattos, J.G.Z., Lawford, R., 2013. Toward global drought early warning capability: Expanding international cooperation for the development of a framework for monitoring and forecasting. *Bulletin of the American Meteorological Society* 94, 776–785. https://doi.org/10.1175/BAMS-D-11-00176.1

Prathumratana, L., Sthiannopkao, S., Kim, K.W., 2008. The relationship of climatic and hydrological parameters to surface water quality in the lower Mekong River. *Environment International* 34, 860–866. https://doi.org/10.1016/j.envint.2007.10.011

Puczko, K., Jekatierynczuk-Rudczyk, E., 2020. Extreme hydro-meteorological events influence to water quality of small rivers in urban area: A case study in Northeast Poland. *Scientific Reports* 10, 10255. https://doi.org/10.1038/s41598-020-67190-4

Rajesh, M., Rehana, S., 2021. Prediction of river water temperature using machine learning algorithms: A tropical river system of India. *Journal of Hydroinformatics* 23, 605–626. https://doi.org/10.2166/hydro.2021.121

Rajesh, M., Rehana, S., 2022. Impact of climate change on river water temperature and dissolved oxygen: Indian riverine thermal regimes. *Scientific Reports* 12, 9222. https://doi.org/10.1038/s41598-022-12996-7

Rehana, S., 2019. River water temperature modelling under climate change using support vector regression, in: Singh, S.K., Dhanya, C.T. (Eds.), *Hydrology in a Changing World: Challenges in Modeling, Springer Water.* Springer International Publishing, Cham, pp. 171–183. https://doi.org/10.1007/978-3-030-02197-9_8

Rehana, S., Dhanya, C.T., 2018. Modeling of extreme risk in river water quality under climate change. *Journal of Water and Climate Change* 9, 512–524. https://doi.org/10.2166/wcc.2018.024

Rehana, S., Mujumdar, P.P., 2011. River water quality response under hypothetical climate change scenarios in Tunga-Bhadra river, India. *Hydrological Processes* 25, 3373–3386. https://doi.org/10.1002/hyp.8057

Rehana, S., Mujumdar, P., 2012. Climate change induced risk in water quality control problems. *Journal of Hydrology* s 444–445, 63–77. https://doi.org/10.1016/j.jhydrol.2012.03.042

Rehana, S., Munoz-Arriola, F., Rico, D.A., Bartelt-Hunt, S.L., 2019. Modelling water temperature's sensitivity to atmospheric warming and river flow, in: Sobti, R.C., Arora, N.K., Kothari, R. (Eds.), *Environmental Biotechnology: For Sustainable Future.* Springer, Singapore, pp. 309–319. https://doi.org/10.1007/978-981-10-7284-0_12

Safavi, H.R., Malek Ahmadi, K., 2015. Prediction and assessment of drought effects on surface water quality using artificial neural networks: Case study of Zayandehrud River, *Journal of Environmental Health Science and Engineering* 13, 68. https://doi.org/10.1186/s40201-015-0227-6

Sagasta, J.M., Burke, J., 2005. *Agriculture and Water Quality Interactions: A Global Overview.* Food and Agriculture Organization of the United Nations (FAO), Rome, Italy.

Salehnia, N., Alizadeh, A., Sanaeinejad, H., Bannayan, M., Zarrin, A., Hoogenboom, G., 2017. Estimation of meteorological drought indices based on AgMERRA precipitation data and station-observed precipitation data. *Journal of Arid Land* 9, 797–809. https://doi.org/10.1007/s40333-017-0070-y

Schewe, J., Levermann, A., 2012. A statistically predictive model for future monsoon failure in India. *Environmental Research Letters* 7, 044023. https://doi.org/10.1088/1748-9326/7/4/044023

Schindler, D.W., Bayley, S.E., Parker, B.R., Beaty, K.G., Cruikshank, D.R., Fee, E.J., Schindler, E.U., Stainton, M.P., 1996. The effects of climatic warming on the properties of boreal lakes and streams at the Experimental Lakes Area, northwestern Ontario. *Limnology and Oceanography* 41, 1004–1017. https://doi.org/10.4319/lo.1996.41.5.1004

Shafer, B.A., Dezman, L.E., 1982. Development of a surface water supply index (SWSI) to assess the severity of drought conditions in snowpack runoff areas. 50th Annual Western Snow Conference, Proceedings of the 50th Annual Western Snow Conference.

Sheffield, J., Wood, E.F., Wood, E.F., 2012. *Drought : Past Problems and Future Scenarios.* Routledge, London. https://doi.org/10.4324/9781849775250

Somville, M., De Pauw, N., 1982. Influence of temperature and river discharge on water quality of the western Scheldt estuary. *Water Research* 16, 1349–1356. https://doi.org/10.1016/0043-1354(82)90213-5

Stagge, J.H., Tallaksen, L.M., Gudmundsson, L., Loon, A.F.V., Stahl, K., 2015. Candidate distributions for climatological drought indices (SPI and SPEI). *International Journal of Climatology* 35, 4027–4040. https://doi.org/10.1002/joc.4267

Stahl, K., Hisdal, H., 2004. Hydroclimatology, in: *Hydrological Drought: Processes and Estimation Methods for Streamflow and Groundwater.* Developments in Water Sciences, 48, pp. 19–51.

Svoboda, M.D., Fuchs, B.A., 2016. *Handbook of Drought Indicators and Indices.* WMO-No. 1173.

Takeuchi, K., 1974. *Regional Water Exchange for Drought Alleviation*, Hydrology paper 70. Colorado State University, Colorado.

Tallaksen, L.M., Madsen, H., Clausen, B., 1997. On the definition and modelling of streamflow drought duration and deficit volume. *Hydrological Sciences Journal* 42, 15–33. https://doi.org/10.1080/02626669709492003

Teegavarapu, R.S.V., 2017. Climate variability and changes in precipitation extremes and characteristics, in: Kolokytha, E., Oishi, S., Teegavarapu, R.S.V. (Eds.), *Sustainable Water Resources Planning and Management Under Climate Change. Springer, Singapore*, pp. 3–37. https://doi.org/10.1007/978-981-10-2051-3_1

The Times of India, 2019. Drought, floods new normal as Karnataka rain pattern changes [WWW Document]. *The Times of India.* URL https://timesofindia.indiatimes.com/city/bengaluru/drought-floods-are-new-normal-in-karnataka/articleshow/69191638.cms

Thomann, R.V., Mueller, J.A., 1987. *Principles of Surface Water Quality Modeling and Control.* Harper & Row, New York.

Udmale, P., Ichikawa, Y., Nakamura, T., Shaowei, N., Ishidaira, H., Kazama, F., 2016. Rural drinking water issues in India's drought-prone area: A case of Maharashtra state. *Environmental Research Letters* 11, 074013. https://doi.org/10.1088/1748-9326/11/7/074013

Van Lanen, H., Fendekova, M., Kupczyk, E., Kasprzyk, A., Pokojski, W., 2004. Flow generating processes. Hydrological drought. *Processes and Estimation Methods for Streamflow and Groundwater* 48, 53–96.

van Vliet, M.T.H., Franssen, W.H.P., Yearsley, J.R., Ludwig, F., Haddeland, I., Lettenmaier, D.P., Kabat, P., 2013. Global river discharge and water temperature under climate change. *Global Environmental Change* 23, 450–464. https://doi.org/10.1016/j.gloenvcha.2012.11.002

van Vliet, M.T.H., Zwolsman, J.J.G., 2008. Impact of summer droughts on the water quality of the Meuse river. *Journal of Hydrology* 353, 1–17. https://doi.org/10.1016/j.jhydrol.2008.01.001

Vicente-Serrano, S.M., Beguería, S., López-Moreno, J.I., 2010. A multiscalar drought index sensitive to global warming: The standardized precipitation evapotranspiration index. *Journal of Climate* 23, 1696–1718. https://doi.org/10.1175/2009JCLI2909.1

Webb, B.W., Clack, P.D., Walling, D.E., 2003. Water–air temperature relationships in a Devon river system and the role of flow. *Hydrological Processes* 17, 3069–3084. https://doi.org/10.1002/hyp.1280

Whitehead, P.G., Wilby, R.L., Battarbee, R.W., Kernan, M., Wade, A.J., 2009. A review of the potential impacts of climate change on surface water quality. *Hydrological Sciences Journal* 54, 101–123. https://doi.org/10.1623/hysj.54.1.101

Wilbers, G.-J., Zwolsman, G., Klaver, G., Hendriks, A.J., 2009. Effects of a drought period on physico-chemical surface water quality in a regional catchment area. *Journal of Environmental Monitoring* 11, 1298–1302. https://doi.org/10.1039/B816109G

Wilhite, D.A., 2000. Drought as a natural hazard: Concepts and definitions, in: *Drought: A Global Assessment*. Drought Mitigation Center Faculty Publications, p. 22.

Wilhite, D.A., Glantz, M.H., 1985. Understanding the drought phenomenon: The role of definitions. *Water International* 10, 111–120. https://doi.org/10.1080/02508068508686328

Willeke, G., Hosking, J.R.M., Wallis, J.R., 1994. *The National Drought Atlas*. U.S. Army Corps of Engineers, Water Resources Support Center, Institute for Water Resources Report 94-NDS-4.

Williams, M.W., Losleben, M., Caine, N., Greenland, D., 1996. Changes in climate and hydrochemical responses in a high-elevation catchment in the Rocky Mountains, USA. *Limnology and Oceanography* 41, 939–946. https://doi.org/10.4319/lo.1996.41.5.0939

Wolford, R.A., Bales, R.C., 1996. Hydrochemical modeling of Emerald Lake watershed, Sierra Nevada, California: Sensitivity of stream chemistry to changes in fluxes and model parameters. *Limnology and Oceanography* 41, 947–954. https://doi.org/10.4319/lo.1996.41.5.0947

Wu, H., Hayes, M.J., Weiss, A., Hu, Q., 2001. An evaluation of the standardized precipitation index, the China-Z index and the statistical Z-score. *International Journal of Climatology* 21, 745–758. https://doi.org/10.1002/joc.658

Yeh, C.-F., Wang, J., Yeh, H.-F., Lee, C.-H., 2015. SDI and Markov Chains for regional drought characteristics. *Sustainability* 7, 10789–10808. https://doi.org/10.3390/su70810789

Zelenhasić, E., Salvai, A., 1987. A method of streamflow drought analysis. *Water Resources Research* 23, 156–168. https://doi.org/10.1029/WR023i001p00156

Zwolsman, J.J.G., van Bokhoven, A.J., 2007. Impact of summer droughts on water quality of the Rhine River: A preview of climate change? *Water Science & Technology* 56, 45–55. https://doi.org/10.2166/wst.2007.535

27 Integrated Drought Management
Moving from Managing Disasters to Managing Risk

Donald A. Wilhite

CONTENTS

27.1 INTRODUCTION

In recent years, concern has grown worldwide that droughts are and will continue to increase in frequency, severity, and duration given changing climatic conditions and documented increases in extreme climate events. This concern is of particular importance given climate model projections showing pronounced warming and decreased precipitation in the coming decades in many regions. The narrowing gap between water supply and demand will result in a dramatic increase in frequency and magnitude of drought impacts and the number of sectors affected. Although agriculture is typically the first and most drought-affected sector, many other sectors, including energy production, tourism and recreation, transportation, urban water supply, public health, and the environment, have experienced significant impacts and these will increase at an accelerating rate as a result of further warming and growing competition for finite water resources.

Historically, the response to drought by governments and other organizations globally have been reactive – poorly coordinated, ineffective, and untimely. This approach is commonly referred to as crisis management. This model leads to the provision of relief or assistance in a post-drought setting to those most affected. I have typically illustrated this post-impact response with the "hydro-illogical cycle" (Figure 27.1). In the early stages of drought development, the initial awareness of drought conditions is followed by concern and panic as drought conditions intensify. Oftentimes, the need for a more proactive approach to drought management that includes the adoption of risk reduction actions in advance of the next drought is considered by policy makers. However, when normal rainfall eventually returns, the government's plan to develop a proactive drought management strategy in order to be better prepared for the next drought often is discarded, i.e., a return to the apathy stage of the cycle. Associated with the crisis management approach is the lack of recognition that drought is a normal part of the climate. To prepare for drought's inevitable recurrence, drought-prone countries should develop a national drought policy and associated preparedness and mitigation plans aimed at risk reduction. This step will require a greater upfront investment in drought planning and

DOI: 10.1201/9781003276548-27

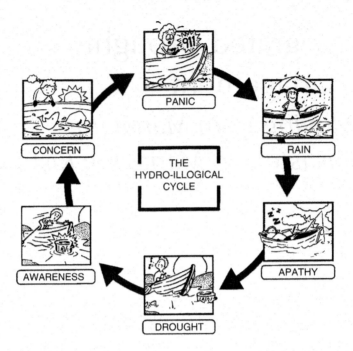

FIGURE 27.1   The hydro-illogical cycle, the traditional crisis management approach.

preparedness, but the cost of action to reduce the risk associated with future drought episodes is far less than the costs associated with inaction (Wilhite and Pulwarty 2018a; Venton et al. 2019).

Research has demonstrated that the reactive or crisis management approach to drought management increases vulnerability to future drought episodes by reducing self-reliance and increasing dependence on governments and donor organizations (Wilhite and Pulwarty 2018b). Crisis management also reinforces the continuation of past water, agricultural, and other management practices that have contributed to current vulnerabilities. As Albert Einstein once said, "We cannot solve our problems with the same level of thinking that created them". Therefore, it is imperative for all drought-prone nations to adopt a new paradigm for drought management based on the principles of risk reduction.

Until recently, despite the ineffectiveness of past drought response efforts, no concerted effort has been made at the global level to initiate a dialogue on the formulation and adoption of national drought management policies that provide a framework for proactive, risk-based drought management. Without a coordinated national drought policy that includes comprehensive monitoring, early warning and information delivery systems, vulnerability and impact assessments, the identification and adoption of appropriate local-level mitigation, and response measures aimed at risk reduction, nations will continue to respond to drought as they have in the past.

A primary stimulus for this changing paradigm for drought management is a direct outcome of the High-Level Meeting on National Drought Policy (HMNDP) held in Geneva, Switzerland, in March 2013 (HMNDP 2013). The principal sponsors of this meeting were the World Meteorological Organization (WMO), the Secretariat of the United Nations Convention to Combat Desertification (UNCCD), and the Food and Agriculture Organization of the United Nations (FAO), in collaboration with a number of other partners. The goal of HMNDP was to provide practical insight into useful, science-based actions to address key drought issues and to identify various strategies that would lead nations to cope more effectively with drought and, therefore, reduce societal impacts. The organizers and key participants of this meeting encouraged national governments to adopt policies that engender cooperation and coordination at all levels of their administration in order to increase their capacity to cope with extended periods of water shortage resulting from drought. The ultimate goal of this effort is to create more drought-resilient societies and ensure food security and the sustainability of natural resource systems at the domestic level. Wilhite and Pulwarty (2018b) highlighted the principles associated with this paradigm shift and provided many case studies of its application in countries and regions.

A series of outcomes were forthcoming from the HMNDP. One such outcome was the launching of the Integrated Drought Management Programme (IDMP) in 2013 by the World Meteorological Organization and the Global Water Partnership (GWP) (http://www.droughtmanagement.info/). The IDMP's objective is to support stakeholders at all levels by providing policy and management guidance and by sharing scientific information, knowledge, and best practices for integrated drought management (IDM). IDM mitigates drought risk and builds drought resilience by addressing multiple components of drought management, including disaster risk reduction, climate adaptation strategies, and national water policies.

The IDMP emphasizes the following tenets:

1. Shift focus [of drought management] from reactive to proactive
2. Foster horizontal integration by collaborating with partners from diverse sectors
3. Facilitate vertical integration on global, regional, national and local levels
4. Improve knowledge base with better access to information
5. Promote projects that demonstrate innovation
6. Build countries' capacity to implement integrated approaches

Following on the recommendations of the HMNDP, the IDMP adopted a three-pillar approach to drought management based on a ten-step planning process (WMO and IDMP 2014). These pillars are illustrated in Figure 27.2.

As illustrated in Figure 27.2, to create a successful drought policy aimed at risk reduction, it is vital for that policy to emphasize a comprehensive monitoring and early warning system, vulnerability and impact assessment and mitigation and response, i.e., the three pillars.

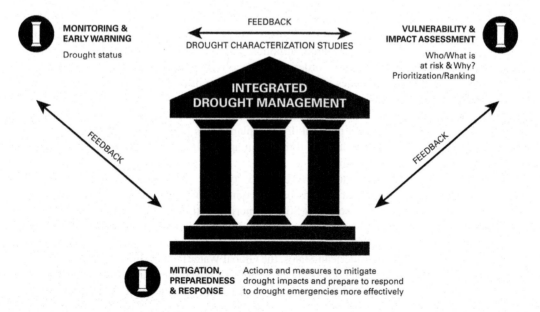

FIGURE 27.2 The three-pillar approach to drought management. (From Integrated Drought Management Programme, World Meteorological Organization and Global Water Partnership.)

Some of the drivers of and barriers to drought risk management are:

Drivers to drought risk management
- Increased financial burden of drought relief costs and their ineffectiveness
- Impact and vulnerability trends in economic, social, and environmental sectors
- Increased water-related conflicts and scarcity
- Increased awareness of efficiency of drought risk management, i.e., risk reduction
- Evidence on co-benefits of mitigation actions and preparedness
- Climate change and increased incidence of extreme events

Barriers to drought risk management
- Path dependency, size of upfront costs for multiyear droughts
- Information failure on the occurrences, impacts, costs/benefits of drought risk management
- Market failure (credit constraints)
- Economic rationality of ex ante action (uncertainty and irreversibility)
- Negative externalities of preparedness plans
- Inertia of government policies and programs hinder institutional and programmatic changes
- Government silos lead to minimal flow of information, coordination and cooperation between ministries
- Lack of political will

Another outcome of the HMNDP was the conduct of six regional capacity-building workshops on drought preparedness and national drought management policies. These workshops were held in Asia, Eastern Europe, Asia, Latin America, and Africa, and were organized through the efforts of the WMO, FAO, UNCCD, UN-Water, and the Convention on Biological Diversity. These workshops included the participation of approximately 75 countries and each workshop stressed the three-pillar approach to drought management and the ten-step process put forward by the IDMP (2014). The basics of the ten-step process is presented later, but a more detailed summary is included in the IDMP report. A summary of the outcomes from each of these workshops is available on the IDMP website, https://www.droughtmanagement.info/.

Following the conduct of these workshops, the UNCCD launched a drought initiative in 2018 to promote the development of national drought policies. The goal of this effort is to mitigate, adapt to, and manage effects of drought in order to enhance resilience of vulnerable populations and ecosystems. The expected outcomes of this initiative are to reduce the vulnerability of ecosystems to drought through sustainable land and water management practices and to increase community resilience to drought episodes. Currently, 73 countries are participating in this initiative and more than 50 countries have completed national drought plans. This initiative stresses the ten-step process developed through the IDMP and the three-pillar approach that has become the centerpiece of drought preparedness efforts worldwide.

Continuing this global and regional effort on development of drought management policies, a project was recently launched that focuses on the countries of southern South America, i.e., the Drought Information System for southern South America, with funding from the Inter-American Development Bank. This effort is partnering with the Regional Climate Center for Southern South America along with the WMO, NOAA (National Oceanic and Atmospheric Administration), UNCCD, the Organization of American States, the Global Water Partnership, and the U.S. National Drought Mitigation Center, among others. The overarching goal of the planned workshop is to help launch and support the nationally led design and implementation of national drought preparedness plans and policies by SISSA member nations (Argentina, Bolivia, Brazil, Chile, Paraguay, and Uruguay). This effort is relying on the ten-step process and the three pillars of drought management promoted by the IDMP and other organizations for developing national drought plans and policies.

At the conclusion of this workshop, plans are for there to be follow-up and ongoing support from the workshop partners to further assist the countries in drought policy development.

These examples represent only a few of the international activities that are currently underway to promote a more proactive approach to drought management aimed at risk reduction. This new paradigm continues to gain widespread momentum at the regional and national level in response to the spiraling impacts of drought and the evolving threat of climate change.

27.2 THE THREE PILLARS OF INTEGRATED DROUGHT MANAGEMENT

Drought, like all natural hazards, has both a natural and social dimension. In most cases, the social dimension is the factor that turns a hazard into a disaster. The risk associated with drought for any region is a product of both the region's exposure to the event (i.e., probability of occurrence at various severity levels) and the vulnerability of society to the event. The natural event (i.e., meteorological drought) is a result of the occurrence of persistent large-scale disruptions in the global circulation pattern of the atmosphere. Exposure to drought varies spatially and there is little, if anything, that can be done to alter drought occurrence. Vulnerability, on the other hand, is determined largely by social factors. Natural disasters are a consequence of the interactions between the weather and climate extremes and the vulnerability of human and natural systems to such extremes.

Vulnerability is "determined by physical, social, economic and environmental factors or processes which increase the susceptibility of an individual, a community, assets or systems to the impacts of hazards" (UNISDR 2017). Vulnerability is dynamic because of societal changes that occur over time that may increase or decrease vulnerability. For example, vulnerability drivers include factors such as population changes, population shifts (regional and rural to urban), demographic characteristics, technology, government policies, environmental awareness and degradation, water-use trends, and social behavior. Vulnerability assessments provide a framework for identifying the social, economic, and environmental causes of drought impacts, i.e., who and what is at risk and why. It bridges the gap between impact assessment and policy formulation by directing policy attention to the underlying or root causes of vulnerability rather than to its result, the negative impacts, which follow triggering events such as drought. Drought impacts cut across many sectors and across normal divisions of government authority, reinforcing the need for cooperation and coordination between government ministries and NGOs (nongovernmental organizations).

27.3 NATIONAL DROUGHT POLICY: MOVING TOWARD INTEGRATED DROUGHT MANAGEMENT

The challenges that nations face in the development of a risk-based national drought management policy are complex. The process requires political will at the highest level possible and a coordinated approach within and between levels of government and with the diversity of stakeholders who must be engaged in the policy development process. A national drought policy could be a stand-alone policy. Alternatively, it could contribute to or be a part of a national policy for disaster risk reduction or climate change adaptation with holistic and multihazard approaches centered on the principles of risk management. Regardless of the path chosen, the policy would provide a framework for shifting the paradigm from one traditionally focused on reactive crisis management to one that focuses on a proactive risk-based approach. This approach will increase the coping capacity of the country, reduce recovery times, and increase resilience to future drought episodes, i.e., the goal of IDM.

To facilitate the development of national drought policies, the IDMP published *National Drought Management Policy Guidelines: A Template for Action* (WMO and GWP) in 2014 that is based on the ten-step planning process developed by Wilhite (1991; Wilhite et al. 2000). These guidelines provide a template for development of a national drought policy based on the principles of drought risk management and following the three-pillar approach. The ten-step process is generic, i.e., nations are encouraged to adapt this process to their national needs and institutional capacity, as exemplified in Central and Eastern Europe (GWP CEE 2015) as well as Brazil, Mexico, and Morocco (WMO and GWP 2014; Wilhite and Pulwarty 2018b).

The formulation of a national drought policy, while providing the framework for a paradigm shift, is only the first step in vulnerability reduction. The development of a national drought policy must include development and implementation of preparedness and mitigation plans at the subnational level, an outcome of the three-pillar approach. These plans will be the instruments for implementing a national drought policy.

If proactive risk management is socially optimal, compared to reactive crisis management, why has the shift from crisis management to risk management been so slow? This question is raised repeatedly in drought management and policy workshops and in discussions with policy makers. In consultation with practitioners and experts from a wide range of organizations, some pointers have emerged through focused discussions between the IDMP, the World Bank, and other experts (WMO and GWP 2017; Venton et al. 2019). They include that the economic argument and assessments can support an integrated approach to drought management, but numbers alone will not lead to action by creating the political will to change the paradigm. The context matters, in which the economic argument must connect to the political economy, including the governance context and what drives political will, in order to create change. In addition, focusing on actions that have socioeconomic co-benefits beyond drought management is important in support of the argument for proactive and integrated drought management.

A national drought policy should establish a clear set of principles or operating guidelines to govern the management of drought and its impacts. It should be consistent; equitable for all regions, population groups, and economic sectors; and consistent with the goals of sustainable development. By following the three-pillar approach, the policy is directed toward reducing risk by developing better awareness and understanding of the drought hazard and the underlying causes of societal vulnerability along with developing a greater understanding of how being proactive and adopting a wide range of mitigation and response measures can increase societal resilience. Risk management promotes many proactive actions that have been discussed previously.

27.3.1 Drought Policy Objectives

The objectives associated with a national drought policy will vary from nation to nation but, in principle, will likely reflect some common themes. These objectives would likely

- Encourage vulnerable economic sectors and population groups to adopt self-reliant measures that promote risk management
- Promote sustainable use of the agricultural and natural resource base
- Facilitate early recovery from drought through actions that reinforce the philosophy of risk management

Drought planning refers to actions taken by individual citizens, industry, government, and others before drought occurs with the purpose of reducing or mitigating impacts and conflicts arising from drought. It can take the following forms: response planning or mitigation planning. The three-pillar approach emphasizes mitigation planning that leads to risk reduction. It is important to note that planning must occur on multiple government levels from national to subnational, and the objectives of these policies at the local, state, or regional levels must reflect the goals of national drought policies. Stakeholder engagement is critical at all levels.

The ten steps recommended for the development of a national drought policy (WMO and GWP 2014) are as follows:

Step 1: *Appoint* a national drought management policy commission
Step 2: *State* or *define* the goals and objectives of a risk-based national drought management policy

Step 3: *Seek* stakeholder participation; *define* and *resolve* conflicts between key water-use sectors, considering also transboundary implications

Step 4: *Inventory* data and financial resources available and *identify* groups at risk

Step 5: *Prepare/write* the key tenets of a national drought management policy and preparedness plans, which would include the following elements: monitoring, early warning, and prediction; vulnerability and impact assessment; and mitigation and response

Step 6: *Identify* research needs and *fill* institutional gaps

Step 7: *Integrate* science and policy aspects of drought management

Step 8: *Publicize* the national drought management policy and preparedness plans, and *build* public awareness and consensus

Step 9: *Develop* educational programs for all age and stakeholder groups

Step 10: *Evaluate* and *revise* national drought management policy and supporting preparedness plans

In brief, steps 1–4 of the ten-step process focuses on making sure the right people and agencies/ministries are brought together, have a clear understanding of the process, know what the drought preparedness plan must accomplish, and are given adequate data to make fair and equitable decisions when formulating and writing the actual drought mitigation plan. Step 5 describes the process of developing an organizational structure or framework for completion of the tasks necessary for the preparedness plan, essentially emphasizing the three pillars of drought management referred to previously. The development of the plan is a process, rather than a discrete event that produces a static document. A vulnerability assessment, completed in conjunction with this step, provides a vulnerability profile for key economic sectors, population groups, regions, and communities. Steps 6 and 7 detail the need for ongoing research and coordination between scientists, ministries, and policy makers. Steps 8 and 9 stress the importance of promoting and testing the plan before drought occurs. Finally, step 10 emphasizes revising the plan to keep it current and making an evaluation of the plan's effectiveness following each drought event. Although the steps are sequential, many of these tasks are addressed simultaneously under the leadership of a drought commission or task force and its complement of committees and working groups.

27.4 SUMMARY

Clearly, there is an urgent need to change the paradigm for drought management in all drought-prone nations from one primarily focused on managing the disaster to one focused on managing risk. Convincing policy makers and natural resource managers to adopt a more proactive approach to drought management has been, to date, a challenging task. However, the growing acceptance of the philosophy of disaster risk reduction and the momentum of recent international efforts suggests that the time is now to adopt this new paradigm.

The growing emphasis on and adoption of the concept of national drought policies and the three-pillar approach to drought is fundamental to this change in the approach to drought management. The three pillars exemplify the importance of developing a comprehensive and effective drought monitoring and early warning system (pillar 1) to provide reliable and timely information to managers and other decision makers, and linking the information derived from this system to vulnerability and impact assessments (pillar 2), and the adoption of appropriate mitigation and response actions (pillar 3). It would be expedient for all drought-prone nations to explore this new paradigm for drought risk management.

ACKNOWLEDGMENTS

The author acknowledges the contributions of Robert Stefanski, Technical Support Unit, Integrated Drought Management Programme (World Meteorological Organization and the Global Water

Partnership); and Frederik Pischke, formerly with the Technical Support Unit of the Integrated Drought Management Programme, Geneva, Switzerland.

REFERENCES

HMNDP. 2013. High-level Meeting on National Drought Policy (Final Declaration). Geneva, Switzerland. http://www.droughtmanagement.info/literature/WMO_HMNDP_final_declaration_2013_en.pdf

Integrated Drought Management Programme (IDMP). *World Meteorological Organization (Geneva, Switzerland) and the Global Water Partnership (Stockholm, Sweden)*. https://www.droughtmanagement .info/.

UNISDR. 2017. Terminology. Geneva, Switzerland. (https://www.unisdr.org/we/inform/terminology#letter-v)

Venton, P., C. Cabot Venton, N. Limones, C. Ward, F. Pischke, N. Engle, M. Wijnen, and A. Talbi. 2019. *Framework for the Assessment of Benefits of Action/Cost of Inaction (BACI) for Drought Preparedness*. World Bank, Washington, DC.

Wilhite, D.A. 1991. Drought planning: A process for state government. *Water Resour. Bull.* 27(1):29–38.

Wilhite, D.A., and R.S. Pulwarty. 2018a. *Drought and Water Crises: Integrating Science, Management and Policy* (541 pp). CRC Press, Boca Raton.

Wilhite, D.A., and R.S. Pulwarty. 2018b. Drought as hazard: Understanding the natural and social context (Chapter 1, pp. 3–22). *Drought and Water Crises: Integrating Science, Management and Policy* (541 pp). CRC Press, Boca Raton.

Wilhite, D.A., M.J. Hayes, C. Knutson, and K.H. Smith. 2000. Planning for drought: moving from crisis to risk management. *J. Am. Water Res. Assoc.* 36:697–710.

World Meteorological Organization (WMO) and Global Water Partnership (GWP). 2014. *National Drought Management Policy Guidelines: A Template for Action* (D.A. Wilhite). Integrated Drought Management Programme (IDMP) Tools and Guidelines Series 1. WMO, Geneva, Switzerland and GWP, Stockholm, Sweden. (http://www.droughtmanagement.info/literature/IDMP_NDMPG_en.pdf)

World Meteorological Organization (WMO) and Global Water Partnership (GWP). 2017. *Benefits of Action and Costs of Inaction: Drought Mitigation and Preparedness: A Literature Review* (N. Gerber and A. Mirzabaev). Integrated Drought Management Programme (IDMP) Working Paper 1. WMO, Geneva, Switzerland and GWP, Stockholm, Sweden. (http://www.droughtmanagement.info/literature/IDMP _BACI_WP.PDF)

28 Is India Ready to Account for Ecological Droughts?

Diptimayee Nayak

CONTENTS

28.1 INTRODUCTION

Droughts can be characterized by different densities and intensities of variables like rainfall, weather, temperature, and resilience, and can have economic, social, and environmental impacts, and hence they do not have a single unanimous definition. Factors that mostly influence the recurring droughts in India are the status of soil moisture, surface, and groundwater, agroclimatic features, cropping choices, cropping patterns, agricultural practices, availability of fodder, and socioeconomic vulnerabilities of the population (GoI, 2016). India experienced 25 major drought years from 1871 to 2015, which is defined as years with the All India Summer Monsoon Rainfall (AISMR) less than one standard deviation below the mean (GoI, 2016). Since its independence, India has been exposed to 15 worst-hit droughts including the recent droughts in Maharashtra and some southern states. Zhang et al. (2017) identified four different types of droughts and their frequencies in India from 1981 to 2013 and stated that India has faced seven drought-years in each of the categories of meteorological, hydrological, soil moisture, and vegetation type droughts (Table 28.1). In ranking the world's most water-stressed countries in 2040, a report published in 2015 by the World Resources Institute, ranks India 28 out of 153 countries. In a recent report published by the same institute in 2019, India's water stress ranking is now 13 out of 164 countries and is 13 out of 17 of the extremely high baseline water stress countries. Around 600 million people in India experience a high to extreme form of water stress, 70% of water is contaminated and three-fourths of households do not have access to drinking water on their premises (NITI Aayog, 2018). These figures show the potential risk of drought as one of the most impending disasters in the coming days and knowing the very nature and typology of drought would help us monitor, mitigate, and manage it in an efficient way.

The Ministry of Jal Shakti (MoJS), Department of Water Resources, River Development and Ganga Rejuvenation, Government of India (GoI) has mentioned three types of droughts: meteorological drought, hydrological drought, and agricultural drought. Meteorological drought is defined

DOI: 10.1201/9781003276548-28

TABLE 28.1

Frequency of Droughts in India

Drought Year	Type of Drought	Affected Regions	Consequences/Affected People (in millions)
1966	—	Bihar, Odisha	50
1969	—	Rajasthan, Gujarat, Tamil Nadu, Uttar Pradesh, Andhra Pradesh, Haryana, Madhya Pradesh, Karnataka	15
1970	—	Bihar, Rajasthan	172
1972	—	Rajasthan, Himachal Pradesh, Uttar Pradesh	50
1979	—	Eastern Rajasthan, Punjab, HP, UP	200
1982	—	Rajasthan, Punjab, HP	100
1983	Hydrological/vegetation	TD, WB, Kerala, Rajasthan, Karnataka, Bihar, Odisha	100
1987	—	Eastern and northern India	300
1992	—	Rajasthan, Odisha, Gujarat, Bihar	—
2000	Meteorological/ hydrological/vegetation	Rajasthan, Gujarat, Odisha, Andhra Pradesh, MP	More than 100
2013	—	Maharashtra	
2015	—	Maharashtra	

Sources: Compiled from Zhang et al. (2017) and GoI (2016).

as rainfall deficiency with respect to the long-term average and can be of three types, viz., normal (when long-term average deficiency is 25% or less), moderate (when long-term average deficiency is between 26% and 50%), and severe (when long-term average deficiency is more than 50%) (MoJS, n.d.). Hydrological drought can be understood as water deficiencies in surface and subsurface water supplies due to average precipitation and high-water usage; whereas, agricultural drought is defined as four consecutive weeks of metrological drought, with weekly rainfall being 50 mm and 80% of crops is planted during kharif season. Agriculture drought can be decomposed into soil moisture drought and vegetation drought (Zhang et al., 2017).

The National Drought Mitigation Center, at the University of Nebraska-Lincoln (United States), added another two types of droughts, i.e., socioeconomic and ecological. Socioeconomic drought is associated with the supply and demand of economic goods like water, fish, forage, and hydro-electric power that have elements of and are impacted by the first three types of droughts (Wilhite and Glantz, 1985). All these droughts are basically the impacts and repercussions of meteorological droughts and are perceived through a human-centric lens.

The recently conceived ecological drought, which was coined by a group of scientists led by ecologist and paleoecologist Shelley D. Crausbay and biologist Aaron R. Ramirez at the University of California in 2017, is ecosystem-centric, emphasizing ecological dimensions, where impacts of droughts on the very ecosystems are highlighted, identified, and measured (Crausbay et al., 2017). They use the example of the Millennium Drought in Australia (2002–2010) that has led to losses of many basic ecosystem services provided by the hydrological ecosystem in the Murray–Darling basin, and the total cost, as estimated by Banerjee et al. (2013), in terms of loss of ecosystem services, replacement cost for reviving ecosystem services, and adaptation cost, was more than AUD$800 million. Although ecosystem-centric in approach, ecological droughts capture the vulnerability of the human population, which depends more on vulnerable ecosystem resources and ecosystem services. Ecological drought poses integrated impacts to both the ecology and human beings, therefore experts try to decrease the vulnerability and differential vulnerability of ecosystem and human populations with a set of well-designed policy interventions and preparedness.

Thus, ecological drought encompasses not only the cost of meteorological droughts that have immense socioeconomic impacts, which is largely due to the loss of ecosystem services, but also the cost to nature, which is reflected in terms of the loss of ecosystems. Ecological drought can be defined as "an episodic deficit in water availability that drives ecosystems beyond thresholds of vulnerability, impacts ecosystem services, and triggers feedback in natural and/or human systems" (Crausbay et al., 2017). Likewise, the National Drought Mitigation Center defines ecological drought as "a prolonged and widespread deficit in naturally available water supplies – including changes in natural and managed hydrology – that create multiple stresses across ecosystems". The basic thrust of the nascent concept is its acknowledgment of the impacts of meteorological drought beyond human and hydrological dimensions, which has been expanded to vegetation and ecosystems (Redmond, 2002) capturing and integrating all the possible impacts of meteorological droughts on nature (ecology, hydrology, climate) and human beings (socioeconomic and cultural aspects). The conceptual framework of the ecological drought has two dimensions, viz., the components of vulnerability like exposure, sensitivity, and adaptive capacity; and a continuum from human to natural factors, which has been produced in Figure 28.1.

The very crux of ecological droughts is that if droughts impact ecosystems, then ecosystem-based responses and solutions can ameliorate the problem; however, these are largely absent in drought planning efforts. Generally, the anthropogenic need for water demand is placed ahead of the needs of ecosystems (Crausbay et al., 2017). Therefore, appropriate planning to address ecological droughts must integrate ecosystem services and vulnerability assessment (Raheem et al., 2019). In this context, when the problem of climate change–driven droughts has spread its effects around the globe, adequate adaptation and mitigation attempts must be reflected in the plans at the micro and meso level not only to address its effects on human beings but also accounting for the impacts on non-humans (McEvoy et al., 2018).

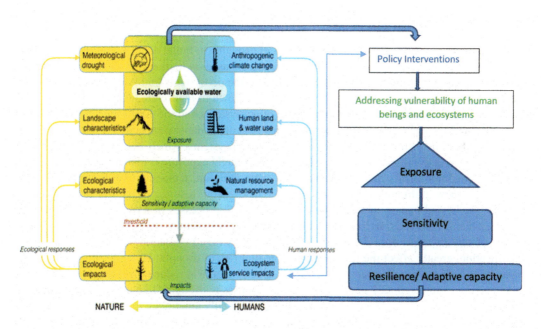

FIGURE 28.1 Ecological drought, ecosystem services and vulnerability. (Adapted and Modified from Crausbay et al., 2017, p. 2544.)

The vital questions that must be answered include: What are the characteristics and criteria of ecological droughts or how do we understand the occurrence of ecological droughts? What are the causes and consequences of ecological droughts? What are the factors affecting ecological droughts? What are the mechanisms to ameliorate ecological droughts? Which ecosystems are adversely affected and what ecosystem services are going to be decreased and lost due to ecological droughts? What institutional mechanisms are required to address the issue? To what extent has India developed plans, programs, mechanisms, and institutional structures to tackle the severity of the existing and potential threat of ecological droughts? Is there any ecosystem-based responses to the drought management plan in India? Is India ready to integrate ecosystem-based approaches to address ecological droughts? These questions can be answered by contextualizing the prevalence of the impacts of ecological droughts and then by analyzing the action plans and policies to find whether ecosystem-based approaches are integrated into the policy documents to address the consequences of ecological droughts.

Thus, the objectives of the study are given as follows: (a) to find the status of ecological droughts and the factors responsible for them by exploring and identifying the existence of scenarios and variables of droughts in India; and (b) to analyze challenges as to how far India is ready to address ecological droughts that can be verified from its drought management plans, policy documents, and plans of various decision-making units.

The rest of the paper is organized as follows: the data and methods section is described in the next section followed by the results and discussion. The final section deals with the overall conclusion with appropriate policy suggestions.

28.2 DATA AND METHODS

28.2.1 CONCEPTUAL FRAMEWORK

The conceptual framework to understand the impacts of ecological drought is shown in Figure 28.2.

28.2.2 DATA AND MATERIALS

For understanding the intertemporal occurrences of droughts in India, data and information on historical droughts must be explored and the very features based on their severity analyzed. Different key factors and variables of droughts that may affect the ecosystems and Anthropocene based on the understanding of the vulnerability framework are shown in Figure 28.2 (Crausbay et al., 2017). The factors that affect the drought vulnerability of ecosystems and the Anthropocene to a reduced water availability depends on their exposure, sensitivity, and adaptive capacity (Glick et al., 2011).

Factors, like the ultimate availability of water to ecosystems, which is known as ecologically available water, determine drought exposure. Again, drought exposure is also impacted by the geographical condition of the area, frequency, intensity, duration, temperature/evaporation, precipitation type (snow, rain) and time, duration of the dry season, features of natural landscapes like topology, soils, and human modifications of hydrological processes in terms of water infrastructure (e.g., dams, reservoirs, irrigation canals, groundwater extractions, river regulations). Likewise, factors of sensitivity, such as the nature of vegetation, human population, and their composition/classification, greatly affect the likelihood of potential risk of drought. However, it is the adaptive capacity of both the natural species and human species that help reduce the impacts of ecological droughts. For example, policy interventions and responses to droughts like natural resources management, and landscape and land use management can greatly ameliorate the impacts of ecological droughts. Likewise, the adaptive capacity of animals in terms of migration to a suitable nearby area may help to address ecological droughts. The relevant data are collected and compiled from the Central Water Commission, Central Groundwater Board, India Meteorological Department, and various literature and policy documents.

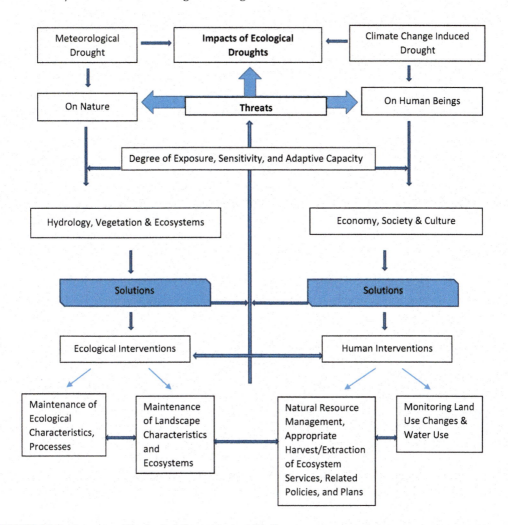

FIGURE 28.2 Impacts of Ecological Droughts And Challenges.

28.3 RESULTS AND DISCUSSION

28.3.1 Ecological Droughts: Benchmarking the Actual and Potential Threats

Drought is a common natural phenomenon in India and every drought cannot be called an ecological drought. Thus, the question that very often comes is how to identify ecological droughts in the Indian context. For this, we must delineate the very characteristics, typology, and drivers of ecological droughts, which can be taken as a standard to evaluate their occurrence in India. The very characteristics of ecological droughts can be extracted from the definition given by Crausbay et al. (2017) as "a disturbance that pushes coupled natural- human systems beyond their adaptive capacity and triggers important socioecological feedbacks". Therefore, ecological drought can be characterized by the following features. First, it does not encompass small-scale and short-term effects of drought within a system's adaptive capacity. For example, during droughts, plants reduce their productivity, which is small-scale and short-term in nature, and if this phenomenon is not beyond the adaptive capacity of the natural system, then it cannot be called ecological droughts. Second, it has ecological and/or social footprints in terms of its impacts on natural systems as well as human beings.

Based on this understanding of features, ecological droughts can be of four types (Crausbay et al., 2017): type I, having moderate impacts and feedback to natural systems only; type II, having moderate impacts and feedback to human systems only in terms of loss of ecosystem services; type III, having impacts and feedback to coupled natural–human systems; and type IV, transformational impacts and feedback to coupled natural–human systems). Among these four types of ecological droughts, the first three can have temporary changes on natural or human systems or both, whereas the last one has extreme impacts and brings persistent change on both natural and human systems. However, drought planning does not generally encompass all the ecological impacts in a landscape like loss of ecosystem services but rather considers only a few (Raheem et al., 2019). The general perception is that ecosystems and ecosystem services are complementary to each other, however, at times, demand for healthy ecosystems and ecosystem services for human well-being can be competing rather than interdependent. Therefore, exploring the drivers of ecological drought impacts may help policymakers understand the trade-offs of water use and land use change for human and ecosystem water needs among many other factors, and if these two can be monitored effectively to reduce the impacts.

28.3.1.1 Key Drivers of Ecological Drought, Impact, and Vulnerability in India

Noticing the frequency of droughts in India in Table 28.1, the most pertinent query is whether India is accounting for the drivers of ecological droughts in the policy and planning to reduce their impacts. The frequency of droughts differs region-wise in India, and the probability of drought across the regions is different. In a study by Shewale and Kumar (2005), the India Meteorological Department, Pune, estimated drought probabilities across Indian regions and found that East Rajasthan has the highest drought probabilities (25%), followed by Saurasthra and Kutch (24%), whereas Assam and Meghalaya and the Gangetic Plains, each has the lowest drought probabilities (1%). They listed 35 subdivisions of drought probabilities across Indian states and accordingly classified these regions into three parts, viz., chronically drought-affected areas, areas having a drought probability of more than 20%, like West Rajasthan and Gujarat; frequently drought-prone areas, areas having drought probability between 10% and 20%, like Uttar Pradesh, Uttaranchal, Haryana, Punjab, Himachal Pradesh, East Rajasthan, West Madhya Pradesh, Marathwada, Vidarbha, Telangana, coastal Andhra Pradesh, and Rayalaseema; and the rest of the regions of India are the least drought-affected areas, or areas having drought probability of less than 10%.

The drought research literature contains the key drivers of drought impacts and vulnerability based on the physical, socioeconomic, and demographic characteristics of the affected region and population. These drivers and variables include precipitation/rainfall, soil moisture, crop sown area, streamflow, ground and surface water levels, reservoirs, temperature/hotter weather/heat waves, vegetation moisture, climatic conditions, Southern Oscillation, marine events like El Niño in the Pacific to some extent, and seasonal climate forecast (GoI, 2016; WMO and GWP, 2014; Shewale and Kumar, 2005). Moreover, cropping patterns, cropping choices and water-use efficiency in agriculture can also affect drought vulnerability. Agriculture consumes around 80% of India's water resources and its water-use efficiency is one of the lowest in the world at 25%–35% in comparison to 40%–45% in Malaysia and Morocco, and 50%–60% in Israel, Japan, China, and Taiwan (Mihir Shah Committee, 2016). Moreover, India's water infrastructure[1] and its very purpose now is debatable for drought management and water scarcity, which calls for revamping the water institutions and the national water policy[2] as well.

The scientific community uses different indices to measure the technicalities of drought than the central and state governments for drought management (GoI, 2009). The government of India and state governments use key indicators like rainfall, reservoirs, surface and groundwater tables, soil moisture, and sowing/crop conditions, whereas scientific researchers use a number of indices like the Aridity Anomaly Index to measure agricultural drought and moisture stress; Standardized Precipitation Index (SPI); Palmer Drought Severity Index that indicates standardized moisture conditions; Crop Moisture Index to measure the degree to which crop moisture requirements are satisfied; Surface Water Supply Index that shows the water availability across a river basin; Normalized Difference Vegetation Index to measures moisture stress in vegetation; Normalized Difference Wetness Index that shows the crop turbidity and health; and the Effective Index and Moisture Adequacy Index to measure intensity, duration, and spatial extent of drought (GoI, 2009).

The indicators/indices that are used for drought forecast and estimation are the rainfall deviation/SPI dry spell; deviation from normal crop sown area; Normalized Differential Vegetation Index (NDVI) and Normalized Difference Wet Index (NDWI) or deviations from normal form of NDVI/NDWI; Streamflow Drought Index (SFDI); and Groundwater Drought Index (GWDI) (GoI, 2016). However, ecological drought indicators/indices are an integrated form of all forms of droughts and can be measured by the ecosystem service approach, which is yet to be recognized and developed both in policy documents and practice (Banerjee et al., 2013). The change In ecosystem services and vulnerability to both ecosystems and human beings due to droughts are the key measurements in capturing the impacts of ecological droughts. Some of the key variables in drought management that may impact ecological droughts are discussed as follows.

28.3.1.1.1   The Status of Water and Water Scarcity in India

The per capita water availability in India has been continuously decreased from 5177 m^3/year in 1951 to 1816 m^3/year in 2001 to 1545 m^3/year in 2015, and a projected 1486 m^3/year in 2021, 1367 m^3/year in 2031, and 1140 m^3/year in 2050, which reveals a water scarcity in India (GoI, 2013).[3] Narasimhan and Gaur (2009)[4] estimate 3840 BCM of annual rainfall in India and as per the analysis of the Ministry of Water Resources, the annual water use is 634 BCM against the supply of 1123 BCM of utilizable water, which is a comfortable situation, whereas, as per the analysis based on worldwide comparisons, the annual use of water of 634 BCM against the utilizable water of 654 BCM is a concern. With 17% of the world's population and 4% of water resources, India must account for factors responsible for drought, its management, and mitigation. Moreover, there is a large disparity in basin-wise per capita availability of water, for example, in the Brahmaputra/Barak basin, this figure is 14059 m^3/year, whereas in the Sabarmati basin, it is only 307 m^3/year (GoI, 2013), and other basins like the Tapi, Pennar, and Mahi are experiencing water stress.

28.3.1.1.2   The Status of Surface Water and Water Abstraction

Factors like rainfall, live storage status of reservoirs, and heat waves affect the availability of surface water in rivers, canals, and ponds. Again, weak rainfall in successive years adversely affects by decreasing recharge in surface and groundwater. During major drought years from 1965 to 2015, it has been observed that the noticeable departure of rainfall from normal rainfall caused droughts years. From 1965 to 2015, rainfall was erratic and deviated from normal, as shown in Table 28.2.

TABLE 28.2

Drought Years and Departure of Rainfall from Normal during 1965–2015 in India

Year	% Departure from Normal	Year	% Departure from Normal
1965	−18	2000	−5
1966	−16	2001	−8
1985	−7	2002	−19
1986	−13	2014	−12
1987	−19	2015	−14
1999	−4		

Source: Adapted from GoI (2016, p. 12).

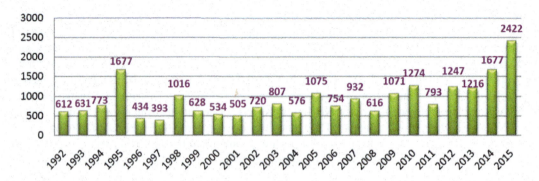

FIGURE 28.3 Heat Wave Deaths In India, 1992–2015. (Adapted from Guleria and Gupta, NDMA, 2018, p. 15.)

Likewise, as "winds bring heat bombs", the frequency of heat waves may add fuel to the fire of increased water abstraction, storage of reservoirs below their capacity, and above all the availability of surface water, and this has been a common phenomenon both in the plains and hill regions of India.

Its impact can be considered in terms of deaths due to heat waves. Heat waves[5] are considered when the normal maximum temperature in the plains is at least 40°C with at least a 4 degree deviation from the normal, and in hills, when the normal maximum temperature is at least 30°C, with a deviation from normal of at least 6 degrees (Guleria and Gupta, 2018). The impact of heat waves can be measured in terms of the number of deaths over the years, which is reflected in Figure 28.3, and during the period 2001–2010, two Indian states, Madhya Pradesh (1030 deaths) and Andhra Pradesh (1210 deaths) are placed in the top ten highest mortality heat events across the globe (Guleria and Gupta, 2018, p. 7).

The data shows that deaths due to heat wave has increased by almost four times over a period of less than 15 years, which would severely affect urban India first, and would have severe economic repercussions in terms of loss of working hours, outdoor labor productivity, and gross domestic product (GDP). As per a report by McKinsey in 2020, the increased climate variability of heat waves, rainfall, and humidity will affect the average number of lost daylight working hours in India and could increase to the point where between 2.5% and 4.5% of the GDP could be at risk annually.

The live water storage status of reservoirs along with hydrometeorological data like river water level, river discharge, sediment flow/siltation, and water quality are also good indicators of surface water, which can also influence drought impacts. Glacial lakes and water bodies can also impact drought scenarios on the low land.

The live water storage data of 123 reservoirs maintained by the Central Water Commission shows that live storage decreased in 2020 corresponding to the same the previous year. For example, live storage available in these reservoirs is 92.916 BCM, which is 54% of the total live storage capacity of these reservoirs. However, in August 2019, the live storage available in these reservoirs was 105.856 BCM and the average of last ten years' live storage was 94.348 BCM. Thus, the live storage available in 123 reservoirs is 88% of the live storage in 2019 and 98% of storage of average of the last ten years (Reservoir Storage Bulletin, August 13, 2020). The same trend has also been depicted in Table 28.3, i.e., the overall storage position is less than the corresponding period of the previous year (2019) in the country and is also less than the average storage of the last ten years during the corresponding period.

Figure 28.4 shows that current reservoir storage in India. The southern regions has less than the designated capacity, depicting water shortages even in a meteorological drought situation. In each region, current storage is even less than the last year's storage and ten years' average storage, except in the southern regions. All the drought-prone states in the northern, western, and central regions

TABLE 28.3

Status of Reservoir Storage in Indian States

State	% Departure from Normal Storage	State	% Departure from Normal Storage
Himachal Pradesh	−18	Nagaland	3
Punjab	−31	Gujarat	−9
Rajasthan	−30	Maharashtra	4
Odisha	−11	Uttar Pradesh	13
Jharkhand	1	Uttarakhand	5
WB	41	Madhya Pradesh	−13
Tripura	36	Chhattisgarh	21
AP & TG	18	Telangana	2
AP	121	Karnataka	4
Kerala	12	Tamil Nadu	25

Source: Central Water Commission, live storage status as of August 13, 2020.

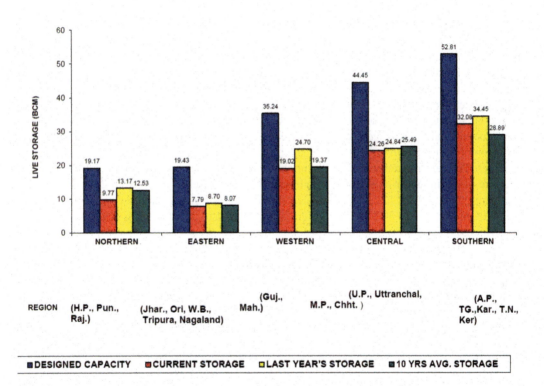

FIGURE 28.4 Region-wise storage position as of August 13, 2020. (Adapted from Reservoir Storage Bulletin of 13.08.2020, page no.6, Central Water Commission, live storage status, 2020.)

of India, experience current storage to be less than their capacity, which increases the likelihood of drought severity and vulnerability of ecosystem and human systems impacting the economy, society, environment, and ecology. All these factors make major portions of India's ecosystems vulnerable to ecological droughts because of its high exposure, sensitivity, and lack of adequate adaptive capacity.

TABLE 28.4

Status of Groundwater Resources of Drought Prone States, 2017

State	Overexploited (% of total blocks)	Critical (% of total blocks)	Semicritical (% of total blocks)	Safe (% of total blocks)	Saline (% of total blocks)
Rajasthan	62.9	11.8	8.9	15.7	0.7
Tamil Nadu	39.46	6.16	14.64	36.72	2.99
Punjab	78.98	1.44	3.62	15.94	0
Maharashtra	3.14	2.57	17.42	76.57	0.28
Gujarat	10.04	2.00	4.41	78.31	5.22
Uttar Pradesh	9.10	6.70	18.33	65.86	0
Uttarakhand	0	0	27.77	72.23	0
Haryana	60.9	2.34	16.40	20.30	0

Source: Data compiled from Block-wise Ground Water Resources Assessment, 2017, CGWB.

28.3.1.1.3 *Status of Groundwater and Water Abstraction*

As far as the demand side of the water is concerned, around 85% of rural domestic requirements, 50% of urban requirements, and 50% of irrigation requirements depend on groundwater resources (GoI, 2013). Because of hydrological change, which occurred due to groundwater development and extraction, there has been great alternation in the natural recharge system also. There has been a galloping trend in groundwater abstraction due to factors like improved technology and electrical and diesel-based equipment for ground water extraction, viable schemes for water resource development, institutional finance agencies, quality seeds, subsidized agricultural inputs, and suboptimal extraction of water resources, which are primarily responsible for overexploitation of groundwater resources. The net irrigated area has gone up thrice in 2008–2009 in comparison to 1950–1951. and the share of groundwater irrigation through wells has immensely shot up from 28% in 1950–1951 to 61% in 2008–2009. It has been estimated that the annual groundwater extraction in 2017 was 248.69 BCM and around 89% has been used for irrigation and the remaining 11% for domestic and industrial use (GoI, 2019).

The status of groundwater assessment of most of the drought probability states is in the categories of overexploited, critical, or semicritical (Table 28.4).

Table 28.4 shows the status of groundwater in potential drought-prone states. Punjab is the most groundwater overexploited state and around 79% of blocks are categorized as overexploited. In the overexploitation of groundwater category, Rajasthan, Haryana, Tamil Nadu Gujarat, Uttar Pradesh, and Maharashtra follow Punjab. For drought-prone states like Punjab, Haryana, Rajasthan and Delhi, current annual groundwater extraction is more than the annual extractable groundwater (GoI, 2019), which may accelerate the impacts of ecological droughts in these states.

However, these scenarios can be improved with suitable policy interventions, like artificial water recharge programs. For instance, in 2004, about 15% of the total assessment units were found overexploited and 3.9% were critical, whereas, in 2009, 13.72% assessments were overexploited and 2.8% were found critical. These relatively better figures were due to large-scale policy intervention of artificial water recharge initiatives. For the first time, in 1994, during the [8]h Five-Year Plan, artificial recharge studies were undertaken in nine states/union territories at the cost of 3.23 crore of rupees (GoI, 2013). Likewise, in different plan periods, projects were undertaken to implement artificial recharge facilities across vulnerable districts and states. However, after the end of the five-year plan period, these figures have started worsening. For example, in the assessment year 2017, about 17% of the total assessed units were found to be overexploited, where the percentage of groundwater extraction was more than 100%; 5% were critical, where the stage of groundwater extraction was between 90% and 100%; 14% were semicritical, in which the groundwater extraction was between 70% and 90% percent; and 63% were safe, where the groundwater extraction was below 70% (GoI, 2019).

28.3.1.1.4 Identification and Quantification of Groundwater and Surface Water Ecosystems and Their Services

Ecosystem services are the benefits that human beings receive from ecosystems. In the typology of ecosystem services, Millennium Ecosystem Assessment reports delineate four categories of ecosystem services, viz., provisioning, regulating, cultural, and supporting (MA, 2005). Although in recent times, there has been the new classification of ecosystem services like the Common International Classification of Ecosystem Services (CICES) developed by the European Union, and nature's contribution to people by the Intergovernmental Science-Policy Platform on Biodiversity and Ecosystem Services (IPBES). Some of the ecosystem services provided by groundwater ecosystems are purification of water, storage, biodegradation of anthropogenic contaminants, nutrient recycling, and mitigation of droughts (Griebler and Avramov, 2015). Both surface water and groundwater ecosystems help in the functions of other groundwater-dependent ecosystems like rivers and wetlands. The implications of droughts on freshwater ecosystems have been rising (Li et al., 2017). The availability of freshwater is an important provisioning service used for multiple purposes such as drinking, irrigation, fishing, recreation, and hydropower production.

In traditional drought impact studies, loss of ecosystem services like decrease in drinking water or scarcity of water in irrigation, or loss of nutrients and the health of concerned ecosystems are generally not accounted. Thus, there is an urgency for the identification and economic valuation of the changed ecosystem services by using ecological tools like the Integrated Valuation of Ecosystem Services and Tradeoffs (InVEST) to account for the repercussions of ecological droughts that largely impact both natural and human systems (Terrado et al., 2014). A set of ecosystem services indices for provisioning services has been developed that includes the freshwater provisioning index, food provisioning index, regulating services, and the Soil and Water Assessment Tool (SWAT), which can be used to quantify ecosystem services and better represent the impact of droughts on ecosystem services (Logsdon and Chaubey, 2013; Li et al., 2017). The concerned ecosystems become vulnerable because of pressure from both natural factors, like hydrological droughts disrupting the water cycle, and anthropogenic factors, like overextraction of surface and groundwater or inappropriate cropping choices, which must be fixed by ecosystem-based approaches.

All these scenarios on rainfall, heat waves, storage capacity, and groundwater divulges are factors of drought exposure making India vulnerable to ecological droughts. Moreover, scenarios like the nature of vegetation and trajectories of socioeconomic conditions (society, demographics, and economics) exposed to these factors of drought exposure make India sensitive and vulnerable to ecological droughts. It has been claimed that heat waves also impact natural habitats, causing increased forest fires, and, as per the Forest Survey of India report, 6.17% of forests are prone to severe forest fires, affecting about 3.73 million hectares of forests annually (Guleria and Gupta, 2018). As far as compositions of human dimensions are concerned, India's estimated population in 2025 will be 1392.086 million, up from 935.744 million in 1995, and the percentage of the urban population demanding water will increase from 27% in 1995 to 45% in 2025 (IPCC, 1997). The income inequality as found by Oxfam India report shows that the top 10% of the Indian population holds 77% of the total national wealth (Oxfam, 2022) which can have poverty consequences. Thus, shared socioeconomic pathways in India make it not only vulnerable, but the differential vulnerability is also high as populations with differential vulnerability with low socioeconomic conditions have low adaptive capacity. It is the policy interventions, drought management plans and programs, preparedness, risk reduction, and mitigation strategies that may help increase the adaptive capacity and hence lessen the degree of vulnerability.

28.3.2 Policy Prospects, Preparedness, and Decision-Making in Drought Challenges

The government of India has made several attempts and instituted reforms to manage, monitor, and mitigate the impacts of droughts for a long time even before its independence. In the report "21st Century Institutional Architecture for India's Water Reforms" (2016), the Shah Committee prescribes a demand-side management of water resources instead of the supply side, and highlights three crucial elements for addressing water scarcity and drought challenges. First, limiting the demand for water by increasing water-use efficiency and reducing waste, with a mandate for *harkhetkopaani*, which means "water for each agricultural field". Second, recognizing nature as an important stakeholder, which has been hitherto ignored in the engineering-tuned water resources, and highlights to maintain required environmental flows (e-flows) and minimize impacts of water developments on ecosystems. Third, understanding the importance of both the intrinsic and instrumental value of water resources and aquifer mapping[6] through participatory management as the key collective action superior to the top-down technocratic management approach. If implemented successfully, they may prove important responses to the risk of ecological droughts. For the said objectives to be achieved, the proposed institutional restructuring of the Central Water Commission (CWC) and Central Ground Water Board (CGWB) to a "national water commission" will guard not only the work of civil engineering and hydrology but provide an interdisciplinary understanding of social sciences, management, ecological economics, and agronomy. These are truly new steps to prepare to avert risks, and manage and mitigate ecological droughts. However, there is always a gap between committee prescriptions and implementations, and this is not an exception in this case.

Two important policy guidelines, i.e., "Manual for Drought Management, 2009", prepared by the National Institute for Disaster Management in association with the Department of Agriculture, Cooperation, and Farmers Welfare; and the revised manual "Manual for Drought Management, 2016", prepared by a team of scientists and policymakers from reputed institutes in association with the Department of Agriculture, Cooperation, and Farmers Welfare, are key to decision makers, planners, and professionals for addressing the manifold issues of drought. After seven years, although the revised version introduced updated scientific knowledge focusing on drought impact mitigation measures, introducing new indices and parameters on hydrological indicators, and prescribing a timeline for the declaration of droughts and follow-up action, it did not cover ecological droughts and their repercussions.

While gathering information for decision-making, both documents focused on meteorological, hydrological, and agricultural information, and on the impacts of droughts on the economy and the human dimensions, they ignored impacts on ecological aspects in terms of loss of ecosystems and ecosystems services, which are vital for policy decisions. Therefore, policies related to drought management must consider and integrate ecosystem services and vulnerability assessment in addressing ecological droughts (Raheem et al., 2019).

28.3.2.1 Institutional Arrangements and Mechanisms

The government of India follows an integrated system of drought management through a set of institutional arrangements at the national and state level. At the central level, two institutional structures have been formed for the overall management of droughts: the Central Drought Relief Commissioner (CDRC) and the Crop Weather Watch Group (CWWG). The CDRC is led by the Additional Secretary in the Department of Agriculture, Cooperation, and Farmers Welfare (DAC&FW) and assisted by the Disaster Management Division and the Drought Management Cell. It is responsible to collate information and coordinate with ministries. The CWWG consists of several expert bodies along with an Additional Secretary, the DAC&FW, and a Central Drought Relief Commissioner to monitor the drought situation. The other partners are Economics and Statistics Adviser, DAC&FW, Agriculture Commissioner, Animal Husbandry, India Meteorological Department, Central Water Commission, Central Ground Water Board, Ministry of Power, Indian Council of Agricultural Research, Indian Space Research Organisation, National Centre for Medium Range Weather

Forecasting, and Mahalanobis National Crop Forecast Centre. The state governments have a state drought monitoring center (DMC) with expert bodies of meteorologists, hydrologists, and agricultural scientists. The DMC is also assisted by the state disaster management departments and greatly supplemented with drought information from the national and state level.

Both the central and state governments prioritize monitoring mechanisms to account for drought risk and ameliorate repercussions of drought, and accordingly provide early warnings and forecasts, and response and mitigation strategies through institutional arrangements by maintaining information on variables at temporal and spatial scales. Records are kept on three important parameters of drought risk, viz., meteorological, hydrological, and agricultural.

Meteorological parameters, like delay in the onset of monsoon, rainfall, dry spells during sowing period, and dry spells during critical crop-growth periods, are recorded on daily basis across spatial scales at the national level, state level, district level, and field agencies. Hydrological parameters, like water availability in reservoirs, are recorded weekly at spatial scales. Water availability in tanks and streamflow data are kept fortnightly across spatial scales. Information on all these parameters is recorded daily at the field agencies level, whereas information on groundwater level across spatial scales is maintained seasonally, pre- and post-rains. Likewise, agricultural data, like delay in showing, sown area, crop vigor, soil moisture deficit, change in cropping pattern, and supply and demand of agricultural inputs, are recorded on a weekly basis at the national, state, district, and field levels, except data on crop vigor and soil moisture deficit, which are recorded fortnightly across spatial scales. With the help of these data, forecast, and monitoring, the state declares drought in specified administrative units.

28.4 CONCLUSIONS

A number of attempts including some controversial decisions have been taken by the Indian government, like the National River Linking Project announced in 2014, the soil health card scheme introduced in 2015, the drip irrigation scheme introduced in 2016 by the Agriculture Ministry to increase water and energy efficiency, the Swachha Bharat Mission to get rid of sewage runoff in local water systems, interest in Israel's water management systems and technologies, and interests to cut down on non-revenue water with the help of Denmark and Japan. However, the key drivers and factors affecting the occurrences of ecological droughts (e.g., rainfall, heat wave, reservoirs status), the overexploited status of groundwater in major drought-prone areas, and failure to recognize and quantify ecosystem services in plans and policies depict a gloomy picture to account for ecological droughts in India.

To efficiently utilize the groundwater and irrigation systems, stakeholders (mainly farmers) must adhere to the policy of "more crop per drop", whereas most regions like Maharashtra's sugar bowl and Punjab's wheat belt are far from choosing the right crops. Moreover, the irrigation system is operating at only 38% efficiency, which is a major challenge in drought impact on water stress areas in India (Rajawat, 2016). As a part of improving disaster preparedness, mitigation, and regional cooperation, the government of India along with all the Bay of Bengal neighboring nations (Sri Lanka, Thailand, Bangladesh, Bhutan, Myanmar, and Nepal) hosted the first Bay of Bengal Initiative for Multi-Sectoral Technical and Economic Cooperation, Disaster Management Exercise in 2017. India has also hosted the South Asian Annual Disaster Management Exercise in 2015 and the Asian ministerial conference for disaster reduction in 2016.

However, it has to be noted that drought management practices in India mostly focus on crisis management over prevention (Choudhury and Sindhi, 2017). Moreover, water is a state subject, and water governance is very weak and challenging due to a lack of quality robust data about water storage, water flows, groundwater, etc., and water monitoring agencies are mostly ill-equipped to handle water resources (Thakkar, 2019; Thakkar and Harsha, 2019) requiring a push for "cooperative and competitive federalism" in water governance. Even the government's top policy institution, NITI Aayog, has stated that most of the water-related data are very limited in their coverage, robustness, and efficiency (NITI Aayog, 2018). Reformation is urgently required in the existing institutions in water governance, but most of the reforms to the CWC have not been implemented and most of the recommendations by the Mihir Shah Committee were also rejected (Parsai, 2016).

In the face of inadequate water-related data to understand water scarcity, risk, and vulnerability, EPW ENGAGE (2019) has explored alternatives to data availability and stated how to manage drought without water-use data. The report stated that understanding the severity of drought in drought-prone areas is the key to address the water crisis along with the frequency, trends, coverage, and patterns of drought-affected areas. In Maratha Wada in Maharashtra, a drought-prone area, the number of drought-hit talukas have been increasing over the years. During 2018–2019, 151 talukas were declared drought-hit in comparison to 138 talukas during 2015–2016. This data states the water crisis in the absence of water-related data. Moreover, there is a need to understand the causes behind this increase in drought areas and the existing water and soil conservation efforts only focus on the supply-side solution ignoring the demand-side factors that create a water crisis. About 53% of India's agriculture is dependent on rainwater only and the cropping choice significantly affects the severity of droughts (NITI Aayog, 2018). It has been estimated that the supply of water in 2008 was almost double the demand for water, whereas by 2030, the demand for water would outweigh the supply of water by twofold.[7] Therefore, a comprehensive approach encompassing both supply- and demand-side factors and solutions thereof is required. Since water is a state subject, state action in water management is the most crucial in addressing the challenges of ecological droughts. Two major states – Maharashtra and Tamil Nadu – have experienced severe droughts recently, however, their Composite Water Management Index (CWMI) has shown deterioration (Maharashtra −1.41 and Tamil Nadu −3.33) during 2016–2017 in comparison to the base year of 2015–2016. Both states are classified as medium performers in the CWMI 2018. Out of nine indicators used for the CWMI, most of the states underperformed in the indicators of "groundwater source augmentation", which constitutes 40% of water supply, and "sustainable on-farm water use practices", and these statuses signify India's underpreparedness for ecological droughts. From the ongoing discussion, India is not adequately prepared to understand, address, and account for the issues of ecological droughts and it must accommodate ecosystem-based solutions to water management and ecological droughts.

NOTES

1. It is ironic that the state of Maharashtra, where about 40% of large dams are located, is one of the hardest drought-hit states in India calling us to revise our thrusts on large dams and irrigation systems, which only can help the cases of droughts and prevent farmers' suicides. For details, follow the Shah Committee report (2016).
2. The central government of India formulated its first national water policy in 1987 and subsequently revised it in 2002, and considered it to be an economic good. In 2012, again the policy was revised to accommodate managing the water resources as common pool resources and finally as per the recommendations of Shah Committee, 2016, India is in the process of getting a revised national water policy in 2020 anticipating accountability and institutionalising mechanisms for equitable water resource allocation.
3. "Per capita availability of water," Ministry of Jal Shakti, https://pib.gov.in/PressReleasePage.aspx?PRID =1604871#:~:text=As%20per%20Ministry%20of%20Housing,to%20higher%20level%20by%20states.
4. "A framework for India's water policy," National Institute for Advanced Studies, Bangalore.
5. "When the daily maximum temperature of more than five consecutive days exceeds the average maximum temperature by 5°C, the normal period being 1961–90" (WMO).
6. An aquifer mapping programme was started during the 12th Five-Year Plan (2012–2017) and continued during the 13th Five-Year Defence Plan (2017–2022).
7. In 2030, the supply of water would be 744 BCM, whereas the demand for water would be 1498 BCM (NITI Ayog, 2018, p. 28).

REFERENCES

Banerjee, O., Bark, R., Connor, J. and Crossman, N.D. 2013. An ecosystem services approach to estimating economic losses associated with drought. *Ecological Economics*, 91, pp. 19–27. https://doi.org/10.1016 /j.ecolecon.2013.03.022.

Choudhury, P.R. and Sindhi, S. 2017. Improving the drought resilience of the small farmer Agroecosystem. *Economic and Political Weekly*, LII(32), pp. 41–46.

Crausbay, S.D., Ramirez, A.R., Carter, S.L., Cross, M.S., Hall, K.R., Bathke, D.J., Betancourt, J.L., Colt, S., Cravens, A.E., Dalton, M.S., Dunham, J.B., Hay, L.E., Hayes, M.J., McEvoy, J., McNutt, C.A., Moritz, M.A., Nislow, K.H., Raheem, N. and Sanford, T. 2017. Defining ecological drought for the twenty-first century. *American Meteorological Society*, 58(12), pp. 2543–2550. DOI: https://doi.org/10.1175/BAMS-D-16-0292.1.

Glick, P., Stein, B.A. and Edelson, N.A. 2011. Scanning the conservation horizon: A guide to climate change vulnerability assessment. National Wildlife Federation Publ., 168 pp., www.nwf.org/~/media/pdfs/global-warming/climate-smart-conservation/nwf scanningtheconservationhorizonfinal92311.ashx.

GoI. 2009. *Manual for Drought Management*. Department of Agriculture and Cooperation, Ministry of Agriculture, New Delhi.

GoI. 2019. *Dynamic Ground Water Resources of India, 2017*. Central Ground Water Board, Department of Water Resources, River Development & Ganga Rejuvenation, Ministry of Jal Shakti, Faridabad.

GoI. 2016. *Manual for Drought Management*. Department of Agriculture, Cooperation and Farmers Welfare, Ministry of Agriculture and Farmers Welfare, New Delhi.

GoI. 2013. *Masterplan for Artificial Recharge to Ground Water in India*. Ministry of Water Resources and Central Ground Water Board, New Delhi.

Griebler, C. and Avramov, M. 2015. Groundwater ecosystem services: A review. *Freshwater Science*, 34(1), pp.355–367.

Guleria, S. and Gupta, A.K. 2018. *Heat Wave in India Documentation of State of Telangana and Odisha, 2016*. National Institute of Disaster Management, New Delhi.

IPCC. 1997. Tropical Asia. In *Special Report on "The Regional Impacts of Climate Change: An Assessment of Vulnerability"*, Chapter 11. Cambridge University Press, Cambridge.

Li, P., Omani, N., Chaubey, I., and Wei, X. 2017. Evaluation of drought implications on ecosystem services: Freshwater provisioning and food provisioning in the upper Mississippi River Basin. *International Journal of Environmental Research and Public Health*, 14, p. 496, https://doi.org/10.3390/ijerph14050496.

Live Storage Status. 2020. *Reservoir Level & Storage Bulletin*. Central Water Commission, India.

Logsdon, R.A. and Chaubey, I. 2013. A quantitative approach to evaluating ecosystem services. *Ecological Modelling*, 257, pp. 57–65.

MA (Millennium Ecosystem Assessment). 2005. *Ecosystems and Human Well-Being: Synthesis*. Island Press, Washington, DC. https://www.millenniumassessment.org/documents/document.356.aspx.pdf

McEvoy, J. et al. 2018. Ecological drought: Accounting for the non-human impacts of water shortage in the Upper Missouri Headwaters Basin, Montana, USA. *Resources*, 7(14), https://doi.org/10.3390/resources7010014.

Mihir Shah Committee. 2016. A 21st Century Institutional Architecture for India's Water Reforms. http://cgwb.gov.in/INTRA-CGWB/Circulars/Report_on_Restructuring_CWC_CGWB.pdf

Narasimhan, T.N. and Gaur, V.K. (2009). *A Framework for India's Water Policy*. National Institute of Advanced Study, Bangalore. http://eprints.nias.res.in/235/1/R4-09.pdf

NITI Aayog. 2018. *Composite Water Management Index: A Tool for Water Management*, NITI Ayog, Governmnet of India.

Oxfam. 2022. *India: Extreme Inequality in Numbers*. https://www.oxfam.org/en/india-extreme-inequality-numbers

Parsai, G. 2016. Central water commission up in arms over report calling for its restructuring. *Wire*, https://thewire.in/agriculture/cwc-restructuring.

Raheem, N., Cravens, A.E., Cross, M.S., Crausbay, S., Ramirez, A., McEvoy, J., Zoanni, D., Bathke, D.J., Hayes, M., Carter, S., Madeleine, R., Schwend, A., Hall, K., Suberu, P. 2019. Planning for ecological drought: Integrating ecosystem services and vulnerability assessment. *WIRES Water*, 6(4), https://doi.org/10.1002/wat2.1352.

Redmond, K. T. 2002. The depiction of drought: A commentary. *Bulletin of the American Meteorological Society*, 83, pp. 1143–1147, https://doi.org/10.1175/1520-0477(2002)083<1143:TD ODAC>2.3.CO;2.

Rajawat, S. 2016. *Drought and Water Security in India*. Strategic analysis paper 21 June 2016, Future Directions International.

Sewale, M. P., Kumar, S. 2005. *Climatological Features of Drought Incidences in India. Meteorological Monograph, Climatology No. 21/2005*. India Meterological Department, GoI.

Terrado, M., Acuña, V., Ennaanay, D., Tallis, H., Sabater, S. 2014. Impact of climate extremes on hydrological ecosystem services in a heavily humanized Mediterranean basin. *Ecological Indicators*, 37, pp. 199–209.

Thakkar, H. 2019. Challenges in water governance: A story of missed opportunities. *Economic and Political Weekly*, 54(15), pp. 12–14.

Thakkar, H. and Harsha, J. 2019. Can outdated water institutes steer India out of water crisis? *Economic and Political Weekly (Engage)*, 54(9), pp. 2–5.

Wilhite, D.A., Glantz, M.H. 1985. Understanding the drought phenomenon: The role of definitions. *Water International*, 10(3), pp. 111–120.

WMO, GWP. 2014. *National Drought Management Policy Guidelines: A Template for Action (D.A. Wilhite). Integrated Drought Management Programme (IDMP) Tools and Guidelines Series 1*. WMO, Geneva, Switzerland and GWP, Stockholm, Sweden.

Zhang, X., Obringer, R., Wei, C., Chen, N., Niyogi, D. 2017. Droughts in India from 1981 to 2013 and implications to wheat production. *Scientific Reports*, 7, p. 44552. https://doi.org/10.1038/srep44552, pp. 1-12.

29 Water Transfer

Saeid Eslamian and Saida Parvizi

CONTENTS

29.1 INTRODUCTION

Water scarcity resulting from economic and population growth is considered as one of the most important threats to human societies and a constraint to sustainable development (UN-Water, 2008). Within the next decades, water may become the most strategic resource, especially in arid and semiarid regions of the world (UN-Water, 2005). Historically, policy makers in these regions have tried to solve water scarcity problems through dam building, groundwater recharge, cloud seeding, desalination, wastewater reuse, and developing massive water transfer projects, among others (Hutchinson et al., 2010). However, there is a growing body of evidence that water scarcity can be created or intensified by unsustainable decisions to meet the increasing water demands (Gleick, 1998; Cai et al., 2003). In arid regions, supply-oriented water management schemes, although promising in the short run, are typically associated with unintended secondary consequences in the long run (Madani and Mariño, 2009). In essence, the failure to develop sustainable water resources solutions at the watershed scale is rooted in the lack of understanding of the interrelated dynamics of different subsystems of complex watershed systems (Mirchi et al., 2010).

Interbasin water transfer from water-abundant regions (donors) to regions with water shortages (recipients) has been recognized as a solution to secure water supply for supporting development in recipient basins (Gupta and van der Zaag, 2008). Thus, numerous water transfer projects have been implemented around the world, including Australia (Wright, 1999), China (Shao et al., 2003), Germany (Schumann, 1999), Iran (Madani and Mariño, 2009), Mexico (Medellin-Azuara et al., 2011), and the United States (Lund et al., 2010; Medellin-Azuara et al., 2011; Madani and Lund, 2012). Worldwide, approximately 14% of global water withdrawal is provided through interbasin water transfer projects, and this portion is expected to rise to 25% by 2025 (ICID, 2005).

Water transfer initiatives have relieved water stress by providing "sufficient" water for different users, enhancing socioeconomic development, and increasing freshwater availability for ecosystem augmentation in the recipient basins (Gohari et al., 2013). However, water transfer may entail negative long-term social, economic, and environmental impacts, raising concern as to its effectiveness as a panacea to water shortage (Yan et al., 2012).

Investigating the reasons for the success or failure of water transfer projects can provide valuable lessons to water resources planners and policy makers, who have historically based their decisions on a simple comparison of water balances in the recipient and donor basins (Andrade et al., 2011).

Therefore, this section discusses water transfer between basins. The advantages and disadvantages of water transfer between basins and models that can simulate it are discussed. In this regard, the experiences of other countries will be used.

29.2 METHODOLOGY

29.2.1 WATER TRANSFER

Water transfer is a water management strategy aimed at reducing the mismatch between water supply and demand by transferring water to augment local supply in water-scarce areas or reduce damage caused by excess water. Water transfer has three dimensions (Gichuki and McCornick, 2010):

- First, water can be transferred from one use/user (donor) to another (recipient). Common examples include the transfer of water from agricultural to urban use and the transfer of water rights from one user to another, either through water trading, at the expiry of the water right duration, or when one user simply takes the water with no compensation to the previous user.
- Second, in the temporal dimension, alternative forms of water storage (groundwater recharge, natural or manmade reservoir) increase water availability in the dry seasons by storing the excess water received during the rainy seasons.
- Third, the spatial dimension involves the transfer of water from one location to another using groundwater pathways, natural waterways, canals and/or pipelines.

These dimensions are not mutually exclusive and in most cases occur in combination. Water transfer requires that there be a social, environmental, political, or economic benefit, which provides the justification to offset:

- The cost of transferring it
- Any compensation demanded by the donor
- Any other costs associated with the negative externalities that the transfer may generate

Generally, water transfer schemes have multiple complementary objectives that include:

- To increase total water benefit by transferring surplus water to a water-scarce basin/region
- To facilitate the reallocation of water from a low- to a high-value use
- To reduce regional inequity by transferring water to promote socioeconomic development in water-scarce regions
- To meet treaties, agreements, or other legal obligations
- To facilitate broader cooperation and promote solidarity between donor and recipient regions
- To restore degraded freshwater ecosystems

As stated earlier, there is a wide range of transfers and a variety of terms associated with them. Water transfers are classified based on the geographic scope as follows:

- Interproject – transfer within a water project.
- Intrabasin – transfer from one subbasin to another in the same basin.

- Interbasin – transfer from one basin to another basin. This is further subdivided into short and long interbasin transfers. In the case of the former, the transfer is to a basin immediately adjacent to the donor basin, whereas, with the latter, it may cross multiple basins.

Other defining characteristics of a water transfer arrangement include:

- Types of water – the transfer may involve surface, ground, wastewater (reclaimed, treated, or untreated), brackish and even saline water. For the sake of completeness, it could also include virtual water, that is, through trade.
- Water transfer route – can be direct (above or below ground pipelines, open or closed canals, and natural waterways) or indirect as in the case of groundwater flow.
- Water transfer duration – permanent, long-term and short-term transfers of a water right.
- Water transfer operation criteria – define the volume, rate, and timing (seasonal, constant, pulsed or combination) of the water to be transferred.
- Planned or unplanned water transfer – although the focus of this chapter is on planned transfers, it is important to recognize that there are also unplanned transfers.

Given the preceding criteria, there is considerable variation in the form of a given transfer, and generalizations of their appropriateness and/or impacts may be misleading. However, matching water transfer purpose, type, and characteristics with the unique site conditions is an important step toward reducing negative impacts.

29.2.2 GLOBAL WATER TRANSFER EXPERIENCES

While water scarcity is a complex issue, unequal distribution of water across the world is definitely a factor. Some regions enjoy abundant access to fresh water, while many others suffer from acute shortages. This inequality of water availability will increase in the future based on climate change scenarios. In the face of rapid urbanization, population growth, and climate change, the organizations tasked with securing our water supply must tackle bigger challenges than ever. Water transfers and interconnections are vital elements of these strategies. Water transfers are large and lengthy undertakings that expend significant resources and can have long-term impacts on large areas.

Water shortages are forecast to affect 50% of the world's population by 2030, impacting developing nations most acutely. To increase water security, there has been a significant increase in interbasin water transfer (IBWT) schemes and engineering projects that redistribute water from one basin to another. A simple example of an interbasin water transfer is shown in Figure 29.1 (Sinha et al., 2020).

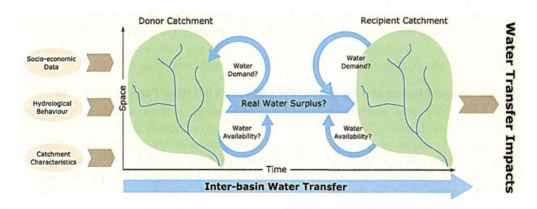

FIGURE 29.1   Interbasin water transfer.

TABLE 29.1

Major Features of International Transfers and Other Relatively Large Water Transfers to Reduce Water Shortage

Water Transfer Project	Year of Construction	Length (km)	Capacity (MCM/Year)	Costs (US$ bn)
California State Water Project (USA)	Early 1960s–1990	715	25	5.2
Colorado River Aqueduct (USA)	1933–1941	392	1603.5	—
Central Arizona Project (USA)	1973–1993	541	1850.2	3.6
National Water Carrier (Israel)	1953–1964	130	1.7	—
Cutzamala System (Mexico)	Late 1970s–1990s	154	2.1	1.3
All American Canal (USA)	1930s	132	64	—
Periyar Project (India)	Commissioned in 1895	—	3.5	—
Indira Gandhi Canal (India)	Since 1958	650	—	—
Telugu Ganga Project (India)	1977–2004	406	10.1	—
Irtysh Karaganda Scheme (Kazakhstan)	1962–1974	450	6.5	—

As mentioned, one of the most ambitious responses to unequal distribution is the large-scale physical transfers of water from one source or basin to another, otherwise known as water transfers. Table 29.1 shows examples of projects that have been done in this field throughout the world. In the following sections, some of these projects will be reviewed and global experiences in this field will be recounted.

29.2.2.1 Two Water Transfer Projects in Iran

The impact of the operation of two water transfer projects – Solegan to Rafsanjan (receiving basin 1) and Koohrang III (receiving basin 2) to Zayandeh-Rud Dam – from the Karoon River are investigated in Karamouz et al. (2010). The model was developed with an economic objective function considering different components of the interbasin water transfer system and it is solved using a genetic algorithm (GA).

The water quality simulation for the sending basin using an artificial neural network (ANN) model is linked with the optimization model. In order to determine the operating rules, a k-nearest neighbor (KNN) model was developed to be used in real-time operation. The main benefit was related to the agricultural production in the Rafsanjan Plain for the high-value pistachio crop. The maximum costs were the loss of agricultural production in the Khuzestan Plain and the environmental costs resulted when headwater quality in the Karoon River was altered.

The results show that if the sending basin receives $0.38 per cubic meter of water transfer, it could affect the loss of agricultural income and environmental costs (Figure 29.2).

29.2.2.2 Major Existing and Proposed IBWT Schemes in Other Parts of the World

IBWT is the "purposeful rearrangement of natural hydrologic patterns via engineering works [...] to move water across drainage divides to satisfy perceived human needs" (Micklin, 1984). IBWT schemes exist or are planned, across all continents, with the exception of Antarctica, in both developed and developing countries (Gupta and Zaag, 2008; Shumilova et al., 2018). Shumilova et al. (2018) identified 76 "mega" IBWT projects costing over US$1 billion, either under construction (25) or in the planning phase (51), which, if all are constructed, will transfer up to 1910 km^3 a^{-1} of water, equivalent to 26 times the average annual flow of the Rhine (Figure 29.3).

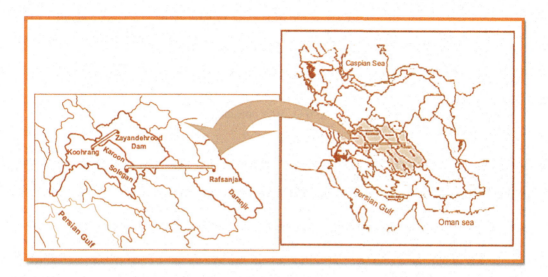

FIGURE 29.2   Iran's major watersheds and the close-up of the three basins, the sending basin, and two receiving basins (Karamouz et. al., 2010).

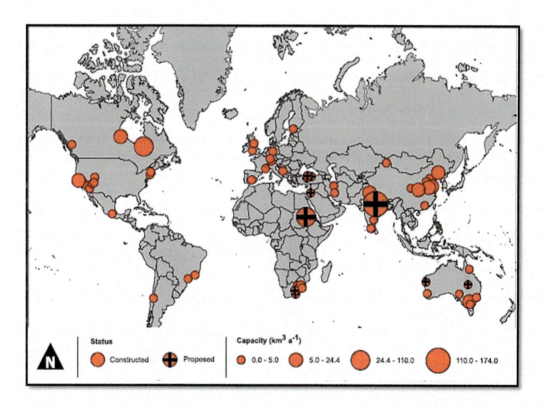

FIGURE 29.3   Major existing and proposed IBWT schemes globally (Sinha et al., 2020).

29.2.2.3   Evaluating Interbasin Water Transfer Schemes

Conflict over IBWT schemes reflects their lack of systematic evaluation, a direct result of their complexity, multiple stakeholders and their myriad of objectives (Wilson et al., 2017). Scholars have proposed standardized approaches, based on specific criteria drawn from the integrated water resources managementprocess, by which schemes can be assessed (Gupta and Zaag, 2008).

Existing criteria sets share two main common requirements (Sinha, 2020):

1. The donor basin must have surplus water availability (WA), taking into account existing and future water demand (WD).
2. The recipient basin must have real WD, after considering all possibilities for WA within the basin.

Although the criteria sets generally contain references to multidisciplinarity or holistic assessment, the use of sound science, and integrated approaches to social, environmental, and economic sustainability, there is no consistency in how this is incorporated or the extent to which this is prioritized.

Early criteria sets were founded on a relatively restrictive set of procedures for assessment and included specific statements on environmental impacts and social and cultural impacts (Cox, 1999). Gupta and Zaag's (2008) criteria are more flexible, recommending only that projects should be "socially, environmentally and economically sustainable".

In contrast, the most recent criteria, proposed by Kibiiy and Ndambuki (2015), contains no specifics on what should be assessed or how to rely instead on the addition of "subbranches" of assessment at different levels to consider different aspects. This transition from mechanistic to flexible sets of criteria may reflect a changing understanding that no two IBWT schemes can be evaluated by the same criteria, and that the complexity of benefits and impacts changes over time.

However, relying on a process that lacks specificity opens the assessment process to potential criticism, given the highly contested and often technocratic nature of IBWT projects (Pasi and Smardon, 2012).

Water surplus/deficit was then calculated by

$$(\text{Water SurPlus/Deficit} = WA_{nat} - WD_{tot}) + R_{dom} + R_{ind} + R_{irrig} \qquad (29.1)$$

where WA_{nat} is the natural water availability; WD_{tot} is the total water demand; and R_{dom}, R_{ind}, and R_{irrig} are regenerated flows from domestic, industrial, and irrigation, respectively. All values are in $mm^3 \ a^{-1}$.

Potential water surplus or deficit is evaluated using the following equation:

$$\text{Water SurPlus/Deficit} = WA_{tot} - WD_{tot}) + TWR \qquad ((29.2)$$

where WA_{tot} is the total water available at 75% dependability, WD_{tot} is total water demand, and TWR is total regenerated flow.

29.2.2.4 Water Transfer Compensation Classification between China, Japan, America, and Australia

The compensation mechanism for agricultural water transfer (CMAWT) is an important method for reducing or eliminating the negative effects of agricultural water transfer. In this section, a classification theory for CMAWT is established to evaluate the CMAWT of China, Japan, America, and Australia (Dai et al., 2016).

As a result of current water resource shortages, agricultural water transfer (AWT) is becoming increasingly significant in China. AWT is the transfer of the rights of irrigation water to others in the agricultural industry or in other industries. The CMAWT is necessary to reduce or eliminate the negative effects of AWT (Liu et al., 2001; Rosegrant and Ringler, 2000). The CMAWT is the mechanism for eliminating the negative effects of AWT by compensating for the cost of agricultural water savings and the losses incurred. There are some compensation examples in China, such as the compensation to farmers in the middle stream of the Hei River Basin (Liu et al., 2005) and to farmers in Miyun County of Beijing (Peisert and Sternfeld, 2004). However, whether the compensation mechanism is efficient remains a concern.

The CMAWT can be analyzed from a vertical and horizontal perspective. The former mainly reflects the relationship of the government with other stakeholders (Dai et al., 2016). The factors that influence the vertical form of CMAWT include property rights, the participation level of stakeholders, and compensation approaches. The vertical form of CMAWT can be classified into three types:

- Bureaucratic
- Autonomous
- Market

In the bureaucratic type, agricultural water users have the right to use agricultural water but not the right to transfer it. Agricultural water users and third parties do not participate in this type of CMAWT. The government determines the amount of water transferred, the terms of transfer, and the compensation amount and method.

In the autonomous type, agricultural water users possess the right to use agricultural water and can transfer those rights under certain conditions. Under the guidance of the government, agricultural water users and third parties can decide the transfer amount, the terms of transfer, and the compensation methods. The government and other stakeholders in AWT pay the compensation fee.

In the market type, agricultural water users possess the complete right to use and transfer agricultural water. AWT becomes an agricultural water rights transfer, and the compensation fee becomes the transfer fee. This transfer price is decided based on negotiations between the transfer participants. Third parties participate in agricultural water rights transfers to different extents. Only the buyer pays the transfer fee. The government only supervises the water rights transfers.

The horizontal form of CMAWT involves compensation methods in AWT, which include fund compensation and material compensation. The former uses cash to decrease or eliminate the losses incurred. The latter uses materials or projects to improve the development capability of the losers. These include investments in agricultural water conservation projects and ecorestoration projects. All CMAWT modes are listed in Table 29.2.

As shown in Table 29.2, CMAWT can be divided into six modes. CMAWT can classify different countries using this approach. However, this method remains imperfect because it cannot quantitatively classify CMAWT modes.

Results showed that China's CMAWT is converting from the bureaucratic to market type. Japan's CMAWT is mainly the autonomous type. The CMAWT of America and Australia are market types. The water transfer market in Australia is more sophisticated among the comparison countries. The governments of China and Japan participated more in the agricultural water transfer than the governments of America and Australia. Farmers participated least in agricultural water transfer in China than in other compared countries.

The CMAWT of China is gradually improving, but in terms of farmer participation, government intervention and adverse impact regulation, a huge gap remains between it and the CMAWT of

TABLE 29.2
CMAWT Model

Type	Bureaucratic	Autonomous	Market
Material compensation	Bureaucratic–material	Autonomous–material	Market–material
Fund compensation	Bureaucratic–fund	Autonomous–fund	Market–fund

Source: Dai et al. (2016).

other typical countries. China's agricultural water rights system should be improved to enable all stakeholders to participle in AWT. A suitable CMAWT for China not only compensates the losses of farmers but also stimulates farmers to save agricultural water.

29.2.2.5   Effective Watershed Management

Limited water resources with uneven distribution and growing demands are the main challenges of water management in Iran. The government has planned several water resources development projects. The complex technical, socioeconomic, and environmental outcomes of these projects require a comprehensive evaluation. To select more adaptive and accountable projects, suitable group decision support systems are needed.

Lake Urmia in northwestern Iran is the largest inland lake in the country and one of the largest saline lakes in the world (613,253 ha). The lake is also one of the most important and valuable aquatic ecosystems in the country. Some 1500 species of vascular plants, including unique *Artemia* sp., are distributed among 85 families and represent 15% of the total number of flora species found in the country. Because of its unique natural and ecological features, the lake has been designated as a national park, Ramsar site, and UNESCO biosphere reserve (Conservation of Iranian Wetland Project, 2008).

The lake basin, as a unique socioecological region, is facing extreme water shortages due to poor water governance and climate change (Alipour, 2006; Zarghami et. al., 2009). Because of the intense agricultural development and rapid urbanization, the groundwater level in some parts of the basin has decreased by up to 16 m (Figure 29.4).

According to Alesheikh et al. (2009), the area of Urmia Lake decreased from 5650 km^2 in August 1998 to 4610 km^2 in August 2001. The water level of the lake is now 2–3 m below its critical level (Figure 29.5). The lake requires a minimum flow of 3 billion m^3/yr to compensate for annual evaporation (Conservation of Iranian Wetland Project, 2008). These decreasing water levels are leading to environmental disasters, shown in Figure 29.5 (Zafarnejad, 2010). The wetlands around the lake are major staging areas for migrating birds, such as flamingos. However, in recent years their number has declined (Zafarnejad, 2010).

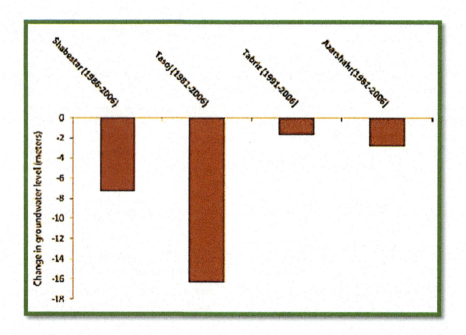

FIGURE 29.4   Declining groundwater levels in some areas of the Lake Urmia basin (Zarghami, 2011).

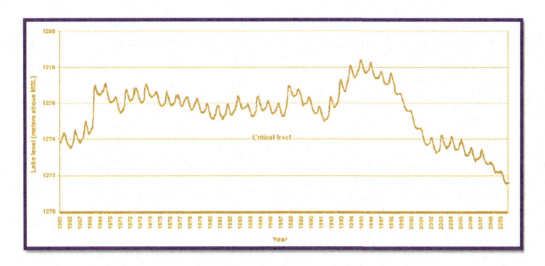

FIGURE 29.5 Annual stage of Lake Urmia (Zafarnejad, 2010).

Zarghami (2011) proposed four routes of water transfer schemes to Urmia Lake in Iran, which is in danger of completely drying out. The routes are evaluated with respect to different criteria. The criteria and weights were obtained from an organization responsible for major water infrastructures in the basin. By using an efficient multicriteria decision-making method of compromise programming, the four alternatives were ranked, and the most robust water transfer route was selected (by using a fuzzy algorithm). This best alternative transfers about 300 million cubic meters per year from another basin in the north of the lake. Results show the importance of using decision support systems for participatory and effective lake management.

According to the case study results, the compromise programming had a number of advantages (Zarghami, 2011):

1. The model considers several criteria in the decision-making process.
2. The criteria weights were obtained by using the fuzzy set theory from different decision makers, facilitating consensus among them instead of considering the preferences of a single decision maker.
3. The manager assigned different power weights for the decision makers.
4. Compromise programming presents the aggregate distance from both nadir and ideal points, and we have the complete rankings of the alternatives.
5. Changing the p value in the compromise programming shows the risk acceptance in the decision-making problem. The sensitivity analysis presents the most robust alternative in uncertain conditions.

29.2.2.6 Water Marketing

From an economic perspective, water resources are composite assets that provide a variety of services for consumptive and productive activities. However, water quality degradation has been an important cause of water scarcity in countries (Wei and Gnauck, 2007a). Water resources management on those problems usually involve interactive and interdependent stakeholders with contradictory or conflicting interests (Van der Veeren and Tol, 2003), goals, and strategies (Wei and Gnauck, 2007a).

Pollutant discharge is an essential but complex issue in water resources management, and this complexity is not only from intricate biochemical processes, but also from different pollutant sources and multipolluters with conflicting aims. Water quality and quantity conflicts are usually caused by (Wei et al., 2010):

- Water scarcity due to uneven precipitation
- Multiple users and pollutant sources discharging waste into water
- Different degrees of upstream pollutions restricting the water use in the downstream catchment
- Interbasin water transfer breaking the long-established balance of water quality and quantity in a basin.

To solve water conflicts caused by water scarcity, some engineering solutions have been proposed in terms of reducing water losses, increasing water-use efficiency and wastewater recycling, water conservation, and water transfer, and some other nonengineering measures. However, methods of using water efficiency and wastewater recycling are not so sufficient for the regions facing extreme water shortage.

In addition, interbasin water diversion involves a multidisciplinary problem (Yevjevich, 2001), which usually brings about fundamental issues and conflicts concerning socioeconomical, environ-ecological, administrative, and legislative problems (Shao and Wang, 2003; Yang and Zehnder, 2005). In addition, different economic and political instruments have been widely used to solve water-use conflicts (Wang et al., 2003).

The water markets approach is one cited frequently in the literature (Bhaduri and Barbier, 2003). Water market methods can provide water users with incentives to allocate water and reduce pollutants discharge efficiently, and such a market exists in some countries, including Australia, USA (California), Chile, India, and Spain.

However, the water market requires defining the original water rights, creating institutional and legal mechanisms, and establishing basic infrastructures for water trade (Wang et al., 2003) before it can operate well.

Waste discharge is a public bad, and every polluter can free-ride others' achievement of treatment (Wei and Gnauck, 2007b). The free-riding problem causes market failure. In the absence of market and property rights, conflicts between multistakeholders competing for water uses are unavoidable (Wei and Gnauck, 2007a). True "markets" for water are rare, and they are not really free markets (Dellapenna, 2000).

Those economic- and political-based water conflict-solving methods can be categorized into two classes: direct regulations and economic instruments (Wei and Gnauck, 2007a). Direct regulation is also known as "command and control" strategies, which usually includes limitation quotas, standards, laws, etc. Economic and political instruments make use of the market mechanism, price incentives, water rights, subsidies, compensation, tradable permits, green taxations, etc.

However, environmental resource problems and their interrelationships with economic activities and the dynamic ecosystem are very complex and cannot be solved with simple policy tools (Wei et al., 2010).

Command and control strategies usually lack incentive, because they are mainly in virtue of legislation, power, or force. Wei and Gnauck (2007a) stated that the existing economic and regulation instruments do not work so well in solving these conflicts. From a technical strategy point of view, multiobjective optimization models have been used early to maximize the overall benefit in order to solve transboundary water conflict in a river basin (Yang and Zeng, 2004).

In recent years, more advanced and popular multiobjective evolutionary algorithms have been used to solve conflict objectives in the watershed (Bekele and Nicklow, 2005; Muleta and Nicklow, 2005). In general, however, those optimization measures neglect the real interests and benefits of the stakeholders in the basin, though they can capture the multiple optimal solutions, sometimes called Pareto optimal solutions.

Game theory is an appropriate approach to model and to solve such water conflicts. It was launched by John von Neumann, a great mathematician, and Oskar Morgenstern in 1944. Game theory modeling concepts and reasoning have been widely applied in economic, commercial, social, political, biological, and other sciences to help people analyze social and behavioral phenomena.

29.3 RESULTS AND DISCUSSION

Water transfer has been and continues to be a complementary water management strategy for promoting socioeconomic development in water-scarce regions. Over 2500 years ago, the Babylonians, the Roman Empire, and the Chinese constructed extensive canal networks, famous aqueducts, and the Grand Canal, respectively, to support human settlement in water-scarce areas. The Anuradhapura Kingdom of Sri Lanka, too, developed major water transfers as far back as 100 AD to support the irrigation the civilization needed to feed a growing population (deSilva, 2005).

In the 20th century, the phenomenal population growth, economic activities, and human settlement in water-scarce regions, advances in science and technology, political will, and availability of resources led to the development of many water transfer projects. The global interbasin water transfer increased from 22 to 56, from 56 to 257, and from 257 to 364 km³/yr during the periods 1900–1940, 1940–1980, and 1980–1986, respectively, and is estimated to increase to 760–1240 km³/year by 2020 (Shiklomanov, 1999). Most of these transfers took place in Canada, the former USSR, India, and the United States .

The interbasin water transfer project is an alternative to balance the nonuniform temporal and spatial distribution of water resources and water demands. Transferring water from an area may cause a variety of negative impacts, including social and environmental impacts. But a water transfer project can be executed if it is environmentally and economically justified.

When water is intended to be used in another basin, water rights could be traded for financial resources. In the national arena, water is equity for all: equity for those who are in need of water and do not have access to water, and those who actually have water rights and may have a surplus that is wasted in a variety of ways.

To analyze the aforementioned issues, tangible and nontangible costs and benefits should be evaluated. Several investigators have emphasized the need for economic and environmental assessment of interbasin water transfer plans. Lund and Israel (1995) presented the application of multistage linear programming for the estimation of the least-cost integration of several water marketing opportunities with water conservation and traditional water supplies.

Draper et al. (2003) developed an economic-based optimization model for California's major water supply system. They noted that optimization models driven by economic objective functions are practical for assessing the development project.

Feng et al. (2007) developed a decision support system (DSS) for assessing the social-economic impact of China's south-to-north water transfer project. The DSS provides decision support through simulation with a water embedded computable general equilibrium model.

Gupta and Zaag (2008) assessed the interbasin water transfers from a multidisciplinary perspective and attempted to answer whether such transfers are compatible with the concept of integrated water resources management and the criteria for assessing such transfers.

Matete and Hassan (2005) developed an analytical framework that can be applied to integrate environmental sustainability aspects into economic development planning in the case of exploiting water resources through interbasin water transfers.

The brief review of existing experiences indicated that the apogee of implementing interbasin water transfer projects has been met in the 19th century. According to the results, 43.75%, 18.75%, 12.5%, and 25% of the projects have been implemented for drinking and agricultural water supply, energy production, environmental objectives, and multipurpose water supply purposes, respectively.

However, in many developed countries, more than 80% of the interbasin water transfer projects were implemented for providing drinking water. According to the data, more than 27% of the global water withdrawal capacity is transferred with interbasin water transfer projects.

In Iran, based on available data, the capacity of interbasin water transfer is 6.35 km^3 per annum mainly transferred for agricultural purposes through the construction of long tunnels. Since interbasin water transfer projects directly affected the management of origin and destination basins, the socioeconomic and environmental conditions were often weakened in one of the basins in the long term.

Consequently, interbasin water transfer projects should be applied only in emergency conditions when no other alternative solution is practicable and when an drinking water supply emergency has emerged. This should also satisfy the integrated recognition of study region conditions and its potentials, having a comprehensive and systematic management approach, and ultimately a consideration of environmental, economic, social, and political dimensions.

29.4   CONCLUSIONS

Water transfer decisions should be based on a holistic view of the problem, which not only includes the hydrological aspects, but also the socioeconomic and environmental concerns. Developing integrated water resources management models can facilitate a holistic understanding of complex watershed systems, leading to sustainable water resources planning and management decisions (Mirchi et al., 2010).

System dynamics models are tools that facilitate understanding of the interactions among diverse but interconnected subsystems that drive the dynamic behavior of the system (Sterman, 2000). These models can facilitate water resources planning and management by identifying problematic trends and their root drivers within an integrated framework (Mirchi et al., 2012), which is critical for the sustainable management of water resources systems (Hjorth and Bagheri, 2006; Madani, 2010).

In general, the following are adopted as basic criteria in the study of interbasin water transfer (Sinha et al., 2020):

1. The donor basin must have surplus water availability after fulfilling all its present and future water deficits.
2. The recipient basin must have a water deficit after tapping all possibilities of water availability within the basin.
3. The completed project must be supported by an integrated, multidisciplinary assessment of potential impacts and benefits, intended to minimize adverse impacts and maximize benefits, and demonstrate equitable distribution among donor and recipient basins.
4. Analyses must use, where possible, data that is freely available within the public domain, and which should be made available for scrutiny.

REFERENCES

Alesheikh, A.A., Ghorbanali, A., Talebzadeh, A. 2009. Generation the coastline change map for Urmia Lake by TM and ETM+ imagery. Available from http://www.gisdevelopment.net/application/nrm/coastal/mnm/ ma04022pf.htm. Accessed 14 Nov 2010.

Alipour, S. 2006. Hydrogeochemistry of seasonal variation of Urmia Salt Lake, Iran. *Saline Syst.* 2(9). https://doi.org/10.1186/1746- 1448-2-9.

Andrade, J.G.P.D., Barbosa, P.S.F., Souza, L.C.A., Makino, D.L. 2011. Interbasin water transfers: The Brazilian experience and international case comparisons. *Water Resour. Manage.* 25 (8), 1915–1934.

Bekele, E.G., Nicklow, J.W. 2005. Multiobjective management of ecosystem services by integrative watershed modeling and evolutionary algorithms. *Water Resources Research* 41, W10406. https://doi.org/10.1029/2005WR004090.

Bhaduri, A., Barbier, E.B. 2003. *Water Transfer and International River Basin Cooperative Management: The Case of the Ganges*. Department of Economics and Finance, University of Wyoming, Laramie, WY.

Cai, X.M., McKinney, D.C., Rosegrant, M.W. 2003. Sustainability analysis for irrigation water management in the Aral Sea region. *Agric. Syst.* 76 (3), 1043–1066.

Conservation of Iranian Wetlands Project. 2008. *Integrated Management Plan for Lake Urmia,*. Department of Environment, National Project Manager, Govt. of Iran, Tehran.

Cox, W.E. 1999. Determining when Interbasin water transfer is justified: Criteria for evaluation. In: Bogardi, J.J., Bruk, S., Vienot, C., de La, J.M., González, F. (Eds.), *Proceedings of the International Workshop: A Contribution to the World Water Vision Consultation Process, Technical Documents in Hydrology (No. 28)*. International Hydrological Programme (IHP-V). UNESCO, Paris, pp. 173–178.

Dai, C., Tan, Q., Lu, W.T., Liu, Y., Guo, H.C. 2016. Identification of optimal water transfer schemes for restoration of a Eutrophic Lake: An integrated simulation – optimization method. *Ecol. Eng.* 95, 409–421.

Dai, X., Han, Y., Zhang, X., Chen, J., Li, D. 2016. Development of a water transfer compensation classification: A case study between China, Japan, America and Australia. *Agric. Water Manage.* 182, 151–157.

De Silva, K.M. 2005. *A History of Sri Lanka*. Penguin Books, India.

Dellapenna, J.W. 2000. The importance of getting names right: The myth of markets for water. *William Mary Environ. Law Policy Rev.* 25, 317–377.

Draper, A., Jenkins, M., Kirby, K., Lund, J., Howitt, R. 2003. Economic-engineering optimization for California water management. *J. Water Resour. Plann. Manage.* 129(3), 155–164.

Feng, S., Li, L., Duan, Z., Zhang, J. 2007. Assessing the impacts of South-to-North Water transfer project with decision support systems. *Decision Support Sys.* 42(4), 1989–2003.

Gichuki, F., McCornick, P.G. *International Experiences of Water Transfers*. Relevance to India, International Water Management Institute, Colombo, Sri Lanka, pp. 345–371.

Gleick, P.H. 1998. Water in crisis: Paths to sustainable water use. *Ecol. Appl.* 8(3), 571–579.

Gupta, J., van der Zaag, P. 2008. Interbasin water transfers and integrated water resources management: Where engineering, science and politics interlock. *Phys. Chem. Earth* 33(1–2), 28–40.

Gohari, A.R., Eslamian, S., Mirchi, A., Abedi-Koupai, J., Massah Bavani, A.R., Madani, K. 2013. Water transfer as a solution to water shortage: A fix that can backfire. *J. Hydrol.* 491, 23–39.

Hjorh, P., Bagheri, A. 2006. Navigation towards sustainable development: A system dynamics approach. *Future* 38(1), 74–92.

Hutchinson, C.F., Varady, R.G., Drake, S. 2010. Old and new: Changing paradigms in arid lands water management. In: Schneier-Madanes, G., Courel, M.F. (Eds.), *Water and Sustainability in Arid Regions*, vol. 3. Springer, New York, pp. 311–332.

ICID, 2005. *Experiences in Inter-basin Water Transfers for Irrigation, Drainage or Flood Management (3rd Draft 15 August 2005)*. Unpublished report. International Commission on Irrigation and Drainage ICID-CIID, New Delhi.

Karamouz, M., ASCE, F., Mojahedi, A., Ahmadi, A. 2010. Interbasin water transfer: Economic water quality-based model. *J. Irrig. Drain.* 136, 90–98.

Kibiiy, J., Ndambuki, J. 2015. New criteria to assess interbasin water transfers and a case for Nzoia-Suam/ Turkwel in Kenya. *Phys. Chem. Earth* 89–90, 121–126.

Liu, H., Cai, X., Geng, L., Zhong, H. 2005. Restoration of pastureland ecosystems: A case study of western Inner Mongolia. *J. Water Resour. Plann. Manage.* 131(6), 420–430.

Liu, J., Jiang, W.L., Ren, T.Z. 2001. Research on the compensation mechanism of agricultural water rights transfer. *J. China Agric. Resour. Reg. Plann.* 22(6), 42–44 (in Chinese).

Lund, J.R., Hanak, E., Fleenor, W.E., Bennett, W.A., Howitt, R.E., Mount, J.F., Moyle, P.B. 2010. *Comparing Futures for the Sacramento–San Joaquin Delta*. University of California Press, Berkeley, CA.

Lund, J.R., Israel, M. 1995. Optimization of transfers in urban water supply planning. ASCE library. *Journal of Water Resources Planning and Management*, 121(1):1–41.

Madani, K. 2010. *Towards Sustainable Watershed Management: Using System Dynamics for Integrated Water Resources Planning*. VDM Verlag Dr. Müller, Saarbrücken, Germany, ISBN 978-3-639-18118-0.

Madani, K., Lund, J.R. 2012. California's Sacramento–San Joaquin Delta conflict: from cooperation to chicken. *Water Resour. Plan. Manage.* ASCE 138(2), 90–99.

Madani, K., Mariño, M.A. 2009. System dynamics analysis for managing Iran's Zayandeh-Rud river basin. *Water Resour. Manage.* 23, 2163–2187.

Matete, M., Hassan, R. 2005. An ecological economics framework for assessing environmental flows: The case of inter-basin water transfers in Lesotho. *Glob. Planet. Change* 47, 193–200.

Medellin-Azuara, J., Mirchi, A., Madani, K. 2011. Water supply for agricultural, environmental and urban uses in California's borderlands. In: Contreras, L.M. (Ed.), *Agricultural Policies: New Developments.* Nova Science Publishers, New York, pp. 201–212.

Micklin, P.P. 1984. Inter-basin water transfers in the United States. *Int. J. Water Resour. Develop.* 2, 37–65.

Mirchi, A., Watkins, D., Madani, K. 2010. Modeling for watershed planning, management, and decision making. In: Vaughn, J.C. (Ed.), *Watersheds: Management, Restoration and Environmental Impact.* Nova Science Publishers, New York.

Mirchi, A., Madani, K., Watkins, D., Ahmad, S. 2012. Synthesis of system dynamics tools for holistic conceptualization of water resources problems. *Water Resour. Manage.* 26(9), 2421–2442.

Muleta, M.K., Nicklow, J.W. 2005. Decision support for watershed management using evolutionary algorithms. *J. Water Resour. Plan. Manage.* 131 (1), 35–44.

Pasi, N., Smardon, R. 2012. Inter-linking of Rivers: A solution for water crises in India or a decision in doubt? (policy analysis). *J. Sci. Policy Govern.* 2, 1–42.

Peisert, C., Sternfeld, E. 2004. Quenching Beijing's thirst: The need for integrated management for the endangered Miyun reservoir. *China Environ.* Series 7, 33–45.

Rosegrant, M.W., Ringler, C. 2000. Impact on food security and rural development of transferring water out of agriculture. *Water Policy* 1(6), 567–586.

Schumann, H. 1999. Water transfer systems for freshwater supply in Germany. In: Proceedings of the International Workshop on Interbasin Water Transfer, 25–27 April, 1999. UNESCO, Paris, pp. 115–122.

Shao, X.J., Wang, H. 2003. Interbasin transfer projects and their implications: A China case study. *Int. J. River Basin Manage.* 1 (1), 5–14.

Sinha, P., Rollason, E., Bracken, L.J., Wainwright, J., and Reaney, S.M. 2020. A new frame work for integrated, holistic, and transparent evaluation of inter-basin water transfer scheme. *Science of the Total Environment,* 721, 1–16.

Shao, X., Wang, H., Wang, Z. 2003. Interbasin transfer projects and their implications: A China case study. *Int. J. River Basin Manage.* 1 (1), 5–14.

Shumilova, O., Tockner, K., Thieme, M., Koska, A., Zarfl, C. 2018. Global water transfer megaprojects: A potential solution for the water-food-energy nexus? *Front. Environ. Sci.* 6:150, 1–11.

Sterman, J.D. 2000. *Business Dynamics, Systems Thinking and Modeling for A Complex World.* McGraw-Hill, Boston.

UN-Water. 2005. A gender perspective on water resources and sanitation. Interagency task force on gender and water. In: The 12th Session of the Commission on Sustainable Development.

UN-Water. 2008. Status report on integrated water resources management and water efficiency plans. In: The 16th Session of the Commission on Sustainable Development.

Van der Veeren, R.J.H.M., Tol, R.S.J. 2003. Game theoretic analyses of nitrate emission reduction strategies in the Rhine river basin. *Int. J. Global Environ. Issues* 3(1), 74–103.

Wang, L.Z., Fang, L., Hipel, K.W. 2003. Water resource allocation: A cooperative game approach. *J. Environ. Inform.* 2 (1), 11–22.

Wei, S.K., Gnauck, A. 2007a. Water supply and water demand of Beijing: A game theoretic approach for modeling. In: Go´mez, J.M., Sonnenschein, M., Muller, M., Welsch, H., Rautenstrauch, C. (Eds.), *Information Technologies in Environmental Engineering.* Springer Verlag, Berlin Heidelberg, pp. 525–536.

Wei, S.K., Gnauck, A. 2007b. Game theoretic approaches to model water conflicts on a river basin scale. In: Gnauck, A. (Ed.), *Modellierung und Simulation von Ökosystemen.* Shaker Verlag, Aachen, pp. 22–40.

Wei, S., Yang, H., Abbaspour, K., Mousavi, J., Gnauck, A. 2010. Game theory-based models to analyze water conflicts in the middle route of the South-to-North water transfer project in China. *Water Research,* 44, 2499–2516.

Wright, G. 1999. Interbasin water transfers: The Australian experience with the snowy mountains scheme. In: *Proceedings of the International Workshop on Interbasin Water Transfer,* 25–27 April, 1999. UNESCO, Paris, pp. 101–105.

Wilson, M.C., Li, X.-Y., Ma, Y.-J., Smith, A.T., Wu, J. 2017. A review of the economic, social, and environmental impacts of China's south–north water transfer project: A sustainability perspective. *Sustainability* 9, 1489.

Yan, D.H., Wang, H., Li, H.H., Wang, G., Qin, T.L., Wang, D.Y., Wang, L.H. 2012. Quantitative analysis on the environmental impact of large-scale water transfer projects on water resource area in a changing environment. *Hydrol. Earth Syst. Sci.* 16, 2685–2702.

Yang, Z.F., Zeng, Y. 2004. Mathematical model for water conflict and coordination in transboundary regions. *Acta Sci. Circums.* 24(1), 71–76.

Yang, H., Zehnder, A. 2005. The South-North water transfer project in China: An analysis of water demand uncertainty and environmental objective in decision making. *Water Int.* 30 (3), 339–349.

Yevjevich, V. 2001. Water diversions and interbasin transfers. *Water Int.* 26(3), 342–348.

Zafarnejad F. 2010 Aug 2. Funeral of the largest lake of the country. *Hamshahri.* Available from http://www .hamshahri.org/news-112867.aspx. Accessed 14 Nov 2010. In Persian.

Zarghami, M. 2011. Effective watershed management: Case study of Urmia Lake, Iran. *Lake Reserv. Manage.* 27, 87–94.

Zarghami M, Hassanzadeh Y, Babaeian I, Kanani R. 2009. Climate change and water resources vulnerability; Case study of Tabriz City, Socio-economic and Natural Sciences of the Environment (SENSE). In: Symposium on Climate Proofing Cities, 1 Dec 2009. Amsterdam.

30 A Compact Policy to Combat Water Scarcity

Chandrashekhar Bhuiyan

CONTENTS

30.1 INTRODUCTION

Water is the elixir of life. Therefore, the sustainability of life and proliferation of civilization is heavily dependent on water resources. History tells us that most ancient civilizations were established and flourished along major rivers such as the Tigris and Euphrates in Mesopotamia, the Nile in Egypt, the Indus and the Ganga in India, the Danube in Europe, and the Huang-He and the Yangtze in China because of easy access and a continuous supply of water. History also reveals that many civilizations were wiped out or migrated due to shifting or dying of rivers, such as the Indus Valley civilization of India. Thus, scarcity of water has posed a direct threat to civilizations across time. Unfortunately, at present, the balance of demand and supply of water resources is jeopardized leading to water scarcity in different parts of the world due to an unprecedented rise in human population followed by rapid industrialization, multi-cropping practice, and uncontrolled urbanization. Ongoing climate change has led to heatwaves, reduction of rainy days and rainfall, and frequent droughts, which have further aggravated the situation. As a consequence of acute water

DOI: 10.1201/9781003276548-30

scarcity, the environment, ecosystems, and human societies are struggling to maintain equilibrium and harmony.

Water scarcity is defined as the inadequacy of water caused due to physical shortage and/or institutional mismanagement in water distribution. Physical water scarcity refers to the inadequacy of water resources due to natural reasons (mainly climate and hydrogeology) as are the cases of Sudan, Yemen, Senegal, Venezuela, Ethiopia, Tunisia, and Cuba. Conversely, water scarcity also develops due to lack of infrastructure or mismanagement in many developing as well as developed countries, such as Afghanistan, Algeria, Australia, Brazil, Canada, China, Cyprus, Egypt, India, Iran, Mexico, Pakistan, Spain, Sri Lanka, and Uzbekistan. In fact, with time and with so-called progress in civilization, gradually more and more places and countries are coming into the grief of water scarcity. Many important cities across the globe, such as Cape Town, Perth, Chennai, Mexico City, and Los Angeles, have witnessed acute water scarcity. It is feared that Perth in Western Australia might become the world's first abandoned ghost metropolis due to acute and persistent water scarcity. Cape Town in South Africa is also waiting for its "day zero" when no water will be left to drink. If water scarcity is not eradicated, many cities will meet the same tragic fate in the near future. The following facts and figures further portray an authentic picture of the present and future states of a water-scarce world.

1. At present, over 2 billion people live in countries that suffer from acute water stress (UN 2018).
2. It is estimated that by 2040, nearly 600 million children under 18 (every one in four) will live in areas of extreme water stress (UNICEF 2017).
3. By the year 2030, 700 million people worldwide could be displaced by intense water scarcity (Hameeteman 2013).
4. Two-thirds of the global population (~4 billion people) and half of the world's large cities are affected by water scarcity, which is going to be the biggest threat in the days ahead (Mekonnen and Hoekstra 2016).
5. Under the ongoing climate change scenario, 24 million to 700 million people of arid and semiarid regions will be displaced by 2030 (UN 2009).
6. One-third of the world's large groundwater systems are already in distress (Richey et al. 2015).
7. By 2050, one-half of the global population (4.8–5.7 billion) will face water scarcity at least for a month. About 73% of the affected people will live in Asia (Burek et al. 2016).

In this background, attainment, restoration, and maintenance of equilibrium between water demand and supply in a sustainable manner, under natural climate variability as well as long-term climate change are big challenges but crucial to local viability and vitality (Morehouse 2000). The challenge is indeed many folds, as there is little choice or capability in controlling population rise, urbanization, industrialization, climate change, and water demand. The immediate need of the hour is to arrest the rise in water demand, delinking water consumption from economic development, and subsequent enhancement of water resources and reserves. These measures need proper planning, strategy building, and policy making aimed at real-time water resources assessment, enhancement, and management. In this chapter, a framework of a compact policy to combat water scarcity is proposed. Causes and impacts of water scarcity, policy development, and their systematic implementation to achieve a water-sufficient system and drought-resilient society are discussed.

30.2   WATER SCARCITY: CAUSES AND IMPACTS

Water scarcity may develop due to natural reasons, inappropriate human practices, or a combination of both. It is a complex problem that may originate due to climatic reasons, but ultimately evolves into a management problem. Whatever might be the cause(s), impacts of water scarcity are diverse, affecting domestic, agricultural, and industrial domains leading to socioeconomic unrest.

30.2.1 CAUSES

There is a subtle difference between drought and aridity, although both are associated with lack of precipitation. Aridity refers to consistent low precipitation, which is the climatic characteristic of a place, whereas drought develops due to an abrupt occurrence of low precipitation due to anomalous climatic behavior. Water scarcity may develop at drought-prone areas, in arid or hyperarid climatic zones, and also in humid and non-drought zones. There is a difference in the nature of water scarcity resulting due to aridity, drought, and mismanagement of water resources.

The primary cause of physical water scarcity is inadequate precipitation. Since precipitation is naturally low in hyperarid, arid, and semiarid regions, water bodies and water supply systems that are dependent on precipitation are extremely sensitive to precipitation variations. The sensitivity of water resources is comparatively less in humid and subhumid regions. However, precipitation is not the sole governing factor for water scarcity. For example, Meghalaya, a state in India, receives the highest rainfall in the world but suffers from water scarcity owing to rapid surface runoff and inadequate recharge of aquifers. The reasons behind high surface runoff and insufficient groundwater recharge are impervious soil, steep land–surface slope, deforestation, and rapid change in land cover. Because of this, Meghalaya is also referred as the "wettest desert in the world". Other important causes of water scarcity include delayed arrival and early retreat of rain, temperature extremes, overextraction of groundwater, rapid urbanization and industrialization, multicropping agricultural practice, cultivation of crops having high water requirements, and unscientific irrigation practices. Overextraction of groundwater during recurrent droughts is found to lower water tables and cause water scarcity (Zektser et al. 2005, Bhuiyan et al. 2009).

30.2.2 IMPACTS

Arid and hyperarid regions receive very low precipitation, and, therefore, ecosystems and people in such regions are naturally acclimatized, adopted, and habituated to living with less water. On the contrary, people living in semiarid to humid regions are adapted to living with adequate to excess water. Therefore, when intense, prolonged, or frequent droughts reduce water availability, people of such regions suffer severely due to abrupt water scarcity.

Impacts of water scarcity are many folds and develop in different spheres simultaneously or in a chain reaction. The immediate impact is visible in the forms of, for example, dried-up water bodies (ponds, lakes, streams, springs, waterfalls, and wells), dry cracked crop fields, low water reserves in dams and barrages. Long queues of people and containers in front of public bore wells, tube wells, and water taps is a daily routine and well-known scene in water-scarce regions (Figure 30.1). Acute water scarcity leads to disrupted water supply for homes, irrigation, and industry. In many parts of India, water scarcity has created unimaginable problems, such as social unrest, human trafficking, prostitution, looting, riots, and mass migration (Bhuiyan 2017, Bandyopadhyay et al. 2020). Several states of India have witnessed large-scale farmers' suicide on a regular basis for which acute water scarcity is one of the reasons (Kale et al. 2014, Hardikar 2017). Similarly, mass migration of people from the Bundelkhand region of Central India has become a summer ritual due to acute water scarcity (UNDP 2012). In rural areas, women are at the receiving end for bearing the primary responsibility to fetch water. In Asia and Africa, in particular, women are worst affected both physically and mentally by water scarcity (Sony 2002). The infamous "water-wife" concept in rural Maharashtra is a burning example (Blakemore 2015, Siddiqui 2015).

30.3 WATER-SCARCE SOCIETY

In combating water scarcity, it is crucial to assess the severity, duration, and frequency of drought and water scarcity in the concerned region. The first step is the assessment of regional vulnerability,

FIGURE 30.1 Empty containers in queue for water in rural Gujarat (India) in April 2007.

probability and risk of water scarcity, and their causes. While vulnerability is inherent in a region, the probability of water scarcity is governed mainly by climatic factors (see Figure 30.2).

30.3.1 VULNERABILITY ASSESSMENT

Vulnerability assessment of a system is important since it plays a critical role in the relationship between hazards and society (Wilhelmi and Wilhite 2002). Although arid and hyperarid regions naturally receive very little rainfall, a region with high rainfall amounts but erratic rainfall patterns are more vulnerable to water scarcity in comparison to a region with lower but consistent rainfall (Bhuiyan 2017). However, precipitation or rainfall is not the sole governing factor for water scarcity. Absence of active surface water bodies (streams, lakes, ponds, springs, and waterfalls), low yield and productivity of aquifers, high population and population density, and skewed water supply compared to demand also increase the vulnerability of water scarcity. Vulnerability of a region to water scarcity is governed by and can be assessed by examining the (1) number and capacity of active surface water bodies; (2) productivity of wells and/or discharge of springs and streams; (3) availability and adequacy of water resource; (4) mean groundwater draft; (5) mean recharge of aquifers and storage in reservoirs; and (6) average water demand compared to availability in the region

Water scarcity vulnerability assessment is and should be carried out through field survey along with the aid of accessory data and information on water resources, reserves, water supply, demography, water extraction, water utilization patterns, and water demand. Such assessment should be seasonal (summer, rainy season, winter, spring) and ground verification is required at least once a year, preferably during the summer. Mapping and microzoning for water scarcity will be extremely useful for identifying the most vulnerable zones for subsequent planning and implementation of remedial measures.

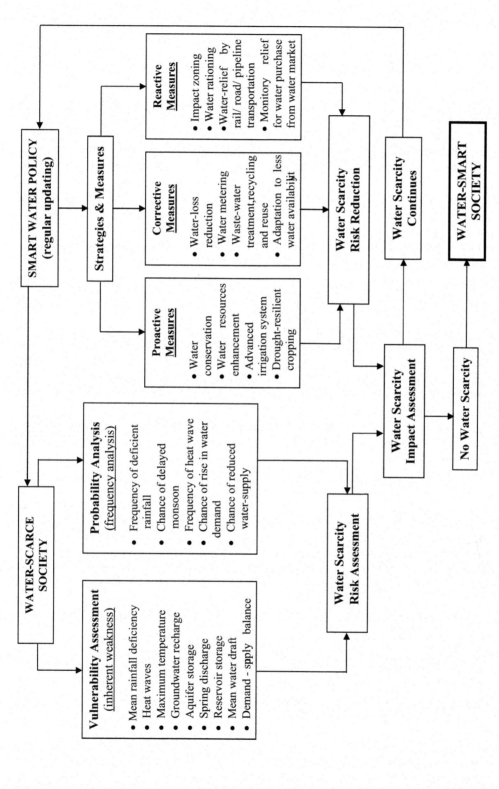

FIGURE 30.2   The flowchart representing the proposed policy for transforming a water-scarce society into a water-smart society.

30.3.2 Probability and Risk Assessment

Probability of development of water scarcity at a region increases with (1) deficiency in rainfall; (2) frequency of deficient rainfall; (3) high temperature; (4) frequency of heat waves; and (5) late arrival and/or early retrieval of monsoon. Since the availability of water resources is governed mainly by precipitation, the probability of occurrence of water scarcity is also dependent on rainfall. The high probability of abrupt low rainfall, and heatwaves leading to drought increases the chance of water scarcity. A region that suffers from recurrent or intense or prolonged droughts also suffers from a perennial water scarcity problem. Frequency of deficient rainfall (i.e., the gap between actual and required rainfall for replenishment of water bodies and aquifers) is also a key factor governing the water scarcity risk. A region of inadequate water resources has a higher frequency and greater chance of suffering from acute water stress. The other important factor determining water-scarcity risk is the sensitivity of a region and its population to rainfall deficiency compared to normal rainfall. A region with a high probability of very low rainfall (hyperarid to semiarid climatic zones) is more prone to water scarcity but is also well adapted to survive with less water. The probability and risk of water scarcity are also linked with factors governing the intrinsic vulnerability of a region (listed in Section 30.3.1). Therefore, for water scarcity risk assessment, these facts and factors should be considered together and not in isolation since water scarcity risk is a result of the combined influence of climatic, hydrological, and socioeconomic parameters.

30.4 A POLICY TO COMBAT WATER SCARCITY

Every policy is framed with the aim of reduction of vulnerability and risk, and enhancement of resilience. Although many countries have a national drought policy, there is hardly any policy dedicated to combat or mitigate water scarcity. Since water scarcity may develop even without drought, it needs a different policy and specific planning. Just as a physician diagnoses a disease by examining symptoms and causes, and prescribed medicines accordingly, so a policy-making body needs to identify the causes of water scarcity in a region through ground verification of water resource infrastructure, water stress severity, impacts, and future risk of water scarcity. The primary goal of such a policy would be transformation of a water-deficient region into a water-sufficient region through appropriate strategies including reactive, corrective, and proactive measures, and adaptation to a lesser quantity of water. Such a policy framework is presented in Figure 30.2.

30.4.1 Reactive Measures

Reactive measures help in crisis management and are implemented on war footing during an emergency. However, reactive measures do not ensure risk reduction since these can provide only short-term relief from the concerned problem (Bandyopadhyay et al. 2020). There are limited reactive measures to manage water scarcity, and restoration of adequate water supply by any means is the only way out.

30.4.1.1 Water Rationing

Water rationing involves the distribution of limited water resources to ensure minimum and essential use. Water rationing may be carried out through controlled water supply to household taps at a definite time for a definite duration, or a fixed quantity of water distribution from community water tanks. Water rationing can only prolong the combat against the water crisis; it is not a solution to the water scarcity problem. To ensure a permanent or long-term solution to water scarcity, corrective and proactive measures must be implemented.

30.4.1.2 Water Relief

Water relief is the ultimate measure to reduce the impact of water scarcity when the system has failed and the water supply is interrupted or disrupted. This is very common in many Asian, African, and

Latin American countries that suffer from recurrent drought, and little water is available in streams, canals, tanks, and wells. Water is transported through railways or roadways or piped networks to the places suffering from water scarcity and then distributed among the residents of the water-scare villages and towns. Again, water relief offers only a short-term relief to the immediate crisis and does not erase or reduce water scarcity vulnerability or risk of the society.

30.4.2 CORRECTIVE MEASURES

Corrective measures reduce the impact of water scarcity by minimizing water loss and prolonging the water supply. Water loss could be due to natural reasons or due to faulty human practices. While arresting the untapped water prevents loss of fresh water, recycling wastewater for irrigation and industrial uses may ensure the conservation of fresh water for drinking purpose.

30.4.2.1 Water-Loss Reduction

A significant amount of water is wasted unutilized both naturally and also due to human negligence. For example, Meghalaya state of India hosts the villages Mawsynram and Cherrapunji, which receive the highest (11,873 mm) and the second-highest (11,777 mm) annual rainfall in the world. People in major parts of the state despite receiving extremely high rainfall (state average: 11,500 mm) suffer from acute water scarcity and walk a long distance to fetch drinking water. In this case, the reasons are quick rainfall runoff due to steep slopes and impervious soil. Similar are the cases in most intermittent and ephemeral streams at mountainous terrains where stormwater quickly runs down through the stream channels. The cumulative volume of untapped stormwater running down the dry channels is large enough to recharge the underlying aquifers. Again, in cities and towns, low infiltration of rainwater and small recharge of aquifers are due to urbanization and concretization. With capturing this untapped water, the severity of water scarcity could be reduced. Although unaccounted, a significant amount of water is also lost regularly through evaporation from reservoirs on impounded rivers. Also, people living in towns and big cities, who receive continuous water supply in house taps, free of cost, have little perception about water scarcity and are least sensitive to wastage of water. To combat water scarcity, water loss and wastage must be prevented through use of smart technology, awareness building, and penalization on violation of norms.

30.4.2.2 Water Metering

In the USA, Canada, Australia, as well as in most of the developed countries in Europe, proper utilization of water is ensured, as well as wastage of water is checked through water metering. People of those countries are well aware of the value and cost of water and are sensitive to wastage of water. Unfortunately, in most other countries, water is a natural resource without any price. Therefore, if and whenever available, its use and misuse are rampant. This causes water scarcity, ultimately leading to interruption in the water supply. Therefore, water metering and water pricing are essential for proper utilization of water and to make people adapted to the limited water supply. This will also make people realize that water is the most essential and most precious natural resource.

30.4.2.3 Water Recycling and Reuse

Reuse of wastewater requires physical, chemical, and biological treatment to remove impurities, pollutants, and harmful microbes. The very first step of water recycling is a physical examination and biochemical analysis of the wastewater. Without proper identification and quantification of the unwanted and harmful materials, treatments plants for wastewater cannot be designed. Wastewater treatment makes impure and polluted water reusable and is carried out probably all over the world. Posttreatment analysis of wastewater is needed to determine the possible use of treated water. Proper monitoring of the adverse effect, if any also should be monitored, assessed, and retreated, if possible. If there is no risk of any adverse effect on ecosystems and human life, the treated water may be recycled and reused. Treated wastewater is mostly used for irrigation and industrial purposes, and are seldom used for drinking purpose.

30.4.3 Proactive Measures

Although reactive and corrective measures can provide temporary relief, these are not sufficient to combat water scarcity. With the progress of time, policies need a pragmatic shift from a reactive, crisis management approach to a proactive, risk management approach (Wilhite 2000). Therefore, a dedicated policy to eradicate water scarcity must include visionary strategies and proactive measures (Bandyopadhyay et al. 2020). Proactive measures as discussed next, have been found useful in minimizing drought impacts and may be effective in reducing the intensity of water scarcity.

30.4.3.1 Water Conservation

With the growing population, industrialization, and eccentricity of climate, water conservation is essential to combat drought and water scarcity. Water conservation aims to increase the storage of excess and/or untapped water. Flowing water in streams and rivers is stored through dams, barrages, and canals. Excess water also can be stored in reservoirs and tanks. However, aboveground storage of water has many disadvantages such as high evaporation loss, the chance of contamination, occupancy of land, costly infrastructure, and adverse environmental impacts particularly in the riparian ecosystems of the impounded rivers. Underground storage of water, also referred to as subsurface water banking, has several advantages over aboveground water storage (Pyne 1995, Bouwer 2002, De Vries and Simmers 2002, Dillon 2005, Gale 2005, Maliva 2014). These include availability of underground aquifers, nonrequirement of over-ground space or expensive storage structures, no evaporation loss, autofiltration and purification in the soil and underlying rock formations, and little chance of contamination. However, site selection for groundwater banking through artificial recharge of aquifers requires detail geological and geophysical survey and is crucial to achieve success in water recovery at the time of need (Bhuiyan 2015).

30.4.3.2 Water Resources Enhancement

Water scarcity cannot be eradicated without enhancing water resources and increasing water reserve facilities. There are numerous techniques and structures to increase surface water as well as groundwater reserve. Appropriate techniques have to be applied and suitable water storage structures constructed depending upon the source of water, infrastructure, prospect, and consequences. In general surface water bodies, such as ponds, lakes, and streams, are naturally abundant in rural areas. Village communities may be trained in the upgradation of existing natural water bodies and construction of new structures such as check dams, percolation tanks, ponds, and dug wells (Figure 30.3). At suitable locations, augmentation of streams and construction of check dams, ditch and furrow structures, trenches may also be useful (Hofkes and Visscher 1986, CGWB 2007). These techniques have been widely used in different parts of the world, and have ensured success in combating water scarcity (Mutiso 2002, Petry et al. 2002, Nissen-Petersen 2006).

Contrary to villages, most of the modern towns and cities are concretized. Thus, water resources enhancement and storage in the urban area needs different approaches and techniques. Rooftop rainwater harvesting is the easiest and one of the most widely used techniques for replenishing aquifers. However, channeling rainwater into the underground soil or rock formation has a high chance of subsurface leakage and lateral movement across the aquifer and, therefore, has high uncertainty of recovery at the time of need. Aquifer storage and recovery (ASR) involves the injection of water into aquifers through wells, and its withdrawal at the time of need. The technique has been found to enhance groundwater resources significantly (Pyne 1995, Bower 2002, Dillon 2005) and is instrumental in combating water scarcity. At appropriate locations and situations, strategies and techniques such as runoff capturing, and dew harvesting could also be effective for water resources enhancement. Desalinization of sea water is another effective technique that can meet

FIGURE 30.3 Self-help: villagers in Gujarat (India) are digging a community well in 2014 to ensure water in the dry summer ahead.

the huge water demand in the agricultural and industrial sectors. This in turn will reduce the stress on aquifers if desalinated water is transported through pipeline or railways from coastal regions to peninsular regions, and will make a significant impact in the war against water scarcity. In spite of high expenditure and adverse environmental risk, both developed and developing countries have now started using desalinated water to meet growing water demand. At present, approximately 16,000 desalination plants are operational across 177 countries, which generate about 95 million m^3/day of fresh water (Jones 2019).

30.4.3.3 Risk Reduction

The risk of water scarcity and its adverse impacts on society can be reduced through proper planning, proactive strategy, preventive and corrective measures, and their timely execution. With proper implementation of the policy, a water-deficient region can be transformed into a water-sufficient region. Risk-reduction strategies and measures could be proactive as well as corrective. Proactive strategies aim at conservation and enhancement of water resources, and recycling of wastewater. Corrective measures ensure minimization of water loss by advanced irrigation facility, cultivation of low water-budget crops, crop rescheduling, and water metering in urban areas. With the reduction of vulnerability, the risk of water scarcity reduces automatically, which subsequently will enhance the resilience of the system and the society.

30.4.3.4 Adaptation to Lesser Water

Human behavior has a direct bearing on water resources management and water scarcity mitigation. Education, upbringing, family practice, and awareness play important roles in water use and conservation. Emphasis is to be given through school education on practicing water conservation, minimizing water loss, and proper use of water resources. Regular practice will make people adapt to living with a lesser amount of water. Decoupling economic growth and water consumption is an important step to achieve this goal, which is already proven by a developed country like Australia. Other countries need to learn the lesson and get adopted to using less water. In other words, less is the new optimum.

30.5 POLICY IMPLEMENTATION

Timely and step-by-step implementation of the policy framework is key to combating water scarcity, and to transforming a water-scarce community into a water-smart society through assessment of vulnerability, impact, and risk of water scarcity, and eradication of water-scarcity impact and risk through different proactive, corrective, and reactive measures (Figure 30.4). Strategies and measures and their order of implementation will vary from place to place (hyperarid, semiarid, or subhumid climatic zones; hard-rock areas or coastal zones; rural or urban areas, etc.), at various seasons and events (summer or winter; drought or flood), and for different types of water scarcity (sudden or prolonged; occasional or frequent, etc.). For example, rural villages contain natural water bodies (ponds, lakes, and streams), which are unlikely in an arid urban setup. Again, in peninsular regions, a water resource is obtained mainly from dug wells and tube wells, whereas at hill stations, springs may be the main source of drinking water. In coastal regions, desalinated water may be the primary source of irrigation water. Therefore, water-scarcity mitigation in different regions involves different plans, measures, and techniques.

The success of a policy depends not only on the policy framework but also on its timely and systematic implementation. Policy making is not an easy task, and a correct and effective policy to combat water scarcity can be framed involving scientists, engineers, managers, government officers, and legislators. A policy becomes efficient and effective through its regular assessment, revision, and upgradation. Implementation of a policy becomes easier with the involvement of nongovernmental organizations (NGOs) and common people. NGOs in particular can be instrumental in creating public awareness through public announcements, door-to-door campaigns, seminars, and field activities. When water scarcity occurs, it affects the whole community or society. Therefore, to combat water scarcity, all-around active participation of all stakeholders is crucial and essential.

30.6 DISCUSSION AND CONCLUSION

Studies and trend analysis have projected a sharp 40% rise in water demand compared to supply by the year 2030, leading to severe to extreme water scarcity in different parts of the world (Boltz 2017). Water scarcity is feared also to be intense with urbanization, industrialization, and economic development. The situation demands immediate attention and actions, beginning with outcome-based decisive policies that involve proper strategy and various proactive, reactive, and corrective measures as discussed in this chapter. A smart policy emerges not only due to the incorporation of effective measures but also with its sincere and timely implementation. People and society play the most crucial roles in curbing problems such as water scarcity since a water-centric policy can be implemented successfully by sincere and enthusiastic participation and cooperation of the people who suffer due to recurrent water scarcity (Mudrakartha 2007). Decoupling economic growth from water use is an important step required to transform water deficiency to water sufficiency. This is already proven possible by some countries, such as Australia where the economy grew by more than 30% in spite of the decline in water consumption by 40% from 2001 to 2009 (UNEP 2016). Enforcement of law in the time-bound implementation of holistic water management plans and policies covering both the natural water cycle (precipitation–runoff–infiltration–evapotranspiration) and man-made water cycle (extraction–distribution–consumption–treatment–recycling–reuse–return to the environment) may be instrumental and effective toward this achievement.

Technology, in this regard, can help significantly in implementing a policy. More accurately, without the incorporation of advanced technology, it is impossible to implement a smart policy to combat water scarcity. Technological innovations such as wireless sensor networks can efficiently

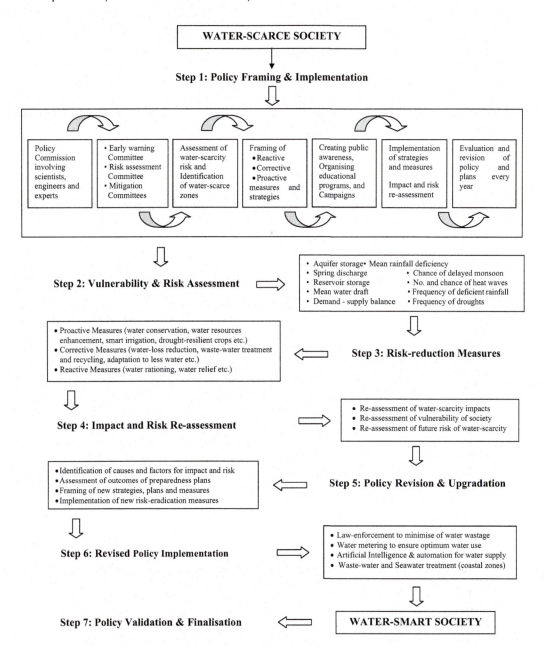

FIGURE 30.4 Step-by-step implementation of the policy to combat water scarcity.

detect leakage in water pipes and valves, and thus can help in minimizing water loss due to leakage. Artificial Intelligent systems and devices may be used extensively for automatic harvesting of rainwater, drip irrigation, water metering, water quality monitoring, detection of pollution, NS timely water supply, as well as the early warning on probable water scarcity. Internet of Things (IOT) and data analytics (DA) may be useful for automated water conservation and distribution through analysis and optimization of user consumption behavior (Merchanta et al. 2013, Savio and Chakraborty 2021). In fact, in the coming days, the application of IOT- and DA-based systems will be the key instruments in smart cities for water resources enhancement, conservation, distribution, recycling, and management.

This is to emphasize that vulnerability and risk of a system can be reduced through the implementation of appropriate policies. Although most countries in the world have some sort of disaster management policy, there is hardly any dedicated policy to curb water scarcity. However, recurrent droughts and acute water scarcity across the globe due to galloping rise in population, urbanization, multicropping agricultural practices, industrial development, and changing climate will force nations to frame and implement focused and dedicated policies to combat water scarcity. The sooner this is done, the better it will be.

REFERENCES

Bandyopadhyay, N., Bhuiyan, C., Saha, A.K. 2020. Drought mitigation: Critical analysis and proposal for a new drought policy with special reference to Gujarat (India). *Progress in Disaster Science* 5, 100049.

Bhuiyan, C. 2015. An approach towards site selection for water banking in unconfined aquifers through artificial recharge. *Journal of Hydrology*, https://doi.org/10.1016/j.jhydrol.2015.01.052

Bhuiyan, C. 2017. Chapter 1: Drought vulnerability. In: S. Eslamian, F. Eslamian (eds.), *Handbook of Drought and Water Scarcity*. Volume 1, Taylor and Francis, CRC Press, Boca Raton.

Bhuiyan, C., Flügel, W.A., Singh, R.P. 2009. Erratic monsoon, growing water demand, and declining water table. *Journal of Spatial Hydrology* 9(1), pp. 1–9.

Blakemore, E. 2015. Water wives: Men in India marry extra women to fetch them water. *Smithsonian Magazine*, June 2015, https://www.smithsonianmag.com/smart-news/water-wives-men-india-marry-extra-women-fetch-them-water-180955511/ (Retrieved 20 November 2020).

Boltz, F. 2017. "How do we prevent today's water crisis becoming tomorrow's catastrophe?" In: World Economic Forum, 23 March 2017. (Retrieved 15 November 2020).

Bouwer, H. 2002. Artificial recharge of groundwater: Hydrogeology and engineering. *Hydrogeology Journal* 10, pp. 121–142.

Burek, P., Satoh, Y., Fischer, G., Kahil, T., Jimenez, L.N., Scherzer, A., Tramberend, S., Wada, Y., Eisner, S., Flörke, M., Hanasaki, N., Magnuszewski, P., Cosgrove, W., Wiberg, D. 2016. *Water Futures and Solution Fast Track Initiative*. Working Paper-WP-16-006, Final Report, Laxenburg, Austria.

CGWB. 2007. *Manual on Artificial Recharge of Ground Water*, Central Ground Water Board, Ministry of Water Resources, Govt. of India, New Delhi, 198p.

De Vries, J., Simmers, I. 2002. Groundwater recharge: An overview of processes and challenges. *Hydrogeology Journal* 10(1), pp. 5–17.

Dillon, P.J. 2005. Future management of aquifer recharge. *Hydrogeology Journal* 13(1), pp. 313–316.

Gale, I. 2005. Strategies for Managed Aquifer Recharge (MAR) in semi-arid areas. UNESCO IHP. 34p. www.iah.org/recharge

Hameeteman, E. 2013. *Future Water (In) Security: Facts, Figures, and Predictions*. Global Water Institute, pp. 15.

Hardikar, J. 2017. With no water and many loans, farmers' deaths are rising in Tamil Nadu. *The Wire*. 21/JUN/2017 (Retrieved 15 October 2020).

Hofkes, E.H., Visscher, J.T. 1986. *Artificial Groundwater Recharge for Water Supply of Medium-size Communities in Developing Countries*. International Reference Centre for Community Water Supply and Sanitation, The Hague, The Netherlands.

Jones, E., Qadir, M., van Vliet, M.T.H., Smakhtin, V., Kang, S. 2019. The state of desalination and brine production: A global outlook. *Science of the Total Environment* 657, pp. 1343–1356.

Kale, N.M., Khonde, S.R., Mankar D.M. 2014. Socio-economic, psychological and situational causes of suicides of farmers in Vidarbha region of Maharashtra. *Karnataka Journal of Agricultural Science* 27, pp. 40–46.

Maliva, R. 2014. Groundwater banking: Opportunities and management challenges. *Water Policy* 16(1). https://doi.org/10.2166/wp.2013.025

Mekonnen, M.M., Hoekstra, A.Y. (2016). Four billion people facing severe water scarcity (PDF). *Science Advances* 2(2). American Association for the Advancement of Science. https://doi.org/10.1126/sciadv.1500323 (Retrieved 13 November 2020).

Merchanta, A., Mohan-Kumar, M.S., Ravindra, P.N., Vyas, P., Manohar, U. 2013. Analytics driven water management system for Bangalore city. 12th International Conference on Computing and Control for the Water Industry, CCWI2013, *Procedia Engineering* 70, pp. 1137–1146.

Morehouse, B.J. 2000. Climate impacts on urban water resources in the southwest: The importance of context, *Journal of the American Water Resources Association* 36(2), pp. 265–277.

Mudrakartha, S. 2007. To adapt or not to adapt: The dilemma between long-term resource management and short-term livelihood. In: Giordano M., Villholth KG (eds.), *The Agricultural Groundwater Revolution: Opportunities and Threats to Development*, CAB International, Wallingford, UK.

Mutiso, S. 2002. The significance of subsurface water storage in Kenya. Seminar Proceedings, Wageningen, The Netherlands, 18–19 December 2002, Netherlands National Committee for the IAH and Netherlands Hydrol. Society, Amsterdam, 29 pp

Nissen-Petersen, E. 2006. *A Handbook on Water from Dry Riverbeds*. English Press, Nairobi, Kenya, 68p.

NITI Aayog-UNDP. 2012. *Human Development Report - Bundelkhand 2012*.

Petry, B., Van Der Gun, J., Boeriu, P. 2002. Coping with water scarcity a case history from Oman. Management of aquifer recharge and subsurface storage. In: IAH Seminar Wageningen, The Netherlands, 18–19 December 2002, pp 55–65.

Pyne, R.D.G. 1995. *Groundwater Recharge and Well: A Guide to Aquifer Storage and Recovery*. CRC Press, Boca Raton.

Richey, A.S., Thomas, B.F., Lo, M.-H., Reager, J.T., Famiglietti, J.S., Voss, K., Swenson, S., Rodell, M. 2015. Quantifying renewable groundwater stress with GRACE. *Water Resources Research* 51, pp. 5217–5238, https://doi.org/10.1002/2015WR017349.

Savio, I., Chakraborty, U. 2021. An IOT based water management system for smart cities. In: Bhuiyan C, Flügel WA, Jain SK (eds.), *Water Security and Sustainability. Lecture Notes in Civil Engineering*, Springer, Cham.

Siddiqui, D. 2015. Drought-hit Maharashtra village looks to 'water wives' to quench thirst. https://in.reuters.com/article/india-waterwives-maharashtra-idINKBN0OK1CI20150604 (Retrieved 12 November 2020).

Sony, J. 2002. Gender dimension of water scarcity: Result of a study in "non-source" villages of four districts in Gujarat. In: IWMI-Tata Water Policy Research Programme Annual Partners' Meet 2002, Vallabh Vidyanagar, Gujarat, India.

UN. 2009. *Water in a changing world. UN World Water Development Report March 2009*, UN, New York.

UN. 2018. *Sustainable Development Goal 6: Synthesis Report 2018 on Water and Sanitation*, UN, New York.

UNEP. 2016. *Half the World to Face Severe Water Stress by 2030 unless Water Use is 'Decoupled' from Economic Growth, Says International Resource Panel*, United Nations Environment Programme Report, 21st March 2016. (Retrieved 15 November 2020).

UNICEF. 2017. Thirsting for a Future: Water and children in a changing climate. UNICEF Publication. https://www.unicef.org/publications/ (Retrieved 20 October 2020).

Wilhelmi, O.V., Wilhite, D.A. 2002. Assessing vulnerability to agricultural drought: A Nebraska case study. *Natural Hazards* 25, pp. 37–58.

Wilhite, D.A. 2000. Chapter 1: Drought as a natural hazard: Concepts and definitions. In: Wilhite, D.A. (ed.), *Drought: A Global Assessment, Natural Hazards and Disasters Series*, Routledge Publishers, London.

Zektser, S., Loáiciga, H.A., Wolf, J.T. 2005. Environmental impacts of groundwater overdraft: Selected case studies in the southwestern United States, *Environmental Geology* 47, pp. 396–404.

31 Water Pricing during Drought Conditions

Saeid Eslamian and Mousa Maleki

CONTENTS

31.1 INTRODUCTION

Water pricing mechanisms are generally seen as one of the most important instruments for water demand management in the context of overextraction of water. The use of water pricing is frequently proposed as a strategic tool for water policy, such as in the Water Framework Directive (European Commission, 2000b) and the Blueprint to Safeguard Europe's Water Resources (European Commission, 2012). Moreover, water pricing is seen by many environmental organizations as a social issue, and even the agricultural subsidies linked to the European Common Agricultural Policy depend on water pricing and the implementation of cost-recovery strategies.

Water pricing is an efficient and effective instrument to manage water resources and associated services, such as improving allocation and increasing revenue. However, water pricing can have many objectives that often conflict. Economic efficiency related to resources improves water allocation but does not consider social affordability or financial and environmental sustainability (Bazrkar et al. 2015). Cost recovery guarantees financial sustainability for service providers but might not be affordable for the public or efficient with respect to resources. The ability of the poor to access water is a social concern. Water shortages, pollution costs, and ecosystem values are reflected in an environmental target. Consequently, water pricing reflects trade-offs among these objectives (OECD, 2010), although it is politically most difficult to implement policies that promote equity, efficiency, and sustainability in the water sector (Dinar, 2000; Rogers, de Silva, and Bhatia, 2002).

31.2 WATER PRICE DEFINITION

A water price represents the unit value of water of specific quality at a certain point of time, hence it varies across quality grades and over time. Optimal prices reflect prices along the optimal water policy. Why should we be concerned with calculating water prices? Can't we just rely on water markets to provide this information? The main problem with water markets is that they are far from ubiquitous and, where they do exist, are prone to fail for various reasons, including (1) common pool externality (when pumping/diverting water from a shared aquifer, reservoir, or stream); (2) returns

DOI: 10.1201/9781003276548-31

to scale associated with the water infrastructure (which constitutes a considerable share of the cost of water allocation and is shared by many users); (3) supply uncertainty (due to stochastic precipitation); and (4) dependence on water ownership rights, allocation rules, and norms. Moreover, water markets take a long time to form and operate properly in any given circumstance, hence rarely exist in young and/or rapidly changing water economies. Absent properly operating water markets, water allocation must be regulated and such regulation is based on one way or another on water prices (Tsur, 2019).

The literature on irrigation reveals a wide range of water pricing methods as well as practical challenges to modifying water prices. Important contributions in this area include Johansson et al. (2002), Tsur et al. (2004), and Tsur (2005). For example, Johansson et al. (2002) provide a comprehensive review of theoretical and practical issues regarding the pricing of irrigation water. In this regard, various methods for allocating irrigation water were reviewed and factors such as physical, social, political, and institutional settings were identified as having an important impact on pricing policies used in different countries. Tsur et al. (2004) emphasized that demand management should be a central point in contemplating water pricing policies to promote the efficient use of water, while Tsur (2005) discussed the economic aspects of irrigation water pricing with an emphasis on supply management and the implication for policy development.

Traditionally, the concept of the full cost of irrigation water was conceptualized simply as the financial costs of irrigation provisions. However, it has been recognized that irrigation may induce costs to other economic actors, including those who gain utility from the environment. The contemporary conceptualization of the full cost of irrigation thus includes consideration of the resource cost, and environmental cost of water provision, although there is some conjecture about the demarcation between each. This approach to the full cost of water was discussed in detail by Rogers et al. (1998), and has subsequently been widely supported by a number of scholars and practitioners (see, for example, Rogers et al., 2002; Ward and Pulido-Velazquez, 2009; Bithas et al., 2014).

Figure 31.1, modified from Rogers et al. (1998), depicts the elements of the full cost of water and its related services. It shows schematically the various components that comprise the full cost. There are three main concepts: (1) the full supply cost, which involves operation and maintenance (O&M) costs and capital costs; (2) the full economic cost, representing the sum of the full supply cost and the opportunity cost associated with the alternate use of the water resource; and (3) the full cost being the combination of the full economic cost and the environmental externalities or public health and ecosystem impacts of water use.

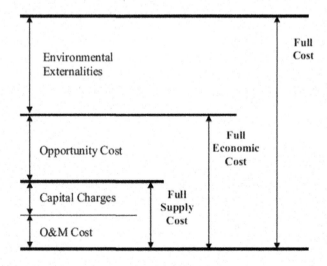

FIGURE 31.1   General principles for cost of water. (Modified from Rogers et al., 1998.)

31.3 WATER PRICING, AN IMPORTANT POLICY FOR DROUGHT MANAGEMENT

The design and implementation of policy instruments to reach desirable outcomes need quantitative support. The Dutch economist Jan Tinbergen (1955) formulated a theory of economic policy that identified the required set of policy instruments to best reach a set of targets. In that work, he recognized that some economic quantities are targets and others are policy instruments. Targets are the economic outcome variables that the policy maker would like to influence. Instruments are the variables over which the policymaker has control. Tinbergen made the important insight that achieving the desired values of a known number of targets requires the policy maker to control an equal or greater number of instruments.

Access to more instruments than targets widens the range of choices for reaching the targets. For this study, an important application of Tinbergen's framework for economic policy is the identification of instruments and targets. Control over the pricing structure for irrigation water is an instrument. Important targets include economic efficiency in the allocation of scarce water, affordable access to irrigation water for family food security, and financial sustainability of the water pricing system. Since there is only a single policy instrument considered here, it will be generally impossible for a single water pricing program to hit all three targets.

Pricing is an important policy instrument in the world's irrigated regions. In those areas, regional water managers face the challenge of managing water-use patterns to raise the productivity of irrigated agriculture (Huang et al., 2009, 2010). The price of irrigation water, where such a price exists, can provide an important signal that guides its use patterns (Griffin, 2001). Where implementing affordable access to enough irrigation water for farm family food subsistence needs is an important social goal, access to that irrigation water will not be limited by water users' ability to pay (UNDP, 2006).

Setting the price of irrigation water needs to achieve a number of goals. A number of recent studies conclude that water prices should recover the full costs of supply (e.g., Brooks, 2006; Kostas, 2008; Rogers et al., 2002). So, water pricing arrangements need to be carefully designed so that water is affordable to the water user while also being financially sustainable for the supplier. Numerous recent published works suggest that the price of irrigation water should promote more widespread access to water for all users (Abu-Zeid, 2001; Ruijs et al., 2008; Whiteley et al., 2008). Volumetric pricing is one important signal that rewards water users who avoid using high-cost water for low-valued uses as they adapt to changing water scarcity (Dinar and Mody, 2004; Easter and Liu, 2005; Tsur, 2005; Ward, 2007). However, the costs of implementing volumetric pricing can be high (Tsur, 2005). Moreover, the more conventional pricing method in which prices are set to the average cost of supply suffers from the widely recognized failure to reflect water's underlying economic scarcity. Average cost pricing (ACP) causes special problems with revenue adequacy in drought periods when supplies fall off without a reduction in the total costs of supply (Nikouei, 2012). Users have less water to buy, so revenues also fall, making it difficult for the supplier to cover costs.

A less commonly used water pricing method is marginal cost pricing (MCP). MCP can improve economic efficiency in agricultural water use (Griffin, 2001; Le Gal et al., 2003; Tardieu and Préfol, 2002). However, MCP can undermine meeting subsistence water needs for food security when the marginal cost of supply rises considerably during all-too-common severe or sustained droughts that occur in the dry parts of the world (Dinar and Mody, 2004; Howe, 2005). The use of two-tiered pricing (TTP) is one recognized method to achieve efficient water pricing as well as securing the subsistence needs for food (Easter and Liu, 2005; Tardieu and Préfol, 2002; Ward and Pulido-Velazquez, 2008). Under this arrangement, the price established at the first tier for subsistence needs can be set at an affordable level. However, the price charged at the second tier for discretionary (nonsubsistence) uses increases in the face of rising water scarcity from drought or climate change (Abu-Zeid, 2001; Tsur, 2005). While the dividing line between subsistence and discretionary use is never clear, smaller landholders typically use their first few parcels of land for food grain security in order to meet subsistence calorie needs. Additional land, if available, is more commonly used to secure income from commercial production for sale to local or export markets.

An analysis of irrigation water pricing reform by Bar-Shira et al. (2006) showed that a TTP arrangement for large-scale farming operations can reduce aggregate water use while having little effect on small farms for which the main use of water is to protect farm family food security. Chohin-Kuper et al. (2003) examined financial cost-recovery objectives in several Mediterranean countries and found similar results. Several recent studies concluded that setting up a TTP arrangement for irrigation water use in a shared water system offers considerable potential to address both economic efficiency while protecting food security (Barberán and Arbués, 2009; Garcia-Valiñas, 2005; Garcia and Reynaud, 2004; Tardieu and Préfol, 2002).

Adopting TTP for irrigation water achieves the aim of securing access to water for subsistence irrigation food production. TTP differentiates between water to support the right to subsistence food production (food security) and the use of water for commercial irrigation. To meet the food security policy objective, the provision of water for subsistence food production would be set at an affordable price, while water would be priced at its marginal cost for its commercial use in irrigated agriculture. Agriculture, the largest user of freshwater resources, encompasses both subsistence farming and commercial use (larger-scale farming). So, the policy challenge is to find an effective pricing mechanism that supports the right to subsistence food production while accurately reflecting the scarcity of water and also recovering the financial costs of the water supplier.

The aforementioned literature has made important advances in the analysis of irrigation water allocation and pricing systems that meet one or more of the three objectives described. Despite these contributions, we have found few studies that examined irrigation water allocation and pricing arrangements that could meet or at least address all three objectives. In light of these gaps, the objective of this chapter is to examine the potential impacts of establishing a TTP arrangement for farm water uses in order to secure greater economic efficiency, more affordable access to irrigation water for the production of food staple subsistence, and financial sustainability in water use. Financial sustainability is defined for the purposes of this chapter as having revenues equal to or exceeding costs for water for the indefinite future.

31.4 VALUATION AND PRICING OF AGRICULTURAL IRRIGATION WATER BASED ON MACRO AND MICRO SCALES

Water is an indispensable natural resource for human survival and development. The continuous development of socioeconomic water demand, water resource shortages, degradation of the water environment, and water security issues are becoming increasingly serious. In stark contrast, the price of water is generally low, which neither objectively reflects the scarcity of water resources nor contributes to solutions for water resource problems (Ren et al., 2018).

The typical methods of pricing include calculating shadow prices for various types of water resources using mathematical models (Shen and Lin, 2017), using the full cost of water resources (Xian et al., 2014), or using farmers' willingness to pay for water (Motta and Ortiz, 2018) as a reference for pricing. As relevant research has progressed, the value of water resources has attracted attention. It was proposed by Jiang et al. (1993) that the essence of value was in the capitalization of water resource rent, upon which the differences in price and value of water resources would be clarified. Scholars have estimated the value of water resources by employing equilibrium pricing (Gan et al., 2012), mosaic of values (Hermans et al., 2008), emergy estimation (Lv et al., 2009), and fuzzy comprehensive evaluation (Cai et al., 2012). In addition, there is clear theoretical and empirical evidence that using the price to manage water demand is more cost-effective than other approaches (Olmstead and Stavins, 2007; Olmstead et al., 2007). Even in the absence of water metering, an incentive water pricing instrument could influence farming behavior toward more efficient use of water (Lika et al., 2017). Therefore, the rational pricing of agricultural irrigation water has become the focus (Bar-Shira et al., 2010; Singh, 2007; Esmaeili and Vazirzadeh, 2009; Ohadi and Nejad, 2014).

31.4.1 DETERMINATION OF MACRO-AGRICULTURAL IRRIGATION WATER PRICE

As early as 1931, Leontief carried out the first economic analysis with input–output tables and mathematical models (Raa, 2009). In this study, we determined the macro price based on an input–output analysis. As shown in Table 31.1, the input–output table for Heilongjiang Province in 2012 was divided into three parts. Part I reflects the intersectoral relationships within an economy (x_{ij}, where i is the sector that inputs the product and j is the sector that uses the product), part II reflects the final demand for the production of sector i, and part III reflects the constitution of the benefit of sector j. An additional row shows the amount of water used in sector j in the study area in 2012 (W2012j), which is an input to part I and completes the water resource input–output table (see Table 31.1). The input–output computable general equilibrium (CGE) model of water resources (Equations 31.1 and 31.2) was obtained via linear programming, and its dual solution was used to determine the shadow price of agricultural water resources, or the macro-agricultural irrigation water price:

$$\max Z = \sum_{i,j=1}^{10} a_{vj} X_i \tag{31.1}$$

$$S.t. \begin{cases} AX_j + Y_i - IM_i = X_i \\ X_i - CX_j \geq V_j \\ X_i^l \leq X_i \leq X_i^h \\ Y_i^l \leq Y_i \\ V_j^l \leq V_j \leq X_i \\ \sum_{i,j=1}^{10} a_{wj} X_i \leq W \end{cases} \tag{31.2}$$

where Z is the total surplus (10^8 yuan) and W is the total amount of water consumed (10^8 m^3). For sector i, Y_i is the final demand and Y_i^l is its lower bound (10^8 yuan), IM_i is import (10^8 yuan), X_i is total output, and X_i^h and X_i^l are its upper and lower bounds (10^8 yuan). For sector j, S_j is surplus (10^8 yuan), $a_{vj} = S_j / X_i$ is the profit ratio coefficient, $a_{wj} = W_j^{2012} / X_i$ is the direct water consumption coefficient, V_j is the added value and V_j^l is its lower bound (10^8 yuan), X_j is the total input (10^8 yuan), $A = [aij], aij = xij / Xj$ is the direct consumption coefficient matrix, and $C = A^T$ is the material consumption coefficient matrix.

31.4.2 DETERMINATION OF MICRO-AGRICULTURAL IRRIGATION WATER PRICE

The European Environment Agency (EEA) uses the principle of full-cost recovery to evaluate water prices in the European Union (European Commission, 2000a; Anderson and Gaines, 2009; Kampas et al., 2012). The full-cost water price is widely believed to best reflect the commercialization of water resources (OECD, 2003). Under the guidance of China's basic national conditions and the theory of sustainable development, Wang proposed that the full-cost water price should be composed of resource cost, project cost, environmental cost, and reasonable profit and tax revenue (Wang et al., 2003). For agriculture, profit and tax revenue are not counted. The water pricing framework in China was constructed in Shen and Wu (2017), clearly showing the category of water price (resource charge, services charge, environmental charge). Based on the theory, we used resource cost (Cr), project cost (Cp), and environmental cost (Ce) to determine the full-cost water price (P) and represent the micro-agricultural irrigation water price in this chapter. Emergy theory was applied to estimate the cost of natural resources that are not transformed by human labor (e.g., natural rainfall,

TABLE 31.1

Water Resource Input–Output Table of Heilongjiang Province in 2012 (in 10^8 yuan)

Input	Output	1	2	3	4	5	6	7	8	9	10	Total	Final Demand (Y_j)	Import (IM_j)	Total Output (X_j)
Intermediate Input	1 Agriculture	387.2	1.1	1611.5	15.5	0.1	0.1	7.0	88.6	0.1	0.0	2111.3	2861.4	1020.3	3952.3
	2 Mining	7.3	261.5	41.6	1015.2	200.2	362.6	88.1	4.9	12.3	18.0	2047.7	2736.1	1368.2	3415.6
	3 Light industry	599.2	14.5	649.6	51.3	18.6	2.4	23.9	181.7	21.4	118.6	1681.3	4344.2	1713.7	4311.8
	4 Petrifaction	455.0	163.0	162.7	727.5	159.6	108.8	97.8	175.9	303.0	297.6	2651.1	1374.8	1205.3	2820.6
	5 Manufacturing	58.1	162.2	63.7	33.8	1239.8	130.4	1415.4	84.9	116.7	96.5	3401.4	4048.7	4555.6	2894.4
	6 Power	30.0	97.4	83.5	50.3	101.8	208.1	26.6	163.1	55.0	62.2	878.0	388.8	8.5	1258.4
	7 Architecture	5.3	1.7	3.8	2.5	4.2	1.0	138.1	59.9	5.8	17.0	239.3	5030.4	2339.7	2930.0
	8 Business	208.0	149.5	317.6	160.0	207.7	56.5	183.6	713.2	253.6	213.1	2463.0	2360.3	275.4	4547.8
	9 Traffic posts	80.5	103.7	190.3	78.2	149.9	33.4	135.5	134.3	304.7	88.3	1298.7	719.4	76.6	1941.6
	10 Services	8.1	54.4	37.5	20.1	48.2	24.2	17.1	67.5	48.6	107.8	433.4	3522.9	1090.7	2865.6
	Total	1838.6	1009.0	3161.6	2190.4	2130.2	927.5	2133.0	1674.3	1121.2	1019.2	17,205.0	27,386.9	13,653.8	30,938.1
$W_j^{2012}(10^8\ m^3)$		294.9	2.1	1.2	1.1	1.4	9.8	10.5	4.1	1.9	4.3	331.2			
Added value	Remuneration	1884.8	343.0	217.3	163.7	269.4	167.9	379.2	525.3	275.4	1192.0	5417.9			
	Net product tax	−157.7	696.2	435.3	183.2	169.8	67.2	185.7	452.4	45.2	30.0	2107.3			
	Depreciation	84.6	280.5	117.9	78.1	138.2	89.8	63.1	359.6	147.2	190.1	1549.1			
	Surplus (S_j)	302.0	1086.9	379.7	205.2	186.8	6.2	169.0	1536.3	352.5	434.3	4658.8			
	Total (V_j)	2113.7	2406.6	1150.2	630.2	764.2	330.9	797.0	2873.6	820.4	1846.4	13,733.1			
Total input (X_j)		3952.3	3415.6	4311.8	2820.6	2894.4	1258.4	2930.0	4547.8	1941.6	2865.6	30,938.1			

Sources: Ren et al. (2018). Economic data are from the China Region Input-Output Table 2012 for Heilongjiang Province, pages 94–105. Water consumption for each sector is from the China Statistical Yearbook 2012, Second Heilongjiang Economic Census Yearbook, and Second China Economic Census Yearbook.

surface water, groundwater). The cost of water supply for engineering of state farms was used to estimate project cost, and environmental cost was determined by economic loss due to water environment degradation as accounted for by econometric theory.

The specific steps are as follows:

Step 1: Determine resource cost.

The resource cost (C_r) reflects the natural value of water resources. The emergy theory proposed by H.T. Odum (1996) provides objective and quantitative criteria for the scientific evaluation of natural resource value, which can truly reflect the cost of natural resources. In this research, we chose the emergy theory to determine the natural value of water resources, integrating the solar emergy in water with currency in reality, so that the natural value of water resources can be shown in the form of price. That is, rainfall was used as the emergy source of natural water resources, and a conversion between emergy and currency was completed by EDR (the ratio of emergy to gross domestic product, GDP) to calculate the resource cost:

$$CE_r = W_r \times G \times \rho, CE_s = W_s \times G \times \rho, CE_g = W_g \times G \times \rho$$

$$SE_r = CE_r \times r_r$$

$$SE'_s = SE_r / W_s, \quad SE'_g = SE_r / W_g \qquad (31.3)$$

$$C'_{rs} = SE'_s / EDR, \quad C'_{rg} = SE'_g / EDR$$

where CE_r, CE_s, and CE_g are the chemical emergy of rainfall, surface water, and groundwater (J), respectively; W_r, W_s, and W_g are the total amount of rainfall, surface water, and groundwater (10^8 m^3), respectively; G is the Gibbs free emergy value (J/g); ρ is the density of water (g/cm^3); SE_r is rainfall solar emergy (sej); r_r is rainfall solar transformity (sej/J) (Odum, 1996); SE'_s and SE'_g are solar emergy per unit surface water and groundwater (sej / m^3), respectively; C'_{rs} and C'_{rg} are the resource cost per unit of surface water and groundwater (yuan/m^3), respectively; and EDR is the ratio of emergy to GDP (sej/yuan) (Tan, 2012).

Step 2: Determine project cost.

The project cost (C_p) was determined by the reasonable expenses incurred from agricultural water supply engineering. In this study, we use a cost accounting method to calculate project cost, including depreciation cost (D_p), overhaul expense (R_p), employee work reward (L_c), employee welfare (F_1), union funds (F_2), and employee education funds (F_3), using Equation 31.4:

$$C_p = (C_p + D_p + L_c + F_1 + F_2 + F_3) / (EIA \times IWQ)$$

$$D_p = \frac{K_0 - K_1}{n} \qquad (31.4)$$

$$K_0 = 85\% \times I, K_1 = 4\% \times K_0$$

where K_0 and K_1 are the fixed assets of the original value and net residual value (10^8 yuan), respectively; values for R_p, F_1, F_2, and F_3 are from "Measures of water supply pricing cost supervision and examination of water conservancy project (Trial)"; n is the number of years depreciated; EIA is the mean of the effective irrigation area of state farms (10^4 hm^2); IWQ is the irrigation water quota (10^4 m^3/hm^2); and I is the fixed assets investment (10^8 yuan).

Step 3: Determine the environmental cost.

In this chapter, the environmental cost (C_e) was determined by economic losses due to water environment degradation. Based on econometric theory (Jiang et al., 2015), economic loss caused by the use of pesticides, fertilizers, and plastic mulching during the cultivation process was determined by Equation 31.5:

$$C_e = \frac{U}{W_a}$$

$$U = \sum_{k=1}^{6} l S_u P Q_k \gamma_k$$

$$l = 1 / \left(1 + e^{3 - \frac{1}{E_n}} \right), S_u = W_d / \lambda W_s \tag{31.5}$$

$$\lambda = \sqrt{\sum_{t=1}^{4} \left[(E_t / E) - 1 \right]^2 / 4}$$

where U is economic loss due to degradation of the water environment in rural areas (10^8 yuan); W_a is the average amount of agricultural irrigation water consumption (10^8 m^3); $k = 1, 2, ..., 6$ refers to water quality classifications; l is the rural socioeconomic development stage coefficient; S_u is the water scarcity index; P is the current agricultural water price (yuan/m^3); Q_k is the amount of water in the specified water quality class (10^8 m^3); γ_k is the ecological function loss rate of different water quality; E_n is Engel's coefficient for rural areas; W_d and W_s are the amount of water demand and water supply (10^8 m^3), respectively; l is the rainfall variation coefficient; and E_t and E are the ratios of t quarter rainfall and mean quarterly rainfall to annual rainfall, respectively.

Step 4: Determine the full-cost water price.

We regarded the sum of the resource cost, project cost, and environmental cost as the full-cost water price as follows:

$$P = C_r + C_P + C_e \tag{31.6}$$

31.5 DRINKING WATER PRICING POLICIES

Several European member states have completed (fully or partially) the development of appropriate water pricing policies according to WFD 2000/60/EC toward the implementation of full water cost recovery. More specifically, 25 out of the 27 (except for Portugal and Greece) have completed this obligation (Kanakoudis and Tsitsifli, 2015). In Greece, the implementation of the WFD 2000/60/EC requirements is being monitored following the implementation of 14 River Basin Management Plans (RBMPs), highlighting problems that occurred and weaknesses identified. The Greek government launched in June 2016 the 14 RBMPs (1st revision) public consultation process, along with the draft Joint Ministerial Decision (JMD) regarding the "Adoption of general water pricing rules and services, methods and procedures for full water cost recovery in several water uses". This JMD is called to incorporate WFD 2000/60/EC requirements, covering all uses of water, including methodologies and tools for the estimation of full water cost, including continuous monitoring and gradual improvement measures of public water services, along with guidelines for a socially fair water pricing policy.

Regarding, the level of environmental and resource costs integration into pricing policies through economic instruments varies not only by member state (DANVA, 2014), but also by region or river basin district (European Environment Agency, 2013). The price of water and wastewater prices per household differs also between member states (Figure 31.1). In Switzerland, for example, water prices are almost five times higher than in Bulgaria. One of the basic parameters is the differences in GDP between countries. The International Benchmarking Network for Water and Sanitation Utilities (IBNET) indicates the differences in water tariffs at a global level. The provision of comparative information and its use in benchmarking is an important management tool for managers and professionals in water and sanitation utilities (IBNET, 2017).

Even in more developed countries, the price of water differs by region, partly because of structural differences and partly due to political and strategic differences. A comparative analysis of water tariffs based on a consumption of 15 m³ per month (average tariffs per country weighted by population served), taking into consideration the GDP per capita of several countries, also highlights also the differences (Figure 31.2). It should be mentioned that the price of water in some countries is about a hundred times more expensive than others (i.e., South Korea compared to Cape Verde).

The type of charging differs also in several countries. In Denmark, for instance, a number of water utilities have chosen to charge a fixed annual charge for water and/or wastewater and a price per cubic meter for water consumed, while others charge only for the water used (DANVA, 2014). Additionally, although water pricing structures in selected European countries (England and Wales, Scotland, Netherlands, France, Germany, Slovenia, Croatia, Serbia, Spain) are a combination of fixed and volumetric charge (European Environment Agency, 2013), in a number of OECD countries, the structure of prices for public water services are volumetric oriented rather than fixed (Organisation for Economic Co-operation and Development, 2010). In Greece, a fixed charge is also used additionally to a volumetric water pricing model. On the other hand, several countries, such as Hungary, Poland, and Czech Republic, have already adopted water pricing policies based only on volumetric pricing with a trend in moving toward increasing block tariffs.

A recent study (Kanakoudis et al. 2014) showed that there is no common pricing policy among Greek water utilities. Each water utility charges different fees and tariffs on their water bills. The mean payable amount does not display great variation between low and high consumption, while high consumption and water wasting are not discouraged (Kanakoudis et al. 2014).

Looking at the nexus between all specific parameters, reliable metering systems of water consumption are also a precondition for the application of efficient water pricing policies. Safeguarding

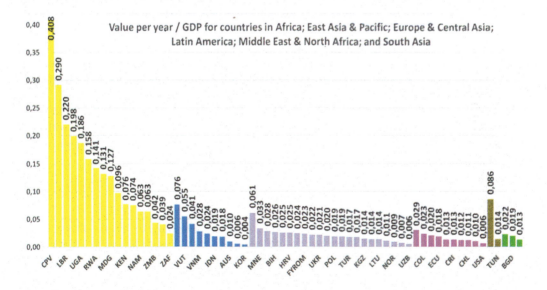

FIGURE 31.2 Drinking water tariffs in relation to GDP for a 15 m³ monthly water use for the year 2015 (average tariffs per country weighted by population served). (The colors refer to Africa, yellow; East Asia and Pacific, blue; Europe and Central Asia, purple; Latin America, magenta; Middle East and North Africa, olive; and South Asia, green) (Tsitsifli et al., 2017).

both transparency and fairness of water pricing policies based on reliable water metering and improved cost-benefit assessments to ensure cost recovery is one of the basic principles of the European Commission.

In conclusion, the development of "appropriate water tariffs" is influenced by a number of factors, such as local characteristics, different geological and climatic parameters, and different institutional and regulatory frameworks (Kanakoudis et al., 2014). In this respect, water utilities should adopt a more strategic approach that could be cost-effective and could be used to signal water scarcity and to create incentives for efficient domestic water use. The overall objective is to implement an appropriate water pricing policy in Greece, better designed tariff structures and targeted measures, to implement European Union guidelines regarding Full Water Cost (FWC) recovery.

31.6   CONCLUSION AND RECOMMENDATIONS

Anything scarce and in demand commands a price; this is one of the basic principles of economics. Water is scarce in some contexts (drought, degraded quality), so water pricing is increasingly seen as an acceptable instrument of public policy. Water-use charges, pollution charges, tradable permits for water withdrawals or release of specific pollutants, and fines are all market-based approaches that can contribute to making water more accessible, healthier, and more sustainable over the long term.

One particular area of water policy that has become increasingly subject to pricing principles is that of public water supply and wastewater services. Efficient and effective water pricing systems provide incentives for efficient water use and for water quality protection. They also generate funds for necessary infrastructure development and expansion, and provide a good basis for ensuring that water services can be provided to all citizens at an affordable price. The metering of water consumption is a prerequisite for the application of efficient water pricing policies. About two-thirds of OECD member countries already meter more than 90% of single-family houses, although universal metering remains a controversial issue in some contexts.

Selective metering is less controversial, particularly if the public knows that new water resources are scarce, or if the metering applies to discretionary water use, like private swimming pools. Metering new homes is also more widely accepted than converting older ones.

In terms of the structure of prices for public water services, there is a clear trend in OECD countries away from fixed charges and toward volumetric charging; in other words, the more you use, the more you pay. Even where fixed charges still exist, the policy of allowing large free allowances is declining. Hungary, Poland, and the Czech Republic, for example, already use pricing systems based solely on volumetric pricing, with no fixed charge element at all.

To encourage conservation, the trend in volumetric charging is also moving away from decreasing block tariffs and toward increasing block tariffs. This means that the charge increases with each additional unit of water used or wastewater treated, rather than providing discounts to high-volume users.

REFERENCES

Abu-Zeid, M., 2001. Water pricing in irrigated agriculture. *Int. J. Water Resour. Dev.* 17 (4), 527–538.

Anderson, K.M., Gaines, L.J., 2009. International water pricing: An overview and historic and modern case studies. In *Managing and Transforming Water Conflicts*, Cambridge University Press, New York, pp. 249–265.

Assessment of Cost Recovery Through Water Pricing. Technical Report No16, European Environment Agency.

Barberán, R., Arbués, F., 2009. Equity in domestic water rates design. *Water Resour. Manage.* 23 (10), 2101–2118.

Bar-Shira, Z., Finkelshtain, I., Simhon, A., 2006. Block-rate versus uniform water pricing in agriculture: An empirical analysis. *Am. J. Agric. Econ.* 88 (4), 986–999.

Bar-Shira, Z., Finkelshtain, I., Simhon, A., 2010. Block-rate versus uniform water pricing in agriculture: An empirical analysis. *Am. J. Agric. Econ.*, 88, 986–999.

Bazrkar, M.H., Zamani, N., and Eslamian, S., 2015, *Evaluation of Socioeconomic Impacts of Urban Water Reuse Using System Dynamics Approach, Urban Water Reuse Handbook*, Ch. 28, Ed. By Eslamian, S., Taylor and Francis, CRC Group, pp. 331–340.

Bithas, K., Kollimenakis, A., Maroulis, G., Stylianidou, Z., 2014. The water framework directive in Greece. Estimating the environmental and resource cost in the water districts of western and Central Macedonia: Methods, results and proposals for water pricing. *Procedia Econ. Fin.* 8, 73–82. doi:10.1016/S2212-5671(14)00065-3

Brooks, D.B., 2006. An operational definition of water demand management. *Int. J. Water Resour. Dev.* 22 (4), 521–528.

Cai, C.; Huang, T.; Li, X.; Li, Y., 2012. Application of fuzzy maths in urban water resources value: A case study of water resources value in Cheng Du region. *Adv. Mater. Res.* 361–363, 1571–1575.

Chohin-Kuper, A., Rieu, T., Montginoul, M., 2003. Water policy reforms: Pricing water, cost recovery, water demand and impact on agriculture. Lessons from the Mediterranean experience. In Water Pricing Seminar (June 30–July 2, 2003). Agencia Catalana del Agua & World BANK Institute, Barcelona.

Dinar, A., 2000. Political economy of water pricing reforms. In Dinar, A. (Eds.), *The political economy of water pricing reforms*. Washington, DC: Oxford University Press, pp. 1–26.

Dinar, A., Mody, J., 2004. Irrigation water management policies: Allocation and pricing principles and implementation experience. *Nat. Resour. Forum* 28 (2), 112–122.

Easter, K.W., Liu, Y., 2005. *Cost Recovery and Water Pricing for Irrigation and Drainage Projects, Agriculture and Rural Development Discussion Paper 26*. The International Bank for Reconstruction and Development/The World Bank, Washington, DC.

Esmaeili, A., Vazirzadeh, S., 2009. Water pricing for agricultural production in the south of Iran. *Water Resour. Manag.* 23, 957–964.

European Commission (EC), 2000a, Directive 2000/60/EC of the European parliament and of the council. *Off. J. Eur. Communities* L327, 12–13.

European Commission, 2000b. Water framework directive. Directive 2000/60/EC of the European Parliament and of the Council of 23 October 2000 establishing a framework for Community action in the field of water policy. Official Journal L 327, 22/12/200, P.0001–0073. Available at: http://eur-lex.europa.eu/legalcontent/ EN/TXT/?uri=celex:32000L0060. Accessed 5 May 2016.

European Commission, 2012. Communication from the commission (COM(2012)673): A blueprint to safeguard Europe's water resources. *European Commission*, Brussels. Available at: http://ec.europa. eu/environment/water/blueprint/index_en.htm. Accessed 5 May 2016.

Gan, H., Qin, C.H., Wang, L., Zhang, X.J., 2012. Study on water pricing method and practice I. Discussion on the connotation of water resources value. *J. Hydraul. Eng.* 39, 289–295. (In Chinese).

Garcia, S., Reynaud, A., 2004. Estimating the benefits of efficient water pricing in France. *Resour. Energy Econ.* 26 (1), 1–25.

Garcia-Valiñas, M.A., 2005. Efficiency and equity in natural resources pricing: A proposal for urban water distribution service. *Environ. Resour. Econ.* 32 (2), 183–204.

Griffin, R.C., 2001. Effective water pricing. *J. Am. Water Resour. Assoc.* 37 (5), 1335–1347.

Hermans, L.M., Halsema, G.E.V., Mahoo, H.F., 2008. Building a mosaic of values to support local water resources management. *Water Policy* 8, 415–434.

Huang, Q.Q., Rozelle, S., Wang, J.X., Huang, J.K., 2009. Water management institutional reform: A representative look at Northern China. *Agric. Water Manage.* 96 (2), 215–225.

Huang, Q.Q., Wang, J.X., Easter, K.W., Rozelle, S., 2010. Empirical assessment of water management institutions in Northern China. *Agric. Water Manage.* 98 (2), 361–369.

Howe, C.W., 2005. The functions, impacts and effectiveness of water pricing: Evidence from the United States and Canada. *Int. J. Water Resour. Dev.* 21 (1), 43–53.

Jiang, Q.X., Zhu, C.H., Fu, Q., Wang, Z.L., Zhao, K., 2015. A study on green GDP of Heilongjiang province based on cost accounting of water resources value. *Water Saving Irrig.* 11, 80–84. (In Chinese).

Jiang, W.L., Yu, L.S., Liu, R.H., Han, G.G., Wang, H.D., 1993. Study on the price upper limit of water resources. *China Water Wastewater* 2, 58–59. (In Chinese).

Johansson, R.C., Tsur, Y., Roe, T.L., Doukkali, R., Dinar, A., 2002. Pricing irrigation water: A review of theory and practice. *Water Policy* 4, 173–199. doi:10.1016/s1366-7017(02)00026-0

Kampas, A., Petsakos, A., Rozakis, S., 2012. Price induced irrigation water saving: Unraveling conflicts and synergies between European agricultural and water policies for a Greek Water District. *Agric. Syst.* 113, 28–38.

Kanakoudis, V., Tsitsifli, S., 2015. River basin management plans developed in Greece based on the WFD 2000/60/EC guidelines. *Desal. Wat. Treat.* 56 (2015), 1231–1239.

Kanakoudis, V., Papadopoulou, A., Tsitsifli, S., 2014. Domestic water pricing in Greece: Mean net consumption cost versus mean payable amount. *Fresenius Environ. Bull.* 23 (2014), 2742–2749.

Kostas, B., 2008. The sustainable residential water use: Sustainability, efficiency and social equity. The European experience. *Ecol. Econ.* 68 (1–2), 221–229.

Le Gal, P.Y., Rieu, T., Fall, C., 2003. Water pricing and sustainability of self-governing irrigation schemes. *Irrig. Drain. Syst.* 17 (3), 213–238.

Lika, A., Galioto, F., Viaggi, D., 2017. Water authorities' pricing strategies to recover supply costs in the absence of water metering for irrigated agriculture. *Sustainability* 9, 2210.

Lv, C.M., Wu, Z.N., Hu, C.H., 2009. Progress and prospect on theory research of water resource value. *Resour. Environ. Yangtze Basin* 18, 545–549. (In Chinese).

Motta, R.S.D., Ortiz, R.A., 2018. Costs and perceptions conditioning willingness to accept payments for ecosystem services in a Brazilian case. *Ecol. Econ.* 147, 333–342.

Nikouei, A., 2012. *Integrated Economic-Hydrologic Modeling of Water Allocation and Use in Zayandeh-Rud River Basin with Emphasis on Evaluation of Environmental and Drought Policies.* Shiraz University, Shiraz, Iran, 271pp.

Odum, H.T., 1996. Environmental accounting: EMERGY and environmental decision making. *Child Dev.* 42, 1187–1201.

Ohadi, N., Nejad, J.K., 2014. Economic pricing of water in pistachio production of Sirjan. *Int. J. Agric. Manag.* 4, 247–252.

Olmstead, S.M., Hanemann, W.M. and Stavins, R.N., 2007. Water demand under alternative price structures. *J. Environ. Econ. Manag.*, 54(2), 181–198.

Olmstead, S.M., Stavins, R.N., 2007. *Managing Water Demand Price vs. Non-Price Conservation Programs.* Pioneer Institute for Public Policy Research, Boston, MA.

Organisation for Economic Co-operation and Development (OECD), 2003. *Improving Water Management: Recent OECD Experience* (Complete Edition). In Source OECD Development; IWA Publishing, London, UK; ISBN 9264099484.

Organisation for Economic Co-operation and Development (OECD), 2010. *Pricing Water Resources and Water and Sanitation Services*, OECD Studies on Water, OECD Publishing, Paris, https://doi.org/10 .1787/9789264083608-en.

Performance Benchmarking in Water and Sewerage Utilities, 2017. International Benchmarking Network for Water and Sanitation Utilities Database (IBNET), http://www.ib-net.org/, last accessed January 2017.

Raa, T.T., 2009. *Input–Output Economics: Theory and Applications.* Social Science Electronic Publishing, Rochester, NY; Volume 19, p. 568.

Ren, Y., Wei, S., Cheng, K., Fu, Q., 2018. Valuation and pricing of agricultural irrigation water based on macro and micro scales. *Water* 10(8), 1044.

Rogers, P., Bhatia, R., Huber, A., 1998. *Water as a Social and Economic Good: How to Put the Principle into Practice. Global Water Partnership.* TAC Background Paper No. 2. 35 Pages.

Rogers, P., de Silva, R., & Bhatia, R., 2002. Water is an economic good: How to use prices to promote equity, efficiency, and sustainability. *Water Policy* 4, 1–17.

Ruijs, A., Zimmermann, A., van den Berg, M., 2008. Demand and distributional effects of water pricing policies. *Ecol. Econ.* 66 (2–3), 506–516.

Singh, K., 2007. Rational pricing of water as an instrument of improving water use efficiency in the agricultural sector: A case study in Gujarat, India. *Int. J. Water Resour. Dev.* 23, 679–690.

Shen, D., Wu, J., 2017. State of the art review: Water pricing reform in China. *Int. J. Water Resour. Dev.* 33, 198–232.

Shen, X., Lin, B., 2017. The shadow prices and demand elasticities of agricultural water in China: A StoNED-based analysis. *Resour. Conserv. Recycl.* 127, 21–28.

Tan, C., Lv, J., 2012. Scenario prediction for emergy of Heilongjiang eco-economic system. *For. Econ.* 4, 39–42. (In Chinese).

Tardieu, H., Préfol, B., 2002. Full cost or "sustainability cost" pricing in irrigated agriculture. Charging for water can be effective, but is it sufficient? *Irrig. Drain.* 51 (2), 97–107.

Tinbergen, J., 1955. *On the Theory of Economic Policy*, second ed. North-Holland, Amsterdam.

Tsitsifli, S., Gonelas, K., Papadopoulou, A., Kanakoudis, V., Kouziakis, C., Lappos, S., 2017. Socially fair drinking water pricing considering the Full Water Cost recovery principle and the Non-Revenue Water related cost allocation to the end users. *Desalin. Water Treat* 99, 72–82.

Tsur, Y., 2005. Economic aspects of irrigation water pricing. *Can. Water Resour. J.* 30, 31–46. doi:10.4296/cwrj300131

Tsur, Y., 2019. Water pricing (November 14, 2019). Available at SSRN: https://ssrn.com/abstract=3486831 or http://dx.doi.org/10.2139/ssrn.3486831

Tsur, Y., Dinar, A., Doukkali, R.M., Roe, T., 2004. Irrigation water pricing: Policy implications based on international comparison. *Environ. Dev. Econ.* 9, 735–755.

UNDP, 2006. Beyond scarcity: Power, poverty and the global water crisis. In: Ross-Larson, B., Coquereaumont, M.D., Trott, C. (Eds.), *United Nations Development Program: Human Development Report 2006.* United Nations Development Programme, New York.

Wang, H., Ruan, B.Q., Shen, D.J., 2003. *Water Price Theories and Practice Facing to Sustainable Development.* Science Press, Beijing, China. (In Chinese).

Ward, F.A., 2007. Decision support for water policy: A review of economic concepts and tools. *Water Policy* 9 (1), 1–31.

Ward, F.A., Pulido-velazquez, M., 2008. Efficiency, equity, and sustainability in water quantity-quality optimization model in the Rio Grande basin. *Ecol. Econ.* 66, 23–37.

Ward, F.A., Pulido-Velazquez, M., 2009. Incentive pricing and cost recovery at the basin scale. *Environ. Manag.* 90, 293–313. doi:10.1016/j.jenvman.2007.09.009

Water in Figures, 2014. *Benchmarking 2014: Process Benchmarking and Statistics*, Danish Water and Waste Water Association (DANVA), Skanderborg, Denmark, p. 1–48.

Whiteley, J.M., Ingram, H., Perry, R. (Eds.), 2008. *Water, Place, and Equity.* The MIT Press, Cambridge, MA; London, UK.

Xian, W., Xu, Z., Deng, X., 2014. Agricultural irrigation water price based on full cost recovery: A case study in Ganzhou District of Zhangye Municipality. *J. Glaciol. Geocryol.* 36, 462–468. (In Chinese).

32 Incorporating Ecosystem Services into Drought Planning
Lessons from Two Place-Based Applications from the Western US

Nejem Raheem and Deborah J. Bathke

CONTENTS

32.1 INTRODUCTION

Severe drought is a recurring problem in the western United States. Since 2000, drought has caused a "billion-dollar weather disaster" in at least part of the West for all but four years (NCEI 2021). These significant economic losses, averaging $8.2 billion a year in damages, are primarily based on US Department of Agriculture (USDA) indemnity payments (Smith & Katz 2013) and do not address damages to natural capital or to ecosystem services (ES), the benefits that people derive from nature (MEA 2003; Tánago et al. 2016).

The close link between human and natural communities means that ecological impacts from drought have serious effects on human communities (Crausbay et al. 2017; Millar & Stephenson 2015). Recent droughts highlight the extent of ecological impacts (NDMC 2021a). For example, prolonged drought and bark beetle infestations caused unprecedented tree die-off in the western United States, creating more hazardous conditions for firefighters, significantly impacting their ability to stop fires and increasing risks to communities. Drought and increased fire danger caused states in the Southwest to close large swaths of state and federal forests during the height of the busy summer season. Drought and heat damaged Christmas tree farms in Oregon, causing a number to close for good. These impacts cascade through other parts of the local economy. Pederson et al. (2006) described how decadal droughts in Glacier National Park, Montana, have affected the availability of water for irrigation, forest-related tourism, and fire control. With droughts projected to get more severe in the West (Dai 2013; Cook et al. 2015; Van Loon et al. 2016), it is clear that drought planning needs to provide clearer instruction on how to address and incorporate ecological impacts of concern, including addressing changes due to climate change and predicted land-use changes. While ecological impacts are often not considered monetarily in estimates of drought effects, they are real; better accounting of these impacts in drought planning may help to reduce the total costs of these events (Multihazard Mitigation Council 2017). Additionally, integrating an environmental economic perspective can also help to more completely account for these costs.

DOI: 10.1201/9781003276548-32

While drought planning research, planning guidance, and drought plans often discuss environmental impacts, they typically do not discuss the interdependence of these impacts with humans (Wilhite 1992; Wilhite et al. 2005; Tánago et al. 2016; McEvoy et al. 2018). Possible explanations for this lack of integration include complexity in relating these impacts directly to human communities or a perceived lack of priority in stakeholders' eyes (Wilhite 1992). Additionally, while extant information on "natural and biological resources" can be found via state and federal agencies, we have found that groups' and individuals' sense of the relative condition and importance of those resources also matters when planning for drought (Raheem et al. 2019). Agency or institutional/regulatory fragmentation is likely another obstacle, as ecological impacts of drought may be addressed in other planning processes (e.g., U.S. Forest Service National Forest Planning) that are rarely integrated with drought planning (McEvoy et al. 2018). Finally, the lack of a shared and comprehensive framework may hinder the holistic inclusion of ecosystems and the full range of benefits of social, cultural, and economic benefits that they provide.

Support in the following three areas should strengthen the integration of ecological impacts into drought planning: (1) structuring and operating stakeholder discussions, (2) ascertaining the salient ecological impacts of a drought in a given region, and (3) integrating knowledge of those impacts into preparedness efforts. One framework that can help address all of these concerns is the ES framework (MEA 2003). This interdisciplinary framework relates outputs of ecological processes to human well-being through a highly structured approach designed to systematically identify and trace the complex links between natural and human systems (MEA 2003; Haines-Young & Potschin 2012). Used to inform natural resource management and environmental policy at various levels (Koontz et al. 2008; Dell'Appa et al. 2015; Schaefer et al. 2015), incorporating this approach into the existing drought planning process can have many benefits.

First, an ES framework can provide a roadmap for drawing up a suite of impacts to consider in any given planning exercise, which can be used widely throughout the drought planning process. The framework allows planners and stakeholders to consider a broad spectrum of impacts, not just those at top of the mind, that may occur when drought damages the environment. Additionally, an ES approach can help to better ascertain the economic impacts of drought, as ES are clearly linked to human values and to economic impacts through a range of methods (Ranganathan et al. 2008). The ES approach allows groups to use existing planning protocols which rely on ES, and can be helpful as a part of a communication or outreach process, as the ES framework is largely anthropocentric, and the impacts resulting from an ES approach should already be salient to individuals or groups in a community.

In this chapter we provide an overview of the ES approach and its use in planning efforts, provide background on drought planning, and describe how drought planning processes typically address ecological impacts. By using two place-based applications from the western United States that used an ES framework to look at landscape-wide impacts, we demonstrate how these findings can enhance drought planning using the widely applied ten-step drought planning process (Wilhite et al. 2005). Finally, we discuss how to move forward to better integrate ecological impacts into drought planning.

32.2   ECOSYSTEM SERVICES APPROACH

Ecosystem services are broadly defined as the benefits that humans derive from nature. Research has led to the development of multiple conceptual frameworks, such as the Common International Classification of Ecosystem Services (CICES) and The Economics of Ecosystems and Biodiversity (TEEB) (Haines-Young and Potschin 2012; Sukhdev et al. 2012) to classify and help reveal the benefits related to the interdependency of human and natural systems. Most frameworks are generally derived from the Millennium Ecosystem Assessment (MEA) and describe four categories: supporting, provisioning, regulating, and cultural (deGroot et al. 2002; MEA 2003; Sukhdev et al. 2018). Table 32.1 presents an overview of the MEA approach, with examples, to categorizing ES.

TABLE 32.1

Classification Categories of Ecosystem Services

Provisioning Services	Regulating Services	Cultural Services
Products Obtained from Ecosystems	Benefits Obtained from the Regulation of Ecosystem Processes	Generally Nonmaterial Benefits People Derive from Ecosystems
Fresh water	Climate regulation	Ecotourism
Food	Disease prevention	Sense of place
Fiber	Pollination	Traditional cultural practices
Medicines	Pest control	Education
Timber	Temperature regulation	Scientific research
	Climate regulation	Creativity

Supporting Services

The services that are necessary for the production of all other ecosystem services: biodiversity, nutrient cycling, primary productivity, photosynthesis.

Source: Adapted from the Millennium Ecosystem Assessment (2005).

Supporting services underpin all other ES; one example is photosynthesis (primary productivity) in grasses, which leads to a suite of other services (e.g., soil retention due to thriving grassland communities). Not all ES frameworks include supporting services, as they are not considered "final" services and do not directly impact humans' values (Haines-Young and Potschin 2012). Provisioning services are directly consumed, and include fresh water and food (e.g., trout fisheries). Regulating services result from mediating processes, and include regulation of hydrologic processes and water quality (e.g., riparian areas trap sediment and reduce their movement into rivers). Cultural services derive from the relationship of spiritual, aesthetic, educational, and recreational (e.g., hiking) values with ecosystems. While cultural ES are often nonmaterial, in some cases they can be material, as when non-timber forest products are used for cultural ceremonies (Sangha and Russell-Smith 2017).

Although ES are not often used in drought planning, they have a history of use in other planning efforts, such as in ecosystem-based management (Slocombe 1993), and in US legislation known as the Farm Bill (Schaefer et al. 2015). Benefit–cost analyses, environmental accounts, conservation planning, and environmental impact assessments have all used ES as a component of their organization (Chan et al. 2006; Pearce et al. 2003, 2006; ONS 2018). An approach called payment for ecosystem services (PES) is used throughout the world to offset the costs of conservation (Dunn 2011).

Using any of the ES frameworks, planning teams can build a local or regional inventory, list, or table for a given landscape. These inventories have been completed for several different areas already (Raheem et al. 2015, 2019; Turkelboom et al. 2014; Sousa et al. 2016). Some ES frameworks, such as the CICES, provide a spreadsheet, which facilitates cataloging and categorizing ES in various cases (https://cices.eu/resources/) (EEU 2021). A simplified version of the CICES v4.0 spreadsheet, adapted for research in Montana (Raheem et al. 2019) is shown in Table 32.2.

32.3 PLACE-BASED APPLICATIONS OF DROUGHT PLANNING AND ES

Drought, in the most basic sense, can be defined as a deficit of expected water availability that results in water shortages for some activity or group. Because droughts are experienced relative to local baselines and sector-specific demands, operational definitions of drought must focus on location-specific characteristics and impacts. Drought definitions traditionally viewed the effects of meteorological drought through the anthropocentric lenses of agricultural, hydrological, and socioeconomic impacts (Wilhite & Glantz 1985). Recently, the concept of "ecological drought" was introduced to better integrate the human and natural dimensions of drought (Crausbay et al. 2017),

TABLE 32.2

Modified CICES V4.0 Spreadsheet Used in Montana Project

Section	Example by Division/Group		
Type of service	Division: type of organism or material (e.g., biomass, water) Group: further subdivision (e.g., cultivated plants, wild plants, raised animals, wild animals)		
Provisioning	Nutrition (elk, beets)	Materials (lodgepole, willows)	Energy (water, fuelwood)
Regulation and maintenance	Mediation of wastes, toxics, and other nuisances (dilution, bioremediation)	Mediation of flows (riparian buffers)	Maintenance of phys/chem/bio conditions (local water temperature regulation)
Cultural	Physical and intellectual interactions with biota, ecosystems, and landscapes/seascapes (hiking, gathering plants)	Spiritual, symbolic and other interactions with biota, ecosystems, and landscapes/seascapes (documentaries, religious activities)	

creating an opportunity to enhance existing planning guidance to more comprehensively address how drought affects ecosystems and the impacts that cascade through communities that depend on these ecosystems for goods and services.

Planning guides have been created to assist local, state, tribal, and national governments develop drought plans (NDMC 2021b). Many of these guides incorporate an adaptable ten-step planning process (Figure 32.1) developed at the National Drought Mitigation Center (Wilhite et al. 2005). Using this ten-step approach as an example, the following paragraphs discuss ways in which the planning process could be supplemented through an ES approach. At a basic level, ensuring an ES approach at the beginning of the planning process should yield benefits all the way through to the end. Incorporating ES will provide (1) better integration of scientific approaches or interdisciplinarity, (2) a greater breadth of ecological detail, (3) a way for stakeholders to work together, and (4) easier communication along the process. For each of the ten steps, we discuss how an ES approach would be beneficial and how it could be implemented. By combining our drought planning experience (New Mexico Drought Task Force 2006; Wickham et al. 2019; Bathke et al. 2019; Bathke & Bernadt 2021) with our ES research (Raheem et al. 2015, 2019), we provide detailed examples when relevant. For the steps where we have not yet been able to conduct research, we discuss the approach and benefits more generally.

32.3.1 RESEARCH CONTEXT

Integrating ES into drought planning can take several forms. In this chapter, we integrate findings from two such projects (Raheem et al. 2015, 2019). One used a group of experts recruited for their expertise and familiarity with the topics, landscape, and culture of what is sometimes called the Upper Río Grande bioregion (Peña 2003). This group drafted a comprehensive table of ES, based on the MEA (2003), that might exist along a particular altitude gradient in northern New Mexico or southern Colorado, using traditional Spanish landform and cultural terminology. This catalog was not based on drought requirements, but instead sought to provide a comprehensive catalog of all possible ES in this landscape to be used as a template for future planning exercises, including drought planning.

The other group convened a larger group of professionals working in water, drought, and landscape planning in the Upper Missouri Headwaters (UMH) basin in southwestern Montana, the location of a National Drought Resilience Partnership (NDRP) demonstration project for innovative

1	Appoint a drought task force or committee
2	State the purpose and objectives of the drought plan.
3	Seek stakeholder input and resolve conflict.
4	Inventory resources and identify groups at risk.
5	Prepare and write the drought plan.
6	Identify research and institutional needs.
7	Integrate science and policy.
8	Publicize the plan and build public awareness
9	Develop education programs.
10	Evaluate and revise the plan.

FIGURE 32.1 The ten-step drought planning process (Wilhite et al. 2005, 2014).

drought planning (White House 2014). The group included regional and watershed drought planners, scientists, and managers from various federal and state land management agencies (e.g., MT
DNRC, US Forest Service), and conservation scientists and practitioners from nonprofits working
in the region (e.g., The Nature Conservancy, the Center for Large Landscape Conservation, the
Wildlife Conservation Society). These participants came from partner organizations involved in
the Montana NDRP demonstration project (Montana Drought Demonstration Partners 2015). This
group's focus was on the ecological impacts of drought. This group organized an ES table using
the CICES (Haines-Young & Potschin 2012). As such, the two place-based applications represent
somewhat different approaches to integrating ES into planning.

32.3.2 DROUGHT PLANNING APPLICATION

Step 1 of the ten-step drought planning process (Figure 32.1) focuses on establishing the operational
and administrative framework for the drought plan. This includes appointing a leadership team
charged with initiating the drought planning process, coordinating plan development, implementing the plan during times of drought, and defining the purpose and scope of the plan (Wilhite et
al. 2005). Task force composition affects the entire planning process. Ensuring that at least one
individual on the task force is fluent with an ES framework can help ensure that the task force better understands and integrates the natural environment and the full scope of social, cultural, and
economic impacts tied to the services it provides. While agencies might not have an ES expert on

staff, faculty at universities throughout the West do this work. Additionally, many agencies already represented in drought planning processes have ES expertise at some level. This includes the US Forest Service (USFS), US Geological Survey (USGS), Environmental Protection Agency (EPA), National Oceanic and Atmospheric Administration (NOAA), and others.

Step 2 is designed to help focus the plan and ensure that it considers the region or organization's capabilities and needs. Any plan's purpose and objectives will vary due to the planning context, the political structure of the government or organization developing the plan, the sectors affected by drought, and other aspects. Involving ES experts and an ES approach in step 2 increases the likelihood that the plan's purpose reflects a broader set of ecological impacts, processes, and values (Collins and Larry 2008).

Step 3 helps ensure that the right people are brought together to understand drought-related concerns and impacts, resolve conflict, and generate collaborative solutions. Failure to involve key stakeholder groups and identify significant sectoral drought impacts leads to gaps in assessing vulnerability as well as in the development of a successful drought plan (Wilhite 2014; Redmond 2002). Published guidance recommends involving public participation and communications experts and provides options (e.g., citizen advisory councils, public meetings, etc.) for planners, but does not always provide the more tactical levels of detail needed to carry out this step (Wilhite et al. 2005, 2014). On the other hand, when tools are provided (e.g., checklists, discussion questions) (Knutson et al. 1998; Svoboda et al. 2011), they stop short of either comprehensively addressing environmental impacts, coupling these impacts with human systems, or connecting impacts to stakeholder values.

An ES framework can provide a roadmap for identifying a wide range of possible ecological impacts of drought, bringing stakeholders together, communicating plan benefits, and earning buy-in. Scarce water resources during drought require that policy makers mandate allocation decisions, forcing them to juggle or make trade-offs between competing social, economic, and environmental interests and values. If developed collaboratively, an ES framework has the potential to help stakeholders define common goals and promote dialogue and collaboration by helping them draw connections among their shared values. While stakeholders often recognize the need for increased sharing of water resources during drought, the concept of "shared sacrifice" is typically invoked as a top-down emergency enactment and is viewed negatively as "less a shared sacrifice than the sacrifice of the many for the benefit of the privileged few" (Anderson et al. 2016, p. 152). When successful sharing occurred, it was based on a voluntary approach that focused on the equitable distribution of water for the collective goal of preserving the health of the entire system (Anderson et al. 2016). An ES framework supports collaborative decision-making, which might help ensure more durable planning outcomes.

In many communities in northern New Mexico, cultural practices and some landscape terminology are often described in traditional Spanish terms (Raheem et al. 2015). For example, common-property irrigation ditches are known as *acequias*, from the Arabic *as-saqiya*, meaning "water-bearer", "irrigation canal", or "conduit" (Rivera 1998; Rodriguez 2006). Spanish settlers, inheriting Roman and Moorish irrigation systems in southern Spain, brought that technology and law to the New World. The terminology of those communities (not to mention the many indigenous communities in New Mexico which might use Tewa, Kerese, Navajo, or a number of other languages) is not always well known by professional planners in the region. Raheem et al. (2015) used traditional Spanish terminology to catalog ES in what Peña (2003) calls the Upper Río Grande bioregion (URGB). The matrix used to present the inventory follows an altitudinal gradient, from headwaters in the mountains to where rivers drain into the mainstem Río Grande, not unlike Postel's (2005) mountains-to-sea framework. Figure 32.2 shows a typical acequia agricultural landscape.

As ES are anthropocentric, it can be useful to involve individuals with a wide range of expertise pertaining to human experience when assembling an inventory. To describe a landscape in terms that matter to local residents and planning or scientific experts, the research group included experts from the fields of economics, cultural history, linguistics, planning, ecology, and hydrology. This sort of multidisciplinarity is essential to understanding the linkages between ecology and culture (Pielke 2007; Bauer et al. 2013). When used to create a common language for planning, ES can help

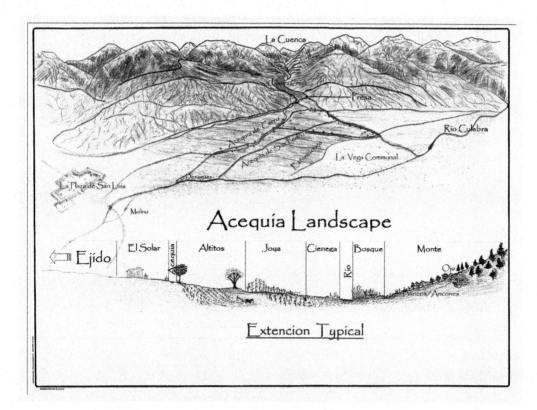

FIGURE 32.2 Schematic of a typical acequia agricultural landscape. From Raheem et al. 2015. (Used with permission from Arnie Valdez.)

incorporate cultural values, foster inclusion, and promote buy-in into the planning process (Sacco et al. 2009). Table 32.3 provides some definitions of the landscape terminology used in Raheem et al. (2015).

Another example of the utility of an ES approach comes from seeking correlation or greater detail on legislation that pertains to water use. The Land Grant and Acequia Traditional Use Recognition and Consultation Act (HR 6487), introduced in 2018, is in discussion in the US House of Representatives. It addresses the importance of acequias and land grants in New Mexico, and recommends some guidance for federal agencies working with those communities. For example, as many of these lands are currently owned by the federal government, the bill recommends waiving fees for certain historical-traditional uses, with the exception of grazing cattle. Under the section defining "historical-traditional uses," the bill discusses the "use of water, religious and cultural use, gathering herbs," and many other cultural and provisioning ecosystem services. The "cultural" and "provisioning" tables in Raheem et al. (2015) show a fairly extensive list of services that correspond to the guidance in HR 6487. Table 32.4 shows selected excerpts from that work.

Breaking out the services along an altitudinal gradient can make it easier to assess where services occur spatially. In these communities, several spiritual and religious values occur in the solares. For example,

> Matachines dances would take place in the solares and near the acequias. La Llorona would exist near the acequias but is not listed in the solar region. La Llorona is typically found near or in water bodies, and she is not invoked as living in people's houses. Christmas services would take place in the solares but not near the acequias, and the blessing of the waters would take place in both locations.

(Raheem et al. 2015; 9)

TABLE 32.3

Summary Descriptions of Principal Landform Terminology

Spanish	English Equivalent
Sierra	Alpine and subalpine life zone
Monte	Forested montane life zone
Dehesa	Unirrigated, sometimes common grazing land
Solar	Land designated for human habitation
Acequia	Irrigation canal or ditch
Altito	Land typically used for orchards
Jolla	"Hollows", high quality irrigated land for row crops
Vega	Unirrigated meadow, used for grazing crops
Ciénaga	Seasonal wetland
Bosque	Typically riparian forest
Rio	River/mainstem river

Source: Adapted from Raheem et al. (2015).

TABLE 32.4

Selected Ecosystem Services in New Mexico

	Services by Category		
Category	**Spiritual and Religious**	**Knowledge Systems**	**Aesthetic Values**
Cultural	Matachines (a traditional dance ceremony), La Llorona (a female spirit who inhabits bodies of water, often luring children to their deaths), feast days, Penitentes (a Catholic sect found in New Mexico and Colorado, formally known as La Sociedad de los Hermanos Penitentes), fiestas (traditional festivals), bultos, Santos (carved saints, usually of wood), blessing of the waters	Indigenous agricultural knowledge, flexible resource management, transmissible knowledge	Social relations, querencia (a term that translates roughly as "love of place"), place-based environmental ethic, cultural heritage values
Category	**Human Habitation**	**Fiber**	**Ornamental Resources**
Provisioning	Building materials (vigas (heavy roof timbers), latillas (lighter cross pieces for roof construction), river rock, flagstone, cal (a kind of lime), adobe, place for building (space), shade (reduced cooling cost, pleasant)	Baskets (willows), wool (sheep)	Tamarisk sticks, willow (furniture), natural dyes, carvings from cottonwoods, roots for katchinas (ceremonial figures or figurines from the Hopi tradition), mica clay, mineral dyes and paints, wildflowers, colonial furniture, ristras (wreaths, often made of dried chiles), Christmas trees, cedar, pine, leather

Source: Adapted from Raheem et al. (2015).

TABLE 32.5

Provisioning Ecosystem Services by Biome in the Missouri Headwaters Basin

	Services by Biome		
Subcategory	Forests	Grasslands	River/Riparian
Cultivated or irrigated crops	Lumber, timber	Alfalfa hay, grass hay, canola, wheat, potatoes, grass seed	Wheat, hay, alfalfa, pasture (native/nonnative), potatoes, seed potatoes, pulses
Livestock	Cattle, some bison, use of domestic animals for invasive/pest species control	Grass-fed livestock, mainly cattle	Cattle, sheep, goats, livestock feed and water often direct from river, meat, wool, llamas
Wild animals and their outputs	Game animals and any kind of hunting, antler sheds, fishes, trapping for furs (multiple species), upland game birds	Forage and habitat for wild game animals, grassland game birds	Hunting, trapping, fishing
Animals from local aquaculture (hatchery)	Not present in these biomes		Fish hatcheries
Surface water for drinking	Forests are a source of clean water	Healthy grassland floodplains improve water storage and groundwater recharge for quantity and quality	Community water supplies – municipal, also cabin supplies. Bozeman reservoir.
Groundwater for drinking	Forests help recharge groundwater; Western juniper encroachment could consume more groundwater than current low-elevation species	Grasslands contribute to ground–surface water interactions; provide groundwater for domestic water use (drinking water)	Rivers contribute to ground–surface water interactions, wells; different types and quantities of riparian vegetation consume groundwater at different rates
Surface water for non-drinking purposes	Water for irrigation, livestock, firefighting, construction, snow-making, hydropower; surface water rights for instream flow/habitat	Wetlands can be sources of water for irrigation, artificial ponds	Irrigation for crops; hydropower; mills; industrial or mining use; breweries; instream flow as a minimum amount for ecological health

Source: Adapted from Raheem et al. (2019).

This context can help align cultural uses and services with the geopolitical boundaries used in drought planning.

Step 4 ensures that the drought task force and policy makers have the data and information needed to evaluate the resources and groups that could be harmed by drought and to assess the underlying causes for the vulnerabilities. Because vulnerability is context- and location-specific, drought vulnerability assessments need to be comprehensive and multidimensional in their approach. Research shows that ecological effects are largely absent from drought planning efforts or that those efforts contain locally important gaps (Tánago et al. 2016; McEvoy et al. 2018).

By constructing an ES inventory, it is possible to tie impacts to relevant biomes and processes, resulting in fewer gaps (Table 32.5). Raheem et al. (2019) provided extensive tables detailing possible ES in the UMH, but for illustrative purposes here we limit examples here to excerpts from the provisioning ES table. An extensive inventory allows stakeholders and experts to double-check what impacts are listed versus what has happened in the past or is likely to occur. This provides greater detail than many current processes and reduces the likelihood of gaps in vulnerability assessments.

Step 5 describes the process of developing the relevant working groups needed to write the drought plan (Wilhite et al. 2005). Drought plans typically focus on three key components: (1) drought monitoring, (2) drought risk and vulnerability assessments, and (3) mitigation and response and mitigation strategies. Drought monitoring is a critical component of drought risk management, essential for providing early warning of impending drought conditions; characterizing the severity, extent, and duration of such conditions; and establishing thresholds for triggering response actions (WMO and GWP 2016). Monitoring drought involves continuous assessment of indicators appropriate to the geographic location, season, impacts of concern, and plan purpose. Incorporating an ES perspective can help ensure that indicators representing ecosystem changes are included in drought monitoring efforts. Additionally, ES literature often focuses on proxy data that can be used to assess ES status.

Drought risk and vulnerability assessments identify the sectors, groups, resources, and regions most at risk from drought and establish who is at risk and why (Wilhite et al. 2005; 2014). As described earlier in step 4, an ES approach offers the potential to reduce the chance that these assessments miss key pieces and result in mitigation and response strategies (step 5) that fail to adequately address drought risk. Mitigation actions are the steps taken before a drought occurs to reduce the effects of future droughts, whereas response actions are activities taken during a drought to help cope with its effects. ES frameworks inherently support these types of management decisions by helping to identify drought-related effects on natural systems and their sociocultural and economic context, set objectives, and assess trade-offs and prioritize actions (Martinez-Harms et al. 2015). Taking an ES perspective in steps 4 and 5 are also likely to highlight research and institutional needs (step 6) since environmental aspects are limited in operational drought monitoring, vulnerability analyses, and plan actions (WMO and GWP 2016; Tánago et al. 206; McEvoy et al. 2018; Raheem et al. 2019; Banerjee et al. 2013).

To address possible concerns that an ES approach might not mesh well with existing planning approaches, we provide one example from past research in Montana and one conducted for this chapter, based in New Mexico. Impact tables are sometimes used to evaluate the vulnerability of various sectors to droughts. In the UMH, local watershed planners drew up a set of impact tables, breaking out impacts by water (e.g., agricultural irrigation, mining) and land-use sectors (e.g., rangelands, and riparian areas). In order to demonstrate the relevance of an ES approach to drought planning, Raheem et al. (2019) conducted a review of these impact tables, reclassified the impacts according to CICES impacts, and aligned them with hydroclimate indicators. Table 32.6 shows the relationship between drought impacts on ES through the CICES framework.

TABLE 32.6

Selected Impacts and Indicators with CICES Equivalents from the Missouri Headwaters Basin

Water or Land Use Sector	Impacts	Indicators	CICES Equivalent
Agricultural irrigation	Impacts to agricultural production: Decreased forage?	Streamflow, snowpack, precipitation	Provisioning > biomass > cultivated terrestrial plants for nutritional purposes > crops by type/quantity
Stockwater	Insufficient water for stock, so stock productivity declines? Grazing restrictions? Impacts to stock	Streamflow, snowpack, precipitation	Provisioning > water > surface/ groundwater used for nutrition, materials, or energy > surface water for drinking

Source: Adapted from Raheem et al. (2019).

We also examined New Mexico's drought plan for 2018 (NMOSE 2018). This plan explicitly mentions ecological drought as a category of drought, which appears to be the first mention of that term in New Mexico drought planning. However, "ecological impacts" are still mentioned as distinct from other impacts. We reviewed Table 32.3 (state and federal agency response) from the drought plan and have mapped the "impact conditions" to an ES framework. By doing this we hope to clarify that using an ES approach can be useful to organize and coordinate drought impact evaluation. Table 32.7 shows impacts from the 2018 New Mexico Drought Plan mapped onto ES.

The fact that in both cases nearly all impacts can be converted into ES suggests that ES might be a useful framing approach. Comparing these tables to the outputs from impact tables in either original report suggests that the lists in this drought plan might not be complete. Examining the tables in both papers suggests a wealth of possible services to be considered that might be impacted by drought.

Steps 7 through 9 focus on using communication and outreach as a means to close the gap between science and policy, clarify the decision-making process for the public, and build trust in drought management actions. The sheer complexity of drought risk management requires a broad range of actions and responses that necessitates effective dialogue between a wide range of policy makers, scientists, and stakeholders (Wilhite et al. 2014). By creating a holistic framework that focuses on things that people care about, ES can be applied as a tool for communication and advocacy (Bull et al. 2016). The anthropocentricity of any ES framework means that a planning process that uses it will result in outcomes that are comprehensible to a wide range of stakeholders, particularly in sectors that are not always well addressed in current planning efforts. Additionally, building an ES inventory through repeated stakeholder consultation should result in broader buy-in across sectors and help different groups identify impacts they might not have previously considered. This cross-fertilization or conversation should help design more effective communication materials.

Finally, step 10 describes the need to keep the drought plan up to date through ongoing and post-drought evaluations (Wilhite et al. 2005). This includes assessing changes in drought risk caused by advances in science and technology; revisions to water laws, policies, and treaties; shifts in political leadership and priorities; and increased demand resulting from factors such as population growth, economic development, shrinking groundwater supplies, and climate change. When repeated and compared with earlier findings, ES frameworks can be used to estimate changes in the value and level of importance of ES and the implications for drought risk management. Additionally, planning frameworks, such as the Drivers–Pressures–State, Ecosystem Services–Response (or DPSER), integrate ES into their evaluations and can be adapted for condition reporting (Office of National Marine Sanctuaries 2015).

32.4 DISCUSSION, POLICY APPLICATIONS, AND NEXT STEPS

Having discussed applications of ES for drought planning purposes, it is essential to cover a few more details about the approach, as it can fulfill several goals. Broadly, it can also serve as a method for implementing and modifying any given ES framework for landscape-specific planning efforts. Using a standardized method such as ES across multiple planning efforts allows for more efficient modification and rapid deployment. Once a state or regional table is created, it could then be modified or subdivided into smaller drought management areas such as watersheds or natural resource districts. Additionally, using a table or catalog approach could allow managers and stakeholders to discuss and plan for impacts ex ante, rather than ex post (Banerjee et al. 2013). Once a regional table is drawn up, the next step could be modifying the table per watershed.

In drought management, allocation and prioritization decisions are complex and often controversial. Using ES can help demonstrate the far-reaching consequences of ecological impacts and clarify their importance to diverse groups and stakeholders. Additionally, it can help identify which ES can be monitored using readily available drought indicators and indices, which in turn can trigger policy measures or on-the-ground projects to mitigate losses during drought. Building an ES inventory

TABLE 32.7

Impacts from New Mexico Drought Plan 2018 Converted to ES Equivalents

Impact by Category as Listed in New Mexico Drought Plan	ES Equivalent
Agriculture	
Production yield/losses and increased costs	Provisioning > food
Increased drought levels	Possibly provisioning > fresh water
Drinking Water	
Public Water Systems (PWS) drought-induced operation system failures (emergency declaration required for emergency funding to PWS)	Engineering, not an ES
Water shortages	Provisioning > fresh water
Increased PWS contaminants and exceeded max levels	Most likely regulating > pollution control
Risk of private wells running dry	Provisioning > fresh water
Water Quantity	
Colorado River and Upper Colorado River Basin compacts: San Juan compact deliveries and water competition	Provisioning > fresh water
La Plata River compact: Water availability severely impacted by reduced flows	Provisioning > fresh water
Pecos River Basin compact: NM delivery obligations, increased pumping likelihood, and Endangered Species Act (ESA) flow requirements	Provisioning > fresh water
Rio Grande Basin, Amended Costilla Creek compact: NM users quantity reduction	Provisioning > fresh water
Rio Grande Basin, San Juan-Chama Project: Supply impairment in both states	Provisioning > fresh water
Wildfire	
Fire danger increases	Regulating > natural hazard
Fire; air quality	Regulating > air quality
Forest and watershed health declines	Slightly vague from an ES perspective but probably map to all categories
Wildlife	
Reduced streamflow effect on endangered species of aquatic plants and animals	Possibly supporting
Forage resource damage	Possibly regulating or provisioning
Reduced grain production at waterfowl preserves	Provisioning
Reduced available wildlife drinking water and food supply	Provisioning > Surface water for non-drinking purposes
Deteriorated fish habitat and reduced fishery productivity	Fishery productivity is managed, not an ES exactly
Loss or impairment of fish and wildlife resources	Could be provisioning or cultural, based on whether for consumption or catch-and-release, or for tribal customary use
Economic Development, Recreation, and Tourism	
Negative effect on tourism from public's negative perception of drought situation in state	Cultural > tourism
Lack of conservation efforts will further deplete water resources in tourism related businesses	Provisioning and cultural
Revenue effects	Most likely cultural > tourism
Fires in NM State Parks	Both regulating > natural hazard and cultural > tourism

also can allow for allocation or prioritization decisions fairly early in the planning process, though of course any such decision is very complex. Tables like these can clarify the importance of ES to different groups, which ES might be most easily or effectively monitored, and which impacts on ES can be mitigated via policy measures or on-the-ground projects.

The use of this framework for decision-making could be supplemented by mapping ES and assigning some kind of value for those services, whether explicitly monetary or otherwise (Pearce et al. 2006). This integration may facilitate comparisons between services provided by, for example, agroecosystems and other, less anthropogenic landscapes. Mapping ecosystem service provisions within a given landscape may be accomplished by delineating landforms into digital GIS coverage. This allows the ES to be represented in a spatially explicit manner and facilitates comparisons at multiple spatial and temporal scales (Burkhard et al. 2012). These ideally community-generated maps can be correlated with biophysical indicators. Spatially explicit information on the land cover – the geographic configuration of different ecosystems –exists for New Mexico and Montana at relatively high resolution (e.g., National Gap Analysis Program Landcover Dataset).

Researchers use a range of methods to score or rank the quality of ES provision. Fleming et al. (2014) evaluated ecosystem services on two farms in the Upper Río Grande in New Mexico. They used an aggregate scoring method from 1 (poor levels of provision) to 4 (excellent levels of provision) for a range of ecosystem services (e.g., vegetation cover vegetation diversity, aquifer recharge, wildlife diversity). Combining approaches such as Fleming's with in-depth interviews, GIS data on hydrology and biology, and economic valuation could help to dramatically improve the quality of information available for decision-making in acequia communities and at the state or county level.

While the UMH and the URGB are culturally quite different, imposing an external analytical frame or language onto a highly local cultural setting can create some confusion. One suggestion coming out of this work is to engage experts and stakeholders familiar with local culture and terminology to help design and check the products or outputs of the ES framework. This approach can help familiarize agencies with terminology and knowledge that enables them to manage systems with a culturally accurate and sensitive approach, which could in turn help to preserve and understand traditional ecological knowledge associated with these terms. Such integration is essential to both good relations and good science (Smith 2013).

While a general ES inventory such as the ones described is useful, it may require additional work to maximize its usefulness in drought planning. Community meetings, combined with expert consultation, give stakeholders an opportunity to identify problems, share concerns and ideas, develop strategies, and make decisions. This type of public involvement is not only good planning practice, it also helps meet many regulations and requirements (IAP2 2016). One possible step would be to classify the vulnerability to drought of any or all of the services through community meetings or expert consultations. Once the most vulnerable services are identified, planners and community members might be more able to prioritize services for anticipatory management, or for measurement and monitoring.

Is an inventory approach "better'" than just interviewing stakeholders individually, as is more often done? Every approach has advantages and drawbacks. One advantage of the fairly formal structure of an inventory is that it works like a checklist; checklist approaches have proven useful in helping participants think carefully through a predetermined set of possibilities (Gawande 2009; Hales & Pronovost 2006), rather than relying on preexisting conceptions they may have about a given landscape (Kahneman 2011).

In conversations before the UMH workshop, some participants noted concerns that the process would not reveal anything new – these were experts, after all. In debriefing discussions at the end of the in-person workshop, several participants indicated that they were surprised by certain services or appreciated having the time to think about less obvious services, particularly in the cultural category. Further, the process of making the modifications was useful itself; discussing and engaging in modifying the original tables resulted in social learning (Reed et al. 2010) that increased collective regional understanding of ecosystem processes and their impacts on human communities, the desired outcome among planners (Koontz 2014).

However, ES is not always how people see nature (Luck et al. 2012). In the UMH work, we found that interviews identified drought impacts on both the natural (e.g., assemblages of macroinvertebrates) and the human world (e.g., people's stress level and overall well-being, perception of the community or Montana) that are not explicitly related to ways humans use nature. Planners could strategically combine interviews and inventory workshops. Complementing quantitative measures with narratives of stakeholders has been suggested as important for capturing the complexity of vulnerability (Schröter et al. 2005). Planners wishing to use some combination of these approaches will have to make some choices. At the least, they will have to choose an ES framework or a mix; decide whether to use an experts-first approach to create the inventory or build up from stakeholders and cross-check; decide how long or "complete" to make any given list; and determine how much time to spend on the process.

In terms of choosing one ES framework over another, we have little to offer. Choosing one and modifying it as needed seems to be the best approach in our experience. Costanza (2008) argued that "any attempt to come up with a single or 'universal' classification system should be approached with caution", due to the complexity of ecosystem processes, and the diversity of decision contexts. Future frameworks for specific regional use might selectively combine aspects of two or more frameworks. In terms of time, we recommend that planners allow adequate time to complete the work. We recognize that this is not easy; staff doesn't get time off in a coordinated way, travel can be frustrating and expensive (the West is a big place), and space can be expensive to rent. The UMH project had a day for the workshop, whereas the URGB project used four. Much of the revision of the UMH data could have been accomplished in a follow-up session, rather than in tracking down participants. We have learned that time is crucial.

Whatever choices planners make, they should always keep in mind how they will actually use the information they collect and how they can coordinate the gathering and using it across complex organizational structures and management. Clearly, whatever recommendations we provide will be incomplete. A brief glance at New Mexico's drought plan shows the complexity of interagency coordination required to address anything related to this. Interagency coordination is difficult and beyond the scope of this chapter. Recognizing this, the National Drought Resilience Partnership was created to address this issue. Additionally, adding new work to already overloaded staff is not simple. Replacing existing frameworks is complicated. Agencies and individuals are liable to fight over territory previously "theirs". However, it is clear that an integrated approach is more likely to solve the clearly integrated problems faced in drought planning communities. We hope this chapter provides some support.

Incorporating ES in the way we are suggesting is broadly applicable to a range of situations, such as marine spatial planning. Once a group becomes familiar with incorporating ES into one planning system, it can likely be integrated into another. As Long et al. (2015) point out, traditional "silo-structured" management "is widely seen as sufficient" and has "failed to protect marine systems from human pressures". As the need for better integration of ecological impacts into many types of related planning becomes more evident, the ES framework offers a way forward for practitioners and experts.

REFERENCES

Anderson B, Ward L, McEvoy J, Gilbertz SJ, & Hall DM. 2016. Developing the water commons? The (post) political condition and the politics of "shared giving" in Montana. Geoforum 74: 147–157.

Banerjee O, Bark R, Connor J, & Crossman ND. An ecosystem services approach to estimating economic losses associated with drought. 2013. *Ecological Economics* 91: 19–27. https://doi.org/10.1016/j.ecolecon.2013.03.022

Bathke D & Bernadt T. 2021. Collaborative drought planning using scenario exercises. National Drought Mitigation Center. http://drought.unl.edu/scenarioguide/

Bathke DJ, Haigh T, Bernadt T, Wall N, Hill H, & Carson A. 2019. Using serious games to facilitate collaborative water management planning under climate extremes. *Journal of Contemporary Water Research and Education* 167: 55–60.

Bauer DM & Johnston RJ. 2013. The economics of rural and agricultural ecosystem services: Purism versus practicality. *Agricultural and Resource Economics Review* 42: v–xv.

Bull JW, Jobstvogt N, Böhnke-Henrichs A, Mascarenhas A, Sitas N, Baulcomb C, ... & Koss R. 2016. Strengths, weaknesses, opportunities and threats: A SWOT analysis of the ecosystem services framework. *Ecosystem Services* 17: 99–111.

Burkhard B, Kroll F, Nedkov S, & Müller F. 2012. Mapping ecosystem service supply, demand and budget. *Ecological Indicators* 21: 17–29.

Chan KMA, Shaw MR, Cameron DR, Underwood EC, & Daily GC. 2006. Conservation planning for ecosystem services. *PLoS Biol* 4(11): e379. https://doi.org/10.1371/journal.pbio.0040379

Cook BI, Ault TR, & Smerdon JE. 2015. Unprecedented 21st century drought risk in the American Southwest and Central Plains. *Science Advances* 1(1): e1400082. https://doi.org/10.1126/sciadv.1400082.

Collins S & Larry B. 2008. Caring for our natural assets: An ecosystem services perspective. In: Deal, R.L., ed. *Integrated Restoration of Forested Ecosystems to Achieve Multi-Resource Benefits: Proceedings of the 2007 National Silviculture Workshop*. Gen. Tech. Rep. PNW-GTR-733. Portland, OR: U.S. Department of Agriculture, Forest Service, Pacific Northwest Research Station: 1–11.

Costanza R. 2008. Ecosystem services: Multiple classification systems are needed. *Biological Conservation* 141(2): 350–352.

Crausbay SD, Ramirez AR, Carter SL, Cross MS, Hall KR, Bathke DJ, Betancourt JL et al. 2017. Defining ecological drought for the twenty-first century. *Bulletin of the American Meteorological Society* 98(12): 2543–2550.

Dai A. 2013. Increasing drought under global warming in observations and models. *Nature Climate Change* 3(1): 52–58.

de Groot RS, Wilson MA, & Boumans RMJ. 2002. A typology for the classification, description and valuation of ecosystem functions, goods and services. *Ecological Economics* 41: 393–408.

Dell'Apa A, Fullerton A, Schwing F, & Brady MM. 2015. The status of marine and coastal ecosystem-based management among the network of US federal programs. *Marine Policy* 60, 249–258.

Dunn H. 2011. *Payments for Ecosystem Services*. DEFRA Paper 4 2011.

European Environment Agency. 2021. *CICES Version 5.1 Spreadsheet*. CICES Resources, European. https://cices.eu/resources/. Accessed January 18, 2021.

Fleming WA, Rivera JA, Miller A, & Piccarello M. 2014. Ecosystem services of traditional irrigation systems in northern New Mexico, USA. *Journal of Biodiversity Science, Ecosystem Services & Management* 10: 1–8.

Gawande A. 2009. *The Checklist Manifesto: How to Get Things Right*. New York: Henry Holt and Co.

Haines-Young R & Potschin M. 2012. *Common International Classification of Ecosystem Services (CICES, Version 4.1)*. Nottingham, UK: Centre for Environmental Management, School of Geography, University of Nottingham, and Copenhagen, Denmark. European Environment Agency 2012.

Hales BM & Pronovost PJ. 2006. The checklist: A tool for error management and performance improvement. *Journal of Critical Care* 21(3); 231–235. https://doi.org/10.1016/j.jcrc.2006.06.002.

IAP2. International Association for Public Participation. 2016. *Planning for Effective Public Participation*, 165 pp., IAP2 Federation.

Kahneman D. 2011. *Thinking Fast and Slow*. New York: Farrar, Strauss, and Giroux.

Knutson C, Hayes M, and Phillips T, 1998. How to reduce drought risk. Western Drought Coordination Council, Preparedness and Mitigation Working Group. https://drought.unl.edu/archive/Documents/NDMC/Planning/risk.pdf

Koontz TM, 2014. Social learning in collaborative watershed planning: The importance of process control and efficacy. *Journal of Environmental Planning and Management* 57(10): 1572–1593.

Koontz TM & Bodine J. 2008. Implementing ecosystem management in public agencies: Lessons from the US Bureau of Land Management and the Forest Service. *Conservation Biology* 22(1): 60–69.

Long RD, Charles A, & Stephenson RL. 2015. Key principles of marine ecosystem-based management. *Marine Policy* 57: 53–60.

Luck G, Kai MA, Chan A, Eser U, Gómez-Baggethun E, Matzdorf B, Norton B, & Potschin MB. 2012. Ethical considerations in on-ground applications of the ecosystem services concept. *BioScience* 62: 1020–1029.

Martinez-Harms MJ, Bryan BA, Balvanera P, Law EA, Rhodes JR, Possingham HP, & Wilson KA. 2015. Making decisions for managing ecosystem services. *Biological Conservation* 184: 229–238.

McEvoy J, Bathke DJ, Burkardt N, Cravens AE, Haigh T, Hall KR, Hayes MJ, Jedd T, Podebraská M, & Wickham E. 2018. Ecological drought: Accounting for the nonhuman impacts of water shortage in the upper Missouri headwaters basin, Montana, USA. *Resources* 7: 14. https://doi.org/10.3390/resources7010014.

Millennium Ecosystem Assessment (MEA). 2003. *Ecosystems and Human Well-being: A Framework for Assessment*. Washington, DC: Island Press.

Millar CI & Stephenson NL. 2015. Temperate forest health in an era of emerging mega disturbance. *Science*, 349 (6250); 823–26.

Montana Drought Demonstration Partners. 2015. *A Workplan for Drought Resilience in the Missouri Headwaters Basin: A National Demonstration Project.* http://dnrc.mt.gov/divisions/water/management /docs/surface-water-studies/workplan_drought_resilience_missouri_headwaters.pdf

Multihazard Mitigation Council. 2017. Natural Hazard Mitigation Saves 2017 Interim Report: An Independent Study: Summary of Findings. Principal Investigator Porter, K.; co-Principal Investigators Scawthorn, C.; Dash, N.; Santos, J.; P. Schneider, Director, MMC. National Institute of Building Sciences, Washington. https://www.fema.gov/sites/default/files/2020-07/fema_ms2_interim_report_2017.pdf

NCEI. NOAA National Centers for Environmental Information. 2021. U.S. Billion-Dollar Weather and Climate Disasters. https://www.ncdc.noaa.gov/billions/. https://doi.org/10.25921/stkw-7w73

NDMC. National Drought Mitigation Center. 2021a. *Drought Impact Reporter.* Retrieved January 18, 2021 from https://droughtreporter.unl.edu/

NDMC. National Drought Mitigation Center. 2021b. *Planning Processes.* Retrieved January 18, 2021 from http://drought.unl.edu/droughtplanning/AboutPlanning/PlanningProcesses.aspx

NMDTF. New Mexico Drought Task Force. 2006. *New Mexico Drought Plan Update: 2006.* Office of the State Engineer. https://www.nmdrought.state.nm.us/DroughtPlan/2006-NM-Drought-Plan.pdf

NMOSE. New Mexico Office of the State Engineer. 2018. *New Mexico Drought Plan.* Albuquerque, NM. Office of the State Engineer. https://www.ose.state.nm.us/Drought/droughtplan.php

ONS (UK Office of National Statistics). (n.d.). UK natural capital: Ecosystem service accounts, 1997 to 2015. https://www.ons.gov.uk/releases/uknaturalcapitallandandhabitatecosystemaccounts, accessed January 2018.

Pearce D. 2003. *Conceptual Framework for Analysing the Distributive Impacts of Environmental Policies.* Paris, France. OECD.

Pearce D, Atkinson G, & Mourato S. 2006. *Cost-benefit Analysis and the Environment: Recent Developments.* Paris, France. OECD.

Pederson GT, Gray ST, Fagre DB, & Graumlich LJ. 2006. Long-duration drought variability and impacts on ecosystem services: A case study from Glacier National Park, Montana. *Earth Interactions* 10(1): 1–28.

Peña DG. 2003. The watershed commonwealth of the upper Río Grande. In: Boyce K, Shelley BG, eds. *Natural Assets: Democratizing Environmental Ownership.* Covelo, WA: Island Press; 169–186.

Pielke RA. 2007. *The Honest Broker: Making Sense of Science in Policy and Politics.* Cambridge, MA: Cambridge University Press.

Postel S. 2005. From headwaters-to-sea: The critical need to protect freshwater ecosystems. *Environment* 47: 8–21.

Raheem N, Archambault S, Arellano E, Gonzales M, Kopp D, Rivera J, Guldan S, Boykin K, Oldham C, Valdez A, Colt S, Lamadrid E, Wang J, Price J, Goldstein J, Arnold P, Martin S, & Dingwell E. 2015. A framework for assessing ecosystem services in acequia irrigation communities of the Upper Río Grande watershed. *WIREs Water* 2015. https://doi.org/10.1002/wat2.1091

Raheem N, Cravens AE, Cross MS, Crausbay S, Ramirez A, McEvoy J, Zoanni D, Bathke DJ, Hayes M, Carter S, Rubenstein M, Schwend S, Hall K, & Suberu P. 2019. Planning for ecological drought: Integrating ecosystem services and vulnerability assessment. *WIREs Water.* https://doi.org/10.1002/wat2.1352

Ranganathan J, Raudsepp-Hearne N, Lucas N, Irwin F, Zurek M, Bennett K, Ash N, & West, P. 2008. *Ecosystem Services: A Guide for Decision Makers.* Washington, DC: World Resources Institute.

Redmond K. 2002. The depiction of drought: A commentary. *Bulletin of the American Meteorological Society* 83(8): 1143–1147.

Reed MS, Evely AC, Cundill G, Fazey I, Glass J, Laing A, Newig J, Parrish B, Prell C, Raymond C, & Stringer LC. 2010. What is social learning? *Ecology and Society* 15(4): r1. [online] http://www.ecologyandsociety.org/vol15/iss4/resp1/

Rivera JA. 1998. *Acequia Culture: Water, Land and Community in the Southwest.* Albuquerque, NM: University of New Mexico Press.

Rodríguez S. 2006. *Acequia: Water Sharing, Sanctity, and Place.* Santa Fe, NM: School for Advanced Research.

Sacco PL, Blessi GT & Nuccio M, 2009. Cultural policies and local planning strategies: What is the role of culture in local sustainable development? *The Journal of Arts Management, Law, and Society* 39(1): 45–64.

Sangha K & Russell-Smith J. 2017. Towards an indigenous ecosystem services valuation framework: A North Australian example. *Conservation and Society* 15: 255–269.

Schaefer M, Goldman E, Bartuska A, Sutton-Grier A, & Lubchenco J. 2015. Nature as capital: Advancing and incorporating ecosystem services in United States federal policies and programs. *PNAS* 112(24): 7383–89.

Schröter D, Polsky C, & Patt AG. 2005. Assessing vulnerabilities to the effects of global change: An eight step approach. *Mitigation and Adaptation Strategies for Global Change* 10(4): 573–595.

Slocombe DS. 1993. Implementing ecosystem-based management. *Bio Science* 43(9): 612–622.

Smith KR. 2013. Economic science and public policy. *Agric Resour Econ Rev* 42: 90–97.

Smith AB & Katz RW. 2013. US billion-dollar weather and climate disasters: Data sources, trends, accuracy and biases. *Natural Hazards* 67(2): 387–410.

Sousa LP, Sousa AI, Alves FL, & Lillebø AI. 2016. Ecosystem services provided by a complex coastal region: Challenges of classification and mapping. *Scientific Reports* 22782. https://doi.org/10.1038/srep22782

Sukhdev P, Wittmer H, Schröter-Schlaack C, Nesshöver C, Bishop J, ten Brink P, Gundimeda H, Kumar P, & Simmons B. 2012. Mainstreaming the economics of nature. A synthesis of the approach, conclusions, and recommendations of TEEB. http://doc.teebweb.org/wpcontent/uploads/Study%20and%20Reports /Reports/Synthesis%20report/TEEB%20Synthesis%20Report%202010.pdf. Accessed January, 2018.

Svoboda MD, Smith K, Widhalm M, Woudenberg DL, Knutson CL, Sittler M, Angel J, Spinar M, Shafer M, McPherson R, & Lazrus H. 2011. Drought-ready communities: A guide to community drought preparedness. Lincoln, NB. National Drought Mitigation Center. https://drought.unl.edu/droughtplanning/ AboutPlanning/PlanningProcesses/Drought-ReadyCommunities.aspx

Tánago IG, Urquijo J, Blauhut V, Villarroya F, & de Stefano L. 2016. Learning from experience: A systematic review of assessments of vulnerability to drought. *Natural Hazards* 80(2): 951–973.

Turkelboom F, Raquez P, Dufrêne M, Raes L, Simoens I, Jacobs S, Stevens M, De Vreese R, Panis JAE, Hermy M, Thoonen M, Liekens I, Fontaine C, & Dendoncker N. 2014. CICES going local: Ecosystem services classification adapted for a highly populated country. In Jacobs S, Dendoncker N, and Keune H, eds. *Ecosystem Services Global Issues, Local Practices*. Amsterdam: Elsevier.

Van Loon AF, Gleeson T, Clark J, Van Dijk, AIJM, Stahl K, Hannaford J, et al. 2016. Drought in the Anthropocene. *Nature Publishing Group* 9(2): 89–91. http://doi.org/10.1038/ngeo2646

White House. 18 July 2014. *Building Drought Resilience in Montana*. https://obamawhitehouse.archives.gov/ blog/2014/07/18/building-drought-resilience-montana

Wickham, ED, Bathke, D, Abdel-Monem, T, Bernadt T, Bulling D, Pytlik-Zillig L, Stiles, C & Wall, N. (2019). Conducting a drought-specific THIRA (Threat and Hazard Identification and Risk Assessment): A powerful tool for integrating all-hazard mitigation and drought planning efforts to increase drought mitigation quality. *International Journal of Disaster Risk Reduction* 39: 101227.

Wilhite, DA. 1992. *Drought. Drought Mitigation Center Faculty Publications*; 64. http://digitalcommons.unl .edu/droughtfacpub/64

Wilhite DA. 2014. *World Meteorological Organization; Global Water Partnership; and National Drought Mitigation Center, National Drought Management Policy Guidelines: A Template for Action*. Drought Mitigation Center Faculty Publications. 83. http://digitalcommons.unl.edu/droughtfacpub/83

Wilhite DA & Glantz MH. 1985. Understanding: The drought phenomenon: The role of definitions. *Water International* 10(3): 111–120. https://doi.org/10.1080/02508068508686328

Wilhite DA, Hayes MJ & Knutson CL. 2005. Drought preparedness planning: Building institutional capacity. *Drought and Water Crises: Science, Technology, and Management Issues*: 93–135. CRC Press. Boca Raton, FL.

Wilhite DA, Sivakumar MV, & Pulwarty R. 2014. Managing drought risk in a changing climate: The role of national drought policy. *Weather and Climate Extremes* 3: 4–13.

WMO and GWP. World Meteorological Organization and Global Water Partnership. 2016. *Handbook of Drought Indicators and Indices (M. Svoboda and B.A. Fuchs). Integrated Drought Management Programme (IDMP), Integrated Drought Management Tools and Guidelines Series 2*. Geneva. https:// www.droughtmanagement.info/literature/GWP_Handbook_of_Drought_Indicators_and_Indices _2016.pdf

33 Drought Tolerance Agronomic Management Strategies for Rainfed and Irrigated Maize Crop in Eastern India

R.K. Srivastava, R.K. Panda, Arun Chakraborty, and Swayam Prava Singh

CONTENTS

33.1 INTRODUCTION

The global population is estimated to increase by 50% by 2050 and thereby the global grain demand is expected to double. Doubling grain production with environmental sustainability is a major challenge to be faced. Sustainable grain production can be achieved by ecofriendly farming techniques of crop production without using toxic agrichemicals. Agricultural crop production is highly sensitive to climatic variables in different regions of the world.

The key factor of agriculture crop production is temperature, which causes drought conditions. It directly affects crop growth rate and development. Rises in greenhouse gases will increase the yearly air temperature by 1°C to 6°C globally by the end of this century (Tubiello and Ewert, 2002). The changes in temperature will be about 2°C–3°C in the next upcoming decades 2030–2050 (Hatfield and Pruger, 2015), which was also suggested by the Intergovernmental Panel on Climate

DOI: 10.1201/9781003276548-33

Change (IPCC, 2007). Additionally, extreme temperature events will be more intense and more frequent. Increased temperature affects crop development by increasing the requirement for water, which reduces the economic yield (Meehl et al., 2007). Heat stress is one of the primary agricultural problems globally. Many pieces of literature also state that warmer temperatures raise the rate of phonological development in a crop, with the least effect on leaf area, leaf weight or plant matter, and water demand (Hatfield and Pruger, 2015).

Sowing date is an important management practice used to adjust the timing and duration of crop phenological phases according to environmental conditions for crop development. Varying sowing dates can alter crop growth and the length of crop phenological phases, which affects potential grain yield and its components. Delayed sowing dates can reduce the synchronization of peak solar radiation and maximum leaf area index for the maize hybrid varieties. The most critical growth stage of maize is flowering, which affects its biomass, grain yield, and plant height.

A crop simulation model has the capability to estimate the yield and progress of a crop by applying a set of genetic coefficients, with suitable initial soil, weather variables, and agronomic management. The Decision Support System for Agrotechnology Transfer (DSSAT) Version 4.5 contains 10 separate models to simulate 29 different crops growth and yield, for example, CERES-Maize, CERES-Rice, Substor-Potato, and CROPGROW-Peanut. Presently, DSSAT is used broadly for evaluating the impact of climate variability and change on agricultural production (Daccache et al., 2011).

Maize is a vital crop cultivated across the globe under varied soil and climatic conditions. It is cultivated in 6.4 m ha area with a production of 11.5 million tonnes (Panda et al., 2004). The crop is mainly affected by two constraints, viz., biotic and abiotic factors, such as soil, drought, pests, and change in weather variables (that is, air temperature (maximum and minimum), relative humidity solar radiation, and precipitation). These factors are well-known factors that affect crop production and thus decline yield efficiency. Both water and nitrogen play a vital role in agriculture production, of which water is the most limiting factor for reduced productivity of the crop, while nitrogen is regarded as the most authoritative factor for crop production and grain quality.

Adaptation of appropriate nitrogen fertilization levels with sowing dates management can maximize agricultural crop production without compromising on the environmental front. Keeping the aforesaid fact in mind, the reported study was undertaken with the major objective of analyzing sowing dates to cope with increased temperature due to climate change for maize crops under rainfed and irrigated conditions in eastern India. Sowing dates evaluation was done under nutrient and water stress conditions for maize crops. Field experiments were conducted to evaluate the crop model, i.e., CERES-Maize v4.5, under different sowing dates and nitrogen fertilization levels, then model was run to evaluate the sowing dates. The objectives of the study were threefold: (1) validate the crop model performance for the predominant cultivar in the eastern region of India, (2) assess the effect of temperature on maize crop under two CO_2 concentration levels, and (3) evaluate the different sowing dates under different crop management in rainfed and irrigated conditions to cope with drought due to increased temperature.

33.2 MATERIALS AND METHODOLOGY

33.2.1 FIELD EXPERIMENTAL SITE

Field experiments were organized at the experimental farm of the Agricultural and Food Engineering Department, Indian Institute of Technology Kharagpur, India. The study region's soil is classified as the lateritic type (Table 33.1) (Srivastava et al., 2020).

Volumetric water content in soil ranged between 0.094–0.12 m^3/m^3 at the wilting point and 0.22–0.24 m^3/m^3 at field capacity.

The climate of the study area is classified as tropical with an average temperature ranging between 21°C–32°C. The area receives an average rainfall of 1450 mm annually. Weather data

TABLE 33.1

Layer-Wise Soil Characteristics of Experimental Site

Soil	Soil Depth (cm)	Clay (%)	Silt (%)	Sand (%)	Organic Carbon (%)	pH of Water	Total Nitrogen (%)
Medium sandy	0–5	14.3	26.2	59.5	0.52	6.1	0.053
loam	5–20	14.3	26.2	59.5	0.39	6.1	0.047
	20–40	27	20.2	52.8	0.12	6.5	0.011
	40–60	26.6	19.2	52.2	0.09	6.5	0.013

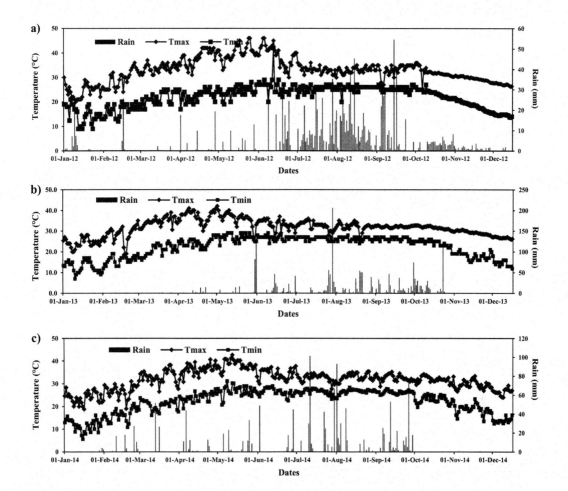

FIGURE 33.1 Daily rainfall (mm) and temperature (T_{max} and T_{min}) at Kharagpur for the years (a) 2012, (b) 2013, and (c) 2014.

was measured using an automated weather station located at the experimental farm at the institute. The daily recorded weather variables were temperature, rainfall, wind speed, relative humidity, and sunshine hours. Figures 33.1a–c show the temporal values of precipitation and temperature (maximum and minimum) during the years 2012, 2013, and 2014 respectively. The daily weather variables, such as rainfall and temperature (maximum and minimum), were obtained from the India Meteorological Department, Pune, India for the period 1982–2012.

33.2.2 Field Experimental Details

Four field experiments on maize crop were conducted for two sowing dates (timely and delayed), and four nitrogen fertilization levels were maintained under rainfed and irrigated conditions. The field experiment was a randomized split-plot design with three replicates, and the crop was sown at a depth of 5 cm with 50 cm × 30 cm spacing. Urea was used as the nitrogen fertilizer and applied in three splits: 50% at sowing time, 25% after 25 days of sowing, and 25% after 50 days of sowing.

Two field experiments were conducted under the rainfed condition of maize (variety Bio-22027) during the years 2012 and 2014 sown on June 10 and 25, and subsequently harvested on September 8 and 23. Nitrogen fertilizer levels were done at 0 (N_0), 60 (N_{60}), 80 (N_{80}), and 100 (N_{100}) kg ha^{-1}.

Two field experiments on maize crops were conducted under irrigated conditions in the years 2013 and 2014. The maize seed was sown on January 5 and 25 in the years 2013 and 2014, respectively, and subsequently, harvested on April 5 and 26, respectively. Maize hybrid (Tx367) (medium duration) was selected for the experiment. Nitrogen fertilizer levels were maintained as 0 (N_0), 75 (N_{75}), 100 (N_{100}), and 125 (N_{125}) kg ha^{-1}.

Irrigation scheduling was done based on effective root zone depth's soil water content. Daily soil water content was measured with a time domain reflectometry (TDR) by setting the instrument in a manual mode. Figure 33.2 represents the applied quantum of irrigation water datewise along with rainfall and volumetric soil moisture (VSM) for the irrigated maize crop for the years 2013 and 2014.

The different crop parameters were recorded at varying crop growth stages, i.e., initial, vegetative, maturing, and harvesting for both rainfed and irrigated conditions. The plant height, plant population, grain yield, and aboveground biomass were recorded. The plants were further separated into different parts similar to the sampling during the growing season and then oven-dried for 48 hours at 70°C.

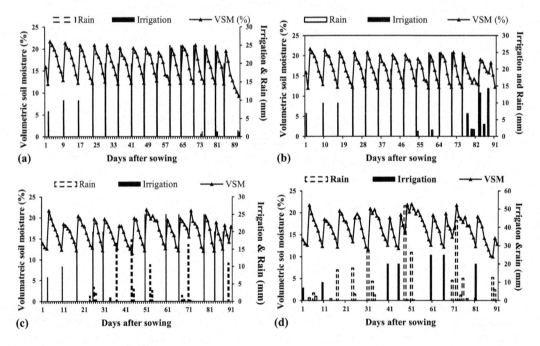

FIGURE 33.2 Variation of volumetric soil moisture (VSM), irrigation (mm), and precipitation (mm) during the maize crop growth period with respect to days after sowing in 2013 for (a) January 5 and (b) January 25, and in 2014 for (c) January 5 and (d) January 25.

33.2.3 CERES-Maize Model

The physiologically based CERES-Maize model is cultivar-specific and simulates crop response to various environmental conditions and predicts productivity more accurately. The genotype coefficients were calibrated for cultivars of P1, P5, G2, and PHINT.

33.2.3.1 Calibration and Validation of the Model

The CERES-Maize model was calibrated with results of the maize crop growth, yield, and yield attributes during the years 2012 under rainfed conditions and 2013 under irrigated conditions. The genetic coefficients were developed by the trial-and-error approach (Hunt et al., 1993) of DSSAT. The calibration of the CERES-Maize was done using nitrogen level N_{80} for the rainfed condition during the year 2012 and N_{100} for the irrigated condition during the year 2013 data sets, and was validated for the same treatment during the year 2014 for both rainfed and irrigated maize. During the model calibration, crop parameters were used such as anthesis day, physiological maturity, aboveground dry matter, and grain yield obtained during the harvest. The aboveground dry matter weight [kg (dm)/ha] was also taken at different maize crop stages for irrigated as well as rainfed maize. After calibration, a set of genetic coefficients were derived.

33.2.4 Effect of Increased Temperature under Two CO_2 Levels

The CERES-Maize was done under different temperature scenarios in order to understand the impact of systematic changes in air temperatures (maximum and minimum), and CO_2 levels on the crop yield in ambient and elevated CO_2 conditions. The scenarios were developed by combining changes in the different climate variables. The model was used for four levels of maximum and minimum temperatures by adding temperature increases of 1°C, 2°C, 3°C, and 4°C, with two CO_2 levels – one at ambient, that is, 380 ppm, and the other at elevated (doubling) CO_2, that is, 760 ppm – under rainfed and irrigated conditions using the 30 years of weather data (from 1982 to 2012) used in this study.

33.2.5 Determination of Sowing Dates

The determination of sowing dates was done using the CERES-Maize model by simulating the seasonal yield from the year 1982 to 2012 using historical weather data (obtained from the Regional Meteorology Department, Kolkata) for the West Medinipur district of West Bengal, India. After validation of CERES-Maize, based on the field experiment, five appropriate sowing dates were considered, such as 10 days and 20 days before and after the appropriate sowing date determined through the field experiment. Sowing dates chosen for the study were May 30, June10, June 20, June 30, and July 10 for rainfed conditions; and December 25, January 5, January 15, January 25, and February 4 for irrigated conditions. Sowing dates evaluations were done for different N levels of 60, 80, 100, and 120 kg/ha for rainfed conditions; and 75, 100, 125, and 150 kg/ha for irrigated conditions. Moreover, sowing dates were also evaluated for varying maximum allowable depletion levels based on the availability of water content in the soil, for example, 10%, 40%, 70%, and 90% for irrigated conditions.

33.2.6 Statistical Method

For determining the accuracy of the validated model, statistical methods such as R^2, the mean relative error (Equation 33.1), normalized roots mean square error (RMSEn) (Equation 33.2), and index of agreement (d-stat) (Equation 33.3) indices were calculated using the following equations:

$$MRE = \frac{1}{N} \sum_{i=1}^{n} \frac{X_i - Y_i}{Y_i} \times 100 \tag{33.1}$$

$$RMSEn = 100 \times \frac{\sqrt{\frac{1}{N} \sum_{i=1}^{N} (X_i - Y_i)^2}}{\overline{\overline{Y}}} \tag{33.2}$$

$$\text{d-stat} = 1 - \frac{\sum_{i=1}^{N} (X_i - Y_i)^2}{\sum_{i=1}^{N} \left(|X_i - \overline{Y}| + |Y_i - \overline{Y}| \right)^2} \tag{33.3}$$

where X_i is the estimated value obtained from different models, Y_i is the estimated value obtained from the crop model, and $\overline{Y}$ is the observed mean value.

33.3 RESULTS AND DISCUSSION

33.3.1 CROP GROWTH AND YIELD SIMULATION

33.3.1.1 Calibration of the CERES-Maize Model

The simulated time series of aboveground dry matter weight (kg ha^{-1}) were in the range of standard deviation of observed values right through the growing season in the course of calibration for the rainfed and irrigated conditions (Figure 33.3).

The statistical analysis with RMSE$_n$ (3.92, 2.42%), and d-stat (0.99) values between the simulated and observed time series of aboveground dry matter weight exhibited that the calibrated model functioned well in both rainfed and irrigated conditions (Figure 33.3).

The comparison between the observed and simulated values is illustrated in Table 33.2 along with the standard deviation. The simulated values of anthesis date and date of physiological maturity were within ± 2 and ± 3 days with their respective observed values for the irrigated and rainfed conditions (Table 33.2). The simulated values of grain yield, tops weight, kernel number per cob, and leaf area index were found to be within the standard deviation of the measured value. The derived genetic coefficients of irrigated and rainfed maize are presented in Table 33.3.

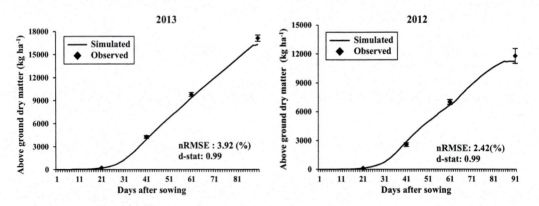

FIGURE 33.3 Comparison of observed and simulated aboveground dry matter at different stages for rainfed maize for the year 2012 and irrigated maize for the year 2013.

TABLE 33.2

Comparison of Observed and Simulated Values for the Rainfed Condition (2012) and Irrigated Maize (2013) Maize during Calibration of CERES-Maize Model

	Irrigated Maize		Rainfed Maize	
	Cultivar Tx 369		Cultivar BIO 22027	
Variable	Observed	Simulated	Observed	Simulated
Anthesis day	54	55	55	55
Physiological maturity day	87	90	89	90
Yield at harvest maturity (kg ha^{-1})	5633 ($\pm$119)	5575	5500 ($\pm$142)	5514
Tops wt at maturity (t[dm] ha^{-1})	19.42 ($\pm$0.79)	18.99	11.15 ($\pm$0.96)	11.23
Number at maturity (no/cob)	265 ($\pm$3.93)	262	320 ($\pm$3.54)	317
Leaf area index (maximum)	4.3	4.4	3.8	3.9

TABLE 33.3

Maize Cultivars TX369 and Bio22027 Genetic Coefficients

	Genetic Coefficient				
Parameter	P1 (°C day)	P5 (°C day)	G2	G3 (mg/day)	PHINT (°C day)
Cultivar Tx369	205.00	650.0	630.0	9.50	35.00
Cultivar Bio22027	250.00	750.00	820.00	9.00	45.00

33.3.1.2 CERES-Maize Model Validation

The model validation was done on the basis of the results of field experiment results conducted during the year 2014 for rainfed and irrigated maize crops. The simulated time series aboveground dry matter (kg ha^{-1}) was compared with the observed data for rainfed and irrigated conditions as illustrated in Figure 33.4, which reveals a close matching between the simulated and observed data. The

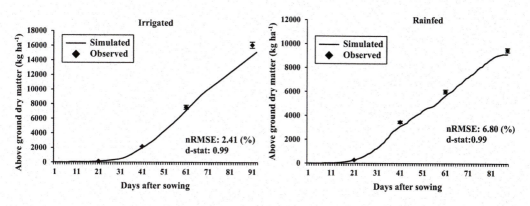

FIGURE 33.4 Observed and estimated values of aboveground dry matter (kg ha^{-1}) at crop growth stages for rainfed and irrigated conditions in 2014 for validation of the model.

TABLE 33.4

Results of the Validation of CERES-Maize Model through Comparison between Observed and Simulated Values for the Rainfed and Irrigated Conditions for Maize for 2014

| | Irrigated Season | | Rainfed Season | |
| | Cultivar Tx369 | | Cultivar BIO22027 | |
Variable	Observed	Simulated	Observed	Simulated
Anthesis day	54	52	53	52
Physiological maturity day	90	91	89	87
Yield at harvest maturity (kg ha^{-1})	4836	4714	4827	4514
Tops wt at maturity (kg[dm] ha^{-1})	17.42	14.85	11.01	9.16
Number at maturity (no./cob)	245	248	280	266
Leaf area index (maximum)	4.3	4.6	4.1	3.8

statistical analysis showing RMSE$_n$ of 2.41, 6.80%, and d-stat of 0.99 values between the simulated and observed time series of aboveground dry matter weight (kg ha^{-1}) indicates that the calibrated model performed well in both rainfed as well as irrigated conditions. The comparison between simulated and observed crop parameters such as anthesis date, grain yield (kg ha^{-1}), physiological maturity, time series tops weight, kernel number per cob, and leaf area index (maximum) revealed good agreement with each other (Table 33.4), thereby indicating the good performance of the model during the validation.

33.3.2 EFFECT OF INCREASED TEMPERATURE UNDER AMBIENT AND ELEVATED CO_2 ON GRAIN YIELD

Figure 33.5a indicates that an increase in maximum temperature by 1°C, 2°C, 3°C, and 4°C decreased the yield by 1.55%, 3.17%, 8.91%, and 12.92%, respectively; while an increase in minimum temperature by 1°C, 2°C, 3°C, and 4°C decreased the yield by 0.88%, 1.84%, 6.78%, and 9.15%, respectively, under ambient CO_2 (380 ppm) for the rainfed condition. Figure 33.5a indicates that an increase in maximum temperature by 1°C, 2°C, 3°C, and 4°C decreased the yield by 0.45%, 0.85%, 5.45%, and 7.74%, respectively; while an increase in minimum temperature 1°C, 2°C, 3°C, and 4°C decreased the yield by 0.06%, 0.20%, 2.55%, and 5.06%, respectively, under elevated doubling CO_2 (760 ppm) for the rainfed condition.

Under irrigated condition (Figure 33.5b), increase in maximum temperature decreased the yield by 1.89%, 8.13%, 14.80%, and 18.10% respectively, and increase in minimum temperature decreased the yield by 0.92%, 5.07%, 10.09%, and 13.82% respectively under ambient CO_2 (380 ppm). Similarly under elevated CO_2 (760 ppm), increase in maximum temperature decreased the yield by 0.67%, 3.56%, 7.69%, and 10.92% respectively, and increase in minimum temperature decreased the yield 0.29%, 1.66%, 4.79%, and 8.22% respectively (Figure 33.5b).

Thus, an increase in temperature tends to have a negative effect on the maize crop yield. An increase in maximum temperature causes more effect on the maize grain yield in comparison to the minimum temperature for both CO_2 levels (380 ppm and 760 ppm) for both rainfed and irrigated conditions. Lizaso et al., (2018) stated that heat stress during the silking period (maturing stage) cause more yield reduction, because high temperatures accelerate development level, which reduce the span of the vegetative and reproductive stages (Hatfield and Prueger, 2015).

However, elevated CO_2 enhanced the grain yield as compared to the ambient CO_2 scenario. The yield reduction is more in irrigated than rainfed condition because soil becomes water deficient

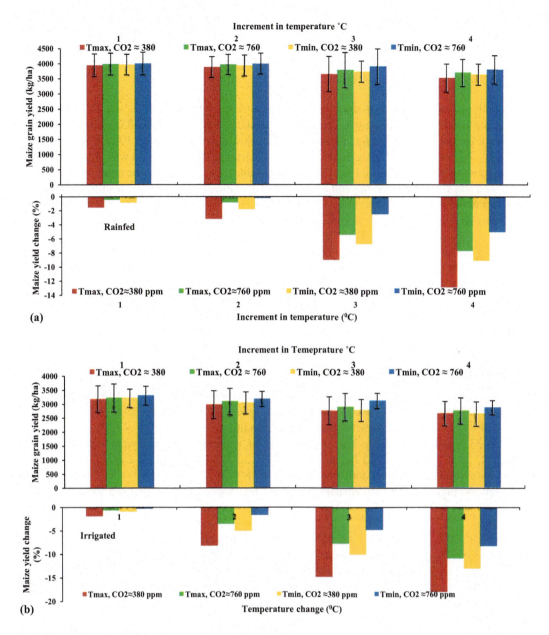

FIGURE 33.5 Individual effects of increased temperature (maximum and minimum) under two CO_2 scenarios at 380 ppm and 760 ppm for the (a) rainfed condition and (b) irrigated condition.

due to the increase in temperature (Mittler, 2006). Moreover, an increase in minimum temperature reduces the biomass (Rosenzweig and Tubiello, 1996), which also affects the maize crop yield.

33.3.3 DETERMINATION OF APPROPRIATE SOWING DATE

To determine the appropriate sowing date for the study region, the maize grain yield was simulated using 30 years of historical weather data from the year 1982 to 2012 under irrigated and rainfed conditions. Sowing dates of May 30, June 10, June 20, June 30, and July 10 for rainfed conditions; and sowing dates of December 25, January 5, January 15, January 25, and February 4 for irrigated conditions were considered. Statistical parameters, like coefficient of variation (CV) and standard deviation with mean yield were considered as the major criteria for determining the most appropriate sowing dates for both rainfed and irrigated conditions.

33.3.3.1 Rainfed Condition

Evaluation of sowing dates was done for the nitrogen levels of 60, 80, 100, and 120 kg/ha under the rainfed conditions to determine the appropriate date of sowing. Table 33.5 indicates that the lowest coefficient of variation for maize grain yield was obtained on June 10 (7.33%–7.77%) and June 20 (7.74%–10.23%). The highest coefficient of variance for grain yield was obtained for July 10 (11.97%–14.79%) and May 30 (10.47%-10.60%) (Table 33.5). Although, the planting date of June 30 showed less variation in grain yield, the standard deviation was quite high, leading to nonacceptance of that sowing date (Table 33.5).

Sowing dates of June 10 and 20 gave higher yields over the remaining sowing dates (Table 33.5), which might be due to the non-water stress condition at the silking stage. The earlier sowing date of May 30 was affected by the drought due to high temperature during the silking stage, while the late sowing dates of June 30 and July 10 faced severe water stress conditions during the sensitive anthesis-silking stage, which is one of the major factors liable for the reduction of the maize yield (Tsimba et al., 2013). Otegui and Bonhomme (1998) reported that maize grain yield is highly correlated with kernel number, which is very sensitive to temperature and water stress conditions. Grassani et al. (2009) reported that rainfed maize crops are severely affected by water stress during the reproductive growth period, which results in reduction of the grain yield.

33.3.3.2 Irrigated Condition

Sowing dates were evaluated for the varying nitrogen levels of 75, 100, 125, and 150 kg/ha and also evaluated for varying levels of maximum allowable depletion (MAD) of available soil water (ASW), that is, 10%, 40%, 70%, and 90% for the irrigated condition.

Table 33.6 indicates that the lowest coefficient of variation for grain yield was obtained for January 15 (9.59%–13.91%) and January 25 (10.91%–13.78%). The highest coefficient of variance for the grain yield was obtained for December 25 (16.28%–18.73%) and January 5 (12.10%–13.81%). Although the planting date of February 5 showed less variance, the simulated yield (maximum and minimum) was quite less in all the nitrogen-level treatments, hence the sowing date was not acceptable.

Similarly, Table 33.7 indicates that the low coefficient of variation for grain yield was obtained for January 15 (8.52%–32.25%) and January 25 (8.83%–32.62%). The highest coefficients of variation for grain yield were obtained for December 25 (16.38%–45.79%) and January 5 (11.49%–37.19%). Although, the planting date of February 5 showed a low coefficient of variation, the simulated yield (maximum and minimum) was low for maximum allowable depletion (MAD) treatments, hence the sowing date was not acceptable.

Under irrigated conditions, the grain yield depends on solar radiation and temperature during the silking stage. The earlier sowing dates of December 25 and January 5 produced less average yield in both the nitrogen and water stress conditions due to less solar radiation, while late sowing dates were affected by high temperatures and water stress conditions during the kernel growth stage, which ultimately reduced the biomass and the grain yield. Comparable findings were found by Grassani et al. (2009). Climatic factors such as solar radiation and temperature have more impact on the grain yield, which is quite useful in the prediction of maize yield potential and its variation (Otegui and Bonhomme, 1998).

33.4 CONCLUSIONS

This study focused on the determination of appropriate sowing dates as an effective management strategy to reduce the effect of an increase in temperature on maize grain yield under rainfed and

TABLE 33.5
Simulation of Grain Yield for Different Sowing Dates with Varying Nitrogen Levels for Rainfed Conditions

N60

Sowing Date	Yield (kg ha⁻¹)				
	Mean (kg ha⁻¹)	Minimum (kg ha⁻¹)	Maximum (kg ha⁻¹)	STD (kg ha⁻¹)	CV (%)
30-May	3409	2711	4450	333.6	10.60
10-Jun	3547	2989	4004	267.9	7.55
20-Jun	3498	2862	4416	358	10.23
30-Jun	3629	2697	4340	366.2	10.09
10-Jul	3596	2482	4451	532.2	14.79

N80

Sowing Date	Yield (kg ha⁻¹)				
	Mean (kg ha⁻¹)	Minimum (kg ha⁻¹)	Maximum (kg ha⁻¹)	STD (kg ha⁻¹)	CV (%)
30-May	4200	3364	5447	439.9	10.47
10-Jun	4273	3676	4949	320.8	7.51
20-Jun	4206	3607	5243	381.9	9.08
30-Jun	4352	3409	5317	394.7	9.07
10-Jul	4302	3120	5364	570.5	13.26

N100

Sowing Date	Yield (kg ha⁻¹)				
	Mean (kg ha⁻¹)	Minimum (kg ha⁻¹)	Maximum (kg ha⁻¹)	STD (kg ha⁻¹)	CV (%)
30-May	4810	3959	6180	507.6	10.55
10-Jun	4875	4184	5606	357.4	7.33
20-Jun	4777	4146	5867	389.6	8.16
30-Jun	4912	4076	5977	381.7	7.77
10-Jul	4904	3629	6276	597.7	12.19

N120

Sowing Date	Yield (kg ha⁻¹)				
	Mean (kg ha⁻¹)	Minimum (kg ha⁻¹)	Maximum (kg ha⁻¹)	STD (kg ha⁻¹)	CV (%)
30-May	5296	4527	6740	561.7	10.6
10-Jun	5369	4440	6248	413.7	7.7
20-Jun	5216	4585	6417	403.8	7.74
30-Jun	5371	4337	6654	409.9	7.63
10-Jul	5373	3926	6914	643.2	11.97

TABLE 33.6
Simulation of Grain Yield for Different Sowing Dates with Varying Nitrogen Levels for the Irrigated Condition

N_{75}

Sowing Date	Grain Yield (kg ha^{-1})				
	Mean (kg ha^{-1})	Minimum (kg ha^{-1})	Maximum (kg ha^{-1})	STD (kg ha^{-1})	CV (%)
25-Dec	2672	1643	3366	435.2	16.28
05-Jan	2658	2062	3231	326.1	12.27
15-Jan	2802	2289	3436	341.8	12.19
25-Jan	2948	2483	3882	375.8	12.75
05-Feb	3016	2290	3483	306.1	10.15

N_{100}

Sowing Date	Grain Yield (kg ha^{-1})				
	Mean (kg ha^{-1})	Minimum (kg ha^{-1})	Maximum (kg ha^{-1})	STD (kg ha^{-1})	CV (%)
25-Dec	3603	2095	4714	635.7	17.64
05-Jan	3589	2716	4567	495.7	13.81
15-Jan	3756	2990	4723	503	13.39
25-Jan	3952	3267	5345	544.8	13.78
05-Feb	3987	2990	4789	441.5	11.07

N_{125}

Sowing Date	Grain yield (kg ha^{-1})				
	Mean (kg ha^{-1})	Minimum (kg ha^{-1})	Maximum (kg ha^{-1})	STD (kg ha^{-1})	CV (%)
25-Dec	4511	2561	6087	845.1	18.73
05-Jan	4514	3450	5564	612.7	13.57
15-Jan	4714	3732	6053	582.5	12.35
25-Jan	4740	4047	6000	578.7	12.20
05-Feb	4900	3658	5866	561.7	11.46

N_{150}

Sowing Date	Grain yield (kg ha^{-1})				
	Mean (kg ha^{-1})	Minimum (kg ha^{-1})	Maximum (kg ha^{-1})	STD (kg ha^{-1})	CV (%)
25-Dec	5268	3009	7085	888.0	16.86
05-Jan	5222	4102	6355	632.0	12.10
15-Jan	5378	4406	6580	515.8	9.59
25-Jan	5395	4659	6294	508.3	10.91
05-Feb	5385	4357	6253	596.2	11.07

TABLE 33.7

Simulation of Grain Yield for Different Sowing Dates with Varying Nitrogen Levels for the Irrigated Condition

MAD10%

Sowing Date	Mean (kg ha⁻¹)	Minimum (kg ha⁻¹)	Maximum (kg ha⁻¹)	STD (kg ha⁻¹)	CV (%)
25-Dec	3567	2278	4849	623.77	17.49
05-Jan	3882	3042	4797	445.94	11.49
15-Jan	4211	3434	5039	358.84	8.52
25-Jan	4145	3542	4770	366.13	8.83
05-Feb	3715	2594	4617	438.51	11.80

MAD70%

Sowing Date	Mean (kg ha⁻¹)	Minimum (kg ha⁻¹)	Maximum (kg ha⁻¹)	STD (kg ha⁻¹)	CV (%)
25-Dec	2966	259	4840	1198	40.38
05-Jan	3267	1138	5086	961	29.40
15-Jan	3489	1793	4675	744	18.45
25-Jan	3307	2383	4449	616	18.62
05-Feb	3355	2166	4335	552	16.45

MAD40%

Sowing Date	Mean (kg ha⁻¹)	Minimum (kg ha⁻¹)	Maximum (kg ha⁻¹)	STD (kg ha⁻¹)	CV (%)
25-Dec	3811	2332	4930	624.47	16.38
05-Jan	3696	2559	5094	548.40	14.84
15-Jan	3866	2961	4691	466.38	12.06
25-Jan	3661	3100	4519	443.7	12.11
05-Feb	3412	2394	4423	497.21	14.57

MAD90%

Sowing Date	Mean (kg ha⁻¹)	Minimum (kg ha⁻¹)	Maximum (kg ha⁻¹)	STD (kg ha⁻¹)	CV (%)
25-Dec	2524	259	4840	1155.61	45.79
05-Jan	2866	1138	5086	1065.99	37.19
15-Jan	2892	1793	4675	932.92	32.25
25-Jan	2802	2383	4449	905.69	32.62
05-Feb	3009	2166	4335	757.83	25.19

irrigated conditions. Results clearly conclude that an increase in maximum temperature causes more effect on the maize grain yield in comparison to the minimum temperature in both CO_2 levels (380 ppm and 760 ppm) for both rainfed and irrigated conditions. Moreover, the yield reduction is more in irrigated than rainfed conditions as the soil becomes water deficient due to an increase in temperature. Also, the elevated CO_2 enhanced the grain yield in comparison to the ambient CO_2 scenario in both rainfed and irrigated conditions.

The combination of sowing dates with nitrogen fertilizer and water management not only can reduce the use of these but can also reduce the effect of drought due to increased temperature. Sowing dates have a significant effect on maize growth parameters starting from plant emergence to maturity as well as on the yield parameters such as kernel number and cob number per unit area. The sowing dates of June 10 and 20 are suitable dates for rainfed conditions, while January 15 and 25 are suitable for irrigated conditions. Thus, the appropriate date of sowing is of prime importance for any climatic region to cope with the drought situation that may occur due to increased temperature.

Thus, optimum sowing dates can alter negative effects of increased temperature due to climate change as a no-cost adaptation strategy. Further, the study suggests that crop modeling can be an effective tool to assess different adaptation measures under different environmental conditions to reduce the climate change impact on crop growth.

REFERENCES

Daccache, A., Weatherhead, E.K., Stalham, M.A., and Knox, J.W. 2011. Impacts of climate change on irrigated potato production in a humid climate. *Agricultural and Forest Meteorology*, 151(12), pp.1641–1653.

Grassini, P., Yang, H. and Cassman, K.G. 2009. Limits to maize productivity in Western Corn-Belt: A simulation analysis for fully irrigated and rainfed conditions. *Agricultural and Forest Meteorology*, 149(8), pp.1254–1265.

Hatfield, J.L. and Prueger, J.H. 2015. Temperature extremes: Effect on plant growth and development. *Weather and Climate Extremes*, 10, pp.4–10.

Hunt, L.A, Pararajasingham, S., Jones, J.W., Hoogenboom, G., Imamura, D.T., and Ogoshi, R.M. 1993 GENCALC: Software to facilitate the use of crop models for analyzing field experiments. *Agronomy Journal*, 85, pp.1090–1094.

IPCC (Intergovernmental Panel on Climate Change). 2007. *Climate Change Impacts, Adaptation and Vulnerability, Report of Working Group II to the Fourth Assessment Report of the IPCC*, Cambridge University Press, Cambridge: UK and New York.

Lizaso, J.I., Ruiz-Ramos, M., Rodríguez, L., Gabaldon-Leal, C., Oliveira, J.A., Lorite, I.J., Sánchez, D., García, E. and Rodríguez, A. 2018. Impact of high temperatures in maize: Phenology and yield components. *Field Crops Research*, 216, pp.129–140.

Meehl, G.A., Arblaster, J.M., and Tebaldi C. 2007. Contributions of natural and anthropogenic forcing to changes in temperature extremes over the U.S. *Geophysical Research Letter*, 34(19), https://doi,org/10.1029/2007GL030948

Mittler, R. 2006. Abiotic stress, the field environment and stress combination. *Trends in Plant Science*, 11(1), pp.15–19.

Otegui, M.E. and Bonhomme, R. 1998. Grain yield components in maize: I. Ear growth and kernel set. *Field Crops Research*, 56(3), pp.247–256.

Panda, R.K., Behera, S.K., and Kashyap, P.S. 2004. Effective management of irrigation water for maize under stressed conditions. *Agricultural Water Management*, 66, pp.181–203.

Rosenzweig, C., and Tubiello, F.N. 1996. Effects of changes in minimum and maximum temperature on wheat yields in the central US A simulation study. *Agricultural and Forest Meteorology*, 80(2–4), pp.215–230.

Srivastava, R.K., Panda, R.K. and Chakraborty, A., 2020. Quantification of nitrogen transformation and leaching response to agronomic management for maize crop under rainfed and irrigated condition. *Environmental Pollution*, 265, pp.114866.

Tsimba, R., Edmeades, G.O., Millner, J.P., and Kemp, P.D. 2013. The effect of planting date on maize grain yields and yield components. *Field Crops Research*, 150, pp.135–144.

Tubiello, F.N., and Ewert, F. 2002. Modeling the effects of elevated CO2 on crop growth and yield: A review. *European Journal of Agronomy*, 18, pp.57–74.

34 Life Despite Drought in the Brazilian Semiarid

Juliana Espada Lichston, Rebecca Luna Lucena,
Virgínia Maria Cavalari Henriques,
Raimunda Adlany Dias da Silva, and
Magda Maria Guilhermino

CONTENTS

34.1 THE BRAZILIAN SEMIARID

Brazil is a country of great territorial dimension, being the fifth largest country in the world, with approximately 94% of its territory located in a tropical climate zone. Because it is located in a low latitude zone, and is bathed by the Atlantic Ocean, and has extensive rivers in its interior, much of the country has a hot and humid climate that favors extensive forest cover of the equatorial and tropical types (rain forest). Brazil has a richness of six biomes, namely, Amazon, Atlantic Forest, Cerrado, Caatinga, Pantanal, and Pampa. The only biome with a semiarid climate is the Caatinga biome. Only a small portion of the Brazilian territory, the south and part of the southeast, is located in a subtropical and temperate zone (Alvares et al., 2013).

Within this humid tropical universe, right in the northeast portion of the Brazilian territory, there is a semiarid zone, surrounded by dry and transition subhumid areas, with an extension of 1,128,697 km², representing 18% of the national territory. The Brazilian semiarid region currently comprises 1262 municipalities in the states of Maranhão (MA), Piauí (PI), Ceará (CE), Rio Grande do Norte (RN), Paraíba (PB), Pernambuco (PE), Alagoas (AL), Sergipe (SE), Bahia (BA), and Minas Gerais (MG) (Figure 34.1). The states of Ceará, Paraíba, Pernambuco and Rio Grande do Norte have more than 80% of their territories in this semiarid zone, where 27,870,241 inhabitants live, with approximately 62% in the urban area and 38% in the rural area (IBGE, 2019).

The criteria used in the delimitation of the semiarid were the average annual rainfall equal to or less than 800 mm; the Thornthwaite Aridity Index equal to or less than 0.50; and the daily percentage of water deficit equal to or greater than 60%, considering all days of the year (IBGE, 2018). The authority to set technical and scientific criteria for delimiting the semiarid region was given to the Deliberative Council of the Northeast Development Superintendence (SUDENE). However, the semiarid area did not always have the current dimensions. According to information from SUDENE,

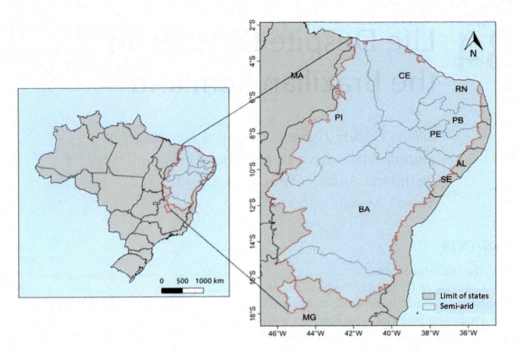

FIGURE 34.1 Brazilian map with the delimitation of the Brazilian semiarid region. (From SUDENE, 2018; Lopes et al., 2019.)

the first official delimitation took place in 1989, when the semiarid region was defined by only one criterion: a region with average rainfall equal to or less than 800 mm/year.

The Brazilian semiarid has two important characteristics: it is located between the tropical Atlantic Ocean and a large equatorial forest, being far from any desert; and it is the most populated semiarid region on Earth. This region also has another particularity since all of its hydrographic basins are exorheic and, therefore, do not present major problems of salinization of waters and soils, very common in other semiarid regions of the world. Only in the flattened areas of the northern coast of the state of Rio Grande do Norte, due to the base level of the estuaries, is there an area of greater salinization that allows the exploitation of sea salt, making this state the main salt producer in all of Brazil (Ab'Saber, 2003). The semiarid climatic type of Northeast Brazil has as its main characteristics high temperatures, and high rates of evapotranspiration associated with the temporal and spatial instability of rains, which contribute to periodic drought events (Nimer, 1979; Cabral Junior and Bezerra, 2018; Cunha et al, 2019). Table 34.1 shows the climatological norms of some municipalities located in the Brazilian semiarid (National Institute of Meteorology).

In the Brazilian semiarid region, rainfall is concentrated annually and it is possible to identify two well-defined seasons: rainy and dry. The rainy season begins at the summer solstice (late December) and ends in late autumn (May). During this period, the Intertropical Convergence Zone (ITCZ) and the high-level cyclonic vortexes operate over the region, the main systems responsible for deep convection clouds that operate in the region during the rainy season. The rains are poorly distributed since the convective clouds generate strong storms, but cover (horizontally) a very restricted area, generally less than a dozen kilometers. Thus, it is common to have an average rainfall in a municipality but in the neighborhood there is a drought.

Seasonal and interannual drought periods are a natural part of the semiarid climatic/rainfall behavior, and there is a high interannual variability, where dry and rainy years follow between "usual" years (Kayano and Andreoli, 2009; Marengo et al., 2017). Drought, both meteorological and hydrological, are common phenomena in the region. Drought can be defined as a prolonged

TABLE 34.1

Climatological Standards of Municipalities Located in the Brazilian Semiarid Region (1961–2010)

Municipality (State)	Temperature (°C)	Precipitation (mm)	PET (mm)*	AI*
Paulo Afonso (BA)	25.9	561.3	2,080.2	0.29
Quixeramobim (CE)	26.9	785.3	2,179.1	0.39
Monteiro (PB)	24.1	682	1,592.8	0.45
Petrolina (PE)	26.9	530	2,399.9	0.24
Picos (PI)	27.5	815.3	2,748.7	0.30
Macau (RN)	27.3	512.8	2,279.2	0.22
Average value	**26.4**	**647.8**	**2,213.3**	**0.32**

Source: INMET data.
*Period 1981–2010.
Note: The data refer to the average annual values of temperature, precipitation, potential evapotranspiration (PET), and Aridity Index (AI).

period of rainfall deficiency, greater than an unusual period of drought (dry spells) (months/season), that is, when precipitation is less than the expected value over a long period (years). Especially considering differences between the dry period of the year, interspersed with the rainy season from longer periods of drought caused by natural phenomena such as El Niño. However, human activities can accentuate the possible impacts of drought, since these "impacts" result from the interaction between the natural event and the demand of people and their activities (Marengo et al., 2017). Another important issue to note is that, even if the rains occur within the "usual", or even if they are above average in a given year, the very marked seasonality means that during the dry season, there is always an absence of rain that, associated with high temperatures (~30°C), surface waters evaporate with sufficient intensity to lower the levels of large water reservoirs and make rivers disappear. The climate and the environment of the semiarid promote a transformation in the incredible landscape, where flora, fauna, and man can and must adapt and not remain subject to a policy that for a long time treated the semiarid as a hostile and inhospitable environment.

Covering the Caatinga and Cerrado biomes, and transition areas, popularly called "wild" or "dry forests", it is in the Caatinga that the driest and most representative areas of the Brazilian semiarid are found.

34.2 CAATINGA, AN EXCLUSIVELY BRAZILIAN BIOME

The Caatinga makes up 90% of the Brazilian semiarid, an endemic biome, and has an indigenous name given to its vegetation cover, which means "white forest", due to the gray aspect it presents during dry periods when trees lose their leaves in adaptation to restrictions water resources. The Caatinga is characterized by maximum plant adaptation to water shortages and, therefore, there are many succulent and deciduous plants. According to Ab'Saber (1977, 2003), there is no better thermometer to delimit the semiarid region than the extremes of the Caatinga vegetation itself, rich in shrub-tree and cactus species. In Figure 34.2, some images reflect the environmental characteristics prevalent in the semiarid region in the domain of Caatinga.

The Caatinga soils have, in general, natural fertility between medium and high, since rainfall totals prevent leaching processes, making them suitable for agriculture where the soil and geomorphological conditions are favorable, however, because of the climatic conditions and ecological factors present, agricultural activities based on monoculture are not adapted to this environment,

FIGURE 34.2 Mosaic of images showing aspects of soils, vegetation, and landforms in the Caatinga domain: (a–d) dry season, (e–f) rainy season.

causing severe negative environmental and social impacts. The removal of vegetation cover and the loss of biodiversity, where fauna and flora suffer retraction and extinction, culminate in the concentration of land and the exploitation of labor, linked to soil degradation and poverty of the local population.

It is important to highlight that in the Caatinga, there is also climatic and environmental diversity, since in the mountain areas, where the altitude is higher than 500 m, there are milder temperatures and the affection of the orography promotes greater humidity in the soil and the air, through the presence of clouds, fog, and orographic dew. It is in these mountain areas where survival fruit cultivation is practiced and where the mainsprings of rivers in the region spring up. In this context, the Brazilian semiarid region is a region with a particular climatic condition, which generated a specific biome, the Caatinga, rich in endemic species, plants, and animals, where precipitation rates modify the landscape, encouraging the "backwoods" peoples to adapt, who must be focused on peaceful coexistence, conservation, and preservation of the Caatinga.

34.3   FAMILY FARMING IN THE BRAZILIAN SEMIARID REGION

The Caatinga and family farming are fused into a historic, complex, diverse, and intense lifestyle and tradition. The Caatinga has been inhabited by indigenous people since prehistory, and droughts have been reported since 1587, described with the Indians going to the coast in search of food. The indigenous peoples were almost wiped out, and the remnants were brutally colonized along with enslaved people who came to work in the sugarcane monocultures of Brazil's ruling class, the European whites. These three ethnicities – whites, Indians, and blacks – formed the "Brazilian" population, and the miscegenation was historically unfair since the predominance of privileges and benefits was always for the whites, and the Indian and the black people were always relegated to a condition of ethnic and economic inferiority imposed by the current society.

The large sugarcane plantations were located mainly in the coastal areas of the Northeast region of Brazil. In the interior of the region, in the so-called *sertões*, cattle ranching was developed to meet the needs of meat, milk, leather, tallow, cargo, and firewood transportation, and to move the mills in the sugar cane mill. Country cattle ranching reached one of the most promising historical moments during the 18th century, but with the great drought that occurred in the years 1877 to 1879, herds were wiped out, motivating the development of mining and cotton culture.

With the decline of cotton farming at the end of the 20th century, beef and dairy farming grew extensively and predatory, and the process for cheese production using firewood together with other product segments, such as mining and the production of ceramics, established themselves as economic alternatives based solely on the exploitation of the environment and cheap labor.

Due to the prolonged droughts and the economic problems, governmental recommendations regarding the climate of the Caatinga, from the 19th century onward, were based on the understanding of "combating drought". In this sense, to meet the interests of the ruling class, dams were built, along with investments in large irrigation projects. For the majority of the population, the economically poor, emergency measures were implemented, such as the opening of work fronts and the distribution of water in containers, measures that never relieved the population from suffering from the prolonged drought period. For this poor population, subsistence agriculture remained the basis of the mini-farm, where they planted crops such as cassava, beans, potatoes, and corn for the family's food.

This "model" of "development" caused a huge gap in the access of the poor population to the dignities of life, such as access to education, information, and health care, and promoted the deforestation of native vegetation and areas of riparian forest. It promoted burning, the salinization of soils and the loss of fertility, the silting of water bodies, the irrational use of water, the rural exodus, the growth and maintenance of poverty, and subservience to politicians in bad faith and the owners of the latifúndios, who perpetuated power. Still, the lack of adequate public policies resulted in the formation of an excluded poor population, the deforestation of more than 50% of the Caatinga, and the production of areas in the process of severe desertification (PAN, 2004).

From 1980, with an understanding of the concept of sustainability in which development must take into account the dependence and the relationship between the economy, society, and the environment, the Semi-Arid Articulation Network (ASA) was created. This network was created through the critical debate of organized society with social movements, and governmental and nongovernmental agents, a fact that brought a new paradigm for family farming and the residents of the Caatinga. From this, the peaceful coexistence with the semiarid environment based on precepts, values, and practices of agroecology, popular and solidary economy, contextualized education, popular communication, food and nutrition security, conservation, sustainable use and environmental recomposition of natural resources, and the breaking of the monopoly of access to land, water, and other means of production in the direction of sustainable development.

Thus, the implementation of public policies such as the construction of cisterns to collect rainwater for human consumption, reuse of wastewater, sidewalk cisterns to collect water for agriculture, methodologies for the conservation of fodder for animal feed, purchase of family-farmed products to be used to feed local populations, among others, have been implemented. However, these public policies are still very much based on assistance with a focus only on production and the product itself; they are not sufficient to meet the broad role that family farming can play.

In Brazil, family farming was known as subsistence farming and only started to be recognized in 2006, when the Family Agriculture Law (Law No. 11.326, of July 24, 2006) was enacted considering that the family farmer and/or rural family entrepreneur practices activities in the rural environment, simultaneously meeting the following requirements: to hold, in any capacity, an area of up to 4 (four) fiscal modules (defined in Brazilian law); use at least half of the family workforce in the production and income-generation process; obtain at least half of the family income from economic activities in your establishment or enterprise; and be the manager of the strictly family establishment or enterprise.

The family farmer, due to the characteristics peculiar to this category, will always have difficulties in reaching the production scale for a given product. To overcome this limitation, individual and collective efforts, technical assistance companies, and rural extension and public policies must move in the direction of the diversity of production associated with the quality of production (ecological production), the processing and protection of productions, as also in the search for the multifunction of family units. In addition to rural agricultural activities, they should seek to associate nonagricultural activities, such as proposing different types of tourism (contemplation, adventure, environmental education, among others), local handicrafts, and ecological services of the rural unit to allow the expression of the singularity of each unit, adding respect to the culture and traditions of each people and community.

It is important to remember that awareness campaigns by society about the role of family farming as well as the healthy form of production, the valorization of products and services, the purchase of products directly from producers or the production unit is important for life in the countryside to be perennial.

Public welfare policies must be resized to allow family farming to play its full role in the molds of its form of production, service, and lives of integrity and valorization of the rural family society, the rural landscape, tradition, and culture. To do so, they should aim at the quality of life (housing, access to water and sanitation, energy, health, education, security, mobility, and leisure) of farmers in all communities (traditional, riverside, indigenous, quilombolas), of all genders and ages (young, adults, the elderly); agroecological production and the adoption of machinery and equipment, techniques, and technologies appropriate to each family unit that meet their own characteristics; the valorization and commercialization prioritizing local and regional trades; digital inclusion; access to education and training for rural society; technical assistance and constant rural extension and quality; the guarantee of their customs and traditions; and the promotion of a full life with dignity, spirituality and leisure.

Family farming is entitled to technical assistance and quality rural extension, constant in guiding individual work, in building the collective, and to supporting and developing activities (agricultural and nonagricultural). Public policies that involve the entire rural society are vital for the establishment of activities, which should leverage family units to promote good living and that everyone can be proud of their work. Family farming should not be a replica of the production method based on the logic of capital, in which capitalist property is exempt from the social and environmental function, in a process of social exclusion and in the standardization of farming landscapes, where hegemonic power determines the inputs and productivity, destroying the native plant and animal genetic heritage of countries in the search for technological dependence and political subservience.

In the capital logic, the animal and plant "genetic resources" used in production systems should only produce and achieve productivity goals as the sole objective of their existence. The large corporations that dominate the financial market, the labor market, and the social-environmental market, justify the use of inputs to maximize productivity with the sophistry of food production for a growing world population and hide that what they produce are only export goods, "commodities" that contribute to the composition of the "gross domestic product" as a development index that does not reflect social exclusion and environmental exhaustion.

Rich countries benefit from the political and social fragility, and exploit the Caatinga through mining, monoculture renewable energy parks, destroying "unproductive" lands of the Caatinga peoples who are naturally family farmers, exploring and destroying their cultures and traditions, cruelly deforesting this biome, and subjecting the local population to the false provision of ecosystem services.

For this reason, we believe that family farming based on agroecology has the ability to break this paradigm of destruction, with the support of a society and a government that is moving toward restoring social and environmental justice, with the help of the use of technology. In favor of human development and not of corporations, and public policies that promote sustainable development to achieve the Sustainable Development Goals (SDGs) of the United Nations (UN), in the promotion of life and good living.

The current reality of family farming in Brazil is complex and diverse, and in the Caatinga, it is no different because many people inhabit it and not a single people and a single way of living. In this exclusively Brazilian biome are 1,446,842 family farmers, who represent 79% of rural producers, the vast majority of whom are male representatives (ISPN, 2023) and are in traditional rural communities made up of the people of the Caatinga, such as the sertanejos and the indigenous, quilombola, and artisanal fishermen, and also by farmers in rural settlements.

However, the peoples of the Caatinga, seen as a unique people, the sertanejo, mostly formed by whites and browns, with the minority being black and indigenous (Censo Agropecuário, 2017), are known for their resilience and resistance, which did not allow the extinction of original knowledge, its forms of food, health care, nature, and their art and values.

In the Caatinga, there are currently 45 indigenous peoples with a population of around 90,000 inhabitants, spread over 36 indigenous lands, which occupy an area of almost 140,000 hectares, most of them in the São Francisco River Basin, the largest perennial river in the region. Among the indigenous peoples are the Kambiwá, Xukuru, Pankararé, Atikum, Fulni-ô, Jenipapo-Kanindé, Jiripancó, Kariri-Xokó, Kantaruré, Kiriri, Kaimbé, Kambiwá, Pankararu, Pitaguary, Potiguara, Pipipan, Tingui Botó, Trem, Truká, Tumbalalá, Tuxá, Xakriabá, Xukuru, Kariri, and Xocó (FUNAI, 2018). They developed their strategies for survival and living in the Caatinga; are guardians of knowledge about the management of plants, their properties and medicinal uses, and about the ancient technique of searching for groundwater with pitchforks.

Quilombolas (a word that comes from the term *kilombo*, from the language of the Bantu peoples, originally from Angola, which means the place of landing or camp) is the name given to communities formed by descendants of enslaved blacks who escaped between the 16th century and the year 1888, the year of the abolition of slavery in Brazil, and by free poor blacks. In Brazil there are 2847 communities in 24 Brazilian states, 60% of them in the Northeast region (1724 communities). The quilombolas of Conceição das Crioulas and the communities of Fundo e Fechos de Pasto are rural communities that still have their way of life based on the use of common grazing areas for raising cattle, goats, and/or sheep, and extracting food and medicinal plants.

Indigenous and quilombola farmers in the Caatinga live in small areas, in endogamous families, with serious health problems; suffer great violence, and neglect by society and public authorities; and are at the mercy of attacks by land grabbers, loggers, and prejudiced people, causing deaths, health, poverty, ignorance, and loss of knowledge, lifestyle, and cultural and artistic wealth.

Artisanal fishermen and riverside peoples are also family farmers, present in rivers and reservoirs in the Caatinga, as well as extractive communities.

There are many ways of production when we reverence what the universe has given us in a wise and balanced way, and agroecology comes to the rescue and makes us think about other ways of producing, respecting the cycles of nature and living. Agroecology through the encounter between modern knowledge and traditional knowledge can revive and rescue forgotten knowledge and create new concepts, actions, and rural practices in connection with social, environmental (ecosystems and biodiversity), economic, and political aspects. This science is in constant movement and construction, and can be understood as a philosophy of life that values healthy eating and good living.

Family farming and agroecology add up in the search for the valorization of the human being, redirecting him to a prominent place in the universe in his fundamental role of taking care of the existing (keeping) and working to produce healthily (cultivating) and his responsibility with recovery and maintenance of good living, recovering the intrinsic value of everything that was created for the present and the future.

Therefore, the scarcity of water must be understood not only concerning the dry season of the year and the bad relationship with agriculture but also as a reality that it is part of the life of all people everywhere in the world. The construction and implementation of developmental public policies that favor the cycles of nature, local knowledge, and the adoption of proper and appropriate technologies without international dependence, and the training and qualification of the people of the Caatinga, the farmers with their specificities, must value life in the countryside and the conservation and preservation of nature (Figure 34.3).

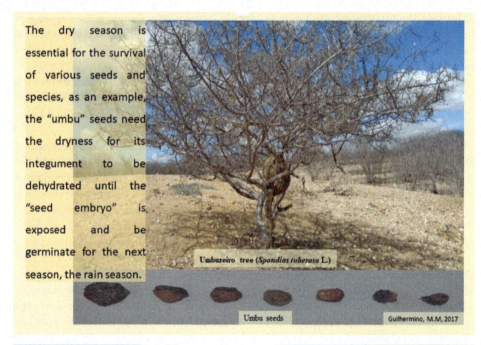

The dry season is essential for the survival of various seeds and species, as an example, the "umbu" seeds need the dryness for its integument to be dehydrated until the "seed embryo" is exposed and be germinate for the next season, the rain season.

Umbuzeiro tree (*Spondias tuberosa* L.)

Umbu seeds   Guilhermino, M.M, 2017

Umbuzeiro tree (*Spondias tuberosa* L.) in the rain season in Caatinga region

FIGURE 34.3   The resilience of nature and its people making the Caatinga an unique biome.

34.4 PRODUCTION POTENTIAL IN THE BRAZILIAN SEMIARID

The Brazilian semiarid region is a region with edaphoclimatic conditions that limit different cultures. In the region, there is also the occurrence of saline soils and prolonged droughts that directly affect agricultural production. The Northeast has been suffering from the increase in areas affected by desertification, which fall into the very serious, severe, and moderate classification (Figure 34.4, right).

Even in the midst of these edaphoclimatic adversities, family farming, as previously stated, is extremely important for the development of Northeast Brazil. The sector is responsible for much of the production of food that is supplied to the urban centers of Brazil (Azevedo and Nunes, 2013). Family farming is highly represented in the productive context of the Northeast region and actively participates in Brazilian agribusiness (Oliveira, Dorner, and Shikida, 2017).

For this sector to boost its productivity, investment in technology, in the development and improvement of techniques and machinery that enable farmers to overcome some difficulties, such as low water availability and the others mentioned earlier, is essential. One of the main needs and concerns of farmers in the semiarid region is related to water supply (Sousa et al., 2019). In this context, it is evident that one of the main limitations faced in the semiarid regions is the low water availability, hindering production in the field by small and large producers, and technology to enhance crops. It is essential to develop and apply technologies aimed at access, the use of water with more efficient irrigation systems, and the reuse of residual water, strategies that can increase production in the region (Coelho et al., 2020). In addition, machinery is needed to facilitate the process from planting to harvesting crops, enabling farmers to increase the final value of their products (Embrapa, 2017).

Another important point is the improvement and selection of crops more adapted and tolerant to semiarid conditions (Lima et al., 2019). Sousa et al. (2019) add that government incentives are necessary, as a way to make production feasible, access to seed, training for handling the crop, and selling products to make agriculture in the Brazilian semiarid region viable.

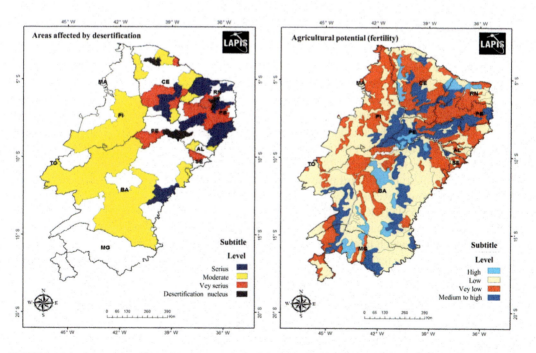

FIGURE 34.4 Maps of the Brazilian Northeast. (Right) Semiarid areas affected by desertification. (Left) Agricultural potential in terms of semiarid fertility. (From SIMA, 2020.)

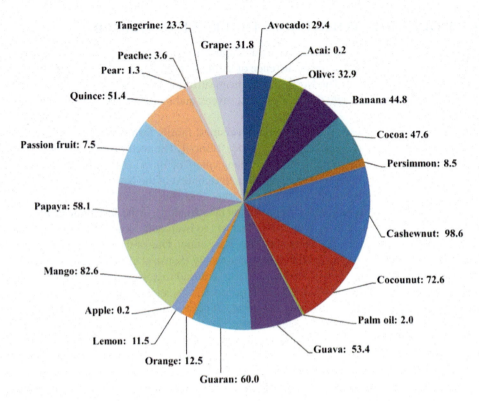

FIGURE 34.5 Total fruit production, in percentage, in 2019 of the states that make up the Brazilian semi-arid. (From IBGE, 2019.)

The Northeast has areas that are classified as high to medium agricultural potential (Figure 34.4, left). The semiarid region has an elevated productive potential, which is responsible for the production of a range of crops in the fruit, horticulture, and bioenergy sectors.

In 2018, Brazil allocated 5,271,748 hectares for fruit production, of which 53.9% of this area refers to production in the states that make up the semiarid region (IBGE, 2018). Among the 21 main crops highlighted in production in the semiarid are cashew nuts, with the production of 98.6% of national production, followed by mango (82.6%), coconut (72.7%), guarana (60.0%), and guava (53.4) (Figure 34.5) (IBGE, 2019).

In the semiarid region, for the horticulture sector, 49 crops are grown, of which 8 stood out in productivity in 2017. In the first position, tomatoes with an approximate production of 261,866 t; in 2019 the Northeast region produced 78,880,761 t of tomatoes (IBGE, 2019); in the second position, peppers with 201,092 t; in the third position, corn with 198,152 t; in the fourth position, coriander with 139,453 t; followed by chayote with 136,114 t, 114,329 t carrots, sweet potatoes with 110,402 t, and lettuce with 105,559 t (IBGE, 2017).

One of the sectors that contributed to the economic development of the Brazilian semiarid region is the textile industry, which has cotton as the main raw material. The cotton production chain, to supply the demand of the textile industry, when compared to other agricultural chains, is shown as one of the most complex and longest (Santos, Silva, and Almeida, 2017). According to the Brazilian Association of Cotton Producers, Brazil ranks first in cotton production in rainfed systems, and remains among the five largest world producers, alongside countries such as China, India, and USA (Abrapa, 2020). In the last five years, Brazilian production has increased from 1,288,800 tons in 2015 to 2,891 tons in 2019, in the same years the Northeast had production from 283,600 tons to 648,400 tons. In 2019, the region contributed about 22.44% of Brazilian production (Abrapa, 2020).

Another crop with high industrial potential is sisal. The semiarid region is the only producer of sisal or *agave* (fiber) in Brazil. Its production is concentrated in the northeast of the country, mainly in the states of Ceará, Rio Grande do Norte, Paraíba, and Bahia (IBGE, 2018). In addition to sisal fiber, mucilage can be used to feed animals (Embrapa, 2017). The sisal production process started in 1940 with a production of 234 t, 12 years later with a production of 49.937 t, the crop became the economic base of a group of municipalities in the semiarid region of the state of Bahia (Santos e Silva, 2017). In the automobile industry, the crop has been used satisfactorily in the production of plastic parts, with the combination of fiber to reinforce polypropylene (Júnior et al., 2020) in upholstery, and it can also be used to reinforce natural rubber (Júnior et al., 2020). Ford, the automaker, replaced the use of plastic obtained through oil, in a percentage of 30% of its vehicle line, with material based on sisal fiber. The crop is highly profitable and requires small areas for planting and harvesting in 2.5 years (ASCOOB, 2008).

34.4.1 RAW MATERIAL FOR WOOD PRODUCTION

The Brazilian semi-arid region provides wood for energy purposes for the whole country. The raw material must be obtained from handling units, taking into account some characteristics such as wood volume production, energy density, basic density, fuel value, and ash content, which must be as small as possible to increase the efficiency (Carvalho et al., 2020). Among the main species with energetic potential in the semiarid are *Mimosa tenuiflora*; *Piptadenia stipulacea*; *Croton sonderianus*; *Aspidosperma pyrifolium* Mart.; *Anadenanthera colubrina* (Vall) Brenan.; *Poincianella pyramidalis* [Tul.] L.P.; *Bauhinia cheilantha* (Bong.) Steud.; *Piptadenia moniliformis*; *Sandy Mimosa* (Willd.) Poir. var. (Calumbi); *Combretum leprosum* Mart.; *Ziziphus joazeiro* Mart.; *Caesalpinia leiostachya* Benth.; *Senna spectabilis*; and *Zanthoxylum stelligerum and Erythroxylum pungens* (Carvalho et al., 2020; Santos et al., 2020).

34.4.2 RAW MATERIAL FOR BIOFUELS PRODUCTION

Among the crops widely studied for the production of biofuels, which can be produced in semiarid regions are palm oil, which originates olive oil and palm oil (Brazílio et al., 2012). The contribution of this crop to biodiesel in the Northeast region went from 3.17% in January 2019 to 24.7% in January 2020. In the same period, the production of cotton oil went up from 3.8% to 6.68% (ANP, 2019, 2020), with oil content in the seeds between 15.56% and 30.15% (Khan et al., 2010). In 2019, the Northeast region produced 1,013,376 t with a contribution of approximately 15% of Brazilian production (IBGE, 2020). The sunflower has approximately 50.66% of oil in its seeds, and of this 26.2% is composed of fatty acid of the oleic type (Gama, Gil, and Lachter, 2010), important for the biofuels industry.

Castor bean (*Ricinus communis* L.), with production in the Northeast region of 21,569 t in 2019, representing 76% of Brazilian production (IBGE, 2020), has about 45% to 50% of oil in its seeds (Beltrão, 2003). The faveleira (*Cnidoscolus quercifolius*), a species native to the semiarid region, has approximately 60% oil in its seeds, comprising 50% to 90% linoleic acids (Noberto, 2013). *Jatropha curcas* L. has a high oil content and can be used medicinally (Arruda et al., 2004). Moringa (*Moringa oleífera* L.) has a 40% oil yield in its seeds (Oliveira et al., 2012). These have representative concentrations of protein, calcium, phosphorus, magnesium, potassium, sodium, 17 fatty acids, vitamin C, and carotenoids (Moyo et al., 2011). Licuri (*Syagrus coronata*) has 40% oil (La Salles et al., 2010) and can be used in human food, with application in the production of fermented milk (Souza et al., 2020), crafts, medicine, and biofuel (Rufino et al., 2008). Babaçu has high added value and oil with a high lauric acid content (Machado, Chaves, and Antoniassi, 2006), also present in crops such as *Orbignya especiosa* (57.5%) and

Cocos nucifera, potential crops for application in the advanced fuel industry as biokerosene (Albuquerque 2017).

In addition to the aforementioned oilseeds, as a potential for biofuel production in the Brazilian semiarid, *Carthamus tinctorius* L. (safflower) is a great alternative for the region. Small-scale experimental crops are already being developed in the northeastern hinterland to obtain data on crop development and productivity. *C. tinctorius* has shown good results in the field and recently, in 2019, it was included in the group of potential matrices for biofuels. The crop has been gaining prominence in the energy sector due to the high concentration of oil in its seeds, ranging from 29% to 40% (Yeilaghi et al., 2012), and composition of linoleic fatty acid from 72% to 78%, oleic acid 16% to 20 %, palmitic 6% to 8%, and stearic 2% to 3% (Hall III, 2016; Junior et al., 2017).

Safflower has been studied since 2010, in field cultivation in the northeastern semiarid by the team of the Research Laboratory of Energy Crops (LIMVE – Federal University of Rio Grande do Norte). The team's focus is the evaluation of the development and production of safflower in semiarid edaphoclimatic conditions, aiming at the production of renewable fuels such as biodiesel and aviation biokerosene. The results obtained by the research group have been encouraging, as they found that the culture significantly reduced its reproductive cycle, from 130–150 days, as mentioned by Galant, Santos, and Almeida-Silva (2015), to 75 days, benefiting regional farmers who can obtain four cultivation cycles per year (Lichston et al., 2016). After cultivation, vegetable biomass can be used for fertilizing the soil, also contributing to the reduction of expenses during cultivation, with the crop being less demanding in terms of nutrition. Another fact observed in the cultivation of safflower in the semiarid region was the absence of herbivory in young and adult plants, contributing significantly to the reduction of expenses by farmers with inputs for combating herbivores.

As for productivity, safflower reaches 6000 kg ha^{-1} a year in the Brazilian semiarid. In 2018, three different varieties were grown – IMAMT 1750, IMAMT 894, IMAMT 946 – with drip irrigation systems (Figure 34.6), with 34% oil yield in their seeds, higher than the report by Yeilaghi et al. (2012) (29% to 33%). The yield obtained in Brazil was even higher than that of soybeans (20%), the main crop used for biodiesel in the country. Safflower oil is of excellent quality for the production of biofuels with a favorable acidity, saponification, and composition index for this purpose.

These crops have potential for the production of biodiesel and advanced fuels such as aviation biokerosene (BioQAV) through the hydrotreating/hydrocracking route (HEFA), resulting in a drop-in fuel (MME, 2018). BioQAV is gaining prominence, due to the aviation sector's entry in the search to reduce greenhouse gas emissions, with the development and use of environmentally compatible biofuels. For the production of this biofuel, both oilseed crops indicated for biodiesel and those indicated for power generation can be used, as well as pruning residues and bark from crops grown in fruit growing, making the semiarid a potential region for the production of renewable fuels.

34.4.3 Aquaponic Cultivation

The historical search for a natural resource for rural production and the development of cities occurs to the present day. Water has been a protagonist in the life of civilizations. The difficulty of migratory life in prehistory led man, around 10,000 years ago, to find new alternatives in search of food. Until then, humans were essentially a hunter-gatherer (Feldens, 2018).

Due to the discovery of agriculture, the first sedentary population clusters emerged. With the expansion of cultivation fields, there was a need to bring water to distant regions. It is known that in 4000 BC, Sumerian peoples in ancient Mesopotamia were the first to create technologies that could make the best use of water. Due to the irregular and violent floods of the Tigris and Euphrates Rivers and to control the power of the water and make better use of it, dams, settlements, wells, aqueducts, dikes, reservoirs, and irrigation channels were built that transported the water to the dry regions, enabling agriculture, animal husbandry, and, in turn, economy and trade. The Egyptians, in the year 5000 BC, built a huge and developed network of canals, which took advantage of seasonal floods to supply themselves with water, thus sustaining extremely fertile agriculture on initially

FIGURE 34.6 Safflower cultivation in the Brazilian northeastern semiarid in 2018: (a) initial stage of development, (b–c) flowering, (d) fruiting.

poor land (sandy and dry). In this way, irrigation provided agricultural efficiency, also allowing the cultivation of hydric-sensitive land.

Over the years, all of these technologies have endured, serving as a foundation for several areas. It is clear that, even with greater access to these new technologies, there is currently no significant gain in terms of their massification. Although in other parts of the world these technologies have been perfected and well disseminated, this reality is not the same in Brazil, however very necessary, given that there are many difficulties in living in regions where rigorous drought periods as in the northeastern semiarid and there are no permanent coping policies for these scenarios.

It is known that humanity has been interacting with ecosystems in the Brazilian semiarid in a degrading way. Barbosa (2008) states that during the period of colonization of Brazil, the Northeast region was one of the first to suffer from the process of European occupation. These peoples were faced with a territory that had a dynamic balance in the Caatinga biome, due to the low degree of anthropism and the habits of the natives who made use of techniques with low environmental impact. In this way it was possible to find fertile lands, water offerings, and an abundance of biomasses. Assessing the current reality, it is possible to consider that humans have caused the evident degradation of natural resources. In turn, the disorderly growth of cities combined with the wrong patterns of economic development add to the composition of irrational use of natural resources and situations of vulnerability (Barbosa, 2008).

The "drought polygon", as a territorial area of coverage of semiarid conditions, is an area of paramount importance for the design of public policies related to the management of the adverse economic, cultural, and social effects of the drought process (Troleis and Silva, 2018). According to Carvalho (2012), drought as a multisectoral phenomenon can categorized as three types: meteorological drought, hydrological drought, and agricultural drought. The different types of drought give rise to a situation of scarcity that can compromise the conditions of survival, to meet both basic and economic demands.

Family farming in the semiarid region is fundamentally characterized by rain-dependent production systems that integrate animal husbandry, plant production, and extraction. Intending to obtain food to last through the year and the weather, the man from the northeastern countryside and, in particular, the countryman, always filled his day seeking to generate food and conserve it in the best possible way. The usual cuisine included manioc flour, beiju, coconut candy, cheeses, bottle butters, dried meat paçocas (Melo and Voltolini, 2019). The main crops historically grown in rain-dependent areas are beans (*Phaseolus vulgaris*), manioc (*Manihot esculenta* Crantz), corn (*Zea mays*), sugar cane (*Saccharum officinarum*; in areas restricted to shallows, humid spaces or in underground dams), peanuts (*Arachis hypogaea*), sorghum (*Sorghum bicolor*), and cotton (*Gossypium hirsutum* L.) (Melo and Voltolini, 2019). As for animal production, popular are cattle, goat, and poultry.

The demand for innovative agricultural production techniques is essential to meet the growing demand for food and reduce the rate of depletion of water resources (Boyd et al., 2020) as well as to enable food production with food security for rural producers. Offering new sources of protein and vegetables or even traditional sources using conventional production systems will not meet the growing demand, of an increasing the world population estimated to be 9.7 billion in 2050 (UN, 2019). In addition, some traditional production techniques are not adapt to the current conditions of natural resources. Conventional terrestrial meat production would also generate dramatic increases in the amounts of greenhouse gases that contribute to climate change (Boyd et al., 2020). Worldwide, capture fisheries are not the solution, as their sustainable limits have already been realized and cannot be significantly expanded (Boyd et al., 2020). Thus, new sources of protein cannot be produced using conventional technologies and management practices, or through commercial overfishing (Boyd et al., 2020).

The interaction between agriculture and aquaculture certainly increases efficiency on a small rural property, especially when the availability of a given resource is limited, as in the case of water (Melo and Voltolini, 2019). Aquaculture offers animal protein of high nutritional value. Integrating crops in the aquaculture system, according to the FAO (2020), means the production of aquatic organisms based on sharing from agriculture, wastewater agroindustries, power plants, and other activities. Integrated systems and biotechnologies will be more appropriate for the semiarid region.

According to Carneiro et al. (2015), the principle for integrated aquaculture production is based on the maximum use of natural resources around a given product, optimizing production with the use of waste generated in the production process, thus reducing the release of agents harmful to the environment. It is based on a concept of high efficiency because it is possible to produce aquatic animals at high densities in a different structure from other animals and plants. The important thing is to have in the system a combination of species of different trophic levels produced in a symbiosis system. Sustainable aquaculture technologies play an important role in meeting the growing demand for food in the coming years. Next, we will highlight successful productions in the northeastern semiarid region as well as other activities suitable to the climatic conditions of the region, such as aquaponics and microalgae production.

Aquaponics is an integrated aquaculture technology suitable for food production in the Brazilian semiarid region. The combination of integrated aquaculture, which includes open, terrestrial, and aquaponic systems, aims to design food production systems providing biomitigative services for the ecosystem and improving economic production through the co-cultivation of complementary species (Chopin et al., 2012). In this way, making it possible to use the same water for more than one product, the residual outlet from one crop serves as an entrance to another in the same system and distributes production costs. Thus, it enables the production of animal and vegetable protein with a higher production volume, and soil and water savings compared to conventional techniques.

The term *aquaponics* originates from the combination of the words *aquaculture* (production of aquatic organisms) and *hydroponics* (production of plants without soil), referring to the integration between the cultivation of aquatic organisms (fish, shrimp) and the cultivation of hydroponic vegetables. Despite being a new technology in Brazil, aquaponics has been tested and validated in several countries in the last 20 years and has proven accessible from a technical and economic point of view (Rakocy et al., 2006; Graber and Junge, 2009).

Aquaponics and other sustainable production techniques were selected by the Sustainable Aquaculture and Biotechnology Laboratory of the Federal University of Rio Grande do Norte (LASBIO/UFRN) among many areas of aquaculture, since it is a modality of food production with low water consumption and high utilization of the organic waste produced, being an alternative for producing fish and vegetables in the same system, generating less impact on the environment (Tyson et al., 2011). Thus, it is suitable for agroecological production for family farming.

LASBIO/UFRN plays the role of improving, expanding, and multiplying these techniques based on the local reality to provide a better quality of life, well-being, convenience, and health through a program that has the main objective of social transformation, and generation of income and work through the development of various sustainable and innovative food production technologies from aquaculture and agriculture in the semiarid in Rio Grande do Norte in the municipality of São José de Campestre. In addition, it aims to reconcile the problems faced with drought and the preservation of natural resources to guarantee the development and citizenship of these communities and, thus, collaborate with the country by building multiplier agents to continue the actions implemented. The program is based on the guarantee of technologies that enable the best use of water through interconnected sustainable systems such as aquaponics, drip irrigation with reuse of the effluent water from shrimp production, and use of the extract and biomass of macroalgae for improvement and optimization of aquaculture production. The program proposal is based on the guidelines of the UN 2023 agenda, which advises to create conditions for sustainable, inclusive and economically sustained growth, shared prosperity and decent work for all, taking into account the different levels of development and national capacities. The actions of the program meet objectives 2, 12, and 15.

For two years, several production tests of more suitable species were carried out at the producer's site where the LASBIO-UFRN program was developed. The following vegetables were produced: cauliflower, basil, cherry tomatoes, okra, taioba, strawberry, chives, lettuce, leaf cabbage, peppers, hibiscus, and bougainvillea. Fertigation is carried out with wastewater from fish farming to drip irrigate a passion fruit plantation. As for the animal species, tilapia is grown in an intensive system with 40 animals per cubic meter (Figure 34.7).

Currently, the producer's goal is to develop a pesto sauce with basil and one with cherry tomatoes to add value and increase shelf life. That way, farmers can sell over months while producing more raw material and increasing revenue entries with a more elaborate product. All plant species continue to be produced for the family's food, and basil and cherry tomatoes are on the production schedule in greater volume for commercialization in the surrounding municipalities. In this way, the program can meet the biggest challenge of food security, that is, access to adequate and healthy food, which is given permanently and sustainably, as clearly articulated by the construction of the Food and Nutrition Security Policy (SAN).

The built structure is of family size. Two types of cultivation structures were built for vegetables with a productivity of 80 feet for the Grow-Bed (gravel bed) and 100 feet for the Nutriente Film Technique (gutters with water). The production of the fish is done in boxes of 1,000 l. Every two months it is possible to harvest 180 feet of basil, which produces 360 pots of pesto sauce with extra virgin olive oil, spices, and nuts. The gross percentage of profit is 50%, obtaining an income with this product of R$1800 or US$360 per month, a percentage of 72% higher than Brazil's minimum wage. The fish production obtained for each six-month cycle is 56 kg. Thus, the per capita

FIGURE 34.7 Integrated sustainable aquaculture system at the site of the LASBIO-UFRN program.

consumption value of fish protein for a family of three is 37 kg. The world average in 2018 was 20.5 kg per capita and in developing countries 19.5 kg per capita (FAO 2020), which is 86% higher than the average for developing countries. The family can sell part of the fish production to increase revenue; a kilo of tilapia in the region is approximately US$2 (Figure 34.8).

The semiarid region has ideal climatic conditions to produce microalgae and to mitigate environmental problems, producing, alternatively, with the use of wastewater from domestic sewage treatment plants. Microalgae are microscopic photosynthetic organisms with a wide variety of species producing biomass from the absorption of sunlight, the capture of CO_2, and nutrients from the aquatic environment, in addition to contributing to the production of atmospheric oxygen. The high temperatures of 30°C and a photoperiod of 12 to 14 hours of light practically the whole year make it possible to produce microalgae with a focus on generating products and income for rural or urban producers.

It is possible to use areas that are not conducive to agriculture. In addition, microalgae are not major consumers of water, as their loss occurs only through photosynthesis and evaporation. The greater solar intensity allows a greater depth of the cultivation media concerning the cultivation of temperate regions and greater productivity. They can be produced in lagoons, raceways, or bioreactors. They are of great biotechnological importance. They are used for food production because they produce various substances, such as vitamins, minerals, pigments, lipids and fatty acids, pigments, and other high-value biomolecules, which may have antimicrobial, antiviral, antibacterial, antioxidant, anti-inflammatory, anti-tumor, and immunomodulatory properties. They present different biotechnological applications, in drugs, cosmetics, human and animal nutrition, biofertilizers, and biofuels. Thus, it becomes an important productive opportunity for the Brazilian semiarid region.

FIGURE 34.8 Production of the sustainable integrated aquaculture system.

34.5 FINAL CONSIDERATIONS

The semiarid region of Brazil has a territorial extension greater than many countries in the world, with 1,128,697 km², representing 18% of the Brazilian territory. This region has peculiar edaphoclimatic characteristics, where rainfall is concentrated annually and it is possible to identify two well-defined seasons (rainy and dry), high temperatures throughout the year, and high rates of evapotranspiration associated with the temporal and spatial instability of the rains, which contributes to periodic events of prolonged drought. However, many plant species have adapted and are tolerant to these environmental conditions, allowing the cultivation of various crops, valuing the family farming of the Brazilian people with agroecological cultivation practices combined with the use of modern technologies.

The role of family farming in Brazil goes beyond food security in the healthy production, naturally without pesticides, of conventional and/or unconventional foods. Family farming also contributes to the appreciation of family work; recovery, conservation, and preservation of native and traditional plant and animal genetic heritage; recovery, conservation and preservation of the Caatinga biome; preservation of the traditions and culture of the people; mitigation of the effects of global warming; decreased hegemonic power over populations; and fair division of land and promotion of social justice.

The economic potential of the Brazilian semiarid region is evident. However, the country needs adequate public policies that allow family farming to fulfill its full potential in the form of production, service, and living, for the integrity and valorization of society, the rural landscape, tradition, culture, leisure, and good living. Public policies should aim at the quality of life (housing, access to water and sanitation, energy, health, education, security, mobility, and leisure) of the inhabitants; the security of commercialization of agricultural and livestock production; access to technologies, agroecology, and constant quality technical assistance and rural extension; digital inclusion; access to education and training for rural society; the preservation of local culture; and the promotion of a dignified and full life for each family in the Brazilian semiarid region.

REFERENCES

Abrapa, Associação Brasileira dos Produtores de Algodão. Algodão no Brasil 2020. Available at: https://www .abrapa.com.br/Paginas/Dados/Algod%C3%A3o%20no%20Brasil.asp Access on 10/08/2020.

Ab'Saber, A.N. 1977. *Problemática da Desertificação e da Savanização no Brasil Intertropical.* São Paulo: IGEOG/USP.

Ab'Saber, A.N. 2003. Caatingas: O domínio dos sertões secos. *In: Os domínios de natureza no Brasil: Potencialidades paisagísticas.* 1ª Ed. São Paulo: Ateliê Editorial.

Albuquerque, M.C.G. 2017. Avaliação físico-química dos óleos de babaçu (Orbignya speciosa) e coco (Cocos nucifera) com elevado índice de acidez e dos ácidos graxos (C6 a C16). *ScientiaPlena*, 13(8). https://doi .org/10.14808/sci.plena.2017.085301

Alvares, C.A., Stape, J.L., Sentelhas, P.C., Gonçalves, J.L.M., Sparovek, G. 2013. Köppen's climate classification map for Brazil. *Meteorologische Zeitschrift*, 22(6), pp. 711–728.

Anp, Agência Nacional do Petróleo, Gás Natural e Biocombustíveis, PERCENTUAL DAS MATÉRIAS-PRIMAS UTILIZADAS PARA PRODUÇÃO DE BIODIESEL - REGIÃO NORDESTE. 2019. Available at: http://www.anp.gov.br/ Access on 06/08/2020.

Anp, Agência Nacional do Petróleo, Gás Natural e Biocombustíveis, PERCENTUAL DAS MATÉRIAS-PRIMAS UTILIZADAS PARA PRODUÇÃO DE BIODIESEL - REGIÃO NORDESTE. 2020. Available at: http://www.anp.gov.br/ Access on 06/08/2020.

Arruda, F.P., Beltrão, N.E.D.M., Andrade, A.P., Pereira, W.E., Severino, L. S. 2004. Cultivo de pinhão manso (*Jatropha curca* l.) como alternativa para o semi-árido nordestino. *Revista Brasileira de Oleaginosas e Fibrosas, Campina Grande*, 8(1), pp. 789–799.

Ascoob, Associação das Cooperativas de Apoio a Economia Familiar, Ford utiliza sisal na fabricação de automóveis 28 deagosto de 2008. Available at: http://www.sistemaascoob.com.br/noticia/34/ford-utiliza -sisal-na-fabricacao-de-automoveis Access on 12/08/2020.

Azevedo, M.B.A., Nunes, E.M. 2013. As feiras da agricultura familiar: Um estudo na rede Xique Xique nos territórios Açu-Mossoró e Sertão do Apodi (RN). *Revista GeoTemas*, 3(2), pp. 59–74.

Barbosa, M.P. 2008. *Desertificação no Estado da Paraíba*. Campina Grande: UFCG/CTRN, 2006. 37 (Apostila). Desertificação. UFCG/CTRN: Campina Grande, (Apostila), pp. 62.

Beltrão, N.E.M. 2003. *Informações sobre o biodiesel, em especial feito com o óleo de mamona*. Embrapa Algodão-Comunicado Técnico (INFOTECA-E).

Brazilio, M., Bistachio, N.J., Cillos Silva, V., Nascimento, D.D. 2012. O Dendezeiro (Elaeis guineensis Jacq.) Revisão. *Bioenergia em Revista: Diálogos* (ISSN: 2236-9171), 2(1), pp. 27–45.

Boyd, C.E.,D'Abramo L.R., Glencross, B.D., Huyben., D.C., Juarez, L.M., . Lockwood, G. S., McNevin, A.A., Tacon, A. G.J., Teletchea, F., Tomasso Jr, J.R., Tucker, C.S., Valenti, W.C. 2020. Achieving sustainable aquaculture: Historical and current perspectives and future needs and challenges. *Journal of the World Aquaculture Society*, 51(3), pp. 578–633.

Carneiro, P.C.F., Morais, C.A.R.S., Nunes, M.U.C., Maria, A.N., Fujimoto, R.Y. 2015. *Produção integrada de peixes e vegetais em aquaponia.* Aracaju: Embrapa Tabuleiros Costeiros, 2015a. (Embrapa Tabuleiros Costeiros. Comunicado Técnico, 189), pp. 23.

Carvalho, O. 2012. A seca e seus impactos. In: *A questão da água no Nordeste. Agência Nacional de Águas.* Brasília, DF: Agência Nacional de Águas. Centro de Gestão e Estudos Estratégicos. BRASIL.

Carvalho, A.C., Santos, R.C., Castro, R.C.V.O., Santos, C.P.S., Costa, S.E.L., Carvalho, A.J.E., Pareyn, F.G.C., Vidaurre, G.B., Junior, A.F.D., Almeida, M.N.F. 2020. Produção de energia da madeira de espécies da Caatinga aliada ao manejo florestal sustentável. *Scientia Forestalis*, 48(126), pp. 3086. https://doi.org/10 .18671/scifor.v48, n 126.08

Cabral Júnior, J.B., Bezerra, B.G. 2018. Análises da evapotranspiração de referência e do índice de aridez para o Nordeste do Brasil. *Revista de Geociências do Nordeste*, 4(1), pp. 71–89.

Chopin, T., Cooper, J.A., Reid, G., Cross, S., Moore, C. 2012. Open-water integrated multi-trophic aquaculture: Environmental biomitigation and economic diversification of fed aquaculture by extractive aquaculture. *Reviews in Aquaculture*, 4(4), pp. 209–220.

Cunha, A.P.M.A., Zeri, M., Leal, K.D., Costa, L., Cuartas, L.A., Marengo, J. A., Tomasella, J., Vieira, R. M., Barbosa, A. A., Cunningham, C., Garcia, J. V. C., Broedel, E., Alvalá, R., Ribeiro-Neto. G. 2019. Extreme drought events over Brazil from 2011 to 2019. *Atmosphere*, 10(11), pp. 642. https://doi.org/10 .3390/atmos10110642

Coelho, E.F., Coelho Filho, M.A., Simões, W.L., Coelho, Y.S. 2020. Irrigação em citros nas condições do nordeste do Brasil. *Citrus Research & Technology*, 27(2), pp. 297–320.

Embrapa, Empresa Brasileira de Pesquisa Agropecuária. Agricultura familiar Convivência com a Seca, Culturas tolerantes à seca e máquinas para agricultura familiar serão apresentadas no Semiárido Show, 2017. https://www.embrapa.br/busca-de-noticias/-/noticia/29523871/culturas-tolerantes-a-seca-e -maquinas-para-agricultura-familiar-serao-apresentadas-no-semiaridoshow

FAO, Food and Agriculture Organization. 2020. *The State of World Fisheries and Aquaculture 2020. Sustainability in Action*. FAO, Rome.

Feldens, F. 2018. *O Homem a Agricultura a História*. 1ª edição. Lajeado/RS. Editora.

FUNAI - CARTA DOS POVOS INDÍGENAS DO CERRADO E DA CAATINGA. 2018. Avaliable in http:// www.funai.gov.br/index.php/comunicacao/noticias/5058-carta-dos-povos-indigenas-do-cerrado-e-da -caatinga-desafios-para-a-gestao-ambiental-e-territorial-das-terras-indigenas?tmpl=component&amp ;print=1&layout=default&page= Access on 04/10/2023.

Galant, N.B., Santos, R.F., Almeida-Silva, M. 2015. Melhoramento de cártamo (*Carthamus tinctorius L.*). *Acta Iguazu*, 4, pp. 14–25.

Gama, P.E., Gil, R.A.S.S., Lachter, E.R. 2010. Produção de biodiesel através de transesterificação in situ de sementes de girassol via catálise homogênea e heterogênea. *Química Nova*, 33(9), pp. 1859–1862.

Graber, A., Junge, R. 2009. Aquaponic systems: Nutrient recycling from fish wastewater by vegetable production. *Desalination*, 246, pp. 147–156.

Hall, C. 2016. *Overview of the Oilseed Safflower (Carthamus tinctorius L.)*. North Dakota State University, Fargo, ND, USA. https://doi.org/10.1016/B978-0-08-100596-5.00030-5

IBGE, Instituto Brasileiro de Geografia e Estatística, Censo agropecuário de horticultura 2017, 2017. Available at: https://sidra.ibge.gov.br/pesquisa/censo-agropecuario/censo-agropecuario-2017#horticultura Access on 04/08/2020

IBGE, Instituto Brasileiro de Geografia e Estatística, 2018a. *Produção Agrícola Municipal*, Available at: https://www.ibge.gov.br/estatisticas/economicas/agricultura-e-pecuaria/9117-producao-agricola -municipal-culturas-temporarias-e-permanentes.html?=&t=resultados Access on 06/08/2020

IBGE, Instituto Brasileiro de Geografia e Estatística. 2018b. Semiárido brasileiro. Available at: https://www .ibge.gov.br/geociencias/cartas-e-mapas/mapas-regionais/15974-semiarido-brasileiro.html?=&t=sobre Access on July 2020.

IBGE, Brazilian Institute of Geography and Statistics, Semiárido Brasileiro. 2019. Avaliable in https://www .ibge.gov.br/geociencias/organizacao-do-territorio/estrutura-territorial/15974-semiarido-brasileiro .html?t=acesso-ao-produto Access on 04/10/2023.

IBGE, Instituto Brasileiro de Geografia e Estatística Diretoria de Pesquisas, 2020a. *Coordenação Agropecuária, Levantamento Sistemático da Produção Agrícola Levantamento Sistemático da Produção Agrícola – LSPA, junho de 2020*.

IBGE, Instituto Brasileiro de Geografia e Estatística, Diretoria de Pesquisas, 2020b. *Coordenação Agropecuária, Levantamento Sistemático da Produção Agrícola Produção de Cereais, Leguminosas e Oleaginosas - Comparação Entre as Safras 2019 e 2020 - Brasil e Grandes Regiões*. Available in: https://www.ibge.gov.br/estatisticas/economicas/agricultura-e-pecuaria/9201-levantamento-sistematico -da-producao-agricola.html?=&t=resultados Access on 07/08/2020.

ISPN – Institute, Society, Population and Nature. 2023. Avaliable in https://ispn.org.br/biomas/caatinga/povos -e-comunidades-tradicionais-da-caatinga/ Access on 04/10/2023.

Júnior, I.B., Belini, U.L., Ellenberger, A., Keinert, A.C. 2020. FIBRAS NATURAIS E COMPÓSITOS NAS INDÚSTRIAS DE MOBILIDADE. *MIX Sustentável*, 6(4), pp. 129–138, https://doi.org/10.29183/2447 -3073.MIX2020.v6.n4.129-138

Junior, L.A.Z., Paschoal, T.S., Pereira, N., Araujo, P.M., Secco, D., Santos, R.F., Prior, M. 2017. Seed productivity, oil content and accumulation of macronutrients in safflower (*Carthamus tinctorius L.*) genotypes in subtropical region. *Australian Journal of Crop Science*, 11(10), pp. 1254.

Kayano, M.T., Andreoli, R.V. 2009. *Clima da região Nordeste do Brasil*. In: Cavalcanti, I. F. de A. et al (Org). *Tempo e clima no Brasil*. São Paulo: Oficina de Textos, pp. 213–233.

Khan, N.U., Marwat, K.B., Hassan, G., Farhatullah, S.B., Makhdoom, K., Ahmad, W., & Ullah, K.H. 2010. Genetic variation and heritability for cotton seed, fiber and oil traits in *Gossypium hirsutum L*. *Pakistan Journal of Botany*, 42, pp. 615–625.

Khan, A.S., Souza, J.S. 2020.Taxa de retorno social do investimento em pesquisa na cultura da mandioca no Nordeste. *Revista de Economia e Sociologia Rural*, 29(4), pp. 411–426.

La Salles K.T.S., Meneghetti S.M.P., La Salles W.F., Meneghetti, M.R., SantosI.C.F., Silva J.P.V., Carvalho, S.H.V., Soletti,J.I. 2010. Characterization of Syagrus coronata (Mart.) Becc. oil and properties of methyl esters for use as biodiesel. *Industrial Crops and Products*, 32(3), pp. 518–521. https://doi.org/10.1016/j.indcrop.2010.06.026

Lichston, J.E, Moreira, F.G.L., Medeiros L.R. 2016. Agregando valor ao cultivo e cártamo (*Carthamus tinctorius* L.) na cadeia energética do semiárido nordestino. In: MENEZES, R. S. (org) Biodiesel no Brasil: Impulso tecnológico, Lavras, UFLA, v. 1 ISBN: 9788565615013, p. 242.

Lima, É.R., Silva, R.A.D., Sousa, E.A.M., Amurim, A.C.I.L., Lichston, J.E. 2019. Perfil dos agricultores familiares da agrovila canudos, Ceará-Mirim/RN, e aceitação do Carthamus tinctorius L. – oleaginosa promissora para biodiesel. *Nature and Conservation*, 12(3), pp. 17–24. https://doi.org/10.6008/CBPC2318-2881.2019.003.0003

Lopes, J.R.F.; Dantas, M.P.; Ferreira, F.E.P. 2019. Identificação da influência da pluviometria no rendimento do milho no semiárido brasileiro. *Revista Brasileira de Agricultura Irrigada*, 13(5), pp. 3610–3618.

Machado, G.C., Chaves, J.B.P., Antoniassi, R. 2006. Composição em ácidos graxos e caracterização física e química de óleos hidrogenados de coco babaçu. *Revista Ceres*, 53(308).

Marengo, J.A., Torres, R.R., Alves, L.M. 2017. Drought in Northeast Brazil: Past, present, and future. *Theoretical and Applied Climatology*, 129, pp. 1189–1200.

Melo, R.F., Voltolini, T.V. 2019. Agricultura familiar dependente de chuva no Semiárido. *Embrapa Semiárido-Livro técnico (INFOTECA-E)*.

MME, Ministério de Minas e energia. 2018. Organização da Aviação Civil Internacional, Available at: http://www.epe.gov.br/sites-pt/publicacoes-dados-abertos/publicacoes/PublicacoesArquivos/publicacao-402/An%C3%A1lise_de_Conjuntura_Ano%202018.pdf Access on 04/ 05/ 2020

Moyo, B., Masika, P.J., Hugo, A., Muchenje, V. 2011. Nutritional characterization of Moringa (Moringa oleifera Lam.) leaves. *African Journal of Biotechnology*, 10(60), pp. 12925–12933.

Nimer, E. 1979. *Climatologia do Brasil*. Rio de janeiro: IBGE, pp. 422.

Noberto, M.N.S. 2013. *Efeito dos substratos rejeito de vermiculita, fibra e pó de coco verde no enraizamento de alporques de faveleira (Cnidoscolus quercifolius Pohl)*. Dissertação (Mestrado – Curso de Ciências Florestais), Universidade Federal da Paraíba, Patos, pp. 64.

Oliveira, F.D.A.D., Medeiros, J.F.D., Oliveira, F.R.A.D., Freire, A.G., Soares, L.C.D.S. 2012. Produção do algodoeiro em função da salinidade e tratamento de sementes com regulador de crescimento. *Revista Ciência Agronômica*, 43(2), pp. 279–287.

Oliveira, T.J.A., Dorner, S.H., Shikida, P.F.A. 2017. A agricultura familiar e o desenvolvimento rural no nordeste do Brasil: Uma análise comparativa com a região sul. *Acta Tecnológica*, 10(2), pp. 59–74.

Programa de Ação Nacional de Combate à Desertificação e Mitigação dos Efeitos da Seca. 2004 (PAN-Brasil) Avaliable in https://www.mma.gov.br/estruturas/sedr_desertif/_arquivos/pan_brasil_portugues.pdf Access on 04/10/2023.

Rakocy, J.E., Losordo, T.M., Masser, M.P. 2006. Recirculating aquaculture tank production systems: Aquaponics - Integrating fish and plant culture. *Southern Reg. Aquaculture Center Publications*, 454.

Rufino, M.U.D.L., Costa, J.T.D.M., Silva, V.A.D., Andrade, L.D.H.C. 2008. Knowledge and use of ouricuri (Syagrus coronata) and babaçu (Orbignya phalerata) in Buíque, Pernambuco State, Brazil. *Acta Botanica Brasilica*, 22(4), pp. 1141–1149.

Santos, C.P.S., Santos, R.C., Carvalho, A.J.E., Castro, R.V.O., Costa, S.E.L., Lopes, L.I., Pareyn, F.G.C., Júnior, A.F.D., Trugilho, P.F., Carvalho, N.F.O., Magalhães, M.A. 2020. Estoque de energia da madeira em áreas sob manejo florestal no Rio Grande do Norte. *Scientia Forestalis*, 48(126), pp. 3080. https://doi.org/10.18671/scifor.v48n126.06

Santos, E.M.C., Silva, O.A. 2017. Sisal na Bahia-Brasil. *Mercator (Fortaleza)*, 16.

Santos, M.O., Silva, O.M., Almeida, F.M. 2017. Uma análise das restrições comerciais no mercado internacional de algodão. *Revista de Estudos Sociais*, 19(38), pp. 67–85.

Sima caatinga, Sitema de monitoramento e alerta para a cobertura vegetal da catinga. Available at: http://lapismet.com.br/SIMACaatinga/maps_ref.php# Access on 06/09/2020.

Sousa, E., Silva, R.A.D., de Morais, F.C., de Lima, É.R., Lichston, J. E. 2019. Perfil dos agricultores de uma cooperativa de Apodi/RN, receptividade ao cultivo de cártamo e percepção sobre agrotóxicos e alternativas. *Nature and Conservation*, 12(3), pp. 25–36 https://doi.org/10.6008/CBPC2318-2881.2019.003.0004

Souza, J.V., de Oliveira, A.P.D., Silva Ferrari, I., Miyasato, I.F., Carrijo, K. F., Schwan, R. F., Dias, F. S. 2020. Autochthonous and commercial cultures with functional properties in goat milk supplemented with licuri fruit. *Food Bioscience, 35*, 100585. https://doi.org/10.1016/j.fbio.2020.100585

SUDENE – Superintendência do Desenvolvimento do Nordeste. 2018. Delimitação do Semiárido. Avaliable in http://antigo.sudene.gov.br/delimitacao-do-semiarido Access on 04/10/2023.

Troleis, A.L., Silva, B.L. 2018. *Do polígono das secas à vulnerabilidade ao colapso hídrico: Uma análise do território do Rio Grande do Norte.* Cajazeiras–PB 3, (5)–jan./jun. 2018 ISSN 2525-5703, 3(5).

Tyson, R.V., Treadwell, D.D., Simonne, E.H. 2002. Opportunities and challenges to sustainability. *Hortscience,* 21, 6–13, 2011 VALENTI, W. C. Aquicultura sustentável. In: Congresso de Zootecnia, 12o, Vila Real, Portugal. 2002, Vila Real: Associação Portuguesa dos Engenheiros Zootécnicos. Anais, pp.111–118.

Yeilaghi, H., Arzani, A., Ghaderian, M., Fotovat, R., Feizi, M., Pourdad, S.S. 2012. Effect of salinity on seed oil content and fatty acid composition of safflower (*Carthamus tinctorius* L.) genotypes. *Food Chemistry,* 3(130), pp. 618–625. https://doi.org/10.1016/j.foodchem.2011.07.085

35 Gender-Responsive Solutions for Managing Drought in the Hindu Kush Himalaya

Karishma Khadka, Subha Khanal, Madhav Prasad Dhakal, Sanjeev Bhuchar, and Nand Kishor Agrawal

CONTENTS

35.1 INTRODUCTION

The Hindu Kush Himalaya (HKH) region stretches over 4 million square kilometers covering the mountain areas of eight countries: Afghanistan, Bangladesh, Bhutan, China, India, Myanmar, Nepal, and Pakistan (Figure 35.1). It has rich biodiversity and diverse ecosystems, and is home to 240 million people. Almost 1.9 billion people in the 10 river basins of the HKH have access to natural resources, and about 3 billion people depend on them (Sharma et al., 2019). Because of its wealth of water resources, the region is often referred to as the "water tower" of Asia. However, the region is also considered to be a highly fragile mountain system due to tectonic activity and the great variation of climates, hydrology, and ecology (Bajracharya & Shrestha, 2011). As a result, it is prone to hazards such as landslides, floods, earthquakes, and droughts.

The effects of climate change are felt acutely in the HKH region, which is extremely prone to natural disasters (Wang et al., 2019). Its high human population, overexploitation of natural resources, depleting ecosystem services, lack of proper infrastructures, and inadequate implementation of rules and regulations have further increased disaster risks (ISDR & UNEP, 2007). Effects

FIGURE 35.1 Map of the Hindu Kush Himalaya region and ten major river basins. (From Sharma et al., 2019.)

of climate change such as downpour in the monsoon and lack of precipitation during winter escalate water-induced calamities such as floods and droughts (Lutz et al., 2019).

Climate change has a direct impact on social and economic life in the HKH region. Rapid urbanization, increasing demand for water for agriculture and industry, and rising pollution have created many challenges for people in the region (Shrestha et al., 2014). Likewise, inconsistent water supply has affected people's livelihoods. Changes in precipitation patterns have forced farmers to rely heavily on river water for irrigation, leading to the diversion of water from major rivers. The increased pressure on water resources will eventually lead to competition for water resources needed for "agriculture, industry, and human consumption" (Hanjra & Qureshi, 2010). The region is vulnerable to even the slightest changes in precipitation patterns, as almost 60% of the agricultural system is directly dependent on rainwater (World Bank, 2016). Seventy percent of the population directly or indirectly relies on farming (Tiwari, 2000). Extreme events such as flood and drought will seriously affect farm yields (Bruinsma, 2003) and threaten people's livelihoods.

To help communities develop adequate coping mechanisms, it is important to assess risks and vulnerabilities during and after a disaster. Vulnerability to hazards may be explained as "the characteristics of a person or group and their situation that influences their capacity to anticipate, cope with, resist, and recover from the impact of a natural hazard" (Donner & Rodríguez, 2011). Drought events have increased vulnerabilities, especially among women and marginalized communities in the HKH. This chapter gives a general overview of climate change and drought from a gendered perspective. It seeks to analyze drought-related risks and their effects and provides some examples of gender-responsive drought preparedness and drought risk reduction in HKH countries, including a few suggestions for scaling them up.

35.2 METHODOLOGY

35.2.1 REVIEW METHODOLOGY

We conducted a literature review to understand the frequency and impacts of drought in the HKH region. The sources we reviewed included journal articles, gray literature, case studies, and reports of various government agencies and international organizations, such as the International Centre for Integrated Mountain Development (ICIMOD), agencies of the United Nations, the Centre for Research on the Epidemiology of Disasters (CRED), International Federation of Red Cross and Red Crescent Societies (IFRC), World Bank, and Watershed Organization Trust (WOTR). We used two major academic databases (Scopus and Google Scholar) using key search strings ("drought", "impacts", "management", "gender" and "HKH"). Data on drought events in the HKH region were extracted from the Emergency Events Database (EM-DAT) website (www.emdat.be). The review aimed to identify good drought management practices in the HKH region. Our field experiences related to drought and gender also shaped our analysis of the findings of the literature review.

35.2.1.1 What Does It Mean to be Gender Responsive?

According to the United Nations Development Programme (UNDP), gender responsiveness reflects understanding of gender-differentiated needs, capacities, and vulnerabilities of women and men of different social groups. To achieve gender responsiveness, it is important to ensure equal participation of men and women, and equitable distribution of benefits (UNDP, 2019). UNICEF (2018) states that

> gender responsive programming refers to programs where gender norms, roles and inequalities have been considered, and measures have been taken to actively address them. Such programs go beyond increasing sensitivity and awareness and actually do something to narrow or remove gender inequalities.

35.2.2 RESEARCH OBJECTIVE

The overall objective of this chapter is to analyze the risk of drought and its adverse effects on the lives and livelihoods of people in the HKH region, and to illustrate how a gender-responsive strategy for drought preparedness and drought risk reduction can help build resilience.

35.2.2.1 Research Questions

The review aimed to address the following key questions:

- What is the drought occurrence and its impacts in the HKH?
- How are women more vulnerable to the impacts of drought?
- What kinds of drought management practices and solutions are being adopted in the HKH?
- Are these practices gender responsive?
- What could be done to improve gender integration and scale up good practices?

35.3 RESULTS AND DISCUSSION

35.3.1 MULTIHAZARD SCENARIO IN THE HKH

The HKH is one of the youngest mountain systems in the world and among the most vulnerable and hazard-prone regions. Due to its fragile geological features, tectonic activity, varied topography, and variable climatic conditions, the region is exposed to different natural disasters such as floods, landslides, earthquakes, avalanches, droughts, and glacial lake outburst floods (GLOFS). In a mountainous setting like the HKH, disasters are closely interrelated; a primary disaster event can trigger subsequent secondary disaster events (Vaidya et al., 2019). For example, an earthquake can trigger a landslide, which ultimately leads to an outburst flood due to the entry of landslide masses into the reservoir area.

According to the EM-DAT, Asia experienced a greater frequency of weather-related disasters between 1995 and 2015 compared to other continents and 3 billion people, the highest number among all continents, were affected (CRED & UNISDR, 2015). Asia was the most impacted in

2019, accounting for 40% of natural disaster events, 45% of deaths, and 74% of the total number of people affected globally (CRED, 2020). Among those affected, the highest percentage was affected by storms (35%), followed by floods (33%) and droughts (31%). Drought accounts for less than 5% of all natural hazards reported globally during the period of 1995–2015.Yet the number of people affected by drought is more than 1 billion, which is more than 25% of all people affected by different hydrological, meteorological, and climatological disasters (CRED & UNISDR, 2015).

Natural hazard events along with associated economic and human impacts are increasing in the HKH. A decadal analysis of disasters in the HKH between 1980 and 2010 shows an increasing trend in the number of disaster events, the number of people killed or affected, and economic losses (Vaidya et al., 2019). Most HKH countries are in the "high risk" category of the INFORM risk index 2021, which is based on three dimensions of risk – hazard and exposure, vulnerability, and lack of coping capacity – measured in 191 countries (INFORM, 2020). The HKH region is already experiencing different natural hazards. The number of extreme events in the region is expected to increase in the future owing to climate change. The HKH region will experience higher temperatures than the global average (Krishnan et al., 2019). Due to elevation-dependent warming, the temperature increase is projected to be around 1.8°C, even if global warming is limited to 1.5°C. Rising temperatures and changing precipitation patterns are likely to cause and intensify climate-induced disasters, including drought. Higher temperatures and prolonged dry periods can increase the frequency of drought, which can have devastating impacts on people's livelihoods and the environment and ecosystems.

35.3.2   DROUGHTS AND THEIR IMPACTS IN THE HKH

Drought is a complex natural hazard resulting from a deficiency of precipitation over an extended period of time, and it has significant impacts on the environment, society, and economy (Wilhite, 1992, 2000). It can lead to water scarcity, crop failure, famine, heat waves, and forest fires. Droughts are frequent in the HKH region. Climate change is likely to increase droughts and other extreme events in the region, further threatening water and food security and affecting human health, the environment, and ecosystems (Rasul, 2021). A survey of local people's perceptions conducted in four major river basins in the HKH indicates an increase in the frequency of drought events (Hussain & Qamar, 2020). According to the EM-DAT global database, among the HKH countries, China and India have faced the highest number of major drought events (39 in China and 16 in India) between 1990 and 2020 (Figure 35.2). While China faced the highest number of drought events, the number of people affected was higher in India.

Drought in the HKH region is a result of low rainfall, rainfall variability, prolonged dry season, and high temperatures. According to the SAARC Disaster Management Centre (2010), about 16% of the total geographical area of India is highly prone to drought. In 2015–2016, India experienced a devastating drought that affected more than 300 million people in ten states (World Bank, 2020). Droughts are also frequent in western Nepal due to the delayed monsoon and low precipitation during winter months (World Bank Group, 2020). According to the World Bank Group (2011), precipitation in Nepal was 16% below normal during the monsoon in 2006–2007, which led to a 21%–30% decrease in rice cultivation. In China, one of the sectors hardest hit by drought is agriculture. The average area affected by drought in China is approximately 21.59 million ha leading to grain losses of up to 10 billion kg annually (He et al., 2011). Pakistan receives less than 200 mm of rainfall annually in 60% of the country's area (World Bank, 2020). A severe drought was experienced throughout the country from 1999 to 2002. During this period, the total flow of water in the major rivers decreased by around 34% below the monthly average leading to water scarcity and crop failure (Ahmad et al., 2004). The prolonged drought affected food security and livelihoods of about 2 million people and resulted in the deaths of 1.76 million livestock in the Balochistan province of Pakistan (Hussain & Qamar, 2020). The impact of drought varies spatially. A study conducted in

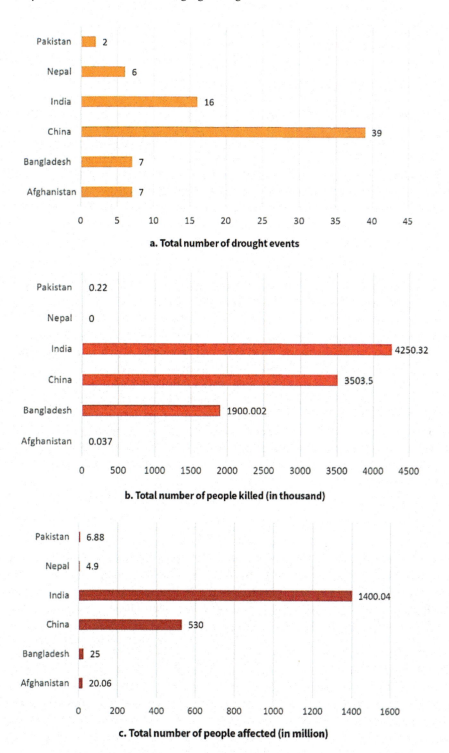

FIGURE 35.2 Graphs representing (a) the total number of major drought events, (b) the total number of people killed, and (c) the total number of people affected by droughts in HKH countries between 1900 and 2020. (From EM-DAT: The Emergency Events Database, www.emdat.be.)

the Koshi River Basin showed that it has more severe impacts on downstream communities than on upstream communities (Wu et al., 2019).

35.3.3 DROUGHT AND GENDERED VULNERABILITIES

Societies in the HKH region are dominated by patriarchal norms, and gender inequalities are visible in all spheres of life (Goodrich et al., 2017). Climate change and disasters in the HKH region have exacerbated the vulnerabilities of women. A gendered perspective on disaster is crucial, as disasters such as drought tend to affect women more severely. Goodrich et al. (2017) suggest that due to preexisting structural inequalities, during disasters women are exposed to greater physical, mental, social, and psychological harm such as sexual abuse and gender-based violence. According to the World Disaster Report, cases of sexual harassment rise during disasters, such as drought (IFRC, 2007).

The HKH region is currently witnessing the effects of climate change and climate-induced disasters, coupled with economic problems such as poverty and unemployment. The construction of roads in rural areas has enabled people, mostly men, to migrate in search of jobs and opportunities (Speck, 2017). Increased outmigration of men has compelled women to take on additional tasks besides performing their traditional gender roles, increasing their workload (Holmelin, 2019). Yet this has also provided women an opportunity to make decisions. More women than men are now involved in agriculture. As a result, women stand on the frontlines of climate change (Nellemann et al., 2011). Although women work longer hours and have a greater workload than men, their roles related to agriculture, water, livestock, food, care, and communal activities are often unaccounted for and invisible (Grassi et al., 2015). These auxiliary roles are strenuous and tedious as women lack access to practical tools, technologies, and services. From a gendered lens, water scarcity has different impacts on men and women. Women are more likely to be affected by water shortages, as women spend a lot of time collecting and managing water for the household. They have to walk additional miles to collect water (Khadka & Verma, 2012). As water is essential for household chores and farming purposes, drought and the resulting water shortage have serious impacts on the lives of women. Water scarcity, commonly faced during and after disasters, is a burden that women are disproportionately tasked with trying to solve.

Intersectionality is an important concept when it comes to assessing the impacts of drought on women. It can be defined as "the intersecting of social differentiations and identities based on class, caste, ethnicity, age, and other factors with gender" (Goodrich et al., 2017). When poverty intersects with gender discrimination, it increases vulnerability (Vaidya et al., 2019). For example, in India, a Dalit woman faces harsher consequences of drought than a woman belonging to an upper caste from the same village, as the former will more likely be barred from touching the main sources of water, such as hand pumps and tube wells (Johns, 2012). People may live in the same community, but the impacts they face vary according to their specific circumstances. Women belonging to poor families are more susceptible to the effects of disaster, as they are highly dependent on available natural resources for their daily survival (Su et al., 2017). Moreover, such families face food shortages during disasters, and often the women of the household receive the smallest portions of food. This not only affects their health, it also disrupts the education of girls and increases their chances of being married off at a very young age (Dilshad et al., 2019). A study conducted in the Tharparkar district in Pakistan showed that in times of food scarcity, the women of the household get less food than other family members and are more undernourished (Memon et al., 2018). The Hindu Kush Himalaya Assessment report shows that around 50% of people in the HKH region face malnutrition, and women and children are especially vulnerable (Rasul et al., 2019).

35.3.4 DROUGHT MANAGEMENT PRACTICES IN THE HKH REGION

Drought management practices in the HKH region can be categorized as technological interventions, adapted approaches, and policy environment at the local to national level. Responses include

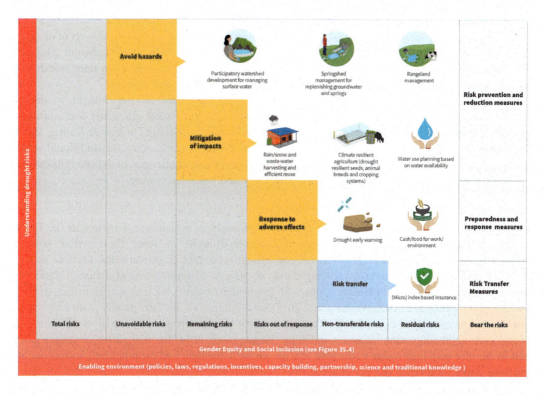

FIGURE 35.3 Risk staircase of possible measures to reduce and prevent drought risks in the Hindu Kush Himalaya. (Adapted from SDC 2018; (https://www.shareweb.ch/site/DRR/Documents/Types%20of%20activity/SDC_Guidelines_on_DRR_April_2018.pdf)

risk prevention, risk reduction and mitigation, managing residual risks through preparedness and response, and risk transfer measures, such as microinsurance (Figure 35.3). Depending on the level and type of drought risks and local context, the interventions can be applied either as stand-alone measures or as a package of practices.

35.3.4.1 Technological Interventions for Reducing Drought Risks

35.3.4.1.1 Water and Watershed Management

There are number of common practices that help mountain communities to partly deal with water stress. Technical interventions such as water storage and conservation facilities for farmland, water-saving irrigation systems, pumping water from the river, and community-managed water tanks and ponds are some common measures for coping with drought (Zhang et al., 2018; Mishra et al., 2019). Different practices are common in different countries depending on the type of local resources found there, existing government schemes, and culture. For example, digging ponds, building water tanks and reservoirs, buying pumps, construction of wells, and efficient use of available water through sprinklers are the most common measures that local people use in China (Jin et al., 2016; Su et al., 2017; Zhang et al., 2018). "Ice stupa" and "artificial glaciers" in Ladakh, India, are unique examples of a water reservoir used for irrigation (Clouse, 2016; Sharma, 2019a). Similarly, roof rainwater harvesting is practiced in Myanmar to cope with drought (Mishra et al., 2017). Rainwater harvesting, water harvesting ponds, and afforestation programs for water conservation are common in Nepal (Vaidya et al., 2019). Rainwater harvesting for groundwater recharge is practiced in Bangladesh and India (Rabbani et al., 2009; Padmaja et al., 2020). Reviving drying springs, which are part of groundwater systems, is another example of drought management in Nepal (Adhikari, 2018). Irrigation systems suited to specific climatic zones are also found in the region. For example, in the arid and semiarid regions of Afghanistan and Pakistan, the karez irrigation system (by a cascade of wells) and spate (canal) irrigation represent some of the finest traditional methods of drought management (Scott et al., 2019). Rehabilitation of hand pumps and wells in Afghanistan (Qureshi & Akhtar, 2004), and irrigation that relies on surface and groundwater sources in Bangladesh and India (Sugden et al., 2014) are measures that have been practiced for generations.

There are other examples of good practices in the HKH countries. During drought, mountain communities in Afghanistan store and reuse wastewater, and share water efficiently and equitably through a traditional water distribution system called the mirab system (Thomas & Ramzi, 2011; Iqbal et al., 2018). As climate and other factors have increased water scarcity, the traditional water management systems are no longer sufficient for addressing these challenges, and more innovations are being tried out, particularly for water use efficiency and water harvesting.

35.3.4.1.2 Agronomic

A major sector that is vulnerable to the impacts of drought is agriculture. Farmers in the HKH region follow different adaptation strategies to deal with drought. In Nepal, these include sowing drought-resistant seeds; cultivation of heat- and drought-resistant crops varieties, such as millet, soybean, and black gram; altering sowing dates; sowing in different plots at different dates for partial risk distribution; and planting crops with low water demand, such as maize, millet, ginger, and turmeric (Adhikari, 2018; Wester et al., 2019). In Afghanistan, farmers mostly plant crops with low water demand (Iqbal et al., 2018). Planting drought-tolerant crops and crop diversification are common drought adaptation measures in China (Huang et al., 2014; Jin et al., 2016). Farmers plant half of the plot with high-value cash crops that are prone to drought and the other half with drought-resistant staple food crops (Zhang et al., 2018).

Other drought adaptation measures include dry seed bed and direct seeding for rice, and mixing local maize varieties with improved varieties (in Nepal), and conservation of drought-resistant crop varieties (in China, Bangladesh, and India) (Mishra et al., 2019). In parts of the Himalayas, women manage the seeds and know which ones are more resistant to drought. For example, women of the Garo hills in Northeast India know which species adapt better to drought (ICIMOD, 2010). In Bhutan, farmers engage in climate-resilient agriculture (Mishra et al., 2019), integrating cereals with legumes such as maize and beans in the same plot, a practice that is also common in Nepal (Goodrich et al., 2017). Drought-tolerant rice varieties, non-rice variety winter crops, and horticultural crops are grown by Bangladeshi farmers (Al-Amin et al., 2019).

35.3.4.2 Management Measures

Some HKH countries have adopted drought management measures such as index-based insurance for drought, establishment of institutions that focus on drought, and programs for building the resilience and adaptive capacity of communities (Vaidya et al., 2019). China increased investment in irrigation and purchased weather index insurance for wheat (Jin et al., 2016). Water management committees were established to improve distribution, and rules for efficient water allocation were developed for irrigation sequence and maintenance of water infrastructures (Zhang et al., 2018). In addition, drought early warning systems were established (Wester et al., 2019). In India, community radio was used to spread awareness about drought risks (Resurrección et al., 2019), and drought-related education was provided in the state of Maharashtra (Joshi, 2019). In Nepal, a drought monitoring system was established as a timely response to food shortages (Mishra et al., 2019).

Communal pooling/storage of food grain and water is an age-old practice in most HKH countries (Vaidya et al., 2019). Grain storage is used in times of drought in Pakistan and Afghanistan (Memon et al., 2018; Iqbal et al., 2018). In China, some farmers shifted from agriculture to forestry and animal husbandry in order to cope with drought (Su et al., 2017). Longan was replaced by mango, as it is more drought resistant (Zhang et al., 2018). In Northeast India, rice was replaced by cotton (Sam et al., 2020), and in Afghanistan, apples were replaced by pomegranates (Qureshi & Akhtar, 2004).

35.3.4.2.1 Good Drought Management Practices

35.3.4.2.1.1 Springshed Management for Drought Risk Reduction

Springs are a major source of water for millions of people in the mid hills of the HKH. However, there is increasing evidence that springs across the region are drying up due to a combination of anthropogenic and climatic factors. While springs are a critical resource during drought, an extended period of dryness could lead to a decline of the groundwater level in the aquifers, which ultimately reduces spring flows. In the Indian Himalaya, about 50 million inhabitants depend on springs for water services (NITI Aayog, 2018). Drying springs and reduced spring discharge are causing water insecurity in mountain communities.

Some governments and institutions in the HKH have initiated programs to manage and revive the drying springs. For example, Dhara Vikas, an innovative program of the government of Sikkim, India, has adopted a springshed management approach to revive springs, focusing on drought-prone areas of the state. Although the fundamental aspects of watershed management and springshed management appear similar, the springshed management approach differs significantly from the watershed development approach in terms of scale, cost, duration, and treatment methods, as well as success indicators; the most important factor is the inclusion of a geohydrological approach (Tambe et al., 2012).

Based on lessons and experiences from drought-prone areas of Sikkim, ICIMOD, through its different projects and partners, initiated a springshed management approach to deal with water scarcity and drought challenges in Bhutan, India, and Nepal. Together with partners and based on the work done by other organizations in the HKH region, ICIMOD has developed a six-step protocol for reviving springs in the HKH (Shrestha et al., 2018), which includes hydrogeological, social, and governance aspects for action research and implementation. Springshed management measures have been applied in some springsheds of Bhutan, India, and Nepal through ICIMOD-coordinated programs. In western Nepal, spring flow was seen to increase after a recharge intervention was carried out in 2016 (Khadka et al., 2019). Similarly, experiences from Sikkim already showed that it is possible to recharge a spring aquifer by establishing a rainwater harvesting structure in the recharge area. That can help ensure the water security of communities in rural areas (Tambe et al., 2012).

35.3.4.2.1.2 Drought-Resilient Watershed Management

Landscapes and climate are interlinked and together form natural systems that provide many ecosystem services – provisioning, regulating, cultural, and supporting (Millennium Ecosystem Assessment, 2005). Degradation of landscapes is known to reduce rainwater or snowmelt infiltration capacity, which is important for replenishing groundwater resources. This intensifies drought impacts, leading to food insecurity and even famines in some countries. Watershed management, which focuses on surface and groundwater management, landscape restoration, and improved local livelihoods, has proven to be a successful nature-based solution for mitigating drought risks in many countries across the world, including the HKH region. The community-led watershed development program in Kaluchi Thakarwadi, a remote settlement in Maharashtra, India, by the Watershed Organization Trust (WOTR) is a case in point (Box 35.1).

BOX 35.1 DROUGHT-RESILIENT WATERSHED MANAGEMENT IN KALUCHI THAKARWADI, INDIA

Kaluchi Thakarwadi is situated in the rain shadow region of India's western mountains where rainfall is unreliable. In the past, villagers here experienced chronic water scarcity, with recurring food and fodder shortages and high rates of migration. In 2006, the WOTR initiated a community-based watershed development program in Kaluchi Thakarwadi, jointly with a local NGO, focusing on three main areas: (i) soil and water conservation treatments such as hillside trenches, check dams, conservation ponds, afforestation, and others at the farm to landscape level; (ii) capacity building and training of women and men on watershed management technologies and approaches; (iii) local institution building for collective and equitable watershed management with equal representation of women and men and wealth classes including the landless (WOTR, 2012).

An evaluation of the watershed interventions in Kaluchi Thakarwadi six years later (WOTR, 2012) found that the program had several positive outcomes. The period of water availability increased from 8 months to 12 months. Area under irrigation increased from 6.2 ha to 30 ha, resulting in more crop production and income. The seasonal outmigration reduced from 70% and 6–7 months to 20% and 3–4 months. Cereal, fodder, and milk yields increased significantly. Women organized into self-help groups and took up enterprises like producing and selling compost, and building and distributing energy-efficient cooking stoves. Lessons learned from Kaluchi Thakarwadi and other watershed programs are helpful in designing gender equality and social inclusion (GESI) responsive adaptation and mitigation strategies for current and future drought risks in the HKH.

35.3.4.2.1.3 Incentives for Ecosystem Services (IES) for Drought Risk Management in Palampur, India Drinking water shortage was a major challenge for people living in Palampur, a town in India (Patterson et al., 2017). Palampur's water supply is connected to a spring located at Dhauladhar Mountain in the Neugal River catchment. The spring source was owned by the upstream community, and users of spring water belonged to the downstream community. To increase water availability for downstream users, interventions had to be carried out in upstream areas. To link upstream and downstream communities, Incentives for Ecosystem Services (IES) schemes were established in phases, adding "pieces of the puzzle" over time. Multiple stakeholders such as municipal members and youth were engaged in the process and later became powerful advocates for IES schemes.

Awareness programs on upstream conservation and its impact downstream were organized. The Palampur community agreed on actions to be carried out for upstream protection, and based on this, an IES proposal was developed. The Palampur Municipal Council provided annual payments to upstream watershed management committees for carrying out activities, such as reforestation in upstream areas, land-use management, water conservation efforts, and spring protection. In addition to upstream conservation actions, efforts were made to manage water demand at the household level by conserving water, saving water, and improving water quality.

The case study of the IES for spring water management in Palampur showed that building youth awareness is critical, as young people can act as "change agents" in households and the community. It is also important to enhance the awareness of staff in local and municipal offices in order to ensure smooth negotiation with upstream communities.

35.3.4.2.1.4 Index-Based Instruments as a Decision Support System Lack of access to formal risk insurance is one of the causes of low drought risk resilience among small and marginal farmers. Informal risk management methods often diminish agricultural productivity and provide only limited coverage. The weather poses the most significant risk to agriculture. Given the rapid advancement in remote-sensing technologies and low-cost automated weather stations, pilot projects on weather index-based crop insurance schemes, complemented by financial literacy, climate-resilient practices, and land rights for women, can help increase agricultural production, income, and savings, and hence increase investment in education and health (Cole et al., 2012; Raju et al., 2016).

However, weather index-based insurance has its own challenges and limitations. Designing such insurance schemes would require good remote-sensing techniques, a network of weather stations, and research and expertise. Fortunately, technology is improving rapidly and satellites now capture high-resolution data. The cost of aerial photography is decreasing, and technologies such as radar sensing and drones have become increasingly accessible (Sandmark et al., 2013).

35.3.4.2.1.5 Fostering Public–Private Partnerships for Resilience to Drought, Baoshan, China The public–private partnership (PPP) approach allows farmers, government and private entities to come up with drought adaptation alternatives (Zhang et al., 2018). This approach has been successfully adopted in China's Yunnan province to widen funding sources, reduce costs, and risks, and increase benefits.

The PPP mechanism has helped widen the funding sources. Private companies in Yunnan have provided funding for the inception phase of the drought management program, which eased the monetary load for governments and farmers. The governments contributed through subsidies in electricity charges and by sharing the construction costs, which reduced the financial burden on farmers and private companies. The farmers contributed by paying for the use of irrigation services. Combining these funds, pumping stations were built that served as alternative water sources for farmers.

The mechanism allowed for effective coordination and cooperation among the three actors, and they were able to share the risks and benefits. The farmers received reliable water supply for irrigation, which improved their yield and household income; the governments fulfilled their obligation to provide public services; and the private companies benefited by collecting land rent or irrigation fees. These kinds of financial gains encouraged private entities to get involved in providing irrigation services.

Adoption of this PPP in Yunnan also helped reduce uncertainties surrounding legal procedures, government supervision, and social connections. The private institutes involved were encouraged and supported by the government – with government funds invested in water conservancy infrastructure, and government backing provided to private and public sector policies such as the "Decision on Accelerating Water Conservancy Reform and Development" policy.

The project combined scientific and local knowledge. Private companies have more access to information and technical expertise than farmers. They also have greater access to resources. In Yunnan, the private sector funded and provided technical support to the construction of an irrigation system in one location and provided technical support in another. The community's active participation and contribution of local knowledge strengthened the project and contributed to its success.

35.3.4.2.1.6 Water Sharing in the Balkh River Basin of Afghanistan Proportional sharing of water enables communities to ward off poverty in times of drought and water shortages (Ward et al., 2013). During drought, precipitation decreases; this in turn, may cause reduction in the streamflows at headwaters. In such a situation, available water in the river basin can be shared proportionally.

Canals allow the water to be distributed evenly. In times of drought, a 25% overall water shortage produces a 25% reduction of each canal's customary full allotment. This approach allows water shortage risk to be distributed proportionally among all canals rather than placing a disproportionate burden on any single canal or group of canals. It rests on the principle of "some for all rather than more for some".

In Afghanistan this approach has been successfully used to prevent food insecurity and income loss. It enables people to adapt to unpredictable changes in water supply. It is a simple, cost-effective technique and is widely considered to be fair and equitable.

All of the aforementioned projects have made huge positive impacts on the lives of people. However, they have not been able to ensure gender integration.

35.3.5 POLICY ENVIRONMENTFOR DROUGHT MANAGEMENT: PROGRESS AND GAPS

Globally, disaster risk management (DRM), including drought risk management, is still evolving and moving from a reactive and crisis management approach to a more comprehensive risk reduction and proactive approach. The international Sendai Framework 2015–2030 for Disaster Risk Management is playing an important role in advancing DRM with a focus on prevention, mitigation, response, and rehabilitation (UNISDR, 2015).

Afghanistan is one of the most drought-impacted countries in the HKH. In 2018, drought affected nearly 10.5 million people, with direct impacts on agriculture, natural resources, water, livelihoods, health, sanitation, protection, and the broader development sectors (FAO & MAIL, 2019). There are three kinds of policy instruments for drought management in Afghanistan. These include the overarching policy for natural resources management and development, such as the Afghanistan National Peace and Development Framework (ANPDF) of 2018; sector-specific strategies; and specific drought management instruments. However, a full-fledged policy for drought management is lacking. A drought management policy was drafted in 2008, while the National Drought Management Strategy and Policy was formulated in 2010–2011, but both are yet to be officially endorsed/adopted (FAO & MAIL, 2019).

China has a policy for regulating weather modifications and disseminating early warning information on drought. The central government allocates special funds to cope with big droughts (Li et al., 2013). Agricultural insurance with subsidy and drought insurance for crops, such as sugarcane, coffee, and tobacco, is provided by the government (Zhang et al., 2018). Yunnan's provincial government formulated a number of policies for drought relief, mitigation and adaptation, and guidelines to steer large-scale water conservancy projects (Pradhan et al., 2017).

In India, the drought relief policy still focuses on providing financial assistance (post-drought) for livelihood and survival. The central government transfers funds to the states for mild and moderate drought incidences (Sharma, 2019b).

In Nepal, the National Policy for Disaster Risk Reduction 2018 and Disaster Risk Reduction and Management Act 2017 have clear provisions for DRM – this includes drought – at the national, provincial, and local level, with a special focus on women, children, the elderly, and marginalized and differently abled people (Government of Nepal, 2018). DRM activities are the sole responsibility of the Ministry of Home Affairs, and this is seen as a policy gap because it leaves other ministries in the role of passive partners (Nepal et al., 2018).

The preceding examples indicate that many HKH countries are formulating and implementing national strategies for DRM in line with the Sendai Framework, with increasing roles for local governments and communities in reducing and managing disaster risks, with a focus on gender equity and social inclusion. However, there are gaps that need to be filled in order to make drought risk reduction efforts more effective.

35.4   A WAY FORWARD

35.4.1 GENDER RESPONSIVENESS IN BUILDING RESILIENCE IN THE HKH

The HKH region's' economy is largely dependent on agriculture, with water being a major natural resource (Su et al., 2017). Due to climate change, the frequency and intensity of drought are projected to increase in the region. The impact of disasters is especially severe on women because of their preexisting vulnerabilities. Therefore, it is important to ensure women's participation in drought response plans and programs. Their knowledge of their surroundings and their long experience coping with disasters could be highly useful in developing effective drought response strategies. Although laws and policies emphasize equal participation of men and women, when it comes to implementation, women's views and experiences are often overlooked (Goodrich et al., 2017). Existing plans and policies do not address the specific issues of marginalized communities, especially those related to gender (Parikh et al., 2012). An assessment of climate change impacts conducted by India's National Action Plan on Climate Change (NAPCC) shows that gender is an important aspect in tackling the impacts of climate change and that it is important for interventions to be designed with gender issues in mind. However, the NAPCC has been ineffective in recognizing the solutions that enable the adaptation to take place and provides very few guidelines on gender-centric measures (Ahmed & Fajber, 2009; Parikh et al., 2012). According to Brauw et al. (2008), in Yunnan Province, China, women are deprived of their basic right to earn the same income as men from the sale of crops and are still excluded from decision-making processes. Although women play a critical role in collecting and managing water for the household and farm, policies and programs lack gender integration (Caizhen, 2009; Tong et al., 2016).

However, government officials and researchers acknowledge that marginalized groups are especially vulnerable to the impacts of climate change (Goodrich et al., 2017). Nellemann et al. (2011) assert that women can play a huge role in climate adaptation efforts because of their ability to use and manage natural resources, juggle multiple tasks for the farm and household, come up with sustainable solutions, and lead community activities through women's groups and cooperatives (Figure 35.4). Given the high rate of male outmigration in the HKH, women are increasingly compelled to take up the decision-making role in their household and community. If women are provided the right platform, resources, and training, they can significantly contribute in combating climate change impacts through adaptation and resilience building. As Achim Steiner of the UNDP states, "Women play a much stronger role than men in the management of ecosystem services and food security" (Nellemann et al., 2011). MacGregor (2010) warns that adaptation efforts that do not address gender inequalities will be inadequate and biased.

It is thus important to mainstream gender in national drought management plans and policies. Women should be involved in the process from the early planning phase to the end of implementation and beyond. They could work in collaboration with government bodies and other local/development organizations from the very beginning. There are many successful examples of gender-responsive adaptation in the HKH region, and the situation can further be improved with more research on the links between gender and disaster. Gender-responsive programs should be properly assessed in order to determine the extent to which they brought "gender related changes, such as, agency, decision-making, equality and empowerment" (UNICEF, 2018).

35.4.2 SCALING UP

Although people have been adopting various drought response measures, minimizing the negative impacts of drought remains a major challenge for communities and governments in the HKH region. This challenge is likely to grow bigger with increasing risks of climate change. It is projected that the HKH region will have more rainfall, but the variability and uncertainty will increase, leading to frequent extreme events, such as flood and drought. In order to deal with the projected uncertainty, a range of flexible and diverse solutions would be needed at every level. While a number of solutions are already being practiced in the region, scaling them up beyond the project areas remains a major challenge. To address this challenge, ICIMOD has initiated a regional initiative called Resilient Mountain Solutions (RMS), which promotes simple, affordable, and replicable solutions that communities can implement without much external support. Combining scientific and technical knowledge with local practices, the initiative promotes solutions that local communities, particularly women, understand very well and can implement on their own. One example is a farm pond lined with soil and cement and connected to a local tap or hand pump. This simple and low-cost solution allows people to reuse water for small-scale irrigation.

On the institutional level, it is imperative that nongovernment organizations align their programs and innovations with government priorities and schemes in order to reach the wider population. Due to increasing decentralization in the region, local governments such as municipalities have much more authority, resources, and willingness to deal with local issues. But they struggle due to a lack

FIGURE 35.4 Combating the impacts of droughts through gender-responsive drought management.

of new knowledge, technologies, and approaches. Local governments need support from external agencies to scale up these solutions and make a real impact.

Another important means for scaling up is digital technology. Digital technologies can enhance the design of these solutions, as well as their use in local planning and implementation. A mobile application–based decision support system on applicability and use of a particular solution in the local language could help farmers and communities to replicate the solution very quickly. For example, the RMS initiative is working with a technology company to develop a user-friendly app for mountain farmers and local municipalities to help them in resource assessment, planning, access to inputs, and marketing. Such solutions could go a long way in ensuring widespread replication of different water management and drought solutions.

As mentioned earlier, climate-induced disasters such as drought have severely impacted communities in the HKH region. Droughts have led to water shortages; threatened the food and nutritional security of rural communities; and affected human health, the environment, and ecosystems. The most affected are those who belong to marginalized and vulnerable groups, specifically women. Water scarcity can severely impact women, who bear the responsibility of fetching water for the household and farm. Therefore, response to drought must take differentiated gender needs into account. Solutions introduced by the RMS initiative have made positive impacts on the lives of people in the mountains and hills. As drought is a natural disaster, seeking a permanent solution to it would not be entirely fruitful. Instead, we should promote sustainable and resilient approaches that will bring positive changes in the lives of mountain communities.

ACKNOWLEDGMENTS

ICIMOD gratefully acknowledges the support of its core donors: the governments of Afghanistan, Australia, Austria, Bangladesh, Bhutan, China, India, Myanmar, Nepal, Norway, Pakistan, Sweden, and Switzerland. The authors acknowledge the support of the Resilient Mountain Solutions (RMS) initiative of ICIMOD. We are grateful to Suman Bisht for reviewing this chapter and providing suggestions from a gendered perspective. We would like to thank Anil Kumar Jha and Nisha Wagle for helping us with the literature search. We are also grateful to the Knowledge Management Committee (KMC) team of ICIMOD, especially Rachana Chettri for providing editorial assistance, Sudip Kumar Maharjan for preparing the graphics, and Samuel Thomas for reviewing the manuscript.

DISCLAIMER

The views and interpretations in this chapter are those of the authors and are not necessarily attributable to ICIMOD.

REFERENCES

Adhikari, S. (2018). Drought impact and adaptation strategies in the Mid-Hill farming system of Western Nepal. *Environments*, 5(9), 1–12. https://doi.org/10.3390/environments5090101

Ahmad, S., Hussain, Z., Qureshi, A. S., Majeed, R., & Saleem, M. (2004). *Drought Mitigation in Pakistan: Current Status and Options for Future Strategies*. International Water Management Institute (IWMI).

Ahmed, S., & Fajber, E. (2009). Engendering adaptation to climate variability in Gujarat, India. *Gender & Development*, 17(1), 33–50. https://doi.org/10.1080/13552070802696896

Al-Amin, A. K. M. A., Akhter, T., Islam, A. H. M. S., Jahan, H., Hossain, M. J., Prodhan, M. M. H., Mainuddin, M., & Kirby, M. (2019). An intra-household analysis of farmers' perceptions of and adaptation to climate change impacts: Empirical evidence from drought prone zones of Bangladesh. *Climatic Change*, 156(4), 545–565. https://doi.org/10.1007/s10584-019-02511-9

Bajracharya, S. R., & Shrestha, B. (eds). (2011). The status of glaciers in the Hindu Kush–Himalayan region. *Mountain Research and Development*. https://doi.org/10.1659/mrd.mm113

Brauw, A. De, Li, Q., Liu, C., Rozelle, S., & Zhang, L. (2008). Feminization of agriculture in China? Myths surrounding women's participation in farming. *China Quarterly*, *194*, 327–348. https://doi.org/10.1017/S0305741008000404

Bruinsma, J. (ed). (2003). *World Agriculture: Towards 2015/2030: An FAO Perspective* (J. Bruinsma (ed.)). Earthscan.

Caizhen, L. (2009). Water policies in CHINA: A critical perspective on gender equity. *Gender, Technology and Development*, *13*(3), 319–339. https://doi.org/10.1177/097185241001300301

Clouse, C. (2016). Frozen landscapes: Climate-adaptive design interventions in Ladakh and Zanskar. *Landscape Research*, *41*(8), 821–837. https://doi.org/10.1080/01426397.2016.1172559

Cole, S., Bastian, G. G., Vyas, S., Wendel, C., & Stein, D. (2012). *Systematic Review: The Effectiveness of Index-based Micro-insurance in Helping Smallholders Manage Weather-related Risks*. London: EPPI-Centre, Social Science Research Unit, Institute of Education, University of London. ISBN: 978-1-907345-35-7.

CRED. (2020). *Natural Disasters 2019*. https://cred.be/sites/default/files/adsr_2019.pdf

CRED & UNISDR. (2015). The human cost of weather related disaster 1995–2015. In *Construction and Building Materials*.

Dilshad, T., Mallick, D., Udas, P. B., Goodrich, C. G., Prakash, A., Gorti, G., Bhadwal, S., Anwar, M. Z., Khandekar, N., Hassan, S. M. T., Habib, N., Abbasi, S. S., Syed, M. A., & Rahman, A. (2019). Growing social vulnerability in the river basins: Evidence from the Hindu Kush Himalaya (HKH) Region. *Environmental Development*, *31*(August 2018), 19–33. https://doi.org/10.1016/j.envdev.2018.12.004

Donner, W., & Rodríguez, H. (2011). *Disaster Risk and Vulnerability: The Role and Impact of Population and Society*. PRB. https://www.prb.org/disaster-risk/

FAO & MAIL. (2019). *Afghanistan Drought Risk Management Strategy*.

Goodrich, C. G., Mehta, M., & Bisht, S. (2017). Status of gender, vulnerabilities and adaptation to climate change in the Hindu Kush Himalaya: Impacts and implications for livelihoods, and sustainable mountain development. http://Lib.Icimod.Org/Record/32618.

Government of Nepal. (2018). *National Policy Reform for Disaster Risk Reduction*.

Grassi, F., Landberg, J., & Huyer, S. (2015). *Running Out of Time: The Reduction of Women's Work Burden in Agricultural Production*.

Hanjra, M. A., & Qureshi, M. E. (2010). Global water crisis and future food security in an era of climate change. *Food Policy*, *35*(5), 365–377. https://doi.org/10.1016/j.foodpol.2010.05.006

He, B., Lü, A., Wu, J., Zhao, L., & Liu, M. (2011). Drought hazard assessment and spatial characteristics analysis in China. *Journal of Geographical Sciences*, *21*(2), 235–249. https://doi.org/10.1007/s11442-011-0841-x

Holmelin, N. B. (2019). Competing gender norms and social practice in Himalayan farm management. *World Development*, *122*, 85–95. https://doi.org/10.1016/j.worlddev.2019.05.018

Huang, J., Jiang, J., Wang, J., & Hou, L. (2014). Crop diversification in coping with extreme weather events in China. *Journal of Integrative Agriculture*, *13*(4), 677–686. https://doi.org/10.1016/S2095-3119(13)60700-5

Hussain, A., & Qamar, F. M. (2020). Dual challenge of climate change and agrobiodiversity loss in mountain food systems in the Hindu-Kush Himalaya. *One Earth 3*, *3*(5), 539–542. https://doi.org/10.1016/j.oneear.2020.10.016

ICIMOD. (2010). Gender perspectives in mountain development: New challenges and innovative approaches. *Sustainable Mountain Development*, *57*, 68.

IFRC. (2007). World disaster report 2007: Focus on discrimination. In *Disasters*. http://www.ifrc.org

INFORM. (2020). *INFORM Risk Index 2021*. The European Commission Disaster Risk Management Knowledge Centre. https://drmkc.jrc.ec.europa.eu/

Iqbal, M. W., Donjadee, S., Kwanyuen, B., & Liu, S. y. (2018). Farmers' perceptions of and adaptations to drought in Herat Province, Afghanistan. *Journal of Mountain Science*, *15*(8), 1741–1756. https://doi.org/10.1007/s11629-017-4750-z

ISDR & UNEP. (2007). *Environment and Vulnerability: Emerging Perspectives*. United Nations Environment Programme.

Jin, J., Wang, W., & Wang, X. (2016). Adapting agriculture to the drought hazard in rural China: Household strategies and determinants. *Natural Hazards*, *82*(3), 1609–1619. https://doi.org/10.1007/s11069-016-2260-x

Johns, H. (2012). Stigmatization of Dalits in access to water and sanitation in India. In *It was submitted at the Human Rights Council in September*. https://idsn.org/wp-content/uploads/user_folder/pdf/New_files/UN/HRC/Stigmatization_of_dalits_in_access_to_water_sanitation.pdf

Joshi, K. (2019). The impact of drought on human capital in rural India. *Environment and Development Economics*, *24*(4), 413–436. https://doi.org/10.1017/S1355770X19000123

Khadka, K., Pokhrel, G., Dhakal, M., Desai, J., & Shrestha, R. B. (2019). *Springshed Management: An Approach to Revive Drying Springs in the Himalayas. Proceedings of the seminar on "Leaving No one Behind"*. Nepal Academy of Science and Technology (NAST). https://nast.gov.np/documentfile/Proceedings.pdf

Khadka, M., & Verma, R. (eds). (2012). Gender and biodiversity management in the Greater Himalayas: Towards equitable mountain development. *Current Opinion in Environmental Sustainability*, *6*(January), 9–16.

Krishnan, R., Shrestha, A. B., Ren, G., Rajbhandari, R., Saeed, S., Sanjay, J., Syed, A., Vellore, R., Xu, Y., You, Q., & Ren, Y. (2019). The Hindu Kush Himalaya assessment. In *The Hindu Kush Himalaya Assessment*. Springer International Publishing. https://doi.org/10.1007/978-3-319-92288-1_3

Li, H., Gupta, J., & Van Dijk, M. P. (2013). China's drought strategies in rural areas along the Lancang River. *Water Policy*, *15*(1), 1–18. https://doi.org/10.2166/wp.2012.050

Lutz, A. F., ter Maat, H. W., Wijngaard, R. R., Biemans, H., Syed, A., Shrestha, A. B., Wester, P., & Immerzeel, W. W. (2019). South Asian river basins in a 1.5 °C warmer world. *Regional Environmental Change*, *19*(3), 833–847. https://doi.org/10.1007/s10113-018-1433-4

MacGregor, S. (2010). "Gender and climate change": from impacts to discourses. *Journal of the Indian Ocean Region*, *6*(2), 223–238. https://doi.org/10.1080/19480881.2010.536669

Memon, M. H., Aamir, N., & Ahmed, N. (2018). Climate change and drought: Impact of food insecurity on gender based vulnerability in district Tharparkar. *The Pakistan Development Review*, *57*(3), 307–331. https://doi.org/10.30541/v57i3pp.307-321

Millennium Ecosystem Assessment. (2005). *Ecosystems and Human Well-being*. World Resources Institute.

Mishra, A., Agrawal, N. K., & Nishikant, G. (2017). Building mountain resilience: Solutions from the Hindu Kush Himalaya. In *ICIMOD*. ICIMOD.

Mishra, A., Appadurai, A. N., Choudhury, D., Regmi, B. R., Kelkar, U., Alam, M., Chaudhary, P., Mu, S. S., Ahmed, A. U., Lotia, H., Fu, C., Namgyel, T., & Sharma, U. (2019). The Hindu Kush Himalaya assessment. In *The Hindu Kush Himalaya Assessment* (pp. 457–490). Springer International Publishing. https://doi.org/10.1007/978-3-319-92288-1_13

Nellemann, C., Verma, R., & Hislop, L. (eds). (2011). *Women at the Frontline of Climate Change : Gender Risks and Hopes. A Rapid Response Assessment*. https://www.researchgate.net/search/publication?q=Women+at+the+Frontline+of+Climate+Change%3A+Gender+Risks+and+Hopes

Nepal, P., Khanal, N. R., & Sharma, B. P. P. (2018). Policies and institutions for disaster risk management in Nepal : A review. *The Geographical Journal of Nepal*, *11*, 1–24.

NITI Aayog. (2018). *Report of Working Group I: Inventory and Revival of Springs in the Himalayas for Water Security*. https://niti.gov.in/writereaddata/files/document_publication/doc1.pdf

Padmaja, R., Kavitha, K., Pramanik, S., Duche, V. D., Singh, Y. U., Whitbread, A. M., Singh, R., Garg, K. K., & Leder, S. (2020). Gender transformative impacts from watershed interventions: Insights from a mixed-methods study in the Bundelkhand Region of India. *American Society of Agricultural and Biological Engineers*, *63*(1), 153–163. https://doi.org/10.13031/trans.13568 153

Parikh, J., Upadhyay, D. K., & Singh, T. (2012). Gender perspectives on climate change & human security in India: An analysis of national missions on climate change. *Cadmus*, *1*(4), 180–186.

Patterson, T., Bhatta, L. D., Alfthan, B., Agrawal, N. K., Basnet, D., Sharma, E., & Oort, B. van (eds). (2017). *Incentives for Ecosystem Services (IES) in the Himalayas: A 'Cookbook' for Emerging IES Practitioners in the Region*. https://www.grida.no/publications/399

Pradhan, N. S., Su, Y., Fu, Y., Zhang, L., & Yang, Y. (2017). Analyzing the effectiveness of policy implementation at the local level: A case study of management of the 2009–2010 drought in Yunnan Province, China. *International Journal of Disaster Risk Science*, *8*(1), 64–77. https://doi.org/10.1007/s13753-017-0118-9

Qureshi, A. S., & Akhtar, M. (2004). *A Survey of Drought Impacts and Coping Measures in Helmand and Kandahar Provinces of Afghanistan*. IWMI Internal Report, December.

Rabbani, M. D. G., Rahman, A. A., & Mainuddin, K. (2009). Women's vulnerability to water-related hazards: Comparing three areas affected by climate change in Bangladesh. *Waterlines*, *28*(3), 235–249. https://doi.org/10.3362/1756-3488.2009.025

Raju, K., Naik, G., Ramseshan, R., Pandey, T., Joshi, P, Anantha, K., Kesava, R. A., Moses, S. D., & Kumara, C. D. (2016). *Transforming Weather Index-Based Crop Insurance in India: Protecting Small Farmers from Distress. Status and a Way Forward*. Research Report IDC-8. Patancheru, *502*, 324.

Rasul, G. (2021). Twin challenges of COVID-19 pandemic and climate change for agriculture and food security in South Asia. *Environmental Challenges*, 2. https://doi.org/10.1016/j.envc.2021.100027

Rasul, G., Saboor, A., Tiwari, P. C., Hussain, A., Ghosh, N., & Chettri, G. B. (2019). The Hindu Kush Himalaya assessment. In *The Hindu Kush Himalaya Assessment* (pp. 301–338). Springer International Publishing. https://doi.org/10.1007/978-3-319-92288-1_9

Resurrección, B. P., Goodrich, C. G., Song, Y., Bastola, A., Prakash, A., Joshi, D., Liebrand, J., & Shah, S. A. (2019). The Hindu Kush Himalaya assessment. In *The Hindu Kush Himalaya Assessment* (pp. 491–516). Springer International Publishing. https://doi.org/10.1007/978-3-319-92288-1_14

Sam, A. S., Padmaja, S. S., Kächele, H., Kumar, R., & Müller, K. (2020). Climate change, drought and rural communities: Understanding people's perceptions and adaptations in rural eastern India. *International Journal of Disaster Risk Reduction*, 44, 101436. https://doi.org/10.1016/j.ijdrr.2019.101436

Sandmark, T., Debar, J.-C., & Clémence, T.-J. (2013). *The Emergence and Development of Agriculture Microinsurance: A Discussion Paper*. Microinsurance Network.

Scott, C. A., Zhang, F., Mukherji, A., Immerzeel, W., Mustafa, D., & Bharati, L. (2019). The Hindu Kush Himalaya assessment. In *The Hindu Kush Himalaya Assessment* (pp. 257–299). Springer International Publishing. https://doi.org/10.1007/978-3-319-92288-1

SAARC Disaster Management Centre. (2010). SAARC Workshop on Drought Risk Management in South Asia.

Sharma, A. (2019a). Giving water its place: Artificial glaciers and the politics of place in a high-altitude Himalayan village. *Water Alternatives*, 12(3), 993–1016.

Sharma, A. (2019b). Drought management policy of India: An overview. *Disaster Advances*, 12(11), 51–62.

Sharma, E., Molden, D., Rahman, A., Khatiwada, Y. R., Zhang, L., Singh, S. P., Yao, T., & Wester, P. (2019). The Hindu Kush Himalaya assessment. In *The Hindu Kush Himalaya Assessment* (pp. 1–16). Springer International Publishing. https://doi.org/10.1007/978-3-319-92288-1

Shrestha, A., Sada, R., & Melsen, L. (2014). Adapting to peri-urban water insecurity induced by urbanization and climate change. *Hydro Nepal: Journal of Water, Energy and Environment*, 14, 43–48. https://doi.org/10.3126/hn.v14i0.11259

Shrestha, R. B., Desai, J., Mukherji, A., Dhakal, M., Kulkarni, H., Mahamuni, K., Bhuchar, S., & Bajracharya, S. (2018). *Protocol for Reviving Springs in the Hindu Kush Himalayas: A Practitioner's Manual 2018/4* (Issue September).

Speck, S. (2017). "They moved to city areas, abroad": Views of the elderly on the implications of outmigration for the middle hills of Western Nepal. *Mountain Research and Development*, 37(4), 425–435. https://doi.org/10.1659/MRD-JOURNAL-D-17-00034.1

Su, Y., Bisht, S., Wilkes, A., Pradhan, N. S., Zou, Y., Liu, S., & Hyde, K. (2017). Gendered responses to drought in Yunnan. *Mountain Research and Development*, 37(1), 24–34. https://doi.org/10.1659/MRD-JOURNAL-D-15-00041.1

Sugden, F., Silva, S. De, Clement, F., Maskey Amatya, N., Ramesh, V., Philip, A., & Bharati, L. (2014). *IWMI Working Paper 159: A Framework to Understand Gender and Structural Vulnerability to Climate Change in the Ganges River Basin: Lessons from Bangladesh, India and Nepal*. Colombo, Sri Lanka: International Water Management Institute (IWMI). 50p. (IWMI Working Paper 159).). https://doi.org/10.5337/2014.230, ISBN 978-92-9090-806-7.

Tambe, S., Kharel, G., Arrawatia, M. L., Kulkarni, H., Mahamuni, K., & Ganeriwala, A. K. (2012). Reviving dying springs: Climate change adaptation experiments from the Sikkim Himalaya. *Mountain Research and Development*, 32(1), 62–72. https://doi.org/10.1659/MRD-JOURNAL-D-11-00079.1

Thomas, V., & Ramzi, A. M. (2011). SRI contributions to rice production dealing with water management constraints in northeastern Afghanistan. *Paddy and Water Environment*, 9(1), 101–109. https://doi.org/10.1007/s10333-010-0228-0

Tiwari, A. K. (2000). *Infrastructure and Economic Development in Himanchal Pradesh*. Indus Publishing.

Tong, Y., Fan, L., & Niu, H. (2016). Water conservation awareness and practices in households receiving improved water supply: A gender-based analysis. *Journal of Cleaner Production*, 141, 947–955. https://doi.org/10.1016/j.jclepro.2016.09.169

UNDP. (2019). *Gender Responsive NDC Planning and Implementation*. http://www.ndcs.undp.org/content/ndc-support-programme/en/home/our-work/focal/cross-cutting-gender.html

UNICEF. (2018). *Gender Responsive Communication for Development* (Issue May).

UNISDR. (2015). *Sendai Framework for Disaster Risk Reduction 2015–2030*.

Vaidya, R. A., Shrestha, M. S., Nasab, N., Gurung, D. R., Kozo, N., Pradhan, N. S., & Wasson, R. J. (2019). The Hindu Kush Himalaya assessment. In *The Hindu Kush Himalaya Assessment*. Springer International Publishing. https://doi.org/10.1007/978-3-319-92288-1_11

Wang, Y., Wu, N., Kunze, C., Long, R., & Perlik, M. (2019). The Hindu Kush Himalaya assessment. In *The Hindu Kush Himalaya Assessment* (pp. 17–56). Springer International Publishing. https://doi.org/10.1007/978-3-319-92288-1_2

Ward, F. A., Amer, S. A., & Ziaee, F. (2013). Water allocation rules in Afghanistan for improved food security. *Food Security, 5*(1), 35–53. https://doi.org/10.1007/s12571-012-0224-x

Watershed Organisation Trust (WOTR). (2012). *Kaluchi Thakarwadi: Rejuvenated landscape, Rejuvenated lives. Farming Matters.* Centre for Information on Low External Input and Sustainable Agriculture (ILEIA). https://www.ileia.org/2012/12/23/kaluchi-thakarwadi-rejuvenated-landscape-rejuvenated-lives/

Wester, P., Mishra, A., Mukherji, A., & Shrestha, A. B. (eds). (2019). *The Hindu Kush Assessment Himalaya Mountains, Climate Change, Sustainability and People.* Springer Open. https://doi.org/https://doi.org/10.1007/978-3-319-92288-1

Wilhite, D. A. (1992). Drought. *Encyclopedia of Earth System Science, 2,* 81–92. https://doi.org/10.4324/9781315298917-26

Wilhite, D. A. (2000). Chapter 1: Drought as a natural hazard. *Drought: A Global Assessment, 1,* 3–18.

World Bank. (2016). *World Bank Data.* https://data.worldbank.org/indicator/

World Bank. (2020). *Managing Groundwater for Drought Resilience in South Asia.* https://data.worldbank.org/country/bangladesh

World Bank Group. (2011). Vulnerability, risk reduction and adaptation to climate change. In *Ecological Complexity* (Vol. 20). https://doi.org/10.1016/j.ecocom.2014.11.002

World Bank Group. (2020). *Climate Change Knowledge Portal for Development Practitioners and Policy Makers.* https://climateknowledgeportal.worldbank.org/country-profiles

Wu, H., Xiong, D., Liu, B., Zhang, S., Yuan, Y., Fang, Y., Chidi, C. L., & Dahal, N. M. (2019). Spatio-temporal analysis of drought variability using CWSI in the Koshi River Basin (KRB). *International Journal of Environmental Research and Public Health, 16*(17). https://doi.org/10.3390/ijerph16173100

Zhang, L., Hu, J., Li, Y., & Pradhan, N. S. (2018). Public-private partnership in enhancing farmers' adaptation to drought: Insights from the Lujiang Flatland in the Nu River (Upper Salween) valley, China. *Land Use Policy, 71,* 138–145. https://doi.org/10.1016/j.landusepol.2017.11.034

36 Conventional and Advanced Irrigation Scheduling Techniques to Mitigate Drought

Navsal Kumar, Arunava Poddar,
Rohitashw Kumar, and Vijay Shankar

CONTENTS

DOI: 10.1201/9781003276548-36

36.1 INTRODUCTION

Irrigation in agriculture is necessary to increase world food production. Irrigated agriculture, which accounts for 20% of the land for cultivation, provides 40% of the world's food supply (Garces-Restrepo et al. 2007). Limited freshwater resources, an increasing population accompanied by changing climate, and severe drought conditions pose a serious threat to global food security by severely damaging agricultural production. These changes threaten the freshwater resources available worldwide for irrigation. To achieve sustainability, the challenge is to meet the rising demand for food by increasing productivity while optimizing the use of water (Behmann et al. 2014). This points toward the efficient utilization of existing water resources, which can be accomplished by developing optimal irrigation schedules, i.e., application of a controlled volume of water to crops at required intervals. This requires an improved and detailed understanding of soil–crop–weather interactions and plant response to various abiotic stresses (Ihuoma and Madramootoo2017).

Conventional methods used for scheduling irrigation rely on meteorological parameters and in-situ soil moisture observations to estimate water loss from the plant–soil system (González-Dugo et al. 2006). Methods based on soil moisture observations provide a fair idea regarding crop water requirements, however, they are based on the assumption that the soil has uniform water-holding capacity. Hence, moisture observations are taken at few points and are considered as a true representation of the soil moisture status in the field. Methods relying on meteorological parameters are relatively simple to use, but assume uniform crop density and the same transpiration rate for the entire field. Other methods based on observations of plant water status are reliable but are destructive, labor-intensive, and not suitable for automation, as they do not consider the heterogeneity of the soil and crop canopy.

Conventional methods that are soil-based, plant-based, and weather-based give us measurements at a few points as an overall representation of the field, which are a poor indication of overall field status. Moreover, these methods are labor-intensive, time-consuming, and might involve expensive instrumentations. To have a sustainable approach for optimal use of water, advanced irrigation scheduling techniques should be implemented that can detect water stress early and accurately to avoid any irreversible damage to crops causing a detrimental effect on crop yield due to water stress.

Recently, many studies have focused on implementing canopy and remote sensing techniques to be used as an alternative to these conventional methods, as they provide data of temporal and spatial variability of crop water status in the field (Suárez et al. 2010; Rossini et al. 2013; Zarco-Tejada et al. 2013; Panigadaet al. 2014; Kumar et al. 2014; Dangwal et al. 2016; Poddar et al. 2021a). Canopy sensing methods are based on the application of infrared thermometry to estimate canopy temperatures and to use it as an indicator of crop water status. Remote sensing techniques employ narrow-band indices based on red-edge and visible regions of the electromagnetic spectrum for water stress detection in crops (Berni et al. 2009; Zhao et al. 2015; Dadich et al. 2016; Kumar et al. 2017). Data on these techniques are obtained from sensors with high resolutions that are mounted on small unmanned aircraft systems (sUAS). These methods are widely used in precision agriculture for monitoring crop water status and developing irrigation schedules.

The present chapter presents a brief description of these conventional and advanced techniques used for irrigation scheduling. Use and application of the methods, principles, advantages, and disadvantages are also discussed.

36.2 SOIL WATER ATMOSPHERE CONTINUUM

The soil water atmosphere continuum (SWAC) is defined as a pathway for the movement of water from the soil to the atmosphere through plants (Philip 1966). The movement of water molecules from soil to plant and plant to atmosphere takes place through roots and leaves, respectively, which act as exchange surfaces. The water from soil enters plant roots through the process of osmosis, is transported to the leaf through the xylem, and transpires via leaf stomata to the atmosphere (Robert 2007). The atmosphere has low water potential and plant leaf has relatively high water potential which causes a pressure gradient across stomatal pore causing water to transpire in vapor form to the atmosphere. The amount of water extracted by a plant depends on atmospheric demand and stomatal conductance but is governed by the maximum supply rate through roots and soil moisture availability (Jackson et al. 2000).

36.2.1 SOIL WATER BALANCE

Water balance has been considered an important tool in most hydrological and hydrogeological studies. The soil water balance is simply a detailed statement of the law of mass conservation applied to the crop root zone of soil. The fundamental principle behind the soil water balance is to account for all quantities of water added to, removed from, and stored within an assumed volume of soil for a given duration For studies related to agricultural water management and irrigation sciences, soil water balance is usually described by the following equation (Bandyopadhyay and Mallick 2003; Shankar 2007):

$$(P + I + CR) - (RO + DP + ET) = \Delta S \tag{36.1}$$

where P is precipitation; I is irrigation; ET is crop evapotranspiration; RO is surface runoff; DP is deep percolation to the groundwater table; CR is capillary rise from the shallow water table; and ΔS is change in soil moisture content.

Figure 36.1 shows the various components of the soil water balance. The water balance equation can be used to compute an unknown component when measurements of all other terms are available. While conducting field crop experiments, the soil water balance is generally applied to lysimeters by isolating the crop root zone from the surrounding environment and controlling processes that are difficult to measure (Kosugi and Katsuyama 2004). Lysimeters are generally used to estimate crop ET from other easily measurable components of water balance. Fluxes like capillary rise from and deep percolation to the water table are difficult to assess for short durations and hence the soil water balance approach is applicable for long periods (Allen et al. 1998; Poddar et al. 2018a; Poddar et al. 2021b).

36.2.2 SOIL MOISTURE DYNAMICS IN THE CROP ROOT ZONE

The knowledge of moisture movement in soil is necessary to determine soil moisture depletion and soil moisture extraction. Studying water movement in the soil–root system quantitatively helps in modeling root uptake by crops and provides key information regarding optimum irrigation scheduling and agricultural water management (Kumar et al. 2013). The upper layer of the soil shows

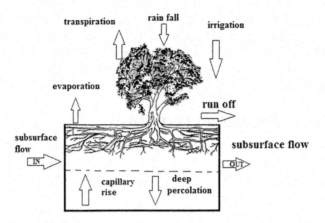

FIGURE 36.1   Soil water balance of the root zone.

the most activity. How the water moves in upper soil determines the rate of soil evaporation, plant transpiration, runoff, and groundwater recharge.

Most of the processes involving soil moisture flow in the crop root zone occur while the soil is unsaturated, i.e., soil voids are partly filled with water and partly with air (Sharma et al. 2016; Kumar et al. 2020c). The zone is generally known as unsaturated and extends from the water table up to the ground surface. Soil moisture flows in the unsaturated zone is an intricate process and difficult to describe quantitatively, since it involves complex relations among variables like suction head, soil moisture content, and hydraulic conductivity (Kumar et al. 2013; Poddar et al. 2017).

Moisture flow in the unsaturated zone is generally simulated by solving Richards's equation (Richards1931). Richards's equation is a highly nonlinear partial differential equation (Feddes et al. 1978)that was obtained by combining Darcy's equation with the mass conservation equation. In the case of cropped soils, the root water uptake is represented by incorporating a sink term in Richards's equation (Govindraju et al. 1992). The equation is given as (Celia et al. 1990)

$$\frac{\partial \theta}{\partial t} = \frac{\partial \left[K(\psi) \dfrac{\partial}{\partial z}(\psi + z) \right]}{\partial z} - S(z,t) \tag{36.2}$$

where θ is the moisture content,ψ is the pressure head,t is the time coordinate,z is the vertical coordinate taken positive upward, K is the unsaturated hydraulic conductivity of the soil, and $S(z, t)$ represents the root water uptake expressed as volume of water per unit volume of soil per unit time.

The solution of Richards's equation (Equation 36.2) is usually obtained by employing a numerical method (finite difference/finite element/finite volume) coupled with a suitable iteration technique (Picard's/Newton's iteration) and constitutive relationships among K, θ, ψ variables (Shankar et al. 2012). Numerous root water uptake models are developed based on the extraction patterns such as constant, linear, nonlinear, exponential, and logarithmic (Ojha et al. 2009; Poddar et al. 2020).

36.2.3   ENERGY EXCHANGE AT LEAF SURFACE

Energy exchange occurs where two mediums interact. In the case of plants, these are leaf surfaces interacting with the atmosphere around them (Gates 1962). This energy exchange causes transpiration through the stomatal pores and also is the driving phenomena for the onset of crop water stress. This energy balance through a crop canopy is generally expressed by the following equation (Monteith & Unsworth 2007):

$$R_n = G + H + \lambda E \tag{36.3}$$

Where R_n is the net radiation (W/m^2), G is the heat flux below the canopy (W/m^2), H is the sensible heat flux (W/m^2) from the canopy to the air, λE is the latent heat flux to the air (W/m^2), and λ is the heat of vaporization.

This energy exchange between leaf and atmosphere occurs when there is a vapor pressure deficit in the air and a difference in crop canopy temperature causes the flow of energy and mass to form an equilibrium between crop canopy and atmosphere (Philip 1966).

36.3 IRRIGATION SCHEDULING OVERVIEW

Irrigation scheduling is defined as planning and decision-making activity regarding the application of water in the field, which the agriculturist, practitioner, or operator of an irrigated farm is involved in during the crop growth season (Jensen 1981). It helps in determining the correct frequency and amount of irrigation events. The main objective of irrigation scheduling is to ensure the optimum amount of water in the plant's root zone as well as a suitable time interval between each consecutive watering such that the crop productivity is not adversely affected (Ojha and Rai 1996; Shankar et al. 2012; Kumar et al. 2019a). The important parameters for assigning irrigation schedules are crop water stress and soil moisture depletion (Martin 1990; Kumar et al. 2020d). The soil moisture is generally maintained at a level so that air may also pass through soil ensuring proper root growth (Unger and Kaspar 1994). The allowable soil moisture depletion is dependent on plant available water, which is the difference between soil moisture at field capacity and the permanent wilting point.

36.4 SOIL-BASED TECHNIQUES

Conventional methods of developing irrigation schedules rely on the precise estimation of *in situ* soil moisture status. Irrigation events are then scheduled based on a target moisture deficit or potential. Estimation of soil moisture content is generally performed using different instruments employing different principles. Brief descriptions of some of the commonly adopted techniques are mentioned in the following sections.

36.4.1 GRAVIMETRIC METHOD

The gravimetric method is the most conventional technique used for estimating soil moisture content (Reynolds 1970). Soil samples are obtained from desired depths for the determination of soil moisture. The samples are weighed and then kept in the oven for drying for a day (24 hours) at 105°C. After drying, the soil sample is weighed again and the water loss is determined. Although the method is reliable, it is not practical for developing irrigation schedules, since it is destructive and takes a full day to determine the soil moisture. For instance, in the case of sandy soils that dry quickly, irrigation might be needed before the results of soil moisture measurements come. Still, this method is most useful when calibrating other instruments that are used for *in situ* soil moisture measurement.

36.4.2 SENSOR-BASED METHODS

The *in situ* soil moisture measurements for scheduling irrigation is generally performed using various instruments such as a tensiometer, capacitance probe, neutron probe, time domain reflectometer

(TDR), and electrical resistance blocks. These techniques are based on direct measurement of either soil moisture content or soil matric potential. The instruments are commercially available in the market, but few farmers have the experience, time, money, or expertise to use them effectively (Shankar et al. 2017). Therefore, there is a need for advisory services related to irrigation scheduling to assist the farmer.

36.4.2.1 Soil Moisture Content

36.4.2.1.1 Time Domain Reflectometer (TDR)

This technique is based on measuring the dielectric constant of soil, which is used to estimate the soil moisture content (Cole 1977; Topp and Davis 1985). The TDR comprises an electronic meter connected to two parallel rods inserted into the soil at the desired depth (Figure 36.2). An electrical signal is sent through the soil with the rods serving as the transmitter and receiver. The rate of conductance of electrical signals through the soil is directly related to soil moisture content (Noborio 2001). The TDR is very easy to use, and the results are reliable and have been extensively used in many studies (Evett 2003; Huebner et al. 2005; Sinha 2017).

36.4.2.1.2 Neutron Probe

A neutron probe uses a radiation source to estimate the amount of soil moisture (Bell 1987). The neutron probe is the most prevalent method currently employed by farmers in the USA and UK (Shankar et al. 2017). A neutron probe is lowered into an aluminum access tube that is installed in the field, and soil moisture measurements are automatically recorded at specified intervals up to the maximum depth of the probe. Although quick and portable, they remain expensive and are bound by legal controls due to the emission of the radioactive source. This approach has become popular with the farming community and continues to have a strong impact. However, *in situ* calibration of neutron probes is rarely done by the manufacturer, which may result in inaccurate readings. The probe is also not reliable for estimating soil moisture changes in the topsoil surface (20 cm).

36.4.2.1.3 Capacitance Probe

The use of the capacitance probe is becoming popular for scheduling irrigation at various places all over the world (Tallon and Si 2015). A capacitance probe is used for measuring the dielectric permittivity of a surrounding soil medium. When soil water content changes in the soil, the capacitance probe measures the change in capacitance due to the change in dielectric permittivity and is

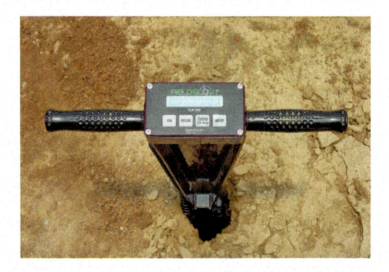

FIGURE 36.2 Soil moisture measurements being done using a TDR soil moisture meter.

FIGURE 36.3   Soil moisture measurements using a capacitance probe.

directly related to a change in soil moisture content (Dean et al. 1987; Bell et al. 1987). The use and operation of the capacitance probe are similar to that of the neutron probe. The probe is inserted in the access tube and the reading is taken and recorded on the display unit (Figure 36.3). After proper calibration, the capacitance probe is reliable, easy to use, and accurate. However, due to its high cost, the capacitance probe is not a practical option compared to other low-cost methods for on-farm use. It might be more practical for operators/farmers with a large area of irrigated land (Evans and Sneed 1991). The major drawback of portable probes is their small sampling range, increasing the sensitivity of measurements to soil disturbance caused by access tube installation.

36.4.2.2   Soil Matric Potential

36.4.2.2.1   Tensiometers

Tensiometer measures soil matric potential or tension in the unsaturated zone. It generally measures the suction that the plant root exerts to pull soil moisture into the plant body (Richards 1965; Wang et al. 1998). It comprises of an airtight plastic tube and a porous ceramic tip at the bottom. A vacuum gauge (to indicate suction pressure)or a resealable rubber bung (through which a portable vacuum meter can be inserted)is generally provided at the top. The tensiometer is filled with water and partially buried in the soil until the desired depth during the irrigation season (Figure 36.4). Water in the tensiometer will come to equilibrium with water in the soil. Readings on the gauge indicate the availability of water in the soil. If used properly, it enables efficient and better irrigation management by accurately determining when water should be applied to maintain optimum crop growth (Stannard 1992; Taber et al. 2002).

36.4.2.2.2   Electrical Resistance Blocks

An electrical resistance block comprises two electrodes embedded in a porous material (gypsum) and installed at different depths in the root zone of soil (Spans and Bakers 1992), as shown in Figure 36.5a. Since the blocks are porous, the water in the gypsum block reaches equilibrium with water in the soil. Observations are recorded by connecting a portable soil moisture meter (Figure 36.5b) to the electrode, which gives a reading from 0 to 199 centibars. The electrical resistance between the electrodes is determined and is related to soil moisture content as a function of soil matric potential (Irmak et al. 2014; Liakos et al. 2015). A fully saturated soil gives a reading of 0 and fully dry soil gives a reading of 199. Proper installation of blocks and good contact with soil is essential for reliable readings. Methods based on electrical resistance are suitable for most soils since the readings cover a wide range of soil moisture applicable to irrigation management. The electrical resistance

FIGURE 36.4 Tensiometer installation in the crop field.

FIGURE 36.5 (a) Electrical resistance (gypsum) block and (b) soil moisture meter.

blocks are inexpensive tools available commercially and can be used by the majority of farm irrigators. The efficiency of blocks tends to deteriorate with time, and it is recommended not to use them for more than two to three seasons (Lieb et al. 2003; Thompson et al. 2006).

36.5 WEATHER-BASED TECHNIQUES

36.5.1 CROP WATER REQUIREMENTS

A crop water requirement (CWR) is defined as the amount of water needed by crops to recoup from water loss occurring through evapotranspiration (Todorovic 2005). In other words, this is water needed by different crops for their optimal growth. CWR is always taken in the context of any crop growing under its optimal conditions, i.e., uniform and healthy (disease-free), completely covering the ground by its canopy, and optimum soil conditions (water and nutrients) (Allen et al. 1998). CWR depends on the following factors:

- The climate of the region
- Type of crop under consideration
- Growth stage of the crop

36.5.2 EVAPOTRANSPIRATION

Evapotranspiration (ET) is defined as a process of water transfer through evaporation and transpiration from the plant and soil system.ET has been utilized as an important parameter in hydrological and climatological studies (Xu and Singh 2002; Poddar et al. 2018b). It is very useful in the planning and management of irrigation in agricultural farms. When dealing with the management of water resources and environmental studies, quantification of ET becomes necessary, which involves the determination of reference ET (ET_0). The rate of evapotranspiration is generally expressed in units of depth (mm) per unit of time. ET is governed by weather parameters, crop characteristics, management, and environmental conditions (Jensen and Allen 2016).

36.5.3 REFERENCE EVAPOTRANSPIRATION

ET_0 represents the rate of evapotranspiration from a uniform grassed area of 10–15 cm height, growing actively, shading the ground completely by its canopy, and having adequate water supply in its root zone (Allen et al. 1998; Tabari 2011). ET_0 estimation is governed by climatic parameters including air humidity, wind speed, temperature, and solar radiation (Kumar et al. 2020b). Precise estimation of ET_0 is necessary for accurate assessment of crop evapotranspiration and improving the efficient use of water resources.

36.5.4 CROP EVAPOTRANSPIRATION

Crop evapotranspiration (ET_c) is defined as the evapotranspiration taking place from well-fertilized and disease-free crops, grown in large fields, having optimum soil moisture conditions, and attaining full productivity under the given climatic conditions (Hargreaves and Samani 1982; Allen et al. 1998). Although the terms ET_c and CWR are theoretically identical, CWR is the amount of water that needs to be supplied to the crop, whereas ET_c indicates the amount of water that is lost through evapotranspiration. The irrigation water requirement essentially represents the difference between CWR and effective precipitation (Allen et al. 1998). Accurate estimates of ET_c can be obtained through water balance studies conducted using a lysimeter.

36.5.5 CROP COEFFICIENT

The crop coefficient (K_c) is an important parameter for scheduling irrigation and allocation of available water resources (Kang et. al. 2003; Kumar et al. 2020e). The value of K_c varies with the crop growth stage, i.e., initial K_c ($K_{c\ ini}$), mid-season K_c ($K_{c\ mid}$), and end-season K_c ($K_{c\ end}$) (Doorenbos and Pruitt 1977). Standard K_c values for different crops are provided in FAO-56. These values are modified according to the local climatic conditions of the concerned area. There are two crop coefficient methods used to calculate ET_c: single crop coefficient method and dual-crop coefficient. The single K_c approach is given by (Allen et al. 1998)

$$ET_0 * K_C = ET_C \tag{36.4}$$

36.5.6 METHODS TO ESTIMATE REFERENCE EVAPOTRANSPIRATION

Several temperatures, radiation, panevaporation–based, and combination-type methods are generally used to estimate ET_0. The methods are mentioned in Table 36.1.

36.5.6.1 Combination-Type Methods

Combination-type methods rely on a large set of input climatic parameters to estimate ET_0 values. These methods are generally known to provide accurate ET_0 estimates. The FAO-56

TABLE 36.1

Details of Reference Evapotranspiration Methods

Type	Method	Basic Equation	Acronym	Reference	Eq. No.
Combination-type methods	FAO-56 Penman–Monteith	$ET_0 = \dfrac{0.408\Delta(R_n - G) + \gamma \dfrac{900}{\overline{T}+273} u(e_s - e_0)}{\Delta + \gamma(1+0.34u)}$	PEN-M	Allen et al. (1998)	5
	FAO-24 Corrected Penman	$ET_0 = c[WR_{nf} + (1-W)0.27(1+0.01U)(e_a - e_d)]$	C-PEN	Doorenbos and Pruitt (1977)	6
Radiation-based methods	FAO-24 Radiation	$ET_0 = C(WR_s)$	RAD	Doorenbos and Pruitt (1977)	7
	Priestley-Taylor	$ET_o = \beta \dfrac{\Delta}{\Delta + \gamma}(R'_n)$	PT	Priestley and Taylor (1972)	8
	Makkink	$ET_0 = 0.61* \dfrac{\Delta}{\Delta+\gamma} * \dfrac{R_s}{58.5} - 0.12$	MK	Makkink (1957)	9
	Jensen-Haise	$ET_0 = \dfrac{C_T(T_a - T_x)*R_s}{\lambda}$	JH	Jensen and Haise (1963)	10
	Irmak	$ET_0 = -0.611 + 0.149*R_s + 0.079*T_a$	IK	Irmak et al. (2003)	11
	Turc	$ET_o = 0.31\left(\dfrac{\overline{T}}{\overline{T}+15}\right)(R'_s + 2.09)\left(1 + \dfrac{50 - RH_{mean}}{70}\right)$ For RH<50 $ET_o = 0.31\left(\dfrac{\overline{T}}{\overline{T}+15}\right)(R'_s + 2.09)$ For RH>50	TC	Turc (1961)	12

(Continued)

TABLE 36.1 (CONTINUED)

Details of Reference Evapotranspiration Methods

Type	Method	Basic Equation	Acronym	Reference	Eq. No.
Temperature-based methods	FAO-56 Hargreaves-Samani	$ET_o = 0.0023(\bar{T}+17.8)(T_{max}-T_{min})^{0.5}R_a'$	HAR	Hargreaves and Samani (1985)	13
	FAO-24 Blaney Criddle	$ET_o = a_b + b_b\left[p(0.46\bar{T}+8.13)\right]$	BC	Doorenbos and Pruitt (1977)	14
	Thornthwaite	$ET_o = 16*\left(\dfrac{10*T_i}{H}\right)^A$	TW	Thornthwaite (1948)	15
	Linacre	$ET_o = \dfrac{\dfrac{500*(T+0.006*A)}{100-\phi}+15*(T-T_d)}{80-T}$	LIN	Linacre (1977)	16
Pan evaporation–based methods	Frevert	$ET_o = k_p * E_{pan}$ $K_{pan} = 0.475 - 2.4*10^{-4}*U$ $+5.16*10^{-3}*RH+1.18*10^{-3}*F-1.6*10^{-5}*RH^2$ $-1.01*10^{-6}*F^2-8*10^{-9}*RH^2*U-1*10^{-8}*RH^2*F$	FV	Frevert et al. (1983)Cuenca (1989)	17
	Allen and Pruitt	$K_{pan} = 0.108 - 0.00031*U + 0.0422*\ln F + 0.1434*\ln RH - 0.00063\ln F^2*\ln RH$	AP	Allen and Pruitt (1991)	18
	Snyder	$K_{pan} = 0.482 + 0.024*\ln F - 0.000376*U + 0.0045*RH$	SD	Snyder (1992)	19
	Modified Snyder	$K_{pan} = 0.5321 - 0.000321*U + 0.0249*\ln F + 0.0025*RH$	MS	Snyder (1992)	20
	Orang	$K_{pan} = 0.51206 - 0.000321*U + 0.002889*RH + 0.031886*\ln F - 0.000107*RH*\ln F$	OG	Orang (1998)	21

Penman–Monteith (PEN-M) method is used as a standard for evaluating other methods, and this method was developed by Allen et al. (1998) considering local calibration of the wind function to achieve satisfactory results. FAO-24 corrected Penman (C-PEN) is another combination-type method and was developed by Doorenbos and Pruitt (1977).

36.5.6.2 Radiation-Based Methods

Six radiation-based methods – FAO-24 radiation (RAD) (Doorenbos and Pruitt 1977), Priestley–Taylor (Priestley and Taylor 1972), Makkink (Makkink 1957), Jensen–Haise (Jensen and Haise 1963), Irmak (Irmak et al. 2003), and Turc (Turc 1961) – are commonly used for estimating ET_0 in various regions worldwide. These methods use solar radiation and other climatic parameters for estimating ET_0. But the accuracy and reliability of these methods depend largely on the radiation data of the study area under consideration.

36.5.6.3 Temperature-Based Methods

Temperature-based methods are generally considered as possible alternatives to combination-type methods due to the fact that temperature data (T_{max}, T_{min}, and T_{mean}) is readily and easily available in nearly all locations. Hargreaves (Hargreaves and Samani 1985), Thornthwaite (Thornthwaite 1948), Blaney–Criddle (Blaney and Criddle 1962), and Linacre (Linacre 1977) are commonly used temperature-based methods for ET_0 estimation purposes.

36.5.6.4 Pan Evaporation–Based Methods

Determination of ET from pan evaporation (E_{pan}) depends on the precise estimation of the pan coefficient (K_{pan}). For the accurate determination of K_{pan}, estimation of different parameters like mean daily wind speed, upwind fetch of the low-growing crop, and relative humidity is essential (Irmak et al. 2002). Numerous researchers have evaluated K_{pan} methods under different climates (Rahimikhoob 2009; Assari and Modaberi 2014; Sabziparvar et al. 2010; Kumar et al. 2020f) and suggested that location influences the values of K_{pan}. Therefore, it is suggested that K_{pan} for the agroclimate of the study area should be determined before estimating ET_0 values. Frevert (Frevert et al. 1983; Cuenca 1989), Allen and Pruitt (Allen and Pruitt 1991), Snyder (Snyder 1992), modified Snyder (Snyder 1992), and Orang (Orang 1998) are the most popular pan evaporation–based methods. Detailed mathematical descriptions are given in Table 36.1.

36.6 PLANT-BASED TECHNIQUES

The following plant-based approaches for scheduling irrigation are some of the conventional methods used in various places by agronomists and researchers.

36.6.1 PLANT RESPONSE TO WATER STRESS

Water stress negatively influences many characteristics of the crop physiology, especially photosynthetic capacity. If stress continues, crop growth and crop yield are damaged. Crops have "evolved complex physiological and biochemical adaptations to adjust and adapt to a variety of environmental stresses" (Osakabe et al. 2014). The plant water status is generally determined using various indicators.

36.6.2 METHODS TO ESTIMATE PLANT WATER STATUS

36.6.2.1 Leaf Water Potential

The leaf water potential (LWP) is an indicator of the plant water status during the day. This method enables measuring of the short-term (hourly basis) hydrophytic response of the plant in response to variation in the leaf transpiration and the root water absorption (Deloire and Heyns 2011). LWP is a major driving force for the movement of water through the plant (Jarvis 1976). LWP is measured by placing the leaf inside a fully sealed chamber. Then pressurized gas is slowly added to the chamber. With an increase in pressure inside the chamber, water inside the xylem will come out and will be visible at the cut end of the stem. The corresponding reading is noted. The values of the LWP observed in the field must be interpreted in relation to volume flux of the water flowing through the plant and the water transfer's pathway characteristics from soil to the leaf.However, LWPfor scheduling irrigation is not recommended, because of the high variability between measurements.

36.6.2.2 Relative Water Content

The relative water content (RWC) relates the maximum water content at full turgor of leaf with the water content of a leaf and can be reflected as an indicator of vegetation status. RWC is obtained from experimental observations of leaf turgid weight (TW) and leaf weight by using the following expression (Colombo et al. 2008):

$$RWC = \frac{FW - DW}{TW - DW} * 100 (\%) \tag{36.5}$$

where FW is the fresh weight, DW is the dry weight, and TW is the turgid weight.

This method is a reliable indicator of the plant water status, and less complex equipment is required for this technique. But the method is time-consuming and destructive, and thus is generally not advised for regular use in irrigation scheduling (Ihouma and Madramootoo 2017).

36.6.2.3 Stomatal Conductance

Stomatal conductance is defined as a "measure of the rate of passage of carbon dioxide (CO_2) entering, or water vapor escaping through the stomata of a leaf" (Taiz & Zeigler1991). Stomatal conductance is a good measure of plant water stress. Measurement of the stomatal conductance is performed by using leaf porometers (Shimshi 1977). Porometers are of different types as described next:

a) Steady-state porometers: They use a sensor with a fixed diffusion path to leaf and measure the concentration of the vapor at two locations in the diffusion path (Schulze et al. 1982). They then calculate vapor flux from vapor concentration measurements and known conductance of the diffusion path using the given equation:

$$\frac{C_{vL} - C_{v1}}{R_{vs} + R_1} = \frac{C_{v1} - C_{v2}}{R_2} \tag{36.6}$$

where C_{vL} is the vapor concentration at the leaf,C_{v1} and C_{v2} are the concentrations at two sensor locations, R_{vs} is the stomatal resistance, and $R_1$and $R_2$are the resistances at two sensors.

If R_1 and R_2 are equal, then vapor concentration is substituted with relative humidity (h), then

$$R_{vs} = \frac{1 - h_1}{h_2 - h_1} R_2 - R_1 \tag{36.7}$$

Stomatal conductance (g_{vs}) is defined as the reciprocal of resistance:

$$g_{vs} = \frac{1}{R_{vs}} \tag{36.8}$$

b) Dynamic porometers: They measure the time it takes for humidity to increase from a specified value to another value in a closed chamber clamped to a leaf (Monteith et al.1988). The resistance (R) is obtained using Equation 36.7:

$$\Delta t = \frac{(R+A)l\Delta h}{1-h} \qquad (36.9)$$

c) Null balance porometers: These conserve a constant humidity in a closed chamber by regulating the dry airflow through the chamber and look for stomatal resistance (Beardshell et al. 1972):

$$R_{vs} = \frac{A}{f}\left(\frac{1}{h}-1\right) - R_{va} \qquad (36.10)$$

where R_{vs} is the stomatal resistance, R_{va} is the boundary layer resistance, f is the flow rate of dry air, h is the chamber humidity, and A is the leaf area.

36.6.2.4 Sap Flow Measurement

The sap flow measurement technique assesses the transpiration rate of the crop by sending the heat pulse through the stem and then measuring the rate at which the sap ascends the stem. Generally, short heat pulses are applied in the crop stem, and by estimating the heat pulse velocity along the stem, sap mass flow can be determined. Changes in the stomatal opening cause the change in the transpiration rate. Singh et al. (2010) utilized sap flow sensors for scheduling irrigation, but this approach only gave indirect estimations for the conductance change. It happened because the sap flow also depends on humidity and atmospheric conditions. Thus, sap flow changes can also happen without any variations in the stomatal opening.

This method is sensitive to water deficit and closing of the stomatal aperture. It is adapted generally for controlling irrigation systems and automatic recording. The main disadvantage of this method is the system must be calibrated for each crop and it is very difficult to duplicate. Also, this method only tells when to irrigate the crop and does not provide any information regarding the amount of water to be applied. This method also requires complex instrumentation.

36.7   ENVIRONMENTAL CANOPY SENSING TECHNIQUES

36.7.1   INFRARED THERMOMETRY

This method uses infrared thermometers to measure canopy temperature, which helps in determining the water stress in the crop. These methods, based on the detection of water stress via canopy temperature measurements, are widely used as an alternative to conventional methods (Berni et al. 2009). This method gives very reliable results. One of the major advantages of this method is it is a nondestructive test. Algorithms based on canopy temperature correlate strongly to crop outputs, including water use efficiency, seasonal evapotranspiration, crop yield, midday leaf water potential, and irrigation rates (Collaizzi et al. 2012; Kumar et al. 2019b). However, these methods do not account for soil and crop heterogeneity.

36.7.2   CROP WATER STRESS INDEX

Crop water stress index (CWSI) is a widely adopted technique used to estimate crop water stresses (Alderfasi and Nielsen 2001; Kumar et al. 2020a). For estimating crop water stress, it adopts crop canopy and air temperature difference. Canopy temperature act as a crop water stress indicator due to depletion of the soil water, stomatal closure, and energy exchange between leaf and atmosphere ceases (Jackson et al. 1981). In return it causes plant temperature to be increased and water uptake decreases, which causes stress to develop in the plant. This method has a few advantages like being sensitive to stomatal closure and crop water deficit. But the estimation of crop water stress is influenced if there is cloud cover. Also, this method requires different baselines to be developed for different crops.

The widely adopted method to derive the CWSI for crops is based on canopy temperature and helps in irrigation scheduling studies as well as indicates crop water status (Gontia and Tiwari 2008). The principle on which the CWSI is based is that due to transpiration the leaf surface cools and with the depletion of soil moisture in the root zone, transpiration through the leaf and the stomatal conductance decreases, which causes the stomatal opening to close and rise in leaf temperature.

Idso et al. (1981) witnessed a linear relationship between canopy air temperature differential and vapor pressure deficit, and an empirical method was then developed for crop water stress estimation. The empirical formula for CWSI shown in Equation 36.11 uses two baselines. The difference between canopy temperature (T_c) and air temperature (T_a) of anoptimum-watered crop that transpires at maximum potential rate represents the lower baseline, and (T_c–T_a) of a non-transpiring crop represents the upper baseline.

$$\text{CESI} = \frac{\left[\left(T_c - T_a\right) - \left(T_{nws} - T_a\right)\right]}{\left[\left(T_{dry} - T_a\right) - \left(T_{nws} - T_a\right)\right]} \tag{36.11}$$

where T_c is canopy temperature (°C), T_a is air temperature (°C), T_{nws} is the non-water-stressed canopy temperature (°C), and T_{dry} is water-stressed canopy temperature (°C).

36.7.3 OTHER INDICES

In recent studies, various other indices have been used to detect crop water stress which requires infrared thermometry and less information input as compared to the CWSI. These indices are the ratio of canopy temperature, degree above non-stressed, and degree above canopy threshold.

36.7.3.1 Ratio of Canopy Temperature

Canopy temperature ratio (T_c ratio), i.e., the ratio of estimated canopy temperature of a crop grown under fully irrigated soil profile (optimum) to that measured under water-stressed soil profile (non-optimum), is analternative for water stress coefficient. The canopy temperature ratio circumvents the uncertainty ofroot depth and soil properties to assess root zone depletion and total available soil water. This method was inspired by practices used for in-season evaluation of crop nitrogen level where the total target area was compared to reference area, which was provided with adequate nitrogen to ease nitrogen deficiency (Schepers et al. 1992; Bausch and Duke 1996).

36.7.3.2 Degrees above Non-Stressed (DANS)

To avail an index that includes a wider range of values and can be predicted by simple subtraction, the difference between stressed and non-stressed canopy temperatures was used. The degrees above non-stressed (DANS) index can be projected with a single type of observation, i.e., canopy temperature.

36.7.3.3 Degrees above Canopy Threshold (DACT)

The degrees above canopy threshold (DACT) index is suggested as an appropriate index that necessitates a single canopy temperature measurement for measuring water stress (DeJonge et al. 2015). DACT is correlated with CWSI at higher temperatures and can be effectively used as the CWSI without requiring additional inputs that are needed by the CWSI.

36.8 REMOTE SENSING TECHNIQUES

The canopy temperature's usefulness measured using infrared thermometry is established in many studies for monitoring and estimating crop water stress. Other indices that are based on the red end and visible spectrum are important for the detection of crop water stress (Zhao et al. 2015; Dangwal et al. 2016; Kumar et al. 2020g). Remote sensing techniques use airborne-based sensors or satellite to fetch data for a specified area. Data collection with remote sensing approaches can be both active and passive. For active remote sensing, radar energy is released and the resultant signal that is reflected is calculated, whereas with passive remote sensing, spectral imagers identify radiation that is reflected by the area under observation.

Remote sensing analysis can afford regular data on hydrological and agricultural conditions for vast areas. The ability of remote sensing to detect and observe crop growth and other biophysical variables have had major advancements in the last 20 years. Now, remote sensing is an essential research tool and has its application in the management of irrigated agricultural sectors both at the local and national scale.

36.8.1 Vegetation Indices/Spectral Reflectance

Vegetation indices can provide meaningful information about plant health, water content, environmental stress, and other important characteristics. Water stress estimation by plant-based techniques use leaf temperature as the main basis, which directly represent transpiration from crops, and do not use other parameters (e.g., non-stomatal reductions) (Zarco-Tejada et al. 2012). Spectral vegetation indices that are correlated with the plant water stress are

- Xanthophyll indices
- Structural indices
- Water indices

36.8.1.1 Xanthophyll Indices

Xanthophyll indices use chlorophyll fluorescence and stomatal conductance as an indicator of plant water stress. It comprises two different indices. These indices are the

- Photochemical Reflectance Index (PRI)
- Normalize Photochemical Reflectance Index (PRI_{norm})

36.8.1.1.1 Photochemical Reflectance Index (PRI)

The Photochemical Reflectance Index (PRI), along with the solar-induced chlorophyll fluorescence emission, acts as a previsual indicator of water stress, which indirectly estimates plant water stress (Gamon et al. 1997; Flexas et al. 2002; Moya et al. 2004). The PRI is sensitive to the efficiency of photosynthesis and xanthophyll cycle pigment's peroxidation state (Suárez et al. 2010). The functional requirement of the PRI is being sensitive to fast changes in carotenoids due to deep oxidation of the xanthophyll pigments (Magney et al. 2016), and also to an increasing amount of heat dissipation in water stress conditions (Panigada et al. 2014).

Xanthophyll cycle pigment interconversion can be identified in leaves in the form of changes in reflectance through 531 nm. The empirical formula for calculating the PRI is (Gamon et al. 1997)

$$PRI = \frac{R_{570} - R_{531}}{R_{570} + R_{531}} \tag{36.12}$$

Here R_{570} and R_{531} are reflectances of light having a wavelength of 570 and 530 nm, respectively.

36.8.1.1.2   Normalized Photochemical Reflectance Index (PRI$_{norm}$)

The Normalized Photochemical Reflectance Index (PRI$_{norm}$) is a method that detects xanthophyll pigment changes with the change in water stress, and normalizes the canopy leaf area reduction and chlorophyll content level initiated by water stress. Berni et al. (2009) stabilized the PRI using the Renormalized Difference Vegetation Index (RDVI). The RDVI is an index that is sensitive to the canopy structure and a red-edge index that is sensitive to chlorophyll content (R_{700}/R_{670}). The empirical formula for PRI$_{norm}$ is

$$PRI_{norm} = \frac{PRI}{RDVI * \left(\dfrac{R_{700}}{R_{600}} \right)} \tag{36.13}$$

where R_{700} and R_{600} arereflectances of light having a wavelength of 700 and 600 nm, respectively.

This method is a modification of the PRI and also possess a better capacity for detecting water stress level when compared to other structural and greenness indices (Gago et al. 2015). Rossini et al. (2013) revealed spectral vegetation indices to be capable of mapping water stress, by explaining the advantage of high spatial resolution capabilities that are found more difficult in thermal areas. PRInorm was also exposed to the potential applicability of remote sensing data in precision agriculture to improve irrigation scheduling.

36.8.1.2   Structural Indices

Structural indices work in the range of the visible spectrum and near-infrared spectrum. The indices measure reflectance within these spectral ranges to indicate changes in the plant canopy that occur due to water stress. Structural indices consist of the following indices:

* Normalized Difference Vegetation Index (NDVI)
* Renormalized Difference Vegetation Index (RDVI)
* Optimized Soil Adjusted Vegetation Index (OSAVI)

36.8.1.2.1   Normalized Difference Vegetation Index (NDVI)

Th NDVI is a widely known index that is used as an indicator for the greenness of vegetation (Rouse et al. 1974). It is used in irrigation studies to map the cropped area to estimate the crop coefficient (K_c) for use in the FAO56 conventional method. The NDVI can be calculated as

$$NDVI = \frac{R_{800} - R_{670}}{R_{800} + R_{670}} \tag{36.14}$$

Here R_{800} and R_{670} are the reflectances of light having a wavelength of 800 and 670 nm, respectively.

36.8.1.2.2   Renormalized Difference Vegetation Index (RDVI) and
Optimized Soil Adjusted Vegetation Index (OSAVI)

When derived empirically the NDVI product comes out to be unstable, as it is affected by sun view geometry and the soil reflectance. So RDVI and OSAVI were developed to improve the NDVI (Roujean and Breon 1995; Haboudane et al. 2002). They were developed with the concept of reducing soil brightness, which is found to influence spectral vegetation indices using near-infrared and red wavelengths. Also, these indices help in reducing variable active photosynthetic radiation due to different non-photosynthetic materials being present. The following equations are used to compute these indices:

$$RDVI = \frac{R_{800} - R_{670}}{\sqrt{R_{800} + R_{670}}} \tag{36.15}$$

$$OSAVI = \frac{(1+0.16)R_{800} - R_{670}}{(R_{800} + R_{670}) + 0.16} \qquad (36.16)$$

Here R_{800} and R_{670} reflectances of light having a wavelength of 800 and 670 nm, respectively. These indices are non-destructive with high spectral and temporal resolution.

36.8.1.3 Water Indices

Normally, the water absorption bands in the 1300–2500 nm spectral region show the greatest sensitivity to leaf water concentration (Carter 1991). Although water absorption in the spectral region is strong, it gives inadequate results. Hence, a reflectance trough in the near-infrared region at 950–970 nm, which corresponds to weak absorption of the water band, has shown to be effective for representing the total crop or canopy moisture content (Peñuelas et al. 1997). Whenever crops are found water-stressed, the 970 nm range of the reflectance spectrum inclines to disappear as it shifts toward lower wavelengths (Peñuelas and Filella 1998). Using this concept, the Water Index (WI) and Simple Ration Water Index (SRWI) were developed to measure reflectance in the near-infrared region, which is used to represent plant moisture content. These are calculated using the following equations:

$$WI = \frac{R_{900}}{R_{970}} \qquad (36.17)$$

$$SRWI = \frac{R_{860}}{R_{1240}} \qquad (36.18)$$

Here R_{900}, R_{970}, R_{860}, and R_{1240} are reflectances of light with a wavelength of 900, 970, 860, and 1240 nm, respectively.

36.9 SUMMARY

Conventional techniques of irrigation scheduling use climatic data, soil moisture observations, and physiological measures of the crop for the assessment of crop water stress. These approaches, though reliable and widely used, comprise expensive equipment, sensors, and costly installation. Collection and analysis of data under conventional techniques are time-consuming and labor-intensive. This is especially more common in the case of heterogeneous soil and plant canopies. Some of the methods even use indicators that are destructive and are not suitable for automation. On the other hand, irrigation scheduling techniques based on canopy sensing and remote sensing are quite suitable for automation, and provide non-destructive, reliable, and rapid estimates of the water status of an entire field. Although new techniques provide automation, results obtained might not be precise with single indicators.

Integrative data management techniques for combining conventional and advanced approaches of irrigation scheduling should be developed. They should be less time-consuming, less labor-intensive, provide more reliable and adequate results, and involve less complexity for the prediction of plant water stress. The combination of both techniques has far bigger potential to have many accurate results for a far wide spectrum of crops and soils.

REFERENCES

Alderfasi, A. A., & Nielsen, D. C. (2001). Use of crop water stress index for monitoring water status and scheduling irrigation in wheat. *Agricultural Water Management*, *47*(1), 69–75.

Allen, R. G., & Pruitt, W. O. (1991). FAO 24 reference evapotranspiration factors. *Journal of Irrigation and Draining Engineering*, *117*(5), 758–773.

Allen, R. G., Pereira, L. S., Raes, D., & Smith, M. (1998). *Crop Evapotranspiration-Guidelines for Computing Crop Water Requirements-FAO Irrigation and Drainage Paper 56*. FAO, *300*(9)

Bandyopadhyay, P. K., & Mallick, S. (2003). Actual evapotranspiration and crop coefficients of wheat (Triticum aestivum) under varying moisture levels of humid tropical canal command area. *Agricultural Water Management, 59*(1), 33–47.

Bausch, W. C., Duke, H. R. (1996). Remote sensing of plant nitrogen status of corn. *Transactions of the ASAE, 39*, 1869–1875.

Beardsell, M. F., Jarvis, P. G., & Davidson, B. (1972). A null-balance diffusion porometer suitable for use with leaves of many shapes. *Journal of Applied Ecology, 9*(3), 677–690.

Behmann, J., Steinrücken, J., & Plümer, L. (2014). Detection of early plant stress responses in hyperspectral images. *ISPRS Journal of Photogrammetry and Remote Sensing, 93*, 98–111.

Bell, J. P. (1987). *Neutron Probe Practice. 3rd edition*. Wallingford, Institute of Hydrology, 51pp. (IH Report no. 19) .

Bell, J. P., Dean, T. J., & Hodnett, M. G. (1987). Soil moisture measurement by an improved capacitance technique, Part II. Field techniques, evaluation and calibration. *Journal of Hydrology, 93*(1–2), 79–90.

Berni, J. A. J., Zarco-Tejada, P. J., Sepulcre-Cantó, G., Fereres, E., & Villalobos, F. (2009). Mapping canopy conductance and CWSI in olive orchards using high resolution thermal remote sensing imagery. *Remote Sensing of Environment, 113*(11), 2380–2388.

Berni, J. A., Zarco-Tejada, P. J., Suárez Barranco, M. D., & Fereres Castiel, E. (2009). *Thermal and Narrow-band Multispectral Remote Sensing for Vegetation Monitoring from an Unmanned Aerial Vehicle*. Institute of Electrical and Electronics Engineers.

Blaney, H. F., & Criddle, W. D. (1962). *Determining Consumptive Use and Irrigation Water Requirements* (No. 1275). US Department of Agriculture.

Carter, G. A. (1991). Primary and secondary effects of water content on the spectral reflectance of leaves. *American Journal of Botany, 78*(7), 916–924.

Celia, M. A., Bouloutas, E. T., & Zarba, R. L. (1990). A general mass conservative numerical solution for the unsaturated flow equation. *Water Resources Research, 26*, 1483–1496

Colaizzi, P. D., Kustas, W. P., Anderson, M. C., Agam, N., Tolk, J. A., Evett, S. R., ... & O'Shaughnessy, S. A. (2012). Two-source energy balance model estimates of evapotranspiration using component and composite surface temperatures. *Advances in Water Resources, 50*, 134–151.

Cole, R. H. (1977). Time domain reflectometry. *Annual Review of Physical Chemistry, 28*(1), 283–300.

Colombo, R., Meroni, M., Marchesi, A., Busetto, L., Rossini, M., Giardino, C., & Panigada, C. (2008). Estimation of leaf and canopy water content in poplar plantations by means of hyperspectral indices and inverse modeling. *Remote Sensing of Environment, 112*(4), 1820–1834.

Cuenca, R. H. (1989). *Irrigation System Design: An Engineering Approach*, Prentice Hall.

Dadich, M. S., Srivastva, R. K., Dadich, H., KumarR., & Sharma, P. K. (2016). Application of Irrigation model under deficit irrigation. *Indian Journal of Ecology, 43*(2), 650–654

Dangwal, N., Patel, N. R., Kumari, M., & Saha, S. K. (2016). Monitoring of water stress in wheat using multispectral indices derived from Landsat-TM. *Geocarto International, 31*(6), 682–693.

Dean, T. J., Bell, J. P., & Baty, A. J. B. (1987). Soil moisture measurement by an improved capacitance technique, Part I. Sensor design and performance. *Journal of Hydrology, 93*(1–2), 67–78.

DeJonge, K. C., Taghvaeian, S., Trout, T. J., & Comas, L. H. (2015). Comparison of canopy temperature-based water stress indices for maize. *Agricultural Water Management, 156*, 51–62.

Deloire, A., & Heyns, D. (2011). The leaf water potentials: Principles, method and thresholds. Wynboer, *265*, 119–121.

Doorenbos, J., & Pruitt, W. O. (1977). *Guidelines for Predicting Crop Water Requirements*. Irrigation and Drain Paper No. 24, Food and Agricultural Organization.

Evans, R. O., & Sneed, R. E. (1991). *Measuring Soil Water for Irrigation Scheduling: Monitoring Methods and Devices*. AG-North Carolina Agricultural Extension Service, North Carolina State University (USA).

Evett, S. R. (2003). Soil water measurement by time domain reflectometry. *Encyclopedia of Water Science*, 894–898.

Feddes, R. A., Kowalik, P. J., & Zaradny, H. (1978). *Simulation of Field Water Use and Crop Yield. Simulation Monographs*. PUDOC.

Flexas, J., Escalona, J. M., Evain, S., Gulías, J., Moya, I., Osmond, C. B., & Medrano, H. (2002). Steady-state chlorophyll fluorescence (Fs) measurements as a tool to follow variations of net CO2 assimilation and stomatal conductance during water-stress in C3 plants. *Physiologia Plantarum, 114*(2), 231–240.

Frevert, D. K., Hill, R. W., & Braaten, B. C. (1983). Estimation of FAO evapotranspiration coefficients. *Journal of Irrigation and Drainage Engineering, 109*(2), 265–270.

Gamon, J., Serrano, L., & Surfus, J. (1997). The photochemical reflectance index: An optical indicator of photosynthetic radiation use efficiency across species, functional types, and nutrient levels. *Oecologia112*, 492–501.

Garces-Restrepo, C., Vermillion, D., & Munoz, G. (2007). *Irrigation Management Transfer. Worldwide Efforts and Results*. FAO. pp. 62.

Gates, D. M. (1962). Leaf temperature and energy exchange. *Archiv für Meteorologie, Geophysik und Bioklimatologie, Serie B, 12*(2), 321–336.

Gontia, N. K., & Tiwari, K. N. (2008). Development of crop water stress index of wheat crop for scheduling irrigation using infrared thermometry. *Agricultural Water Management, 95*(10), 1144–1152.

González-Dugo, M. P., Moran, M. S., Mateos, L., & Bryant, R. (2006). Canopy temperature variability as an indicator of crop water stress severity. *Irrigation Science, 24*(4), 233.

Govindraju, R. S., Or, D., Kavvas, M. L., Rolston, D. E., & Biggar, J. (1992). Error analyses of simplified unsaturated flow models under large uncertainty in hydraulic properties. *Water Resources Research, 28* (11), 2913–2924.

Haboudane, D., Miller, J. R., Tremblay, N., Zarco-Tejada, P. J., & Dextraze, L. (2002). Integrated narrow-band vegetation indices for prediction of crop chlorophyll content for application to precision agriculture. *Remote Sensing of Environment, 81*(2–3), 416–426.

Hargreaves, G. H., & Samani, Z. A. (1982). Estimating potential evapotranspiration. *Journal of the Irrigation and Drainage Division, 108*(3), 225–230.

Hargreaves, G. H., & Samani, Z. A. (1985). Reference crop evapotranspiration from temperature. *Applied Engineering in Agriculture, 1*(2), 96–99.

Huebner, C., Schlaeger, S., Becker, R., Scheuermann, A., Brandelik, A., Schaedel, W., & Schuhmann, R. (2005). Advanced measurement methods in time domain reflectometry for soil moisture determination. In *Electromagnetic Aquametry* (pp. 317–347). Springer.

Idso, S. B., Jackson, R. D., Pinter Jr, P. J., Reginato, R. J., & Hatfield, J. L. (1981). Normalizing the stress-degree-day parameter for environmental variability. *Agricultural Meteorology, 24*, 45–55.

Ihuoma, S. O., & Madramootoo, C. A. (2017). Recent advances in crop water stress detection. *Computers and Electronics in Agriculture, 141*, 267–275.

Irmak, S., Allen, R. G., & Whitty, E. B. (2003). Daily grass and alfalfa-reference evapotranspiration estimates and alfalfa-to-grass reference evapotranspiration ratios in Florida. *Journal of Irrigation and Draining Engineering, 129*(5), 360–370.

Irmak, S., Haman, D. Z., & Jones, J. W. (2002). Evaluation of class A pan 415 coefficients for estimating reference evapotranspiration in humid location. *Journal of Irrigation and Drainage Engineering, 128*(3), 153–159.

Irmak, S., Payero, J. O., VanDeWalle, B., Rees, J., & Zoubek, G. (2014). Principles and operational characteristics of Watermark granular matrix sensor to measure soil water status and its practical applications for irrigation management in various soil textures.

Jackson, R. B., Sperry, J. S., & Dawson, T. E. (2000). Root water uptake and transport: Using physiological processes in global predictions. *Trends in Plant Science, 5*(11), 482–488.

Jackson, R. D., Idso, S. B., Reginato, R. J., & Pinter, P. J. (1981). Canopy temperature as a crop water stress indicator. *Water Resources Research, 17*(4), 1133–1138.

Jarvis, P. G. (1976). The interpretation of the variations in leaf water potential and stomatal conductance found in canopies in the field. *Philosophical Transactions of the Royal Society of London B, 273*(927), 593–610.

Jensen, M. E., & Allen, R. G. (Eds.). (2016, April). *Evaporation, Evapotranspiration, and Irrigation Water Requirements*. American Society of Civil Engineers.

Jensen, M. E., & Haise, H. R. (1963). Estimating evapotranspiration from solar radiation. *Proceedings of the American Society of Civil Engineers, Journal of the Irrigation and Drainage Division, 89*, 15–41.

Jensen, M. J. (1981). Summary and challenges. pp. 225–231. In Proceedings of the ASAE's Irrigation Scheduling Conference on, Irrigation Scheduling for Water & Energy Conservation in the80's, ASAE Publ. 23-81, American Society of Agricultural Engineer, St. Joseph, MI.

Kang, S., Gu, B., Du, T., & Zhang, J. (2003). Crop coefficient and ratio of transpiration to evapotranspiration of winter wheat and maize in a semi-humid region. *Agricultural Water Management, 59*(3), 239–254.

Kosugi, K., & Katsuyama, M. (2004). Controlled-suction period lysimeter for measuring vertical water flux and convective chemical fluxes. *Soil Science Society of America Journal, 68*(2), 371–382.

Kumar, N., Poddar, A., & Shankar, V. (2019a, August). Optimizing irrigation through environmental canopy sensing–A proposed automated approach. In AIP Conference Proceedings (Vol. 2134, No. 1, p. 060003). AIP Publishing LLC.

Kumar, N., Poddar, A., Dobhal, A., && Shankar, V. (2019b). Performance assessment of PSO and GA in estimating soil hydraulic properties using near-surface soil moisture observations. *Compusoft, 8*(8), 3294–3301.

Kumar, N., Adeloye, A. J., Shankar, V., & Rustum, R. (2020a). Neural computing modelling of the crop water stress index. *Agricultural Water Management, 239*, 106259.

Kumar, N., Maharshi, S., Poddar, A., & Shankar, V. (2020b, July). Evaluation of artificial neural networks for estimating reference evapotranspiration in Western Himalayan Region. In 2020 International Conference on Computational Performance Evaluation (ComPE) (pp. 163–167). IEEE.

Kumar, N., Poddar, A., & Shankar, V. (2020c). Nonlinear regression for identifying the optimal soil hydraulic model parameters. In *Numerical Optimization in Engineering and Sciences* (pp. 25–34). Springer.

Kumar, N., Poddar, A., Shankar, V., Ojha, C. S. P., & Adeloye, A. J. (2020d). Crop water stress index for scheduling irrigation of Indian mustard (Brassica juncea) based on water use efficiency considerations. *Journal of Agronomy and Crop Science, 206*(1), 148–159.

Kumar, N., Shankar, V., & Poddar, A (2020e). Investigating the effect of limited climatic data on evapotranspirationbased numerical modeling of soil moisture dynamics in the unsaturated root zone: A case study for potato crop. *Modeling Earth Systems and Environment, 6*, 2433–2449. https://doi.org/10.1007/s40808-020-00824-8

Kumar, N., Shankar, V., & Poddar, A. (2020f). Agro-hydrologic modelling for simulating soil moisture dynamics in the root zone of Potato based on crop coefficient approach under limited climatic data. *ISH Journal of Hydraulic Engineering, 28*(sup1), 310–326.

Kumar, N., Shankar, V., Rustum, R., & Adeloye, A. J. (2020g). Evaluating the performance of self-organizing maps to estimate well-watered canopy temperature for calculating crop water stress index in Indian mustard (Brassica juncea). *Journal of Irrigation and Drainage Engineering, 147*(2), 04020040.

Kumar, R., Jat, M. K., & Shankar, V. (2013). Soil moisture dynamics modelling enabled by hydraulic redistribution in multi-layer root zone. *Current Science, 105*(10), 1373–1382.

Kumar, R., Shankar, V., & Jat, M. (2014). Evaluation of nonlinear root uptake model for uniform root zone vis-à-vis multilayer root zone. *Journal of Irrigation and Drainage Engineering, 140*(2), 04013010. https://doi.org/10.1061/(ASCE)IR.1943-4774.0000655.

Kumar, R., Majid, M., Mir, S., Shahzad, M., & Ahmed, L. (2017). Temporal analysis of drought using standard precipitation index (SPI) Method. *Indian Journal of Soil Conservation, 45*(3), 348–350.

Leib, B. G., Jabro, J. D., & Matthews, G. R. (2003). Field evaluation and performance comparison of soil moisture sensors. *Soil Science, 168*(6), 396–408.

Liakos, V., Vellidis, G., Tucker, M., Lowrance, C., & Liang, X. (2015). A decision support tool for managing precision irrigation with center pivots. In Precision Agriculture'15 (pp. 713–720). Academic Publishers.

Linacre, E. T. (1977). A simple formula for estimating evaporation rates in various climates, using temperature data alone. *Agricultural Meteorology, 18*(6), 409–424.

Magney, T. S., Vierling, L. A., Eitel, J. U., Huggins, D. R., & Garrity, S. R. (2016). Response of high frequency Photochemical Reflectance Index (PRI) measurements to environmental conditions in wheat. *Remote Sensing of Environment, 173*, 84–97.

Makkink, G. F. (1957). Testing the Penman formula by means of lysimeters. *Journal of the Institution of Water Engineerrs, 11*, 277–288.

Martin, D. L., Stegman, E. C., & Fereres, E. (1990). Irrigation scheduling principles. In: *Management of Farm Irrigation Systems* (Vol. 1990, pp. 155–203). American Society of Agricultural Engineers.

Modaberi, H., & Assari, M. (2014). Estimation of reference Evapotranspiration, the best pan coefficient and rice plant coefficient using the pan evaporation data in rice growth period (Case study: Mordab plain region in Guilan province, Iran). *International Journal of Engineering Sciences, 3*(9), 75–84.

Monteith, J. L., & Unsworth, M. (2007). *Principles of Environmental Physics*. Academic Press.

Monteith, J. L., Campbell, G. S., & Potter, E. A. (1988). Theory and performance of a dynamic diffusion porometer. *Agricultural and Forest Meteorology, 44*(1), 27–38.

Moya, I., Camenen, L., Evain, S., Goulas, Y., Cerovic, Z. G., Latouche, G., & Ounis, A. (2004). A new instrument for passive remote sensing: 1. Measurements of sunlight-induced chlorophyll fluorescence. *Remote Sensing of Environment, 91*(2), 186–197.

Noborio, K. (2001). Measurement of soil water content and electrical conductivity by time domain reflectometry: A review. *Computers and Electronics in Agriculture, 31*(3), 213–237.

Ojha, C. S. P., & Rai, A. K. (1996). Nonlinear root-water uptake model. *Journal of Irrigation and Drainage Engineering, 122*(4), 198–202.

Ojha, C. S. P., Prasad, K. S., Shankar, V., & Madramootoo, C. A. (2009). Evaluation of a nonlinear root-water uptake model. *Journal of Irrigation and Drainage Engineering, 135*(3), 303–312.

Orang, M. (1998). "Potential Accuracy of the Popular Non-linear Regression Equation for Estimating Pan Coefficient Values in the Original and FAO-24 Tables." Unpublished, California Department of Water Resources Report, Sacramento, CA.

Osakabe, Y., Osakabe, K., Shinozaki, K., & Tran, L. S. P. (2014). Response of plants to water stress. *Frontiers in Plant Science, 5*, 86.

Panigada, C., Rossini, M., Meroni, M., Cilia, C., Busetto, L., Amaducci, S., & Marchesi, A. (2014). Fluorescence, PRI and canopy temperature for water stress detection in cereal crops. *International Journal of Applied Earth Observation and Geoinformation, 30*, 167–178.

Peñuelas, J., & Filella, I. (1998). Visible and near-infrared reflectance techniques for diagnosing plant physiological status. *Trends in Plant Science, 3*(4), 151–156.

Peñuelas, J., Pinol, J., Ogaya, R., & Filella, I. (1997). Estimation of plant water concentration by the reflectance water index WI (R900/R970). *International Journal of Remote Sensing, 18*(13), 2869–2875.

Philip, J. R. (1966). Plant water relations: Some physical aspects. *Annual Review of Plant Physiology, 17*(1), 245–268.

Poddar, A., Sharma, A., & Shankar, V. (2017, December). Irrigation scheduling for potato (solanum tuberosum l.) based on daily crop coefficient approach in a sub-humid sub-tropical region. In Proceedings of Hydro-2017 International, L. D. College of Engineering.

Poddar, A, Kumar, N, & Shankar, V (2018a, February). Effect of capillary rise on irrigation requirements for wheat. In: Proceedings of International Conference on Sustainable Technologies for Intelligent Water Management (STIWM-2018), IIT Roorkee, India.

Poddar, A., Gupta, P., Kumar, N., Shankar, V., & Ojha, C. S. P. (2018b). Evaluation of reference evapotranspiration methods and sensitivity analysis of climatic parameters for sub-humid sub-tropical locations in western Himalayas (India). *ISH Journal of Hydraulic Engineering, 27*(3), 336–346.

Poddar, A., Kumar, N., & Shankar, V. (2021a). Evaluation of two irrigation scheduling methodologies for potato (Solanum tuberosum L.) in north-western mid-hills of India. *ISH Journal of Hydraulic Engineering, 27*(1), 90–99.

Poddar, A., Kumar, N., & Shankar, V. (2021b). Performance evaluation of four models for estimating the capillary rise in wheat crop root zone considering shallow water table. In: Pandey, A., Mishra, S., Kansal, M., Singh, R., Singh, V. P. (eds) *Hydrological Extremes. Water Science and Technology Library* (Vol. 97), 423-434. Springer. https://doi.org/10.1007/978-3-030-59148-9_29

Poddar, A., Kumar, N., Kumar, R., Shankar, V., & Jat, M. K. (2020). Evaluation of non-linear root water uptake model under different agro-climates. *Current Science, 119*(3), 485.

Priestley, C. H. B., & Taylor, R. J. (1972). On the assessment of surface heat flux and evaporation using large-scale parameters. *Monthly Weather Review, 100*(2), 81–92.

Rahimikhoob, A. (2009). An evaluation of common pan coefficient equations to estimate reference evapotranspiration in a subtropical climate (north of Iran). *Irrigation Science, 27*(4), 289–296.

Reynolds, S. G. (1970). The gravimetric method of soil moisture determination. *Journal of Hydrology, 11*, 258–273.

Richards, L. A. (1931). Capillary conduction of liquids through porous mediums. *Physics, 1*(5), 318–333.

Richards, S. J. (1965). Soil suction measurements with tensiometers. *Methods of Soil Analysis. Part 1. Physical and Mineralogical Properties, Including Statistics of Measurement and Sampling* (methodsofsoilana), 153–163.

Roberts, K. (Ed.). (2007). *Handbook of Plant Science* (Vol. 1). John Wiley & Sons.

Rossini, M., Fava, F., Cogliati, S., Meroni, M., Marchesi, A., Panigada, C., & Colombo, R. (2013). Assessing canopy PRI from airborne imagery to map water stress in maize. *ISPRS Journal of Photogrammetry and Remote Sensing, 86*, 168–177.

Roujean, J. L., & Breon, F. M. (1995). Estimating PAR absorbed by vegetation from bidirectional reflectance measurements. *Remote Sensing of Environment, 51*(3), 375–384.

Rouse, J., Haas, R. H., Schell, J. A., & Deering, D. W. (1974). *Monitoring Vegetation Systems in the Great Plains with ERTS*.

Sabziparvar, A. A., Tabari, H., Aeini, A., & Ghafouri, M. (2010). Evaluation 459 of class A pancoefficient methods for estimation of reference crop evapotranspiration in cold semi-arid and warm arid climates. *Water Resources Management, 24*(5), 909–920.

Schepers, J.S., Francis, D.D., Vigil, M.F., Below, F.E. (1992). Comparison of corn leaf nitrogen concentration and chlorophyll meter readings. *Communications in Soil Science and Plant Analysis, 23*, 2173–2187.

Schulze, E. D., Hall, A. E., Lange, O. L., & Walz, H. (1982). A portable steady-state porometer for measuring the carbon dioxide and water vapour exchanges of leaves under natural conditions. *Oecologia, 53*(2), 141–145.

Shankar, V. (2007). *Modelling of Moisture Uptake by Plants*. Ph.D. dissertation, Department of Civil Engineering, Indian Institute of Technology.

Shankar, V., Hari Prasad, K. S., Ojha, C. S. P., & Govindaraju, R. S. (2012). Model for nonlinear root water uptake parameter. *Journal of Irrigation and Drainage Engineering*, *138*(10), 905–917.

Shankar, V., Ojha, C. S., Govindaraju, R. S., Prasad, K. H., Adebayo, A. J., Madramoottoo, C. A., ... & Singh, K. K. (2017). Optimal use of irrigation water. In *Sustainable Water Resources Management* (pp. 737–795).

Sharma, V., Poddar, A., & Shankar, V. (2016, September). Performance evaluation of data-driven and statistical rainfall runoff models for a mountainous catchment. In: Proceedings of National Conference: Civil Engineering Conference–Innovation for Sustainability (CEC–2016), vol *1*, pp. 247–261.

Shimshi, D. (1977). A fast-reading viscous flow leaf porometer. *New Phytologist*, *78*(3), 593–598.

Singh, A. K., Madramootoo, C. A., & Smith, D. L. (2010). Water balance and corn yield under different water table management scenarios in Southern Quebec. In 9th International Drainage Symposium held jointly with CIGR and CSBE/SCGAB Proceedings, 13-16 June 2010, Québec City Convention Centre, Quebec City, Canada (p. 1). American Society of Agricultural and Biological Engineers.

Sinha, S., Norouzi, A., Pradhan, A., Yu, X., Seo, D. J., & Zhang, N. (2017). A field soil moisture study using time domain reflectometry (TDR) and time domain transmissivity (TDT) Sensors. *DEStech Transactions on Materials Science and Engineering*. 2017 International Conference on Transportation Infrastructure and Materials (ICTIM 2017) ISBN: 978-1-60595-442-4.

Snyder, R. L. (1992). Equation for evaporation pan to evapotranspiration conversions. *Journal of Irrigation and Draining Engineering*, *1186*, 977–980.

Spaans, E. J., & Baker, J. M. (1992). Calibration of watermark soil moisture sensors for soil matric potential and temperature. *Plant and Soil*, *143*(2), 213–217.

Stannard, D. I. (1992). Tensiometers: Theory, construction, and use. *Geotechnical Testing Journal*, *15*(1), 48–58.

Suárez, L., Zarco-Tejada, P. J., Berni, J. A. J., González-Dugo, V., & Fereres, E. (2010, August). Orchard water stress detection using high-resolution imagery. In XXVIII International Horticultural Congress on Science and Horticulture for People (IHC2010): International Symposium on 922 (pp. 35–39).

Tabari, H., Marofi, S., Aeini, A., Talaee, P. H., & Mohammadi, K. (2011). Trend analysis of reference evapotranspiration in the western half of Iran. *Agricultural and Forest Meteorology*, *151*(2), 128–136.

Taber, H. G., Lawson, V., Smith, B., & Shogren, D. (2002). Scheduling microirrigation with tensiometers or Watermarks. *International Water and Irrigation*, *22*(1), 22–26.

Taiz, L., & Zeiger, E. (1991). Plant physiology. The Benjamin. *Cummings Redwood City*, 565.

Tallon, L. K., & Si, B. C. (2015). Representative soil water benchmarking for environmental monitoring. *Journal of Environmental Informatics*, *4*(1), 31–39.

Thompson, R. B., Gallardo, M., Agüera, T., Valdez, L. C., & Fernández, M. D. (2006). Evaluation of the Watermark sensor for use with drip irrigated vegetable crops. *Irrigation Science*, *24*(3), 185–202.

Thornthwaite, C. W. (1948). An approach toward a rational classification of climate. *Geographical Review*, *38*(1), 55–94.

Todorovic, M. (2005). Crop water requirements. *WaterEncyclopedia*, *3*, 557–558.

Topp, G. C., & Davis, J. L. (1985). Measurement of soil water content using time-domain reflectometry (TDR): A field evaluation 1. *Soil Science Society of America Journal*, *49*(1), 19–24.

Turc, L. (1961). Estimation of irrigation water requirements, potential evapotranspiration: A simple climatic formula evolved up to date. *Annales Agronomiques*, *12*(1), 13–49.

Unger, P. W., & Kaspar, T. C. (1994). Soil compaction and root growth: A review. *Agronomy Journal*, *86*(5), 759–766.

Wang, D., Yates, S. R., & Ernst, F. F. (1998). Determining soil hydraulic properties using tension infiltrometers, time domain reflectometry, and tensiometers. *Soil Science Society of America Journal*, *62*(2), 318–325.

Xu, C. Y., & Singh, V. P. (2002). Cross comparison of empirical equations for calculating potential evapotranspiration with data from Switzerland. *Water Resources Management*, *16*(3), 197–219.

Zarco-Tejada, P. J., González-Dugo, V., & Berni, J. A. (2012). Fluorescence, temperature and narrow-band indices acquired from a UAV platform for water stress detection using a micro-hyperspectral imager and a thermal camera. *Remote Sensing of Environment*, *117*, 322–337.

Zarco-Tejada, P. J., González-Dugo, V., Williams, L. E., Suárez, L., Berni, J. A., Goldhamer, D., & Fereres, E. (2013). A PRI-based water stress index combining structural and chlorophyll effects: Assessment using diurnal narrow-band airborne imagery and the CWSI thermal index. *Remote Sensing of Environment*, *138*, 38–50.

Zhao, T., Stark, B., Chen, Y., Ray, A. L., & Doll, D. (2015). A detailed field study of direct correlations between ground truth crop water stress and normalized difference vegetation index (NDVI) from small unmanned aerial system (sUAS). Unmanned Aircraft Systems (ICUAS). 2015 International Conference on IEEE. 520–525.

37 Water Resources, Uses, and Their Integrated Management in the United Arab Emirates

Ahmed Sefelnasr, Abdel Azim Ebraheem,
Mohsen Sherif, and Mohamed Al Mulla

CONTENTS

DOI: 10.1201/9781003276548-37

37.1 INTRODUCTION

The lack of renewable freshwater resources in the United Arab Emirates (UAE) is a major challenge for its sustainable water resources development. Since the 1960s, the growth in population, higher standards of living, and expansion of the agricultural, forestry, and industrial sectors have created a huge demand for more fresh water. Initially, demand was met from fresh groundwater resources, but, those are being rapidly depleted. Increased reliance on nonconventional water supplies is required to maintain economic growth in the UAE. The major challenges are to balance water supply and demand as efficiently as possible given that the per capita consumption of freshwater is among the highest in the world and new water supplies are expensive. The water resources in the UAE can be grouped into two categories:

- Conventional resources: composed of surface water and groundwater
- Nonconventional resources: represented by desalination and treated wastewater

The current supply from groundwater is about 51%, mostly for irrigation uses, but some limited quantities are used for potable uses, particularly in the Northern Emirates. The desalinated water supply is about 37%, mainly for potable water uses, and in some places also used for irrigation purposes. Reclaimed water supply (treated sewage effluent) accounts for 12% and is mainly used for irrigating amenity areas. The agricultural sector water use remains the largest consumer with 34% of total water. The domestic and industrial water sector consumes 32%, the forestry sector uses 15%, and amenities utilize 11%. The remaining 8% is considered as loss. The largest portion of water is consumed in Abu Dhabi Emirate (61%), followed by Dubai Emirate (18%), and the Sharjah and the Northern Emirates use about 21% of total water consumption (Hydro Atlas, 2015; Sherif et al., 2018).

37.2 CONVENTIONAL WATER RESOURCES

37.2.1 SURFACE WATER

Arid conditions form the major constraint for development in the UAE. The entire region is mostly dry throughout the year; however, surface water runoff may happen to take place during rainy days/seasons. A thorough understanding of the regional and temporal variations in precipitations is vital for water resources planning and hydrologic studies. The rainfall patterns may significantly affect agriculture productivity, as the cycles of maximum or minimum precipitation may coincide with the peak growing seasons in a particular area. Precipitation varies spatially and data from several representative gauges should be used to estimate the average precipitation for an area and to evaluate its reliability. This is especially important in mountainous areas where the variation tends to be large. Time variations in rainfall intensity are extremely important in the rainfall–runoff process, particularly in urban areas. The locations of climatological stations in the UAE are shown in Figure 37.1. On the other hand, Figures 37.2, 37.3, and 37.4 present the variation of the annual rainfall in the period 1934–2010, the average annual rainfall departures from the normal rainfall over the duration 1965–2010, and the spatial distribution of average annual rainfall, respectively. The lowest rainfall is observed in the desert foreland, and the highest rainfall is observed in the mountains and east coast regions (Sherif et al., 2018).

The runoff generation from rainfall varies from one wadi to the other depending on topography and geomorphology. In mountains, floods result from heavy rainfall where a small part of the rainwater infiltrates in the wadi beds and thus produces large discharge and frequent flash floods. Most

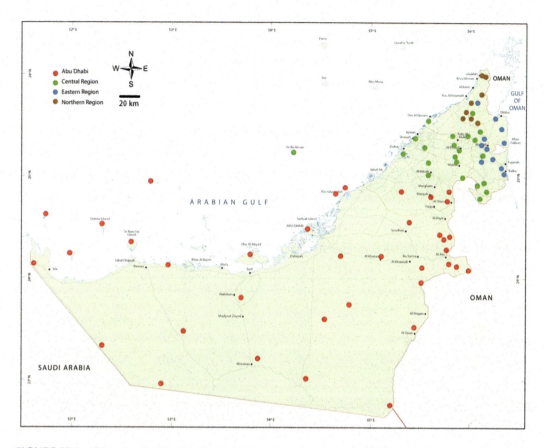

FIGURE 37.1 Climatological stations in the UAE. (From Hydro Atlas, 2015.)

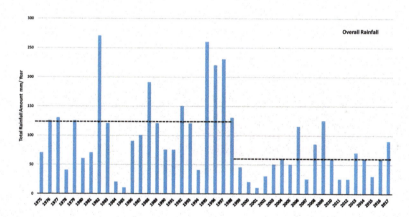

FIGURE 37.2 Total and average annual rainfall.

of the annual rainfall, therefore, reaches the plain, which, in some cases, corresponds to a relatively high annual runoff. In the piedmont areas, both infiltration and runoff are observed. In the permeable zone (gravel plain), a large part of the annual rainfall flows out into the dune region. In such a region where runoff is not generated, the only water loss is due to evaporation. The water loss is directly proportional to the number of days of rainfall. Flood flow takes place generally in winter months. However, isolated and heavy flows have been observed in the other months, especially by the end of summer (Sherif et al., 2014, 2017).

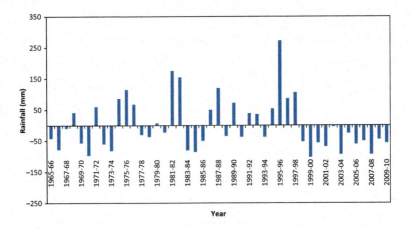

FIGURE 37.3 Annual rainfall departures from the normal rainfall.

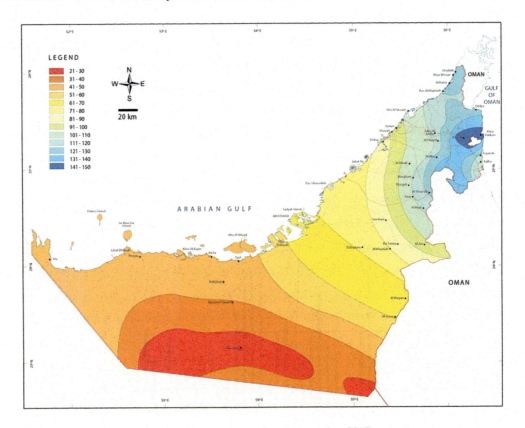

FIGURE 37.4 Annual average precipitation of the UAE (Hydro Atlas, 2015).

Sixty surface water catchments were defined in the UAE, of which 48 are located in the Northern Emirates and 12 in the Abu Dhabi Emirate. The areas of these catchments range from 3 km² at Khatt to 1362 km² at Idayah (Table 37.1). The drainage network of the northern part of the UAE is presented in Figure 37.5. The Al Hajar Mountain group in the eastern region divides the watersheds into two main orientations. One of those orientations drains toward the Gulf of Oman in the east, and the other one drains to the desert and Arabian Gulf in the west. The flow in the western region is limited by the relatively low rainfall, and relatively flat and permeable terrain (Brook et al., 2006a). The minimum rainfall required to produce runoff was 75 mm in the Oman basin and 90 mm in the

TABLE 37.1

Annual Discharges of Selected Springs

Year	Khatt North	Khatt South	Siji	Bu Sukhanah
1984	0.54	1.04	0.09	0.96
1985	0.44	0.60	0.06	1.10
1986	0.23	0.13	0.09	1.42
1987	0.22	0.30	0.13	1.58
1988	0.67	0.76	0.13	1.51
1989	0.35	0.35	0.13	1.45
1990	0.51	0.63	0.13	1.58
1991	0.51	0.69	0.13	2.50
1992	0.25	0.35	0.13	1.20
1993	0.35	0.45	0.13	1.30
1994	0.45	0.52	0.13	1.21
1995	0.60	0.75	0.13	1.42
1996	0.52	0.65	0.13	1.54
1997	0.51	0.65	0.13	1.38
1998	0.54	0.75	0.13	1.56
1999	0.37	0.48	0.13	1.42
2000	0.38	0.47	0.13	1.24
2001	0.50	0.65	0.13	1.27
2002	0.45	0.51	0.13	1.35
2004	0.06	0.07		
2005	0.06	0.07		
2006	0.06	0.07		
2007	0.21	0.26		
2008	0.216	0.27		

Jabel Hafit basin (Rizk and Al Sharhan, 2003). However, a study by the Ministry of Environment and Water (MOEW) and United Arab Emirates University (MOEW-UAEU, 2005) showed that a 30 mm rainfall event will produce runoff in the Oman mountains in the eastern region.

Several studies estimated the average annual surface water flow in the eastern region. Rizk and Al Sharhan (2003) estimated the runoff percentage from rainfall to range from 3% in the Jabal Hafit basin to 18% in the Northern Oman Mountains. Based on these estimated percentages, the potential average annual surface water flow was estimated between 2% and 20%. Considering the two extremes of the runoff coefficient, the potential annual surface runoff estimated is evaluated between 14,771,352 and 147,713,524 m^3 (Hydro Atlas, 2015).

Dams and reservoirs have been used successfully in collecting, storing, and managing surface water resources. To manage wadi flows and floods, dams and diversion structures were built along the main wadis in the different emirates. The main objectives of these dams are to recharge the groundwater aquifer, protect the downstream areas from flood damages, meet local agricultural demand, and conserve the damage from soil erosion. The locations, capacity, and affected area are presented in Figure 37.5.

The majority of these multipurpose dams are distributed in Northern Emirates where significant flash flood events occur and from watersheds and originating from the mountain series. However, Abu Dhabi Emirate is less prone to flash floods, and dams only serve as a recharge pond. There are 150 rainwater storage dams in the UAE with a total storage capacity of 150 MCM – about 118 MCM in the Northern Emirates including Dubai and 32 MCM in Abu Dhabi. These figures show the

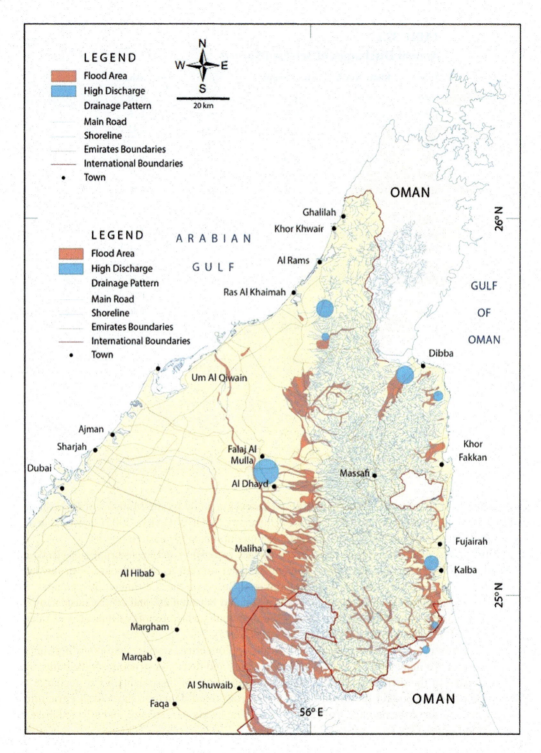

FIGURE 37.5 Location of major geomorphological units with flooding and high discharge areas (Ghonim, 2008).

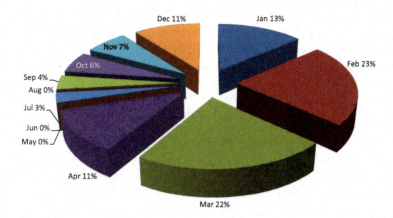

FIGURE 37.6 Percentage of monthly floods occurrences (flood frequency).

high-capacity dams that are available in the Northern Emirates. Flood events take place generally in the months of November through April with the maximum number of floods occurring during the months of February and March. The average percentage of flood occurrence on a monthly basis is presented in Figure 37.6.

37.2.1.1 Falajs and Springs

The strict definition of a spring is a point where groundwater flows out of the ground and is thus where the groundwater level meets the ground surface. In the UAE, springs seem to discharge from local and intermediate groundwater flow systems (Rizk and El-Etr, 1997). Discharge of the springs shows wide variation during the period from 1984 to 2008. Table 37.1 summarizes the discharge of the main four springs in the UAE. After 2004, the discharge of springs decreased dramatically as compared to the precedent years. The quality of the springs' water changed, and the total dissolved solids (TDS) increased by about 10% since 1984. The TDS ranges from 1414 mg/L to 10228 mg/L. The concentration of major ions and cations varies from one spring to another according to their local geological and hydrogeological conditions (MOEW, 2010). A falaj is an old traditional irrigation system through which water is diverted from the groundwater by gravity to be distributed among farms (Brook et al., 2006a; MOEW, 2010). A total of forty falajs were active until 1999. Most of the falajs dried up by 2005 due to the absence of rainfall and overexploitation of groundwater nearby falajs. The falajs' annual discharges were measured to be about 31.2 million m³. The TDS of falajs water varies from 400 mg/L to 6700 mg/L. Many of these falajs are not functional now due to the declined groundwater table, and they became dry, especially in the Al Ain area.

37.2.2 Groundwater

The groundwater system in the UAE consists of four main aquifers: (1) Quaternary aquifer flanking the eastern mountain, (2) limestone aquifer, (3) fractured ophiolite aquifer, and (4) sand dune aquifer in the south and west. The Quaternary aquifer system is the main aquifer system in the UAE. Field measurements indicate that the depth of groundwater ranges from 5 m to more than 100 m below ground level. Excessive groundwater abstraction from this aquifer system results in a local cone of depressions due to the limited annual replenishment.

The Water Balance Model (WBM) developed by the National Water Center, UAEU, and the former MOEW in 2015 was used to estimate the groundwater reserves in this aquifer and included two new components: groundwater recharge and abstraction for irrigation. The hydrologic budget of the Quaternary aquifer indicated that the total annual recharge is about 200 million m³. The estimated fresh groundwater reserve in the Quaternary aquifer across the UAE is less than 10 km³ while the estimated groundwater storage of the slightly brackish to slightly saline water is 350 km³ (Sherif et al., 2020). The flow system in this aquifer depends on local topography, aquifer boundary and extent, groundwater abstraction, and recharge rate. The hydraulic conductivity of this system ranges from 10 to 40 m/day and the TDS ranges from 500 up to 150,000 mg/L in the sabkha areas along the Al Ain–Abu Dhabi main road (Sherif et al., 2020).

The limestone aquifer in the northern part of the UAE is mainly composed of fractured limestone and dolomite. The hydrologic budget of the limestone aquifer in the northern part of the UAE indicated that the total annual recharge is about 9.57 million m^3 and the annual losses to the sea are around 1.47 million m^3. The estimated fresh to slightly brackish groundwater reserve is 14 km^3 (IWACO, 1986; NDC and USGS, 1996; Ebraheem et al., 2014). In Jabal Hafit in the Al Ain area, this aquifer is composed of 1500 m thick limestone and marl interbedded with gypsum and dolomite and evaporates formations of the Lower Eocene to Miocene age. The estimated slightly brackish to saline groundwater reserve in the limestone aquifer in this area is estimated at 350 km^3 (Sherif et al., 2020). The ophiolite aquifer system sequence is a jointed and fractured system, and it was subjected to faulting. The main fault system runs in a northwest–southeast direction. The aquifer has very limited groundwater potential. Sand dunes cover about 74% of the total area of the UAE and gradually increase in elevation from sea level at the western coast to reaching 250 m (amsl) near Liwa. These sand dunes represent a very good shallow aquifer system. However, the available data is insufficient to characterize this aquifer. The available drilling data is not enough to calculate accurate hydraulic conductivity.

37.3 NONCONVENTIONAL WATER RESOURCES

37.3.1 DESALINATION

Domestic water demands based on desalination in the UAE increased dramatically during the last decade due to the continuous increase in total water consumption reaching about 6.0 km^3 in 2019 (Figure 37.7). The UAE is now a world leader in the application of desalination technology, brought about by a rapid and very comprehensive program of construction of new plants and a detailed research and development campaign. The total desalinated seawater in the UAE per year in 2019 was around 2000 MCM and contribute to about 43% of the total water budget of the country (4620 MCM per year in 2011). The contribution percentage is steadily growing from less than 10% in 1995 to 21% in 2005 (Brook et al., 2005a, 2005b) to 40% in 2011 and is now around 43%. The distribution of the desalination plants is shown in Figure 37.8. The majority of these installations are in the Abu Dhabi Emirate with about 67% of the total plants in the country. The remaining plants are distributed as 18% in Dubai, 10% in Sharjah, and 5% in the Northern Emirates. All these plants use both thermal distillation and membrane technologies (Brook and Dawoud, 2005).

37.3.2 RECLAIMED OR TREATED WATER

Reclaimed water is considered an important water resource for certain areas of usage. More than 60 wastewater treatment plants are present in the UAE. Most of the major treatment plants in the UAE use conventional treatment methods including tertiary treatment. Currently, about 780 MCM water is being reclaimed and 450 MCM is being reused for gardening purposes mainly in Abu Dhabi and Dubai Emirates (Brook et al., 2006a). Dubai is reclaiming a large portion of its wastewater as compared to all the emirates. Out of 260 MCM of wastewater, around 215 MCM (82.5%) is reused for landscaping and also used for cooling towers in industries in Dubai. Abu Dhabi Emirate utilizes 120 MCM of its 250 MCM of treated wastewater to irrigate landscapes, parks, and gardens. In Sharjah Emirate, 95 MCM of treated water (around 65% of the available treated wastewater) is reused (Brook et al., 2005b). In Ajman Emirate, limited amounts (18%) of the available treated water are used and the remaining part is further processed using microfiltration and reverse osmosis and then supplied to the industry. In the Ras Al Khaimah Emirate, only 35% of treated water is utilized and the remaining part is discharged to surface ponds to recharge the aquifer. The Fujairah Emirate uses

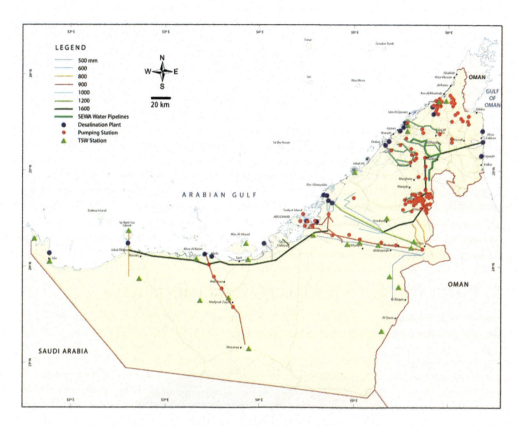

FIGURE 37.7   Abu Dhabi's desalinated water distribution network (Hydro Atlas, 2015).

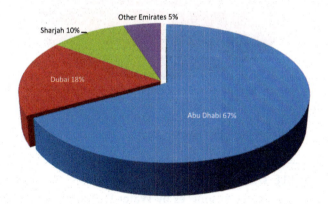

FIGURE 37.8   Percentage of desalination installation in the UAE.

about 33% of reclaimed water out of 20 MCM, for irrigation purposes and landscaping. Overall, at the country level, about 39% of treated water is not being used for any purposes (UAE Water Conservation Strategy, 2010; UAE Hydro Atlas, 2015).

37.4   WATER USES IN THE UAE

Desalinated water is mainly used to meet domestic and industrial demand. A significant portion is also used for amenity and gardening purposes. Agricultural and forestry water use is mainly met from groundwater resources. Water for amenities is predominantly supplied from treated water resources. The agricultural sector is the largest user with about 46% of total water resources (Figure 37.9). The domestic and industrial water sector accounts for 35%, the forestry sector uses 4%, and amenities consume 9%. The remaining 6% is considered for losses. The water consumption in Abu Dhabi Emirate represents about 61% of the total water use in the country followed by Dubai Emirate, which consumes 17% of the water budget. Sharjah and the Northern Emirates account for 22% of total water use.

The use of water resources by different sectors is summarized in Table 37.2 and Figure 37.9. The total groundwater used in the year 2019 was about 3.22 km^3 including 2.68 km^3 (83%) for agriculture, 0.25 km^3 (7.7%) for forestry, 0.22 km^3 (6.8%) for domestic, and the remaining 78 MCM (2.5%) for urban area amenity purposes. The desalinated water was used mostly for potable water supplies (Figure 37.10). The total desalinated water used for all purposes was about 2.2 km^3. The water used for the urban water supply is about 1.82 km^3 (83%), about 120 MCM (5%) is used for agricultural purposes, 100 MCM (4.5%) is used for amenities and landscaping, and 160 MCM (7.5%) is considered losses in the network. Reclaimed water is only used for amenities and landscaping, about 0.78 km^3.

37.5   INTEGRATED WATER RESOURCES MANAGEMENT (IWRM)

Integrated water resources management (IWRM) requires the recognition of a number of principles. The first principle emphasizes the key role of freshwater in maintaining all forms of life and

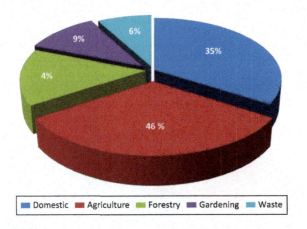

FIGURE 37.9   Water use distribution by demand sectors.

TABLE 37.2
Use of Water Resources by Water Demand Sectors

	Type of Resource (MCM)		
Water Use	Groundwater	Desalination	Reclaimed
Domestic	216	1820	120
Agriculture	2680	120	78
Forestry	248	0	0
Gardening	78	100	383
Waste	0	160	207
Supply	3220	2200	788
Total supply			6208

Source: Updated from Hydro Atlas (2015).

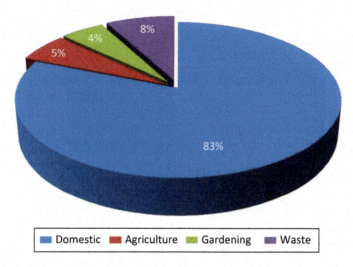

FIGURE 37.10   Desalinated water distributions.

its necessity for socioeconomic development. Despite their abundance in certain parts of the world, freshwater resources have limited physical quantities and can be adversely affected by human activities that not only make them unfit for human consumption, but also disturb the ecological balance. The second principle is failing in encouraging public engagement in decision-making, and restraining accountability would lead to lopsided water allocation and services. The third principle brings into the spotlight women's roles as a main stakeholders and water users. The fourth principle underscores the economic value of water, which tends to be overlooked in setting water management policies in the UAE. Over the last few decades, the expansion of agriculture with a view to creating employment, protecting the rural heritage, and making the UAE less dependent on imported food has driven the demand for water to unsustainable levels.

In 2011, water demand for farms and landscape irrigation accounted for 60% of water consumption in the UAE, but agriculture was just responsible for less than 2% of GDP and for less than 15% of the fruits and vegetables available in local markets (Brook et al., 2006b; MOEW, 2010). Sherif et al. (2018, 2020) indicated that current abstraction rates exceed 15 times the average groundwater recharge rate. At current consumption patterns, fresh groundwater will be depleted within less than eight years. This also affects groundwater quality, as salinity is rapidly increasing. In 2012, 80% of the groundwater was saline, 17.4% brackish, and only 2.6% was fresh. Setting proper pricing policies can motivate consumers to appreciate the real value of water and hence implement conservation principles and measures. Despite being universally endorsed, the aforementioned principles provide only a general sketch of IWRM without offering a clear definition of how it could be implemented. There is a general agreement that water resources management should achieve a balance among economic efficiency, social equity, and environmental sustainability. The World Bank adopted this approach in a water resources management policy paper published in 1993 and reconfirmed its commitment to IWRM (World Bank, 2004).

37.5.1   Issues to be Addressed by IWRM

Managing water resources is a delicate balancing act between meeting demand and maintaining the viability of the resource for future use without jeopardizing the integrity of the environment. Meeting basic human demands has the first priority in utilizing water resources. Although several advanced societies have achieved near completion in meeting these demands, several others are still lagging in securing access to clean water. The following sections are based on the widely accepted framework proposed by the Global Water Partnership (GWP) initiative, depicted in Figure 37.11. The three key objectives of IWRM, namely, economic efficiency, social equity, and ecological sustainability, are interactively connected with the implementation components in order to underscore the importance of keeping them incorporated in all implementation decisions. The first objective emphasizes the necessity to optimize water usage, particularly under conditions of water scarcity (Sherif et al., 2014; World Bank, 2004). To avoid placing less influential and poor groups at a disadvantage due to the lack of representation or inability to pay for services, the second objective calls for special provisions or compensation for these groups. Any implementation strategy or tool has to observe rules established by the third objective (Figure 37.11). As elaborated by Tuinhof et al. (2010), the following key issues for groundwater supply management should be understood:

- Aquifer systems are susceptible to negative impacts under abstraction stress interactions between groundwater and surface water, such as abstraction effects (on river base flow and some wetlands) and recharge reduction effects (due to surface water modification).
- Social development goals greatly influence water use, especially where agricultural irrigation and food production are concerned, thus management can only be fully effective if cross-sector coordination occurs. Regulatory interventions and economic tools become more effective if they are not only encoded in water law but implemented with a high level of user participation.
- Other generic principles can emerge. Both hydrogeologic and socioeconomic conditions tend to be somewhat location-specific and thus no simple blueprint for integrated groundwater management can be readily provided.

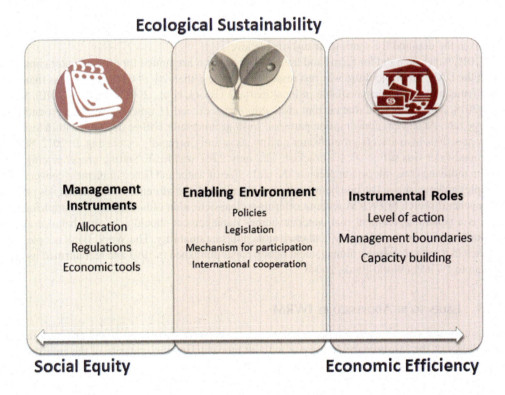

FIGURE 37.11   Global Water Partnership IWRM framework. (Modified after Tuinhof et al., 2010.)

37.5.2 THE ENABLING ENVIRONMENT OF IWRM

IWRM requires a system of policies, laws, regulations, platforms, and mechanisms to support its activities and players. It requires a culture that facilitates and encourages communication and participation of all stakeholders, particularly marginalized groups that tend to be the most affected by it yet have the least say over water resources management decisions.

37.5.3 WATER LEGISLATION FOR FACILITATING IWRM

Legislation establishes the powers, responsibilities, and rights of different stakeholders in water resources management. In particular, it gives authority to the government to take action to implement and enforce water regulations. It also clarifies the role of different stakeholders and sets the rules for managing water resources. The first element represents a compendium of policies, legislation, and regulations available for stakeholders. The second element delineates the roles of different institutional players and stakeholders. The third element constitutes a wider range of tools for regulation, economic optimization, and monitoring.

37.5.4 ROLES AND RESPONSIBILITIES IN WATER RESOURCES DEVELOPMENT AND INVESTMENT

There is generally no well-defined delineation between the contribution of public authorities versus the private sector in water resources development and investment. Generally, however, governments are expected to be responsible for developing and managing infrastructure that offers public goods and services such as storage and transfer facilities to manage water variability, correcting uneven distribution, and offering protection from floods and extended droughts. Moreover, infrastructure projects, such as dams, have long life and cost recovery times, and thus it is difficult to attract private funding to finance their construction. Likewise, in the case of the UAE, the construction of desalination plants, storage and recharge dams, and aquifer storage and recovery (ASR) projects may have a strategic nature, and hence should be constructed and financed by the government. Water plays an important role in various sectors, including energy, housing, tourism, and commerce. Consequently, overall planning for water resources should involve different ministries to ensure an optimal allocation of water resources, coordinate public spending on water resource development, and avoid conflicting policies. For example, in the UAE ministries responsible for urban development, irrigation, and environmental protection, as well as municipalities, and electricity and water authorities should coordinate their policies and activities to ensure an optimal socioeconomic and environmental allocation of water resources. An atmosphere of counterproductive competition among different ministries may result in unsustainable management of water resources.

37.5.5 MANAGING WATER ACROSS NATIONAL BOUNDARIES

Integrated watershed/aquifer management has merits in coordinating efforts at the national level. However, as water crosses national boundaries, national sovereignty arises as a major obstacle to productive collaboration. Although international water laws exist for resolving conflicts among riparians, countries are not obliged to abide by them and may choose not to do so if potential resolutions are likely to undermine their current privileges.

37.5.6 KEY CHALLENGE FOR GROUNDWATER RESOURCES MANAGEMENT IN THE UAE

Groundwater resources management in the UAE deals with balancing the exploitation of a complex resource (in terms of quantity, quality, and surface water interactions) with the increasing demands of water and land users. This section deals mainly with the quantitative, essentially resource-related, issues of groundwater management, and touches marginally upon groundwater pollution protection. Calls for groundwater management in the UAE did not arise until a decline in well yields and/or quality affected the stakeholder groups. It is already becoming obvious that any further uncontrolled pumping would result in the development of a "vicious circle" (Figure 37.12) and damage to the resource as a whole (complete aquifer dewatering and severe saltwater intrusion). Despite having one of the highest water scarcity indexes in the world, the UAE also has one of the highest per capita water consumption rates. This large demand is driven by a water policy largely based on supply-side management, rather than demand-side management, that is, on large infrastructure developments to ensure water supply rather than on initiatives to use water resources more efficiently. For example, the UAE has the third largest capacity for desalination after the Kingdom of Saudi Arabia and the United States. Abu Dhabi generates

Current Management

Supply-driven groundwater development leading to a vicious circle

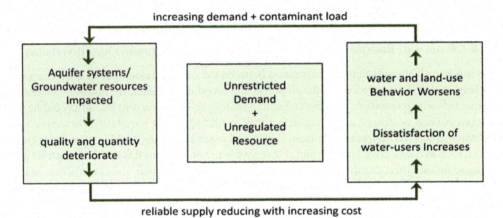

Target Management

Integrated groundwater resource management leading to a virtuous circle

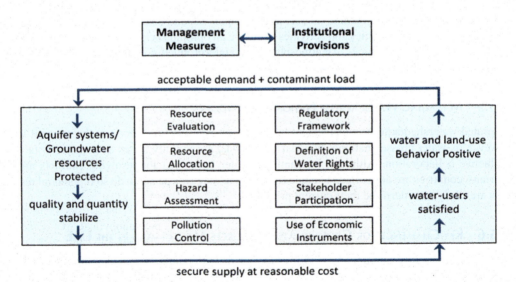

FIGURE 37.12 Current supply-driven versus target integrated groundwater resources management system. (After Tuinhof et al., 2010.)

100% of desalinated water from combined cycle power and desalination plants fueled primarily by natural gas. However, the UAE's growing dependency on desalinated water for domestic consumption has a high economic cost and an even higher environmental impact. The annual policy brief of the Environment Agency–Abu Dhabi (EAD, 2016) pointed out that if the current water supply-driven groundwater development in the UAE has not changed to integrated groundwater resources management (Figure 37.11), the following immitigable (or hardly avoidable) implications may occur.

- Environmental implications
  - Usable groundwater resources will decrease leading to its depletion. In the absence of groundwater, agriculture will depend on desalinated water.
  - The volume of brine and cooling water disposed into the gulf will be more than double, intensifying the impact on the marine ecosystem and local fisheries.
  - CO_2 emissions from desalination will be further doubled, exacerbating climate change.
- Economic implications
  - A new desalination capacity will be needed, which requires large capital expenditures.
  - Fuel consumption will contribute to natural gas depletion.
  - Subsides of AED 217 billion will be incurred in 20 years.
- Social implications
  - Overabstraction of groundwater will increase its salinity and thus reduce the availability of freshwater. The developed UAE WBM in this study indicated that freshwater resources in the UAE have decreased from 42 km^3 in 2005 to less than 10 km^3 in 2012. This will have an adverse impact on the agriculture sector and threaten food security in the country as well as create many social problems.
  - The volume of brine and cooling water disposed into the gulf will increase sea temperature and intensify coral bleaching and adversely affect fisheries that represent a source of employment and income.
  - Higher emissions of CO_2 will increase global warming, which will decrease precipitation and affect groundwater levels and marine fisheries.

Figure 37.12 indicates that the socioeconomic dimension (demand-side management) is as important as the hydrogeological dimension (supply-side management), and integration of both is always required (Tuinhof et al., 2010; Lawrence and Morris, 1997). In general, water-use sectors in the UAE are facing dramatic changes, as the management focus shifts from sectorial approaches toward more integrated, intersectoral management principles. This has far-reaching consequences for how the sectors are being organized and for managing implementation frameworks. IWRM will not take place if the legal framework is not adapted and the necessary institutional arrangements are not made (Tuinhof et al., 2010). The enabling conditions for creating a strong IWRM framework in the UAE include an enabling institutional framework, including the legal roles and responsibilities of institutions and their interrelationships; mechanisms to strengthen the role of women and other stakeholders; and ensuring the principles of fairness, affordability, and protection of the low-income segment of the society. The proposed UAE national water policy must be completed with clear government objectives pertaining to IWRM, including closer collaboration among the competent authorities, the private sector, and civil society to ensure more rational use of this precious resource to meet basic vital needs. Water legislation will naturally take different positions on a number of water topics: water rights, water privatization or concession, and public or private water use, among others. Legislation ought to emphasize the principles or elements in support of the proposed UAE IWRM, such as the polluter pays principle, equity, social justice, allocation, economic efficiency, financial feasibility, and licensing, as well as the river basin management approach.

37.5.7   How Should Integrated Groundwater Management Be Practiced?

In most situations, groundwater management will need to be kept in reasonable balance. The condition of excessive and unsustainable abstraction (unstable development), which is widely occurring, is shown in Figure 37.13. For this case, the total abstraction rate (and usually the number of production water wells) will eventually fall significantly as a result of a nearly irreversible degradation of the aquifer system itself.

The framework provided in Table 37.3 can be used as a diagnostic instrument to assess the adequacy of existing groundwater management arrangements for a given level of resource development (both in terms of technical tools and institutional provisions). The MOEW is currently developing an IWRM plan with several performance indicators.

By working down the levels of development of each groundwater management tool, these performance indicators are calculated and compared to the target stage of resource development to

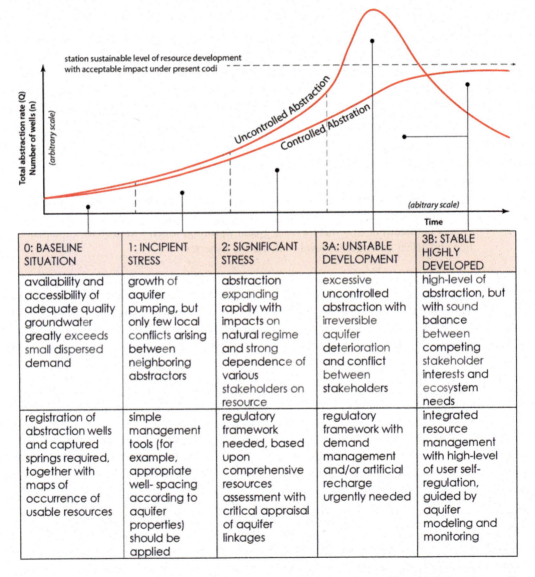

0: BASELINE SITUATION	1: INCIPIENT STRESS	2: SIGNIFICANT STRESS	3A: UNSTABLE DEVELOPMENT	3B: STABLE HIGHLY DEVELOPED
availability and accessibility of adequate quality groundwater greatly exceeds small dispersed demand	growth of aquifer pumping, but only few local conflicts arising between neighboring abstractors	abstraction expanding rapidly with impacts on natural regime and strong dependence of various stakeholders on resource	excessive uncontrolled abstraction with irreversible aquifer deterioration and conflict between stakeholders	high-level of abstraction, but with sound balance between competing stakeholder interests and ecosystem needs
registration of abstraction wells and captured springs required, together with maps of occurrence of usable resources	simple management tools (for example, appropriate well-spacing according to aquifer properties) should be applied	regulatory framework needed, based upon comprehensive resources assessment with critical appraisal of aquifer linkages	regulatory framework with demand management and/or artificial recharge urgently needed	integrated resource management with high-level of user self-regulation, guided by aquifer modeling and monitoring

FIGURE 37.13   Stages of groundwater resource development in a major aquifer and their corresponding management needs. (After Tuinhof et al., 2010.)

TABLE 37.3

Levels of Groundwater Management Tools, Instruments, and Interventions Necessary for a Given Stage of Resource Development

Groundwater Tools and Management Instruments	Level of Development of Corresponding Tool or Instrument (According to Hydraulic Stress Stage)			
	1	2	3	4
Technical Tools				
Resource assessment	Basic knowledge of aquifer	Conceptual model based on field data	Numerical model(s) operational with simulation of different abstraction scenarios	Models linked to decision-support and used for planning and management
Quality evaluation	No quality constraints experienced	Quality variability is issue in allocation	Water quality processes understood	Quality integrated in allocation plans
Aquifer monitoring	No regular monitoring program	Project monitoring, ad hoc exchange of data	Monitoring routines established	Monitoring programs used for management decisions
Institutional Instruments				
Water rights	Customary water rights	Occasional local clarification of water rights (via court cases)	Recognition that societal changes override customary water rights	Dynamic rights based on management plans
Regulatory provisions	Only social regulation	Restricted regulation (e.g., licensing of new wells, restrictions on drilling)	Active regulation and enforcement by dedicated agency	Facilitation and control of stakeholder self-regulation
Water legislation	No water legislation	Preparation of groundwater resource law discussed	Legal provision for organization of groundwater users	Full legal framework for aquifer management
Stakeholder participation	Little interaction between regulator and water users	Reactive participation and development of user organizations	Stakeholder organizations co-opted into management structure (e.g., aquifer councils)	Stakeholders and regulator share responsibility
Awareness and education	Groundwater is considered an infinite and free resource	Finite resource (campaigns for water conservation and protection)	Economic good and part of an integrated system	Effective interaction and communication between stakeholders
Economic instruments	Economic externalities hardly recognized (exploitation is widely subsidized)	Only symbolic charges for water abstraction	Recognition of economic value (reduction and targeting of fuel subsidies)	Economic value recognized (adequate charging and increased possibility of reallocation)

(Continued)

TABLE 37.3 (CONTINUED)

Levels of Groundwater Management Tools, Instruments, and Interventions Necessary for a Given Stage of Resource Development

Groundwater Tools and Management Instruments	Level of Development of Corresponding Tool or Instrument (According to Hydraulic Stress Stage)			
	1	2	3	4
Management Actions				
Prevention of side effects	Little concern for side effects	Recognition of (short- and long-term) side effects	Preventive measures in recognition of *in situ* values	Mechanism to balance extractive uses and *in situ* values
Resource allocation	Limited allocation constraints	Competition between users	Priorities defined for extractive use	Equitable allocation of extractive uses and *in situ* values
Pollution control	Few controls over land use and waste disposal	Land surface zoning but no proactive controls	Control over new point source pollution and/or siring of new wells in safe zones	Control of all point and diffuse sources of pollution; mitigation of existing contamination

Source: After Tuinhof et al. (2010).

indicate priority aspects for urgent attention. Through this type of approach, the necessary management interventions for a given hydrogeological setting and resource development situation can be agreed (Custodio and Dijon, 1991; Salman, 1999). The proposed MOEW IWRM plan adopted the following guiding principles from the International Bottled Water Association as the foundation for executing a comprehensive groundwater resources management policy and plan. First is scientific documentation. The primary effort of protecting and managing groundwater resources must be based on a full understanding of the hydrogeological setting. The entire aquifer system must be viewed within the context of science supported by empirical data.

37.5.7.1   Water Management Instruments

Agarwal et al. (2000) provided an extensive overview of established methods and approaches that could be employed in water resources management. They comprised a wide range of instruments used in assessing water resources, regulation, economic management, conflict resolution, communication, and new technology. This section covers some of these methods. Developing an information base is an important prerequisite for the reliable assessment of water resources. This task was particularly challenging in the UAE before the development of the central water resources information system in the MOEW in the year 2010. Understanding and management of the social and environmental implications of water resources development are at the core of the proposed UAE's IWRM.

37.5.7.2   Regulatory Methods

Depending on the method of application, Agarwal et al. (2000) categorized regulatory instruments into three types: direct controls, economic instruments, and self-regulation. Direct controls are those applied by public authorities to dictate or influence the use of water. Governments resort to executive regulations to enforce certain rules such as restrictions on groundwater extraction or discharge of wastewater. Under certain circumstances, especially when dealing with impending crises, executive regulations can be effective and efficient.

37.5.7.3 Meeting the Challenge of Progressive Scarcity

The World Bank (2005, 2007, 2009) argued that managing water resources under scarcity conditions goes through three stages of policy changes and response. At first, the emphasis will be on securing water supply from resources having the least cost of development. When the most affordable water resources have been developed, the emphasis shifts to developing and strengthening organizations responsible for serving users to optimize the use of water at the user level.

With mounting scarcity, there is a more urgent need to devise policies to achieve a more efficient allocation of water among users. This last and most challenging task requires a transparent institutional system that ensures accountability and creates an atmosphere of trust and confidence between users and policy makers.

As mentioned earlier, present groundwater abstraction in the UAE is more than 14-fold its renewability. This is exhausting the groundwater systems and at current consumption patterns, fresh groundwater will be totally depleted. This is also affecting its quality, as salinity is rapidly increasing. To address this challenge, the UAE government is adopting measures to encourage more efficient use of water, in particular in the agriculture sector. One of the measures is the implementation of the new Agriculture and Food Safety Policy led by the Abu Dhabi Food and Control Authority (ADFCA). To that end, ADFCA has established the Farmer's Services Centre to provide agricultural services and incentives to farmers to adopt best agricultural practices. Another measure will be the adoption of targets for water consumption as part of the UAE Water Conservation Strategy and Abu Dhabi Environment Vision 2030 led by the Environment Agency. The UAE government is also calling for the cooperation of all competent authorities to formulate five-year strategies for the environment to achieve the targets. In October 2012, the Permanent Committee for Water and Agriculture Resources approved the creation of a Water Council that will play a supervisory and coordination role between its member entities (EAD, Abu Dhabi Water and Electricity Authority, ADFCA, and Department of Municipal Affairs) to ensure integrated and coherent water policies in the future

37.6 WATER AS AN ECONOMIC GOOD

The emphasis on water as an economic good reflects a practical reality that the financial viability of water services necessitates full cost recovery. This may, however, not be fully applicable to the case of the UAE, where the government subsidizes several of the basic services that are provided to the community. Nevertheless, it is important to recognize that offering water services below their real cost encourages wasteful usage and imposes on society water opportunities costs. Emphasizing the economic worth of water was surrounded with controversy since its application may result in neglecting water's social and environmental role, particularly in potentially reducing access of less privileged groups to fresh water. A practical policy for water pricing is to increase the unit cost (m^3 in this case) with the increase in water consumption.

37.6.1 DEMAND-SIDE MANAGEMENT INITIATIVES IN THE UAE

In recent years, water policy has attempted to rationalize water consumption through demand-side management initiatives, as these are more cost-efficient options than supply-side alternatives. In 2011, the ADFCA initiated a strategic plan to be implemented by the Farmers' Services Centre (FSC) aiming to achieve a 40% presence of locally produced fruits and vegetables in the markets of the emirate by 2015 from the current share of about 15%. In the meantime, the strategic plan involves a 40% reduction in the consumption of irrigation water by the year 2013. To implement the strategy, the FSC should provide agricultural services and strengthen awareness, in addition to helping farmers to market their produce. With this in mind, the FSC launched the Zera'atona (our agriculture) campaign seeking to ensure economic agricultural sustainability through preserving vital natural resources by motivating farmers to adopt best agricultural practices, deemed essential to preserve their land, improve the agricultural product, protect the environment in general, preserve water in particular, and ensure better income for the farmers.

37.6.2 PHASING OUT OF SUBSIDIES FOR THE CULTIVATION OF CROPS WITH HIGH WATER CONSUMPTION

Regulation No. 7 of 2010 phased out UAE subsidies for the cultivation of crops with high water consumption, particularly Rhodes grass. Rhodes, which was found to consume more that 59% of water irrigation in agriculture, was limited to about 15,500 farms. The FSC is also working with ADFCA and the International Centre for Biosaline Research to introduce feed that has high tolerance to salinity and drought. To compensate farmers' losses, a program for improving farmers' income was activated on April 1, 2012, for the emirate provided the farmers comply with the rules and regulations governing agriculture including the directives and pieces of advice of the FSC regarding irrigation methods and best practices.

37.6.3 RATIONALIZING THE USE OF WATER FOR IRRIGATION OF PALM TREES

ADFCA, through the FSC, is also working on rationalizing the use of water irrigation for palm trees, responsible for almost 34% of water consumption, the second highest water consumer after Rhodes grass. A total of 5443 out of 8373 farms in the western region have been provided with modern irrigation networks, which are expected to help reduce water irrigation consumption in these farms by 50%.

37.6.4 REPLACING DRIP IRRIGATION WITH SUBSURFACE IRRIGATION

EAD and the Abu Dhabi Municipality (ADM) are introducing subsurface irrigation as an alternative to traditional drip irrigation. This will help reduce large water losses resulting from high evaporation.

37.6.5 ADOPTING NEW TECHNOLOGIES FOR GREENHOUSE AGRICULTURE

ADFCA and EAD have partnered to set up the Abu Dhabi Protected Agriculture Center to promote the use of greenhouses and soilless agriculture, a technique that involves growing plants in hydroponic and soil alternatives. The process uses 90% less water for the same volume of crops produced.

37.6.6 USING TREATED WASTEWATER IN AGRICULTURE

Recycled water is used well below its potential. In 2009, only 55% of treated wastewater water was reused for irrigation. The remaining 45% was discharged to the Arabian Gulf due to the lack of proper infrastructure for transmission. ADFCA tends to increase the use of treated wastewater in agriculture and is currently implementing a project to irrigate 216 farms with treated wastewater in AI Nahda and AI Wathba areas.

37.7 GROUNDWATER RESOURCES GOVERNANCE

Water governance can be defined as the range of political, social, economic, and administrative systems that are in place to regulate the development and management of water resources and the provision of water services at different levels of society, while recognizing the role played by environmental services (Rogers and Hall, 2003; Abdel-Dayem and Odeh, 2010). The relatively weak and unclear water governance system is reflected in the wasteful use of water, increasing pollution trends, lack of transparency, and inconsistent water services. Although there is no agreed-upon definition of good water governance, issues and basic principles for its achievement include the participation of all stakeholders (government, private sector, and civil society), transparency, promotion of water legislation, and regulations. This section defines the governance challenge and reviews the water governance discourse that can achieve water sustainability goals in the UAE. It summarizes the progress achieved in improving the governance system, potential desired outcomes of water governance reform, and their impacts and implementation possibilities. It provides, where possible, some case studies to highlight success stories and/or lessons learned in water management. Furthermore, there is a broad discussion of what may constitute the core of water legislation. The IWRM concept depicted in Figure 37.14 is a comprehensive approach to the development and management of water resources, addressing its management both as a resource and as a group of water services (WWAP, 2006). Water governance provides the framework in which the IWRM approach can be applied, taking into consideration the ways governments and social organizations interact, how they relate to the public, how decisions are taken, and how accountability is rendered (Graham and Plumptre, 2003).

37.7.1 WATER GOVERNANCE CHALLENGES IN THE UAE

The UAE is a federal nation and its constitution of 1971 defines, under Article 23, that natural resources are the property of individual emirates. As a result, although federal ministries hold some strategic and coordinating responsibilities, institutions, legislation, and regulations governing natural, and more recently nonconventional, water resources are to be vested with the individual emirate level. Day-to-day operations and management also take place at the local level. In addition, there are increasing inputs to water (and energy) management and regulations from Gulf Cooperative Council (GCC) agreements. Water governance becomes "effective" or "good" when conditions of equity, accountability, participation, transparency, predictability, and responsiveness prevail (Tiihonen, 2004). Kooiman's (2003) governance is a complex function of social–political interactions in which various societal actors are interdependent, and no one agent, group, or sector has all the knowledge and facts required to set policy, make decisions, or take actions (Kooiman, 2003). Because of this complexity, good governance does not just appear but is the culmination of multifaceted, long-term processes that have to be carefully planned and nurtured (Rhodes, 1996; Kooiman, 2003; Tiihonen, 2004). For good governance to emerge, contextually appropriate conditions must exist and an enabling environment must be cultivated (Tiihonen, 2004). Parties concerned must be open to committing to collective decision-making; effective and functional institutions need to be developed; and policy, legal, and political frameworks must be suitable to the goals that are being pursued for the common good (Rhodes,1996; Kooiman, 2003). In addition to the water scarcity challenges in the UAE, water governance in the country has the following challenges.

FIGURE 37.14 The IWRM–governance nexus (Abdel-Dayem and Odeh, 2010).

37.7.2 CHALLENGES

For the last few decades, water management in the UAE was mostly based on a supply approach driven where local and federal governments developed infrastructure to capture and distribute water to users without much emphasis on demand management. Water policy and management paradigms that dominated in that period can no longer be sustained in an era of water scarcity. The ability to overcome the challenge of managing scarce water resources with the least ecological and social costs will depend on the ability to create the water governance structure needed for sustainable water allocation and management (Saleth and Dinar, 2004). In the UAE, there is a crucial need to adopt a water policy that can achieve a balance between the cost of delivering water for agricultural use and the revenue associated with crop cultivation. They should also be adaptable to changes in the global markets, while efforts should be expended to raise the efficiency of water and production inputs and achieve higher revenue per unit of water (Abdel-Dayem and Odeh, 2010). The current system of water governance within the Emirate of Abu Dhabi has reasonably clear lines of demarcation between the entities responsible for each type of water:

- Groundwater for agriculture, forestry, and amenities is managed by ADFCA, EAD, and Department of Municipal Affairs (DMA).
- Seawater for potable water supply for residential and commercial customers is controlled by the Regulation and Supervision Bureau (RSB) and Abu Dhabi Water and Electricity Authority (ADWEA).
- Wastewater collection and treatment is controlled by the Abu Dhabi Sewage Services Company (ADSSC).

However, in the area of groundwater management, there are overlaps and gaps between the activities of the various federal- and emirate-level environmental organizations, such as establishing regulations and controlling resource use, and collecting and managing data. Many of the overlaps and gaps result from organizations operating as silos with limited communication between the various management and user groups. It is imperative that planning involves input and knowledge from these various groups. To address these issues, the Abu Dhabi government has made significant progress to provide an effective governance framework to bring greater clarity as to the roles and responsibilities of the entities managing underground water and better coordination. A number of key initiatives and associated policy instruments have been defined and are currently under review. Unsurprisingly, a priority area for new water policies is water demand management, particularly in agriculture. This would achieve both sustainable development and significant reductions in related future investments in production capacity and infrastructure. Another key area is the coordination of water governance among the different emirates, which would foster increased consistency and transparency in areas such as technical, economic, and environmental regulations and standards. This could bring greater scales of economy in future infrastructure development as well as more effective regulatory control (MOEW, 2010).

37.7.3 WATER LAW AND WATER LEGISLATION

Water law is made up of all the provisions that one way or another govern the various aspects of water management, e.g., water conservation, use and administration, the control of the harmful effects of water, and water pollution. In the UAE, control of water pumping and extraction is either absent or very limited. Enforcement of water regulations is limited due to acquired water rights, political attitudes, and the inflexibility and resistance of most farmers; lack of accurate data and information about water resources, quantities, and quality; and inadequately qualified technicians and maintenance personnel in the water sector. In the last few decades, IWRM has become a central feature of many water legislations. Laws, regulations, standards, and their enforcement are an important part of any governance system ensuring the protection of human and environmental health as well as economic efficiency. They give direction, transparency, and clarity in many areas, such as in responsibilities, roles, and standards for a particular environment or sector. It is therefore important to understand the current systems in place in the UAE. Although water is not mentioned explicitly in the constitution, by implication of some of its provisions (Articles 23, 120, 121, 122), water resources and their regulation fall within the remit of the individual emirates. As a result, legislation and regulations governing the management, development, protection, conservation, and use of natural water resources engage mostly the emirate level of legislative, executive, and judicial authority, with the federal-level MOEW retaining a national policy/strategy, coordination, and standard-setting authority (UAE Water Conservation Strategy, 2010). The legal status of nonconventional water resources, like desalinated water and reclaimed water, is not explicitly defined in the constitution (unlike electricity).

There is no clear indication as to responsibilities for water demand management from the constitution, except in Article 23, which states that the federal government is "responsible for the protection and proper exploitation of such natural resources and wealth for the benefit of the national economy". Thus, there is a burden on the federal legislature and executive to protect water resources and to ensure sustainable use of groundwater. However, the introduction of practical measures has to date emanated from emirate-level organizations (MOEW, 2010). The balance between the roles and responsibilities of the emirate and federal governments is further affected by agreements made at the GCC level. There is an increasing number of agreements on legal and regulatory standards and operations for various aspects of water, particularly desalinated water, that are then to be translated into action at the federal and then emirate level. The various organizations involved in water governance in the UAE are bound by a number of water and environmental laws, regulations, and standards, summarized in Table 37.4. Arguably the most influential law is Federal Law No. 24 of 1999 for the Protection and Development of the Environment, which covers various areas that affect water management including:

- Requirements for environmental assessments of developments
- Various aspects of environmental protection
- Environmental monitoring
- Emergency and disaster planning
- Protection of the marine environment from oil industries and transport
- Polluted water discharges
- Protection of drinking water quality from storage tanks
- Control of air emissions such as from vehicles, the burning of soil and liquid wastes, as well as from the oil extractive industries
- Handling of dangerous substances
- Natural reserves

At the federal level, the MOEW's Water Resources and Nature Conservation section is responsible for developing strategic policies and plans for water, establishing national standards in certain areas, and for coordinating activities with the emirates and other federal organizations. The vision of the MOEW is "conserving the environment and the natural resources for sustainable development", and this reflects the functions and organizational structure given under Ministerial Resolution No. 21 of 2009. In 2009, the Ministry of Environment and Water assumed the various roles of the Federal Environment Agency, set up under Federal Law No. 7 of 2003 and abolished with Federal Law No. 7 of 2009.

While water is a part of many aspects of Resolution No. 21, a number of the specific functions under Article 3 are particularly relevant:

- Developing plans, strategies and policies in the field of water
- Developing programs that ensure output in various sectors including water to ensure food security
- Ensuring environmental protection in economic and social plans for the country
- Evaluating water resources and determining programs and means that ensure good management and conservation
- Proposing legislation to support the ministry's functions

The day-to-day operations and management of natural water resources takes place at the emirate level. These competent organizations, usually a department within the municipality, are responsible for managing the natural water resources, including the implementation of articles of federal laws. The development of protection measures, controls, and management practices for natural water resources vary remarkably among the competent authorities (MOEW, 2010). The competent water authorities in the UAE include:

TABLE 37.4
Agreements and Laws Affecting the Environment and Water in the UAE

Legal jurisdiction	Date of ratification and legal instruments in place
International agreements	1990 Basel Convention on the Control of Transboundary Movements of Hazardous Wastes and their Disposal (1989)
	1995 United Nations Framework Convention on Climate Change (1992)
	1998 United Nations Convention to Combat Desertification (1994)
	1999 Convention on Biological Diversity (1992)
	2002 Convention on Persistent Organic Pollutants (2001)
	2002 Prior Informed Consent Procedure for Certain Hazardous Chemicals and Pesticides in International Trade (PIC Convention) (1998)
	2005 Kyoto Protocol (1997)
	2007 Ramsar Convention (1971
Implementing regional agreements	1979 Kuwait Regional Convention for Cooperation on the Protection of the Marine Environment from Pollution (1978).
	2003 Convention on Conservation of Wildlife and its Natural Habitats in the GCC countries
Federal level	Federal Decree No. 77 of 2005 regarding the Protocol of Control of Marine Cross-Border Transport and Disposal of Hazardous and Other Wastes
	Federal Law No. 23 of 1999 on Exploitation, Protection & Development of Living Aquatic Resources in Waters of the State of UAE
	Federal Law No. 24 of 1999 regarding the Protection and Development of the Environment
	Ministerial Declaration No. 24 of 1999 System for Assessment of Environmental Impacts
	Cabinet Resolution/Decision No. 37 of 2001 regulating Federal Law No. 24 of 1999 concerning Protection and Development of the Environment
	Cabinet Resolution No. 3 of 2002 concerning the National Environmental Strategy and Environmental Action Plan for UAE
	Federal Law No. 11 of 2006 amending certain provisions of Federal Law No. 24 of 1999 concerning the Protection and Development of the Environment
	Cabinet Resolution No. 12 of 2006 concerning Air Pollution System
	Federal Decree No. 9 of 2007 regarding the Law (regulations) on Fertilizers and Agricultural Conditioners of the GCC Member States
	Federal Decree No. 10 of 2007 regarding the GCC Pesticides Law (Regulations)

Source: MOEW (2010).

- Abu Dhabi: Environment Agency–Abu Dhabi (EAD)
- Dubai: Dubai Municipality (DM), Department of Environment
- Sharjah: Sharjah Electricity and Water Authority (SEWA) and Sharjah Environment and Protected Areas Authority (SEPA)
- Umm Al Quwain: Umm Al Quwain Water Authority (UAQWA)
- Ajman: Ajman Municipality, Department of Public Health and Environment (ADPHE)
- Ras Al Khaimah: Ras Al Khaimah Municipality Authority of Environment and Nature (RAKMAEN)
- Fujairah Municipality (FM)

The protection of groundwater resources from overabstraction or pollution comes under the scope of authority of the emirates. The federation has limited authority on that matter (Articles 120, 121, and 122 of the constitution). However, the existing federal environmental protection regulation (i.e., Federal Law No. 24 of 1999 for the Protection and Development of the Environment) does address the protection of groundwater resources directly (Article 39) and indirectly. In this, the need for facilities for the disposal and treatment of waste, including wastewater, of municipal and industrial origin, and from desalination plants, are subject to the prior environmental clearance requirements of Law 59. As a result of this federal regulation, all relevant projects must undergo an environmental impact screening prior to undergoing any separate licensing process under federal or emirate laws.

The following laws have been passed for controlling and managing groundwater:

- UAE Water Conservation Strategy, 2010
- Abu Dhabi Law No. 6, 2006, Regulation of Well Drilling followed by Executive Rules No. 6, 2007
- Dubai Law No. 15, 2008, Protection of Groundwater in the Emirate of Dubai
- Proposed Sharjah Law, 2007 Protection of Water in Sharjah Emirate
- Umm Al Quwain, Law No. 2, 2008, Protection of Groundwater and Regulation of Well Drilling in Umm Al Quwain Emirate
- Ajman Law No. 4, 2009, Regulation of Well Drilling and Consumption of Groundwater
- Ras Al Khaimah Law No. 5, 2006, Regulation of Well Drilling
- Sharjah Law No. 6, 2012, Regulation of Well Drilling

Most of these laws establish provisions for new wells or those that need further development. In a number of laws, there is a provision to establish protection zones for groundwater. Some include details on the actual control measures (metering, pumps, etc.), and data to be provided by well owners, such as the water use and estimated rate of abstraction (Table 37.5). The most difficult area is in regulating existing wells. These wells, and traditional groundwater rights (falajs), which predate regulatory legislation,

TABLE 37.5

Natural Water Resources Legal and Regulatory Provisions

	EAD	DM	SEWA	SM	UAQWA	ADPHE	RAKMAEN	FM
Information gathering/inventorying	✓	✓	✓		✓	✓	✓	
Licensing of wells and their further development	✓	✓	✓		✓	✓	✓	✓
Regulating drilling	✓	✓	✓		✓	✓	✓	✓
Regulating abstraction and rates	✓	✓	✓		✓	✓	✓	✓
Regulating use		✓	✓		✓	✓	✓	
Monitoring and enforcement	✓	✓	✓	✓	✓	✓	✓	✓
Pollution protection	✓		✓		✓	✓	✓	

play a significant role in much of the country's rural areas. Their use is grounded on the deep and diffuse conviction of the owners that, regardless of what the constitution says about all-natural resources, in general, being the property of the emirates, groundwater is the property of the landowner, for him to extract and dispose of as he sees fit. A further difficult area is the actual implementation, administration, and enforcement by the different competent authorities of their water legislation.

In light of the clear-cut constitutional division of powers and responsibilities between the federal and the emirate governments in relation to water resources in general, there is no doubt that the matter of enforcement falls under the exclusive remit of the emirates. This tends to be problematic for many jurisdictions, and the emirates of the UAE is no exception, with large variations in actions. A number of reasons may be put forward to explain this, including a lack of both numbers and adequately trained law enforcement of the regulatory authorities. Inspectors do exist but their numbers and mandate vary greatly across the different emirates. Thus, groundwater is regulated and managed by seven different organizations at the emirate level and one at the federal level (Table 37.5). Different laws, regulations, and practices are currently managing this resource. There is a need to agree at the national level on various standards and regulations for water resources management and information collection, and these should then be implemented by the competent authority within the emirates. This harmonization would bring a greater understanding and protection of the resource. The Abu Dhabi Government has recently established a new governance framework for agriculture endorsing ADFCA to lead the development of the agriculture policy framework and create the Farmers' Services Centre to implement the policy. The Abu Dhabi government has also introduced new laws and policies to make the agriculture sector more competitive by increasing and diversifying production while promoting more rational use of water for irrigation in line with Abu Dhabi's Environment Vision 2030.

37.7.3.1   Law No. 9 of 2007, Establishing the Department of Municipal Affairs

In 2007, Section 5 of the Law No. 9 established the Department of Municipal Affairs (DMA) and hence transferred powers and mandates around agriculture to ADFCA, which became the competent authority for agriculture. Accordingly, ADFCA developed the agricultural policy and prepared the plans for achieving sustainable agricultural growth, while mitigating the harmful effects of certain improper agricultural practices on the environment. The new agricultural policy was expected to take into consideration certain imperatives:

  (i) Restructure the agricultural sector with a view to make it more sustainable
 (ii) Reduce harmful effects on the environment and the pressure on natural resources
(iii) Ensure fair income for the farmers and increase their competitiveness in the market
 (iv) Focus on products that Abu Dhabi has a competitive edge on
  (v) Improve the quality of agricultural products
 (vi) Strengthen national productivity for better food security

37.7.3.2   Law No. 4 of 2009, Pertaining to the Setting Up of the Farmers' Services Centre in the Emirate of Abu Dhabi

In 2009, Law No. 4 established the Farmers' Services Centre with responsibility for implementing Abu Dhabi's agricultural policy by engaging farmers to adopt best agricultural practices.

37.7.3.3   Agriculture and Food Safety Policy (2011)

The Agricultural Water Use Policy has the objective to maximize efficiency and support sustainability. It covers the following aspects: (i) barriers to efficient agricultural water use; (ii) water targets for use; (iii) Water Use Impact Assessment addresses economic, social, and environmental factors in reaching decisions on agriculture activities using water; (iv) data for water impact assessments; and (v) liaison with other departments and agencies.

- In 2011, ADFCA embarked on an ambitious program of policy development that expanded responsibility for the entire food chain from farm to fork, including the safety of imported food.
- The new Agriculture and Food Safety Policy for the Emirate of Abu Dhabi consists of 11 general policies and 15 agriculture policies. The latter covers agriculture production.
- Production relates to the activity of growing, rearing, and producing raw material for entry into the food chain. There are four agriculture policies related to production: Agricultural Land Use Policy, Agricultural Water Use Policy, Production Choice, and Economic Sustainability. The agriculture policy recognizes the challenge of supporting agriculture growth in the context of water scarcity and addresses the potential environmental concerns in the first three policies.

Institutional arrangements, councils, and commissions need to be addressed. These arrangements should incorporate powers, mandates, and responsibilities as well as rights, obligations, and roles of stakeholders. Regulatory approach, demand management, dispute resolution, and well-defined water infractions as well as transitional and final provisions should be made. The main components of the national water legislation are described in Box 37.1.

BOX 37.1 SUGGESTED CORE ISSUES IN A NATIONAL NATURAL WATER RESOURCES LEGISLATION/ACT

Introduction (statement of motives, with a brief summary of underlying water policy principles and priorities)

Part I: General Provisions
- Definition of general terms used in the act
- Authorities responsible for enforcement of the act

Part VI: Water Allocation
- Domestic water and right to water
- Agricultural, aquaculture, coastal management, and industrial water
- Water permits, licenses, and authorizations
- Dam control
- Water trade/allocation/transfer

Part II: Ownership of water resources/Classification of water
- Surface water (such as wadis, storage at dams)
- Groundwater

Part VII: Dispute Settlement
- Courts and tribunals
- Arbitration and alternative disputes resolution
- Aquifer Development and Recharge (ADR) techniques

Part III: Conservation and Protection of Water Resources
- Groundwater conservation and protection
- Flood control
- Ecosystem protection and environmental sustainability

Part VIII: Infractions to Water Resources and Sanctions
- Police powers of water services officers
- Procedures
- Penalties

Part IV: Management of Water Resources
- Institutional arrangements (basin management, catchment management agencies, surface water augmentation and protection)
- Powers, mandates, and responsibilities
- Rights, obligations, and roles of the stakeholders (water users associations, gender role)

Part IX: Transitional and Final Provisions
- Existing water rights and entitlements of indigenous communities
- International cooperation on shared water courses

Part V: Regulation of Water Services
- Water pricing
- Public–private partnerships (PPP), concessions and privatization

37.8   DROUGHT IMPACT ON WATER RESOURCES IN THE UAE

Drought is a natural phenomenon that occurs because of low precipitation conditions. Droughts have major negative impacts on the environment, economy, agriculture, and society. Several researchers estimated the duration and frequency of drought in the UAE (e.g., Sherif et al., 2013; Yilmaz et al., 2020). The results of these studies indicated that climate change and drought could have implications for the UAE's development objectives. Direct impacts of extreme weather events and sea level rise could disrupt the daily functioning of transport and infrastructure, and affect the value of the real estate and tourism industry. As shown in Table 37.6, the range of drought duration in the UAE is from 1 to 7 years, with an average of 2.8 years. Meanwhile, 4.5 years is the average frequency of drought. The Paris Agreement governs climate action from 2020 onward, and involves the commitment of both developing and developed countries alike to hold the global temperature increase to below 2 degrees Celsius compared to preindustrial levels.

The UAE has been a strong supporter of the Sustainable Development Goals (SDGs) that underpin the 2030 Agenda, which was adopted at the 2015 United Nations Sustainable Development Summit. In light of the Paris Agreement and the SDGs, the UAE developed a climate plan to strengthen the UAE's regional and global leadership in climate action by minimizing risks and improving adaptive capacity. The adaptation measures include (MOCCAE, 2017):

- Increasing the UAE's economic diversification through innovative solutions
- Managing greenhouse gas (GHG) emissions
- Increasing climate resilience

TABLE 37.6
Characteristics of Droughts in Different Regions in the UAE

	Drought Events	Drought Duration (years)	Drought Severity Index (DSI)	Average DSI	Average Duration (years)	Frequency (years)	Drought Years (%)
East Coast	10	1–7	0.16–3.36	1.40	2.9	4.5	64
Mountain	10	1–12	0.16–7.02	1.49	2.8	4.5	62
Gravel Plain	10	1–12	0.07–6.57	1.42	2.8	4.5	62
Desert Foreland	10	1–7	0.16–4.61	1.50	2.7	4.5	60
UAE	10	1–12	0.21–6.46	1.41	2.8	4.5	62

Source: After Sherif et al. (2013).

TABLE 37.7
Khatt Spring Characteristics Before and After the Drought

North Spring Situation 1979–1998	North Spring Situation 1999–2005
Outflow 10–50 l/s	Outflow 1–10 l/s
Rainfall rate 60–360 mm/year	Rainfall rate 20–66 mm/year
Water temperature 39°C–40°C	Water temperature 30°C
Irrigation falajs are working without pumping	Irrigation falajs are only working with pumping from hand-dug wells

Source: Wycisk et al. (2008).

Droughts in the UAE are affecting water resources (surface and groundwater in shallow and deep aquifers), coastal areas, marine and terrestrial ecosystems, urban areas, agriculture and food security, and public health. The effect of drought on water resources could be recognized from several observations, among those:

- The development of a cone of depression in the water table in Khatt, Ras Al Khaimah, due to the drought from 1999 to 2005
- The decrease in the spring outflow rates and temperature of Khatt Spring (Table 37.7)
- The increase of the evapotranspiration rates in the shallow aquifer due to the increase in air temperature
- The increase in saltwater intrusion and spread of soil salinization in coastal areas

37.9 CONCLUSIONS

In the UAE, water is supplied from three main sources:
1. Groundwater supply – mostly for irrigation uses, but some limited quantities are used for potable uses, particularly in the Northern Emirates
2. Desalinated water supply – mainly for potable water uses, and in some places also used for irrigation
3. Reclaimed water supply (also known as treated sewage effluent) – mainly used for irrigating amenity areas

The increase in total water consumption is a growing problem in the UAE, reaching about 6 km^3 in 2019. The current supply from groundwater is about 51%, mostly for irrigation uses, but some limited quantities are used for potable uses, particularly in the Northern Emirates. The desalinated water supply is about 37%, mainly for potable water uses, and in some places also used for irrigation purposes. Reclaimed water supply (treated sewage effluent) accounts for 12%, and is mainly used for irrigating amenity areas. For decades, water resources management in the UAE has been mainly based on supply management, and this created severe water resources problems including groundwater depletion, groundwater salinization, geotechnical problems in coastal urban areas, and brine discharge environmental problems.

Integrated water resources management based on demand management is the most feasible alternative in which more experience and institution-strengthening measures are needed to expand public–private partnership (PPP) capacity so that it can make a real contribution to meeting the region's growing urban water service needs. Participation must not be understood as an end in itself with the rise of organized water user groups as the final objective. Participation has to be a means of achieving joint responsibility at all levels of decision-making processes, where actors form part of the problem as well as the solution. Four major regulatory development gaps need to be addressed in order to consolidate the progress achieved in improving water governance:

1. The capacity gap to raise the technical expertise and competencies of staff
2. The policy gap that can establish regulatory agencies' autonomy and independence from the executive power
3. The information gap to reduce the asymmetry of information between the regulator, the operator, and the user
4. The participation gap to allow real citizen involvement in the work of the regulatory agencies

Although these are globally recognized gaps, they aptly apply to the UAE as well. More effort should be mobilized to close gaps in policy and institutional reforms, build capacities and skills, disclose information, raise awareness, and allow broader participation of stakeholders. In the UAE, extraordinary measures of cooperation are required at local, national, and regional levels to improve the existing patterns of water governance in an area where different interests as well as dissimilar values and norms prevail. Therefore, the way forward is to recognize the importance of planning and implementing frameworks for good water governance that takes account of differing social, economic, environmental, and cultural contexts, including the introduction of processes of interaction between state and non-state actors and mechanisms for recognizing mutual responsibilities for governance. The UAE government's insight needs to give priority to developing the capacity to generate credible and relevant water research. This requires a national science policy, a locally accountable research agenda, political commitment, outstanding research management and leadership, sustainable funding mechanisms, and career development incentives to attract and retain young and senior talent.

REFERENCES

Abdel-Dayem, S., Odeh, N. (2010). Water governance. In El-Ashry, M.; Saab, N.; Zeitoon, B. (eds.), *Arab Environment: Water: Sustainable Management of Scarce Resources. Report of the Arab Forum for Environment and Development*, 171–188.

Agarwal, A., de los Angeles, M. S., Bhatia, R., Chéret, I., Davila-Poblete, S., Falkenmark, M., Gonzalez-Villarreal, F., Jønch-Clausen, T., Aït Kadi, M., Kindler, J., Rees, J. A., Roberts, P., Rogers, P., Solanes, M., Wright, A. (2000). Integrated Water Resources Management [TAC Background Papers, no. 4]. http://hdl.handle.net/10535/4986.

Brook, M. C., Dawoud, M. (2005). Coastal water resources management in the coastal zones of United Arab Emirates. Symposium on Integrated Coastal Zone Management, Abu Dhabi, UAE.

Brook, M. C., Al Houqani, H., Dawoud, M. (2005a). The opportunities for water resources management in the emirate of Abu Dhabi, United Arab Emirates. In Proceedings of the 7th Gulf Water Conference, Kuwait, 2005.

Brook, M. C., Al Houqani, H., Dawoud, M. (2005b). The water management challenges faced by the various water sector users in the Emirate of Abu Dhabi, United Arab Emirates. *Water – Resources, Technologies and Management in the Arab World*, 8–10 May, 2005, University of Sharjah, Sharjah, United Arab Emirates.

Brook, M., Al Houqani, H., Al Mugrin, A. (2006a). The current status and future requirements of water resources management in the Arabian Peninsula. In Mustafa, A. K., Benno, B., Michael, B., *Zafar*, Miguel, A. C. G., Walid, S. (Eds.), *Policy Perspectives for Ecosystem and Water Management in the Arabian Peninsula*. Hamilton: United Nations University and UNESCO.

Brook, M. C., Al Shoukri, S., Amer, K. M., Boer, B. Krupp, F. (2006b). Physical and environmental setting of the Arabian Peninsula and surrounding seas. In Amer, K. M. Boer, B., Brook, M. C., Adeel, Z. Clusener-Godt, M. and Saleh, W. (Eds.), *Policy Perspectives for Ecosystem and Water Management in the Arabian Peninsula*.

Custodio, E., Dijon, R. (1991). Groundwater overexploitation in developing countries. UN Interregional Workshop Report, UN-INT/90/R43.

EAD (2016). *Economic Valuation of Groundwater in the Abu Dhabi Emirate*, Unpublished report. Abu Dhabi, Environment Agency - Abu Dhabi.

Ebraheem, A. M., Al Mulla, M. M., Sherif, M. M., Awad, O., Akram, S. F., Al Suweidi, N. B., Shetty, A. (2014). Mapping groundwater conditions in different geological environments in the northern area of the UAE using 2D earth resistivity imaging survey. *Environmental Earth Sciences*, 72(5), 1599–1614. https://doi.org/10.1007/s12665-014-3064-5.

Ghoneim, E. (2008). Optimum groundwater locations in the northern Unites Arab Emirates. *International Journal of Remote Sensing*, 29(20), 5879–5906.

Global Water Partnership (GWP) (2000). *Towards Water Security: A Framework for Action*. Stockholm, Sweden, ISBN 91-630-9202-6, p. 1–10.

Graham, J., Amos, B., Plumptre, T. (2003). Governance principles for protected areas in the 21st century. The Vth IUCN World Parks Congress. Durham, South Africa. Available at: https://www.files.ethz.ch/isn/122197/pa_governance2.pdf.

Hydro Atlas (2015). *UAE HYDROATLAS*. Internal Report, National Water Center and Ministry of Energy and Industry, UAE.

IWACO (1986). *Groundwater Resources. Drilling of Deep-Water Wells at Various locations in the UAE, Groundwater Development in the Northern Agricultural Region*. Internal Report, Vol. 70, Ministry of Environment and Water, Dubai.

Kooiman, J. (2003). *Governing as Governance.* Sage Publications, London.

Lawrence, J. F., Morris, B. (1997). *Groundwater in Urban Development: Assessing Management Needs and Formulating Policy Strategies.* World Bank Technical Paper, 390.

MOCCAE (2017). *National Climate Change Plan of the United Arab Emirates 2017–2050.* Internal Report. UAE Ministry of Climatic Change & Environment (MOCCE), Dubai, UAE.

MOEW (2010). *UAE Water Conservation Strategy.* Internal Report, Ministry of Environment and Water, Dubai, UAE.

MOEW-UAEU (2005). Assessment of the effectiveness of Al Bih, Al Tawyean and Ham Dams in groundwater recharge using numerical models. Internal report, Ministry of Environment and Water.

NDC and USGS (1996). *Groundwater Resources of Abu Dhabi Emirate.* Internal report, National Drilling Company (NDC) and US Geological Survey.

Rhodes, R. A. W. (1996). The new governance: Governing without government. *Political Studies,* 44, 652–667.

Risk, Z., Alsharhan, A. S. (2003). Water resources in the United Arab Emirates. In Alsharhan, A. S., Wood, W. W. (Eds), *Water Management Perspective: Evaluation, Management and Policy.* Elsevier Science, 245–264.

Rizk, Z. S., El-Etr, H. A. (1997). Hydrogeology and hydrogeochemistry of some springs in the United Arab Emirates. *The Arabian Journal for Science and Engineering* 22(1C), 95–111.

Rogers, P., Hall, A. (2003). Effective water governance. *Technical Background Paper,* No. 7.

Saleth, R. M., Dinar, A. (2004). The institutional economics of water: A cross-country analysis of institutions and performance. Washington, DC: World Bank and Cheltenham, UK, Edward Elgar Publishing Ltd. https://openknowledge.worldbank.org/handle/10986/14884.

Salman, M. A. (1999). *Groundwater: Legal and Policy Perspectives.* World Bank Technical Paper, no. 456, Washington, DC.

Sherif, M., Almulla, M., Shetty, A. Chowdhury, R. K. (2014). Analysis of rainfall, PMP and drought in the United Arab Emirates. *International Journal of Climatology,* 34, 1318–1328.

Sherif, M., Ebraheem, A., Shetty, A. (2017). *Groundwater Recharge from Dams in United Arab Emirates.* World Environmental and Water Resources Congress 2017, 139–146, American Society of Civil Engineers, California. DOI: 10.1061/9780784480618.014.

Sherif, M., Ebraheem, A., Al Mulla, M., Shetty, A. (2018). New system for the assessment of annual groundwater recharge from rainfall in the United Arab Emirates. *Environmental Earth Sciences,* 77(11), 412. DOI: 10.1007/s12665-018-7591-3.

Sherif, M., Ebraheem, A., Al Mulla, M., Sefelnasr, A. (2020). Variations of groundwater storage during the last fifty years in UAE. *Environmental Earth Sciences* (Submitted).

Sherif, M., Mohamed, A., Shett, A., Chowdhury, R. K. (2013). Analysis of Rainfall, PMP and Drought in United Arab Emirates. *International Journal of Climatology.* https://doi.org/10.1002/joc.3768

Tiihonen, S. (2004). *From Governing to Governance: A Process of Change.* Tampere University Press, 323p.

Tuinhof, A., Dumar, C., Foster, S., Kemper, K., Garduño, H., Nanni, M. (2010). *Sustainable Groundwater Management: Concepts and Tools. World Bank Group, Global Water Partnership Associate Program. GW-MATE, Briefing Note Series,* Note no.1, World Bank, Washington, D.C., USA, p. 1–6.

UAE Water Conservation Strategy (2010). *Internal Report.* Ministry of Environment and Water, Dubai, UAE.

World Bank (2004). *Water Resources Sector Strategy: Strategic Directions for World Bank Engagement.* The World Bank, Washington, DC, Vol. 1, Report no. 28114, DOI: 10.1596/0-8213-5697-6.

World Bank (2005). *A Water Sector Assessment Report on the Countries of the Cooperation Council of the Arab States of the Gulf.* The World Bank, Washington, DC, Vol. 1, Report no. 32539-MENA.

World Bank (2007). Making the most of scarcity: Accountability for better water management in the Middle East and North Africa. *Mena Development Report.* The World Bank, Washington, DC, Water P-Notes, Vol. 1, no. 40.

World Bank (2009). *Water in the Arab World: Management Perspective and Innovations. Middle East and North Africa Region.* Edited by Jagannathan, N. V., Mohamed, A. S., Kremer, A. The World Bank, Washington, DC, Vol. 1, Report no. 49593.

World Water Assessment Program (WWAP) (2006). *Water: A Shared Responsibility; the United Nations World Water Development Report 2.* UNESCO-WWAP, New York, 601p.

Wycisk, P., Al Assam, M., Akram, S., Al Mulla, M., Schlesier, D., Sefelnasr, A., Al Suwaidi, N. B., Al Mehrizi, M. S., Ebraheem, A. (2008). Three dimensional geological and groundwater flow modeling of drought impact and recharge potentiality in Khatt Springs area, Ras Al Khaimah, UAE. Proceedings of the 7th International Conference on Gulf Area Water Resources, March 3–5, 2008. Manamah, Bahrain.

Yilmaz, A., Najah, A., Hussein, A., Khamis, A., Kayemah, N., Atabay, S. (2020). Climate change effects on drought in Sharjah, UAE. *International Journal of Environmental Science and Development,* 11(3), 116–122.

38 Droughts, Distress, Impact, and Mitigation
A Case Study of Jammu and Kashmir

F.A. Shaheen

CONTENTS

38.1 INTRODUCTION

Drought results from long-continued dry weather and/or insufficiency of rain, which causes exhaustion of soil moisture, depletion of underground water supply, and reduction of streamflow. Drought is frequently defined according to disciplinary perspective. The four types of droughts, namely (i) meteorological drought, (ii) surface water drought, (iii) groundwater drought, and (iv) soil-water drought. He argues that the various forms of droughts get generated independently but are inseparable and linked to each other through the water cycle. The National Commission on Agriculture in India defines three types of droughts, namely, meteorological, agricultural, and hydrological. Meteorological drought is defined as a situation when there is a significant decrease from normal precipitation over an area (i.e., more than 25%). Agricultural drought occurs when soil moisture and rainfall are inadequate during the growing season to support healthy crop growth to maturity and causes crop stress and wilting. Hydrological drought may be a result of long-term meteorological droughts that result in the drying up of reservoirs, lakes, streams, and rivers, and a fall in groundwater level. Many others have also included economic or socioeconomic factors. Social drought relates to the impact of drought on human activities, including indirect as well as direct impacts. An

DOI: 10.1201/9781003276548-38

economic drought is "a meteorological anomaly or extreme event of intensity, duration (or both), outside the normal range of events that enterprises and public regulatory bodies have normally taken into account in their economic decisions and that, therefore, results in unanticipated (usually negative), impacts on production and the economy in general". The concept of drought varies from place to place depending upon normal climatic conditions, available water resources, agricultural practices, and the various socioeconomic activities of a region. The various approaches taken by scientists and nonscientists to define drought demonstrate its complex and interdisciplinary nature. At the same time, although most definitions emphasize the physical aspects of drought, the social aspects are closely related.

Drought thus has many definitions and meanings for different stakeholders. For farmers, it simply means a lack of adequate moisture for normal crop growth. Indian agriculture still largely depends upon monsoon rainfall where about two-thirds of the arable land lack irrigation facilities and is termed as rainfed. The effect is manifested in the shortfalls of agricultural production in drought years. History is replete with examples of a serious shortfall in cultivated areas and drop in agricultural productivity. For example, there once was such a severe shortage of food grains that the country had to resort to importing food grains to save poor people from hunger and starvation. However, India has since been able to build a buffer stock of foodgrains and the threat from droughts is not as serious as it used to be before the Green Revolution. It is worth mentioning here that the shortfall in agricultural production may be the direct impact of meteorological droughts, but the succeeding hydrological and agricultural droughts have a long-range and far-reaching impact on agriculture. This impact may be in the form of changes in the cropping patterns and impoverishment in cattle.

India's northern most state of Jammu and Kashmir (J&K) also faces the paradoxical situations of floods and droughts of moderate to severe degrees with environmental, economic, and social ramifications. Despite the state being surrounded by glaciers with perennial rivers, the region has a replete history of droughts. With climatic aberrations, the state experienced floods or flood-like situations and droughts alternatively of various degrees during the last one and half decades. Even in the current year (2020), the state faced severe drought with no rains from mid-April to the end of July, thus making the crucial crop growth period one of the worst drought seasons of the last 40 years. This chapter explores the droughts in the subsequent past and their impact on the overall economy, in general, and agriculture, in particular, along with mitigating strategies in the J&K.

38.2 PHYSIOGRAPHIC FEATURES OF THE STUDY AREA

The total geographical area of the erstwhile state of Jammu and Kashmir is 222,236 km^2 and extends between 32°17′ and37°″north parallels of latitude and 73°2″ and 80°-3″ east of meridians of longitudes and 81°east of Greenwich. The J&K is located almost in the middle of three climatic regimes of Asia. On its southern border lies the weak monsoon zone of Punjab. On the northeast, the state is bordered by the vast arid plateau of Tibet, while the northwest border areas face the eastern limits of the Mediterranean climatic region.

The J&K falls in the northwestern complex of the Himalayan ranges with marked relief variation, snow-capped summits, antecedent drainage, complex geological structure, and rich temperate flora and fauna. The state has mostly a mountainous area and occupies a central position in the continent of Asia. Out of 3.5 million ha of the mountainous area of India, nearly two-thirds(2.3 million ha) are found exclusively in J&K. In India, J&K lies in the extreme north of the Himalayas and constitute about 67.5%of the northwest Himalayan region. The state on the basis of physiography may be divided into three main divisions: (i) the outer Himalayas, which comprises of whole Jammu province; (ii) the lesser Himalayas, which comprises of whole Kashmir Valley; and (iii) the inner Himalayas, which comprises of whole Ladakh province. Agroclimatically, the state has been divided into four zones keeping in view the altitude, rainfall, temperature humidity, and topography. The salient features of the four agroclimatic zones are given in Table 38.1 and described briefly in the following sections.

TABLE 38.1

Physiographic Features of Different Agroclimatic Zones of J&K

	Jammu		Kashmir	Ladakh
Particulars	**Subtropical**	**Intermediate**	**Temperate**	**Cold Arid**
Geographical distribution	Jammu district, lower parts of Udhampur, Rajouri, Kathua, Poonch districts	Doda district; all outer hills of Jammu division; and parts of Poonch, Rajouri, Udhampur, and Kathua	All ten districts of Kashmir valley: Anantnag, Pulwama, Srinagar, Budgam, Baramulla, Kupwara, Bandipore, Kulgam, Shopian, Ganderbal	Two districts of Ladakh (Leh, Kargil)
Principal crops/fruits	Paddy, maize, wheat, oats	Maize, wheat, barley, paddy, oats, oilseeds	Paddy, maize, oilseeds, temperate fruits almond, saffron	Barley, wheat, alfalfa, apricot
Major livestock	Cross and local cow, buffalo, sheep and goat	Local cow, buffalo, crossbred cow	Crossbred and local cow, sheep and goat	Local and crossbred cow, yak, pashmina goat, sheep
Average land holdings (ha)	0.99	0.93	0.53	1.08
Net irrigated area (%)	36	10	62	100
Major rivers	Ravi, Tawi	Chenab	Jhelum	Indus, Shyok
Altitude (masl range)	300–1350	800–1500	2400–3000	3500–8400
Annual rainfall (mm)	1069	1649	789	83
Temperature (°C)				
Maximum	32.1	31.4	24.5	17.4
Minimum	13.6	11.5	1.2	−7.0
Thermal index	Mild	Mild	Cold	Very cold
Hydric index	Humid	Humid	Humid	Arid

38.2.1 Cold Arid Zone

The cold arid zone is confined mostly to the districts of Leh and Kargil of the Ladakh region. It covers the cold arid belt in the Indus Valley interwoven with the complex network of Himalayan ranges. It is one of the loftiest inhabited regions of the world. It is a mountainous country, mostly a desert of bare crags and granite dust with vast arid tableland at high elevation. The soils are coarse and sand content usually shows an increasing trend from Drass to Leh. Ladakh gets very little rainfall. The temperature in Ladakh falls below the freezing point for about five months in a year. Ladakh has the distinction of having the second coldest inhabited place (Drass) in the world. Because of an extremely cold climate, the plants become dormant during a major part of the year and the growth of the trees is stunted. The climate also affects the growth of crops like food grains and because of this, the major food crops in Ladakh are barley and other millets. However, some areas are now also covered by wheat. Most of the area in Ladakh is monocropped. No crop can be grown except where water for irrigation is available. Irrigation and the harsh climate are thus two limiting factors for crop husbandry. Some progress has, however, been made in the trench and protected vegetable cultivation. The extreme climate of Ladakh, however, has bestowed the region with unique livestock-based Karakul sheep and a pashmina goat–based production system along with apricots in fruits.

38.2.2   MID- TO HIGH-ALTITUDE TEMPERATE ZONE

The temperate zone constitutes the whole of the Kashmir Valley and the higher reaches of the Poonch and Doda districts. The altitude of the zone varies between 1500 and 2500 m above mean sea level. The normal precipitation is 650 mm. The Kashmir Valley experiences a moderate climate. The temperature during summer is much lower than that prevalent in the plains of India as well as in the plains of Jammu. The rains are experienced mostly during March–April and the valley is usually not affected by the summer monsoons to the extent experienced in the rest of the country. The winter is quite cold and is often divided into a few parts indicating the extremity of the winter. From March to October, the temperature remains between 21°C and 33°C. Whenever the temperature goes above 33°C, the valley often experiences rainfall bringing the temperature down. It is this period when the annual Amar Nath Ji Yatra is also held, which attracts lakhs of people from the whole of the country. With the fertile soil, and well-developed irrigation facilities (more than 50% of the net area sown is irrigated), the yield rates of major crops in the Kashmir Valley are much higher than those in the rest of the state. The valley has also created a niche in temperate fruits like apples, pears, walnut, cherry, peach, and plum. The area under horticulture has registered a substantial increase since the 1960s in the valley. The production of vegetables has also registered a lot of growth with the result that the valley is now able to supply vegetables to Jammu and other parts of the country during summer.

38.2.3   MID- TO HIGH-ALTITUDE INTERMEDIATE ZONE

The intermediate zone consists of sub mountainous soils and comprises parts of the districts of Doda, Poonch and Rajouri and Udhampur. The average altitude is about 750 m. It experiences snowfall in its higher reaches during winter. The mountains are cut by ravines and valleys. The area encompasses rugged terrains, steep slopes, and lofty hills with very little area available for cultivation. The zone has severely eroded soils. and these are shallow to moderately deep. The hilly region, mostly in the Jammu division, receives fairly good rainfall. Most of the land is undulating and only a limited area is irrigated. Quite a large part of the forest area is in this zone. Because of the nature of the soil, terrain, and other factors, the yield rate of agricultural crops is relatively low. The area is congenial for the development of horticulture, however, efforts have not been pursued on commercial lines mostly because of the difficulty of transportation of the horticultural products as well as the inferior quality of fruit trees in this region.

38.2.4   LOW-ALTITUDE SUBTROPICAL ZONE

The subtropical zone comprises the outer plains and outer hills of the Jammu province and includes the districts of Kathua, Jammu, and parts of Udhampur. The altitude ranges from 215 to 360 m above mean sea level. The normal annual precipitation is 1000 mm. Shiwalik and lower belt soils are observed in this zone. The plains of Jammu are quite similar to Punjab in terms of temperature and rainfall. This is the area that has been covered by irrigation systems on a large scale. In addition, the command area development program, which was originally started to cover the command of the Ravi-Tawi irrigation system, has been implemented in this part. The land is fertile and the yield rates are also quite high. This area produces more food grains than the demand of the local people and, therefore, has become a granary of the state. Most of the surplus food grains are sold to other areas of the state, and some of the food grains are even sent outside the state. The RS Pura Basmati is known for its quality. The plains of Jammu are irrigated by the Ravi-Tawi Irrigation Canal, Kathua Canal, and Ranbir Canal.

38.3   HISTORY OF DROUGHTS IN JAMMU AND KASHMIR

India is experiencing widespread drought conditions more frequently, thus making around 600 million Indians face extreme drought stress. The repercussions are in the form of crop failures and

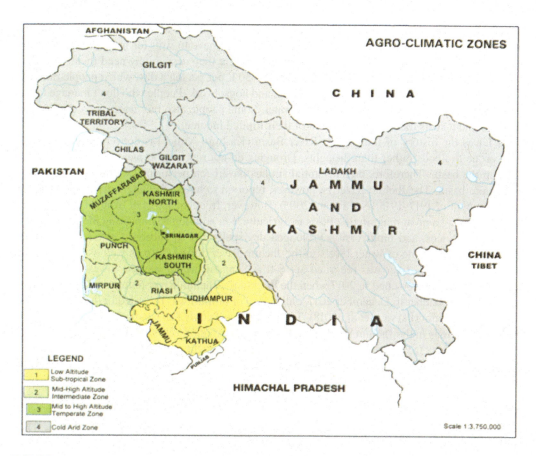

FIGURE 38.1 Agroclimatic Zones of Jammu and Kashmir (India)

struggling farmers resulting in rising cases of suicides in certain circumstances. It will be soon when 40%of country's population may have no access to drinking water. This is because groundwater is being depleted at an alarming rate; some 85%of the country's drinking water comes from aquifers, according to Water Aid. What is alarming is that this also has serious implications for India's health. Currently, nearly 200,000people die every year, due to scarce access to safe water. With 70%of India's water contaminated, the country ranks 120th among 122 countries in a global water quality index. To make matters worse, water levels in India's 90major reservoirs have fallen to 20% of the total capacity, since 2019. The effects of drought are more visible in rural areas than in urban areas. About 300,00 Indian farmers have killed themselves in the past 25 years. Many have deserted their crops and moved to cities in search of livelihood, leaving behind the elderly. Experts suggest that the present situation could also inflame water conflicts between farms, cities, and industries. In states, such as Maharashtra and Bihar, petty crimes, clashes, and even murders have happened over drought crises. The crimes that were taking place over water distribution were mainly between castes, communities, villages, and between urban and rural areas, due to inequitable water distribution. The water crisis disputes, in 2018, were almost double the number in 2017, according to National Crime Records Bureau. By the end of May 2019, 43% of India was experiencing drought, with failed monsoon rains seen as the primary reason. The country has seen widespread drought every year since 2015, with the exception of 2017. According to the Ministry of Agriculture, the frequent droughts in the last four years not only affected the kharif and rabi crop, but also destroyed kharif supplementary crops.

The state of water affairs of J&K is no different than the rest of the country despite having abundant water resources. The distribution of water resources across the state and its availability over the calendar year varies, thus making a large chunk of agriculture area as well as the population devoid of the resource. In rural areas, villagers sometimes have to wait for hours, in hot temperatures, before government tankers carry water trucks where they desperately need them. But, these trucks only provide one bucket, roughly equal to 25–30 L per person a day, which people ration for everything including drinking, cooking, bathing, and house work. People who don't manage to wait for the water supply tanker simply don't get water. Some people also buy water according to their finances, and it costs them a good amount of their limited income.

As far as the history of the Jhelum River Basin (Kashmir Valley) is concerned, the region is more prone to floods rather than droughts. Droughts do not represent typical extreme events that occur in the basin. Therefore, no substantial studies have been performed on drought analysis in the Jhelum basin. The Jhelum River is, however, reported to have completely dried up during the severe drought of 1917–1918 (Nagarajan, 2003). However, frequent spells of drought during recent years is creating a great magnitude of loss to agriculture in addition to impacting the environment, ecosystem services, and biodiversity. Prolonged dry spells and the absence of adequate rainfall resulted in the lowering of water levels in the Jhelum River during 2016 as well as 2017. In 2016, the water level in the river was recorded at –6 ft at Sangam, which was the lowest in the last 55 years. The situation worsened in 2017 when the level of the Jhelum River plunged to its lowest in 61 years. From September to December 2017, only 5 mm of rainfall was recorded in the valley against a normal of 100 mm (Parvaze et al., 2018). Thus, the present study has been undertaken to study the drought conditions in the Jhelum basin of J&K over a period of 38 years. In this study, the Standard Precipitation Index (SPI)has been selected to assess the drought condition in the Jhelum basin of Kashmir Valley. The SPI is the most commonly used drought index because of its simplicity, the ability to represent drought on multiple timescales, and is based on probability. The SPI for the basin was calculated on the basis of precipitation data from seven meteorological stations located across the basin.

Droughts were identified and characterized by Parvaze et al, (2018) using drought indices. Their study deals with the potential of using precipitation-based SPI to duplicate observed meteorological droughts in the Jhelum River Basin of Kashmir Valley. Historical droughts that occurred from 1980 to 2017 were examined by them. The SPI analysis shows a good agreement with the recorded historical drought events. The SPI was also used to study the time-based pattern of drought occurrence and its severity. The results of the SPI reveal that droughts in the Jhelum basin occurred during the years 1999, 2000, 2001, 2007, and 2016. The intensity of these droughts was found to be severe to extreme. The most prolonged droughts in the basin occurred during three successive years: 1999, 2000, and 2001. The SPI values were less than –1 for these years on 6-month, 9-month, and 12-month timescales. The 12-month SPI values for these years were –2.37, –2.39, and –2.46, respectively. However, the SPI values suggest moderate dryness in place of acute dryness during years of severe and extreme drought. Moreover, the SPI trendlines suggest a decreasing nature of the SPI on all scales. This implies that the frequency of droughts in the basin has increased during the last two decades. The decreasing trend of precipitation further increases the vulnerability of the basin to droughts.

38.3.1 DRIEST WINTER IN RECENT TIMES

The winter of 2017–2018 was one of the driest in recent memory. In January, a normally snowy month in the valley, Kashmir saw a nearly 90%drop in precipitation. The valley received 3.2 mm of rain and snowfall against the average of 65.3 mm for the month. It was followed by several dry months, which reduced water levels in the Jhelum River to 2ftin the valley. At Srinagar before the rain started, the water level was 1.94 ft against the normal average of 8ft. The winter was declared as one of the driest in the recent past, and the government was forced to declare an emergency with orders in place before the onset of the spring season to not plant paddy crops and other water-dependent irrigated agriculture in the ensuing summer season as the irrigation department would be in a position to supply assured water for crop growth.

A number of water supply schemes for household consumption also failed, which made the Public Health and Engineering (PHE) Department formulate a contingency plan to supply water through its fleet of tankers. The PHE Department stipulates a per capita supply of 135 L in the cities, 70 L in towns, and 40 L in villages. Rural Kashmir suffers most during summers, as the department currently supplies 75 million gallons a day (MGD) against a requirement of 85 MGD. The PHE Department declared scores of areas "drought affected" in the summer of 2018 and started rationing water supply to the entire valley. For the first time, the PHE Department had prepared a drought action plan to deal with the water crisis. The plan, which was formulated in coordination with all 24 PHE divisions of Kashmir, was estimated to cost Rs 700 crore. Under this plan, the need to upgrade the existing system of distribution and manage water resources was apriority.

38.3.2 Driest Summer in Recent Times

The state of J&K experienced its driest summer of the last 40 years in 2020. At 35.7°C, Srinagar recorded the hottest August day in 39 years (MeT, Srinagar), with 73%below normal rainfall between April to July, the peak crop growth period in which moisture availability determines the yield and quality of agricultural produce. The south Kashmir was severely affected with a 94%rainfall deficit. The rainfed areas are severely affected by such an unprecedented drought condition relative to irrigated agriculture where some supplemental canal and other sources of irrigation water is available. A number of paddy areas were found to harvest the standing unmatured crop in order to use it as fodder for cattle, as irrigation was not available due to drought in the summer 2020. The rainfed areas were most hit. Similarly, it also severely impacted the fruit industry both through quality as well as yield through deficient rainfall.

On the other hand, the state's PHE Department, which supplies water to households, faces problems with both water quality and availability. With low precipitation in 2020, some 39 water supply schemes in the valley have been hit. In north Kashmir's Baramulla district alone, 16 water supply schemes face severe shortages at the source. Similar problems have been reported in Srinagar, Budgam, Anantnag, Bandipora, Kulgam, Shopian, Kupwara, and Ganderbal.

Rising temperatures are threatening Kashmir's glaciers, the main source of water in the valley. These glaciers are shrinking as much as 15–18 ma year, and in the past 70 years Kashmir has lost three major glaciers. Data analysis of the impact of changing weather patterns in the Himalayan region showed that the glaciers in the Lidder Valley in south Kashmir have shrunk by 17%. The total glaciated area of nine benchmark glaciers had fallen from 29 km^{2}in 1980 to 23.81 km^{2}in 2013(Romshoo et al., 2020). The study revealed that smaller glaciers were more vulnerable to recession than major ones. If this trend of the recession continues into the next few decades, it may pose a serious threat to the availability of water for irrigation, hydropower, horticulture, and recreational use.

38.4 AGRICULTURE SECTOR IN THE STATE

Agriculture continues to be a major source of livelihood for the majority of people in J&K. The share of agriculture and animal husbandry in the net state domestic product (NSDP) has declined from 37.6%in 1980–1981 to 16.5%in 2018–2019. However, still, about 50% of the working population in the state is dependent on agriculture and allied activities for its living.

The changes in the cropping pattern presented in Table 38.2 showed that the area under different foodgrains has increased consistently over the years from 836 to 917 thousand hectares. Rice area has shown a decline of 30 thousand hectares from 1990–1991 to 2000–2001 and then onward its area grew again to 283 thousand hectares up to 2016–2017. However, the data reflected by the Statistical Digest with regard to the increase in rice area does not conform with the ground reality

TABLE 38.2

Shifts in Cropping Pattern of J&K (in thousand hectares)

Crop	1980–1981	1990–1991	2000–2001	2010–2011	2016–2017
Rice	264.0	274.5	244.1	261.4	283.4
Bajra	15.0	16.1	13.3	16.6	13.8
Maize	275.0	294.9	330.2	308.2	295.2
Wheat	202.0	245.1	281.0	290.7	290.3
Barely	12.0	8.1	8.9	13.3	6.7
Jowar/millets	19.0	13.5	14.9	4.55	9.1
Pulses	48.6	41.3	27.5	28.9	18.9
Foodgrains	835.6	893.5	919.9	923.65	917.4
Condiments and spices	1.9	1.4	2.1	2.4	2.4
Fruits and vegetables	51.1	60.4	67.3	87.2	132.07
Sugar plus other food crops	1.3	0.5	3.3	5.1	8.7
Total food crops	889.8	955.9	992.5	1018.4	1060.7
Oilseeds	52.7	67.6	73.9	64.6	54.7
Fibers	1.6	0.8	0.4	0.2	0.03
Dyes and tanning material	0.2	4.0	2.8	2.7	3.7
Drugs, narcotics, and plantation crops	1.0	0.1	0.1	0.1	0.02
Fodder crops	25.0	37.9	43.9	21.8	45.8
Other nonfood crops	2.9	0.5	1.5	32.1	11.9
Total nonfood	83.4	110.8	122.5	121.5	116.2
Total area sown	973.1	1066.7	1115.0	1139.8	1177.1
Net sown area	715.3	730.7	748.3	731.6	757.03
Cropping intensity (%)	**136.0**	**146.0**	**149.0**	**156**	**155**

Source: Digest of Statistics, (various issues), DE & S, Govt. of Jammu and Kashmir.

as vast areas of paddy lands were either converted to orchards or came under urbanization and other infrastructure projects. The area under maize, wheat, and bajra, however, has significantly increased over the period with a slight decline during the last decade. The area under pulses, jowar, and other cereals and millets witnessed a decrease of varying degree since 1980s. Among food crops, there has been an expansion of 81 thousand hectare area under fruits and vegetables. The oilseed crops area has remained almost stagnant in the state over the years. Figures documented in the table revealed that the area available for cultivation in the state has increased with multiple cropping and has intensified the cropping system of the state.

38.4.1 IRRIGATION DEVELOPMENT

Irrigation water is one of critical inputs required for better performance of crops. Robustness of irrigation infrastructure and storage capacity in terms of water bodies, tanks, reservoirs, sars, on-farm ponds, and other water harvesting and conservation structures act as drought-proofing and mitigating assets. It increases the land use/cropping intensity, provides incentives for use of more inputs, and thus results in higher agricultural output. The proportion of cultivated area irrigated in the state has remained almost stagnant at 42% of the net area sown in J&K. Regarding sources of irrigation, canal irrigation constituted around 90% of net irrigated area in J&K, though its proportion has shown a little decline over the years (Table 38.3). Tanks and other sources of irrigation are gradually gaining importance and would yet be more prominent in climate change regimes. The stagnation of irrigation capacities coupled with the frequent advent of drought hampered productivity gains

TABLE 38.3

Area Irrigated by Source (in thousand hectares)

Source	1974–1975	2006–2007	2007–2008	2008–2009	2009–2010	2010–2011	2011–2012	2016–2017
Canals	279	286.64	285.78	287.77	289.80	288.48	285.40	285.35
	(94.58)	(92.64)	(92.77)	(91.72)	(90.77)	(91.88)	(89.40)	(87.78)
Tanks	Neg.	4.24	4.21	4.85	5.11	5.02	7.10	8.02
		(1.37)	(1.37)	(1.55)	(1.60)	(1.60)	(2.22)	(2.47)
Wells	3.00	1.04	0.99	3.80	4.33	6.21	7.42	10.42
	(1.02)	(0.34)	(0.32)	(1.21)	(1.36)	(1.98)	(2.32)	(3.20)
Other sources	13.00	17.52	17.05	17.32	20.03	14.28	19.33	21.29
	(4.41)	(5.66)	(5.53)	(5.52)	(6.27)	(4.54)	(6.06)	(6.55)
Total	**295**	**309.4**	**308.04**	**313.74**	**319.27**	**313.99**	**319.26**	**325.08**

Source: Financial Commissioner (Rev.), J&K.

Note: Figures in brackets indicate percentage.

that may otherwise accrue due to the adoption of water-intensive technologies. Although, the state has abundant surface water in the form of perennial rivers, the availability of water for irrigation at the critical stages of crop growth is an issue. During peak summers with low rainfall seasons, there is tough competition among the users, and tail-end farmers sufferer most during such hard times. Most of the canal irrigation schemes in the state are gravity systems, but a sizable portion of them is lifted irrigation schemes where water is drawn from the river and put in the canal that flows through gravity system. The functioning of these lift irrigation schemes is hampered due to many reasons, which ultimately affect the crop production system.

38.4.2   PRODUCTION OF FOOD GRAINS

Foodgrain production of a state or region defines its food security status at the regional level. Table 38.4 shows the region-wise area, production, and yield of major field crops in J&K state. Rice is grown in the Jammu and Kashmir regions only, with no area in the Ladakh region. The Jammu region has a major (53%) area under rice, whereas the Kashmir region with a lesser proportion (37%) under area commands a major contribution (62.6%) to the total state rice, which can be mainly attributed to relatively higher yields (24.96 q/h) than the Jammu region (13.32 q/h). In the case of maize, both area and production are majorly contributed by the Jammu region. On the yield front also, the Jammu region has an edge over the Kashmir region. Wheat is mostly confined to the

TABLE 38.4

Region-Wise Area, Production, and Yield of Major Field Crops in J&K State

Crop		Jammu	Kashmir	Ladakh	J&K
Rice	Area (ha)	160848	143655	—	304503
		(52.8)	(47.2)		(100)
	Production (qtls)	2143000	3585000	—	5725000
		(37.4)	(62.6)		(100)
	Yield (q/h)	13.32	24.96	0.00	18.80
Maize	Area (ha)	210699	83159	—	293858
		(71.7)	(28.3)		(100)
	Production (qtls)	4336000	1075000	—	5411000
		(80.0)	(20.0)		(100)
	Yield (q/h)	20.58	12.93	—	18.41
Wheat	Area (ha)	277013	1069	3788	281870
		(98.2)	(0.4)	(1.4)	(100)
	Production (qtls)	5422000	24000	39000	5485000
		(98.8)	(0.4)	(0.8)	(100)
	Yield (q/h)	19.57	22.45	10.30	19.46
Fodder	Area (ha)	16988	23150	5732	45870
		(37.0)	(50.5)	(12.5)	(100)
	Production (qtls)	—	—	—	—
	Yield (q/h)	—	—	—	—
Oilseeds	Area (ha)	14971	39646	97	54714
		(27.4)	(72.5)	(0.1)	(100)
	Production (qtls)	—	—	—	285000
	Yield (q/h)	—	—	—	5.21

Source: Digest of Statistics, (2016-17), DE & S, Govt. of Jammu and Kashmir.
Note: Figures in parentheses represent percentages to total.

Jammu region with a very minuscule area under the Ladakh region. In Ladakh, wheat is grown as a kharif crop. Fodder and oilseed crop areas majorly fall in the Kashmir region. Ladakh has also a sizable area under fodder crops, particularly alfalfa, to augment the livestock economy of the region.

38.4.3 PERFORMANCE OF HORTICULTURE SECTOR

J&K state has natural niches in production of quality fruits and has been exclusively considered a horticultural state. Accordingly, the state has made rapid strides in area expansion and production of fruits. The major fruits grown in the state include apple, pear, cherry, walnut, and almond. The other fruit crops grown in the state include fruits such as peach, pear, plum, and apricot, as well as sub-tropical fruit crops such as mango, citrus, and litchi. Area under all fruits in the state has increased more than four times from 82.5 thousand hectares in 1974–1975 to 338.53 thousand hectares by 2016–2017 (Table 38.5). The corresponding production has gone up ten times from 2.16 lakh metric tonnes to 22.3 lakh metric tonnes during the same period. However, the large gains in this production have come from the area as the productivity increased marginally from 2.6 metric tonnes per hectare in 1974–1975 to 6.6 metric tonnes per hectare in 2016–2017. The share of apples to the total area under fruits and consequent production was 46.2 thousand hectares and 190.5 metric tonnes, respectively, in 1974–1975. The percent share of different fruits in the area and production show that over 56%of area and 88%of the total fruit production was contributed by apple, which was an important fruit crop of the state in 1974–1975. In the same year, walnut and almond together constituted over 27%of the total fruit area, but contributed around 5%to total fruit production in the state. Although the area and production of all crops increased over years in absolute terms, their

TABLE 38.5
Area, Production, and Yield of Different Fruits in J&K

Fruit		1974–1975	1980–1981	1990–1991	2000–2001	2016–2017
Apple	Area (000 ha)	46.2	60.3	68.7	88.1	162.97
	Production (000mt)	190.5	536.3	658.2	757.6	1726.83
	Yield (mt/ha)	4.1	8.9	9.6	8.6	10.59
Pear	Area (000 ha)	2.3	5.5	8.1	9.2	14.53
	Production (000mt)	7.7	3.2	16.7	31.3	88.33
	Yield (mt/ha)	3.3	0.6	2.1	3.4	6.08
Cherry	Area (000 ha)	0.7	1.1	1.4	2.4	2.83
	Production (000mt)	0.5	0.5	4.2	5.3	8.28
	Yield (mt/ha)	0.8	0.5	2.9	2.2	2.93
Walnut	Area (000 ha)	13.2	26.7	40.9	59.8	89.34
	Production (000mt)	10.1	15.0	38.6	83.4	266.28
	Yield (mt/ha)	0.8	0.6	0.9	1.4	2.98
Almond	Area (000 ha)	9.4	16.3	19.2	18.1	7.11
	Production (000mt)	1.5	1.9	2.2	10.9	6.26
	Yield (mt/ha)	0.2	0.1	0.1	0.6	0.88
Others	Area (000 ha)	10.7	21.1	37.9	41.4	61.75
	Production (000mt)	5.5	6.1	50.2	49.6	142.0
	Yield (mt/ha)	0.5	0.3	1.3	1.2	2.30
Total	Area (000 ha)	82.5	131.0	176.3	219.0	338.53
	Production (000mt)	216.2	563.0	769.9	938.1	2234.98
	Yield (mt/ha)	2.6	4.3	4.4	4.3	6.6

Source: Digest of Statistics, (various issues), DE & S, Govt. of Jammu and Kashmir.

proportion in total fruit area and production in the state has shown a varying pattern. Presently, apple has a major share both in area (48%) and production (77%) to that of total fruit.

38.5 IMPACT

Drought affects various components of ecosystems and the environment. The availability of water level in the reservoirs, amount of rainfall, soil moisture, and groundwater depths are some of the environmental indicators of drought. There may be permanent and temporal impacts on biodiversity depending on the duration, intensity and scale of drought. Reduction in water may alter the food supply to different life forms, which may subsequently alter the food web. The drought brings a food crisis, which has a number of social and economic ramifications. It affects people's health and safety, in addition inviting conflicts regarding the use of water resources. Drought impacts the economic conditions of nations at a local to global scale. It may affect not only the people living in the drought-stricken areas but also those who live outside the drought areas, as they may also depend on drought-stricken areas for their livelihoods. The agriculture sector is most influenced by drought.

Apparently, drought affects the agriculture sector the most, though it also has serious implications on the environment, ecology, ecosystem services, biodiversity, and carrying capacity of natural resources that support the agroecosystems. Through agriculture, it affects the agricultural value chains, the industries dependent on agriculture supplies, and employment. This results in lower farm incomes, thereby deafening the purchasing power and effective demand, impacting the overall economy of the region/country.

The long-term impact of continuous droughts from 1999 to 2004 in Kashmir Valley resulted in changes by farmers switching from paddy to apple cultivation. So robust conversion of paddy areas to fruit crops by farmers during the early 2020s was due to the cumulative effect of both climate change as well as market forces (as apple gives better returns per unit land area than paddy). In order to assess the impact of the summer drought of 2020 on the agriculture/ farm economy, a Rural Rapid Appraisal was conducted in all zones (North, South, and Central) of the Kashmir Valley taking into consideration both irrigated and rainfed areas at different altitudes in order to better dissect the impact of deficient rainfall, which varied from 54% to 94% across the Kashmir Valley.()

The farmers were asked about the effect of drought on their crops. The loss was observed on account of both yields, as well as the quality of the products across all major crops in the Kashmir Valley. However, the magnitude of loss was found to vary from place to place and crop to crop. The rainfed areas usually with maize and pulses as mixed cropping systems in higher altitudes were most affected. The loss in terms of value was more observed in the case of fruit crops, particularly apples due to their high value of nature.

38.6 CLIMATE CHANGE, FARMERS PERCEPTION, AND ADAPTIVE MEASURES

Climate change has added an important dimension to the management aspect of the agriculture sector with regard to varietal adoption, technology use, resource use options, and management practices. Farmers have evolved and strategized their own adaptation and mitigation strategies over time through their experience and trails, as well as the information from extension agencies. Farmers' perceptions with regard to climate change and its anticipated impact/effect is presented in Table 38.7. Farmers say they are experiencing an increase in temperature. Likewise, the occurrence of floods as well as droughts and the incidence of pests/diseases have increased and are now becoming a common phenomenon with the Kashmir Valley context. In the recent past, some years even experienced paradoxical situations of drought and floods in the same year. The changes in climate variables are also increasing the incidence of disease and pests, and farmers are incurring more costs to manage it. Most of the sampled farmers were found affected by these climate-induced changes and are being affected in terms of crop loss, livestock diseases, etc. as shown in Table 38.7.

TABLE 38.6

Extent of Loss in Yield by Farmers Due to Drought in 2020 in Kashmir

Crop/Fruit	South Kashmir		Central Kashmir		North Kashmir	
	Yield Loss (%)	Quality Loss (%)	Yield Loss (%)	Quality Loss (%)	Yield Loss (%)	Quality Loss (%)
Rice	33	27	37	25	41	25
Maize	78	55	63	47	60	43
Pulses	65	43	57	40	62	38
Fodder	56	32	46	28	62	37
Apple	38	42	33	37	38	37
Pear	35	27	39	27	43	35
Cherry	25	15	20	27	30	25
Walnut	40	37	43	40	45	40
Almond	42	33	45	32	38	27

Source: Rapid Rural Appraisal Survey 2020.

TABLE 38.7

Perception about Climate Change and Effect on Farm Households

Particulars	Have These Things Changed in the Last 10 Years			Are You Affected by This in the Last 10 Years	
	Increased (%)	Decreased (%)	No Change (%)	Yes	%
Temperature	90.14	1.81	8.06	640	88.89
Precipitation	55.0	35.28	9.72	642	89.17
Drought	75.83	0.28	23.89	544	75.56
Flood	62.08	10.56	27.36	425	59.03
Pest/disease	67.36	0.56	32.08	427	59.31

Source: Field Survey Data, 2018–2019 (IFPRI Project).

Adaptation is an important tool to mitigate the climate risks, and range from adjusting the timing of cultural operations to resource use as well as switching to off-farm avenues. The mitigation strategies practiced/adopted by farmers to reduce climate risks are depicted in Table 38.8. Farmers were found to rely more on weather advisories, as they are an important information tool adopted in the region with regard to decisions about farming, particularly in the case of apples. About 46% of farmers have also switched their livelihood options toward labor work and other sources of income. Changes in resource use as well as the timing of cultural operations have also emerged as important adaptation measures as perceived by the farmers.

With advancements in science and technology as well as the economy, information and associated technologies are becoming famous with an easy approach to cater to the needs of stakeholders/beneficiaries. The majority of farmers were found to benefit from information shared on radio, TV, newspapers, and the internet as 88% of farmers used these sources (Table 38.9). However, public sector agencies, viz., agriculture and allied departments, Sher-e-Kashmir University of Agricultural Sciences and Technology–Kashmir (SKUAST-K) and its associated research stations, KVKs, and colleges are making efforts through demonstrations, FLDs, farm camps, participatory breeding programs, and farmer fairs to share information about the improved technologies and varieties,

TABLE 38.8

Farm Adaptation Strategies to Reduce Climate Risks

Adaptation Strategies to Reduce Climate Risks	Yes	Percent
Change timing	177	24.58
Change crop	144	20.00
Introduce/change varieties	158	21.94
Crop Insurance	101	14.03
Keep land fallow	140	19.44
Leased-out	45	6.25
Increase the use of fertilizers	345	47.92
Increase the use of pesticides	273	37.92
Increase ground water irrigation	7	0.97
Seasonal migration for job	130	18.06
Labor work/secondary occupation	335	46.53
Increase in livestock	164	22.78
Weather advisory	611	84.86

Source: Field Survey Data, 2018–2019 (IFPRI Project).

TABLE 38.9

Access to Technical Advice by Farmers

Source of Technical Advice	Yes	Percent
Public sector extension agencies	470	65.73
Private commercial agents	28	3.89
Progressive farmers	96	13.33
Radio/TV/newspaper/internet	637	88.47
Veterinary department	242	33.61
NGO	28	3.89

Source: Field Survey Data, 2018–2019 (IFPRI Project).

as perceived by farmers. But there is a lot of scope to further widen the outreach of public sector agencies in order to improve the livelihoods of small farmers through these newer and improved technologies.

38.7   CONCLUSION AND POLICY IMPLICATION

Climate change is a reality and with every passing year, the severity as well as intensity of extreme events in the form of weather aberrations are going to occur. The pace with which the states are responding to this reality and on-ground preparedness is lacking. There should be a multipronged strategy in dealing with climate-induced changes, particularly droughts. The government must strengthen and increase its outreach of the national flagship program of watershed management in the state, as it has proved the best bet program for drought-proofing (Shaheen et al., 2010). The state government should also provide incentives to farmers through various development programs for on-farm water conservation structures and practices. Moreover, the government, through the rural development department, should focus on creating water infrastructure in terms of rejuvenating tanks, sars, community ponds, springs, etc. The investment in water infrastructure projects, particularly at local village levels throughout the length and breadth of the state will be a great strategy for drought-proofing and building resilience in communities to combat the adverse effects of climate change.

The future strategy should be more focused on improving climate-resilient crops and technologies in order to sustain the food and nutritional security of the region. Intensive efforts at the research and development level are the need of the hour for coming up with location-specific crop technologies. Consumer preference, farmers' choice, and market demand, economic viability, and conserving local gene pool diversity should also figure in when promoting improved crop variety schemes and programs.

Outreach of research and development institutes and development departments has a lot of scope to improve the livelihood of farming communities in J&K, as a good number of farmers are ignorant about the new and emerging climate-resilient technologies and farm practices. Robust and farmer-centric agricultural extension policies should create a favorable environment for collaboration and provision of services to rural people through pluralistic extension approaches in order to adopt and acclimatize the farming community with climate-resilient improved crop varieties and management practices in addition toother *in situ* water conservation practices.

REFERENCES

Nagarajan, R. (2003), *Drought: Assessment, Monitoring, Management and Resource Conservation*, Capital Publishing Company, New Delhi.

Parvaze, S., Parvaze, S. and Ahmad, L. (2018), Meteorological drought quantification with standardized precipitation index for Jhelum Basin in Kashmir Valley, *International Journal of Advance Research in Science and Engineering*, 7(4):688–697.

Romshoo, S. A; Fayaz, M., Mehraj, G. and Bahuguna, I. (2020), Satellite-observed glacier recession in the Kashmir Himalaya, India from 1980 to 2018, *Environmental Monitoring and Assessment*, 192(597):1–17.

Shaheen, F. A.; Wani, M. H. and Baba, S. H. (2010), Sustainable hill agricultural practices through watershed development programs and their impact in Himalayan States, *Indian Journal of Agricultural Economics*, 65(3):344–366.

39 Different *In Situ* Moisture Conservation Options in Rainfed Agroecologies of Odisha

S.K. Behera and D.K. Bastia

CONTENTS

DOI: 10.1201/9781003276548-39

39.1   INTRODUCTION

Uneven and erratic rainfall in arainfed area creates moisture stress conditions during critical growth stages of crops, resulting in severe yield reduction. Even when the rainfall is high, it is often lost as runoff, when the surface of the soil is not suitably formed. Moisture conservation, therefore, plays a key role in successful crop production in the rainfed agroecologies of Odisha, a state in eastern India.

Aberrant rainfall in time and space, often spanning over two weeks or more, coupled with a low water-retention capacity of soil results in moisture stress conditions that culminate in sparse crop stand and hence productivity. Thus, the major constraint for establishing a crop in such areas is the lack of adequate moisture in the root zone (Hadda et al., 2000; Bhatt et al., 2004). The area, therefore, requires the adoption of location-specific *in situ* soil moisture conservation technologies by which the area could be ecologically rehabilitated and its production potential could be realized on a sustained basis. The collection of rainwater directly in the field where it falls is generally termed *in situ* moisture conservation. The collection can be localized in the crop land itself, usually by reshaping the soil surface. The water will be stored in or close to the root zone, ensuring a high degree of utilization. The techniques describe in this section involve localized surface storage and require some reshaping of the soil surface, and hence require substantial inputs of energy. In many cases, it is essential to use human labor, and animal-drawn and tractor-drawn implements. There are different methods used for *in situ* rainwater conservation (Katiyal et al., 1995; Gouda, 2011). Summer plowing, plowing across the slope, increasing bund height, broad bed furrow system and ridge furrow system, organic mulching, cropping system, alternate land use system and use of bulky organic manures are the most important practices, among various cultural, mechanical, and agronomic measuresthathave been reported to reduce soil erosion and increase *in situ* soil moisture storage and improve the productivity of crops (Hadda et al., 2005; Bhatt et al., 2004).

In the drylands of Odisha, the practice of sole cropping is predominant but is risky and often results in low yields or sometimes even crop failure due to erratic monsoon rainfall. In such areas, intercropping is a feasible option to harness efficient utilization of natural resources both spatially and temporally besides minimizing risk in crop production, ensuring reasonable returns at least from the intercrop, and improving soil fertility as well. Intercropping of complementary and supplementary crops is more profitable and is a key drought coping strategy (Behera et al., 2012; Behera et al., 2005). This chapter elaborates on efficient and adoptable *in situ* moisture conservation techniques and suitable cropping systems appropriate for rainfed agroecologies in Odisha.

39.2   DIFFERENT *IN SITU* MOISTURE CONSERVATION OPTIONS IN RAINFED SITUATIONS

Different *in situ* moisture conservation methods in rainfed conditions are essential for *in situ* conservation of soil and water. The main aim of these practices is to reduce or prevent either water erosion or wind erosion, while achieving the desired moisture for sustainable production. The suitability of any *in situ* soil and water management practices depends greatly upon the soil, topography, climate, cropping system, and farmers' resources. Based on past experiences, several *in situ* soil and water conservation measures have been found promising for the various rainfall zones in India (Table 39.1).

39.2.1   SUMMER PLOWING

Summer plowing is defined as plowing the field across the slope during the hot summer with the help of specialized tools. The primary objective is opening the soil crust accompanied by deep plowing and simultaneously overturning of the soil underneath to disinfect it with the help of piercing sun rays. Usually, after the harvest of rabi crops the field remains fallow until the arrival of

TABLE 39.1

Different *In Situ* Soil Moisture Conservation Practices for Various Rainfall Zones of India

	Seasonal Rainfall		
<500 mm	500–750 mm	750–1000 mm	>1000 mm
• Contour cultivation or cultivation across the slope with conservation furrows • Mulching • Summer tillage	• Contour cultivation or cultivation across the slope with conservation furrows • Mulching • Summer tillage • Broad bed furrow (BBF) system • Field bund or increase bund height	• Broad bed furrow (BBF) system • Conservation furrow system • Sowing across the slope • Summer tillage • Field bund or increase bund height • Vegetative bund or barriers	• Broad bed furrow (BBF) system • Ridge furrow (RF) system • Field bund or increase bund height • Vegetative bund or barriers

monsoon. About 15 days before the arrival of monsoon, farmers begin land preparation by spreading the farm yard manure (FYM) and plowing the soil to mix the FYM in soil. Thereafter, the sowing or transplanting starts. There is hardly any time gap in which the plowed soil gets exposed to sharp rays of hot sun. But summer plowing is an operation of plowing that is done well ahead of the arrival of monsoon, in hot summer (April and May) with a specific purpose. A moldboard plow is the most appropriate tool for opening the hard crust of the soil thathas been left fallow from the harvest of the rabi crop.

39.2.1.1 Benefits of Summer Plowing

The major benefits of summer plowing are as follows:

(i) The summer plowing breaks the hard crusted upper layer of the soil, and the deep plowing-increases the infiltration capacity and permeability of the soil as well as increases *in situ* moisture conservation. Consequently, plant roots get more moisture with less effort.

(ii) Summer plowing improves soil structure due to alternate drying and cooling.

(iii) Tillage improves soil aeration, which helps in the multiplication of microorganisms. Organic matter decomposition is hastened, resulting in higher nutrient availability to the plants.

(iv) Since the capacity to absorb rainwater increases, atmospheric nitrate mixed with water enters the soil, and it increases soil fertility and yield of the crops.

(v) Multitudes of insects and pests hibernate underneath the soil crust or stubbles during the hot summer season. Due to the overturning of the soil in summer plowing, the sharp rays of the sun enter the soil and kill the eggs, larvae, and pupae of soilborne insects and pests, thereby the hazards of insects and pests on a subsequent crop is reduced. Consequently, the farmer's expenditure in procuring insecticides and pesticides decreases.

(vi) A lot of harmful bacteria spores and fungal microbes die due to exposure to the heat of summer. Farmers can get relief from purchasing fungicides and pesticides because of the inhibition of plant diseases due to summer plowing.

(vii) Deep plowing and overturning uproots the weeds. Consequently, the roots and stems of the weeds become desiccated and die. As a result, weed control and less weedicide applications are one of the major advantages of summer plowing. As a result of it, competition between the crops and weeds for the same plant nutrients is reduced, thereby the productivity increases.

39.2.1.2   Limitations of Summer Plowing

(i)   Summer plowing is not advisable in sandy or excessive wet soils. After the rain, the farmer has to wait until proper soil condition (tilth) is obtained. The soil after summer plowing should be left as clods.

(ii)   Repeated plowing should not be done because it will increase the cost of cultivation and destroy the soil structure, and the soil will be prone to erosion by air in summer season.

(iii)   The operation of summer plowing should be done properly and timely so that it will reduce the operation costs of applying insecticides, pesticides, and weedicides, and saving in fertilizers and manures. Subsequently it will increase the crop yield and maintain the fertility of the soil.

39.2.1.3   Case Study

The case study of Phulbani block of the Kandhamal district in Odisha state on summer plowing is given in Table 39.2. The operation of summer plowing was tested for the rice crop and maize + cowpea (2:2) intercropping system. The increase in yield and rainwater-use efficiency (RWUE) of the rice crop and maize + cowpea (2:2) intercropping system due to summer plowing is summarized. The increase in yield of 26%–28% was found to be the case of the rice crop, whereas an increase in yield of 15% was found in the case of the maize + cowpea (2:2) intercropping system. The RWUE of rice was found to be 2.43 and 2.53 kg/ha-mm, whereas for the maize + cowpea (2:2) intercropping system the RWUE was 4.50 kg/ha-mm.

39.2.2   Plant Establishment across the Slope

The property of water is to flow down the slope. The greater the slope, the greater the velocity of runoff, and hence the higher energy of water to scour and carry the valuable and nutritive top soil. Similarly, the longer the length of a sloped land, the greater the velocity, and hence the energy of flow to carry soil particles. If the length of a slope is doubled, the energy increase varies with the square of the increase of length. It results in two major losses to the farmers of this region. First, water has less time to seep into the soil. Second, the top soil containing organic matter and applied nutrients is eroded, thereby the farmer wastes money in inputs and has reduced production as compared to expectations. These problems can be solved if the length of run in the sloped land is reduced and some obstruction is placed before the flow to increase its time to infiltrate the soil. This can be achieved by plowing the sloped land across the slope and sowing or planting crop across the slope.

TABLE 39.2

Effect of *In Situ* Moisture Conservation Practice of Summer Plowing on Yield and RWUE of Rice and Maize + Cowpea Intercropping System

| Crop | Variety (duration) | Yield (kg/ha) | | % Increase in Yield | RWUE (kg/ha-mm) |
		With Intervention	Without Intervention		
Direct seeded rice	Sahabhagi (110 days)	2420	1920	26.04	2.43
	Naveen (125 days)	2540	1980	28.28	2.55
Maize + Cowpea (2:2)	Maize (SA 701) + Cowpea (Gomti)	MEY: 4480 kg/ha	MEY: 3880 kg/ha	15.46	4.50

MEY, maize equivalent yield.

39.2.2.1 Benefits

The main benefits of plowing across the slope are as follows:

(i) It increases moisture conservation in the soil.
(ii) It prevents washout of fertilizers and plant nutrients applied by the farmer.
(iii) Production of crops increases because of increased availability of water and plant nutrients.
(iv) Rabi pulses can be grown by pyra-cropping with residual moisture available in the soil.
(v) Energy of flow water is reduced by this method of cultivation, as well as erosion of top soil is drastically reduced.
(vi) With multiple cropping, the net economic returns of the farmer will be enhanced.

39.2.2.2 Management Options

There may be certain difficulties in plowing the soil with heavy machines at higher slopes. Power tillers may be given preference over tractors at these places. The depth of plowing should be as deep as possible depending upon the availability of soil depth. Ridge and furrow methods of cultivation should be given priority over flat beds. The ridges and furrows should run across the slope irrespective of peripheral bunds. Sowing/planting should be done in furrows for more water and nutrient availability.

39.2.2.3 Case Study

Maize, pigeonpea, and cowpea were cultivated across the slope in the Phulbani block of the Kandhamal district in Odisha. Increases in yield of the crops are well documented due to cultivation across the slope rather than cultivation along the slope. The yield increase is attributed to the availability of soil water, retention of organic matter, and applied nutrients in the cultivated soil across the slope (Table 39.3).

39.2.3 INCREASE IN BUND HEIGHT

Bunds are among the most common techniques used in agriculture to collect surface runoff, increase water infiltration, and prevent soil erosion. It is recommended for agricultural fields with permeable soils and land slopes of <6%. Bunding is the construction of small embankments across the slope of the land to prevent erosion and promote better utilization of rainwater. All through the years, field bunding has proved to be one of the most effective interventions in conserving moisture, and in the long term, reducing land degradation. Field bunding has a lot of advantages in terms of conserving

TABLE 39.3

Effect of *In Situ* Moisture Conservation Practice of Plant Establishment across the Slope on Yield and RWUE of Different Crops

Crop and Variety	Water Content (w/w)		Yield (kg/ha)		% Increase in Yield	RWUE (kg/ha-mm)
	During Saturation	At 5 Days of Dry Spell	With Intervention	Without Intervention		
Maize (SA 701)	15.2	9.4	2580	1980	30.3	2.59
Pigeonpea (NTL 724)	15.3	9.0	880	610	44.26	0.88
Cowpea (Gomti)	15.2	10.2	2870	1860	54.3	4.42

moisture content. But, due to improper design criteria and lack of proper maintenance, it sometimes does not solve the actual problem. Therefore, it is necessary to plan, design, and maintain the field bunding properly to get the best results from it. The main advantages of field bunding and an increase in bund height are to reduce the velocity of the runoff water and allow less water to flow down at non-erosive velocity. It also allows more water to be absorbed in the soil profile for crop use and prevents nutrient loss from the soil.

39.2.3.1 Case Study

Bund height of the rice field has been increased for the study in the Phulbani block of the Kandhamal district of Odisha. With the increase in bund height, an increase in the yield of rice up to 25%–27% was recorded (Table 39.4).

39.2.4 Broad Bed Furrow System and Ridge Furrow System

39.2.4.1 Broad Bed Furrow System

The broad bed furrow (BBF) system consists of broad beds of about 100 cm wide separated by sunken furrows of about 50 cm wide. The depth of the furrow is about 15 cm or more. The preferred slope along the furrow is between 0.4% and 0.8% on vertisols (heavy, black, clay soils, sometimes called cotton soils). Two to four rows of a crop can be grown on the broad bed. The bed width and crop geometry can be varied to suit the cultivation and planting equipment. It encourages moisture storage in the soil profile. Deep vertisols may have soil moisture storage up to 250 mm, which is sufficient to support plants through mid-season or late-season dry spells.

39.2.4.1.1 Advantages of Broad Bed Furrow System

Advantages of the BBF system include:

(i) It helps in safely disposing of the surplus surface runoff without causing soil erosion.
(ii) It provides more soil aeration for better plant growth.
(iii) It provides better drained and more easily cultivated soil in the beds. It also encourages double cropping by means of intercropping or sequential cropping.
(iv) The BBF system is particularly suitable on deep black soils in areas with dependable rainfall averaging 750 mm or more. It has not been as productive in areas of less dependable rainfall, or on alfisols or shallower black soils, although in the latter case more productivity is achieved than with traditional farming methods.
(v) Other methods, with more emphasis on storage and irrigation within a package that includes BBF, are more likely to be viable for alfisols. It is recommended that the BBF

TABLE 39.4

Effect of *In Situ* Moisture Conservation Practice of Increase in Bund Height on Yield of Rice Crop

Crop	Variety (duration)	Yield (kg/ha) With Intervention	Without Intervention	% Increase in Yield	RWUE (kg/ha-mm)
Rice	Sahabhagi (110 days)	2400	1910	25.65	1.90
	Naveen (125 days)	2510	I1970	I27Y41	1.98

system should not be considered in isolation, but only as part of an improved farming systems package.

39.2.4.2 Ridge and Furrow System

The ridge and furrow (RF) system is used in row crops with furrows developed between the crop rows. The size and shape of the furrow depend on the crop grown, the equipment used, and the spacing between crop rows. Water is applied by running small streams in furrows between the crop rows. Water infiltrates the soil and spreads laterally to irrigate the areas between the furrows The length of time the water is to flow in the furrows depends on the amount of water required to replenish the root zone, the infiltration rate of the soil, and the rate of lateral spread of water in the soil. Both large and small irrigation streams can be used by adjusting the number of furrows irrigated at any one time to suit the available flow. In areas where surface drainage is necessary, the furrows can be used to rapidly dispose of the runoff from the rainfall. The ridge and furrow system of irrigation is suitable to most soils, except sands that have a very high infiltration rate and provide poor lateral distribution of water between furrows. The ridge furrow system helps to conserve water when there is lower rainfall by holding water in ridges and when there is higher rainfall. This helps to drain the excess water, thus avoiding the water logging situation as depicted by the periodic soil moisture data.

39.2.4.3 Case Study

A study was conducted for cauliflower, tomato, and okra crops with the BBF and RF system at Phulbani block of the Kandhamal district in Odisha. Results show that the RF system is more suitable than the BBF system in rainfed situations with the red lateritic soils of Odisha. The water content during dry spells is more in the case of the RF system than the BBF system in all vegetable crops (Table 39.5).

39.2.5 CONSERVATION FURROW SYSTEM

The conservation furrow is a simple and low cost *in situ* soil and water conservation practice for rainfed areas with moderate slopes. The conservation furrow system is suitable for alfisols and its associated soils. This system is very much suitable for sloping land of 1%–4% with rainfall of 400–900 mm. This practice is highly suitable for soils with severe problems of crusting, sealing, and hard setting. Due to these problems, early runoff is quite common on these soils. In the conservation furrow system, series of furrows are opened on a contour or across the slope at 3–5 m apart. The spacing between the furrows and its size can be chosen based on the rainfall, soils, crops, and

TABLE 39.5

Yield and RWUE of Crops under Broad Bed Furrow (BBF) System and Ridge Furrow (RF) System

	Treatment	Water Content (w/w)			
Crop	Irrigation Method	At Saturation	At 5 Days of Dry Spell	Yield (q/ha)	RWUE (kg/ha-mm)
Cauliflower	BBF	15.3	9.2	58.5	11.16
	RF	15.4	9.8	61.5	11.74
Tomato	BBF	15.2	9.0	112.5	21.47
	RF	15.3	9.6	120.5	23.00
Okra	BBF	15.2	9.0	55.6	10.61
	RF	15.3	9.4	59.2	11.30

topography. The furrows can be made either during planting time or during interculture operations using a country plow. Two to three passes in the same furrow may be needed to obtain the required furrow size. These furrows harvest the local runoff water and improve the soil moisture in the adjoining crop rows, particularly during the period of water stress. The practice has been found to increase the crop yields by 10%–25% and it costs around Rs 250–350 ha^{-1}. To improve its further effectiveness, it is recommended to use this system along with contour cultivation or cultivation across the slope (Ram Mohan Rao et al., 1981).

39.2.5.1   Benefits

Benefits of the conservation furrow systems are as follows:

  (i) Conservation furrows harvest the local runoff and increase the soil moisture for adjoining crop rows.
 (ii) It helps to reduce runoff and soil loss.
(iii) Conservation furrow is a simple and low-cost system.
 (iv) This system is easy to adopt and can be implemented using traditional farm implements.
  (v) Adoption of this system increased crop yields (10%–25%).

39.2.5.2   Case Study

A study was conducted with conservation furrows for maize, pigeonpea, and groundnut crops in the Phulbani block of the Kandhamal district in Odisha. Soil water content was found to be more at five days of dry spell for each crop. The increases in yield were found to be 23.2%, 20.6%, and 15.8% for maize, pigeonpea, and groundnut, respectively, by adopting the intervention of conservation furrows. The rainwater-use efficiency was found to be increased by adopting conservation furrow in all crops (Table 39.6).

39.2.6   Bulky Organic Manures

Manures are the organic materials derived from animal, human and plant wastes that contain plant nutrients in complex organic forms. The art of collecting and using wastes from animal, human, and vegetable sources for improving crop productivity is as old as agriculture. Naturally occurring or synthetic chemicals containing plant nutrients are called fertilizers. Manures with low nutrient content per unit quantity have longer residual effects besides improving soil physical properties, compared to fertilizer with high nutrient content. Manures are grouped as bulky organic manures and concentrated organic manures based on the concentration of the nutrients. Bulky organic

---

TABLE 39.6

Yield, Water Content, and RWUE as Affected by the Intervention of Conservation Furrows in Maize, Pigeonpea, and Groundnut Crops

Crop	Treatment	Water Content (w/w)		Yield (kg/ha)	% Increase in Yield	RWUE (kg/ha-mm)
		At Saturation	At 5 Days of Dry Spell			
Maize	With intervention	15.6	10.2	2440	23.2	2.44
	Without intervention	15.4	9.2	1980		1.98
Pigeonpea	With intervention	15.2	10.4	820	20.6	0.82
	Without intervention	15.1	9.5	680		0.68
Groundnut	With intervention	15.0	10.3	880	15.8	0.88
	Without intervention	15.2	9.4	760		0.76

manures contain a small percentage of nutrients and they are applied in large quantities. Farmyard manure (FYM), compost, and green manure are the most important and widely used bulky organic manures.

39.2.6.1 Advantages
The advantages of bulky organic manures include:

(i) Uses of bulky organic manures supply plant nutrients including micronutrients.
(ii) It improves soil physical properties such as structure and water-holding capacity.
(iii) It increases the availability of nutrients.
(iv) Carbon dioxide released during decomposition acts as a CO_2 fertilizer.
(v) Plant nematodes and fungi are controlled to some extent by altering the balance of microorganisms in the soil.

39.2.6.2 Case Study
A study was conducted on rice, maize, pigeonpea, and cowpea by using bulky organic manures as the source of nutrients. The average yield results of all the crops under organic manures were compared with the application of the recommended dose of chemical fertilizers over five years. The effect of use of bulky organic manures on yield and RWUE of rice, maize, pigeonpea and cowpea are given in Table 39.7. Results on use of bulky organic manures on rice, maize, pigeonpea, and cowpea are encouraging.

39.2.7 ORGANIC MULCHING

Mulching is one of the most important ways to maintain healthy plants. Mulches conserve moisture by restricting evaporation. As organic mulches slowly decompose, they provide organic matter, which helps keep the soil loose. This improves root growth, increases the infiltration of water, and also improves the water-holding capacity of the soil. Mulch is any material applied to the soil surface for protection or improvement of the area covered with mulches. Nature produces large quantities of mulch all the time with fallen leaves, twigs, pieces of bark, spent flower blossoms, fallen fruit, and other organic parts. But, how these can be most effectively applied and used in agriculture needs great attention. A 2- to 3-inch layer of leaves provides good weed control. Leaves are usually easy to get, attractive as mulch, and they improve the soil once they decompose. After the leaves decompose, dig them into the soil and add a new layer of mulch on top.

TABLE 39.7

Effect of Use of Bulky Organic Manures on Yield and RWUE of the Crops

Crop	Variety	Water Content (w/w)		Yield (kg/ha)		% Increase in Yield	RWUE (kg/ha-mm)
		At Saturation	At 5 Days of Dry Spell	With Intervention	Without Intervention		
Rice	Sahabhagi (110 days)	15.5	10.6	2450	2120	14.25	2.43
	Naveen (110 days)	15.5	10.5	2580	2180	16.70	2.55
Maize	Maize (SA 701)	15.4	9.2	2880	2160	33.33	2.59
Pigeonpea	NTL 724	15.3	9.0	880	710	23.94	0.88
Cowpea	Gomti	15.4	9.4	2870	2360	21.61	4.42

39.2.7.1 Benefits of Organic Mulching

The main benefits of mulching are as follows:

(i) It prevents the loss of water from the soil by evaporation.
(ii) It reduces the growth of weeds by the smothering effect.
(iii) It keeps the soil cooler in summer and warmer in the winter, thus it maintains even soil temperature.
(iv) It improves soil structure as the mulch decays, and the decaying mulch adds nutrients to the soil.

39.2.7.2 Case Study

A study was undertaken on the use of organic mulching in different vegetable crops such as cauliflower, tomato, and okra. Reports promulgate that use of organic mulching results in more yield and higher water-retention capacity than without mulching in the case of vegetable crops. The results of the effect of organic mulching on the yield of cauliflower, tomato, and okra crops and water content in the soil are given in Table 39.8.

39.2.8 CROPPING SYSTEM

The selection of crops for rainfed uplands is of paramount importance for yield and rainwater utilization. The crops must be of short duration, low duty, or deep rooted to extract soil moisture from deeper soil layers during dry spells. Some of the promising crops for rainfed upland rice areas are maize, blackgram, greengram, pigeonpea, cowpea, and groundnut. These may be grown as pure or intercrops such as maize + cowpea (2:2), maize + pigeonpea (2:2), and pigeonpea + radish (2:2); and maize–horsegram/niger as double crops. Intercropping systems increase the use efficiency of natural resources like land and water, besides augmenting *in situ* moisture conservation and abating soil erosion.

39.2.8.1 Case Study

A study was undertaken on a different intercropping system such as maize + cowpea (2:2), maize + pigeonpea (2:2), and pigeonpea + radish (2:2) in Phulbani block of the Kandhamal district in Odisha. The demonstrations on different intercropping systems revealed a great advantage can be obtained as regards to yield, economics, and natural resources–use efficiency. Increase in yield for maize + cowpea (2:2), maize + pigeonpea (2:2), and pigeonpea + radish (2:2) are reported to be 99.6%, 61.9%, and 97.6% higher, respectively, than sole cropping (Table 39.9).

TABLE 39.8

Effect of Organic Mulching on Yield and RWUE of Different Vegetable Crops

Crop	Treatment	Water Content (w/w)		Yield (q/ha)	RWUE (kg/ha-mm)
		At Saturation	At 5 Days of Dry Spell		
Cauliflower	Organic mulching	15.2	11.6	64.5	12.31
	Without mulching	15.2	7.3	60.2	11.49
Tomato	Organic mulching	15.3	10.5	126.5	24.14
	Without mulching	15.2	7.4	118.0	22.52
Okra	Organic mulching	15.3	10.0	65.2	12.44
	Without mulching	15.3	7.5	58.3	11.13

TABLE 39.9
Yield and RWUE of Different Promising Intercropping Systems

Normal Crop	Normal Variety (duration)	Yield (kg/ha) With Intervention	Without Intervention	% Increase in Yield	RWUE (kg/ha-mm)
Maize + Cowpea (2:2)	Maize (P-3501) + Cowpea (Gomti)	Maize (2550 kg grain) Cowpea (2240 kg green pods) MEY-4790 kg	Sole maize 2400 kg	99.6	3.96
Maize + Pigeonpea (2:2)	Maize (P-3501)+ Pigeonpea NTL 30	Maize (2350 kg grain) Pigeonpea (640 kg seeds) MEY-3886 kg	Sole maize 2400kg	61.9	3.21
Pigeonpea+ Radish (2:2)	Pigeonpea (NTL 30) +Radish (Pusa Chetki)	Pigeonpea (780 kg seeds) + Radish (10580 kg) PEY-1661 kg	Sole Pigeonpea 840kg	97.6	1.37

MEY, maize equivalent yield.

39.2.9 ALTERNATE LAND USE SYSTEM

Aberrant monsoon behavior in rainfed agroecologies results in poor crop yields and makes crop production unstable and, many times, uneconomical. These soils are invariably low in organic matter and poor in nitrogen, phosphorus, and potassium contents. Moreover, these soils have low water-holding capacity and are prone to erosion. Poor management of marginal lands, particularly the top most soil tier of the undulating topography, results in land degradation. For profitable cultivation, some alternate land use systems need to be advocated for these lands.

Land use systems that are alternatives to crop production are called alternate land use systems. They are the effective economic utilization of land without harming the natural resource structure based on land capability. An alternate land use system involves the addition of a perennial component that has drought tolerance, can withstand the aberrations of monsoon, and imparts stability to production. Various models of alternate land use systems are as follows:

(i) Agri-horticulture: Practice of cultivating crops along with a horticulture component. It is suitable in semiarid regions where annual crop production is low and highly unstable.

(ii) Agri-silviculture: Production technique that combines the growing of agricultural crops with simultaneously raised and protected forest crops.

(iii) Alley cropping: An alternate land use system where crops are grown in alleys formed by hedgerows of trees or shrubs. It is more profitable to plant hedgerows on contours with 10 to 20 m spacing. The essential feature of the system is that hedge rows are kept pruned during cropping to prevent shading and to reduce competition with food crops.

(iv) Ley farming: Ley farming aims to generate *in situ* fertility through the rotation of legume forages with cereals. It is important for increased fodder production, a critical component in rainfed farming.

(v) Silvi-pasture: Marginal drylands, generally shallow and poor in nutrients, can also yield better returns through fuel wood and fodder. After a six- to eight-year rotation even arable crops can be grown on built-up soil fertility. This system is most preferable where fodder shortages are experienced frequently and fallow land can be spared for this system.

The alternate land use system not only helps in generating much-needed off-season employment in monocropped drylands but also utilizes off-season rains that may otherwise go to waste as run-off, prevents degradation of soils, and restores ecological balance. Different land use systems so developed should be used in better ways for increased and stabilized production in drylands. A fruit-based agri-horti system, i.e., fruit trees with agricultural crops, could be an alternate land use system for rainfed or dryland conditions. Under this system, fruit trees can be grown successfully with pulses such as greengram, blackgram, and cowpea; or with oilseed crops such as groundnut and spices such as turmeric as bonus crops and economic alternative systems for marginal and submarginal lands. It is observed that dry land horticultural fruit trees like mango, papaya, and jackfruits integrated with short-duration arable crops like legumes and spices are the most profitable practice among different agri-horti systems.

39.2.9.1 Case Study

Arable crops cultivated in alleys or the interspaces of fruit trees such as mango, jackfruit, and papaya as compared to sole fruit crops recorded a positive effect of intercropping on tree productivity and overall equivalent yield. A higher value of organic carbon, and available soil nitrogen, phosphorus, and potassium are also observed in the alternate land use system as compared to control, which indicates the improvement in soil fertility. The income from crops and fruits, and gross income under each treatment during the study for four consecutive years are given in Table 39.10.

39.2.10 Vegetative Barriers

Vegetative barriers or vegetative hedges or live bunds are effective in reducing soil erosion and conserving moisture. In several situations, the vegetative barriers are more effective and economical than the mechanical measures, viz., bunding. The vegetative barriers are suitable for alfisols, vertisols, vertic inceptisols, and associated soils. These are suitable for an area of slope of more than 2.5% with a rainfall of 400–2500 mm. Vegetative barriers can be established either on contour or

TABLE 39.10

Yield and Economics of Different Agri-Horti-Based Alternate Land Use Systems

System	Yield (kg/ha)	Income from Crop (Rs./ha)	Income from Fruits (Rs./ha)	Cost of Cultivation (Rs./ha)	Gross Income (Rs./ha)	Net Return (Rs./ha)	B:C Ratio
Mango + Greengram	GG: 650	39000	0	20000	39000	19000	1.95
Mango + Blackgram	BG: 680	34000	0	18000	34000	16000	1.89
Mango + Groundnut	GN: 1240	49600	0	25000	49600	24600	1.98
Mango + Turmeric	M: 4400 T: 4800	240000	88000	110000	328000	218000	2.98
Jackfruit + Turmeric	J: 8500 T: 4750	237500	85000	110000	322500	212500	2.93
Papaya + Greengram	P: 8400 GG: 660	39600	84000	35000	123600	88600	3.53
Papaya + Cowpea	P: 8600 CP: 2820	56400	86000	35000	142400	107400	4.06
Papaya + Groundnut	P: 8400 GN: 1220	48800	84000	40000	132800	92800	3.32
Papaya + Blackgram	P: 8500 BG: 670	33500	85000	33000	118500	85500	3.59

on a moderate slope. In this system, the vegetative hedges act as barriers to slow the runoff velocity, resulting in the deposition of eroded sediments and increased rainwater infiltration. It is advisable to establish the vegetative hedges on small bunds. This increases their effectiveness, particularly during the first few years when the vegetative hedges are not so well established. The key aspect of the design of vegetative hedges is the horizontal distance between the hedge rows, which mainly depend on rainfall, soil type, and land slope. Species of vegetative barriers to be grown, the number of hedge rows, plant-to-plant spacing, and the method of planting are very important and should be decided based on the main purpose of the vegetative barrier. If the main purpose of the vegetative barrier is to act as a filter to trap the eroded sediments and reduce the velocity of runoff, then grass species such as vetiver, sewan (*Lasiurus sindicus*), sania (*Crotolaria burhia*), and kair (*Capparis aphylla*) can be used as vegetative barriers. But if the purpose of vegetative hedges is to stabilize the bund, then plants such as *Glyricidia* or others can be effectively used. *Glyricidia* plants grown on bunds not only strengthen the bunds while preventing soil erosion, but also provide nitrogen-rich green biomass, fodder, and fuel. In areas with long dry periods, vegetative hedges may have difficulty surviving. In very low rainfall areas, the establishment, and in high rainfall areas, the maintenance, could be the main problem. Proper care is required to control pests, rodents and diseases for optimum growth and survival of both vegetative hedges and main crops.

39.2.10.1 Benefits

The benefits of vegetative barriers include the following:

(i) Once the vegetative barriers are properly established, then the system is self-sustaining and almost maintenance-free.
(ii) Land under the vegetative hedge is used for multiple purposes, viz., nitrogen-rich biomass, fodder, and fuel.
(iii) They can be successfully used under a wide range of rainfall (400–2500 mm) and topography.
(iv) Vegetative barriers are very economical and often more effective than other erosion control measures.

39.2.11   CONSERVATION TILLAGE METHODS

Conservation tillage or reduced tillage is a practice of minimizing soil disturbance and allowing crop residue or stubble to remain on the ground instead of throwing it away or incorporating it into the soil. The cover of crop residue helps prevent soil erosion by water and air, thus conserving valuable top soil. Soil structure is also improved, because heavy machinery (which causes soil compaction) is not used and soil tilth does not tamper artificially. Another important environmental effect of reduced tillage is the reduction in the use of fossil fuels on farms. Successful conservation tillage practices depend upon the careful selection of crops that are grown in selected locations. Crops that are fast-growing in nature suppress weed growth, and competition for nutrients is minimized. Large-seeded vegetable crops have a very fast growth rate as compared to small-sized vegetable crops. Appropriate planting equipment is also very important in conservation tillage.

39.2.11.1 Benefits

The benefits of conservation tillage are as follows:

(i) Conservation tillage helps to reduce the fuel consumption of tractors and power tillers.
(ii) It reduces soil compaction.
(iii) In conservation tillage methods, farmers have flexibility in planting and harvesting crops.
(iv) It helps to improve soil health.

(v) In conservation tillage, times can be saved, as the fields can be planted quickly in a single pass. The farmers need fewer tractors or tractor time to get the work done.

(vi) There is less wear on tractors and other implements in conservation tillage methods.

39.3 CONCLUSIONS

The study results in the following conclusions.

1. Summer plowing, plowing across the slope, and increasing the bund height for *in situ* moisture conservation can register higher yield and hence net returns, and should be strongly advocated.

2. The use of bulky organic manures improves soil physical properties such as structure and water-holding capacity, in addition to providing plant nutrients including micro nutrients. The uses of bulky organic manures are imperative for sustainable crop production in rainfed agroecologies.

3. The ridge furrow system is more suitable than the broad bed furrow system in rainfed upland situations. The ridge furrow system helps to conserve water when there is lower rainfall by holding water within ridges, and when there is higher rainfall, furrows help to drain the excess water thus avoiding waterlogging.

4. The use of conservation furrows across the slope harvests the local runoff and increases the soil moisture for adjoining crop rows. Adoption of this system increased crop yields 10%–25%.

5. Organic mulching is found to be more pertinent in dryland situations. Organic mulching conserves moisture by restricting evaporation as well as improves root growth, increases the infiltration of water, and also improves the water-holding capacity of the soil.

6. Intercropping systems such as maize + cowpea (2:2), maize + pigeonpea (2:2), and pigeonpea + radish (2:2) have shown to be more promising than sole cropping in utilizing the natural resources of the region.

7. In rainfed marginal lands, pulses such as greengram, blackgram and cowpea; and oilseed crops such as groundnut and spices such as turmeric can be intercropped with fruit trees as bonus crops and economic alternative systems.

8. Vegetative barriers or vegetative hedges or live bunds are effective management practices in reducing soil erosion and conserving *in situ* soil moisture.

9. Conservation tillage helps in improving soil health, reducing cost of cultivation and *in situ* moisture conservation in rainfed conditions.

REFERENCES

Behera, B., Mohanty, S. K., and Pal, A. K.2005. Integrated nutrient supply system for pigeon pea + rice intercropping system. In: *Resource-conserving Technologies for Social Upliftment*, ed. V. N. Sharda et al., 226–229.

Behera, B., Sankar, G. R. M., Sharma, K. L., Mishra, A., Mohanty, S. K., Mishra, P. K., Rath, B. S., Grace, J. K. 2012. Effects of fertilizers on yield, sustainability and soil fertility under rainfed pigeon pea + rice system in subhumid oxisol soils. *Communications in Soil Science and Plant Analysis*, 43: 2228–2246.

Bhatt, R., Khera, K.L., Arora, S. 2004. Effect of tillage and mulching on yield of corn in the submontaneous rainfed region of Punjab, India. *International Journal of Agriculture and Biology*, 6: 26–28.

Gouda, R. K. 2011. *Soil and Water Conservation Measures in Arable lands*. A Bulletin produced by Water Technology Centre for Eastern Region (WTCER), Bhubaneswar, Odisha, pp 176.

Hadda, M.S., Khera, K.L., Kukal, S.S., 2000. Soil and water conservation practices and soil productivity in North-Western sub-montaneous tract of India: A review. *Indian Journal of Soil Conservation*, 28: 187–192.

Hadda, M. S., Arora, S., Khera, K. L., 2005. Evaluation of different soil and nutrient management practices on moisture conservation, growth and yield of maize in rainfed sub-montaneous tract of Punjab. *Journal of Soil Water Conservation*, 4: 101–105.

Katiyal, K. C., Sharma, S., Padmanabham, M. V., Das, S. K., Mishra, P.K., 1995. *Field Manual on Watershed Management*. Central Research Institute for Dryland Agriculture, Santoshnagar, Hyderabad, India. 165 pp.

Ram Mohan Rao, M. S., Chittaranjan, S., Selvarajan, S., Krishnamurthy, K., 1981. Proceedings of the Panel Discussion on Soil and Water Conservation in Red and Black Soils, 20 March 1981, UAS, Bangalore, Karnataka: Central and Soil and Water Conservation Research and Training Institute, Research Center, Bellary, Karnataka and University of Agricultural Sciences, Bangalore, India. 127 pp.

Index

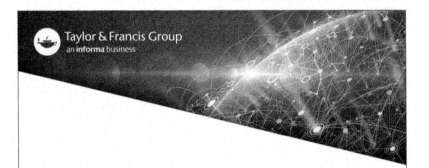